Runmath Series

Fundamentals of Functions and Graphs

Book 3 : The Conic Section 1 (Line & Parabolas)

Seong Ryeol Kim

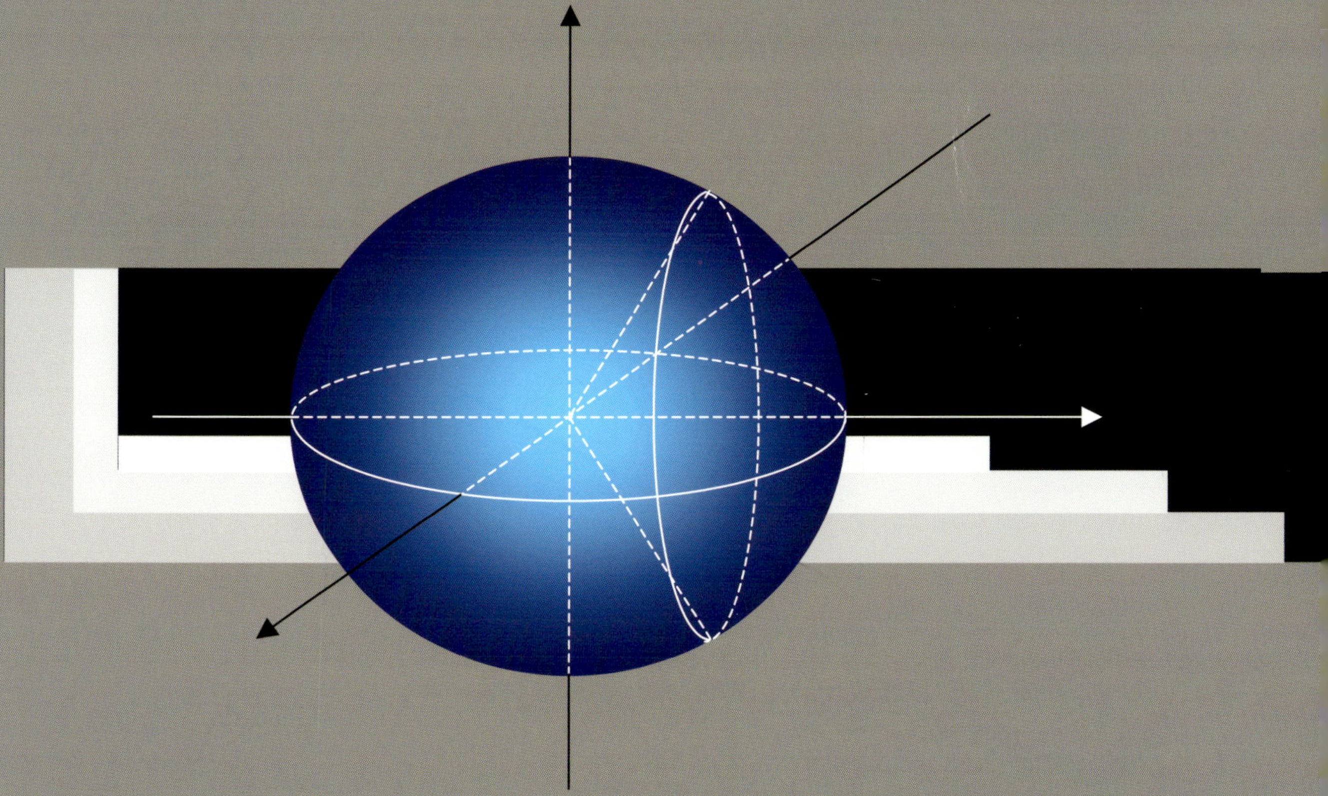

Runmath

Kyo Woo Sa

Fundamentals of Functions and Graphs

Book 3: Conic Sections 1 (Lines and Parabolas)

Seong Ryeol Kim

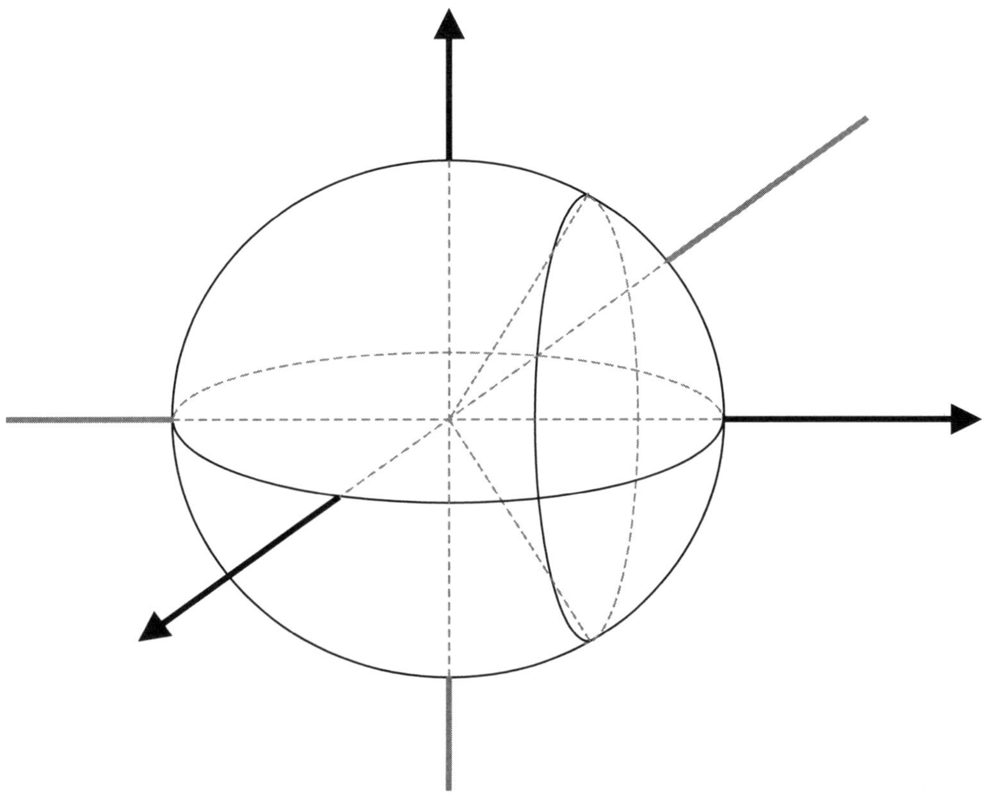

Runmath

Printed in Korea

Fundamentals of Functions and Graphs
Runmath

Preface

This book is the third book in the series, **Fundamentals of Functions and Graphs**. To begin with, in the Book 1: **What Functions?**, discussions on what basic functions are, how they work are covered. Next, in the Book 2: **Graph Operations**, discussions on how to operate the graphs of basic functions and equations. If we know what functions are, how they get constructed, and how to use them, and also, can understand how to construct, use, and manipulate the graphs, then we have properly covered the most part of the fundamentals of functions and graphs. In this book, the discussions will be heavily on straight lines and parabolas which are two of the kinds in basic curves called conic sections. We will get to understand what we can do about the individual graphs of the curves and how we can do it in a concrete manner. Covered in the rest of the series are circles, ellipses, and hyperbolas.

As shown above, conic sections are cross sections of double circular cones meeting at the vertices. A straight line is one of the basic curves called conic sections. Now, when we say a line in math, it is a straight line. Suppose that two right circular cones meet each other at the vertices and the circular bottoms are parallel to each other. Suppose also, that a plane cuts through the cones and includes the point where the vertices meet. Then, the cross section is made up of two lines crossing each other. However, if the plane is tangent to the sides of the cones and includes the point, only one line gets created. Suppose this time, that the plane is parallel to the side of the cone but is not tangent to the side and does not include the vertex either. Then, the cross section is a parabola.

Lines and parabolas are basic ones of the basic curves. Since they are such basic, they are very significant curves. The more basic, the more significant. In fact, of the two kinds in curves, a line is more basic and therefore, more important. Of course, the most basic is a point, though. In fact, a point too, can be taken for one of the conic sections. That's because the plane can include the vertex only. Rigorously therefore, we can say that a point too, is a conic section. However, it is not quite appropriate to call a point a curve, is it?

Without a line, there is not much we can do in general math. It means that without proper knowledge of lines, it is often impossible to take care of not only basic geometry but basic general algebra where we work on basic equations. As all the other objects in math, a line is an idea, too. Therefore, we need to know what we mean by a line. Without the knowledge, we get to fall down working on many problems, but the knowledge can give us tremendous power.

Next, parabolas too, are important curves that show up quite frequently in general math. In our ordinary lives, too, many parabolas are actually observed. A lot of natural phenomena follow parabolas. For instance, the pathways baseballs fly in the air are parabolas. The shape of satellite dish is basically a parabola, too. Parabolas are widely used when we collect waves like lights, sounds, etc. and concentrate them to a particular position and also, send efficiently those waves emitting from a specific position.

Now, we work on ideas in math. Any object in math is an idea and has its definition. In real life, any physical object is a 3-D object no matter how small it may be. That is, no matter what small point we may put on paper, it is a lump of ink or lead. Objects in math are quite different from those in real world since those in math are ideas. In real world, we have thick or thin lines, but in math, lines don't have thickness since they are ideas. Besides, a line cannot be perfectly straight in real world, but a line has to be perfectly straight in math. The same is true for a plane, triangle, quadrangle, circle, etc. Any thing in real worlds is merely an approximation to the idea. That is, it only looks like the idea.

We often put simple geometric objects in mathematical expressions. For instance, we put a line or parabola in an equation. Such an equation is a convenient tool we can use when we work on many problems. However, such an object as a line is an idea, and equations alone are often a little too abstract when we explain ideas or relationship between the ideas. That's because an idea itself is not visible. Therefore, visual presentation of ideas that we want to convey to others is a very important skill. Such a skill is the ability of construction of graphs, and the curves in the graphs are such objects. Then, we can make visible the actual content of such an equation by means of a graph.

In math, we usually use equations or graphs when we want to show objects themselves or how the objects behave. That is to say that when we explain ideas or phenomena, we want to show them systematically or physically rather than in words only. The objects can be for instance, amounts of productions or savings, consumptions or expenditures, distances, inputs, outputs, ratios, prices, demands, supplies, temperatures, volumes, etc. Also, the phenomena tell us how such objects change, for instance.

Therefore, we should be able to construct and work with both of the tools. With equations, we can show and describe ideas or phenomena in analytic manner. In other words, it is convenient for us to explain the ideas or phenomena based on relevant rules or principles by means of equations. Also, we can describe visually the ideas or phenomena by means of the curves of the equations.

What about functions, then? Functions are different from equations. In some cases, the curves of a function and an equation can be the same. Basically, a function explains correlation between groups of data and show how the individual data in one group are correlated to those individual ones in the other. So to speak, a function can be said to indicate or explain a relation between cause and effect. Now, in general math or Calculus, a particular function can produce the same one effect with many different causes, but cannot produce two or more different effects with one cause only. An equation of a geometric object such as a parabola or circle explains the relation between coordinates of points that constitute the object. In other words, the equation is the connective expression between the variables which are used in the coordinate system where the object as the curve is located. However, we can directly use some equations for functions. For instance, we can use for a function an equation of a single line or an equation of some parabola.

When we construct a graph of a function or equation, we usually use a perpendicular coordinate system, which is often called Cartesian coordinate system. In accordance with the number of the variables in the function or equation, the number of the axes is determined, and the axes are perpendicular to each other. In basic general math, we normally use an x-y system. Then, the variables are **x** and **y**, and the x-axis and the y-axis are perpendicular to each other. Besides, the plane made up of the axes is called an x-y plane.

Now, let's find out about how to put in equations simple and basic geometric objects and how to put such equations in graphs. The basic geometric objects are lines, parabolas, circles, ellipses, and hyperbolas. Those are basic curves in general. In particular, they are called conic sections. That's because they are cross sections of double circular cones cut with a plane.

Now, let's begin with lines.

Contents

0. What is a line?

A line is a geometric object which is a collection of points. Of course, a collection of points can be some other object. Normally, we mean a straight line when we say a line. However, if a line is curved, we call it a curved line. Now, whether the line is curved or not, we generally call it a curve. In fact, no matter what an object in a graph may be, we just call it a curve. That is, a curve is a common designation of all the geometric objects in graphs. For instance, a curve in a graph can be a line, a parabola, a curved plane, or a sphere. A line in math is a straight line which is one of the curves in a graph.

A line is one of the basic curves called conic sections. Suppose that two right circular cones meet at their vertices and the circular bottoms are parallel to each other. Suppose also, that a plane cuts through the cones and includes the point where the vertices meet. Then, the cross section is made up of two lines crossing each other. However, if the plane is tangent to the sides of the cones and includes the point, only one line gets created.

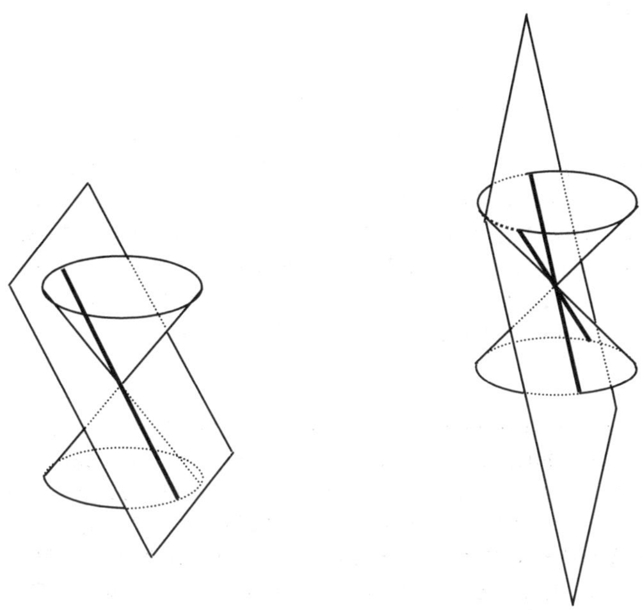

Now, a line has no thickness or width. That's because a line in math is an idea. Basically, a line has an indefinite length. Normally, a line proceeds infinitely forward and backward, that is, in both directions which are opposite of each other. In other words, both ends proceed indefinitely. Of course, it can proceed in one direction only. In that case, it is called a ray. What is a line segment, then? It is a part of a line and has a finite length. In general, when we say a line segment, we mean a straight line segment, and a length in math means a length of such a line segment. For instance, we indicate the distance between two positions by means of the length of the line segment connecting the two positions. Then, what do we call a line segment that is curved? It can be called a curved line segment. A line in math is a straight line which is an idea like any other object in math.

Now, what do we mean by a line? Although we use a part of a line that is, a line segment to indicate a length, what's often important in a line is a direction. That is, a line is used to indicate a direction rather than a length. In analytic geometry, a line frequently means a direction. Analytic geometry can be said to be geometry where we mainly use graphs in coordinate systems. In a line, each of all the points constituting the line has the same direction. Therefore, there is no bending, turning, or twisting in a line. A line or a line segment is an object that is composed of points each of which has the same direction. The direction is indicated by an angle the line makes with one of the axes in the coordinate system where the line is put. The axis is normally the horizontal axis. For instance, it is the x-axis in the x-y coordinate system. Therefore, a line has its direction, and the direction of a point in a plane is the same as the direction of the line including the point. However, we need to note that all the lines parallel to each other have the same direction.

Now, how do we take the measurement of such an angle? Suppose that we have a line in the x-y system. Then, we take the measurement of the angle by means of the slope of the line. What slope? A slope is the value of the ratio of the height to the base of a right triangle. Now, a right triangle is made up of both of the axes and the line. Then, a part of the x-axis is the base of the triangle, a part of the y-axis is the height, and a part of the line takes up the hypotenuse. So, the ratio of the height to the bottom is the slope of the line. Therefore, the slope tells us how the line is slanted to or inclined with the x-axis. In fact, the slope is not a ratio but a rate. We will get to see the detail in the next section.

The curve of a function, too, can be a line. That is, if we construct the graph of some function, the curve is a line. However, we may not express every line in terms of a function. That is to say that we cannot put some line in a function. For instance, there is no function that has a line parallel to the axis of output variable such as the y-axis in the x-y system. That is, we cannot put in a function any line perpendicular to the x-axis. When we work on a line, we usually put the line in an equation. It is not an equation for an unknown but an equation of a geometric object. If the object is a line, such an equation is called an equation of line. We normally put one line in one equation. However, we can put many lines in one equation, too. When we put a line in an equation, we often set to 0 one of the two sides of the equal sign. We don't have to in many cases, however. We can choose the way we put the equation in accordance with our needs.

A line is a set of points, and a part of a line can have a value of a length or distance. In addition, all the points constituting the line have the same direction which can be indicated by the slope of the line. Therefore, we use a line when we measure a distance or talk about a direction. Of course, it has other uses, too. Particularly in analytic geometry, we use a line to explain a direction of a point. For instance, when we find the direction of a point in a curved line, we find and use the slope of a line that is tangent to the curved line at the point, and such a line is called a tangent line which is covered in Calculus. Normally, we indicate a line by means of the equation of the line. Now, let's begin with an equation of line that indicates a single line.

0.0. Single Lines

A single line is just a line. Let's find out about how to express a line, and talk about what we can do about it and how we can do that. First of all, in math, where do we put a line? We put it in a graph. Then, how do we describe a line in a graph by means of mathematical expressions? We put it into an equation. Then, the equation is an equation of line. That is, the equation indicates a line. Now, let's find out about how to put a line into an equation.

Wait a minute! How about using a function?
We can express most lines in functions but cannot put every line in a function. For instance, it's not possible to put in a function a line parallel to the y-axis in the x-y system.

What is a line made up of? It is composed of points. That is, a line is a collection of points. Therefore, a line is said to be basically a set of points. However, a set of points is not necessarily a line. In fact, every geometric object in a graph is composed of points. Suppose now, that a line is in a plane. Then, depending on what points in the plane we put together, the shape of the geometric object is determined. Then, we need to know what points get together to form a line. However, we already know what points need to get together. If we put those points together, all the points have the same direction. Now, what we have to do is that we present the points in a specific manner. How? We put them into an equation of line. How? That's exactly what we are getting at now.

A line is put in a graph. Then, where is the graph? It is in a particular coordinate system. Then, each of the points constituting a line has coordinates. Now, suppose that a line is in a graph in the x-y system. By the way, a graph in the x-y system is also called an x-y plane. Then, each of the points in a line in the x-y plane has a pair of coordinates. Let's describe the points, then. However, since the points are uncountable, we cannot present each and every point. What should we do, then? We can make and use a point representing all the points. We call the point a general point. Also, it is called a variable point or a point variable. A general point can use the applicable variables for its coordinates. Since the line is in the x-y system, the variables are **x** and **y**. Then, the general point can be put in **(x, y)**. We know that a line is a set of points. Now, we can put a line called **L** in an x-y plane in such a way as follows:

L = {(x, y)| ax + by + c = 0 where **a**, **b**, and **c** are constant, but **a** and **b** cannot be 0 concurrently.**}**

In the set expression above, we can say that **ax + by + c = 0** is a connective expression between the variables. The connective expression is the very equation of line. In particular, such an equation of line is said to be in a **general form**. Now, using the general form, we put a line **L** in the x-y system in such a way as follows:

L(x, y) = ax + by + c = 0 where **a**, **b**, and **c** are constant, but **a** and **b** cannot be 0 simultaneously.

Then, the equation **L(x, y) = 0** is an equation of line indicating a single line.
Then, the **name** of the equation of line is **L(x, y)**, and the **equation** is **ax + by + c = 0**.
Also, the **L** in **L(x, y)** can be said to be the equation designator. What is the designator? It is a simple name to call the equation. Then, **L(x, y)** is a complete name. Therefore, the equation can be called either 'the equation **L**', 'the equation **L(x, y)**', or 'the equation **L(x, y) = 0**'.

Besides, we often put a line into an equation in such a form of as follows: **y = ax + b** where **a** and **b** are constant. However, we need to note that the **a** and **b** are different from the **a** and **b** used earlier. Such an equation is called a **standard form** in which we can put nearly all lines. Why not **all** but **nearly all**? That's because we cannot put in the standard form the lines parallel to the y-axis. However, it can give us advantages, too. It is easier to use the standard form when we take a line for a curve of a function rather than an equation. Now, what's more important is that in such a standard equation, we can readily see the slope or the direction of the line. In the standard equation, the **a** is the slope. In addition, we can immediately see where the line meets the y-axis. How do we know where the line meets the y-axis? If we put 0 into the **x** in the equation, then we get **y = b**. Therefore, the line meets the y-axis at the point **(0, b)**. In other words, the line crosses or cuts the y-axis at **(0, b)**. Therefore, the **b** is called a y-intercept. What about the x-intercept, then? Setting **y = 0** in the equation, we get **x = -b/a**. Then, the line intersects the x-axis at the point **(-b/a, 0)**. Therefore, the –**b/a** is the x-intercept.

Now, let's find out about how the two equations described above get constructed. First, we are going to begin with the standard form. Constructing something, we need to have material to construct with. Then, what do we need to construct the equation with? At least, we need either two points in the line or a point in the line and the slope of the line. What we should first do when we find an equation of line is to find the slope of the line. The slope (or direction) of the line is the most significant of all the elements of the line. In other words, the slope of a line is the most important when we work on a line. In fact, a line is no other than a slope. Of course, we have infinitely many lines that have the same slope. However, of all the lines having the same slope, we have no more that one line that can include a particular point.

Finding the equation of line means that we show in specific manner all the points constituting the line. That is, it is the solution to the question, "What points in the world constitute a line?" However, a line is made up of points uncountable, and therefore, we cannot specify one by one all the innumerable points. What should we do, then? We make and use a general point representing all the points in the line. We can use the variables of the coordinate system to make the general point. That is, the system variables are the coordinates of the general point. Now, suppose that a line is in x-y system and the name of the general point is **P**. Then, the general point is **P(x, y)**. Then, we need to come up with a relation between the coordinates, that is, the variables, and present the relation. That is to say that the connective expression between the variables is the equation of the line. Now, suppose that the line includes a particular point **$P_1(x_1, y_1)$**, and that the slope of the line is **a**. Let's now, put in a graph the general point **P(x, y)** and the point **P_1** as shown in the figure below.

Fig. 0.0.0

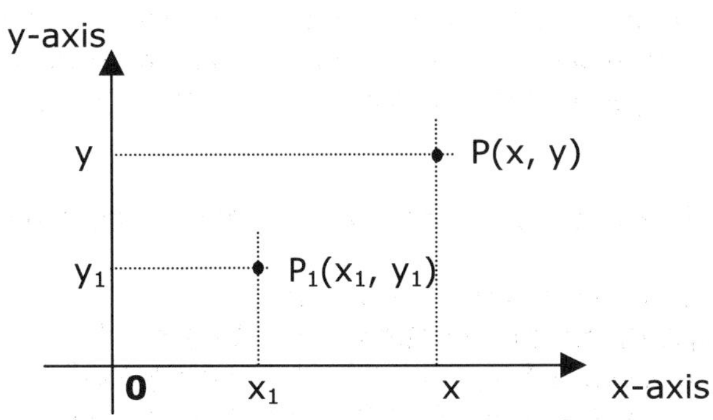

Now, let's put in the graph the line that includes the two points. Usually, in the x-y system, we indicate by **Δx** the change along the x-axis and by **Δy** the change along the y-axis.

First of all, by convention, the value in the y-axis depends on the value in the x-axis.
Since the value in the x-axis changes (increases) in the direction of the x-axis, the line we are after proceeds from the point $\mathbf{P_1(x_1, y_1)}$ to the point $\mathbf{P(x, y)}$.
Then, since the value in the x-axis changes from $\mathbf{x_1}$ to $\mathbf{x}$, the value in the y-axis changes from $\mathbf{y_1}$ to $\mathbf{y}$.
Then, we get $\Delta\mathbf{x = x - x_1}$ and $\Delta\mathbf{y = y - y_1}$.

In the figure below, we can see that the amount of $\Delta\mathbf{x}$ $\mathbf{(= x - x_1)}$ is the change in the direction of the x-axis, and that the amount of $\Delta\mathbf{y}$ $\mathbf{(= y - y_1)}$ is the change in the direction of the y-axis.

Now, let's talk about the slope.

The slope of a line says how the line is slanted to the horizontal axis, that is, the x-axis. More specifically, it explains in what amount and manner the line is slanted to the x-axis.

In the figure below, we can see that there is a right triangle in the graph.

Fig. 0.0.1

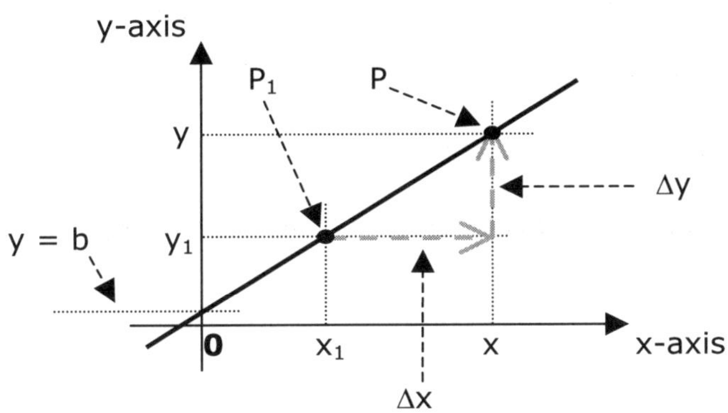

In the triangle, the $\Delta\mathbf{x}$ is the base, the $\Delta\mathbf{y}$ is the height, and the line segment between $\mathbf{P_1}$ and $\mathbf{P}$ is the hypotenuse.
Then, the slope of the line is the ratio of the height to the base of the triangle.

However, the slope is not simply a ratio but actually a rate.
How come?
The reason will be explained later in this section.

Therefore, we can define the slope **a** in such a way as follows: $\mathbf{a} = \Delta\mathbf{y}/\Delta\mathbf{x} = \mathbf{(y - y_1)/(x - x_1)}$.

Now, can we set $\mathbf{a = (y_1 - y)/(x_1 - x)}$, too? Of course, we can. That's because:

$$\mathbf{(y - y_1)/(x - x_1) = (-y_1 + y)/(-x_1 + x) = \{-(y_1 - y)\}/\{-(x_1 - x)\} = (y_1 - y)/(x_1 - x)}.$$

Therefore, we get $\mathbf{a} = \Delta\mathbf{y}/\Delta\mathbf{x} = \mathbf{(y - y_1)/(x - x_1) = (y_1 - y)/(x_1 - x)}$.

On the other hand, a line can be put as shown in the figure below.
In this case, the slope **a** gets a negative value.

Fig. 0.0.2

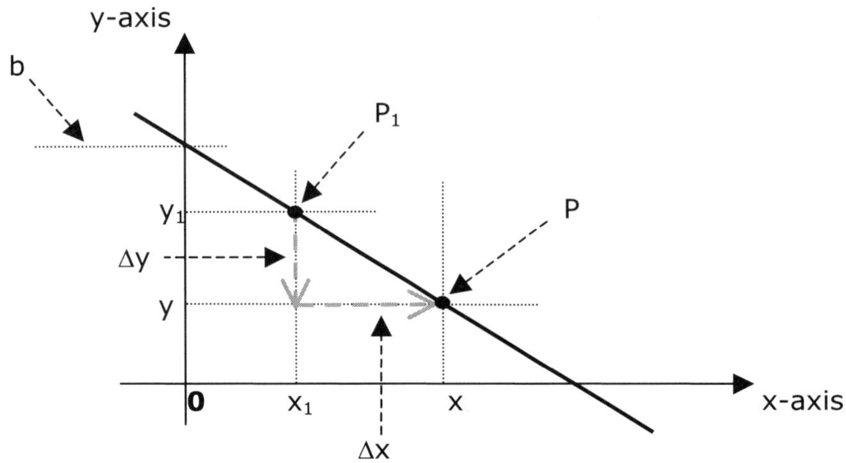

Now, how does the slope of a line indicate the direction of the line?
The value of the slope can specify the direction of the line. The value is the rate of the height to the base of the right triangle. Since it is the value of a rate, we can adjust it. For instance, if we take for 1 the length of the base, the height is the value of the rate.

height/base = $\Delta y/\Delta x = (\Delta y/\Delta x)/1 \Rightarrow$ base = 1 and height = $\Delta y/\Delta x$.

Now, if the value of the slope is positive, the triangle is the one on the left in the figure below, and if the slope is negative, the triangle is the one on the right.

Fig. 0.0.3

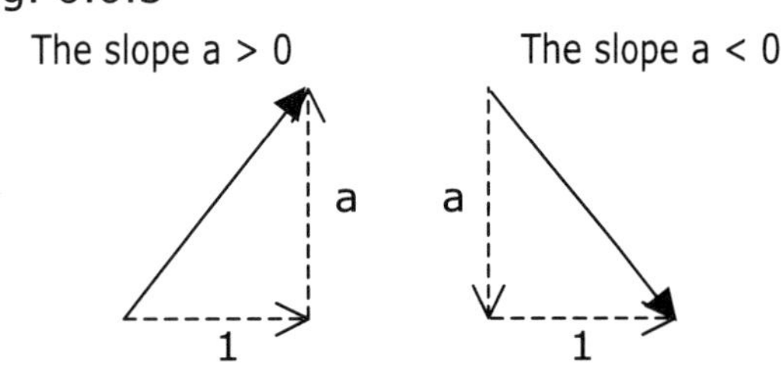

Then, the hypotenuse along with the base and the height can specify the direction.
All the lines parallel to each other have the same slope. Therefore, all the lines parallel to each other have the same direction.

Now, we've got the slope **a** in the standard equation of line $\mathbf{y = ax + b}$.
Then, we need to find the value of the y-intercept **b**.

Let's now, put $\mathbf{a = (y - y_1)/(x - x_1)}$ in such a way as follows: $\mathbf{y - y_1 = a(x - x_1)}$.
Then, $\mathbf{y - y_1 = a(x - x_1) \Rightarrow y = ax - ax_1 + y_1}$.

The **a** is constant. Therefore, $\mathbf{-ax_1 + y_1}$ is constant, too.
Then, we can see that the **b** in $\mathbf{y = ax + b}$ is $\mathbf{-ax_1 + y_1}$.
The expression $\mathbf{y = ax - ax_1 + y_1}$ is a connective expression between the coordinates of the general point.

Then, the connective expression is the expression that is satisfied by each of all the points in the line. Therefore, the equation of the line that includes a particular point (x_1, y_1) in the x-y plane and has the slope of **a** is $y = ax + b$ where $b = -ax_1 + y_1$.

Now, let's move on to the general equation of line. The general form is $ax + by + c = 0$.

The standard equation is $y = ax + b$.
Then, we can set $ax - y + b = 0$, but in this equation, the coefficient of the **y** is a particular number, that is, -1. Then, it does not have complete generality.
How can we give the generality to it, then?
We can change the coefficient of the **y** with a constant since unlike a number, a constant has generality.
For instance, we can use a letter **q** for the constant.
In this case however, the slope is not the **a**, and the y-intercept is not the **b**, either.
The slope of the line is $-a/q$, and the y-intercept is $-b/q$.

$$ax + qy + b = 0 \Rightarrow y = (-a/q)x - b/q.$$

However, by convention, we use **a**, **b**, and **c** for the constants.
Now, the general equation of a line is as follows:

$ax + by + c = 0$ where **a**, **b**, and **c** are constants but **a** and **b** cannot be 0 at the same time.

However, we need to note that in the general form above, the **a** is not the slope of the line, and the **b** is not the y-intercept, either.

Now, why bother to have such a general form?
The standard equation of line cannot indicate every line.
It cannot indicate any line parallel to the y-axis (or perpendicular to the x-axis).
For instance, the equation of a line parallel to the y-axis is as follows: $x = d$ where **d** is constant.
However, the general form can indicate such lines, too.
Suppose that $b = 0$ in the general equation of line $ax + by + c = 0$.
Then, $ax + c = 0 \Rightarrow x = -c/a$ where $-c/a$ is constant.
Therefore, the equation $x = -c/a$ indicates a line parallel to the y-axis (or perpendicular to the x-axis).

In addition, we get to see another use of the general equation in the next section.

Is there any form other than the standard form and the general form?
Yes, there is. We can put a line in $x/p + y/q = 1$ where **p** and **q** are nonzero constants.
In this case, the **p** is the x-intercept, and the **q** is the y-intercept. However, such a line cannot include the origin.

This time, let's take a line for a curve of a function, review the process of finding the standard equation, and also, find out how come the slope is a rate.
In the standard equation $y = ax + b$, the **a** is the slope.
Suppose now, that a line is the curve of a function $y = f(x) = ax + b$ where **x** is real, of course. Suppose also, that the line includes a particular point (x_1, y_1), and that (x, y) is the general point representing all the points in the line.
Now, we are going to find the connective expression between the coordinates **x** and **y**.

Usually, in the x-y system, we indicate by Δx the change (increase) in input and by Δy the change in output.
In a function, the output depends on the input.
Since the input changes (increases) in the direction of the x-axis, the line we are after proceeds from the point (x_1, y_1) to the point (x, y).
Then, since the input changes from x_1 to x, the output changes from y_1 to y.
Then, we get $\Delta x = x - x_1$ and $\Delta y = y - y_1$.

Now, let's talk about the slope.
The slope of a line says how the line is slanted to the horizontal axis or the axis of input variable, that is, the x-axis. More specifically, it explains in what amount and manner the line is slanted to the axis.
In the figure below, we can see that there is a right triangle in the graph.

Fig. 0.0.4

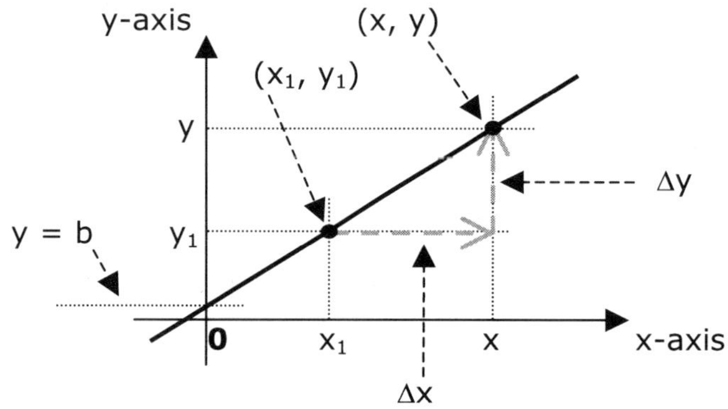

In the triangle, the Δx is the base, the Δy is the height, and a part of the line is the hypotenuse.
Then, the slope of the line is the ratio of the height to the base of the triangle.
However, the slope is not simply a ratio but actually a rate.
How come?

First of all, the input and output are amounts usually different from each other in nature. For instance, it can be the case where the input is time and the output is distance, or the case where the output is temperature when the input is volume, etc.

Next, the base of the triangle is the Δx which is the change (increase) in input, and the height is the Δy which is the change in output. That is, the amounts in the two kinds are amounts that change.
In other words, the slope specifies the change in output with respect to the change (the amount of increase) in input. Therefore, the value of the slope can be either positive, negative, or zero since the slope specifies how much the output either increases, decrease, or remains.

Then, it is convenient to understand the slope if we specify the change in output with respect to a unit change in input. Therefore, the **slope** indicates the **value of the output with respect to a unit of the input**, and is a **rate**. A unit of the input is an amount usually corresponding to 1. For instance, 1 meter, 1 centimeter, 1 kilogram, 1 gram, 1 liter, etc.

Also, such a rate is more specifically called a **rate of change**.

Now, the slope **a** tells us how rapidly or slowly (in fact, big or small) the output changes as the input changes. In turn, it can tell us how the line is slanted to the axis of the input (the horizontal axis) in the graph. Also, the slope **a** tells us to which direction the line is slanted: if the **a** is positive, the line is slanted to the left, and if the **a** is negative, the line is slanted to the right.

Now, the slope **a** itself can be either positive, 0, or negative. Therefore, the **slope a** can be taken for a **velocity**. Then, the **magnitude** of the slope **a** can be said to be a **speed**.

If the slope is positive, the output increases as the input increases. Then, the velocity is positive, too. On the other hand, if the slope is negative, the output decreases as the input increases. Then, the velocity is negative, too. What if the slope is 0? Then, though the input changes, there is no change in output; in other words, the output is constant. Then, what kind of line is it? It is a horizontal line which is parallel to the axis of input variable, that is, the x-axis in the x-y system, for instance.

Now, a slope is a rate, and also, can be taken for a velocity.
In addition, such a rate is more specifically called a **rate of change**.
Then, if the rate of change increases, the velocity increases. Consequently, the slope increases, too.
On the other hand, if the rate of change decreases, the velocity decreases, and consequently, the slope decreases, too. Therefore, a rate of change, a velocity, and a slope can be the same as each other.

Suppose now, that $\mathbf{b = 0}$ in the standard equation $\mathbf{y = ax + b}$. Then, the point $(\mathbf{x_1, y_1})$ is at the origin where the two coordinates are 0. Then, $\mathbf{x_1 = 0}$ and $\mathbf{y_1 = 0}$. Then, the line includes the origin. Then, those students who have studied the **Book 2: Graph Operations** can readily see that $\mathbf{y - y_1 = a(x - x_1)}$ is the result of the translation (parallel transformation) of the equation of the line $\mathbf{y = ax}$. Also, we can say that it is the equation of a line that includes a point $(\mathbf{x_1, y_1})$ and has the slope of **a**.

Fig. 0.0.4

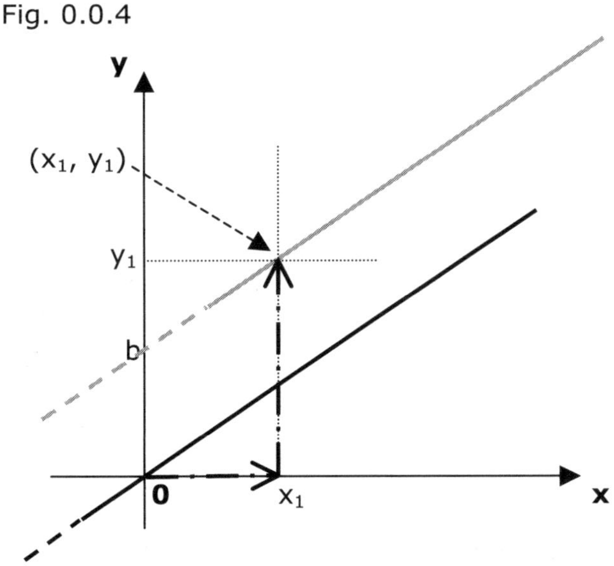

Now, what do we need when we find the equation of a particular line? At least, we need either two points in the line or a point in the line and the slope of the line. Now, why do we need the two points? That's because the slope can be determined by two points. Then, when we work on problems with finding equations of line, we should be able to somehow extract such information from the problem descriptions. Sometimes, the information is explicit, but some other times, some part or all of the information is implicit or hidden somewhere in the problem description. Let's take some examples, now.

Examples

in Single Lines

0.Suppose that the three points $P_1(-2k - 1, 5)$, $P_2(1, k + 3)$, and $P_3(k + 1, k - 1)$ are in a line, and then, find the value of the **k**.

1. Suppose that **s** and **t** are nonzero constants, and then, find the equation of a line including two points **(s, 0)** and **(0, t)** in the x-y plane.

2. Find the equation of a line that includes two points as follows: **(1, 2)** and **(3, 5)**.

3. Suppose that a line includes a point **(-2, 9)**, and that the slope is 3, and then, find the equation of the line.

4. Suppose that a line **L** includes two points **(2, 1)** and **(3, 5)**, and that a line **N** includes a point **(1, 1)** and is perpendicular to the line **L**, and then, find the equation of the line **N**.

5. Suppose that **a** and **b** are nonzero constants, and that:
The equation of a line **A** is $3x + y/a - 2 = 0$,
The equation of a line **B** is $5x - by + 3 = 0$, and
The equation of a line **C** is $2x + y/(1 - 2b) - 5 = 0$.

Suppose also, that the line **A** is perpendicular to the line **B** and is parallel to the line **C**.

Then, find the values of the **a** and the **b**.

6. Suppose that a line includes two points **(1, 2)** and **(3, 5)**, and then, find the distance from the line to a point **(-5, 3)**.

7. Suppose that a line includes a point **(1, 3)** and the slope is 2. Suppose also, that one of the two points symmetric about the line is **(3, 2)**. Then, find the other point.

8. Suppose that a line includes a point at **(2, 5)** and the slope is 2. Then, find the point **(s, t)** from which the distance is 5 to the line.

9. Find a point **(s, t)** the same distance away from a point **(-2, 1)** as away from a point **(3, 2)**.

A. Suppose that:
We have six points $\mathbf{P_1(1, 1)}$, $\mathbf{P_2(1, 3)}$, $\mathbf{P_3(2, 1)}$, $\mathbf{P_4(4, 1)}$, $\mathbf{P_5(s, t)}$, and $\mathbf{P_6(u, v)}$ in the x-y plane.
The $\mathbf{P_5}$ and the $\mathbf{P_6}$ are above the $\mathbf{P_3}$ and the $\mathbf{P_4}$, and also, on the right of the $\mathbf{P_2}$.
We make a triangle $\mathbf{P_5P_1P_2}$ and another triangle $\mathbf{P_5P_3P_4}$ so that the sum of the areas taken by the two triangles is 3.

A.0. Specify the point $\mathbf{P_5}$.

A.1. Suppose also that:
$\mathbf{D_1}$ is the distance between the $\mathbf{P_6}$ and the $\mathbf{P_4}$.
$\mathbf{D_2}$ is the distance between the $\mathbf{P_6}$ and the $\mathbf{P_2}$.
$\mathbf{D_1^2 - D_2^2 = 5}$.

Now, specify the point $\mathbf{P_6}$.

B. Suppose that the distance between two points $\mathbf{P_1}$ and $\mathbf{P_2}$ in the x-y plane is 5.
Suppose also that:
$\mathbf{D_1}$ is the distance between the $\mathbf{P_1}$ and another point $\mathbf{P_3}$.
$\mathbf{D_2}$ is the distance between the $\mathbf{P_2}$ and the $\mathbf{P_3}$.
$\mathbf{D_1^2 - D_2^2 = 10}$.

Then, specify the point $\mathbf{P_3}$.

C. Suppose that we have 4 points $\mathbf{P_1(-2, 3)}$, $\mathbf{P_2(-2, -1)}$, $\mathbf{P_3(2, 2)}$, and $\mathbf{P_4\ (-2, 0)}$ in the x-y plane and make a triangle **T** connecting the three points $\mathbf{P_1}$, $\mathbf{P_2}$, and $\mathbf{P_3}$. Suppose also, that **A** is a line connecting the $\mathbf{P_1}$ and $\mathbf{P_2}$, and that another point $\mathbf{P_5}$ is in the **A**.

C.0. If we draw a line that includes the $\mathbf{P_5}$ and is perpendicular to the line **A**, then the perpendicular line divides the triangle into two equal parts in area. Now, find the position of the $\mathbf{P_5}$.

C.1. If we draw a line **B** that includes the $\mathbf{P_4}$, then the line **B** divides the **T** into two equal parts in area. Now, find the equation of the line **B**.

In the next several examples, we are going to practice putting lines in a graph. Graphs are helpful when we solve problems with equations and functions, and are crucial in many problems.

D. Suppose that we have an equation of a line: $\mathbf{y = (t/4 + 1)x + t + 4}$ where **t** is constant.
Then, find the values of the **t** such that the $\mathbf{y > 0}$ for the $\mathbf{x > -3}$.

E. Suppose that we have an equation of a line: $\mathbf{y = (t/4 + 1)x + t + 5}$ where **t** is constant.
Then, find the values of the **t** such that the $\mathbf{y > 0}$ for the $\mathbf{x > -3}$.

F. Suppose that we have an equation of a line: $\mathbf{y = (t/4 + 1)x + t + 4}$ where **t** is constant.
Then, find the values of the **t** such that the $\mathbf{y \geq 0}$ for the $\mathbf{x > -3}$.

G. Suppose that we have an equation of a line: $y = (t/4 + 1)x + t + 5$ where **t** is constant. Then, find the values of the **t** such that the $y \geq 0$ for the $x > -3$.

H. Suppose that we have an equation of a line: $y = (t/4 + 1)x + t + 3$ where **t** is constant. Then, find the values of the **t** such that the $y \geq 0$ for the $x > -3$.

I. Suppose that we have an equation of a line: $y = (t/4 + 1)x + t + 3$ where **t** is constant. Then, find the values of the **t** such that the $y > 0$ for $-2 < x < 3$.

J. Suppose that we have an equation of a line: $y = (t/4 + 1)x + t + 4$ where **t** is constant. Then, find the values of the **t** such that the $y > 0$ for $-2 < x < 3$.

K. Suppose that we have an equation of a line: $y = (t/4 + 1)x + t + 4$ where **t** is constant. Then, find the values of the **t** such that the **y** is positive for $-5 < x < -3$.

L. Suppose that we have an equation of a line: $y = (t/4 + 1)x + t + 4$ where **t** is constant. Then, find the values of the **t** such that the **y** is positive for $-5 \leq x \leq -3$.

M. Suppose that we have an equation of a line: $y = (t/4 + 1)x + t + 4$ where **t** is constant. Then, find the values of the **t** such that the $y \geq 0$ for $-5 < x < -3$.

N. Suppose that we have an equation of a line: $y = (t/4 + 1)x + t + 4$ where **t** is constant. Then, find the values of the **t** such that the $y \geq 0$ for $-5 \leq x \leq -3$.

O. Suppose that we have an equation of a line: $y = (t/4 + 2)x + t + 3$ where **t** is constant. Then, find the values of the **t** such that the **y** is negative, 0, or positive for $-5 \leq x \leq -3$.

P. Suppose that we have an equation of a line: $y = (t/4 + 2)x + t + 3$ where **t** is constant. Then, find the values of the **x** such that the $y > 0$ for $1 \leq t \leq 5$.

Q. Suppose that we have an equation of a line: $y = 3(t - 1)x + 3 - 2t$ where **t** is constant. Then, find the values of the **x** such that the $y > 0$ for $0 < t < 2$.

It is often a good idea to construct graphs when we work on problems. Graphs help us see better. What do they help us see better? We can see better the problem we are working on, and therefore, we get to see better where we are in the problem, where we are going in search of the solution, and in turn, where the solution is likely around. Quite often, we can get across the solution while we are constructing the graph. The more accurate, the better, of course! However, schematic drawing is frequently enough to figure out the problem we are working on and find out about what we need to do about it.

Now, let's put some lines in graphs.

R. Suppose that $\mathbf{y = f(x) = 2x + 1}$ for $\mathbf{-1/2 \le x \le 2}$, that $\mathbf{y = f(x) = 9 - 2x}$ for $\mathbf{2 < x \le 9/2}$, and that, $\mathbf{g(x) = (f \bullet f)(x)}$. Then, construct the graph of the $\mathbf{g(x)}$.

Note:
An absolute value is a magnitude, which can be either 0 or positive. A value can be either positive, 0, or negative. When we want to indicate the magnitude of a value, we use a pair of short vertical bars. Therefore, $\mathbf{|v|}$ is called the absolute value of the $\mathbf{v}$, and the two bars **| |** are called absolute signs. For instance, since $\mathbf{|x|}$ indicates the magnitude of the value of the $\mathbf{x}$, we get $\mathbf{|x| = 1}$ not only when $\mathbf{x = 1}$ but when $\mathbf{x = -1}$. For another instance, since $\mathbf{|x - 3|}$ and $\mathbf{|x - y|}$ indicate the magnitudes of the values of $\mathbf{(x - 3)}$ and $\mathbf{(x - y)}$, if $\mathbf{x = 3}$ and $\mathbf{y = 5}$, we get $\mathbf{|x - 3| = 0}$ and $\mathbf{|x - y| = 2}$. Therefore, absolute values are greater than or equal to 0. Now, we need to consider both the case where $\mathbf{x \ge 0}$ and the case where $\mathbf{x < 0}$.

S. Construct a graph of each of the equations as follows:

S.0. $\mathbf{y = |3 - 2x| - 1}$.

S.1. $\mathbf{y = |2x - 3| + 3x + 1}$.

S.2. $\mathbf{y = x - |3x - 2|/2}$.

S.3. $\mathbf{2x + |y/2| = 2}$.

S.4. $\mathbf{|2x| + y/2 = 2}$.

S.5. $\mathbf{|2x| + |y/2| = 2}$.

S.6. $\mathbf{y = (x^2)^{1/2}} = \sqrt{x^2}$.

S.7. $\mathbf{y = (x^2)^{1/2}/x} = \sqrt{x^2}/x$.

S.8. $\mathbf{y = (x^2)^{1/2}/|x|} = \sqrt{x^2}/|x|$.

S.9. $\mathbf{y = (x^2 - 5x + 25/4)^{1/2}} = \sqrt{x^2 - 5x + 25/4}$.

S.A. $\mathbf{y = 2x\{|x^2 - 1|/(x^2 - 1)\}}$.

S.B. $\mathbf{y = ((2x + 1)^2)^{1/2} + ((2x - 1)^2)^{1/2}} = \sqrt{(2x+1)^2} + \sqrt{(2x-1)^2}$.

S.C. $\mathbf{||2x| - |3y|| = 1}$.

S.D. $\mathbf{y = |2x - 1| + |3x + 2|}$.

S.E. $\mathbf{y = ||2x - 1| - |3x - 1||}$.

T. Suppose that three points $\mathbf{P_1(x_1, y_1)}$, $\mathbf{P_2(x_2, y_2)}$, and $\mathbf{P_2(x_3, y_3)}$ in the x-y plane are in a line, and find the connective expression among the coordinates.

U. Suppose that the distance from a point $\mathbf{P_1(2, 1)}$ to a particular line is the same as the distance from another point $\mathbf{P_2(-2, -3)}$ to the particular line, and find the equation of the particular line.

V. Suppose that:
T: (x, y) ⟶ (3x – y, x – y).
y = f(x) = ax where the **a** is a constant.
T•f = f.

Then, find the value of the **a**.

W. Suppose that:
T: (x, y) ⟶ (3x – 2y, x + y).
y = f(x) = ax where the **a** is a constant.
T•f = f.

Then, find the value of the **a**.

X. Suppose that
T: (x, y) ⟶ (2x + y + 3, 3x + 2y – 2).
y = f(x) = ax + b where the **a** and the **b** are constants.
T•f = f.

Then, find the values of the **a** and the **b**.

Suggestions or Solutions

To the Problem in the Example 0

Suppose that the three points $P_1(-2k - 1, 5)$, $P_2(1, k + 3)$, and $P_3(k + 1, k - 1)$ are in a line, and find the value of the k.

Since all the three points are in a line, they have to satisfy an equation of line.
Then, we need to apply the points to an equation of line. That is, we put the points into the equation.
Then eventually, we get to have an equation for the **k**.

Then, of the general form and standard form which one should we use?
The general form is $ax + by + c = 0$, and the standard form is $y = ax + b$.

If we use the general form, 3 equations get made but we get 4 unknowns which are **a**, **b**, **c**, and **k**.
In this case, we get troubled since the number of equations is less than that of the unknowns.

Let's use the standard form, then.
How do we use it?
By means of the standard equation, we connect the coordinates of the three points to make a connective expression. Then, the connective expression will be the equation for the **k**.

Suppose now, that the equation of the line is $y = ax + b$.
Suppose also, that $x_1 = -2k - 1$, $y_1 = 5$, $x_2 = 1$, $y_2 = k + 3$, $x_3 = k + 1$, and $y_3 = k - 1$.

Then, since the P_1 is in the line, $y = ax + b \Rightarrow y_1 = ax_1 + b \Rightarrow b = y_1 - ax_1$.
So, $y = ax + b \Rightarrow y = ax + y_1 - ax_1 \Rightarrow y - y_1 = a(x - x_1)$.

Also, since the P_3 is in the line, we can see that $y_3 - y_1 = a(x_3 - x_1)$.
Now, let's use the P_1 and the P_2 to find the slope **a** of the line.

Then, $a = (y_2 - y_1)/(x_2 - x_1)$.
Then, $y_3 - y_1 = a(x_3 - x_1) = \{(y_2 - y_1)/(x_2 - x_1)\}(x_3 - x_1) = (y_2 - y_1)(x_3 - x_1)/(x_2 - x_1)$
$\Rightarrow (x_2 - x_1)(y_3 - y_1) = (y_2 - y_1)(x_3 - x_1)$.

Now, we put the coordinate values into the expression above. Then, we can get the equation for the **k**.

$(x_2 - x_1)(y_3 - y_1) = \{1 - (-2k - 1)\}\{(k - 1) - 5\} = (1 + 2k + 1)(k - 6) = (2k + 2)(k - 6)$
$= 2k^2 - 10k - 12$.

$(y_2 - y_1)(x_3 - x_1) = \{(k + 3) - 5\}\{(k + 1) - (-2k - 1)\} = (k - 2)(3k + 2) = 3k^2 - 4k - 4$.

Then, $2k^2 - 10k - 12 = 3k^2 - 4k - 4 \Rightarrow k^2 + 6k + 8 = 0 \Rightarrow (k + 2)(k + 4) = 0$.
Therefore, $k = -2$ or -4.

Putting the value of the **k** into each of the points we can see that:
(3, 5), **(1, 1)**, and **(-1, -3)** are in a line.
(7, 5), **(1, -1)**, and **(-3, -5)** are in a line.

In short:

Suppose that the equation of line is $y = ax + b$.
Suppose also, that $x_1 = -2k - 1$, $y_1 = 5$, $x_2 = 1$, $y_2 = k + 3$, $x_3 = k + 1$, and $y_3 = k - 1$.

Then, since P_1 is in the line, $y = ax + b \Rightarrow y_1 = ax_1 + b \Rightarrow b = y_1 - ax_1$.
Then, $y = ax + b \Rightarrow y = ax + y_1 - ax_1 \Rightarrow y - y_1 = a(x - x_1)$.

Also. since P_3, too, is in the line, we can see that $y_3 - y_1 = a(x_3 - x_1)$.
Now, let's use P_1 and P_2 to find the slope of the line.

Then, $a = (y_2 - y_1)/(x_2 - x_1)$.
Then, $y_3 - y_1 = a(x_3 - x_1) = \{(y_2 - y_1)/(x_2 - x_1)\}(x_3 - x_1) = (y_2 - y_1)(x_3 - x_1)/(x_2 - x_1)$
$\Rightarrow (x_2 - x_1)(y_3 - y_1) = (y_2 - y_1)(x_3 - x_1)$.

$(x_2 - x_1)(y_3 - y_1) = \{1 - (-2k - 1)\}\{(k - 1) - 5\} = (1 + 2k + 1)(k - 6) = (2k + 2)(k - 6)$
$= 2k^2 - 10k - 12$.

$(y_2 - y_1)(x_3 - x_1) = \{(k + 3) - 5\}\{(k + 1) - (-2k - 1)\} = (k - 2)(3k + 2) = 3k^2 - 4k - 4$.

Consequently, $2k^2 - 10k - 12 = 3k^2 - 4k - 4 \Rightarrow k^2 + 6k + 8 = 0 \Rightarrow (k + 2)(k + 4) = 0$.
Therefore, $k = -2$ or -4.

Suggestions or Solutions

To the Problem in the Example 1

Suppose that s and t are nonzero constants, and find the equation of a line including two points (s, 0) and (0, t) in the x-y plane.

(s, 0) is the point where the x-intercept is, and **(0, t)** is the point where the y-intercept is.
Let's put the two points into the standard form and express the slope and the y-intercept in terms of the **s** and **t**.
Suppose now, that the equation of the line is $y = ax + b$ where **a** and **b** are constant.
Then, the slope $a = \Delta y/\Delta x = (t - 0)/(0 - s) = -t/s$.
So, we get $y = (-t/s)x + b$.

Now, let's use the **(s, 0)** to find the **b**.

$0 = (-t/s)s + b \Rightarrow b = t$.

Therefore, the equation is $y = (-t/s)x + t$.

Now, let's make the equation look a little better.
Dividing both sides by the **t**, we get $y/t = -x/s + 1 \Rightarrow x/s + y/t = 1$.
So, the equation has the new look of $x/s + y/t = 1$.

Therefore, the equation $x/s + y/t = 1$ is the equation of the line where the **s** is the x-intercept and the **t** is the y-intercept.

In short:

The standard form is $y = ax + b$ where the **a** and **b** are constant.

The slope $a = \Delta y/\Delta x = (t - 0)/(0 - s) = -t/s$.
Then, we get $y = (-t/s)x + b$.

Since **(s, 0)** is in the line, $0 = (-t/s)s + b \Rightarrow b = t$.

Then, $y = (-t/s)x + t \Rightarrow y/t = -x/s + 1$.
Therefore, the equation of the line is $x/s + y/t = 1$.

Suggestions or Solutions
To the Problem in the Example 2

Find the equation of a line that includes two points as follows: (1, 2) and (3, 5).

The equation of the line that includes the point **(s, t)** and has the slope of **a** is **y − t = a(x − s)**.
How come?
Suppose that the standard form is **y = ax + b**.
Then, **t = as + b ⇒ b = t − as**.
So, **y = ax + t − as = a(x − s) + t ⇒ y − t = a(x − s)**.

The equation of the line that includes the point **(1, 2)** and has the slope of **a** is **y − 2 = a(x − 1)**.

We can use the two points **(1, 2)** and **(3, 5)** to find the slope.
Then, the **a = Δy/Δx = (5 − 2)/(3 − 1) = 3/2**.
Therefore, the equation of the line is **y − 2 = (3/2)(x − 1)**.

Moving the line **y = (3/2)x** by 1 in the direction of the x-axis and by 2 in the direction of the y-axis, we get the line above.
It is called a parallel transformation.

Fig. 2.0

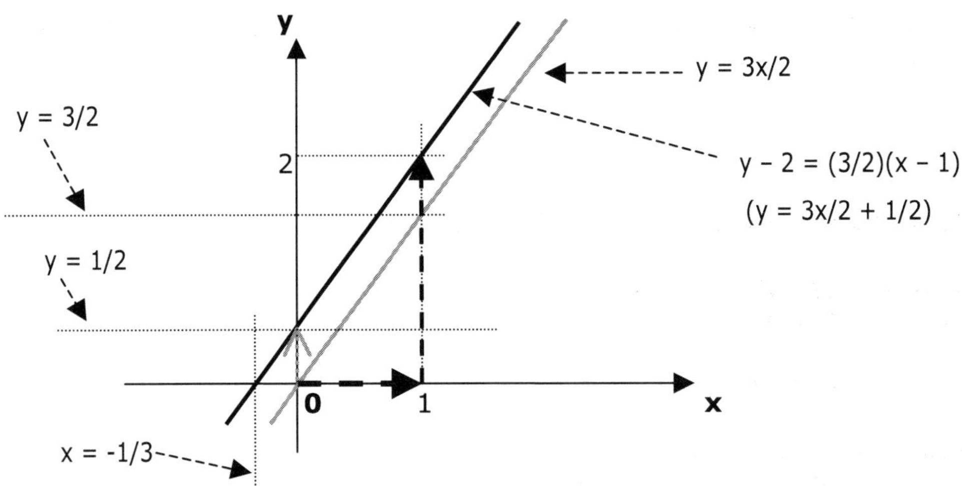

In short:

Since the line includes the point **(1, 2)** and the slope is **a**, the equation of the line is **y − 2 = a(x − 1)**.

Since the line includes **(1, 2)** and **(3, 5)**, **a = Δy/Δx = (5 − 2)/(3 − 1) = 3/2**.

Therefore, the equation of the line is **y − 2 = (3/2)(x − 1)**.

Suggestions or Solutions

To the **Problem** in the Example **3**

Suppose that a line includes a point (-2, 9), and that the slope is 3, and find the equation of the line.

Since the point in the line and the slope are specified, we can readily get the equation using the standard form. Since the slope is 3, suppose that the equation is **y = 3x + b**.
Since the line includes the point **(-2, 9)**, we can see that **9 = 3(-2) + b = -6 + b**.
Then, we get **b = 9 + 6 = 15**.
Therefore, the equation is **y = 3x + 15**, which can be interpreted to be the result of moving the line **y = 3x** by **15** in the direction of the y-axis.

Now, the equation of a line including a point **(s, t)** and having the slope of **a** is **y – t = a(x – s)**.
Then, since the line includes **(-2, 9)** and the slope is 3, the equation is **y – 9 = 3(x – (-2)) = 3(x + 2)**.
Therefore, it can be presented in **y – 9 = 3(x + 2)**, too.
Moving the line **y = 3x** by –2 in the direction of x-axis and by 9 in the direction of y-axis, we get the line above. Expanding the equation and simplifying the result, we will see that it is the same as **y = 3x + 15**.

Fig. 3.0.

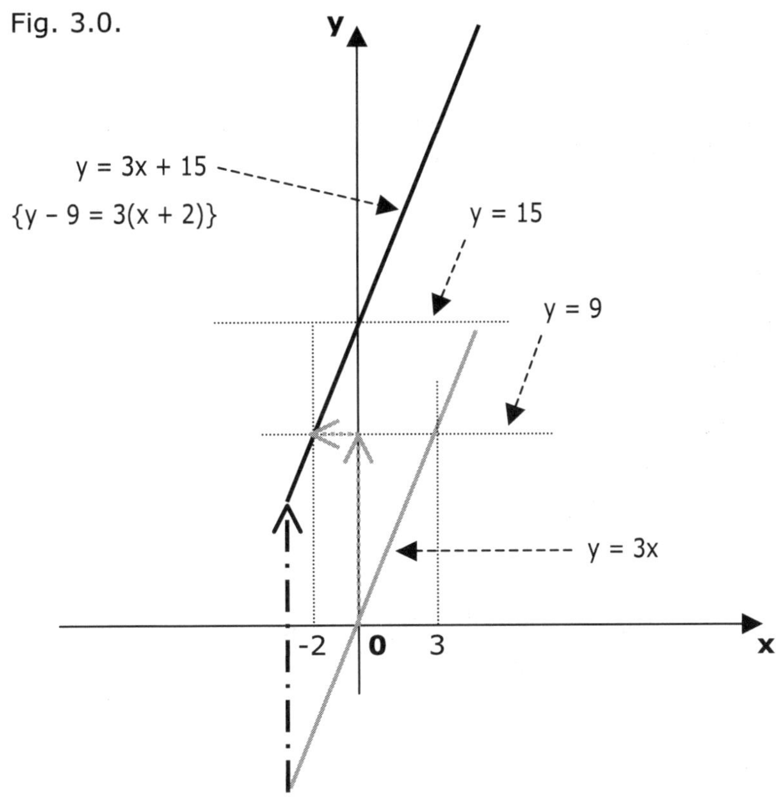

Suggestions or Solutions

To the Problem in the Example 4

Suppose that a line L includes two points (2, 1) and (3, 5), and that a line N includes (1, 1) and is perpendicular to the line L, and find the equation of the line N.

The two lines **L** and **N** are perpendicular to each other. Then, we need to find the condition that two lines get perpendicular to each other. Suppose that the two lines are perpendicular to each other as shown in the graph below.

Fig. 4.0

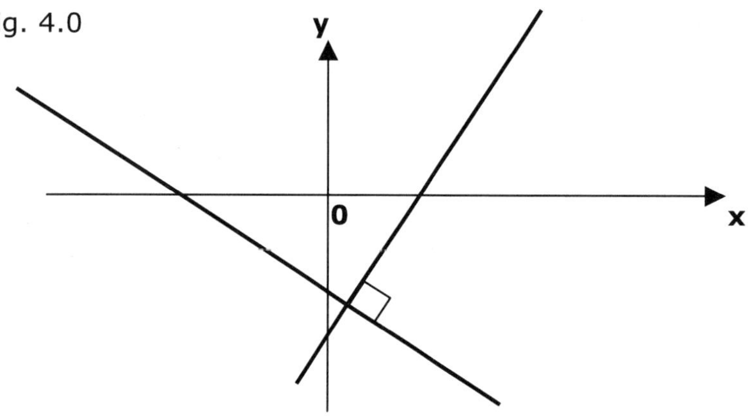

If we move the lines in parallel manner, their directions remain the same, and they still remain perpendicular to each other. Now, we move the lines in such a way shown below.

Fig. 4.1

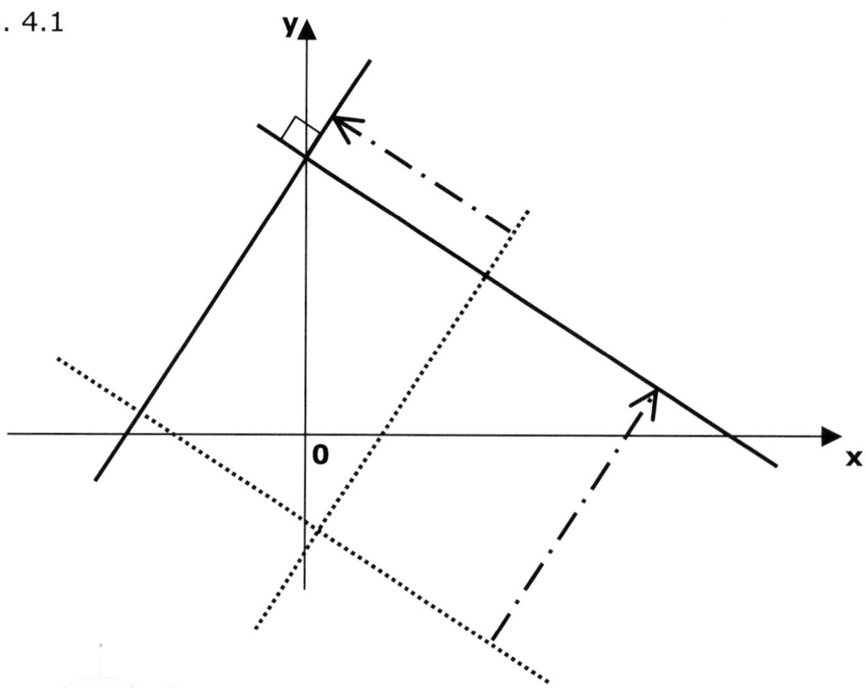

Fig. 4.2

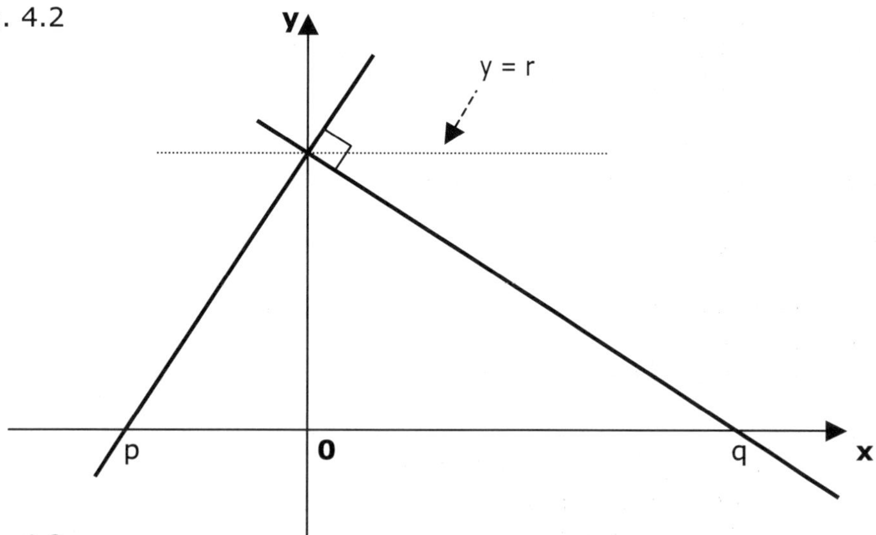

Fig. 4.3

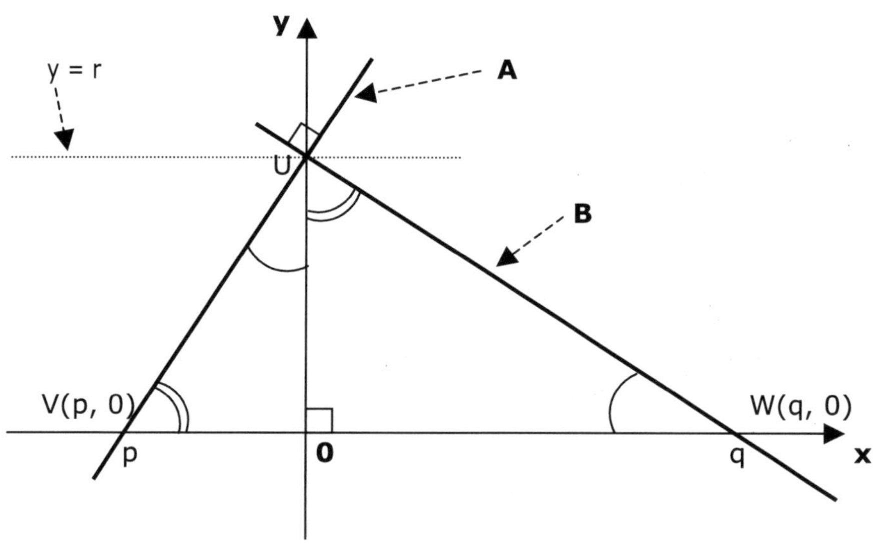

Now, let's fine the slope of each of the lines. We know that the slope is defined to be Δ**y/**Δ**x**.
We can see in the graph above that:
The line **A** meets the x-axis at **V(p, 0)**, and the line **B** meets the x-axis at **W(q, 0)**.
Both of the lines meet each other at **U(0, r)** in the y-axis.

Suppose now, that the origin is **O**. Then, since **OU = r**, **UO = -r**, and also, since **OV = p**, **VO = -p**. So:
The slope of the line **A** is $\Delta y/\Delta x = OU/VO = r/(-p) = -r/p > 0$ since $r > 0$ but $p < 0$.
The slope of the line **B** is $\Delta y/\Delta x = UO/OW = (-r)/q = -r/q < 0$ since both **q** and **r** are positive.
Now, the triangle **UOV** and the triangle **WOU** are similar to each other.
Then, $|OW|/|OU| = |OU|/|OV| = |OU|/|VO|$. (Note: $|x|$ means the magnitude of the **x**.)
So, since $|OW|/|OU| = q/r$, $|OU|/|VO| = q/r$.
Now, **OU/VO** is the slope of the line **A**, which is positive.
Therefore, $OU/VO = q/r$ since $q/r > 0$.

Then, **(OU/VO)(UO/OW) = (q/r)(-r/q) = -1**.
In other words, the product of the slope of the line **A** and the slope of the line **B** is -1.
Therefore, if two lines are perpendicular to each other, then the product of the slopes is -1.

Now, let's find the equation of the line **N**.
Since we know that the point **(1, 1)** is in the **N**, if we find the slope, we can get the equation of **N**.

Since the product of the slope of **N** and the slope of **L** is −1, we can get the slope of **N** from that of **L**.

Since we have two points the **L** includes, we can readily see the slope of the **L**.
The two points are **(2, 1)** and **(3, 5)**.
Then, the slope of the **L** = **$\Delta y/\Delta x$ = (5 − 1)/(3 − 2) = 4/1 = 4**. So, the slope of the **N** = **-1/4**.

We know that the **N** includes **(1, 1)**.
Then, we can see that the equation of the **N** is **y − 1 = (-1/4)(x − 1)**.
If we put it in the standard form, **y = (-1/4)x + 1/4 + 1 = (-1/4)x + 5/4** ⇒ **y = (-1/4)x + 5/4**.
For reference, the equation of the **L** is **y − 1 = 4(x − 2)** ⇒ **y = 4x − 8 + 1 = 4x − 7** ⇒ **y = 4x − 7**.

Fig. 4.4

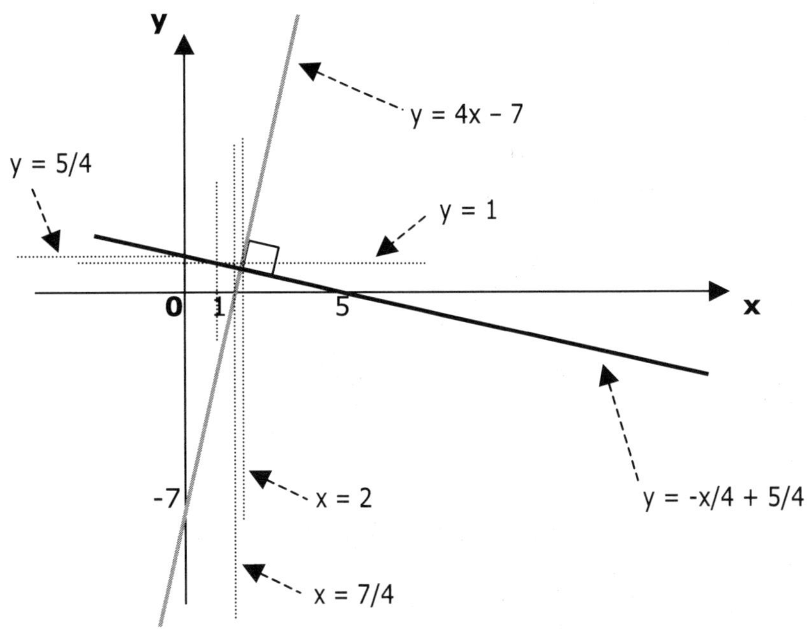

In short:

Since the **N** and the **L** are perpendicular to each other, the product of the slopes of the **N** and the **L** is −1.
The slope of the **L** = **$\Delta y/\Delta x$ = (5 − 1)/(3 − 2) = 4/1 = 4**. So, the slope of the **N** = **-1/4**.
Since the **N** includes **(1, 1)**, the equation of the **N** is **y − 1 = (-1/4)(x − 1)**.

Suggestions or Solutions

To the Problem in the Example 5

Suppose that a and b are nonzero constants, and that:
The equation of a line A is $3x + y/a - 2 = 0$,
The equation of a line B is $5x - by + 3 = 0$, and
The equation of a line C is $2x + y/(1 - 2b) - 5 = 0$.

Suppose also, that the line A is perpendicular to the line B and is parallel to the line C.

Then, find the values of the a and the b.

We know that:
The product of the slopes of two lines perpendicular to each other is -1.
The slopes of the two lines parallel to each other are the same.

Let's examine the slopes, then.
To begin with, let's put the equations given in standard form.

The line **A:** $3x + y/a - 2 = 0 \Rightarrow y = -3ax + 2a$.
The line **B:** $5x - by + 3 = 0 \Rightarrow y = 5x/b - 3/b$.
The line **C:** $2x + y/(1 - 2b) - 5 = 0 \Rightarrow y = 2(2b - 1)x + 5(2b - 1)$.

Then, the slope of the **A** = $-3a$, and the slope of the **B** = $5/b$.
Since the **A** is perpendicular to the **B**, $(-3a)(5/b) = -1 \Rightarrow 15a = b$.

Besides, the slope of the **C** = $2(2b - 1)$, and since the **A** is parallel to the **C**, the **B** is perpendicular to the **C**. Therefore, we can see that $(5/b)2(2b - 1) = -1$.

$(5/b)2(2b - 1) = -1 \Rightarrow 10(2b - 1) = -b \Rightarrow 20b - 10 + b = 0 \Rightarrow 21b = 10 \Rightarrow b = 10/21$.

We have $15a = b$. Then, $a = b/15 = (10/21)/15 = 10/(21X15) =2/(21X3) = 2/63$.
Therefore, $a = 2/63$, and $b = 10/21$.

In short:

Line **A:** $3x + y/a - 2 = 0 \Rightarrow y = -3ax + 2a$.
Line **B:** $5x - by + 3 = 0 \Rightarrow y = 5x/b - 3/b$.
Line **C:** $2x + y/(1 - 2b) - 5 = 0 \Rightarrow y = 2(2b - 1)x + 5(2b - 1)$.
Then, the slope of the **A** = $-3a$, the slope of the **B** = $5/b$, the slope of the **C** = $2(2b - 1)$.
Since the **A** is perpendicular to the **B**, $(-3a)(5/b) = -1 \Rightarrow 15a = b$.
Since the **B** is perpendicular to the **C**, $(5/b)2(2b - 1) = -1 \Rightarrow 10(2b - 1) = -b \Rightarrow 21b = 10 \Rightarrow b = 10/21$.
$15a = b \Rightarrow a = b/15 = (10/21)/15 = 10/(21X15) =2/(21X3) = 2/63$.
Therefore, $a = 2/63$, and $b = 10/21$.

Suggestions or Solutions
To the Problem in the Example 6

Suppose that a line includes two points (1, 2) and (3, 5), and find the distance from the line to a point (-5, 3).

The distance between two points is the length of the line segment connecting the two points.
Then, how do we define the distance from a particular point to a line?
In math, the distance is the shortest distance between the line and the particular point.
What do we mean by the shortest distance?
Suppose that **A** is the line that includes the particular point and is perpendicular to the line.
Suppose also, that **B** is the point where the **A** meets the line.
Then, the shortest distance is the distance between the particular point and the point **B**.

Now, suppose that **L** is the line including the two points **(1, 2)** and **(3, 5)**, and that **P** is the point **(-5, 3)**.
Then, we need to find the line that includes the **P** and is perpendicular to the **L**.
Suppose now, that the perpendicular line is **N**.
Suppose also, that **Q** is the point where the **L** meets the **N**.
Then, the distance between the **L** and the **P** is the distance between the **P** and the **Q**, that is, the length of the line segment connecting the two points **P** and **Q**.

We can find the length by means of the famous distance formula, that is, the Pythagorean Theorem.
The theorem says that in a right triangle, the square of the hypotenuse is the same as the sum of the squares of the other two respective sides.
That is, the square of the distance between two points is the sum of the squares of the differences between the respective **x** and **y** coordinates of the two points.
For instance, if **D** is the distance between two points, then we get $\mathbf{D}^2 = (\Delta \mathbf{x})^2 + (\Delta \mathbf{y})^2$.

Fig. 6.0

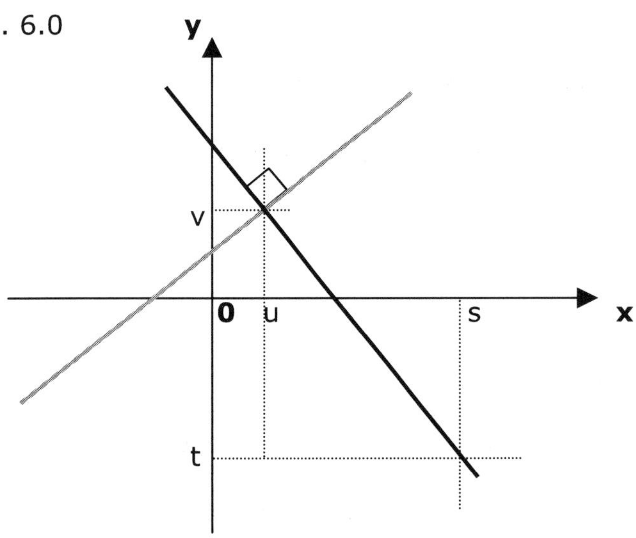

Suppose that in the graph above, **D** is the distance between **(s, t)** and **(u, v)**.
Then, $\mathbf{D}^2 = (\Delta x)^2 + (\Delta y)^2 = (s - u)^2 + (t - v)^2 = (u - s)^2 + (v - t)^2 \Rightarrow \mathbf{D} = \{(s - u)^2 + (t - v)^2\}^{1/2}$.
Therefore, $\mathbf{D} = \sqrt{(s-u)^2 + (t-v)^2}$.

Now, let's find the distance between the **P** and the **Q**.
We need to find the equation of the **L** first. The **L** includes **(1, 2)** and **(3, 5)**.
Then, let's begin with the slope.
The slope of the **L** = $\Delta y/\Delta x = (5 - 2)/(3 - 1) = 3/2$.
Then, since the **L** includes **(1, 2)**, the equation of the **L** is $y - 2 = (3/2)(x - 1)$.
We know that −1 is the product of the slopes of two lines perpendicular to each other.
Suppose that **n** is the slope of the **N**. Then, we get $(3/2)n = -1$.
Then, we get $n = -2/3$.
So, since the **N** includes **(-5, 3)**, the equation the **N** is $y - 3 = (-2/3)\{x - (-5)\} = (-2/3)(x + 5)$.

Now, we need to find the point **Q** where the **L** meets the **N**.
Since the two lines **L** and **N** meet at a particular point the two lines include at the same time, the particular point is the **Q**. Then, the value of the **x** is the x-coordinate of the **Q** when the values of the **y** in the equations of **L** and **N** are the same. Then, if we put the value of the **x** into either of the two equations, the value of the **y** is the y-coordinate of the **Q**.

The equation of **L** is $y - 2 = (3/2)(x - 1) \Rightarrow y = (3/2)(x - 1) + 2 = (3/2)x + 1/2$.

The equation of **N** is $y - 3 = (-2/3)(x + 5) \Rightarrow y = (-2/3)(x + 5) + 3 = (-2/3)x - 1/3$.

Then, $(3/2)x + 1/2 = (-2/3)x - 1/3 \Rightarrow (3/2 + 2/3)x = -1/3 - 1/2$
$\Rightarrow (13/6)x = -5/6 \Rightarrow 13x = -5 \Rightarrow x = -5/13$.

We have $y = (3/2)x + 1/2$.
Then, $y = (3/2)(-5/13) + 1/2 = -15/26 + 1/2 = -2/26 = -1/13$.

Therefore, **Q** is at **(-5/13, -1/13)**. Also, the **P** is at **(-5, 3)**.

We have $D^2 = (\Delta x)^2 + (\Delta y)^2$ where the **D** is the distance.

$D^2 = \{-5/13 - (-5)\}^2 + (-1/13 - 3)^2 = (-5/13 + 5)^2 + (1/13 + 3)^2 = (60/13)^2 + (40/13)^2$
$= (60^2 + 40^2)/13^2 = 3600 + 1600 = 5200/13^2 = (13)(4)10^2/13^2 = 20^2(13)/13^2$
$\Rightarrow D = (20/13)(13)^{1/2} = 20\sqrt{13}/13 = \dfrac{20\sqrt{13}}{13}$.

Therefore, the distance is $\dfrac{20\sqrt{13}}{13}$.

Why not just $\dfrac{20}{\sqrt{13}}$? We normally prefer a rational number in the denominator. However, we can leave the solution in $\dfrac{20}{\sqrt{13}}$, too.

Of course, we can use the formula for the distance between a point and a line.

Suppose that **D** is the distance from a point **(s, t)** to a line $ax + by + c = 0$.
Then, $D = |as + bt + c|/(a^2 + b^2)^{1/2}$.

We need to know how the formula gets constructed, of course. We will find out about it in the next section.

In short:

Suppose that:
L is the line that includes the two points **(1, 2)** and **(5, 3)** and that **P** is the point **(-5, 3)**.
N is the line that includes the **P** and is perpendicular to the **L**.
Q is the point where the **L** meets the **N**.

Then, the slope of the **L** = $\Delta y/\Delta x = (5 - 2)/(3 - 1) = 3/2$.
So, since the **L** includes **(1, 2)**, the equation of the **L** is $y - 2 = (3/2)(x - 1)$.
We know that −1 is the product of the slopes of the two lines perpendicular to each other.
Suppose that the slope of the **N** is **n**. Then, $(3/2)n = -1 \Rightarrow n = -2/3$.
Since the **N** includes **(-5, 3)**, the equation of the **N** is $y - 3 = (-2/3)\{x - (-5)\} = (-2/3)(x + 5)$.

$y - 2 = (3/2)(x - 1) \Rightarrow y = (3/2)(x - 1) + 2 = (3/2)x + 1/2.$
$y - 3 = (-2/3)(x + 5) \Rightarrow y = (-2/3)(x + 5) + 3 = (-2/3)x - 1/3.$

Then, $(3/2)x + 1/2 = (-2/3)x - 1/3 \Rightarrow (3/2 + 2/3)x = -1/3 - 1/2$
$\Rightarrow (13/6)x = -5/6 \Rightarrow 13x = -5 \Rightarrow x = -5/13$.

Consequently, $y = (3/2)x + 1/2 = (3/2)(-5/13) + 1/2 = -15/26 + 1/2 = -2/26 = -1/13$.

Therefore, **Q** is **(-5/13, -1/13)** and **P** is **(-5, 3)**.

Suppose that **D** is the distance between the **P** and the **Q**. Then:

$D^2 = \{-5/13 - (-5)\}^2 + (-1/13 - 3)^2 = (-5/13 + 5)^2 + (1/13 + 3)^2 = (60/13)^2 + (40/13)^2$
$= (60^2 + 40^2)/13^2 = 3600 + 1600 = 5200/13^2 = (13)(4)10^2/13^2 = 20^2(13)/13^2$

$\Rightarrow D = (20/13)(13)^{1/2} = 20\sqrt{13}/13 = \dfrac{20\sqrt{13}}{13}$.

Therefore, the distance is $\dfrac{20\sqrt{13}}{13}$.

Suggestions or Solutions

To the Problem in the Example 7

Suppose that a line includes a point (1, 3) and the slope is 2. Suppose also, that one of the two points symmetric about the line is (3, 2). Then, find the other point.

If two points are symmetric about a line, the distance from one of the two points to the line is the same as the distance from the other to the line. In this case, the line is called the axis of symmetry. Also, the two points are in the line that is perpendicular to the axis of symmetry.

Then, putting the above in proper expressions, we should be able to find the solution.

Also, we can work out this problem in such a way as follows, and let's take the way first.

The midpoint of the two points is in the axis of symmetry.
Then, the coordinates of the midpoint are the respective averages of the x-coordinates and y-coordinates of the two points.

Suppose now, that **P** is the point **(1, 3)**, **L** is the line of slope 2 including the point **P**, and **Q(s, t)** is the point symmetric to the **P** about the **L**. Suppose also, that **R(u, v)** is the midpoint between the **P** and the **Q**. Then, the **R** is in the **L**. Also, the coordinates of the **R** are the **u** and **v**, each of which is in terms of the **s** and **t** since the **R** is the midpoint between the **P** and the **Q**, and the **Q** is **(s, t)**.
Therefore, putting the **R** into the equation of the **L**, we will eventually get an equation for the **s** and the **t**.

However, the number of the unknowns we want to find is 2.
Then, we need another equation.
What equation can it be?
We know that the midpoint **R** is included in the line perpendicular to the **L**, too.
Suppose now, that **N** is the perpendicular line. Then, the **N** includes the **P**.
Also, since the **N** is perpendicular to the **L**, the slope can be found from the slope of the **L**.
We know that –1 is the product of the slopes of two lines perpendicular to each other.
Now, putting the **R** into the equation of the **N**, we get another equation for the **s** and the **t**.

Then, solving the system of the two equations, we will get the **R**.

Now, let's get started with the equation of the **L**.
Since the **L** includes **(1, 3)** and has the slope of 2, the equation is as follows: $y - 3 = 2(x - 1)$.
Next, let's find the equation of the **N**.
Suppose that the slope of the **N** is **n**. Then, $2n = -1 \Rightarrow n = -1/2$. Then, since the **N** includes **P(3, 2)** and the slope is –1/2, the equation is as follows: $y - 2 = (-1/2)(x - 3)$.

Let's now, find the midpoint **R**.
We know that the coordinates of the **R** are the respective averages of the x-coordinates and the y-coordinates of the **P** and the **Q**.
The **P** is at **(3, 2)**, the **Q** is at **(s, t)**, and the **R** is at **(u, v)**.

Then, we get $u = (3 + s)/2$ and $v = (2 + t)/2$.

Now, the **R** has to be in both **L** and **N**.
The equation of the **L** is $y - 3 = 2(x - 1)$, and that of the **N** is $y - 2 = (-1/2)(x - 3)$.

Let's put the **R** into the **L** first. Then, $v - 3 = 2(u - 1) \Rightarrow (2 + t)/2 - 3 = 2\{(3 + s)/2 - 1\}$
$\Rightarrow 2 + t - 6 = t - 4 = 4\{(3 + s)/2 - 1\} = 2(3 + s) - 4 = 2s + 2 \Rightarrow t - 4 = 2s + 2 \Rightarrow 2s - t = -6$.

Let's now, put the **R** into the **N**. Then, $v - 2 = (-1/2)(u - 3)$
$\Rightarrow (2 + t)/2 - 2 = (-1/2)\{(3 + s)/2 - 3\} \Rightarrow 2 + t - 4 = t - 2 = -\{(3 + s)/2 - 3\}$
$\Rightarrow 2t - 4 = -(3 + s) + 6 = 3 - s \Rightarrow 2t - 4 = 3 - s \Rightarrow s + 2t = 7$.

Now, we get the system of equations as follows: $2s - t = -6$ and $s + 2t = 7$.
Let's get rid of the **t** first.
$2s - t = -6 \Rightarrow 4s - 2t = -12$. Then, $(4s - 2t) + (s + 2t) = -12 + 7 \Rightarrow 5s = -5 \Rightarrow s = -1$.
So, $s + 2t = 7 \Rightarrow -1 + 2t = 7 \Rightarrow 2t = 8 \Rightarrow t = 4$.

Therefore, **Q** is at **(-1, 4)**.

Let's put in expressions the statements below this time, and then, find the solution.

If two points are symmetric to each other about a line, the distance from one of the two point to the line is the same as the distance from the other to the line. In this case, the line is called the axis of symmetry. Also, the two points are in the line perpendicular to the axis of symmetry.

Then, we can set up equations with the two facts as below:
The fat that the distance from the given point **P** to the given line **L** is the same as the distance from the symmetric point **Q** to the **L**, and the fact that the **P** and the **Q** are in the **N** perpendicular to the **L**

Using the first fact, we need to find the point where the **N** and the **L** meet each other. In fact, the point is the midpoint **R** found above.

Now, we are going to use the equations or points as are defined or found already above, but start anew the algorithm, that is, the problem solving procedure.

Since the coordinates of the **Q** we are after are two unknowns, we will have to set up a system of two equations for the coordinates.

We need to find the equation of the **L** first, and then, the equation of the **N**.
Then, we find the point where the **L** meets the **N**, that is, the midpoint **R**.
Next, we find the distance between the given point **P** and the **R**. Let **D** be the distance.
Then, we set up an equation saying that the **D** is the same as the distance between the **R** and the symmetric point **Q**. Then, it is one equation.
Now, since the **Q** is the point that is to be in the **N**, we put the **Q** into the **N** to set up the other equation.

The equation of the **L** is as follows: $y - 3 = 2(x - 1)$.
The equation of the **N** is as follows: $y - 2 = (-1/2)(x - 3)$.

Also, naming each of the equations themselves, we can do it as follows:
$L(x, y) = 2(x - 1) - (y - 3) = 0$. $N(x, y) = (-1/2)(x - 3) - (y - 2) = 0$.

Let's now, find the midpoint R where the **L** meets the **N**.
Then, we need to solve the system of equations as follows:
L(x, y) = 0 and **N(x, y) = 0**.

$L(x, y) = 0 \Rightarrow 2(x - 1) - (y - 3) = 0 \Rightarrow y - 3 = 2(x - 1) \Rightarrow y = 2x - 2 + 3 = 2x + 1 \Rightarrow y = 2x + 1.$

$N(x, y) = 0 \Rightarrow (-1/2)(x - 3) - (y - 2) = 0 \Rightarrow y - 2 = (-1/2)(x - 3) \Rightarrow y = -x/2 + 3/2 + 2$
$\Rightarrow y = -x/2 + 7/2.$

Therefore, $2x + 1 = -x/2 + 7/2 \Rightarrow 4x + 2 = -x + 7 \Rightarrow 5x = 5 \Rightarrow x = 1$.
Then, $y = 2x + 1 = 3$.
So, the **R** is at **(1, 3)**.

Let's now, find the distance **D** between the **P** and **R**. The **P** is at (3, 2).
We know that $D^2 = (\Delta x)^2 + (\Delta y)^2$.

Since in calculations of distance between two points, Δx and Δy get squared, it doesn't matter to subtract which of the two points from the other. Suppose we subtract the **R** from the **P**. Then:
$D^2 = (3 - 1)^2 + (2 - 3)^2 = 4 + 1 = 5 \Rightarrow D^2 = 5.$

We now, set up an equation where the **D** is the same as the distance between the **R(1, 3)** and the symmetric point **Q(s, t)**. Then, it is one in the system of equations. Then:
$D^2 = 5 = (s - 1)^2 + (t - 3)^2.$

Now, since **Q(s, t)** is the point that is to be in the **N**, we put the **Q** into the **N** to set up the other equation:
$N(x, y) = (-1/2)(x - 3) - (y - 2) = 0 \Rightarrow y - 2 = (-1/2)(x - 3) \Rightarrow y = -x/2 + 7/2$
$\Rightarrow t = -s/2 + 7/2.$

Then, we get the system of equations as follows: $5 = (s - 1)^2 + (t - 3)^2$ and $t = -s/2 + 7/2$.

Then, $5 = (s - 1)^2 + (-s/2 + 7/2 - 3)^2$, and $-s/2 + 7/2 - 3 = -s/2 + 1/2 = (1 - s)/2$.
So, $5 = (s - 1)^2 + (1 - s)^2/4 = (s - 1)^2 + (s - 1)^2/4 = 5(s - 1)^2/4$
$\Rightarrow 5 = 5(s - 1)^2/4 \Rightarrow (s - 1)^2 = 4.$

Now, wait a sec! Why a quadratic equation? The solution to a quadratic equation is normally composed of two roots. Then, are there two points symmetric to **(3, 2)** about the **L**?
No, but we have only one point symmetric to **(3, 2)** about the **L**. It is correct that a quadratic equation has two different roots (solutions) if the equation is not put in a complete square such as $(cs + d)^2 = 0$ where **c** and **d** are constant.
However, the equation being worked out now should not be put in a complete square.
Why not?
That's because one of the two roots has to be the one for that **P**.
So, the other is the one for the symmetric point **Q**.
Therefore, the system of equations is the system by which we find two points that are in the line **N**, and the points are among the points constituting the circle of radius $5^{1/2}$ centered at **(1, 3)**. Equations of circle are covered in the **Book 4**.

Now, let's get going.

$(s - 1)^2 = 4 \Rightarrow (s - 1 = 2 \text{ or } -2) \Rightarrow (s = 3 \text{ or } -1)$.
We now, need to fund the **t**. We have $t = -s/2 + 7/2$.
Then, when $s = 3$, we get $t = -3/2 + 7/2 = 2$.
So, $(s, t) = (3, 2)$, which is the point **P** given already.
Now, when $s = -1$, we get $t = -(-1)/2 + 7/1 = 1/2 + 7/2 = 4$.
Therefore, we can see that $Q(s, t) = Q(-1, 4)$.

Now, let's have a look at in the graph below the two points **(3, 2)** and **(-1, 4)** symmetric to each other about the line $L(x, y) = 2x - y + 1 = 0$.

Fig. 7.0

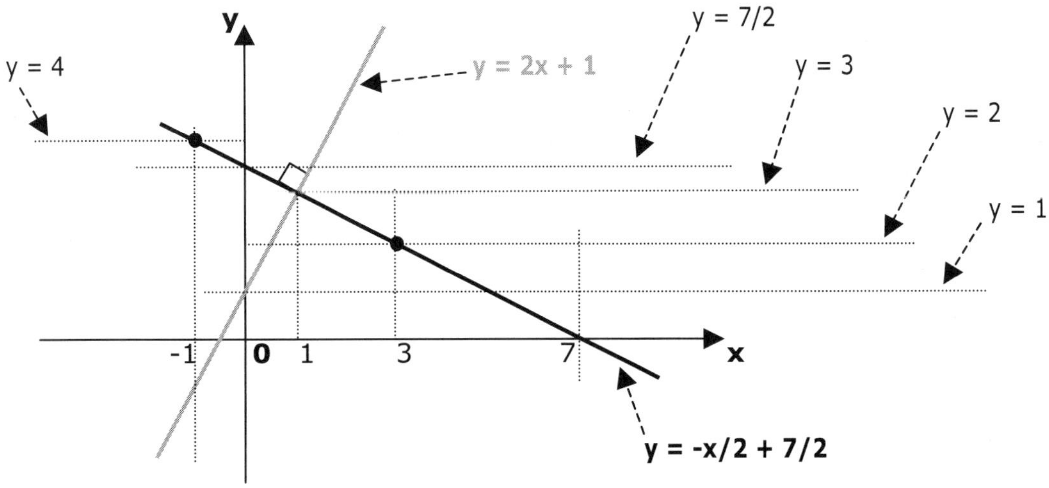

In short:

The first method

Suppose that **L** is the given line, **P** is **(3, 2)**, and **Q(s, t)** is the point symmetric to the **P** about the **L**.
Suppose also, that **R(u, v)** is the midpoint between **P** and **Q**, and **N** is a line perpendicular to the **L**.
Since the **L** includes **(1, 3)** and has the slope of 2, the equation of the **L** is $y - 3 = 2(x - 1)$.
Let **n** be the slope of the **N**. Then, $2n = -1 \Rightarrow n = -1/2$.
Since the **N** includes **P(3, 2)** and has the slope of −1/2, the equation of the **N** is $y - 2 = (-1/2)(x - 3)$.
Let's now, find the midpoint **R(u, v)**.
We know that the coordinates of the **R** are the respective averages of the x-coordinates and the y-coordinates of the **P** and the **Q**.
Then, $u = (3 + s)/2$ and $v = (2 + t)/2$.

First, putting the **R** into the **L**, we get $v - 3 = 2(u - 1) \Rightarrow (2 + t)/2 - 3 = 2\{(3 + s)/2 - 1\}$
$\Rightarrow 2 + t - 6 = t - 4 = 4\{(3 + s)/2 - 1\} = 2(3 + s) - 4 = 2s + 2 \Rightarrow t - 4 = 2s + 2 \Rightarrow 2s - t = -6$.

Next, putting the **R** into the **N**, we get $v - 2 = (-1/2)(u - 3)$
$\Rightarrow (2 + t)/2 - 2 = (-1/2)\{(3 + s)/2 - 3\} \Rightarrow 2 + t - 4 = t - 2 = -\{(3 + s)/2 - 3\}$
$\Rightarrow 2t - 4 = -(3 + s) + 6 = 3 - s \Rightarrow 2t - 4 = 3 - s \Rightarrow s + 2t = 7$.

Now, we get the system of equations as follows: **$2s - t = -6$** and **$s + 2t = 7$**.

Let's, get rid of the **t** first.
$2s - t = -6 \Rightarrow 4s - 2t = -12$. Then, **$(4s - 2t) + (s + 2t) = -12 + 7 \Rightarrow 5s = -5 \Rightarrow s = -1$.**
Then, **$s + 2t = 7 \Rightarrow -1 + 2t = 7 \Rightarrow 2t = 8 \Rightarrow t = 4$**.

Therefore, **Q** is **(-1, 4)**.

The second method

Suppose that **L** is the given line, **P** is **(3, 2)**, and **Q(s, t)** is the point symmetric to the **P** about the **L**.
Suppose also, that **N** is a line perpendicular to the **L**, and **R(u, v)** is the point where **L** meets **N**.

Since the **L** includes **(1, 3)** and the slope is 2, the equation is as follows: **$y - 3 = 2(x - 1)$**.
Let **n** be the slope of the **N**. Then, **$2n = -1 \Rightarrow n = -1/2$**.
Since the **N** includes **P(3, 2)** and the slope is −1/2, the equation is as follows: **$y - 2 = (-1/2)(x - 3)$**.

Let's now find the **R** where the **L** and the **N** meet each other.

$y - 3 = 2(x - 1) \Rightarrow y = 2x - 2 + 3 = 2x + 1 \Rightarrow y = 2x + 1$.
$y - 2 = (-1/2)(x - 3) \Rightarrow y = -x/2 + 3/2 + 2 \Rightarrow y = -x/2 + 7/2$.

Therefore, **$2x + 1 = -x/2 + 7/2 \Rightarrow 4x + 2 = -x + 7 \Rightarrow 5x = 5 \Rightarrow x = 1$**.
Then, **$y = 2x + 1 = 3$**. So, the **R** is **(1, 3)**.

Let's now, find the distance **D** between the **P** and the **R**. The **P** is **(3, 2)**.
$D^2 = (3 - 1)^2 + (2 - 3)^2 = 4 + 1 = 5 \Rightarrow D^2 = 5$.

The **D** is the same as the distance between **R(1, 3)** and **Q(s, t)**.
$D^2 = 5 = (s - 1)^2 + (t - 3)^2$.

Q(s, t) is the point that is to be in the **N**:
$N(x, y) = (-1/2)(x - 3) - (y - 2) = 0 \Rightarrow y - 2 = (-1/2)(x - 3) \Rightarrow y = -x/2 + 7/2$
$\Rightarrow t = -s/2 + 7/2$.

Then, we get the system of equations as follows: **$5 = (s - 1)^2 + (t - 3)^2$** and **$t = -s/2 + 7/2$**.

Then, **$5 = (s - 1)^2 + (-s/2 + 7/2 - 3)^2$**, and **$-s/2 + 7/2 - 3 = -s/2 + 1/2 = (1 - s)/2$**.
Then, **$5 = (s - 1)^2 + (1 - s)^2/4 = (s - 1)^2 + (s - 1)^2/4 = 5(s - 1)^2/4$**.
$\Rightarrow 5 = 5(s - 1)^2/4 \Rightarrow (s - 1)^2 = 4 \Rightarrow s - 1 = 2$ or **$-2 \Rightarrow s = 3$** or **-1**.
We have **$t = -s/2 + 7/2$**.
Then, when **$s = 3$**, we get **$t = -3/2 + 7/2 = 2$**.
So, **(s, t) = (3, 2)** which is the point **P** given already.
Now, when **$s = -1$**, we get **$t = -(-1)/2 + 7/1 = 1/2 + 7/2 = 4$**.
Therefore, **Q(s, t) = Q(-1, 4)**.

Let's take the next example, now.

Suggestions or Solutions
To the Problem in the Example 8

Suppose that a line includes a point at (2, 5) and the slope is 2. Then, find the point (s, t) from which the distance is 5 to the line.

There is not only one point that is a particular distance away from a line. How many are there, then? There can be infinitely many. Nevertheless, it is not true, of course, that those points are anywhere. Those points constitute particular lines. Suppose now, that **L** is the given line. Then, the particular lines are 5 away from the **L**, and they are parallel to the **L**. That is, the particular lines and the **L** are parallel to each other. Then, we can say that each of all the points constituting the particular lines is 5 away from the **L**. How many particular lines are there, then? There are two. There is one each on both sides of the **L**.

Suppose now, that **P** and **Q** are the two parallel lines.
How do we find the **P** and the **Q**, then?
Let's find the **P** only for now. The same is eventually true for the **Q**, too.
First of all, since the **P** is parallel to the **L**, the slope is the same as that of the **L**.
Then, we need to have a particular point that the **P** includes.
The particular point is 5 away from the **L**.
Then, how do we find the particular point?
We know a point in the line **L**. The point is **(2, 5)**.
Then, we need to find a point that is 5 away from the point **(2, 5)**.
Now, is there only one such point?
No, there are infinitely many such. Such are all the points constituting a circle of radius 5 centered at the point **(2, 5)**.
One of all such points is the very particular point.
Where in the circle is the particular point, then?
The particular point is also in a line that includes the point **(2, 5)** and is perpendicular to the line **L**.
In fact, there are two such particular points. One is in the line **P** and the other is in the line **Q**.
That's because the perpendicular line includes the diameter of the circle, and therefore, the **P** and the **Q** have to be respectively at both ends of the diameter.
Suppose now, that **N** is the perpendicular line.
Then, the circle meets the **N** at two points, which are the very particular points.
Then, we need to solve a system of the equation of the circle and the equation of the **N**.
What do we mean by the equation of circle?
An equation of circle indicates all the points that are a particular distance away from the center of the circle, and the particular distance is the radius of the circle. The details on circles are covered in the **Book 4**.
The center of the circle is the point **(2, 5)**, and the radius is 5.
Now, how do call a point representing all the points constituting a geometric object such as a line?
It is called a general point.
Suppose now, that **(x, y)** is the general point in the circle.
Then, the expression stating that 5 is the distance between the point **(2, 5)** and **(x, y)** is the equation of the circle.
Then, how do we find the points where the circle meets the line **N**.
We need to solve the system of the equation of the circle and the equation of the **N**.
Then, the solution says the coordinates of the two particular points.
Then, we can find the two parallel lines **P** and **Q**.

Now, let's get started with the line **N** perpendicular to the **L**.
Suppose that **n** is the slope of the **N**. Then, we get $2n = -1$ since the slope of the **L** is 2, and also, the product of the slopes of two lines perpendicular to each other is −1.
Then, we get $n = -1/2$, and also, since the **N** includes **(2, 5)**, the equation is $y - 5 = (-1/2)(x - 2)$.

Next, let's find the equation of the circle.
The point **(2, 5)** is the center of the circle, and 5 is the distance between the center and the general point **(x, y)**, that is, the radius is 5.
Then, the equation of the circle is as follows: $5^2 = (x - 2)^2 + (y - 5)^2$.

Now, the system of equations is as follows:
$5^2 = (x - 2)^2 + (y - 5)^2$ and $y - 5 = (-1/2)(x - 2)$.

To begin with, let's eliminate the **y** to find the **x**.
Then, let's first put in the standard form the second equation, that is, the equation of the **N**. Then:

$$y - 5 = (-1/2)(x - 2) \Rightarrow y = -x/2 + 1 + 5 = -x/2 + 6.$$

Next, we put $y = -x/2 + 6$ into the first equation, that is, the equation of the circle. Then:

$$5^2 = (x - 2)^2 + (6 - x/2 - 5)^2 = (x - 2)^2 + (1 - x/2)^2$$
$$= (x - 2)^2 + \{(1/2)(2 - x)\}^2 = (x - 2)^2 + (2 - x)^2/4 = (x - 2)^2 + (x - 2)^2/4 = 5(x - 2)^2/4.$$

$$\Rightarrow 25 = 5(x - 2)^2/4 \Rightarrow (x - 2)^2 = 20 \Rightarrow x - 2 = 20^{1/2} \text{ or } -(20^{1/2}) \Rightarrow x = 2 + 20^{1/2} \text{ or } 2 - 20^{1/2}.$$

$$20^{1/2} = 2 \times 5^{1/2} = \sqrt{20} = 2\sqrt{5}$$

Then, $x = 2(1 + 5^{1/2})$ or $2(1 - 5^{1/2})$.

Then:
When $x = 2(1 + 5^{1/2})$, we get $t = -x/2 + 6 \Rightarrow t = -(1 + 5^{1/2}) + 6 = 5 - 5^{1/2}$.
When $x = 2(1 - 5^{1/2})$, we get $t = -x/2 + 6 \Rightarrow t = -(1 - 5^{1/2}) + 6 = 5 + 5^{1/2}$.

Suppose now, that $s_1 = 2(1 + 5^{1/2})$, $t_1 = 5 - 5^{1/2}$, $s_2 = 2(1 - 5^{1/2})$, and $t_2 = 5 + 5^{1/2}$.

Then, the two particular points are (s_1, t_1) and (s_2, t_2).

Suppose now, that the point (s_1, t_1) is in the line **P**, and the point (s_2, t_2) is in the line **Q**.
Since the two line **P** and **Q** are parallel to the **L**, the slopes are 2. Then:
The equation of the **P** is as follows: $y - t_1 = 2(x - s_1)$.
The equation of the **Q** is as follows: $y - t_2 = 2(x - s_2)$.

Now, each of all the points in the two lines **P** and **Q** is 5 away from the **L**.

$s_1 = 2(1 + 5^{1/2}) \approx 6.47$, $t_1 = 5 - 5^{1/2} \approx 2.76$, $s_2 = 2(1 - 5^{1/2}) \approx -2.47$, and $t_2 = 5 + 5^{1/2} \approx 7.24$.

Let's put all those in the graph, now.

Fig. 8.0

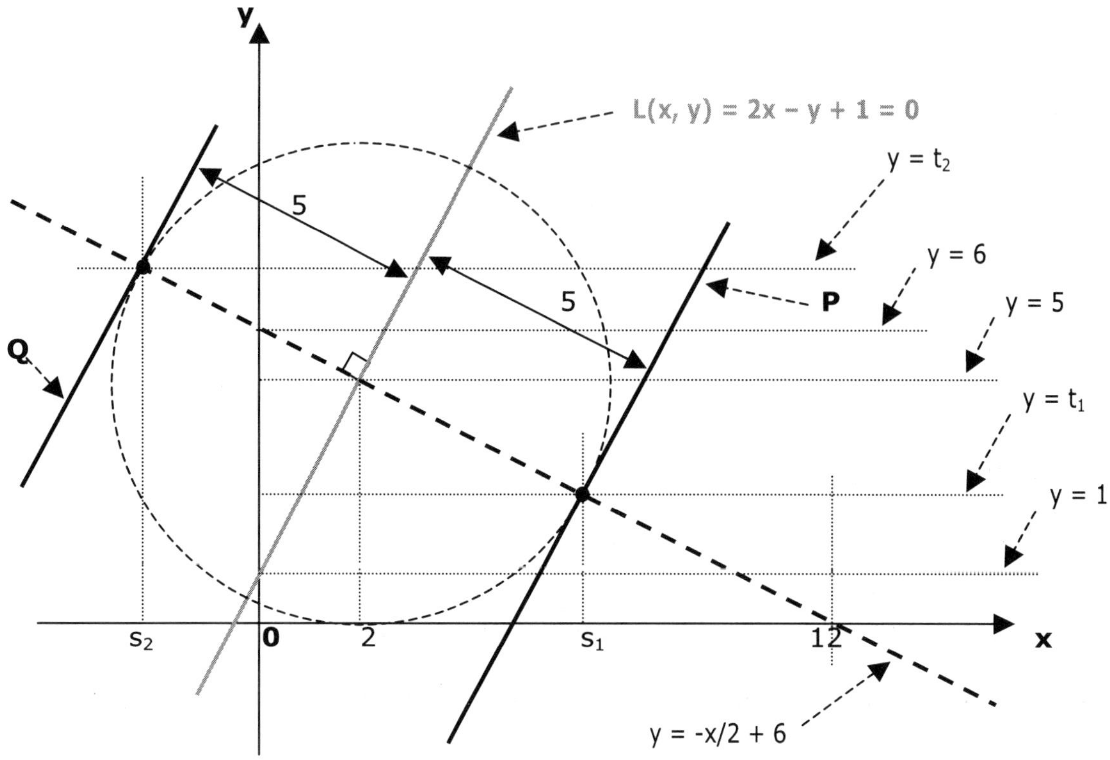

Each of all the points in the black lines (**P** and **Q**) is 5 away from the gray line (**L**).

In short:

Let **L** be the line given, and suppose that two points $\mathbf{R_1}$ and $\mathbf{R_2}$ are 5 away from the point **(2, 5)** in the **L**. Suppose also, that two lines **P** and **Q** are parallel to the **L**, that the **P** includes the $\mathbf{R_1}$, and that the **Q** includes the $\mathbf{R_2}$.
Then, the **P** and the **Q** are the lines that are respectively 5 away from the **L**, and each of all the points in the two lines is 5 away from the **L**.
The two points $\mathbf{R_1}$ and $\mathbf{R_2}$ are in a line **N** which includes the point **(2, 5)** and is perpendicular to the line **L**. Then, the two points are the points where the **N** meets a circle of radius 5 centered at the point **(2, 5)**.

Suppose that **n** is the slope of the **N**. Then, we get $\mathbf{2n = -1}$ since the slope of the **L** is 2 and the product of the slopes of two lines perpendicular to each other is −1. Then, we get $\mathbf{n = -1/2}$, and also, since the **N** includes **(2, 5)**, the equation of **N** is $\mathbf{y - 5 = (-1/2)(x - 2)}$.

Next, the equation of the circle is as follows: $\mathbf{5^2 = (x - 2)^2 + (y - 5)^2}$.

Now, to find the two points $\mathbf{R_1}$ and $\mathbf{R_2}$, we need to solve the system of equations as follows:
$\mathbf{5^2 = (x - 2)^2 + (y - 5)^2}$ and $\mathbf{y - 5 = (-1/2)(x - 2)}$.

To begin with, let's eliminate the **y** to find the **x**.

$y - 5 = (-1/2)(x - 2) \Rightarrow y = -x/2 + 1 + 5 = -x/2 + 6.$

Then:
$5^2 = (x - 2)^2 + (6 - x/2 - 5)^2 = (x - 2)^2 + (1 - x/2)^2$
$= (x - 2)^2 + \{(1/2)(2 - x)\}^2 = (x - 2)^2 + (2 - x)^2/4 = (x - 2)^2 + (x - 2)^2/4 = 5(x - 2)^2/4.$

$\Rightarrow 25 = 5(x - 2)^2/4 \Rightarrow (x - 2)^2 = 20 \Rightarrow x - 2 = 20^{1/2}$ or $-(20^{1/2}) \Rightarrow x = 2 + 20^{1/2}$ or $2 - 20^{1/2}$.

$20^{1/2} = 2 \text{ X } 5^{1/2} = \sqrt{20} = 2\sqrt{5}$

Therefore, $x = 2(1 + 5^{1/2})$ or $2(1 - 5^{1/2})$.

Then:
When $x = 2(1 + 5^{1/2})$, $t = -x/2 + 6 \Rightarrow t = -(1 + 5^{1/2}) + 6 = 5 - 5^{1/2}$.
When $x = 2(1 - 5^{1/2})$, $t = -x/2 + 6 \Rightarrow t = -(1 - 5^{1/2}) + 6 = 5 + 5^{1/2}$.

Now, let's set $s_1 = 2(1 + 5^{1/2})$, $t_1 = 5 - 5^{1/2}$, $s_2 = 2(1 - 5^{1/2})$, and $t_2 = 5 + 5^{1/2}$.

Suppose also, that the $\mathbf{R_1}$ is at (s_1, t_1), and that the $\mathbf{R_2}$ is at (s_2, t_2).

Since the two lines **P** and **Q** are parallel to the **L**, the slopes are 2. Then:
The equation of the **P** is $y - t_1 = 2(x - s_1)$.
The equation of the **Q** is $y - t_2 = 2(x - s_2)$.

Now, each of all the points in the two lines **P** and **Q** is 5 away from the **L**.

Suggestions or Solutions

To the Problem in the Example 9

Find a point (s, t) the same distance away from a point (-2, 1) as away from a point (3, 2).

Of course, we can say that the point **(s, t)** is the midpoint between the two given points.
Then, is the midpoint only the only point that can be the solution?
No, more points can be the solution.

Let **A** and **B** be the two points given. Then, if we use the two points and a particular point to construct an isosceles triangle, the particular point is the same distance away from the **A** and the **B**. That is, the particular point, too, can the **(s, t)**.

Then, what points can be the solution?
How many such isosceles triangles as stated above can we make?
Infinitely many!

Suppose now, that **M** is the midpoint between the **A** and the **B**, and that **L** is the line that includes the **A** and the **B**. Suppose also, that **N** is the line that includes the midpoint **M** and is perpendicular to the **L**.
Then, each of all the points in the line **N** is the same distance away from the **A** and the **B**.
Therefore, each of all the points constituting the **N** can be the **(s, t)**.

Let's now, put in a graph the two lines **N** and **L** along with the two points **A** and **B**, and the midpoint **M**.

Fig. 9.0

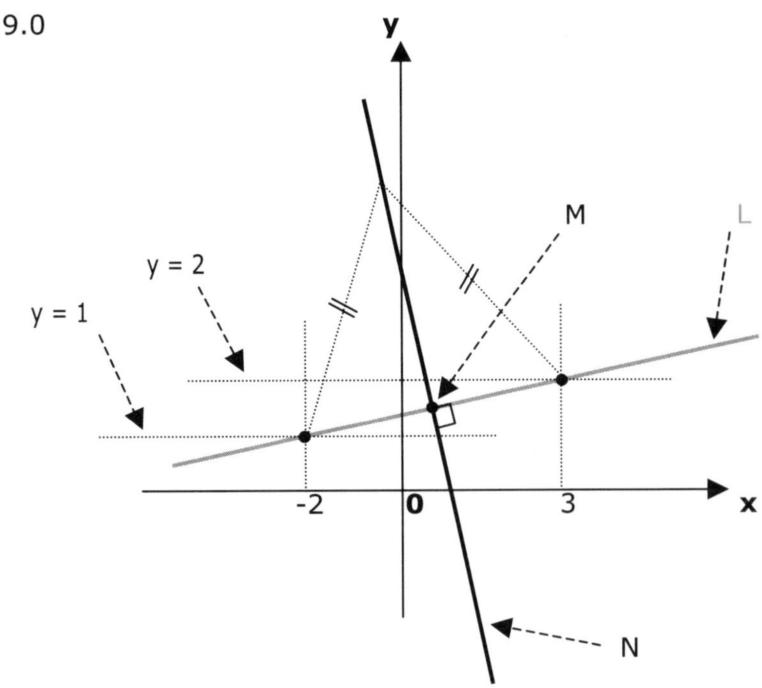

Now, we are going to find the line **N**.

We know that the **N** includes the midpoint **M**. Since the **M** is the midpoint between **(-2, 1)** and **(3, 2)**, the **M** is at **{(-2 + 3)/2, (1 + 2)/2} = (1/2, 3/2)**.
Then, we need to find the slope of the **N**.
The product of the slopes of two lines perpendicular to each other is −1.
Therefore, since the **N** is perpendicular to the **L**, if we find the slope of the **L**, we can readily see the slope of the **N**.
We can see that the slope of the **L** = **Δy/Δx = (2 − 1)/{3 − (-2)} = 1/5**.
Then, the slope of the **N** is −5. Now, the equation of the **N** is **y − 3/2 = -5(x − 1/2)**, and each and every point in the **N** can be the **(s, t)**.

In short:

Let **A** be **(-2, 1)**, **B** be **(3, 2)**, and **M** be the midpoint between the **A** and the **B**.
Then, since the **M** is the same distance away from the **A** and the **B**, the **M** can be the **(s, t)**.
Suppose also, that **L** is the line that includes the **A** and the **B**, and that **N** is the line that includes the **M** and is perpendicular to the **L**.
Then, each and every point in the **N** can be the **(s, t)**.
That's because the **A**, the **B**, and any point in the **N** can make an isosceles triangle where the line segment **AB** is the base.

We know that the **N** includes the midpoint **M**. Since the **M** is the midpoint between **(-2, 1)** and **(3, 2)**, the **M** is at **{(-2 + 3)/2, (1 + 2)/2} = (1/2, 3/2)**.
Then, we need to find the slope of the **N**.
The product of the slopes of two lines perpendicular to each other is −1.
Suppose now, that **n** is the slope of the **N**, and that **m** is the slope of the **L**.
Then, **m = Δy/Δx = (2 − 1)/{3 − (-2)} = 1/5**.
So, **mn = (1/5)n = -1 ⇒ n = −5**.
Therefore, the equation of the **N** is **y − 3/2 = -5(x − 1/2)**, and each and every point in the **N** can be the **(s, t)**.

Suggestions or Solutions

To the Problem 0 in the Example A

Suppose that:
We have six points $P_1(1, 1)$, $P_2(1, 3)$, $P_3(2, 1)$, $P_4(4, 1)$, $P_5(s, t)$, and $P_6(u, v)$ in the x-y plane.
The P_5 and the P_6 are above the P_3 and the P_4, and also, on the right of the P_2.
We make a triangle $P_5P_1P_2$ and another triangle $P_5P_3P_4$ so that the sum of the areas taken by the two triangles is 3.

Then, specify the point P_5.

To begin with, let's put the given points in the x-y plane.

Fig. A.0.0

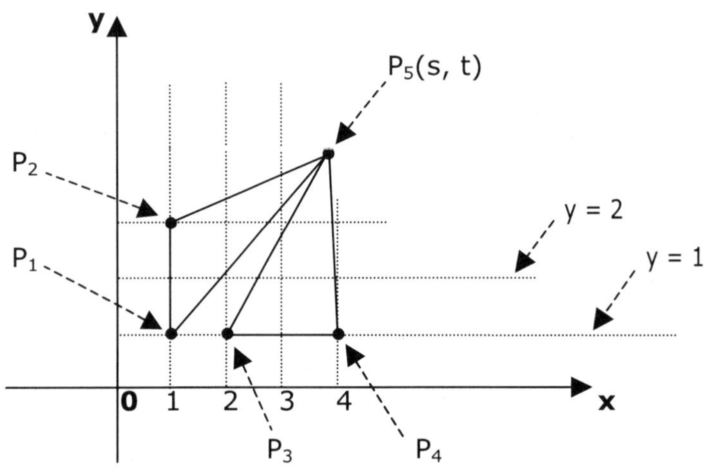

Then, we can notice that if we move the point $\mathbf{P_1}$ to the origin and move the other points in the same manner, then life gets much easier. In other words, we translate all the points under consideration. Then, we can translate the point $\mathbf{P_5}$ back in the opposite direction after finding the $\mathbf{P_5}$. Of course, we can translate the axes, instead. In fact, we are better off translating the axes, and rewrite the coordinate values accordingly. However, from the points' point of view the $\Delta \mathbf{x} = \mathbf{-1}$, and the $\Delta \mathbf{y} = \mathbf{-1}$. Now, after the translations, we get a new graph as follows:

Fig. A.0.1

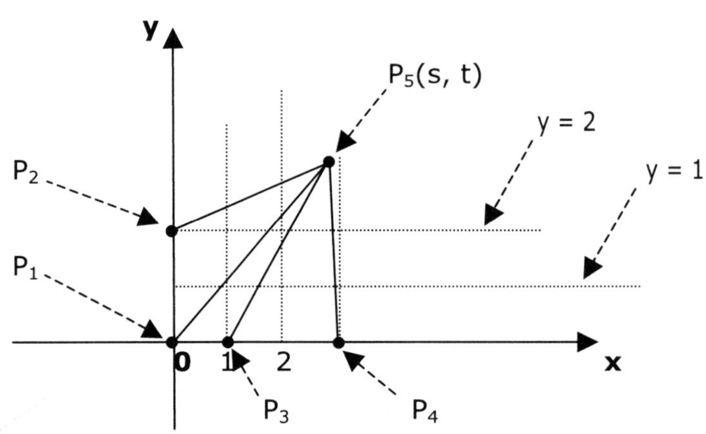

Let's now, put the sum of the areas of the triangles in terms of the coordinates at the point $\mathbf{P_5}$.
Let **A** be the triangle $\mathbf{P_5P_1P_2}$, and **B** be the triangle $\mathbf{P_5P_3P_4}$. Then:
The area of the triangle **A = (1/2)(2)s = s** since we can take the **s** for the height of the triangle.
The area of the triangle **B = (1/2)(2)t = t** for the similar reason to the above.
Then, we can see that **s + t = 3**. Now, we can see that there can be many such points as the $\mathbf{P_5}$ that can make such a pair of triangles of which the sum of the areas is 3. How many, then? In fact, infinitely many! The choices for the **s** and **t** are unlimited.
More precisely, each of all the points in a line can be the **(s, t)**, and the line is **x + y = 3**. In other words, the $\mathbf{P_5(s, t)}$ is now a general point and we choose the names of the coordinates to be **x** and **y** since the general point is in the x-y plane. However, that's not exactly it yet since we need to move the line back to the proper place. We know that the Δ**x** was -1 and Δ**y** was –1. Then, the Δ**x** is now 1 and Δ**y** is now 1 also. Now, the proper equation of the line is **(x – 1) + (y – 1) = 3**.

Now, how come can the equation above be the equation of the line translated in such a reverse manner? Then, let's find this time, the equation in a straight manner.
We know that the point **(1, 2)** satisfies the equation **x + y = 3**.
Then, we translate the point **(1, 2)** in such a manner as follows: Δ**x = 1** and Δ**y = 1**.
Then, after the translation, the point becomes **(2, 3)**.
We know that the slope of the line **x + y = 3** is –1. If we translate a line, the slope remains.
Then, the slope of the line translated is –1, too.
Also, the line translated includes the point **(2, 3)**.
Therefore, the equation of the line translated is **y – 3 = (-1)(x – 2)**.
Now, let's rewrite the equation above in such a manner as follows:
$\mathbf{y - 3 = (-1)(x - 2) \Rightarrow y - 3 = -x + 2 \Rightarrow x - 2 + y - 3 = 0 \Rightarrow x - 1 + y - 1 - 1 - 2 = 0}$
$\mathbf{\Rightarrow x - 1 + y - 1 = 3}$. Putting it in the standard form, we get **y = -x + 5**.

However, that's not the correct solution yet. That's because the general point is supposed to be above the two points $\mathbf{P_3}$ and $\mathbf{P_4}$, and also, on the right of the point $\mathbf{P_2}$. Then, the solution is not a line but a line segment. Now, the equation of the line segment is **(x – 1) + (y – 1) = 3** for **1 < x < 4**, and each of all the points in the line segment can be the **(s, t)**.

Fig. A.0.2

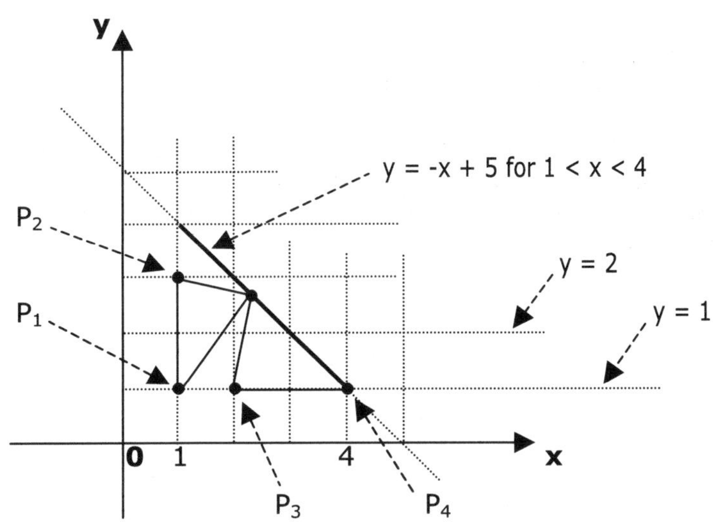

In short:

First, let's translate all the given points in such a manner as follows : $\Delta x = -1$ and $\Delta y = -1$.
After the translation, we get the graph as follows:

Fig. A.0.1

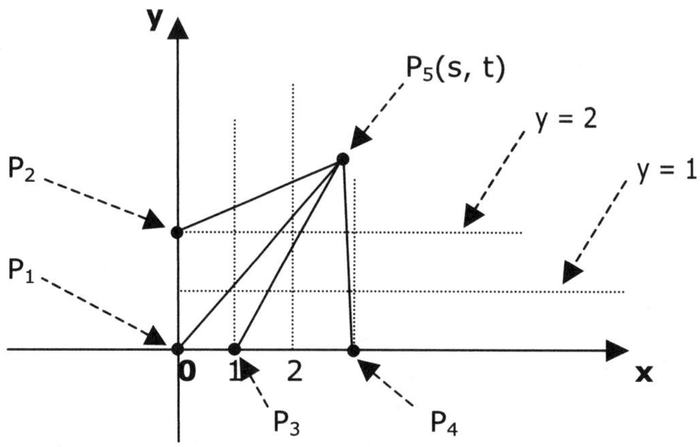

Let **A** be the triangle $P_5P_1P_2$, and **B** be the triangle $P_5P_3P_4$. Then:
The area of the triangle **A** = $(1/2)(2)s = s$ since we can take the **s** for the height of the triangle.
The area of the triangle **B** = $(1/2)(2)t = t$ for the similar reason to the above.
Then, we get $s + t = 3$.
Then, for now, each of all the points in a line can be the **(s, t)**, and the line is $x + y = 3$.
We now need to move the line back to the proper place.
In the first place, the Δx was −1, and the Δy was −1, too. Then, the Δx is now 1, and the Δy is 1 now.
The point **(1, 2)** satisfies the equation $x + y = 3$.
Then, we translate the point **(1, 2)** in such a manner as follows: $\Delta x = 1$ and $\Delta y = 1$.
Then, after the translation, the point becomes **(2, 3)**.
The slope of the line $x + y = 3$ is −1. If we translate a line, the slope remains. Then, the slope of the line translated is −1, too.
Also, the line translated includes **(2, 3)**.
Therefore, the equation of the line translated is $y - 3 = (-1)(x - 2)$.

However, it is not the correct solution yet. That's because the general point is supposed to be above the two points P_3 and P_4, and also, on the right of the point P_2. Then, it is not a line but a line segment.
Now, the equation of the line segment is $(x - 1) + (y - 1) = 3$ for $1 < x < 4$, and each of all the points in the line segment can be the **(s, t)**.

Now, let's move on to the next part (the part 1).

Suggestions or Solutions

To the Problem 1 in the Example A

Suppose that:
We have six points $P_1(1, 1)$, $P_2(1, 3)$, $P_3(2, 1)$, $P_4(4, 1)$, $P_5(s, t)$, and $P_6(u, v)$ in the x-y plane.
The P_5 and the P_6 are above the P_3 and the P_4, and also, on the right of the P_2.
We make a triangle $P_5P_1P_2$ and another triangle $P_5P_3P_4$ so that the sum of the areas taken by the two triangles is 3.

Suppose also that:
D_1 is the distance between the P_6 and the P_4. D_2 is the distance between the P_6 and the P_2.
$D_1^2 - D_2^2 = 5$.

Now, specify the point P_6.

Fig. A.1.0

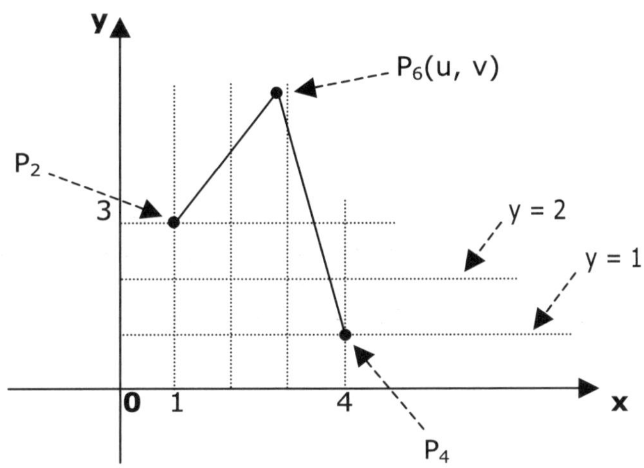

The D_1 is the distance between $P_6(u, v)$ and $P_4(4, 1)$.
Then, $D_1^2 = (u - 4)^2 + (v - 1)^2$.

The D_2 is the distance between $P_6(u, v)$ and $P_2(1, 3)$.
Then, $D_2^2 = (u - 1)^2 + (v - 3)^2$.

Now, $D_1^2 - D_2^2 = 5 \Rightarrow (u - 4)^2 + (v - 1)^2 - (u - 1)^2 - (v - 3)^2 = 5$
$\Rightarrow u^2 - 8u + 16 + v^2 - 2v + 1 - u^2 + 2u - 1 - v^2 + 6v - 9 = 5 \Rightarrow -6u + 4v + 7 = 5$
$\Rightarrow v = 3u/2 - 1/2$.

We can see now, that there can be infinitely many such points as the P_6 that can satisfy the conditions presented in the problem. The choices for the **u** and **v** are unlimited. In other words, the $P_6(u, v)$ is now a general point and we choose the names of the coordinates to be **x** and **y** since the general point is in the x-y plane. Therefore, each of all the points in a line $y = 3x/2 - 1/2$ can be the **(u, v)**.
However, that's not the complete solution yet. That's because the general point is supposed to be above the P_4, and also, on the right of the P_2. Then, the solution is not a line but a ray. A ray (called a half line also) is still a straight line but proceeds in one direction only whereas a normal line stretches itself in both directions. For instance, a normal line proceeds to the left and right, but a ray expands itself either to the left or to the right only.

Therefore, the equation of the ray is **$y = 3x/2 - 1/2$** for **$x > 1$**, and each of all the points of the ray can be the **(u, v)**.

Fig. A.1.2

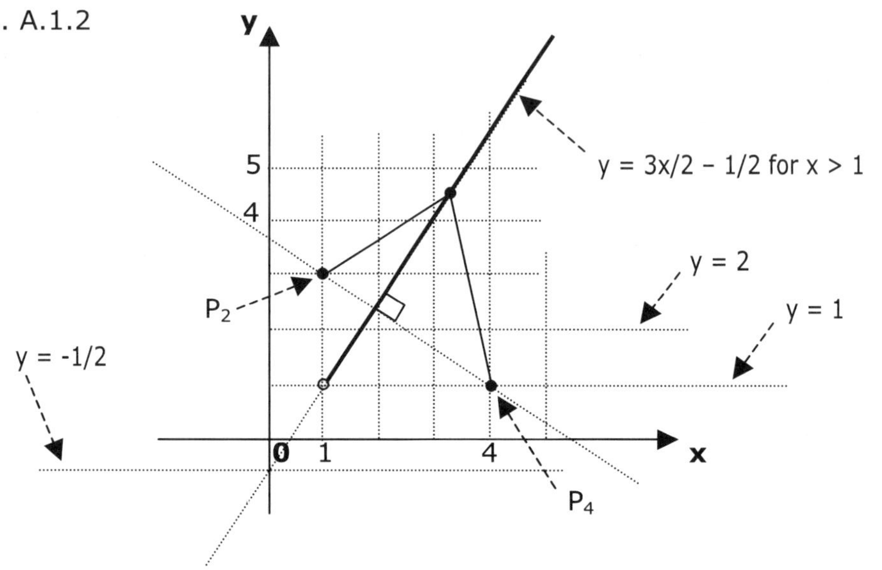

Now, the ray is perpendicular to a line that includes the two points **P_2** and **P_4**.
Then, let's check to see if the product of the slopes of the ray and the perpendicular line is −1.
The slope of the perpendicular line is **$(3 - 1)/(1 - 4) = -2/3$**.
Then, since the product of the slopes is **$(-2/3)(3/2) = -1$**, they are perpendicular to each other.

In short:

Since the **D_1** is the distance between **$P_6(u, v)$** and **$P_4(4, 1)$**, **$D_1^2 = (u - 4)^2 + (v - 1)^2$**.
Since the **D_2** is the distance between **$P_6(u, v)$** and **$P_2(1, 3)$**, **$D_2^2 = (u - 1)^2 + (v - 3)^2$**.

Then, **$D_1^2 - D_2^2 = 5 \Rightarrow (u - 4)^2 + (v - 1)^2 - (u - 1)^2 - (v - 3)^2 = 5$**
$\Rightarrow u^2 - 8u + 16 + v^2 - 2v + 1 - u^2 + 2u - 1 - v^2 + 6v - 9 = 5 \Rightarrow -6u + 4v + 7 = 5$
$\Rightarrow v = 3u/2 - 1/2$.

Then, for now, each and every point of the line **$y = 3x/2 - 1/2$** can be the **(u, v)**.
Now, the **(u, v)** is supposed to be above the point **P_4**, and also, on the right of the **P_2**.
Therefore, all such points constitute a ray, and the equation of the ray is **$y = 3x/2 - 1/2$** for **$x > 1$**, and each of all the points in the ray can be the **(u, v)**.

Now, let's take care of the next example.

Suggestions or Solutions

To the Problem in the Example B

Suppose that the distance between two points P_1 and P_2 in the x-y plane is 5.
Suppose also that:
D_1 is the distance between the P_1 and another point P_3.
D_2 is the distance between the P_2 and the P_3.
$D_1^2 - D_2^2 = 10$.

Then, specify the point P_3.

To begin with, let's put on the x-y plane the points under consideration. Also, let the point **P_3** be **(s, t)**.

Fig. B.0

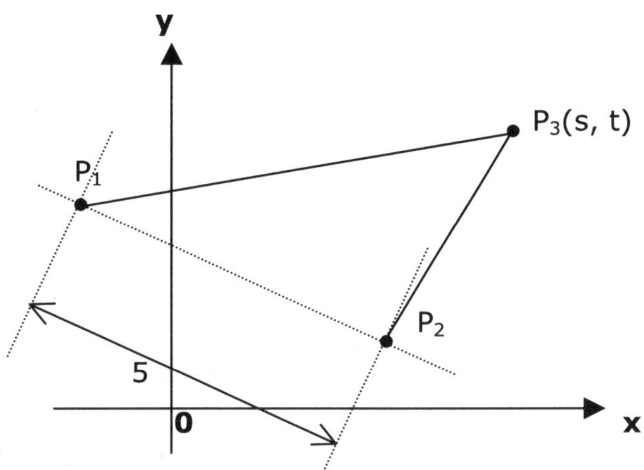

In this example also, let's put again the two points **P_1** and **P_2** at some proper positions where our life gets easier. However this time, we do not have to consider the Δ**x** and the Δ**y**. That's because the constraint is only on the distance between the points applicable. In other words, it doesn't matter wherever those points are as long as the constraint is satisfied. Therefore, what we want to find is the position of the **P_3** relative to the positions of **P_1** and **P_2**. Then, let's put the points in such a way as below.

Fig. B.1

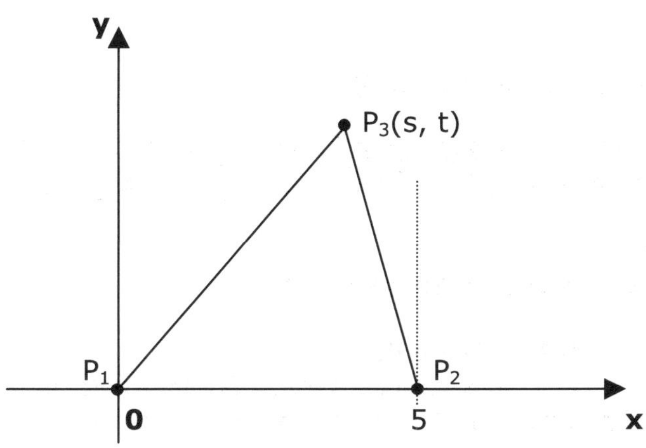

Now, the $\mathbf{P_1}$ is at the origin, and the $\mathbf{P_2}$ is at **(5, 0)**.
Then, we start putting together the facts given to see what we can do to find the solution.

$\mathbf{D_1}$ is the distance between the $\mathbf{P_1}$ and another point $\mathbf{P_3(s, t)}$.
Then, $\mathbf{D_1^2 = (s - 0)^2 + (t - 0)^2 = s^2 + t^2}$.

$\mathbf{D_2}$ is the distance between the $\mathbf{P_2}$ and the point $\mathbf{P_3}$. Then, $\mathbf{D_2^2 = (s - 5)^2 + (t - 0)^2 = (s - 5)^2 + t^2}$.

In addition, we have $\mathbf{D_1^2 - D_2^2 = 10}$. Then, $\mathbf{s^2 + t^2 - \{(s - 5)^2 + t^2\} = 10}$
$\Rightarrow \mathbf{s^2 + t^2 - (s^2 - 10s + 25 + t^2) = 10 \Rightarrow 10s - 25 = 10 \Rightarrow s = 7/2}$.

Then again, there are infinitely many points that satisfy the condition presented in the problem. In this case, each of all the points in a line $\mathbf{x = 7/2}$ can be the $\mathbf{P_3}$. Precisely, the line includes a point that is 7/2 away from the point $\mathbf{P_1}$ and 1.5 away from the $\mathbf{P_2}$, and also, it is perpendicular to a line that includes the points $\mathbf{P_1}$ and the $\mathbf{P_2}$.

Fig. B.2.

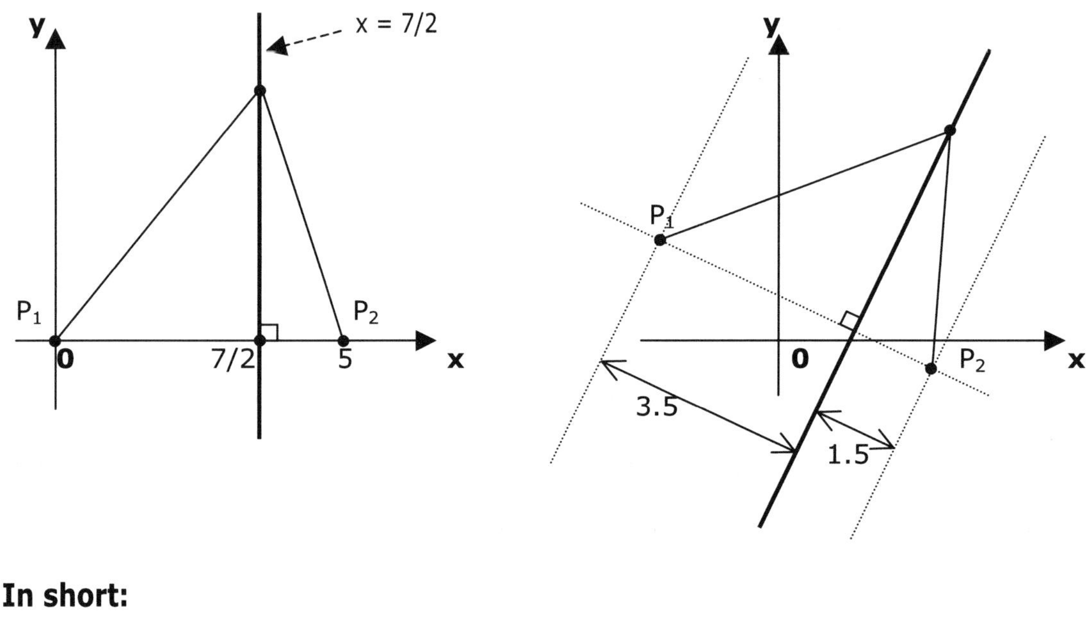

In short:

Suppose that the $\mathbf{P_1}$ is at the origin, and that the $\mathbf{P_2}$ is at **(5, 0)**.
Suppose also, that $\mathbf{P_3}$ is at **(s, t)**.
Since the $\mathbf{D_1}$ is the distance between the $\mathbf{P_1}$ and the $\mathbf{P_3(s, t)}$, $\mathbf{D_1^2 = (s - 0)^2 + (t - 0)^2 = s^2 + t^2}$.
Since the $\mathbf{D_2}$ is the distance between the $\mathbf{P_2}$ and the $\mathbf{P_3}$, $\mathbf{D_2^2 = (s - 5)^2 + (t - 0)^2 = (s - 5)^2 + t^2}$.
In addition, we have $\mathbf{D_1^2 - D_2^2 = 10}$.
Then, $\mathbf{s^2 + t^2 - \{(s - 5)^2 + t^2\} = 10}$
$\Rightarrow \mathbf{s^2 + t^2 - (s^2 - 10s + 25 + t^2) = 10 \Rightarrow 10s - 25 = 10 \Rightarrow s = 7/2}$.

Then, each and every point of the line $\mathbf{x = 7/2}$ can be the $\mathbf{P_3}$.
Precisely, the $\mathbf{P_3}$ can be any point in a line, which includes a point that is 7/2 away from the point $\mathbf{P_1}$ and 1.5 away from the point $\mathbf{P_2}$, and also, is perpendicular to a line that includes the points $\mathbf{P_1}$ and $\mathbf{P_2}$.

Suggestions or Solutions

To the Problem 0 in the Example C

Suppose that we have 4 points $P_1(-2, 3)$, $P_2(-2, -1)$, $P_3(2, 2)$, and P_4 $(-2, 0)$ in the x-y plane and make a triangle T connecting the three points P_1, P_2, and P_3. Suppose also, that A is a line connecting the P_1 and P_2, and that another point P_5 is in the A. If we draw a line that includes the P_5 and is perpendicular to the line A, then the perpendicular line divides the triangle into two equal parts in area. Now, find the position of the P_5.

To begin with, let's put the given points in a graph, and construct the triangle **T**.

Fig. C.0.0

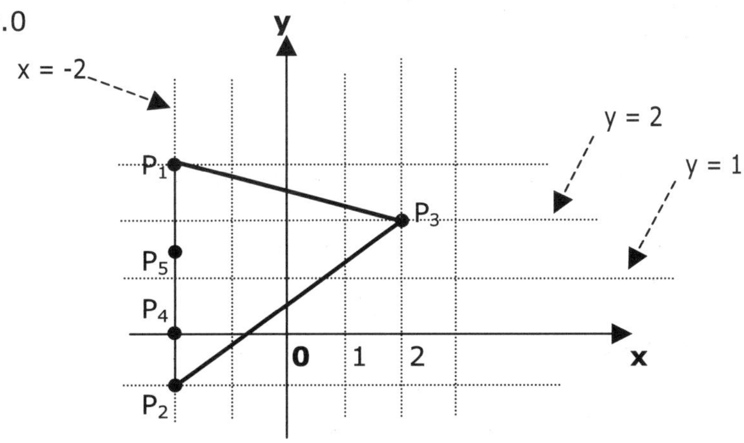

Now, we can notice that if we move the point P_2 to the origin, and move the other points in the same manner, the problem gets much smaller. In this case however, we are better off moving the axes instead. Anyway, from the points' point of view, the **$\Delta x = 2$** and the **$\Delta y = 1$**. Of course, we need to move the solution to the proper location later. Now, the graph looks as follows after the movement.

Fig. C.0.1

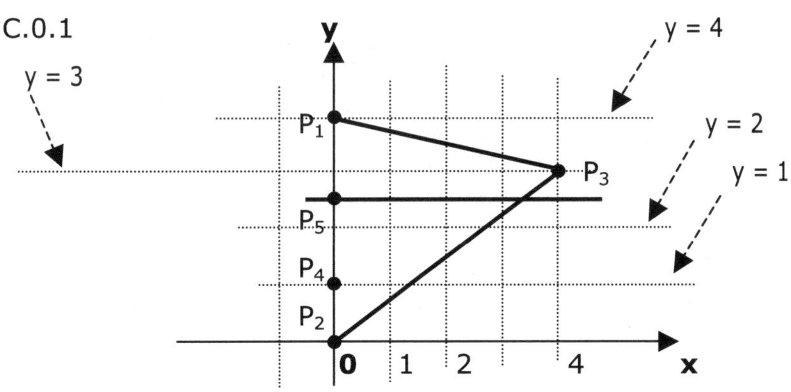

In the graph above, we can see that a line including the P_5 and perpendicular to the y-axis meets the line segment connecting the P_3 and the P_2. Suppose now, that **C** is a line that includes the line segment, and P_6 is the point where the **C** meets the perpendicular line. Suppose also, that **L** is the smaller triangle, and that the P_5 is at **(0, t)**. Then, the equation of the perpendicular line is **y = t**. Then, if we find the P_6, we can express in terms of the **t** the area of the smaller triangle below the perpendicular line. Now, let's begin with the equation of the line **C**. The slope of the **C** is **$\Delta y/\Delta x = (3 - 0)/(4 - 0) = 3/4$**.

So, the equation of the **C** is **$y = 3x/4$** since the line includes the origin.
Let's find the **P_6**, then. We have **$y = t$** and **$y = 3x/4$**. Then, we get **$t = 3x/4 \Rightarrow x = 4t/3$**.
Then, the x-coordinate is **$4t/3$**, and the y-coordinate is **t**. Therefore, **P_6** is at **$(4t/3, t)$**.

Suppose now, that the **t** is the base of the triangle **L**. Then, the height is **$4t/3$**.
Then, the area of the **L** = **$(t)(4t/3)/2 = 2t^2/3$**.
Now, we want the area of the **L** to be half the area of the triangle **T**.
The area of the triangle **T** = **$(4)(4)/2 = 8$**.

Then, **$2t^2/3 = 8/2 = 4 \Rightarrow t^2 = 6 \Rightarrow t = 6^{1/2} \approx 2.45$** since the **$P_5$** is above the origin.
Therefore in this case, the **P_5** is at **$(0, \sqrt{6})$**. We now, need to move the **P_5** to the proper position. The **Δx** was 2, and the **Δy** was 1. Then, **Δx** is now -2, and the **Δy** is -1 now.
Therefore, the **P_5** is at **$(-2, \sqrt{6} - 1)$**.

In short:

If we move the **P_2** to the origin and move the other points in the same manner, the graph looks as below.

Fig. C.0.1

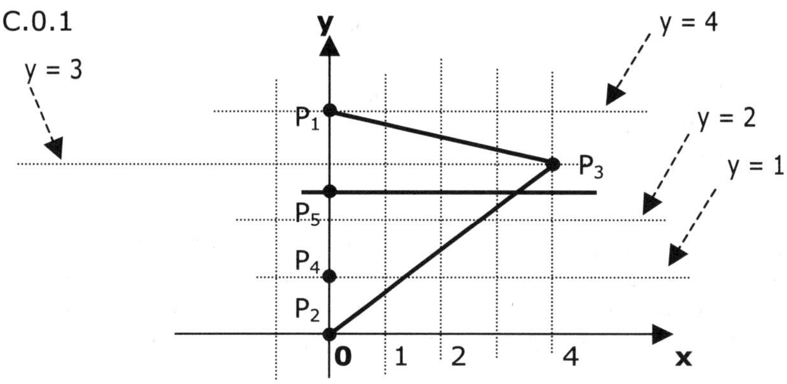

Suppose now, that **C** is a line that includes the **P_2** and the **P_3**, and **P_6** is the point where the **C** meets the horizontal line including the **P_5**. Suppose also, that **L** is the smaller triangle below the horizontal line, and that the **P_5** is at **$(0, t)$**. Then, the equation of the horizontal line is **$y = t$**. Since the **C** includes the origin and the slope is **$\Delta y/\Delta x = (3 - 0)/(4 - 0) = 3/4$**, the equation of the **C** is **$y = 3x/4$**.
Let's find the **P_6**, then. We have **$y = t$** and **$y = 3x/4$**. Then, we get **$t = 3x/4 \Rightarrow x = 4t/3$**.
Then, the x-coordinate is **$4t/3$**, and the y-coordinate is **t**. Therefore, **P_6** is at **$(4t/3, t)$**.

Let's now, take the **t** for the base of the triangle **L**. Then, the height is **$4t/3$**.
Then, the area of the **L** = **$(t)(4t/3)/2 = 2t^2/3$**, which is to be half the area of the triangle **T**.
The area of the triangle **T** = **$(4)(4)/2 = 8$**.

Then, **$2t^2/3 = 8/2 = 4 \Rightarrow t^2 = 6 \Rightarrow t = 6^{1/2} \approx 2.45$** since the **$P_5$** is above the origin.

Therefore in this case, the **P_5** is at **$(0, \sqrt{6})$**. We now, need to move the **P_5** to the proper position. The **Δx** was 2, and the Δy was 1. Then, **Δx** is now -2, and the **Δy** is -1 now.
Therefore, the **P_5** is at **$(-2, \sqrt{6} - 1)$**.
Let's move on to the next part (the part 1) now.

Suggestions or Solutions

To the Problem 1 in the Example C

Suppose that we have 4 points $P_1(-2, 3)$, $P_2(-2, -1)$, $P_3(2, 2)$, and P_4 $(-2, 0)$ in the x-y plane and make a triangle T connecting the three points P_1, P_2, and P_3. Suppose also, that A is a line connecting the P_1 and P_2, and that another point P_5 is in the A. If we draw a line B that includes the P_4, then the line B divides the T into two equal parts in area. Now, find the equation of the line B.

If we move the $\mathbf{P_2}$ to the origin, and move the other points in the same manner, the graph looks as follows after the movement. From the points' point of view, the $\Delta\mathbf{x} = \mathbf{2}$, and the $\Delta\mathbf{y} = \mathbf{1}$.

Fig. C.1.0

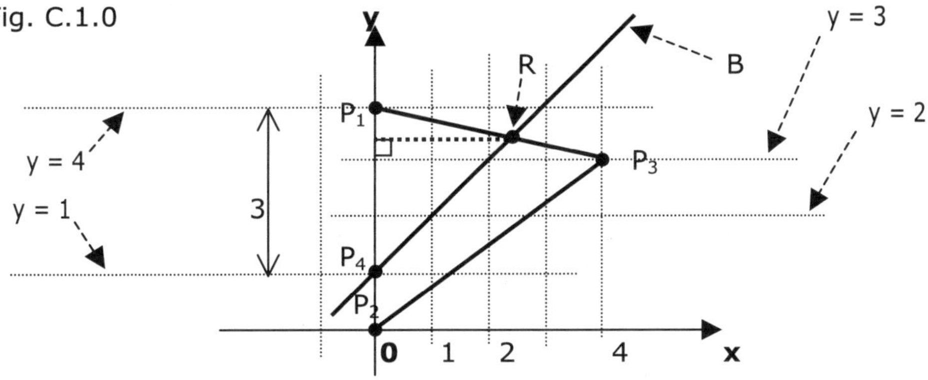

To begin with, let's see which of the two line segments $\mathbf{P_1P_3}$ and $\mathbf{P_2P_3}$ the line **B** needs to meet.
The area of the **T** is 4 X 4 X 1/2 = 8.
Now, if we take for the base the line segment $\mathbf{P_2P_4}$, the area of the triangle $\mathbf{P_2P_3P_4}$ is 1 X 4 X 1/2 = 2.
Then, the **B** needs to meet the line segment $\mathbf{P_1P_3}$.
Suppose now, that **L** is a line that includes the $\mathbf{P_1}$ and $\mathbf{P_3}$, and that **R** is a point where the **L** meets the **B**.
Suppose also, that **H** is the triangle $\mathbf{P_1P_4R}$.
Then, the area of the **H** is to be haft the area of the **T**, that is, 4.

In the graph, we can see that the line segment $\mathbf{P_1P_4}$ is the base of the **H**, and that the length is 3.
Also, we can see that the x-coordinate of the **R** is the height of the **H**.

Now, we need to find the equation of the **B** and that of the **L** to find the x-coordinate of the **R**.
Let's begin with the equation of the **L**.
Since the **L** includes the two points $\mathbf{P_1}$ and $\mathbf{P_3}$, the slope is $\Delta y/\Delta x = (3 - 4)/(4 - 0) = -1/4$. Also, since the **L** includes $\mathbf{P_1(0, 4)}$, the equation of the **L** is $y - 4 = (-1/4)(x - 0) = -x/4 \Rightarrow y = -x/4 + 4$.

Next, suppose that **m** is the slope of the line **B**.
Then, since the **B** includes the point $\mathbf{P_4(0, 1)}$, the equation of the **B** is $y - 1 = m(x - 0) \Rightarrow y = mx + 1$.

Now, let's find the x-coordinate of the **R**.
$(y = -x/4 + 4$ and $y = mx + 1) \Rightarrow -x/4 + 4 = mx + 1 \Rightarrow -x + 16 = 4mx + 4 \Rightarrow (4m + 1)x = 12$
$\Rightarrow x = 12/(4m + 1)$. Therefore, the x-coordinate of the **R** is $12/(4m + 1)$.
Now, the area of the $\mathbf{H} = (1/2)(3)\{12/(4m + 1)\} = 18/(4m + 1)$. Then, $18/(4m + 1) = 4$, which is half the area of the **T**. $18/(4m + 1) = 4 \Rightarrow 18 = 16m + 4 \Rightarrow m = 7/8$.

Then, the equation of the **B** is $\mathbf{y = 7x/8 + 1}$, which however, is not the complete solution yet. We now, need to move the line **B** to the proper position. The $\Delta\mathbf{x}$ was 2, and the $\Delta\mathbf{y}$ was 1. Then, the $\Delta\mathbf{x}$ is now −2, and the $\Delta\mathbf{y}$ is −1 now.

$(y = 7x/8 + 1$, $\Delta x = -2$, and $\Delta y = -1) \Rightarrow y - (-1) = 7\{x - (-2)\}/8 + 1$
$\Rightarrow y + 1 = 7(x + 2)/8 + 1 = 7x/8 + 7/4 + 1.$ Therefore, the equation of the **B** is $\mathbf{y = 7x/8 + 7/4}$.

If we don't know the method above, we can move the **B** in such a way as follows. Since the **B** includes $\mathbf{P_4(0, 1)}$, we move the $\mathbf{P_4}$ first. Then, the original position of the $\mathbf{P_4}$ is $\mathbf{(0 - 2, 1 - 1) = (-2, 0)}$. Due to the translation, the slope of the **B** remains to be 7/8. The equation of a line that has the slope of **a** and includes $\mathbf{(s, t)}$ is $\mathbf{y - t = a(x - s)}$. Therefore, the proper equation of the **B** is $\mathbf{y - 0 = (7/8)\{x - (-2)\} = (7/8)(x + 2) \Rightarrow y = (7/8)x + 7/4}$.

In short:

Let's move the $\mathbf{P_2}$ to the origin and move the other points in the same manner.
Then, $\Delta\mathbf{x} = \mathbf{2}$, and $\Delta\mathbf{y} = \mathbf{1}$.

Fig. C.1.0

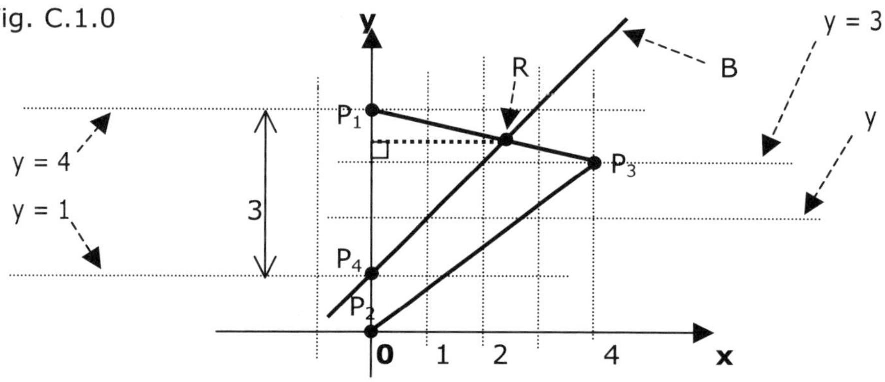

The area of the **T** is 4 X 4 X 1/2 = 8, and if we take for the base the line segment $\mathbf{P_2P_4}$, the area of the triangle $\mathbf{P_2P_3P_4}$ is 1 X 4 X 1/2 = 2. Then, the **B** needs to meet the line segment $\mathbf{P_1P_3}$.
Let **L** be a line including the $\mathbf{P_1}$ and the $\mathbf{P_3}$, **R** be a point where the **L** meets the **B**, and **H** be the triangle $\mathbf{P_1P_4R}$. In the graph, we can see that the line segment $\mathbf{P_1P_4}$ is the base of the **H**, that the length is 3, and also, that the x-coordinate of the **R** is the height of the **H**.
Since the **L** includes the two points $\mathbf{P_1}$ and $\mathbf{P_3}$, the slope is $\Delta y/\Delta x = (3 - 4)/(4 - 0) = -1/4$. Also, since the **L** includes $\mathbf{P_1(0, 4)}$, the equation of the **L** is $\mathbf{y - 4 = (-1/4)(x - 0) = -x/4 \Rightarrow y = -x/4 + 4}$.
Let **m** be the slope of the line **B**.
Then, since the **B** includes the point $\mathbf{P_4(0, 1)}$, the equation of the **B** is $\mathbf{y - 1 = m(x - 0) \Rightarrow y = mx + 1}$.
Now, $\mathbf{-x/4 + 4 = mx + 1 \Rightarrow -x + 16 = 4mx + 4 \Rightarrow (4m + 1)x = 12 \Rightarrow x = 12/(4m + 1)}$.
Therefore, the x-coordinate of the **R** is $\mathbf{12/(4m + 1)}$.
Then, the area of the $\mathbf{H = (1/2)(3)\{12/(4m + 1)\} = 18/(4m + 1)}$. Then, $\mathbf{18/(4m + 1) = 4}$, which is half the area of the **T**. $\mathbf{18/(4m + 1) = 4 \Rightarrow 18 = 16m + 4 \Rightarrow m = 7/8}$.
Then, the equation of the **B** is $\mathbf{y = 7x/8 + 1}$, which however, is not the complete solution yet.
We now, need to move the line **B** to the proper position. The $\Delta\mathbf{x}$ was 2, and the $\Delta\mathbf{y}$ was 1.
Then, the $\Delta\mathbf{x}$ is now −2, and the $\Delta\mathbf{y}$ is −1 now.
Since the **B** includes the $\mathbf{P_4}$, let's move the $\mathbf{P_4}$ first.
Then, the original position of the $\mathbf{P_4}$ is $\mathbf{(0 - 2, 1 - 1) = (-2, 0)}$.
Due to the translation, the slope of the **B** remains to be 7/8. Therefore, the proper equation of the **B** is:
$\mathbf{y - 0 = (7/8)\{x - (-2)\} = (7/8)x + 7/4 \Rightarrow y = (7/8)x + 7/4}$.

Suggestions or Solutions

To the Problem in the Example D

Suppose that we have an equation of a line: y = (t/4 + 1)x + t + 4 where t is constant. Then, find the values of the t such that the y > 0 for the x > -3.

In math, we often put in graphs behaviors of or changes in objects. More precisely, we present such by means of curves in graphs. Among many kinds in curves, the most frequently used are lines. When we work with lines, we usually use equations of line. We have two forms in the equations on of which is the standard form and the other of which is the general form. Of those, we often use the standard form. Beginning with this example through the next several examples, we are going to discuss on what to do and how to do it depending on how the lines in standard equations are put in graphs. Particularly in cases where we use lines for curves of functions, standard equations are appropriate. That's because it is easy to promptly convert such an equation to a function. In fact, we can just use it as is.

Putting an equation or idea in a graph can help us figure out what's going on in the problem so that we can have better estimate on how we should approach the problem. It is naturally the case since we don't just keep thinking about what the problem says but we do it while we keep looking at it. Therefore, the ability to visualize something abstract is very important. In most cases, we should begin the graph with a schematic one. Nevertheless, we should understand the fundamental structure of the curve we want to put in the graph. For instance, a line cannot change the direction once the direction is set; however, a curved line generally keeps changing directions. We should not violate such fundamentals. Once we have started putting the rough curve in the applicable coordinate system, we frequently can notice what's going on in the problem. Then, we keep putting into the graph more information (facts) given or implied in the problem until we get a substantial idea of what's going on. Once we've got it, we refine the graph if necessary. We do not have to be precise all the way from the beginning.

In addition, putting an equation of line in a graph, don't just put one line alone. We should put several. If considered possible, extreme cases should be shown, too. While doing such, we come to the proper understanding of the problem first, and in turn, we frequently get to see whereabouts the solution should be or how it should be constructed. Now, let's work out the problem in this example.

The equation given is as follows: **y = (t/4 + 1)x + t + 4** where **t** is constant.
To begin with, we should reason on what equation it is. Of course, some students can see it at a glance, though. However, in the early stage of working on it, the author couldn't see it very well at a glance. At any rate, it won't hurt to give it a check.
The variables are **x** and **y**, and their exponents are all **1**s. By the way, in **y = 1/x**, the exponent of the **x** is not 1 but −1. Then, we can take **(t/4 + 1)** for the slope of a line, and **(t + 4)** for the y-intercept.
So, we can take the equation for an equation of line.

Now, let's find out about what line the equation is talking about.

The equation presents such a condition as follows: when **x > -3**, **y > 0**. In this case, let's call −3 the critical value of the **x**. A critical value can be said to be a value that we use when we explain the boundary of a particular state. That is, we can say that it shows a turning point. Then, isn't it better to call it a boundary value? We normally use the word, 'boundary value' for some other purpose in math.

Let's now, find out about what the condition stated right above means.

To begin with, let's put some lines in your favorite x-y coordinate system and see what happens. That is, we are going to visualize what the problem is talking about.

Fig. D.0

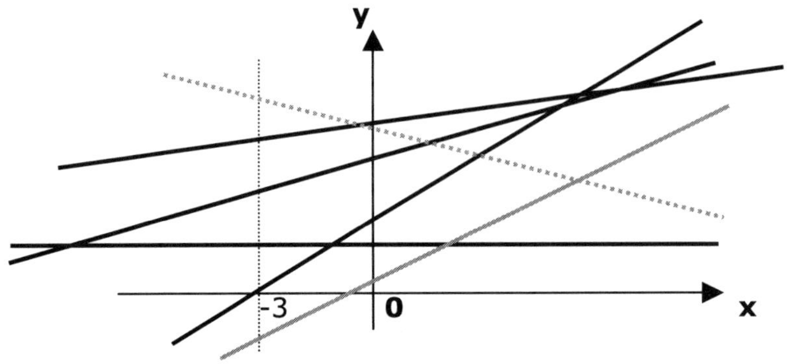

Constructing a graph, we should first show such a crucial item as the critical value mentioned above. Then, putting some lines, we can often notice how the appropriate lines should be placed to meet the specifications (or conditions) stated (or implied) in the problem. That is, we frequently see how the line should be presented to satisfy what the problem is asking. Now, if we want the **y** to have positive values when **x** is greater than -3, the line has to be one of the black ones. We shouldn't neglect the horizontal one. The gray dotted line meets the condition in the problem with the amount shown, but it eventually goes below the x-axis. Therefore, the gray dotted line is not allowed. Now, what can we see?

The solution depends on the slope. First of all, we can see that the slope cannot be negative. That is, the slope needs to be greater than or equal to 0. In other words, we need to consider the two cases as follows: "The slope is positive." and "The slope is 0."

Let's now, make the more specific presentation. Presentations by function make easy the comparisons or presentations of values of variables. Let's put the given equation in a function, then:

Suppose that $\mathbf{y = f(x) = (t/4 + 1)x + t + 4}$ where **t** is constant. The **x** and **t** are real, of course.
Then, the **f** will be determined as for the value of the **t**.
Now, the curve of the **f** that can satisfy "When $\mathbf{x > -3}$, $\mathbf{f(x) > 0}$." is one of the lines less the gray ones in the graph above. The slope of the line is $\mathbf{(t/4 + 1)}$.

Meanwhile, we should follow the fundamental principle on lines. What fundamental principle?
The principle says that once set, the slope (or direction) of a line doesn't change. The slope is constant.
A line keeps going either up or down in the same manner, or is being either horizontal or vertical.
If the slope is positive, the line goes up, and if negative, it goes down. If the slope is 0, the line is horizontal. What about a vertical line, then? In math, a vertical line is said to have no slope. Therefore, we need to note that no slope does not mean the slope of 0.

Suppose now, that in $\mathbf{y = ax + b}$, the **y** is positive for a particular value of the **x**, and afterward, the **y** keeps increasing as the **x** increases. Then, the **y** is always positive for any value of the **x** that is greater than or equal to the particular value.
What makes the **y** keep increasing? That is the slope **a**, which is positive.
Suppose on the contrary, that in $\mathbf{y = ax + b}$, the **y** is negative for a particular value of the **x**, and afterward, the **y** continues increasing as the **x** increases. Then, the **y** is always negative for any value of the **x** that is greater than or equal to the particular value, and what keeps the **y** decreasing is the very slope **a**, which is negative.

Now, the statement, "When **x > -3**, **f(x) > 0**." means the statement as follows: "The slope is positive, and also, **f(-3)** is greater than or equal to 0." or "The slope is 0, and also, **f(-3)** is positive." That is, it means the statement as follows: "**t/4 + 1 > 0** and **f(-3) ≥ 0**." or "**t/4 + 1 = 0** and **f(-3) > 0**."

However, "**f(-3) ≥ 0**." includes "**f(-3) > 0**.", but "**t/4 + 1> 0**" and "**t/4 + 1 = 0**." are two separate stories. Therefore, we first need to check to see if **(t/4 + 1)** can be 0, that is, if the slope of the line can be 0.

Now, what about the y-intercept, **t + 4**? Doesn't it have to be positive, too?
Of course, it does. If the condition above is satisfied, **t + 4** gets a positive value.
In fact, it is included in the case where the slope is 0. We will see why shortly.

Now, let's specify the **t**. To begin with, we should check to see if the slope can be 0.
Then, when the slope, that is, **t/4 + 1** is 0, **f(-3)** has to be positive.

Meanwhile, **t/4 + 1 = 0** ⇒ **f(x) = (t/4 + 1)x + t + 4 = t + 4** ⇒ for all **x**, **f(x) = t + 4**.
The **(t + 4)** is the very y-intercept. Also, since the **t** is constant, **(t + 4)** is constant, too. Therefore, such a function as **f(x) = t + 4** is called a constant function. A constant function gets the same value for all input values.

Now, **t/4 + 1 = 0** ⇒ **t = -4**. Also, **t/4 + 1 = 0** ⇒ **f(x) = t + 4 = -4 + 4 = 0** since **t = -4**.
Then, **f(x) = 0**, which means that the **f** is 0 for all **x**.
So, **f(-3) = 0** which is not positive ⇒ **f(-3)** is not positive. Therefore, the slope cannot be 0.

Therefore, the slope > 0, that is, we need to have **t/4 + 1 > 0**, and also, **f(-3) ≥ 0**.
First, **t/4 + 1 > 0** ⇒ **t > -4**.
Also, **f(-3) = (t/4 + 1)(-3) + t + 4 = -3t/4 – 3 + t + 4 = t/4 + 1 ≥ 0** ⇒ **t ≥ -4**.
Now, we need to consider the two conditions "the slope > 0 and **f(-3) ≥ 0**."
Then, **(t ≥ -4** and **t > -4)** ⇒ **t > -4**.
Therefore, if we choose for the value of the **t** a value greater than –4, we get **y > 0** when **x > -3**.
However, when we write the answer sheet, we also need to explain the case where the slope is negative.

In short:

Since the exponents of the two variables **x** and **y** are 1 respectively and the **t** is constant, it can be an equation of line. Then, **(t/4 + 1)** is the slope.
Suppose that **y = f(x) = (t/4 + 1)x + t + 4** where **t** is constant.
To begin with, in case where the slope is negative, even if **f(-3) > 0**, there is no value for the **t** satisfying the case since the line eventually goes below the x-axis. Therefore, we can have two possible cases as follows:

The first case: **t/4 + 1 = 0** and **f(-3) > 0**. First, **t/4 + 1 = 0** ⇒ **t = -4** and **f(x) = t + 4**.
Then, **t = -4** ⇒ **t + 4 = -4 + 4 = 0**, which is not positive ⇒ **t/4 + 1 ≠ 0**.

The second case: **t/4 + 1 > 0** and **f(-3) ≥ 0**. First, **t/4 + 1 > 0** ⇒ **t > -4**.
Also, **f(-3) = (t/4 + 1)(-3) + t + 4 = -3t/4 – 3 + t + 4 = t/4 + 1 ≥ 0** ⇒ **t ≥ -4**.
Then, **(t ≥ -4** and **t > -4)** ⇒ **t > -4**. Therefore, **t > -4**.

Let's take a similar one in the next example, now.

Suggestions or Solutions

To the **Problem** in the Example **E**

Suppose that we have an equation of a line: y = (t/4 + 1)x + t + 5 where t is constant. Then, find the values of the t such that the y > 0 for the x > -3.

Fig. E.0

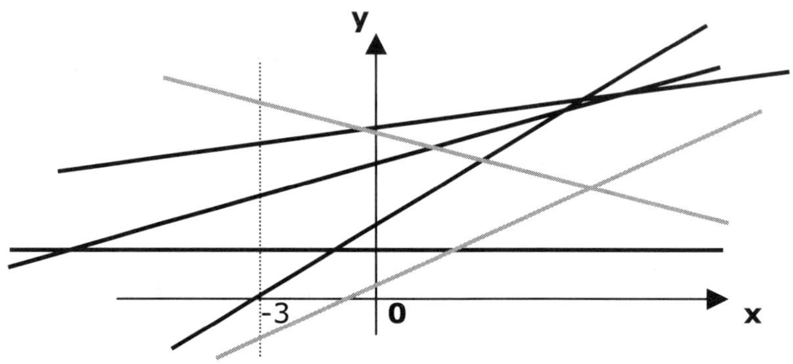

Suppose that **y = f(x) = (t/4 + 1)x + t + 5** where **t** is constant.
Then, when **x > -3**, we need to have **f(x) > 0**. The slope is **(t/4 + 1)**.

A line depends on the slope. A line is no other than a slope, and a slope is no other than a line.

Now, "**f(x) > 0** when **x > -3**." means that after **x = -3**, the value of the **f** has to be always positive.
That is, for any value of the **x** that is greater than –3, the value of the **f** needs to be definitely positive.

The curve of the **f** is now, a line. Going up, a line keeps going up, but going down, it keeps going down, or being horizontal, it constantly proceeds horizontally. Then, as the value of the **x** gets increased, the value of **f(x)** keeps either increasing or decreasing; otherwise, it is constant.

Then, if the value of **f(x)** needs to be positive for all **x** greater than –3, then what does the value of the **f(x)** have to be when **x = -3**, and also, how do the values of the **f** have to be afterwards, that is, for the values of the **x** greater than -3?
That is, if **x > -3** and **f(x) > 0**, then what does the value of **f(-3)** have to be, and also, how do the values of **f(x)** have to be when **x > -3**?

First, the **f(-3)** has to be 0 or greater than 0. That is, we need to have **f(-3) ≥ 0**.
Next, when **x > -3**, the value of **f(x)** needs to keep increasing or **f(x) = c > 0** where **c** is a constant.
In other words, the slope of the **f** is to be positive or 0. That is, we need to have the slope ≥ 0.

Now, let's consider the value of **f(-3)** and the slope altogether.

First, if the slope = 0, we need to have **f(-3) > 0**.
Next, if the slope > 0, we need to have **f(-3) ≥ 0**.

On the other hand, if the slope < 0, even though **f(-3) > 0**, we can have **f < 0** since the line eventually goes below the x-axis.

Now, let's specify the **t**.

There can be such two possible cases as follows:
If the slope = 0, **f(-3) > 0**. If the slope > 0, **f(-3) ≥ 0**.

To begin with, let's see if the slope can be 0.
Then, when the slope is 0, that is, **t/4 + 1 = 0**, **f(3)** needs to be positive.

t/4 + 1 = 0 ⇒ f(x) = (t/4 + 1)x + t + 5 = t + 5
⇒ for all **x**, **f(x) = t + 5**, which means that the **f** is a constant function.
Then, **t/4 + 1 = 0 ⇒ t = -4 ⇒ t + 5 = -4 + 5 = 1 > 0 ⇒ f(-3) > 0**.
So, the slope can be 0. Therefore, we can have **t = -4**.

Next, when the slope is positive, that is, **t/4 + 1 > 0**, we need to have **f(-3) ≥ 0**.

First, **t/4 + 1 > 0 ⇒ t > -4**.
Also, **f(-3) = (t/4 + 1)(-3) + t + 5 = -3t/4 – 3 + t + 5 = t/4 + 2 ≥ 0 ⇒ t ≥ -8**.
Now, we need to consider at the same time the two conditions, "the slope > 0 and **f(-3) ≥ 0**."
Then, **(t > -4** and **t ≥ -8) ⇒ t > -4**.
And yet, since the slope can be 0, the **t** can be -4.
Therefore, **t ≥ -4**.

Now, choosing for the **t** a value greater than or equal to –4, we get **y > 0** when **x > -3**.

In short:

Suppose that **y = f(x) = (t/4 + 1)x + t + 5** where **t** is a constant.

First of all, if the slope < 0, even though **f(-3) > 0**, we can have **f < 0** because the line eventually goes below the x-axis.

Therefore, we can have the two possible cases as follows:
If the slope = 0, **f(-3) > 0**.
If the slope > 0, **f(-3) ≥ 0**.

First, let's see if we can have the slope = 0.
t/4 + 1 = 0 ⇒ t = -4, and also, **f(x) = (t/4 + 1)x + t + 5 = t + 5**.
Then, **t = -4 ⇒ t + 5 = -4 + 5 = 1 > 0 ⇒ f(-3) > 0**.
Consequently, the slope can be 0. Therefore, the **t** can be –4.

Next, we need to have **t/4 + 1 > 0**, and also, **f(-3) ≥ 0**.
To begin with, **t/4 + 1 > 0 ⇒ t > -4**.
Also, **f(-3) = (t/4 + 1)(-3) + t + 5 = -3t/4 – 3 + t + 5 = t/4 + 2 ≥ 0 ⇒ t ≥ -8**.
Then, **(t > -4** and **t ≥ -8) ⇒ t > -4**. In addition, the **t** can be -4.

Therefore, **t ≥ -4**.

Suggestions or Solutions

To the Problem in the Example F

Suppose that we have an equation of a line: y = (t/4 + 1)x + t + 4 where t is constant. Then, find the values of the t such that the y ≥ 0 for the x > -3.

Fig. F.0

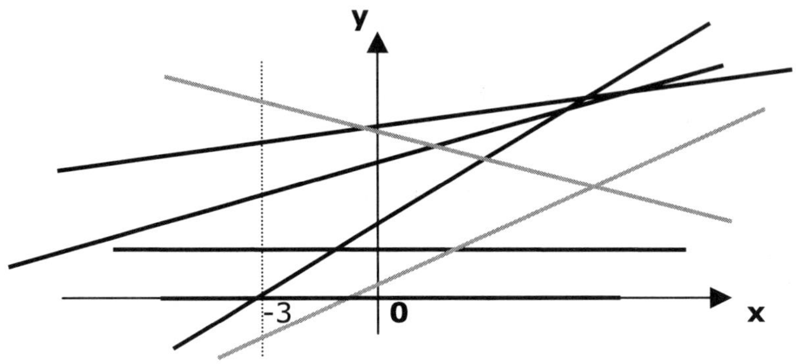

To begin with, suppose that **y = f(x) = (t/4 + 1)x + t + 4** where **t** is a constant.
This time, when **x > -3**, we need to have **f(x) ≥ 0**. The slope is **t/4 + 1**.

What do we mean by a line? A slope! What do we mean by a slope? A line!

Now, "When **x > -3**, **f(x) ≥ 0**." means that after **x = -3**, the value of the **f** has to be always 0 or positive. That is to say that for any value of the **x** greater than −3, the value of the **f** is definitely 0 or positive.

The curve of the **f** is now, a line. Then, as the value of the **x** gets increased, the vale of the **f(x)** keeps either increasing or decreasing; otherwise, it is constant.

Then, if **x > -3** and **f(x) ≥ 0**, what is the value of the **f(-3)** to be, and also, how is the value of the **f(x)** is to be when **x > -3**?

First, **f(-3)** needs to be greater than or equal to 0. That is, we need to have **f(-3) ≥ 0**.
Also, for the values of the **x** greater than −3, the value of the **f(x)** needs to be greater than **f(-3)**, and also, keep increasing; otherwise, we need to have **f(x) = c = f(-3) ≥ 0** where **c** is a constant. That is, the slope of the **f** is positive; otherwise, the slope is 0 and **f(x) = c ≥ 0** where **c** is a constant.

Now, let's consider the value of **f(-3)** and the slope altogether.

First, if the slope = 0, we need to have **f(-3) ≥ 0**.
Next, if the slope > 0, we need to have **f(-3) ≥ 0**.

Then after all, we need to have the slope ≥ 0, and also, **f(-3) ≥ 0**.

On the other hand, if the slope < 0, even though **f(-3) > 0**, we can have **f < 0** since the line eventually goes below the x-axis.

Now, let's specify the **t**.

We need to have the slope ≥ 0 and **f(-3)** ≥ **0**.

Then, **t/4 + 1** ≥ **0** ⇒ **t** ≥ **-4**.
Also, **f(-3) = (t/4 + 1)(-3) + t + 4 = -3t/4 – 3 + t + 4 = t/4 + 1 ≥ 0** ⇒ **t ≥ -4**.
Now, we need to consider at the same time the two conditions, "The slope ≥ 0 and **f(-3) ≥ 0**."
Then, **(t** ≥ **-4** and **t** ≥ **-4)** ⇒ **t** ≥ **-4**.
Therefore, **t** ≥ **-4**.

Now, choosing for the **t** a value greater than or equal to –4, we get **y** ≥ **0** when **x > -3**.

Let's take a look at the case where the slope is 0.

t/4 + 1 = 0 ⇒ **t = -4** ⇒ **t + 4 = -4 + 4 = 0.**
Also, **t/4 + 1 = 0** ⇒ **f(x) = (t/4 + 1)x + t + 4 = t + 4**.
So, **f(x) = 0**.

Therefore, if the slope is 0, the line needs to be the x-axis itself.
If the curve of a function **f** is the x-axis itself, the **f** is also, called a 0-function.

In short:

Suppose that **y = f(x) = (t/4 + 1)x + t + 4** where **t** is a constant.

To begin with, if the slope < 0, even though **f(-3) > 0**, we can have **f < 0** since the line eventually goes below the x-axis.

Then, we can have the two possible cases as follows:
If the slope = 0, **f(-3)** ≥ **0**.
If the slope > 0, **f(-3)** ≥ **0**.

Then after all, we need to have the slope ≥ 0 and **f(-3)** ≥ **0**.
First, **t/4 + 1** ≥ **0** ⇒ **t** ≥ **-4**.
Also, **f(-3) = (t/4 + 1)(-3) + t + 4 = -3t/4 – 3 + t + 4 = t/4 + 1 ≥ 0** ⇒ **t ≥ -4**.
Then, **(t ≥ -4** and **t** ≥ **-4)** ⇒ **t** ≥ **-4**.

Therefore, **t** ≥ **-4**.

Suggestions or Solutions

To the Problem in the Example G

Suppose that we have an equation of a line: $y = (t/4 + 1)x + t + 5$ where t is constant. Then, find the values of the t such that the $y \geq 0$ for the $x > -3$.

Fig. G.0

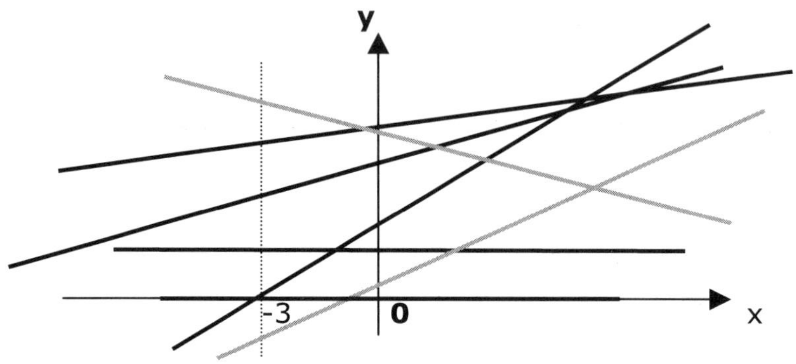

To begin with, suppose that $y = f(x) = (t/4 + 1)x + t + 5$ where **t** is a constant.
This time also, when $x > -3$, we need to have $f(x) \geq 0$. The slope is $t/4 + 1$.
Now, "When $x > -3$, $f(x) \geq 0$." means that after $x = -3$, the **f** is always greater than or equal to 0.

Then, if $x > -3$ and $f(x) \geq 0$, what is the value of the $f(-3)$ to be, and also, how is the value of the $f(x)$ is to be when $x > -3$?
First, we need to have $f(-3) \geq 0$.
Also, the slope of the **f** is positive; otherwise, the slope is 0 and $f(x) = c \geq 0$ where **c** is a constant.

Now, let's consider the value of $f(-3)$ and the slope altogether.
First, if the slope = 0, we need to have $f(-3) \geq 0$. Next, if the slope > 0, we need to have $f(-3) \geq 0$.
Then after all, we need to have the slope ≥ 0, and also, $f(-3) \geq 0$.
On the other hand, if the slope < 0, even though $f(-3) > 0$, we can have $f < 0$ since the line eventually goes below the x-axis.

Now, let's specify the **t**. We need to have the slope ≥ 0 and $f(-3) \geq 0$.
Then, $t/4 + 1 \geq 0 \Rightarrow t \geq -4$.
Also, $f(-3) = (t/4 + 1)(-3) + t + 5 = -3t/4 - 3 + t + 5 = t/4 + 2 \geq 0 \Rightarrow t \geq -8$.
Now, we need to consider at the same time the two conditions, "The slope ≥ 0 and $f(-3) \geq 0$."
Then, $(t \geq -4$ and $t \geq -8) \Rightarrow t \geq -4$. Therefore, $t \geq -4$.

In short:

Suppose that $y = f(x) = (t/4 + 1)x + t + 5$ where **t** is a constant.
To begin with, if the slope < 0, even though $f(-3) > 0$, we can have $f < 0$ since the line eventually goes below the x-axis. Then, we have two possible cases as follows:
The slope = 0 and $f(-3) \geq 0$. The slope > 0 and $f(-3) \geq 0$.
Then after all, the slope ≥ 0 and $f(-3) \geq 0$. First, $t/4 + 1 \geq 0 \Rightarrow t \geq -4$.
Also, $f(-3) = (t/4 + 1)(-3) + t + 5 = -3t/4 - 3 + t + 5 = t/4 + 2 \geq 0 \Rightarrow t \geq -8$.
Then, $(t \geq -4$ and $t \geq -8) \Rightarrow t \geq -4$. Therefore, $t \geq -4$.

Suggestions or Solutions

To the Problem in the Example H

Suppose that we have an equation of a line: $y = (t/4 + 1)x + t + 3$ where t is constant. Then, find the values of the t such that the $y \geq 0$ for the $x > -3$.

Fig. H.0

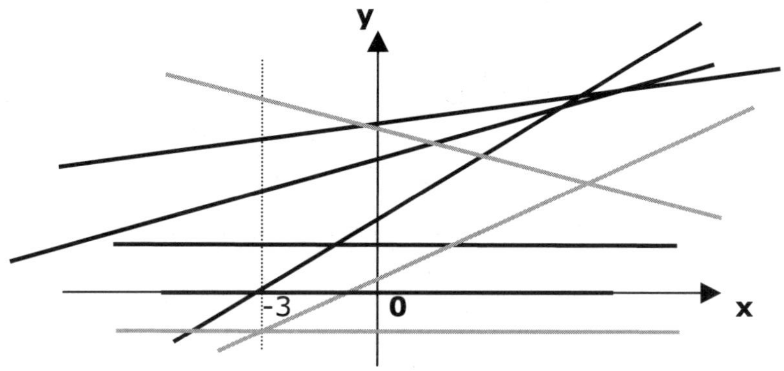

Suppose first, that $y = f(x) = (t/4 + 1)x + t + 3$ where **t** is a constant.
This time also, when $x > -3$, we need to have $f(x) \geq 0$. The slope is $t/4 + 1$.
Now, "When $x > -3$, $f(x) \geq 0$." means that after $x = -3$, the **f** is always greater than or equal to 0.

Then, if $x > -3$ and $f(x) \geq 0$, what is the value of the $f(-3)$ to be, and also, how is the value of the $f(x)$ is to be when $x > -3$?
First, we need to have $f(-3) \geq 0$.
Also, the slope of the **f** is positive; otherwise, the slope is 0 and $f(x) = c \geq 0$ where **c** is a constant.

Now, let's consider the value of $f(-3)$ and the slope altogether.
First, if the slope = 0, we need to have $f(-3) \geq 0$. Next, if the slope > 0, we need to have $f(-3) \geq 0$.
Then after all, we need to have the slope ≥ 0, and also, $f(-3) \geq 0$.
On the other hand, if the slope < 0, even though $f(-3) > 0$, we can have $f < 0$ since the line eventually goes below the x-axis.

Now, let's specify the **t**. We need to have the slope ≥ 0 and $f(-3) \geq 0$.
Then, $t/4 + 1 \geq 0 \Rightarrow t \geq -4$.
Also, $f(-3) = (t/4 + 1)(-3) + t + 3 = -3t/4 - 3 + t + 3 = t/4 \geq 0 \Rightarrow t \geq 0$.
Now, we need to consider at the same time the two conditions, "The slope ≥ 0 and $f(-3) \geq 0$."
Then, $(t \geq -4$ and $t \geq 0) \Rightarrow t \geq 0$. Therefore, $t \geq 0$.

In short:

Suppose that $y = f(x) = (t/4 + 1)x + t + 3$ where **t** is a constant.
To begin with, if the slope < 0, even though $f(-3) > 0$, we can have $f < 0$ since the line eventually goes below the x-axis. Then, we have two possible cases as follows:
The slope = 0 and $f(-3) \geq 0$. The slope > 0 and $f(-3) \geq 0$.
Then after all, the slope ≥ 0 and $f(-3) \geq 0$. First, $t/4 + 1 \geq 0 \Rightarrow t \geq -4$.
Also, $f(-3) = (t/4 + 1)(-3) + t + 3 = -3t/4 - 3 + t + 3 = t/4 \geq 0 \Rightarrow t \geq 0$.
Then, $(t \geq -4$ and $t \geq 0) \Rightarrow t \geq 0$. Therefore, $t \geq 0$.

Suggestions or Solutions

To the Problem in the Example I

Suppose that we have an equation of a line: y = (t/4 + 1)x + t + 3 where t is constant. Then, find the values of the t such that the y > 0 for -2 < x < 3.

Fig. I.0

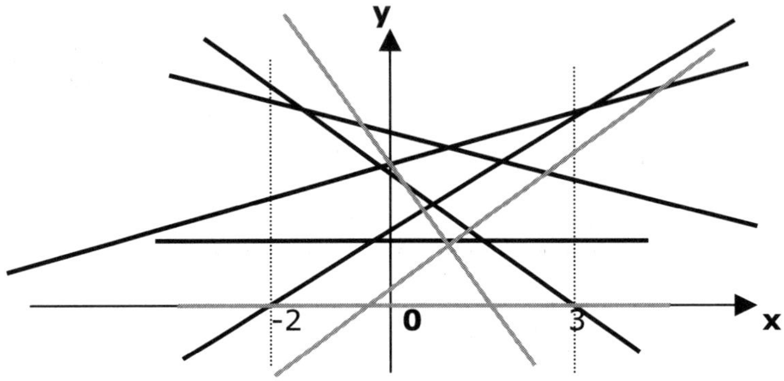

Suppose first, that **y = f(x) = (t/4 + 1)x + t + 3** where **t** is a constant.
Then, since **y > 0** when **-2 < x < 3**, we have **f(x) > 0** for **-2 < x < 3**. Also, **t/4 + 1** is the slope.

Now, what is the difference between this example and the previous examples?
In the previous ones, the slope of the line could not be negative. However, in this, the slope can be 0. Therefore, it seems that we don't have to reason too hard on the slope of the line. Nevertheless, we still need to reason on the slope. Is there any other we need to reason, then? They are the values of the **f** for the critical values. We have two critical values. One is –2 and the other is 3.

Let's first, reason on the slope.

Then, "When **-2 < x < 3, f(x) > 0.**" means that when the value of the **x** is between –2 and 3, the value of the **f** needs to be always positive.

Then, if **-2 < x < 3** and **f > 0**, what is the value of **f(-2)** to be and how are the values of the **f** to be when **-2 < x < 3**?

First, we need to have **f(-2) ≥ 0.**
Also, for the value of the **x** greater than –2, the value of the **f** is to be greater than **f(-2)** and keep increasing; otherwise, **f(x) = c = f(-2) > 0** where **c** is a constant.
Therefore in this case, the slope ≥ 0. However, that's not all.

We can have another situation as follows.
First, **f(-2) > 0.**
Also, for the values of the **x** greater than –2, the value of the **f** is less than **f(-2)** and keeps decreasing but needs to stay positive until the **x** is 3.
That is to say that first, **f(-2) > 0**, and then, when **x > -2, f(x) < f(-2)** but **f(3) ≥ 0.**
In this case, the slope < 0.

Now, the slope can be negative, positive, or 0.

If we reason on every occasion of the slope to solve the problem in such a manner as above, we are likely to bother onerous calculations as shown above.

Now, examining the graph, we can see how the lines should be put.

"When $-2 < x < 3$, $f(x) > 0$." means that when the value of the **x** is between –2 and 3, the value of the **f** needs to be always positive.

Then, when the value of the **x** is between –2 and 3, the line is to be above the x-axis.
Then, we need to have $f(-2) \geq 0$ and $f(3) \geq 0$.
However, we need to note that the curve of the **f** can be the x-axis itself since the **f** can be 0 for both of the critical values of –2 and 3. That is to say that the **f** can be a 0-function. If that is the case, because we need to have $f(x) > 0$, we need to exclude the values of the **t** in such a case.

Now, let's specify the values of the **t**.

We have $y = f(x) = (t/4 + 1)x + t + 3$.

To begin with, let's see if the **f** can be a 0-function.
$t/4 + 1 = 0 \Rightarrow f(x) = (t/4 + 1)x + t + 3 = t + 3.$
$t/4 + 1 = 0 \Rightarrow t = -4 \Rightarrow t + 3 = -4 + 3 = -1 \Rightarrow$ the **f** cannot be a 0-function.

Then, $f(-2) = (t/4 + 1)(-2) + t + 3 = -t/2 - 2 + t + 3 = t/2 + 1 \geq 0 \Rightarrow t \geq -2$.
Also, $f(3) = (t/4 + 1)(3) + t + 3 = 3t/4 + 3 + t + 3 = 7t/4 + 6 \geq 0 \Rightarrow t \geq -24/7 = -3 - 3/7$.

Now, we need to consider at the same time the two conditions, "$f(-2) \geq 0$ and $f(3) \geq 0$."
Then, $(t \geq -2$ and $t \geq -24/7) \Rightarrow t \geq -2$.
So, $t \geq -2$.

Therefore, choosing for the **t** a value greater than or equal to –2, when $-2 < x < 3$, we get $f(x) > 0$.

In short:

Suppose that $y = f(x) = (t/4 + 1)x + t + 3$ where **t** is a constant.
Now, "When $-2 < x < 3$, $f(x) > 0$." means that when the value of the **x** is between –2 and 3, the value of the **f** needs to be always positive.
Then, when the value of the **x** is between –2 and 3, the line needs to be above the x-axis.
Then, we need to have $f(-2) \geq 0$ and $f(3) \geq 0$.

However, the **f** should not be a 0-function. Then, let's give it check.
$t/4 + 1 = 0 \Rightarrow t = -4 \Rightarrow t + 3 = -4 + 3 = -1 \Rightarrow$ the **f** cannot be a 0-function.

First, $f(-2) = (t/4 + 1)(-2) + t + 3 = -t/2 - 2 + t + 3 = t/2 + 1 \geq 0 \Rightarrow t \geq -2$.
Also, $f(3) = (t/4 + 1)(3) + t + 3 = 3t/4 + 3 + t + 3 = 7t/4 + 6 \geq 0 \Rightarrow t \geq -24/7 = -3 - 3/7$.

Then, $(t \geq -2$ and $t \geq -24/7) \Rightarrow t \geq -2$. Therefore, $t \geq -2$.

Suggestions or Solutions

To the Problem in the Example J

Suppose that we have an equation of a line: $y = (t/4 + 1)x + t + 4$ where t is constant. Then, find the values of the t such that the $y > 0$ for $-2 < x < 3$.

Fig. J.0

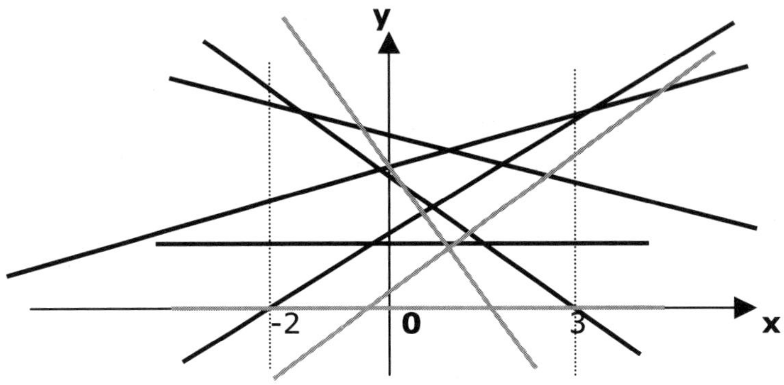

Suppose first, that $y = f(x) = (t/4 + 1)x + t + 4$ where **t** is a constant.
Then, $f(x) > 0$ for $-2 < x < 3$. Also, $t/4 + 1$ is the slope.
This example is not quite different from the previous one.
Now, "When $-2 < x < 3$, $f(x) > 0$." means that when the value of the **x** is between the two critical values –2 and 3, the value of the **f** needs to be always positive.
As in the previous example, examining the graph, we can see how the lines should be put.
Then, when the value of the **x** is between –2 and 3, the line needs to be above the x-axis.
Then, we need to have $f(-2) \geq 0$ and $f(3) \geq 0$.
However, since the **f** should not be a 0-function, let's first check to see if the **f** can be a 0-function.
$t/4 + 1 = 0 \Rightarrow f(x) = (t/4 + 1)x + t + 4 = t + 4$.
$t/4 + 1 = 0 \Rightarrow t = -4 \Rightarrow t + 4 = -4 + 4 = 0 \Rightarrow$ If $t = -4$, the **f** can be a 0-function.
First, $f(-2) = (t/4 + 1)(-2) + t + 4 = -t/2 - 2 + t + 4 = t/2 + 2 \geq 0 \Rightarrow t \geq -4$.
Also, $f(3) = (t/4 + 1)(3) + t + 4 = 3t/4 + 3 + t + 4 = 7t/4 + 7 \geq 0 \Rightarrow t \geq -4$.
Now we need to consider at the same time the two conditions, "$f(-2) \geq 0$ and $f(3) \geq 0$."
Then, ($t \geq -4$ and $t \geq -4$) $\Rightarrow t \geq -4$. However, $t \neq -4$ since if $t = -4$, we get $f(x) = 0$.
So, $t > -4$.
Therefore, choosing for the **t** a value greater than –4, when $-2 < x < 3$, we get $f(x) > 0$.

In short:

Suppose that $y = f(x) = (t/4 + 1)x + t + 4$ where **t** is a constant. Then, when $-2 < x < 3$, $f(x) > 0$.
Then, when the value of the **x** is between –2 and 3, the line needs to be above the x-axis.
Then, we need to have $f(-2) \geq 0$ and $f(3) \geq 0$.
However, since the **f** should not be a 0-function, let's give it a check.
$t/4 + 1 = 0 \Rightarrow t = -4 \Rightarrow t + 4 = -4 + 4 = 0 \Rightarrow$ If $t = -4$, the **f** can be a 0-function.
First, $f(-2) = (t/4 + 1)(-2) + t + 4 = -t/2 - 2 + t + 4 = t/2 + 2 \geq 0 \Rightarrow t \geq -4$.
Also, $f(3) = (t/4 + 1)(3) + t + 4 = 3t/4 + 3 + t + 4 = 7t/4 + 7 \geq 0 \Rightarrow t \geq -4$.
Then, ($t \geq -4$ and $t \geq -4$) $\Rightarrow t \geq -4$. However, $t \neq -4$ since if $t = -4$, $f(x) = 0$.
Therefore, $t > -4$.

Suggestions or Solutions

To the **Problem** in the Example **K**

Suppose that we have an equation of a line: y = (t/4 + 1)x + t + 4 where t is constant. Then, find the values of the t such that the y is positive for -5 < x < -3.

Fig. K.0

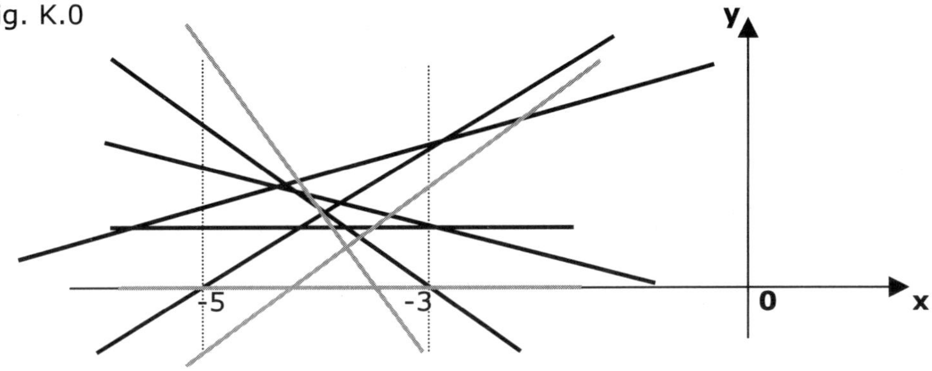

Suppose first, that $y = f(x) = (t/4 + 1)x + t + 4$ where **t** is a constant.
Then, $f(x) > 0$ for $-5 < x < -3$. Also, $t/4 + 1$ is the slope.
This example, too, is not quite different from the previous one.
Now, "When $-5 < x < -3$, $f(x) > 0$." means that when the value of the **x** is between the two critical values −5 and −3, the value of the **f** needs to be always positive.
As in the previous example, examining the graph, we can see how the lines should be put.
Then, when the value of the **x** is between −5 and −3, the line needs to be above the x-axis.
Then, we need to have $f(-5) \geq 0$ and $f(-3) \geq 0$.
However, since the **f** should not be a 0-function, let's give it a check.
$t/4 + 1 = 0 \Rightarrow f(x) = (t/4 + 1)x + t + 4 = t + 4.$
$t/4 + 1 = 0 \Rightarrow t = -4 \Rightarrow t + 4 = -4 + 4 = 0 \Rightarrow$ If $t = -4$, the **f** can be a 0-function.
First, $f(-5) = (t/4 + 1)(-5) + t + 4 = -5t/4 - 5 + t + 4 = -t/4 - 1 \geq 0 \Rightarrow t \leq -4$.
Also, $f(-3) = (t/4 + 1)(-3) + t + 4 = -3t/4 - 3 + t + 4 = t/4 + 1 \geq 0 \Rightarrow t \geq -4$.
Now, we need to consider simultaneously the two conditions, "$f(-5) \geq 0$ and $f(-3) \geq 0$."
Then, $(t \leq -4$ and $t \geq -4) \Rightarrow t = -4$. However, $t \neq -4$ since if $t = -4$, we get $f(x) = 0$.
So, we don't have any value of for the **t** in this case.
Therefore, choosing any value for the **t**, when $-5 < x < -3$, we don't get $f(x) > 0$.

In short:

Suppose that $y = f(x) = (t/4 + 1)x + t + 4$ where **t** is a constant. Then, when $-5 < x < -3$, $f(x) > 0$.
Then, when the value of the **x** is between −5 and −3, the line needs to be above the x-axis.
Then, we need to have $f(-5) \geq 0$ and $f(-3) \geq 0$. Since the **f** should not be 0-function, let's give it a check first. $t/4 + 1 = 0 \Rightarrow t = -4 \Rightarrow t + 4 = -4 + 4 = 0 \Rightarrow$ If $t = -4$, the **f** can be a 0-function.
First, $f(-5) = (t/4 + 1)(-5) + t + 4 = -5t/4 - 5 + t + 4 = -t/4 - 1 \geq 0 \Rightarrow t \leq -4$.
Also, $f(-3) = (t/4 + 1)(-3) + t + 4 = -3t/4 - 3 + t + 4 = t/4 + 1 \geq 0 \Rightarrow t \geq -4$.
Then, $(t \leq -4$ and $t \geq -4) \Rightarrow t = -4$. However, $t \neq -4$ because if $t = -4$, $f(x) = 0$.

Therefore, there is no value for the **t** in this problem.

Suggestions or Solutions

To the Problem in the Example L

Suppose that we have an equation of a line: $y = (t/4 + 1)x + t + 4$ where t is constant. Then, find the values of the t such that the y is positive for $-5 \le x \le -3$.

Fig. L.0

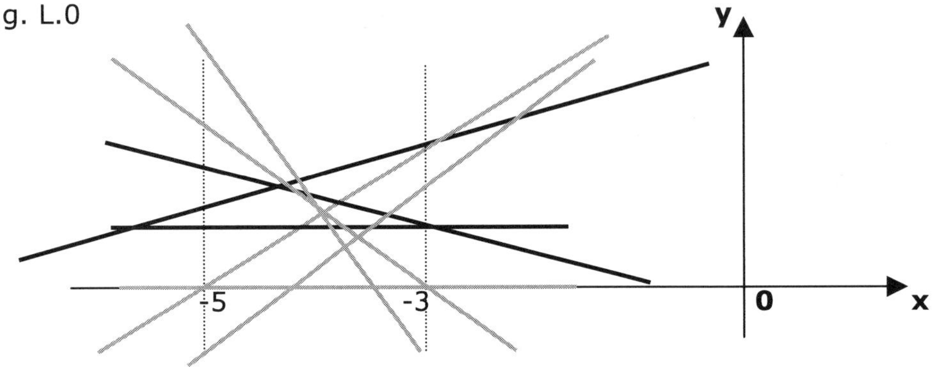

Suppose first, that $\mathbf{y = f(x) = (t/4 + 1)x + t + 4}$ where **t** is a constant.
Then, $\mathbf{f(x) > 0}$ for $\mathbf{-5 \le x \le -3}$. Also, $\mathbf{t/4 + 1}$ is the slope.
This example is not quite different from the preceding one, either. Now, "When $\mathbf{-5 \le x \le -3}$, $\mathbf{f(x) > 0}$."
means that when the **x** takes a value from –5 to –3, the value of the **f** needs to be always positive.
As in the previous example, examining the graph, we can see how the lines should be put.
When, the **x** takes a value from –5 to –3, the line needs to be above the x-axis.
Therefore, for this time, we need to have $\mathbf{f(-5) > 0}$ and $\mathbf{f(-3) > 0}$.

In this case, if the **f** satisfy the conditions above, it cannot be a 0-function.

First, $\mathbf{f(-5) = (t/4 + 1)(-5) + t + 4 = -5t/4 - 5 + t + 4 = -t/4 - 1 > 0 \Rightarrow t < -4}$.
Also, $\mathbf{f(-3) = (t/4 + 1)(-3) + t + 4 = -3t/4 - 3 + t + 4 = t/4 + 1 > 0 \Rightarrow t > -4}$.
Now, we need to consider at the same time the two conditions, "$\mathbf{f(-5) \ge 0}$ and $\mathbf{f(-3) \ge 0}$."
Then, $\mathbf{(t < -4}$ and $\mathbf{t > -4)} \Rightarrow$ There is no value for the **t** in this case.
Therefore, there is no value of the **t** satisfying the problem.
Now, whatever value we choose for the **t**, we can't have $\mathbf{f(x) > 0}$ for $\mathbf{-5 \le x \le -3}$.

In short:

Suppose that $\mathbf{y = f(x) = (t/4 + 1)x + t + 4}$ where **t** is a constant. Then, $\mathbf{f(x) > 0}$ for $\mathbf{-5 \le x \le -3}$.
Then, when the **x** takes a value from –5 to –3, the line needs to be above the x-axis.
Then, we need to have $\mathbf{f(-5) > 0}$ and $\mathbf{f(-3) > 0}$.
First, $\mathbf{f(-5) = (t/4 + 1)(-5) + t + 4 = -5t/4 - 5 + t + 4 = -t/4 - 1 > 0 \Rightarrow t < -4}$.
Also, $\mathbf{f(-3) = (t/4 + 1)(-3) + t + 4 = -3t/4 - 3 + t + 4 = t/4 + 1 > 0 \Rightarrow t > -4}$.
Then, $\mathbf{(t < -4}$ and $\mathbf{t > -4)} \Rightarrow$ There is no value for the **t** in this case.

Therefore, there is no value of the **t** satisfying the problem.

Suggestions or Solutions

To the **Problem** in the Example **M**

Suppose that we have an equation of a line: $y = (t/4 + 1)x + t + 4$ where t is constant. Then, find the values of the t such that the $y \geq 0$ for $-5 < x < -3$.

Fig. M.0

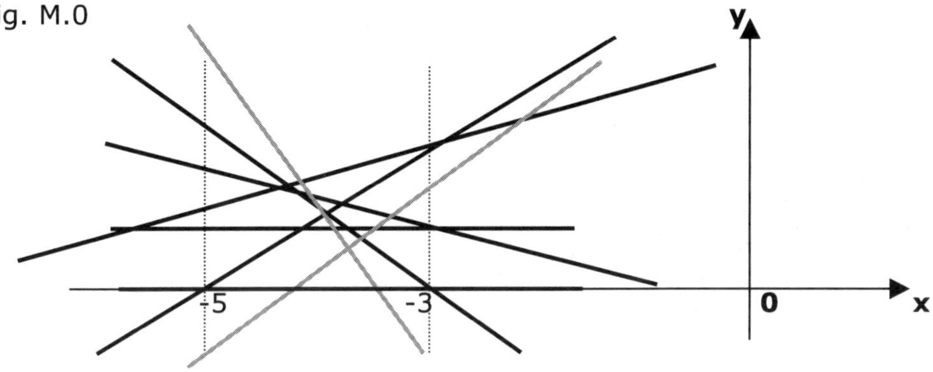

Suppose first, that $y = f(x) = (t/4 + 1)x + t + 4$ where t is a constant.
Then, $f(x) \geq 0$ for $-5 < x < -3$. Also, $t/4 + 1$ is the slope.
The example, too, is similar to the one presented earlier.
Now, "$f(x) \geq 0$ for $-5 < x < -3$." says that when the value of the x is between the two critical values −5 and −3, the value of the f is greater than or equal to 0.
As in the previous example, examining the graph, we can see how the lines should be put.
First, when the value of the x is between −5 and −3, the line needs to be above the x-axis.
Also, we can have $f(-5) = 0$ or $f(-3) = 0$.
Therefore, for this time, we need to have $f(-5) \geq 0$ and $f(-3) \geq 0$.
So, for this time, the f is allowed to be a 0-function, and therefore, we need to do the test.
$t/4 + 1 = 0 \Rightarrow f(x) = (t/4 + 1)x + t + 4 = t + 4$.
$t/4 + 1 = 0 \Rightarrow t = -4 \Rightarrow t + 4 = -4 + 4 = 0 \Rightarrow$ If $t = -4$, the f can be a 0-function.
First, $f(-5) = (t/4 + 1)(-5) + t + 4 = -5t/4 - 5 + t + 4 = -t/4 - 1 \geq 0 \Rightarrow t \leq -4$.
Also, $f(-3) = (t/4 + 1)(-3) + t + 4 = -3t/4 - 3 + t + 4 = t/4 + 1 \geq 0 \Rightarrow t \geq -4$.
Now, we need to consider simultaneously the two conditions, "$f(-5) \geq 0$ and $f(-3) \geq 0$."
Then, $(t \leq -4$ and $t \geq -4) \Rightarrow t = -4$.
Therefore, the value of the t can be −4 only. Then, $f(x) = 0$, that is, the f can be a 0-function only.
Now, choosing -4 for the value of the t, we get $f(x) \geq 0$ for $-5 < x < -3$.

In short:

Suppose that $y = f(x) = (t/4 + 1)x + t + 4$ where t is a constant. Then, $f(x) \geq 0$ for $-5 < x < -3$.
First, when the value of the x is between −5 and −3, the line needs to be above the x-axis.
Also, we can have $f(-5) = 0$ or $f(-3) = 0$. Therefore, we need to have $f(-5) \geq 0$ and $f(-3) \geq 0$.
So, the f is allowed to be a 0-function, and therefore, we need to do the test.
$t/4 + 1 = 0 \Rightarrow t = -4 \Rightarrow t + 4 = -4 + 4 = 0 \Rightarrow$ If $t = -4$, the f can be a 0-function.
First, $f(-5) = (t/4 + 1)(-5) + t + 4 = -5t/4 - 5 + t + 4 = -t/4 - 1 \geq 0 \Rightarrow t \leq -4$.
Also, $f(-3) = (t/4 + 1)(-3) + t + 4 = -3t/4 - 3 + t + 4 = t/4 + 1 \geq 0 \Rightarrow t \geq -4$.
Then, $(t \leq -4$ and $t \geq -4) \Rightarrow t = -4$.

Suggestions or Solutions

To the **Problem** in the Example **N**

Suppose that we have an equation of a line: $y = (t/4 + 1)x + t + 4$ where t is constant. Then, find the values of the t such that the $y \geq 0$ for $-5 \leq x \leq -3$.

Fig. N.0

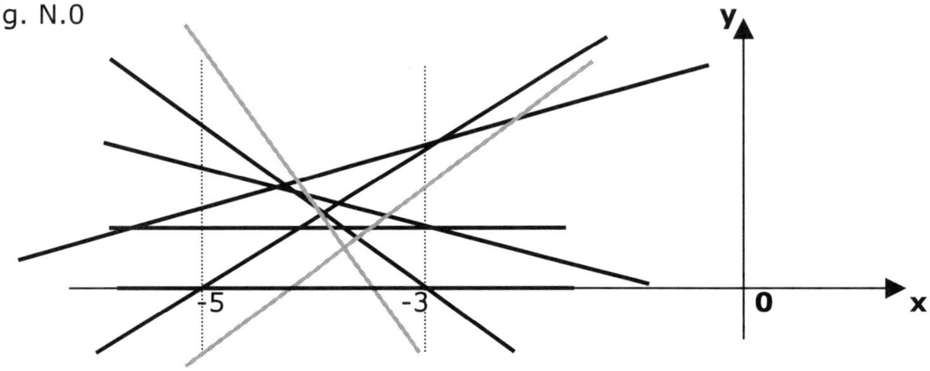

Suppose first, that $y = f(x) = (t/4 + 1)x + t + 4$ where **t** is a constant.
Then, $f(x) \geq 0$ for $-5 \leq x \leq -3$. Also, $t/4 + 1$ is the slope.
The example, too, is similar to the one presented earlier.
Now, "$f(x) \geq 0$ for $-5 \leq x \leq -3$." says that when the **x** takes values from −5 to −3, the value of the **f** is greater than or equal to 0.
As in the previous example, examining the graph, we can see how the lines should be put.
First, when the value of the **x** is between −5 and −3, the line needs to be above the x-axis.
Also, we can have $f(-5) = 0$ or $f(-3) = 0$. Therefore, we need to have $f(-5) \geq 0$ and $f(-3) \geq 0$.
So, the **f** is allowed to be a 0-function, and therefore, we need to do the test.
$t/4 + 1 = 0 \Rightarrow f(x) = (t/4 + 1)x + t + 4 = t + 4$.
$t/4 + 1 = 0 \Rightarrow t = -4 \Rightarrow t + 4 = -4 + 4 = 0 \Rightarrow$ If $t = -4$, the **f** can be a 0-function.
First, $f(-5) = (t/4 + 1)(-5) + t + 4 = -5t/4 - 5 + t + 4 = -t/4 - 1 \geq 0 \Rightarrow t \leq -4$.
Also, $f(-3) = (t/4 + 1)(-3) + t + 4 = -3t/4 - 3 + t + 4 = t/4 + 1 \geq 0 \Rightarrow t \geq -4$.
Now, we need to consider simultaneously the two conditions, "$f(-5) \geq 0$ and $f(-3) \geq 0$."
Then, $(t \leq -4$ and $t \geq -4) \Rightarrow t = -4$.
Therefore, the value of the **t** can be −4 only. Then, $f(x) = 0$, that is, the **f** can be a 0-function only.
Now, choosing -4 for the value of the **t**, we get $f(x) \geq 0$ for $-5 \leq x \leq -3$.

In short:

Suppose that $y = f(x) = (t/4 + 1)x + t + 4$ where **t** is a constant. Then, $f(x) \geq 0$ for $-5 \leq x \leq -3$.
First, when the value of the **x** is between −5 and −3, the line needs to be above the x-axis.
Also, we can have $f(-5) = 0$ or $f(-3) = 0$. Therefore, we need to have $f(-5) \geq 0$ and $f(-3) \geq 0$.
So, the **f** is allowed to be a 0-function, and therefore, we need to do the test.
$t/4 + 1 = 0 \Rightarrow t = -4 \Rightarrow t + 4 = -4 + 4 = 0 \Rightarrow$ If $t = -4$, the **f** can be a 0-function.
First, $f(-5) = (t/4 + 1)(-5) + t + 4 = -5t/4 - 5 + t + 4 = -t/4 - 1 \geq 0 \Rightarrow t \leq -4$.
Also, $f(-3) = (t/4 + 1)(-3) + t + 4 = -3t/4 - 3 + t + 4 = t/4 + 1 \geq 0 \Rightarrow t \geq -4$.
Then, $(t \leq -4$ and $t \geq -4) \Rightarrow t = -4$.

Suggestions or Solutions
To the **Problem** in the Example **O**

Suppose that we have an equation of a line: y = (t/4 + 2)x + t + 3 where t is constant. Then, find the values of the t such that the y is negative, 0, or positive for -5 ≤ x ≤ -3.

Fig. O.0

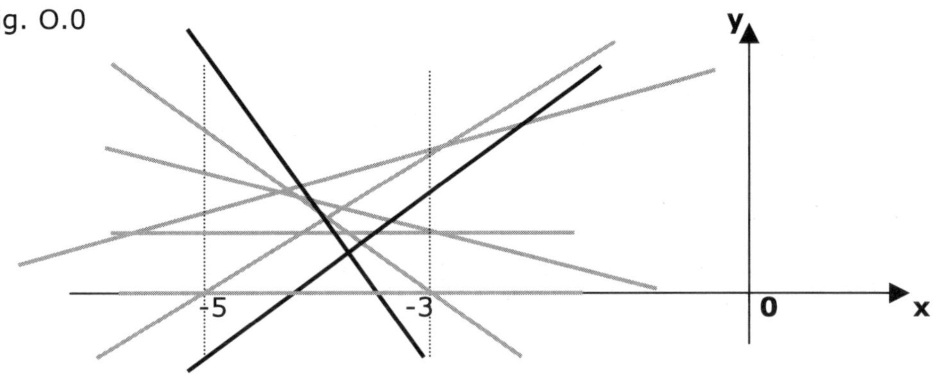

This example looks similar to the previous ones, but is quite different.

Suppose first, that **y = f(x) = (t/4 + 2)x + t + 3** where **t** is a constant.
Then, **f(x)** ≥ **0** or f(**x**) **< 0** for **-5** ≤ **x** ≤ **-3**. Also, **t/4 + 2** is the slope.

Now, "**f(x)** ≥ **0** or **f(x) < 0** for **-5** ≤ **x** ≤ **-3**." means that when the **x** takes values from –5 to –3, the values of the **f** can be positive, 0, and negative. That is, one of the values is 0, and the others are positive values and negative values.

As in the previous examples, examining the graph, we can see how the lines should be put.
Then, the line needs to cross in the x-axis between –5 and -3.
Then, we can have the two cases as follows:

f(-5) < 0 and **f(-3) > 0**.
f(-5) > 0 and **f(-3) < 0**.

Now, let's begin whit the first case: **f(-5) < 0** and **f(-3) > 0**.
We have **f(x) = (t/4 + 2)x + t + 3**.
Then, **f(-5) = (t/4 + 2)(-5) + t + 3 = -5t/4 – 10 + t + 3 = -t/4 – 7 < 0** ⇒ **t > -28**.
Also, **f(-3) = (t/4 + 2)(-3) + t + 3 = -3t/4 – 6 + t + 3 = t/4 – 3 > 0** ⇒ **t > 12**.
Now, we need to consider at the same time the two conditions, "**f(-5) < 0** and **f(-3) > 0**.".
Then, **(t > -28** and **t > 12)** ⇒ **t > 12**.

Next, let's move on the second case: **f(-5) > 0** and **f(-3) < 0**.
Then, **f(-5) = -t/4 – 7 > 0** ⇒ **t < -28**.
also, **f(-3) = t/4 – 3 < 0** ⇒ **t < 12**.
Now, we need to consider the two cases, "**f(-5) > 0** and **f(-3) < 0**."
Then, **(t < -28** and **t < 12)** ⇒ **t < -28**.

Now, we need to put the two cases together.
Then, **$t > 12$** or **$t < -28$**.

Therefore, choosing for the value of the **t** a value greater than 12 or less than −28, we get:
$f(x) \geq 0$ or **$f(x) < 0$** for **$-5 \leq x \leq -3$**.

Now, we can use another method.
We have two cases as follows:
$f(-5) < 0$ and **$f(-3) > 0$**. **$f(-5) > 0$** and **$f(-3) < 0$**.

Then, in both of the cases, **$f(-5)$** and **$f(-3)$** have signs opposite to each other.
Therefore, we can put both of the cases in such a single expression as follows: **$f(-5)f(-3) < 0$**.

We have **$f(-5) = -t/4 - 7$** and **$f(-3) = t/4 - 3$**.

Then, we get **$f(-5)f(-3) = (-t/4 - 7)(t/4 - 3) = -(t/4 + 7)(t/4 - 3) < 0$**.
So, **$-(t/4 + 7)(t/4 - 3) < 0 \Rightarrow (t/4 + 7)(t/4 - 3) > 0 \Rightarrow (t + 28)(t - 12) > 0 \Rightarrow t > 12$** or **$t < -28$**.

In short:

Suppose first, that **$y = f(x) = (t/4 + 2)x + t + 3$** where **t** is a constant.
Then, we have **$f(x) \geq 0$** or **$f(x) < 0$** for **$-5 \leq x \leq -3$**.
Then, the line needs to cross the x-axis between −5 and -3.
Then, we can two possible cases as follows:
$f(-5) < 0$ and **$f(-3) > 0$**.
$f(-5) > 0$ and **$f(-3) < 0$**.

Then, in both of the cases, **$f(-5)$** and **$f(-3)$** have signs opposite of each other.
Therefore, we can put both of the cases in such a single expression as follows: **$f(-5)f(-3) < 0$**.

Then, **$f(-5) = (t/4 + 2)(-5) + t + 3 = -5t/4 - 10 + t + 3 = -t/4 - 7$**.
Next, **$f(-3) = (t/4 + 2)(-3) + t + 3 = -3t/4 - 6 + t + 3 = t/4 - 3$**.

Then, we get **$f(-5)f(-3) = (-t/4 - 7)(t/4 - 3) = -(t/4 + 7)(t/4 - 3) < 0$**.
So, **$-(t/4 + 7)(t/4 - 3) < 0 \Rightarrow (t/4 + 7)(t/4 - 3) > 0 \Rightarrow (t + 28)(t - 12) > 0 \Rightarrow t > 12$** or **$t < -28$**.

Suggestions or Solutions

To the Problem in the Example P

Suppose that we have an equation of a line: $y = (t/4 + 2)x + t + 3$ where t is constant. Then, find the values of the x such that the $y > 0$ for $1 \le t \le 5$.

This time, given the extent of a constant, we need to find a particular extent of a variable.
A constant is different from a variable in nature, but has some property of a variable, too. A variable usually has an extent where it takes a value. The same is true for a constant. Therefore, we also use a constant the way we use a variable but a constant is not a variable. Now, we need to use the extent of the constant **t** to determine the extent of the **x**.

Then, how do we know for what value of the **x** the **y** begins to be positive?
We need to know the x-intercept of the line. However, it is not enough to know the x- intercept only. That because knowing the x-intercept only, we can't see if the **y** gets positive or negative when the value of the **x** gets greater than the x-intercept.
What else do we need know, then?
We need to know the slope of the line, too. However, the slope, too, gets determined depending on the extent of the constant **t**.

Now, we are going to find the extents of the slope and the x-intercept, and let's begin with the slope.
The slope is $(t/4 + 2)$. Suppose $a = t/4 + 2$. We have $1 \le t \le 5$. Then, as the **t** increases, the slope **a** increases, too. That is, as the **t** increases, the line gets steeper. Then, the extent of the **a** is as follows:
$1 \le t \le 5 \Rightarrow 1/4 \le t/4 \le 5/4 \Rightarrow 9/4 \le t/4 + 2 \le 13/4 \Rightarrow 9/4 \le a \le 13/4$.

Next, let's find the extent of the x-intercept. Now, we can use two kinds of methods.
In the first method, we directly find the x-intercept to find the extent.
In the second method, we first find the extent of the y-intercept, and then, find the extent of the x-intercept.

Let's begin with the second method.
The y-intercept is $t + 3$. Suppose that $b = t + 3$. We have $1 \le t \le 5$. Then, as the **t** increases, the y-intercept **b** increases, too. That is, as the **t** increases, the line looks like moving up along the y-axis. Then, the extent of the **b** is as follows: $1 \le t \le 5 \Rightarrow 4 \le t + 3 \le 8 \Rightarrow 4 \le b \le 8$.
How do we find the x-intercept, now?
Since it is the value of the **x** when $y = 0$, we set $y = 0$ in the equation of the line, solve the equation for the **x**, and then, we can take the solution for the x-intercept. Suppose now, that the x-intercept is **s**.
First, when $t = 1$, we get $y = (9/4)x + 4$. Then, $0 = (9/4)x + 4 \Rightarrow x = -16/9 \Rightarrow s = -16/9$.
Next, when $t = 5$, we get $y = (13/4)x + 8$. Then, $0 = (13/4)x + 8 \Rightarrow x = -32/13 \Rightarrow -32/13$.
Therefore, we get $-32/13 \le s \le -16/9$.

Next, the first method is as follows. We have $y = (t/4 + 2)x + t + 3$. Then:
$0 = (t/4 + 2)x + t + 3 \Rightarrow x = -(t + 3)/(t/4 + 2) = -4(t + 3)/(t + 8)$.
$(t + 3)/(t + 8) = (t + 3 + 5 - 5)/(t + 8) = (t + 8 - 5)/(t + 8) = 1 - 5/(t + 8)$.
Therefore, $x = -4\{1 - 5/(t + 8)\} = -4 + 20/(t + 8)$.
Then, we can see that the x-intercept $s = 20/(t + 8) - 4$.
We have $1 \le t \le 5$. Then, as the **t** increases, the **s** decreases. That's because the denominator keeps increasing. Let's now, actually find the extent of the **s**.
When $t = 1$, $s = 20/(t + 8) - 4 = 20/9 - 4 = -16/9 = -1 - 7/9$.
When $t = 5$, $s = 20/(t + 8) - 4 = 20/13 - 4 = (20 - 52)/13 = -32/13 = -2 - 6/13$.
So, we get $-32/13 \le s \le -16/9$.

Now, when **t = 1**, we get **a = 9/4** and **s = -16/9**, and also, when **t = 5**, we get **a = 13/4** and **s = -32/13**. Then, since the slope is always positive when $1 \le t \le 5$, if the value of the **x** is greater than the x-intercept (**s**), the **y** is always positive.
Then, since the biggest x-intercept is −16/9 when $1 \le t \le 5$, if **x > -16/9**, we get **y > 0**.
Now, the graph is approximately as follows:

Fig. P.0

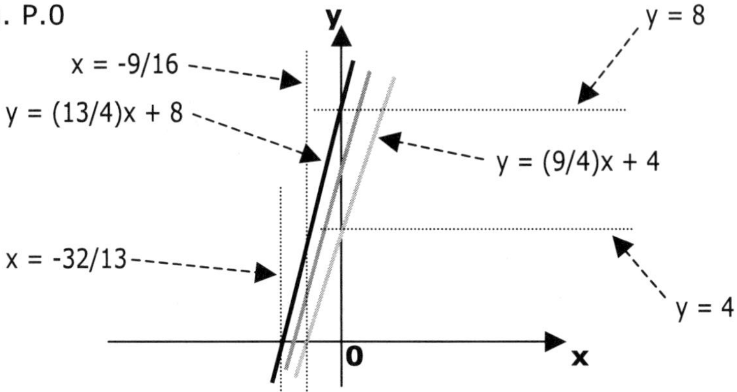

Now, when the **t** is 1, the line is in light gray, and the equation of the line is **y = 9x/4 + 4** where the x intercept is −9/16. Also, if the **t** is 5, the line is in black, and the equation of the line is **y = 13x/4 + 8** where the x-intercept is −32/13. Then, when the **t** changes from 1 to 5, the line changes from the light gray to the black. Now, if the **f** is to be always greater than 0 when $1 \le t \le 5$, the value of the **x** needs to be greater than −16/9, whereas if the **f** is to be always less than 0 when $1 \le t \le 5$, the value of the **x** needs to be less than −32/13.
Therefore, we get **y = (t/4 + 2)x + t + 3 > 0** when $1 \le t \le 5$ and **x > -16/9**.

Now, what if $1 < t \le 5$? Then, the value of the **x** can be −16/9. That's because the light gray line gets made when t = 1, but the **t** cannot be 1 in this case. Therefore, in this case, we get $x \ge -16/9$.
Now, let's approach the solution in this case in the method for the previous examples.
First, we take the **t** for a variable and the **x** for a constant in the given equation, and then, change the form of the equation. Then, we find the extent of the **x**. Of course, a constant is different from a variable. We use the aspect where both a constant and a variable can have extents.
Suppose now, that **y = f(x) = (t/4 + 2)x + t + 3** where **t** is a constant.
Then, we change the **f** to a function **g**. Only their system names are different, so they are the same.
(t/4 + 2)x + t + 3 = xt/4 + 2x + t + 3 = (x/4 + 1)t + 2x + 3.
Suppose now, that **y = g(t) = (x/4 + 1)t + 2x + 3** where **x** is a constant.
Then, we find the extent of the **x** so that we get **g(t) > 0** for $1 < t \le 5$.

Fig. P.1

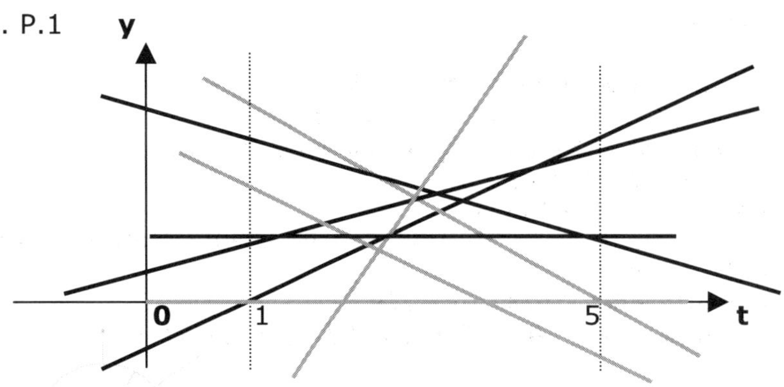

Examining the graph above, we can see how the lines should be put.
Then, when the value of the t is between 1 and 5, the line needs to be above the t-axis.
Also, we can have $\mathbf{g(1) = 0}$, but need to have $\mathbf{g(5) \neq 0}$.
Therefore, we need to have $\mathbf{g(1) \geq 0}$ and $\mathbf{g(5) > 0}$.
Then, $\mathbf{g(1) = (x/4 + 1)(1) + 2x + 3 = x/4 + 1 + 2x + 3 = 9x/4 + 4 \geq 0 \Rightarrow x \geq -16/9}$.
Also, $\mathbf{g(5) = (x/4 + 1)(5) + 2x + 3 = 5x/4 + 5 + 2x + 3 = 13x/4 + 8 > 0 \Rightarrow x > -32/13}$.
Now, we need consider at the same time the two conditions, "$\mathbf{g(1) \geq 0}$ and $\mathbf{g(5) > 0}$."
Then, $\mathbf{(x \geq -16/9}$ and $\mathbf{x > -32/13) \Rightarrow x \geq -16/9}$.
Now, if the value of the **x** is greater than or equal to −16/9, we get $\mathbf{f(x) > 0}$ for $\mathbf{1 < t \leq 5}$.

In short:

Suppose that **a** is the slope and **s** is the x-intercept. Then, $\mathbf{a = t/4 + 2}$. We now, find the **s**.
$\mathbf{0 = (t/4 + 2)x + t + 3 \Rightarrow x = -(t + 3)/(t/4 + 2) = -4(t + 3)/(t + 8)}$.
$\mathbf{(t + 3)/(t + 8) = (t + 3 + 5 - 5)/(t + 8) = (t + 8 - 5)/(t + 8) = 1 - 5/(t + 8)}$.
Therefore, $\mathbf{x = -4\{1 - 5/(t + 8)\} = -4 + 20/(t + 8)}$. Then, $\mathbf{s = 20/(t + 8) - 4}$.

Now, we find the extents of the **a** and the **s**.
First, as the **t** increases, the **a** increases, too.
$\mathbf{1 \leq t \leq 5 \Rightarrow 1/4 \leq t/4 \leq 5/4 \Rightarrow 9/4 \leq t/4 + 2 \leq 13/4 \Rightarrow 9/4 \leq a \leq 13/4}$.
Next, as the **t** increases, the **s** decreases.
When $\mathbf{t = 1}$, $\mathbf{s = 20/(t + 8) - 4 = 20/9 - 4 = -16/9 = -1 - 7/9}$.
When $\mathbf{t = 5}$, $\mathbf{s = 20/(t + 8) - 4 = 20/13 - 4 = (20 - 52)/13 = -32/13 = -2 - 6/13}$.
Therefore, $\mathbf{-32/13 \leq s \leq -16/9}$.

Then:
When $\mathbf{t = 1}$, we get $\mathbf{a = 9/4}$, and the x-intercept is -16/9.
When $\mathbf{t = 5}$, we get $\mathbf{a = 13/4}$, and the x-intercept is -32/13.
When the slope is positive, if the value of the **x** is greater than the x-intercept, the **y** gets positive.
So, when $\mathbf{1 \leq t \leq 5}$ and $\mathbf{x > -16/9}$, we get $\mathbf{y = (t/4 + 2)x + t + 3 > 0}$.
Therefore, $\mathbf{x > -16/9}$.

The other method:

First, we take the **t** for a variable and the **x** for a constant in the given equation, and then, change the form of the equation. Suppose $\mathbf{y = f(x) = (t/4 + 2)x + t + 3}$ where **t** is a constant.
Then, we change the **f** to a function **g** in such a way as follows:
$\mathbf{(t/4 + 2)x + t + 3 = xt/4 + 2x + t + 3 = (x/4 + 1)t + 2x + 3}$.

Suppose now, that $\mathbf{y = g(t) = (x/4 + 1)t + 2x + 3}$ where **x** is a constant.
Then, we find the extent of the **x** so that when $\mathbf{1 \leq t \leq 5}$, we get $\mathbf{g(t) > 0}$.
Then, when the value of the **t** changes from 1 to 5, the line needs to be above the t-axis.
Therefore, we need to have $\mathbf{g(1) > 0}$ and $\mathbf{g(5) > 0}$.
Then, $\mathbf{g(1) = (x/4 + 1)(1) + 2x + 3 = x/4 + 1 + 2x + 3 = 9x/4 + 4 > 0 \Rightarrow x > -16/9}$.
Also, $\mathbf{g(5) = (x/4 + 1)(5) + 2x + 3 = 5x/4 + 5 + 2x + 3 = 13x/4 + 8 > 0 \Rightarrow x > -32/13}$.
Then, $\mathbf{(x > -16/9}$ and $\mathbf{x > -32/13) \Rightarrow x > -16/9}$.
So, if the value of the **x** is greater than −16/9, when $\mathbf{1 \leq t \leq 5}$, we get $\mathbf{f(x) > 0}$.
Therefore, $\mathbf{x > -16/9}$.

Suggestions or Solutions

To the Problem in the Example Q

Suppose that we have an equation of a line: $y = 3(t - 1)x + 3 - 2t$ where t is constant. Then, find the values of the x such that the $y > 0$ for $0 < t < 2$.

For this time, too, given the extent of a constant, we need to find a particular extent of a variable.
A constant is different from a variable, but has a property of a variable, too. A variable usually has an extent where it takes a value. The same is true for a constant. Therefore, we sometimes use a constant like a variable. However, a constant is not a variable.
Now, after all, the extent of the constant **t** gets to determine the extents of the slope, the x-intercept, and the y-intercept.
Then, let's suppose that **a** is the slope, **s** is the x-intercept, and **b** is the y-intercept, and then, find the extents of them.

Now, let's begin with the slope **a**.
We have the slope $a = 3(t - 1)$ for $0 \le t \le 2$. Then, as the **t** increases, the slope **a** increases, too. That is, as the **t** increases, the line gets steeper. Then, the extent of the **a** is as follows:
$0 \le t \le 2 \Rightarrow -1 \le t - 1 \le 1 \Rightarrow -3 \le 3(t - 1) \le 3 \Rightarrow -3 \le a \le 3$, that is, $|a| \le 3$.

Next, we find the extent of the x-intercept **s**.
However, let's first find the extent of the y-intercept **b**, and then, find the extent of the **s**.

We have the y-intercept $b = 3 - 2t$ for $0 \le t \le 2$. Then, as the **t** increases, the y-intercept **b** decreases. That is, as the **t** increases, the line seems to move down along the y-axis. It looks like it, doesn't it? Anyway, the extent of the **b** is as follows: $0 \le t \le 2 \Rightarrow 0 \ge -2t \ge -4 \Rightarrow 3 \ge 3 - 2t \ge -1 \Rightarrow -1 \le b \le 3$.

Let's now, find the extent of the x-intercept. Since the x-intercept is the value of the **x** when $y = 0$, we set $y = 0$ in the equation of the line, solve the equation for the **x**, and then, can take the solution for the x-intercept.
First, when $t = 0$, we get $y = -3x + 3$. Then, $0 = -3x + 3 \Rightarrow x = 1 \Rightarrow s = 1$.
Next, when $t = 2$, we get $y = 3x - 1$. Then, $0 = 3x - 1 \Rightarrow x = 1/3 \Rightarrow s = 1/3$.
Therefore, we get $1/3 \le s \le 1$. Now, is that so, indeed?

Meanwhile, if we directly find the x-intercept **s**, it will be as follows. We have $y = 3(t - 1)x + 3 - 2t$.
Then, $0 = 3(t - 1)x + 3 - 2t \Rightarrow 3x = (2t - 3)/(t - 1) = (2t - 2 - 1)/(t - 1)$
$= (2t - 2)/(t - 1) - 1/(t - 1) = 2 - 1/(t - 1) = 2 + 1/(1 - t)$.
Therefore, $x = (1/3)\{2 + 1/(1 - t)\}$. Then, we get the x-intercept $s = (1/3)\{2 + 1/(1 - t)\}$.

Then, let's see how the **s** changes as the **t** increases. We have $0 \le t \le 2$.
However, we are first going to see how $1/(1 - t)$ changes. Then, suppose that $c = 1/(1 - t)$.
Then, we get $t = 0 \Rightarrow c = 1/(1 - t) = 1/1 = 1$. Then, we get $s = (1/3)(2 + 1) = 1$.
Then, as the **t** increases from 0 toward 1, the **c** increases, too, but up to what is it going to increase? Since the denominator gets close to 0, it increases infinitely.
However, if $t = 1$, we get $1 - t = 0$, but is not allowed since it is the denominator.
Anyhow, if the value of the **t** is nearly the same as 1, the **s** eventually gets to increase infinitely, too.
Then, when $0 \le t < 1$, we get $s \ge 1$.
In fact, if $t = 1$, the line is a horizontal line since the slope $a = 3(t - 1) = 0$, and in this case, we don't have the x-intercept.

Now, we get $t = 2 \Rightarrow c = 1/(1 - t) = 1/(-1) = -1$. Then, we get $s = (1/3)(2 - 1) = 1/3$.

Then, as the **t** decreases from 2 toward 1, the **c** decreases, too, but up to what is it going to decrease? Since the denominator gets close to 0 and is negative, it decreases to be negative infinity.
Then, when $\mathbf{1 < t \le 2}$, we get $\mathbf{s \le 1/3}$.
We know that if the **t** is 1, the line is a horizontal line.

Now, the actual extent of the **s** is as follows: $\mathbf{s \le 1/3}$ or $\mathbf{s \ge 1}$.
Therefore, the **s** cannot have a value between 1/3 and 1. So, $\mathbf{1/3 \le s \le 1}$ found above is wrong. Then, it is not a good idea to find the extent of the x-intercept by means of the extent of the y-intercept only.

Now, for each of the two cases $\mathbf{0 \le t < 1}$ and $\mathbf{1 < t \le 2}$, let's find the extent of the slope $\mathbf{a = 3(t - 1)}$.
Then:
$\mathbf{0 \le t < 1 \Rightarrow -1 \le t - 1 < 0 \Rightarrow -3 \le 3(t - 1) < 0.}$
$\mathbf{1 < t \le 2 \Rightarrow 0 < t - 1 \le 1 \Rightarrow 0 < 3(t - 1) \le 3.}$

Then:
$\mathbf{0 \le t < 1 \Rightarrow -3 \le a < 0}$ and $\mathbf{s \ge 1}$. $\mathbf{1 < t \le 2 \Rightarrow 0 < a \le 3}$ and $\mathbf{s \le 1/3}$.

If the slope is negative and the **x** is less than the x-intercept, the **y** is positive, and also, if the slope is positive and the **x** is greater than the x-intercept, the **y** is positive.
Then, we need to have $\mathbf{s < 1}$ and $\mathbf{s > 1/3}$. Therefore, we get $\mathbf{1/3 < s < 1}$.
Of course, when $\mathbf{t = 1}$, the line is a horizontal line since $\mathbf{a = 0}$.
Also, since $\mathbf{t = 1 \Rightarrow y = 3(t - 1)x + 3 - 2t = 1}$, the **y** is positive for all **x**.
Therefore, when $\mathbf{0 \le t \le 2}$, if $\mathbf{1/3 < x < 1}$, we get $\mathbf{y > 0}$. Now, the graph is approximately as follows:

Fig. Q.0

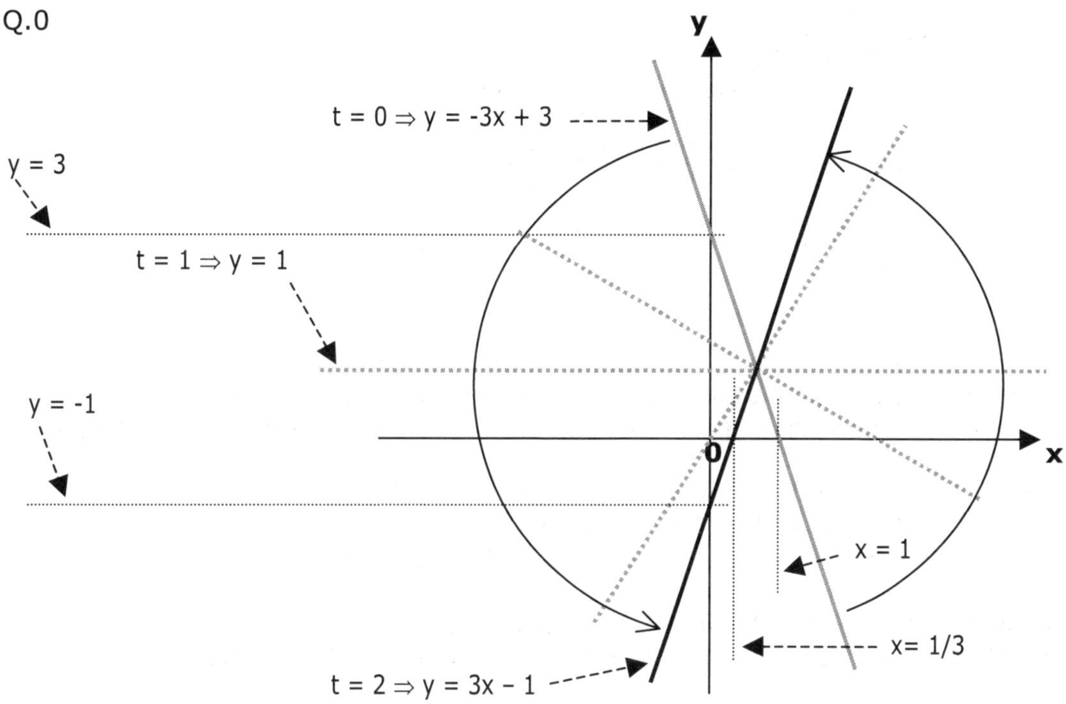

Then, let's find the point where the lines in gray and black meet each other.
Let's begin with the x-coordinate.
$\mathbf{-3x + 3 = 3x - 1 \Rightarrow 6x = 4 \Rightarrow x = 2/3 \Rightarrow y = 3x - 1 = 3(2/3) - 1 = 1.}$

Therefore, the two lines meet each other at the point **(2/3, 1)**. In fact, all the lines that can be expressed in the equation $\mathbf{y = 3(t - 1)x + 3 - 2t}$ where $\mathbf{0 \le t \le 2}$ meet each other at the point (2/3, 1) altogether. How come? We get to see it in the next section.

Therefore, it is not that the line moves down along the y-axis but that it rotates around a point.

Now, what if $\mathbf{0 < t < 2}$? Then, we get $\mathbf{1/3 < x < 1}$. That's because the lines in gray and black get made when the **t** is 0 and 2, but in this case, the **t** cannot be 0 and 2.

Now, let's take care of the problem in the method we used for the previous examples.
First, we take the **t** for a variable and the **x** for a constant in the given equation, and then, change the from of the equation. Then, we find the extent of the **x**. Of course, a constant and a variable are different from each other. We take advantage of the aspect where both a constant and a variable can have extents.
Suppose now, that $\mathbf{y = f(x) = 3(t - 1)x + 3 - 2t}$ where **t** is a constant.
Then, we change the **f** to a function **g**. Only their system names are different, so they are the same.

$$\mathbf{3(t - 1)x + 3 - 2t = 3xt - 3x + 3 - 2t = (3x - 2)t + 3 - 3x.}$$

Suppose now, that $\mathbf{y = g(t) = (3x - 2)t + 3 - 3x}$ where **x** is a constant.
Then, we find the extent of the **x** so that we get $\mathbf{g(t) > 0}$ for $\mathbf{0 < t < 2}$.

Fig. Q.1

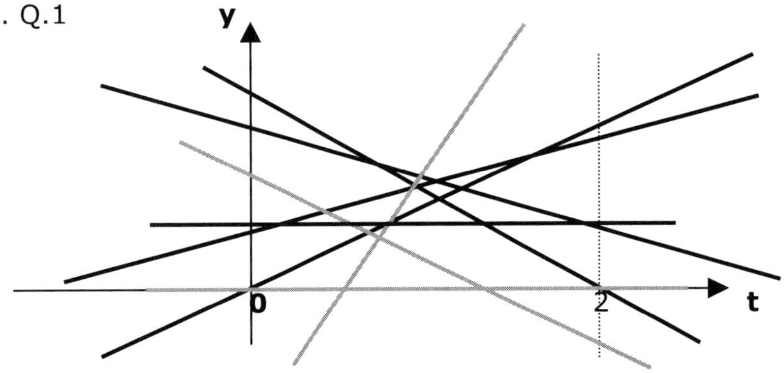

Examining the graph above, we can see how the lines should be put.
Then, when the value of the **t** is between 0 and 2, the line needs to be above the t-axis.
Also, we can have $\mathbf{g(0) = 0}$ and $\mathbf{g(2) = 0}$. Therefore, we need to have $\mathbf{g(0) \ge 0}$ and $\mathbf{g(2) \ge 0}$.

First, since the g should not be a 0-function, we need to do the test. The slope is $\mathbf{3x - 2}$.
$\mathbf{3x - 2 = 0 \Rightarrow g(t) = (3x - 2)t + 3 - 3x = 3 - 3x.}$
$\mathbf{3x - 2 = 0 \Rightarrow x = 2/3 \Rightarrow 3 - 3x = 3 - 2 = 1 \ne 0 \Rightarrow}$ The **g** cannot be a 0-function.

Then, $\mathbf{g(0) = (3x - 2)(0) + 3 - 3x = 3 - 3x \ge 0 \Rightarrow x \le 1}$.
Next, $\mathbf{g(2) = (3x - 2)(2) + 3 - 3x = 6x - 4 + 3 - 3x = 3x - 1 \ge 0 \Rightarrow x \ge 1/3}$.

Now, we need to consider at the same time the two conditions, "$\mathbf{g(0) \ge 0}$ and $\mathbf{g(2) \ge 0}$."
Then, $\mathbf{(x \le 1}$ and $\mathbf{x \ge 1/3) \Rightarrow 1/3 \le x \le 1}$.
Now, if the value of the **x** is greater than or equal to 1/3 and less than or equal to 1, we get $\mathbf{f(x) > 0}$ for $\mathbf{0 < t < 2}$.

In short:

Suppose that **a** is the slope, and that **s** is the x-intercept. Then, $\mathbf{a = 3(t - 1)}$. We now, find the **s**.
$\mathbf{0 = 3(t - 1)x + 3 - 2t \Rightarrow 3x = (2t - 3)/(t - 1) = (2t - 2 - 1)/(t - 1)}$
$\mathbf{= (2t - 2)/(t - 1) - 1/(t - 1) = 2 - 1/(t - 1) = 2 + 1/(1 - t)}$
$\mathbf{\Rightarrow x = (1/3)\{2 + 1/(1 - t)\} \Rightarrow s = (1/3)\{2 + 1/(1 - t)\}}$.

Now, we find the extents of the **a** and the **s**. First, as the **t** increases, the **a** increases, too.
$\mathbf{0 \le t \le 2 \Rightarrow -1 \le t - 1 \le 1 \Rightarrow -3 \le 3(t - 1) \le 3 \Rightarrow -3 \le a \le 3}$.

Next, suppose first, that $\mathbf{c = 1/(1 - t)}$.
Then, $\mathbf{t = 0 \Rightarrow c = 1/(1 - t) = 1/1 = 1 \Rightarrow s = (1/3)(2 + 1) = 1}$.
Then, since the denominator gets close to 0 as the **t** increases from 0 toward 1, the **c** gets to increase infinitely. So, in this case, if the value of the **t** is nearly the same as 1, the **s** eventually gets to increase infinitely, too. Then, when $\mathbf{0 \le t < 1}$, $\mathbf{s \ge 1}$. In fact, the slope $\mathbf{a = 3(t - 1) = 0}$ if $\mathbf{t = 1}$, the line is horizontal, and in this case, we have no x-intercept.

Now, $\mathbf{t = 2 \Rightarrow c = 1/(1 - t) = 1/(-1) = -1}$. Then, $\mathbf{s = (1/3)(2 - 1) = 1/3}$.
Then, since the denominator gets close to 0 as the **t** decreases from 2 toward 1 and it is negative, the **c** gets to decrease to be negative infinity. So, the **s** gets to decrease to be negative infinity, too. Then, when $\mathbf{1 < t \le 2}$, we get $\mathbf{s \le 1/3}$. Also, we know that if the **t** is 1, the line is horizontal.
Now, the extent of the **s** is as follows: $\mathbf{s \le 1/3}$ or $\mathbf{s \ge 1}$.

Then:
$\mathbf{0 \le t < 1 \Rightarrow -1 \le t - 1 < 0 \Rightarrow -3 \le 3(t - 1) < 0 \Rightarrow -3 \le a < 0}$ and $\mathbf{s \ge 1}$.
$\mathbf{1 < t \le 2 \Rightarrow 0 < t - 1 \le 1 \Rightarrow 0 < 3(t - 1) \le 3 \Rightarrow 0 < a \le 3}$ and $\mathbf{s \le 1/3}$.

If the slope is negative and the **x** is less than the x-intercept, the **y** is positive, and also, if the slope is positive and the **x** is greater than the x-intercept, the **y** is positive.
Then, we need to have $\mathbf{s < 1}$ and $\mathbf{s > 1/3}$. Therefore, we get $\mathbf{1/3 < s < 1}$.
Of course, since $\mathbf{a = 0}$ when $\mathbf{t = 1}$, the line is horizontal.
Also, since $\mathbf{t = 1 \Rightarrow y = 3(t - 1)x + 3 - 2t = 1}$, the **y** is positive for all **x**.
So, when $\mathbf{0 \le t \le 2}$, if $\mathbf{1/3 < x < 1}$, we get $\mathbf{y > 0}$. Therefore, $\mathbf{1/3 < x < 1}$.

The other method:

We first take the **t** for a variable and the **x** for a constant in the given equation, and then, change the from of the equation.
Suppose that $\mathbf{y = f(x) = 3(t - 1)x + 3 - 2t}$ where **t** is a constant.
Then, we change the **f** to a function **g** in such a way as follows:
$\mathbf{3(t - 1)x + 3 - 2t = 3xt - 3x + 3 - 2t = (3x - 2)t + 3 - 3x}$.

Suppose now, that $\mathbf{y = g(t) = (3x - 2)t + 3 - 3x}$ where **x** is a constant.
Then, we find the extent of the **x** so that $\mathbf{g(t) > 0}$ for $\mathbf{0 \le t \le 2}$.
Then, when the value of the **t** changes from 0 to 2, the lines needs to be above the t-axis.
Therefore, we need to get $\mathbf{g(0) > 0}$ and $\mathbf{g(2) > 0}$.

Then, $\mathbf{g(0) = (3x - 2)(0) + 3 - 3x = 3 - 3x > 0 \Rightarrow x < 1}$.
Also, $\mathbf{g(2) = (3x - 2)(2) + 3 - 3x = 6x - 4 + 3 - 3x = 3x - 1 > 0 \Rightarrow x > 1/3}$.
Then, $\mathbf{(x < 1}$ and $\mathbf{x > 1/3) \Rightarrow 1/3 < x < 1}$.
Now, if the value of the **x** is greater than 1/3 and less than 1, we get $\mathbf{f(x) > 0}$ for $\mathbf{0 \le t \le 2}$.
Therefore, $\mathbf{1/3 < x < 1}$.

Suggestions or Solutions

To the **Problem** in the Example **R**

Suppose that $y = f(x) = 2x + 1$ for $-1/2 \le x \le 2$, $y = f(x) = 9 - 2x$ for $2 < x \le 9/2$, and $g(x) = (f \bullet f)(x)$. Then, construct the graph of the $g(x)$.

Note: in this book and all the books in the series of **Runmath**, the **expression part** in a presentation of a function is called the **mechanism** of the function. For instance, if $f(x) = 2x^3 + 3/x$, the mechanism of the **f** is $2x^3 + 3/x$. Also, 'f' is the function **designator** and **f(x)** is the **name** of the function.

We often take a line for a curve of a function and use the function. In the function **f**, the domain is in two parts. One is $-1/2 \le x \le 2$, and the other is $2 < x \le 9/2$. Also, the **f** has two mechanisms. Now, the **g** is a simple composite function and is composed of the **f** only. Then, in the function **g** also, the domain is the same as that of the **f**. Let's first find the mechanism of the **g** for each of the parts of the domain.

$y = f(x) = 2x + 1$ for $-1/2 \le x \le 2$
$\Rightarrow g(x) = (f \bullet f)(x) = f\{f(x)\} = 2(2x + 1) + 1 = 4x + 3$ for $-1/2 \le x \le 2$.

$f(x) = 9 - 2x$ for $2 < x \le 9/2$
$\Rightarrow g(x) = (f \bullet f)(x) = f\{f(x)\} = 9 - 2(9 - 2x) = 9 - 18 + 4x = 4x - 9$ for $2 < x \le 9/2$.

Fig. R.0.

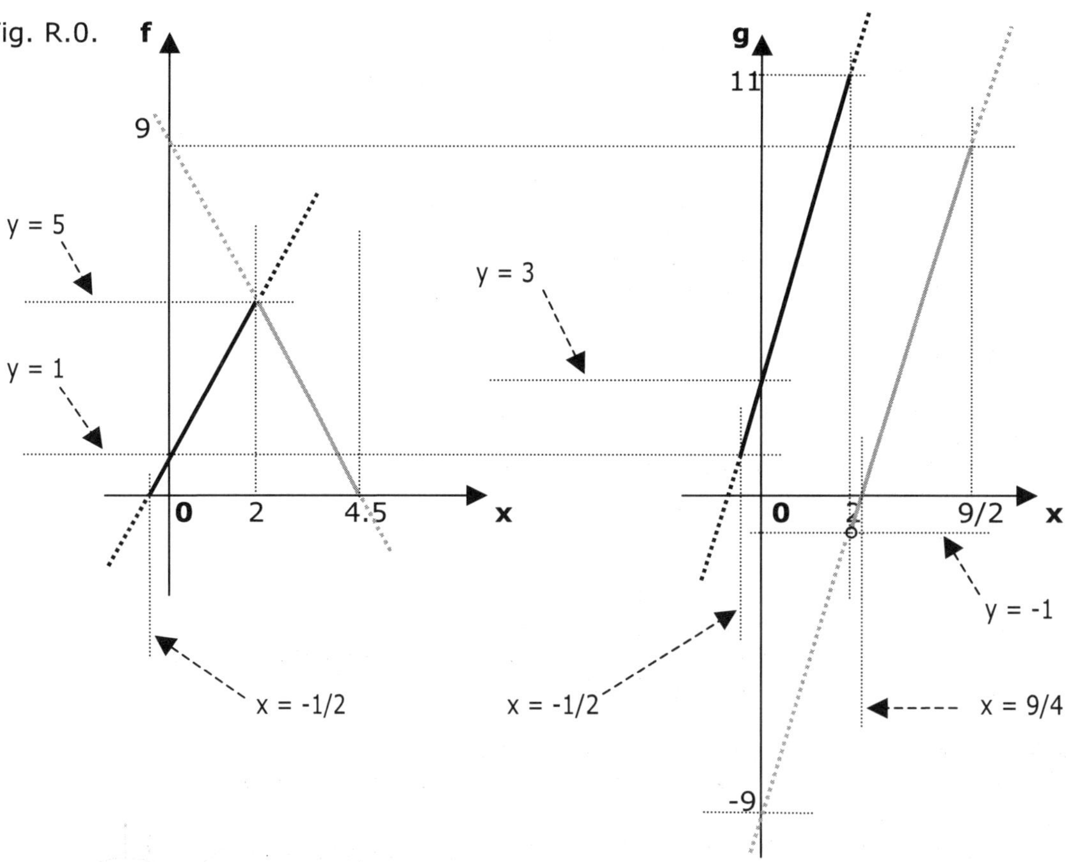

Suggestions or Solutions

To the **Problem 0** in the Example **S**

The equation given is **y = |3 − 2x| − 1**. We can present such an equation for each interval. We can see each of the intervals by means of **|3 − 2x|**. When we take care of **|a|**, we take care of the two cases where **a** ≥ **0** and **a < 0**. Therefore, in terms of the signs (+ or −), we classify the values that the **a** itself can have. We usually take care of 0 along with the positive case. Then, we can have two different cases for the **3 − 2x**.
In the first case, **3 − 2x** is greater than or equal to zero. In the second, **3 − 2x** is less than 0.
Therefore, we can see that without the magnitude sign:

$3 - 2x \geq 0 \Rightarrow 2x \leq 3 \Rightarrow x \leq 3/2 \Rightarrow y = 3 - 2x - 1 = 2 - 2x$ for $x \leq 3/2$.
$3 - 2x < 0 \Rightarrow 2x > 3 \Rightarrow x > 3/2 \Rightarrow y = -(3 - 2x) - 1 = -3 + 2x - 1 = 2x - 4$ for $x > 3/2$.

Then, how come **|3 − 2x| = -(3 − 2x)** for **3 − 2x < 0**?
That's because **|3 − 2x|** is the magnitude of **3 − 2x** and if **3 − 2x** is negative, then we need to make it positive since a magnitude can be 0 or positive.
Therefore, we have **y = 2 − 2x** for **x ≤ 3/2** or 2x − 4 for x > 3/2.

Now, let's put **y = |3 − 2x| − 1** in the graph.

Fig. S.0.0 **y = 2 − 2x** for **x ≤ 3/2** or y = 2x − 4 for x > 3/2

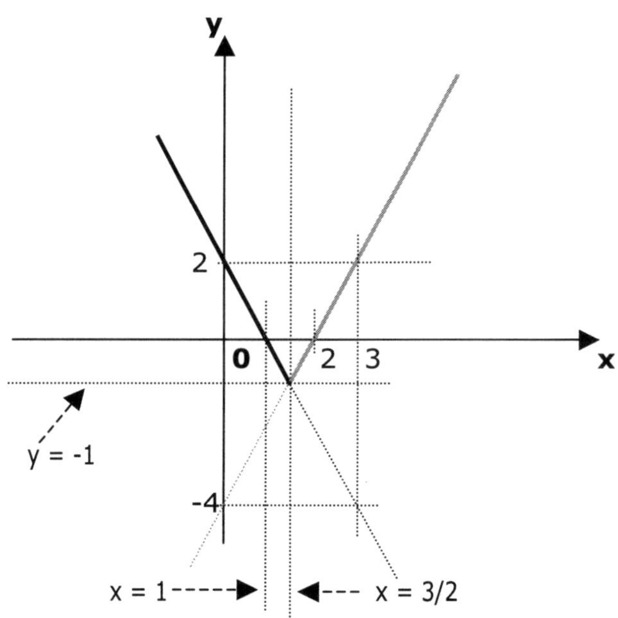

We can see that the lines in gray and black are symmetric to each other about the line **x = 3/2**.
The line in black includes the point **(3/2, -1)**, but the one in gray doesn't include it.
In fact, they are rays.

Suggestions or Solutions

To the Problem 1 in the Example S

The equation given is $\mathbf{y = |2x - 3| + 3x + 1}$.
We can present such an equation for each of the intervals, which we can see from $\mathbf{|3 - 2x|}$.
Then, we classify the values $\mathbf{2x - 3}$ can have in terms of the plus or minus signs. Then, we can have two cases for $\mathbf{2x - 3}$. In the first, $\mathbf{2x - 3}$ is greater than or equal to 0. In the second, $\mathbf{2x - 3}$ is less than 0. Now, let's remove the absolute signs. Then:

$\mathbf{2x - 3 \geq 0 \Rightarrow 2x \geq 3 \Rightarrow x \geq 3/2}$
$\mathbf{\Rightarrow y = 2x - 3 + 3x + 1 = 5x - 2}$ for $\mathbf{x \geq 3/2}$.

$\mathbf{2x - 3 < 0 \Rightarrow 2x < 3 \Rightarrow x < 3/2}$
$\mathbf{\Rightarrow y = -(2x - 3) + 3x + 1 = 3 - 2x + 3x + 1 = x + 4}$ for $\mathbf{x < 3/2}$.

Therefore, $\mathbf{y = 5x + 2}$ for $\mathbf{x \geq 3/2}$, or $\mathbf{y = x + 4}$ for $\mathbf{x < 3/2}$.

Now, let's put in a graph $\mathbf{y = |2x - 3| + 3x + 1}$.

Fig. S.1.0 $\mathbf{y = 5x + 2}$ for $\mathbf{x \geq 3/2}$ or $y = x + 4$ for $x < 3/2$

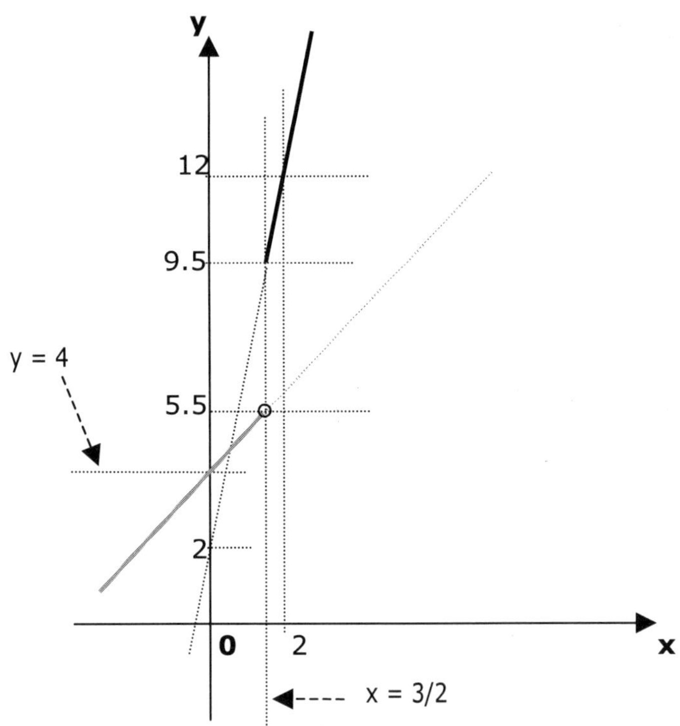

The line in black includes the point **(3/2, 9.5)**, but the one in gray doesn't include the point **(3/2, 5.5)**.

Suggestions or Solutions

To the **Problem 2** in the Example **S**

The equation given is $y = x - |3x - 2|/2$.
We can present such an equation for each of the intervals, which we can see from $|3x - 2|$.
Then, in terms of the signs of + or −, let's classify the values that $3x - 2$ can have.
In the first case, $3x - 2$ is greater than or equal to 0. In the second case, $3x - 2$ is less than 0.
Now, let's remove the absolute signs. Then:

$3x - 2 \geq 0 \Rightarrow 3x \geq 2 \Rightarrow x \geq 3/2$
$\Rightarrow y = x - (3x - 2)/2 = \{2x - (3x - 2)\}/2 = (-x + 2)/2 = 1 - x/2$ for $x \geq 3/2$.

$3x - 2 < 0 \Rightarrow x < 3/2$
$\Rightarrow y = x - \{-(3x - 2)\}/2 = x + (3x - 2)/2 = (5x - 2)/2 = 5x/2 - 1$ for $x < 3/2$.

Therefore, $y = 1 - x/2$ for $x \geq 3/2$, or $y = 5x/2 - 1$ for $x < 3/2$.

Now, let's put in a graph $y = x - |3x - 2|/2$.

Fig. S.2.0 $y = 1 - x/2$ for $x \geq 3/2$, or $y = 5x/2 - 1$ for $x < 3/2$

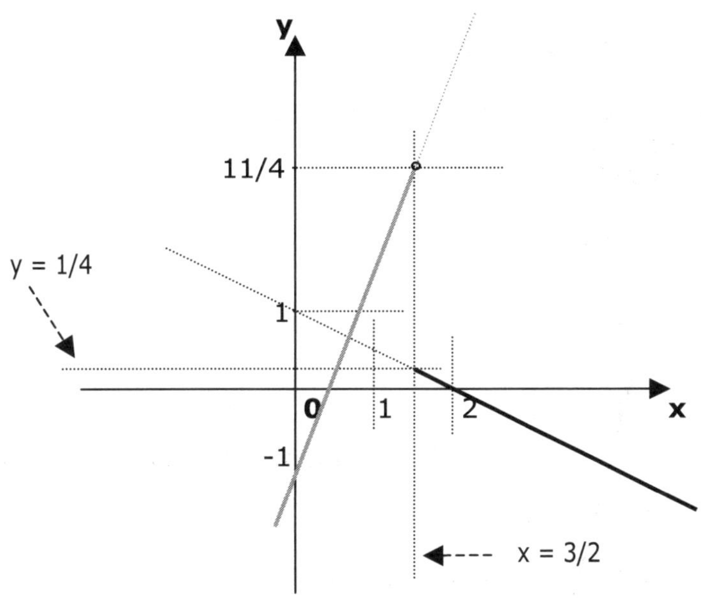

The point **(3/2, 11/4)** belongs to the line in black.

Suggestions or Solutions
To the Problem 3 in the Example S

The equation given is $2x + |y/2| = 2$ where not only the **x** but the **y** is a variable.
Now, whatever the value the **y** may take, $|y/2|$ indicates half the magnitude of the value the **y** has.
Is $|y|/2$ the same as $|y/2|$, then?
Yes, it is. Therefore, $2x + |y/2| = 2$ is the same as $2x + |y|/2 = 2$.
Then, is $|-x|$ the same as $|x|$, too?
Yes, it is. Now, if $x \geq 0$, then $|-x| = |x| = x$; on the other hand, if $x < 0$, then $|-x| = |x| = -x$.

Let's first, begin with the case where $y/2 \geq 0$.
Then, $y/2 \geq 0 \Rightarrow y \geq 0 \Rightarrow 2x + y/2 = 2 \Rightarrow y = 4 - 4x$ for $y \geq 0$.
Then, let's find the extent of the **x**.
$y \geq 0 \Rightarrow 4 - 4x \geq 0 \Rightarrow 4x \leq 4 \Rightarrow x \leq 1$. So, $y = 4 - 4x$ for $x \leq 1$.
Let's put the **x** in terms of the **y**.
Then, $y = 4 - 4x \Rightarrow 4x = 4 - y \Rightarrow x = 1 - y/4$ for $y \geq 0$.

Let's now, move on to the case where $y/2 < 0$.

Then, $y/2 < 0 \Rightarrow y < 0 \Rightarrow 2x + \{-(y/2)\} = 2 \Rightarrow 2x - y/2 = 2 \Rightarrow y = 4x - 4$ for $y < 0$.
Then, let's find the extent of the **x**.
$y < 0 \Rightarrow 4x - 4 < 0 \Rightarrow 4x < 4 \Rightarrow x < 1$. So, $y = 4x - 4$ for $x < 1$.
Let's now, put the **x** in terms of the **y**.
Then, $y = 4x - 4 \Rightarrow 4x = y + 4 \Rightarrow x = y/4 + 1$ for $y < 0$.

Then, we can see that the **y** can be any real number, but the **x** cannot be greater than 1.
We now have $y = 4 - 4x$ for $x \leq 1$, or $y = 4x - 4$ for $x < 1$.
Also equivalently, we have $x = 1 - y/4$ for $y \geq 0$, or $x = y/4 + 1$ for $y < 0$.

Now, let's put in a graph $2x + |y/2| = 2$.

Fig. S.3.0

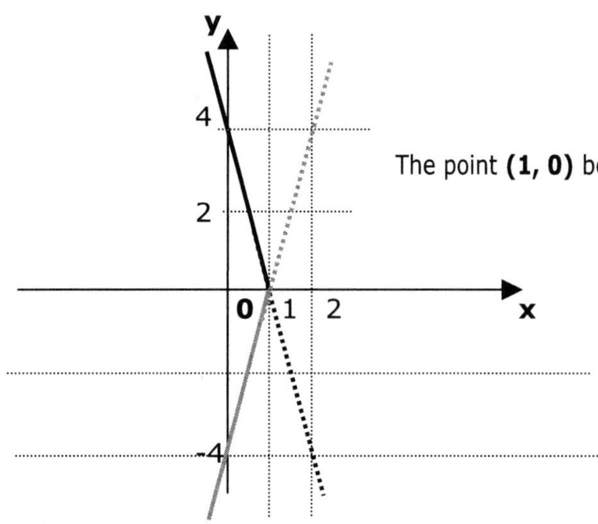

The point **(1, 0)** belongs to the line in black.

We can see that the lines in gray and black are symmetric to each other about the x-axis.

Suggestions or Solutions
To the **Problem 4** in the Example **S**

The equation given is **|2x| + y/2 = 2**, which can be presented for each of the intervals that we can see from **|2x|**.
What do we mean by **|2x|**?
|2x| is the magnitude of the value twice as much as the value the **x** has. Also, if **x** is positive, **2x** is positive, too, and if **x** is negative, **2x** is negative, also.
Therefore, we have **|2x| = 2|x|**.
Then, we have two intervals, in the first of which, the **x** is greater than or equal to 0. In the second, the **x** is less than 0.

First, we can convert the given equation in such a way as follows:
|2x| + y/2 = 2 ⇒ 2|x| + y/2 = 2 ⇒ |x| + y/4 = 1.

If we are more comfortable with the form in **y = bla bla bla**, then we can put it that way. It is similar to the standard form.
Then, **|x| + y/4 = 1 ⇒ y/4 = 1 − |x| ⇒ y = 4 − 4|x|**.

Let's remove the absolute sign.
x ≥ 0 ⇒ y = 4 − 4x. **x < 0 ⇒ y = 4 − 4(-x) = 4 + 4x.**

Now, we have **y = 4 − 4x** for **x ≥ 0**, or **y = 4 + 4x** for **x < 0**.
Then, the graph of **|2x| + y/2 = 2** is as follows:

Fig. S.4.0 **y = 4 − 4x** for **x ≥ 0**, or y = 4 + 4x for x < 0

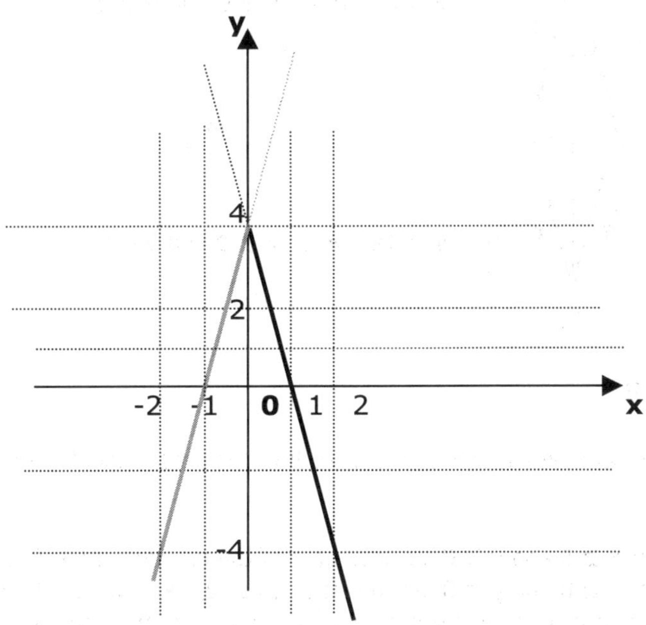

We can notice that the lines in gray and black are symmetric to each other about the y-axis.

Suggestions or Solutions
To the Problem 5 in the Example S

The equation given is $|2x| + |y/2| = 2$. In this case, both the **x** and the **y** have the absolute signs. Let's first, convert the given equation in such a way as follows.
$|2x| + |y/2| = 2 \Rightarrow 2|x| + |y|/2 = 2 \Rightarrow |y| = 4 - 4|x|$.

Then, we need to present the equation for each of the intervals, which we can see from $|x|$ and $|y|$. Since there are two intervals for each of the two variables **x** and **y**, we need to consider 4 cases.

$x \geq 0$ and $y \geq 0 \Rightarrow y = 4 - 4x$ where $x \geq 0$ and $y \geq 0$.
$x \geq 0$ and $y < 0 \Rightarrow -y = 4 - 4x \Rightarrow y = 4x - 4$ where $x \geq 0$ and $y < 0$.
$x < 0$ and $y \geq 0 \Rightarrow y = 4 - 4(-x) = 4 + 4x \Rightarrow y = 4x + 4$ where $x < 0$ and $y \geq 0$.
$x < 0$ and $y < 0 \Rightarrow -y = 4 - 4(-x) = 4 + 4x \Rightarrow y = -4x - 4$ where $x < 0$ and $y < 0$.

Now, let's put in a graph $|2x| + |y/2| = 2$.

Fig. S.5.0

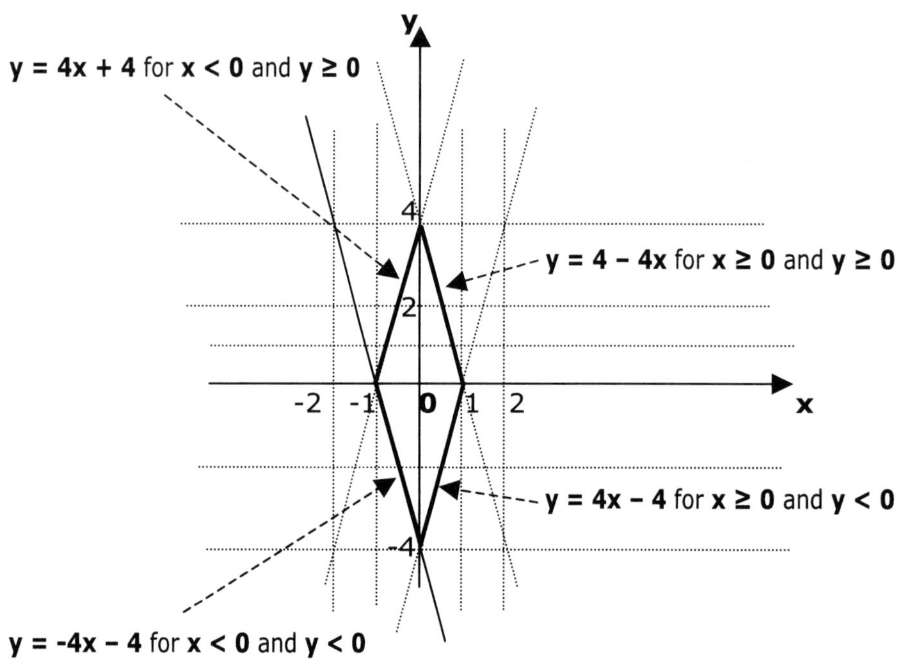

We can see that each of the line segments is symmetric to each of the others about x-axis, y-axis, or the origin.

The line segment $y = 4 - 4x$ for $x \geq 0$ and $y \geq 0$ includes both of the end points **(0, 4)** and **(1, 0)**.
The line segment $y = 4x - 4$ for $x \geq 0$ and $y < 0$ includes the point **(0, -4)** but not **(1, 0)**.
The line segment $y = -4x - 4$ for $x < 0$ and $y < 0$ does not include both of the end points.
The line segment $y = 4x + 4$ for $x < 0$ and $y \geq 0$ includes **(-1, 0)** but not **(0, 4)**.

Suggestions or Solutions

To the Problem 6 in the Example S

The equation given is **y = $(x^2)^{1/2}$ =** $\sqrt{x^2}$. In this case, both the **x** and the **y** do not have the absolute sign. However, we should take care of the square root sign as if it were the absolute sign. For instance, if inside the square root sign, the exponent of a variable is even, then when we remove the square root sign, we do as if we removed the absolute signs.
In general, we have $\sqrt{x^2} \geq$ **0**. Then, let's take such an example as follows.

We know that $\sqrt{4} = \sqrt{2^2} = \sqrt{(-2)^2} = 2$.

Then, if **x = -2** in $\sqrt{x^2}$, we get $\sqrt{x^2} \neq$ **x** since **x = -2** and $\sqrt{x^2}$ **> 0**.

Also, we know that $\sqrt{x^2} = \sqrt{4}$. Therefore, in this case, we get $\sqrt{x^2}$ = -x.

Likewise, we should not just put $\sqrt{(x-2)^2}$ **= x – 2.**

That's because we have $\sqrt{(x-2)^2} \geq$ **0** but we can have **x – 2 < 0.**

Therefore, we should put $\sqrt{(x-2)^2}$ **= x – 2** where **x ≥ 2.**

Now, we have **y = $(x^2)^{1/2}$ =** $\sqrt{x^2}$. Then, we know that since $\sqrt{x^2}$ **≥ 0**, we have **y ≥ 0.**
Therefore, we cannot just put y = x since the **x** can be negative.

Then, we can remove the square root sign in such a way as follows:
y = x where **x ≥ 0**. **y = -x** where **x < 0.**
Let's now, construct the graph of **y = $(x^2)^{1/2}$ =** $\sqrt{x^2}$.

Fig. S.6.0 **y = x** for **x ≥ 0**, or y = -x for x < 0

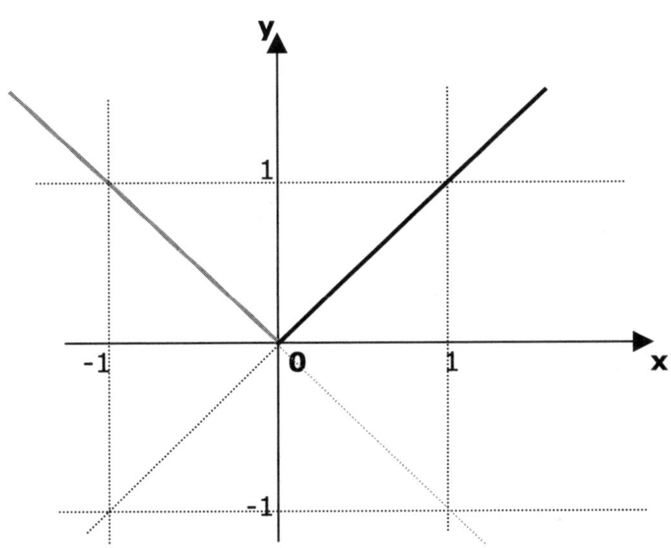

The black line includes the origin, but the gray one doesn't.
We can see that **y = $(x^2)^{1/2}$ = |x|.**

Suggestions or Solutions
To the Problem 7 in the Example S

The equation given is $\mathbf{y = (x^2)^{1/2}/x} = \sqrt{x^2}/x$. In this case also, both the **x** and the **y** do not have the absolute signs. However, we should take the square root sign for the absolute signs. In general, we have $\sqrt{x^2} \geq \mathbf{0}$.

Now, we know that a denominator can never be 0. Therefore, we need to have $\mathbf{x \neq 0}$, and we cannot just put $\mathbf{x \geq 0}$. So, we need to set the extents of the **x** in such a way as follows.

$\mathbf{x > 0 \Rightarrow y = x/x = 1}$ for $\mathbf{x > 0}$.
$\mathbf{x < 0 \Rightarrow y = -x/x = -1}$ for $\mathbf{x < 0}$.

Then, we can present $\mathbf{y} = \sqrt{x^2}/x$ in such a way as follows:
$\mathbf{y = 1}$ for $\mathbf{x > 0}$, or $\mathbf{y = -1}$ for $\mathbf{x < 0}$.

Let's now, construct the graph of $\mathbf{y = (x^2)^{1/2}/x} = \sqrt{x^2}/x$.

Fig. S.7.0 $\mathbf{y = -1}$ for $\mathbf{x > 0}$, or $y = 1$ for $x < 0$

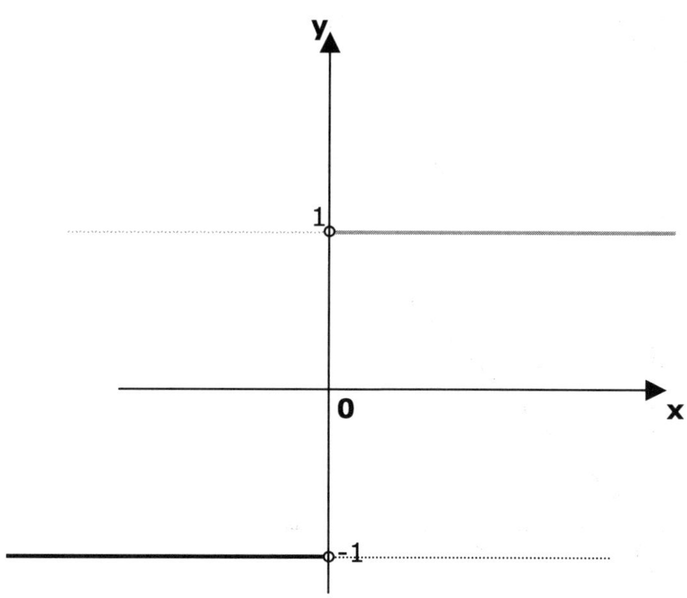

The line in gray doesn't include the point **(0, 1)**, and the one in black doesn't include the point **(0, -1)**.

Suggestions or Solutions
To the **Problem 8** in the Example **S**

The equation given is $\mathbf{y = (x^2)^{1/2}/|x|} = \sqrt{x^2}\ / \mid x \mid$. In this case, the **x** has the absolute signs. We should take care of the square root sign as if it were the absolute sign. In general, we have $\sqrt{x^2} \geq \mathbf{0}$.

Now, we know that a denominator can never be 0. Therefore, we need to have $\mathbf{x \neq 0}$, and we can't just put $\mathbf{x \geq 0}$. So, we need to set the extents of the **x** in such a way as follows:

$\mathbf{x > 0 \Rightarrow y = x/x = 1}$ for $\mathbf{x > 0}$.
$\mathbf{x < 0 \Rightarrow y = -x/(-x) = 1}$ for $\mathbf{x < 0}$.

Then, we can put $\mathbf{y} = \sqrt{x^2}\ / \mid x \mid$ in such a way as follows:
$\mathbf{y = 1}$ for **x** that is any of all real numbers but 0.

Let's put in a graph $\mathbf{y = (x^2)^{1/2}/|x|} = \sqrt{x^2}\ / \mid x \mid$, now.

Fig. S.8.0 $\mathbf{y = 1}$ for **x** that is any of all real numbers but 0.

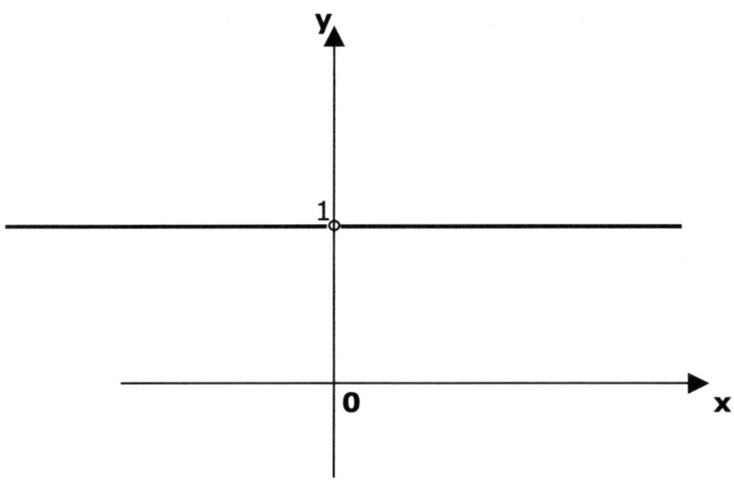

The line above doesn't include the point **(0, 1)**.

Suggestions or Solutions
To the **Problem 9** in the Example **S**

The equation given is $\mathbf{y = (x^2 - 5x + 25/4)^{1/2}} = \sqrt{x^2 - 5x + 25/4}$. We should take care of the square root sign as if it were the absolute sign. For instance, if inside the square root, the exponent of a variable is even, then when we remove the square root sign, we should do it as if we removed the absolute signs. In general, we have $\sqrt{x^2} \geq \mathbf{0}$.

Now, the equation looks like a complete square. Then, let's give it a check to see if it is.
$\mathbf{x^2 - 5x + 25/4 = x^2 - 2x(5/2) + (5/2)^2 = (x - 5/2)^2}$. It is.

Then, we can see that $\mathbf{y = \{(x - 5/2)^2\}^{1/2} = |x - 5/2|}$, which is an equation of line.
Why the absolute sign, then?
That's because the **y** is to be greater than equal to 0 since $\mathbf{(x^2 - 5x + 25/4)^{1/2} \geq 0}$.
We know that $\mathbf{x - 5/2}$ can be negative for some values of the **x**.

Now, let's remove the absolute sign, too.

$\mathbf{x - 5/2 \geq 0 \Rightarrow x \geq 5/2 \Rightarrow y = x - 5/2}$ for $\mathbf{x \geq 5/2}$.
$\mathbf{x - 5/2 < 0 \Rightarrow x < 5/2 \Rightarrow y = -(x - 5/2) = 5/2 - x}$ for $\mathbf{x < 5/2}$.

Therefore, we can put $\mathbf{y = (x^2 - 5x + 25/4)^{1/2}} = \sqrt{x^2 - 5x + 25/4}$ in such a form that:
$\mathbf{y = x - 5/2}$ for $\mathbf{x \geq 5/2}$, or $\mathbf{5/2 - x}$ for $\mathbf{x < 5/2}$.

Let's put in a graph $\mathbf{y = (x^2 - 5x + 25/4)^{1/2}}$, now.

Fig. S.9.0 $\mathbf{y = x - 5/2}$ for $\mathbf{x \geq 5/2}$, or $y = 5/2 - x$ for $x < 5/2$

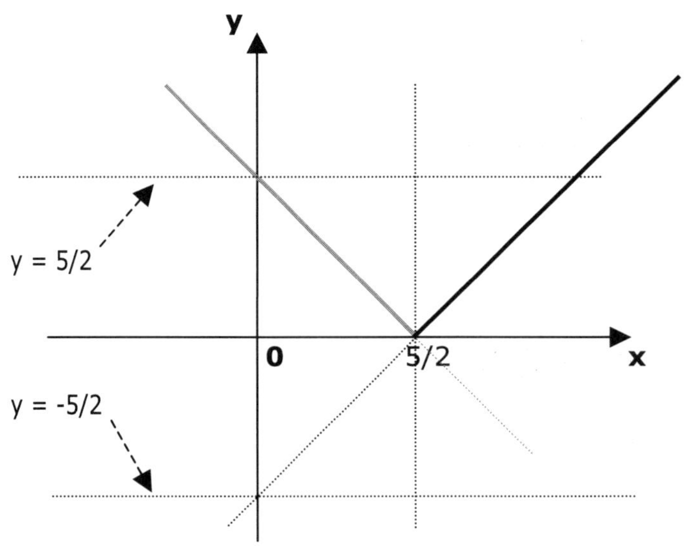

The two lines are symmetric about the line $\mathbf{x = 5/2}$.
The line in gray doesn't include the point **(5/2, 0)**, but the in black does.

Suggestions or Solutions

To the Problem A in the Example S

The equation given is $y = 2x\{|x^2 - 1|/(x^2 - 1)\}$.

Because of the absolute signs, we need to consider two different cases as follows.

First, $x^2 - 1 \geq 0 \Rightarrow (x + 1)(x - 1) \geq 0 \Rightarrow x \geq 1$ or $x \leq -1$
$\Rightarrow y = 2x(x^2 - 1)/(x^2 - 1) = 2x$ for $x \geq 1$ or $x \leq -1$.

Next, $x^2 - 1 < 0 \Rightarrow (x + 1)(x - 1) < 0 \Rightarrow -1 < x < 1$
$\Rightarrow y = 2x\{-(x^2 - 1)\}/(x^2 - 1) = -2x(x^2 - 1)\}/(x^2 - 1) = -2x$ for $-1 < x < 1$.

Therefore, we can put $y = 2x\{|x^2 - 1|/(x^2 - 1)\}$ in such a way as follows:

$y = 2x$ for $x \geq 1$ or $x \leq -1$, or $y = -2x$ for $-1 < x < 1$.

Let's now, put in a graph $y = 2x\{|x^2 - 1|/(x^2 - 1)\}$.

Fig. S.A.0 $y = 2x$ for $x \geq 1$ or $x \leq -1$, or $y = -2x$ for $-1 < x < 1$

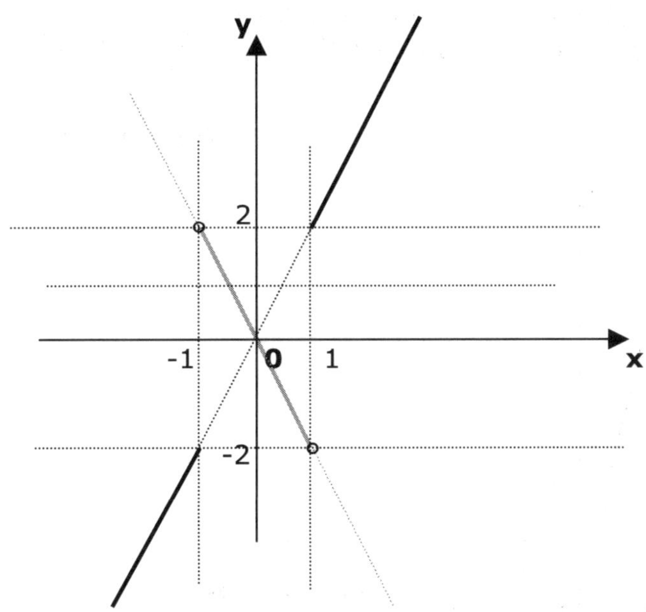

The black line includes the two points **(-1, -2)** and **(1, 2)**, but the gray one doesn't include both of the end points **(-1, 2)** and **(1, -2)**.

Suggestions or Solutions
To the Problem B in the Example S

The equation given is $y = ((2x + 1)^2)^{1/2} + ((2x - 1)^2)^{1/2} = \sqrt{(2x+1)^2} + \sqrt{(2x-1)^2}$.

When we remove the square root sign, we need to make sure the result is greater than or equal to 0. That's because the square root of a real number is greater than or equal to 0. Therefore, we should not just remove the square root sign and the exponent. Since we have two cases for a square root sign and we have two square root signs, we need to consider 4 cases as follows:

$(2x + 1 \geq 0$ and $2x - 1 \geq 0) \Rightarrow (x \geq -1/2$ and $x \geq 1/2) \Rightarrow x \geq 1/2$
$\Rightarrow y = 2x + 1 + 2x - 1 = 4x$ for $x \geq 1/2$.

$(2x + 1 \geq 0$ and $2x - 1 < 0) \Rightarrow (x \geq -1/2$ and $x < 1/2) \Rightarrow -1/2 \leq x < 1/2$
$\Rightarrow y = 2x + 1 + \{-(2x - 1)\} = 2x + 1 - 2x + 1 = 2$ for $-1/2 \leq x < 1/2$.

$(2x + 1 < 0$ and $2x - 1 \geq 0) \Rightarrow (x < -1/2$ and $x \geq 1/2) \Rightarrow$ No existence

$(2x + 1 < 0$ and $2x - 1 < 0) \Rightarrow (x < -1/2$ and $x < 1/2) \Rightarrow x < -1/2$
$\Rightarrow y = -(2x + 1) + \{-(2x - 1)\} = -2x - 1 - 2x + 1 = -4x$ for $x < -1/2$.

Therefore, we can put $\sqrt{(2x+1)^2} + \sqrt{(2x-1)^2}$ in such a form as follows:
$y = 4x$ for $x \geq 1/2$, $y = 2$ for $-1/2 \leq x < 1/2$, or $y = -4x$ for $x < -1/2$.
Now, let's put in a graph $y = ((2x + 1)^2)^{1/2} + ((2x - 1)^2)^{1/2} = \sqrt{(2x+1)^2} + \sqrt{(2x-1)^2}$.

Fig. S.B.0 $y = 4x$ for $x \geq 1/2$, $y = 2$ for $-1/2 \leq x < 1/2$, or $y = -4x$ for $x < -1/2$

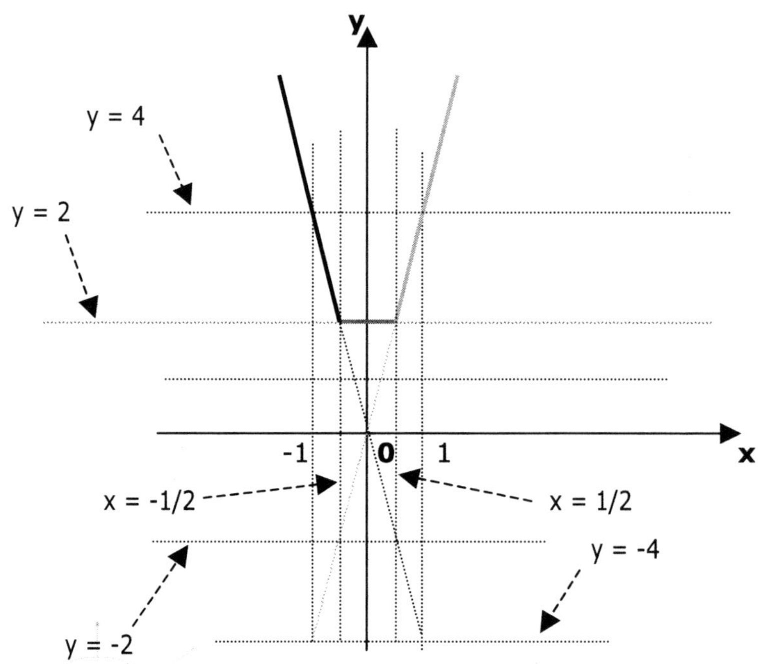

Suggestions or Solutions
To the Problem C in the Example S

The equation given is **$||2x| - |3y|| = 1$**.
We need to consider the case where what's inside the absolute signs is negative and the case where it is greater than or equal to 0.
Therefore, we need to consider not only **$|x|$** and **$|y|$**, but **$|2x| - |3y| \geq 0$** and **$|2x| - |3y| < 0$**.

Then, we need to consider 8 cases as follows.

$(x \geq 0, y \geq 0, \text{and } 2x \geq 3y) \Rightarrow (x \geq 0, y \geq 0, \text{and } y \leq 2x/3) \Rightarrow 2x - 3y = 1.$
$(x \geq 0, y \geq 0, \text{and } 2x < 3y) \Rightarrow (x \geq 0, y \geq 0, \text{and } y > 2x/3) \Rightarrow -(2x - 3y) = 3y - 2x = 1.$

$(x \geq 0, y < 0, \text{and } 2x \geq -3y) \Rightarrow (x \geq 0, y < 0, \text{and } y \geq -2x/3) \Rightarrow 2x - (-3y) = 2x + 3y = 1.$
Now, how come **$-3y$** in "**$2x \geq -3y$**"? That's because **$|3y|$** becomes **$-3y$** since **$y < 0$** if the absolute signs are removed. Let's move on to the next, then.

$(x \geq 0, y < 0, \text{and } 2x < -3y) \Rightarrow (x \geq 0, y < 0, \text{and } y < -2x/3) \Rightarrow -\{2x - (-3y)\} = -2x - 3y = 1.$

$(x < 0, y \geq 0, \text{and } -2x \geq 3y) \Rightarrow (x < 0, y \geq 0, \text{and } y \leq -2x/3) \Rightarrow -2x - 3y = 1.$
$(x < 0, y \geq 0, \text{and } -2x < 3y) \Rightarrow (x < 0, y \geq 0, \text{and } y > -2x/3) \Rightarrow -(-2x - 3y) = 2x + 3y = 1.$

$(x < 0, y < 0, \text{and } -2x \geq -3y) \Rightarrow (x < 0, y < 0, \text{and } y \geq 2x/3) \Rightarrow -2x - (-3y) = 3y - 2x = 1.$
$(x < 0, y < 0, \text{and } -2x < -3y) \Rightarrow (x < 0, y < 0, \text{and } y < 2x/3) \Rightarrow -\{-2x - (-3y)\} = 2x - 3y = 1.$

Now:
$(x \geq 0, y \geq 0, \text{and } y \leq 2x/3) \Rightarrow 2x - 3y = 1 \Rightarrow y = 2x/3 - 1/3.$
$(x \geq 0, y \geq 0, \text{and } y > 2x/3) \Rightarrow 3y - 2x = 1 \Rightarrow y = 2x/3 + 1/3.$
$(x \geq 0, y < 0, \text{and } y \geq -2x/3) \Rightarrow 2x + 3y = 1 \Rightarrow y = -2x/3 + 1/3.$
$(x \geq 0, y < 0, \text{and } y < -2x/3) \Rightarrow -2x - 3y = 1 \Rightarrow y = -2x/3 - 1/3.$
$(x < 0, y \geq 0, \text{and } y \leq -2x/3) \Rightarrow -2x - 3y = 1 \Rightarrow y = -2x/3 - 1/3.$
$(x < 0, y \geq 0, \text{and } y > -2x/3) \Rightarrow 2x + 3y = 1 \Rightarrow y = -2x/3 + 1/3.$
$(x < 0, y < 0, \text{and } y \geq 2x/3) \Rightarrow 3y - 2x = 1 \Rightarrow y = 2x/3 + 1/3.$
$(x < 0, y < 0, \text{and } y < 2x/3) \Rightarrow 2x - 3y = 1 \Rightarrow y = 2x/3 - 1/3.$

Therefore:
$y = 2x/3 - 1/3$ for $x \geq 0, y \geq 0$, and $y \leq 2x/3$; or for $x < 0, y < 0$, and $y < 2x/3$.
$y = 2x/3 + 1/3$ for $x \geq 0, y \geq 0$, and $y > 2x/3$; or for $x < 0, y < 0$, and $y \geq 2x/3$.
$y = -2x/3 + 1/3$ for $x \geq 0, y < 0$, and $y \geq -2x/3$; or for $x < 0, y \geq 0$, and $y > -2x/3$.
$y = -2x/3 - 1/3$ for $x \geq 0, y < 0$, and $y < -2x/3$; or for $x < 0, y \geq 0$, and $y \leq -2x/3$.

Then, let's put in a graph **$||2x| - |3y|| = 1$**.

Fig. S.C.0

$y = 2x/3 + 1/3$ for $x \geq 0$, $y \geq 0$, and $y > 2x/3$; or for $x < 0$, $y < 0$, and $y \geq 2x/3$

$y = -2x/3 + 1/3$ for $x \geq 0$, $y < 0$, and $y \geq -2x/3$; or for $x < 0$, $y \geq 0$, and $y > -2x/3$

$y = 2x/3 - 1/3$ for $x \geq 0$, $y \geq 0$, and $y \leq 2x/3$; or for $x < 0$, $y < 0$, and $y < 2x/3$

$y = -2x/3 - 1/3$ for $x \geq 0$, $y < 0$, and $y < -2x/3$; or for $x < 0$, $y \geq 0$, and $y \leq -2x/3$

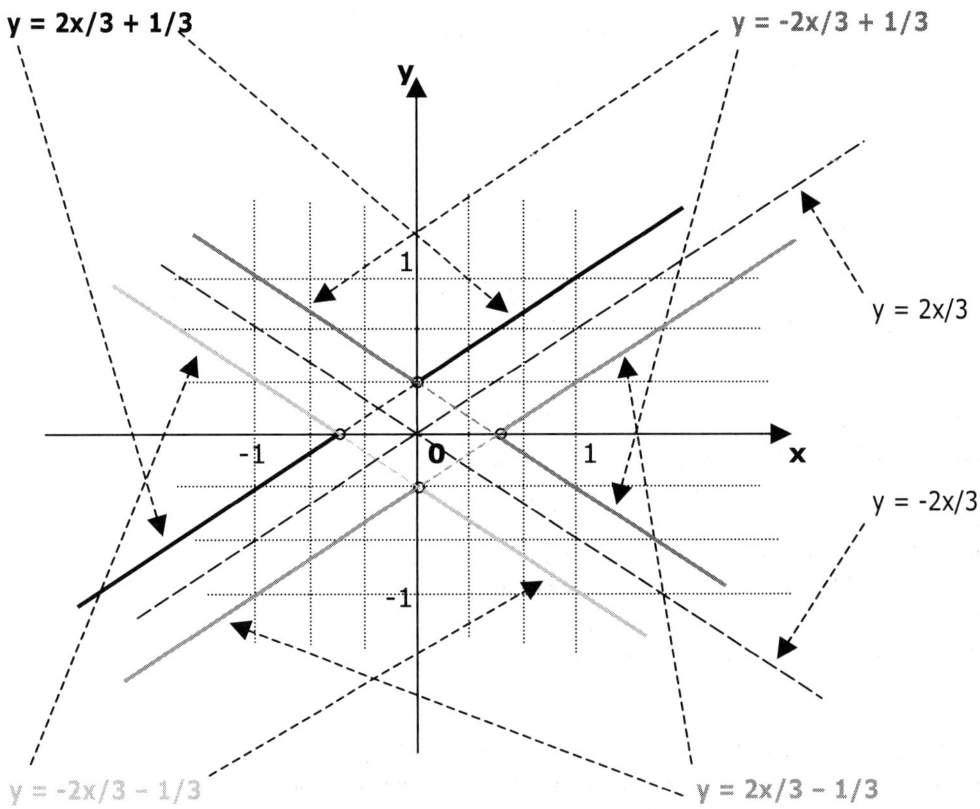

Suggestions or Solutions

To the Problem D in the Example S

The equation given is $\mathbf{y = |2x - 1| + |3x + 2|}$. We need to consider 4 cases.

$(2x - 1 \geq 0 \text{ and } 3x + 2 \geq 0) \Rightarrow (x \geq 1/2 \text{ and } x \geq -2/3) \Rightarrow x \geq 1/2$
$\Rightarrow \mathbf{y = 2x - 1 + 3x + 2 = 5x + 1}$ for $\mathbf{x \geq 1/2} \Rightarrow \mathbf{y = 5x + 1}$ for $\mathbf{x \geq 1/2}$.

$(2x - 1 \geq 0 \text{ and } 3x + 2 < 0) \Rightarrow (x \geq 1/2 \text{ and } x < -2/3) \Rightarrow$ Nonexistence.
$(2x - 1 < 0 \text{ and } 3x + 2 \geq 0) \Rightarrow (x < 1/2 \text{ and } x \geq -2/3) \Rightarrow -2/3 \leq x < 1/2$
$\Rightarrow \mathbf{y = -(2x - 1) + 3x + 2 = -2x + 1 + 3x + 2 = x + 3}$ for $\mathbf{-2/3 \leq x < 1/2}$.
$\Rightarrow \mathbf{y = x + 3}$ for $\mathbf{-2/3 \leq x < 1/2}$.

$(2x - 1 < 0 \text{ and } 3x + 2 < 0) \Rightarrow (x < 1/2 \text{ and } x < -2/3) \Rightarrow x < -2/3$
$\Rightarrow \mathbf{y = -(2x - 1) + \{-(3x + 2)\} = -2x + 1 - 3x - 2 = -5x - 1}$ for $\mathbf{x < -2/3}$.
$\Rightarrow \mathbf{y = -5x - 1}$ for $\mathbf{x < -2/3}$.

Now, let's put in a graph $\mathbf{y = |2x - 1| + |3x + 2|}$.

Fig. S.D.0

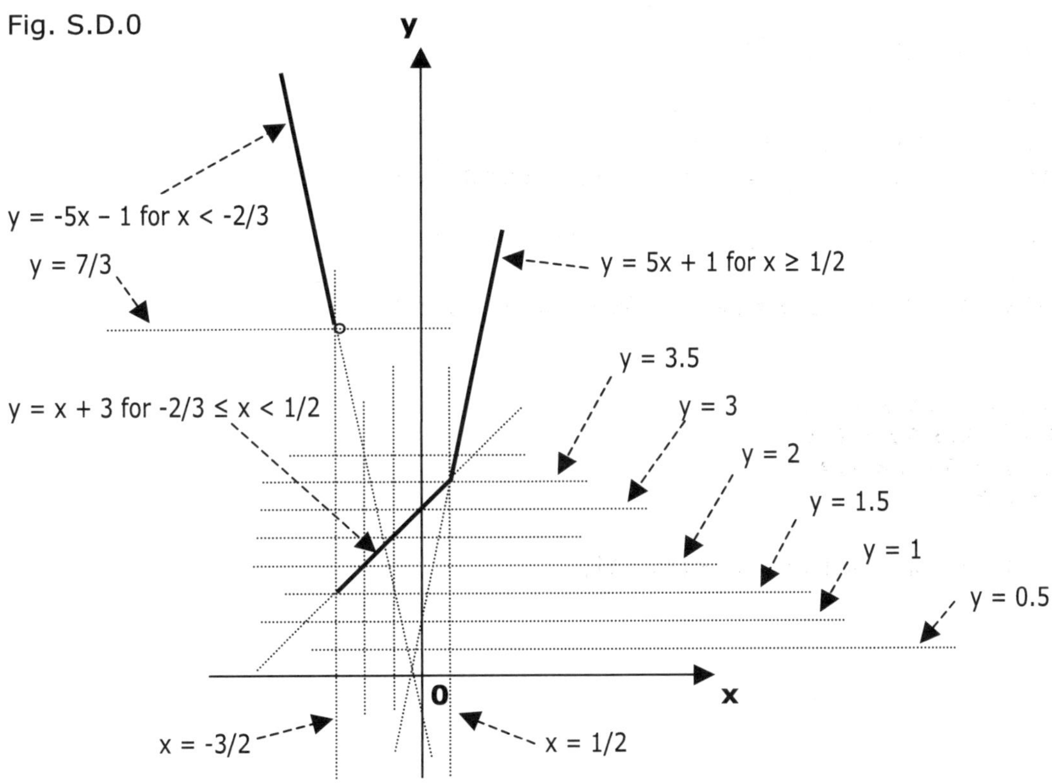

$\mathbf{y = -5x - 1}$ for $\mathbf{x < -2/3}$ includes the point $\mathbf{(-3/2, 7/3)}$.
$\mathbf{y = x + 3}$ for $\mathbf{-2/3 \leq x < 1/2}$ includes the point $\mathbf{(-3/2, 3/2)}$, but not $\mathbf{(1/2, 7/2)}$.
$\mathbf{y = 5x + 1}$ for $\mathbf{x \geq 1/2}$ includes the point $\mathbf{(1/2, 7/2)}$.

Suggestions or Solutions

To the Problem E in the Example S

The equation given is **$y = ||2x - 1| - |3x - 1||$**. We need to consider 8 cases.

$(2x - 1 \geq 0, 3x - 1 \geq 0, \text{ and } 2x - 1 \geq 3x - 1) \Rightarrow (x \geq 1/2, x \geq 1/3, \text{ and } x \leq 0) \Rightarrow$ Nonexistence.

$(2x - 1 \geq 0, 3x - 1 \geq 0, \text{ and } 2x - 1 < 3x - 1) \Rightarrow (x \geq 1/2, x \geq 1/3, \text{ and } x > 0) \Rightarrow x \geq 1/2$
$\Rightarrow$ **$y = -\{2x - 1 - (3x - 1)\} = -2x + 1 + 3x - 1 = x$** for **$x \geq 1/2$**.

$(2x - 1 \geq 0, 3x - 1 < 0, \text{ and } 2x - 1 \geq -(3x - 1) = -3x + 1) \Rightarrow (x \geq 1/2, x < 1/3, \text{ and } x \geq 2/5)$
$\Rightarrow$ Nonexistence.
$2x - 1 \geq 0, 3x - 1 < 0, \text{ and } 2x - 1 < -(3x - 1) \Rightarrow$ Nonexistence.

$(2x - 1 < 0, 3x - 1 \geq 0, \text{ and } -(2x - 1) = -2x + 1 \geq 3x - 1) \Rightarrow (x < 1/2, x \geq 1/3, \text{ and } x \leq 2/5)$
$\Rightarrow 1/3 \leq x \leq 2/5$
$\Rightarrow$ **$y = -(2x - 1) - (3x - 1) = -2x + 1 - 3x + 1 = -5x + 2$** for **$1/3 \leq x \leq 2/5$**.

$(2x - 1 < 0, 3x - 1 \geq 0, \text{ and } -(2x - 1) = -2x + 1 < 3x - 1) \Rightarrow (x < 1/2, x \geq 1/3, \text{ and } x > 2/5)$
$\Rightarrow 2/5 < x < 1/2$
$\Rightarrow$ **$y = -\{-(2x - 1) - (3x - 1)\} = 2x - 1 + 3x - 1 = 5x - 2$** for **$2/5 < x < 1/2$**.

$(2x - 1 < 0, 3x - 1 < 0, \text{ and } -(2x - 1) = -2x + 1 \geq -(3x - 1) = -3x + 1)$
$\Rightarrow (x < 1/2, x < 1/3, \text{ and } x \geq 0) \Rightarrow 0 \leq x < 1/3$
$\Rightarrow$ **$y = -(2x - 1) - \{-(3x - 1)\} = -2x + 1 + 3x - 1 = x$** for **$0 \leq x < 1/3$**.

$(2x - 1 < 0, 3x - 1 < 0, \text{ and } -(2x - 1) = -2x + 1 < -(3x - 1) = -3x + 1)$
$\Rightarrow (x < 1/2, x < 1/3, \text{ and } x < 0)$
$\Rightarrow$ **$y = -\{-(2x - 1) - \{-(3x - 1)\} = 2x - 1 - 3x + 1 = -x$** for **$x < 0$**.

Now:
$y = x$ for **$x \geq 0$**, or **$0 \leq x < 1/3$**.
$y = -5x + 2$ for **$1/3 \leq x \leq 2/5$**.
$y = 5x - 2$ for **$2/5 < x < 1/2$**.
$y = -x$ for **$x < 0$**.

Let's now, put in a graph **$y = ||2x - 1| - |3x - 1||$**.

Fig. S.E.0.

$y = x$ for $0 \le x < 1/3$ or $x \ge 1/2$

$y = -x$ for $x < 0$

$y = 5x - 2$ for $2/5 < x < 1/2$

$y = -5x + 2$ for $1/3 \le x \le 2/5$

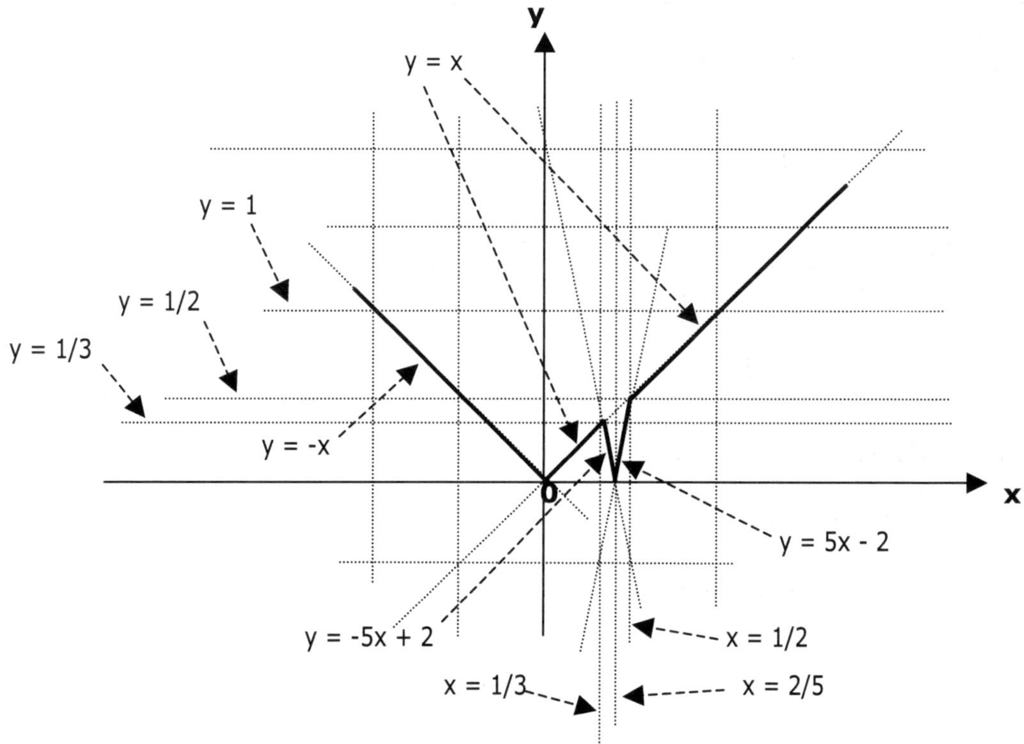

Suggestions or Solutions

To the Problem in the Example T

Suppose that three points $P_1(x_1, y_1)$, $P_2(x_2, y_2)$, and $P_2(x_3, y_3)$ in the x-y plane are in a line, and find the connective expression among the coordinates.

Suppose that the equation of the line is $y = ax + b$ where **a** is the slope and **b** is the y-intercept.
Then, let's use the two points P_1 and P_2 to find the slope **a**.
Then, $a = \Delta y/\Delta x = (y_2 - y_1)/(x_2 - x_1)$.

Now, we use the P_1 to specify the **b**.
$y = ax + b \Rightarrow y_1 = ax_1 + b \Rightarrow b = y_1 - ax_1$.

Then, $y = ax + b \Rightarrow y = ax + y_1 - ax_1 = a(x - x_1) + y_1$.
Then, we can see that $y = a(x - x_1) + y_1 \Rightarrow y - y_1 = a(x - x_1)$.
Since the P_3 is in the line, we can see that $y_3 - y_1 = a(x_3 - x_1)$.

We know that the slope $a = (y_2 - y_1)/(x_2 - x_1)$.
Then, we plug it in $y_3 - y_1 = a(x_3 - x_1)$.

$y_3 - y_1 = \{(y_2 - y_1)/(x_2 - x_1)\}(x_3 - x_1)$
$\Rightarrow y_3 - y_1 = (y_2 - y_1)(x_3 - x_1)/(x_2 - x_1)$
$\Rightarrow (y_3 - y_1)(x_2 - x_1) = (y_2 - y_1)(x_3 - x_1)$.

Now, we get a connective expression as follows:
$(y_3 - y_1)(x_2 - x_1) = (x_3 - x_1)(y_2 - y_1)$.

The connective expression above can be said to be the condition under which three points can be in a line.

For a simple example, the three points $(x_1, y_1) = (1, 1)$, $(x_2, y_2) = (2, 2)$, and $(x_3, y_3) = (3, 3)$ are in a line. The line is $y = x$, of course.
Then, $(y_3 - y_1)(x_2 - x_1) = (x_3 - x_1)(y_2 - y_1) \Rightarrow (3 - 1)(2 - 1) = (3 - 1)(2 - 1) \Rightarrow 2 = 2$.

For another example, let's use the following three points in the line $y = 2x + 1$:
$(x_1, y_1) = (1, 3)$, $(x_2, y_2) = (3, 7)$, and $(x_3, y_3) = (5, 11)$.

Then, $(y_3 - y_1)(x_2 - x_1) = (x_3 - x_1)(y_2 - y_1) \Rightarrow (11 - 3)(3 - 1) = (5 - 1)(7 - 3) \Rightarrow 16 = 16$.

Suggestions or Solutions
To the **Problem** in the Example **U**

Suppose that the distance from a point $P_1(2, 1)$ to a particular line is the same as the distance from another point $P_2(-2, -3)$ to the particular line, and find the equation of the particular line.

A distance in math is a perpendicular distance.
Suppose that the particular line is the line **A**.
Then, the two points $\mathbf{P_1}$ and $\mathbf{P_2}$ are in the line perpendicular to the line **A**.
Also, since the two distances are the same, the point where the two lines perpendicular to each other meet is the midpoint between the two points $\mathbf{P_1}$ and $\mathbf{P_2}$.
Therefore, the midpoint is in the line **A**.
Then, finding the slope of the **A** and the midpoint, we can find the equation of the line **A**.

Since the product of the slopes of two lines perpendicular to each other is −1, we can find the slope of the A from the slope of the line perpendicular to the **A**.
The coordinates of the midpoint are the respective averages of the coordinates of $\mathbf{P_1}$ and $\mathbf{P_2}$.

Now, let's begin with the slope of the line perpendicular to the **A**.

Since the perpendicular line includes $\mathbf{P_1(2, 1)}$ and $\mathbf{P_2(-2, -3)}$:
The slope is $\Delta y/\Delta x = (1 - (-3))/(2 - (-2)) = 4/4 = 1$.
Then, the slope of the A is −1.

Next, the x-coordinate of the midpoint = (2 + (-2))/2 = 0, and also, the y-coordinate = (1 + (-3))/2 = -1.
So, we get the midpoint = **(0, -1)**.

The equation of a line that has the slope of **a** and includes a point **(m, n)** is $y - n = a(x - m)$.

Therefore, the equation of the **A** is $y - (-1) = (-1)(x - 0) \Rightarrow y + 1 = -x \Rightarrow y = -x - 1$.

In short:

Suppose **A** is the particular line and **B** is the line that is perpendicular to the **A**, and includes $\mathbf{P_1}$ and $\mathbf{P_2}$.

Then, since the **B** includes $\mathbf{P_1(2, 1)}$ and $\mathbf{P_2(-2, -3)}$:
The slope is $\Delta y/\Delta x = (1 - (-3))/(2 - (-2)) = 4/4 = 1$.
Then, the slope of the **A** is −1.

Also, the **A** includes the midpoint between $\mathbf{P_1}$ and $\mathbf{P_2}$.
The x-coordinate of the midpoint = (2 + (-2))/2 = 0, and also, the y-coordinate = (1 + (-3))/2 = -1.
So, the midpoint = **(0, -1)**.

Therefore, the equation of the **A** is $y - (-1) = (-1)(x - 0) \Rightarrow y + 1 = -x \Rightarrow y = -x - 1$.

Suggestions or Solutions

To the Problem in the Example V

Suppose that:
T: (x, y) ⟶ (3x − y, x − y). y = f(x) = ax where the a is a constant. T•f = f.

Then, find the value of the a.

We need to determine the value of the **a** so that the function **f** having the curve of an equation **y = ax** does not get changed even if we transform the **f** by the transformation **T**.
Suppose that **g = T•f**, and that **t = g(s)**. Then, we get **g = f**, and also, we have:
T: (x, y) ⟶ (s, t) = (3x − y, x − y).
So, we get **s = 3x − y** and **t = x − y**. Then, we take them for the simultaneous equations for the **x** and the **y**. Then, solving them, we can put each of the **x** and the **y** in terms of the **s** and the **t**.

$s - t = 3x - y - (x - y) = 3x - y - x + y = 2x \Rightarrow s - t = 2x \Rightarrow x = (s - t)/2$. The **x** is now, in terms of the **s** and the **t**. Then, we can put the **x** into **s = 3x − y** to put the **y** in terms of the **s** and the **t**.
$s = 3(s - t)/2 - y = (3s - 3t)/2 - y \Rightarrow y = (3s - 3t)/2 - s = (s - 3t)/2$.
Now, we have the $x = (s - t)/2$, and the $y = (s - 3t)/2$.

Then, we need to apply the expressions above to the original function **f** to get the new function **g**.
The original function is $y = f(x) = ax$. Then, $(s - 3t)/2 = f\{(s - t)/2\} = a\{(s - t)/2\}$.
Then, $(s - 3t)/2 = a(s - t)/2 \Rightarrow s - 3t = a(s - t) \Rightarrow at - 3t = as - s = (a - 1)s$
$\Rightarrow (a - 3)t = (a - 1)s \Rightarrow t = (a - 1)s/(a - 3)$. So, $t = g(s) = (a - 1)s/(a - 3)$.
Now, since the **g** is in the x-y system, we need to change the letters **s** and **t** to **x** and **y** respectively.
Then, we get $y = g(x) = (a - 1)x/(a - 3) = \{(a - 1)/(a - 3)\}x$.

Now, since $f(x) = ax$ and $f(x) = g(x)$, we get $(a - 1)/(a - 3) = a$ where $a \neq 3$.
So, $a - 1 = a(a - 3) = a^2 - 3a \Rightarrow a^2 - 4a + 1 = 0$.
Then, $a^2 - 4a + 1 = a^2 - 4a + 4 - 4 + 1 = (a - 2)^2 - 3 = 0$.
Therefore, $a - 2 = \pm 3^{1/2} \Rightarrow a = 2 \pm \sqrt{3}$.

In short:

Suppose that **g = T•f**, and that **t = g(s)**. Then, we get **g = f**, and also, **s = 3x − y** and **t = x − y**.
$s - t = 3x - y - (x - y) = 3x - y - x + y = 2x \Rightarrow s - t = 2x \Rightarrow x = (s - t)/2$.
$s = 3x - y = 3(s - t)/2 - y = (3s - 3t)/2 - y \Rightarrow y = (3s - 3t)/2 - s = (s - 3t)/2$.
Therefore, $x = (s - t)/2$ and $y = (s - 3t)/2$.
Since $y = f(x) = ax$, $(s - 3t)/2 = f\{(s - t)/2\} = a\{(s - t)/2\}$.
Then, $(s - 3t)/2 = a(s - t)/2 \Rightarrow s - 3t = a(s - t) \Rightarrow at - 3t = as - s = (a - 1)s$
$\Rightarrow (a - 3)t = (a - 1)s \Rightarrow t = (a - 1)s/(a - 3)$. So, $t = g(s) = (a - 1)s/(a - 3)$.
Then, $y = g(x) = (a - 1)x/(a - 3) = \{(a - 1)/(a - 3)\}x$.
Now, since $f(x) = ax$ and $f(x) = g(x)$, we get $(a - 1)/(a - 3) = a$ where $a \neq 3$.
So, $a - 1 = a(a - 3) = a^2 - 3a \Rightarrow a^2 - 4a + 1 = 0$.
Then, $a^2 - 4a + 1 = a^2 - 4a + 4 - 4 + 1 = (a - 2)^2 - 3 = 0$.
Therefore, $a - 2 = \pm 3^{1/2} \Rightarrow a = 2 \pm \sqrt{3}$.

Suggestions or Solutions

To the Problem in the Example W

Suppose that:
T: (x, y) ⟶ (3x − 2y, x + y). y = f(x) = ax where the a is a constant. T•f = f.

Then, find the value of the a.

This time also, we need to determine the value of the **a** so that the function **f** having the curve of **y = ax** does not get changed even if we transform the **f** by the transformation **T**.
Suppose that **g = T•f**, and that **t = g(s)**. Then, we get **g = f**, and also, we have:
T: (x, y) ⟶ (s, t) = (3x − 2y, x + y).
So, we have **s = 3x − 2y** and **t = x + y**.

Then, taking them for a system of equations for the **x** and the **y** and solving it, we can put each of the **x** and the **y** in terms of the **s** and the **t**.
First, **s + 2t = 3x − 2y + 2(x + y) = 3x − 2y + 2x + 2y = 5x ⇒ s + 2t = 5x ⇒ x = (s + 2t)/5**.
Next, **x = (s + 2t)/5 ⇒ s = 3x − 2y = 3(s + 2t)/5 − 2y = (3s + 6t)/5 − 2y**
⇒ 2y = (3s + 6t)/5 − s = (-2s + 6t)/5 ⇒ y = (3t − s)/5.

Now, we have the **x = (s + 2t)/5**, and the **y = (3t − s)/5**.
Then, we need to apply the expressions above to the original function **f** to get the new function **g**.
The original function is **y = f(x) = ax**. Then, **(3t − s)/5 = f{(s + 2t)/5} = a{(s + 2t)/5}**.
Then, **(3t − s)/5 = a(s + 2t)/5 ⇒ 3t − s = a(s + 2t) ⇒ 2at − 3t = -as − s = -(a + 1)s**
⇒ (2a − 3)t = -(a + 1)s ⇒ t = (a + 1)s/(3 − 2a). So, **t = g(s) = (a + 1)s/(3 − 2a)**.
Now, since the **g** is in the x-y system, we need to change the letters **s** and **t** to **x** and **y** respectively.
Then, we get **y = g(x) = (a + 1)x/(3 − 2a) = {(a + 1)/(3 − 2a)}x**.

Now, since **f(x) = ax** and **f(x) = g(x)**, we get **(a + 1)/(3 − 2a) = a** where **a ≠ 3/2**.
So, **a + 1 = a(3 − 2a) = $-2a^2$ + 3a ⇒ $2a^2$ − 2a + 1 = 0**.
Now, **$2a^2 - 2a + 1 = 2(a^2 - a + 1/4 - 1/4) + 1 = 2(a - 1/2)^2 + 1/2$**, which however, can't be 0 since the **a** is supposed to take a real number. Therefore, there is no solution. That is, there is no value of the **a**.

In short:

Suppose that **g = T•f**, and that **t = g(s)**. Then, we get **g = f**, and also, **s = 3x − 2y** and **t = x + y**.
s + 2t = 3x − 2y + 2(x + y) = 3x − 2y + 2x + 2y = 5x ⇒ s + 2t = 5x ⇒ x = (s + 2t)/5.
s = 3(s + 2t)/5 − 2y = (3s + 6t)/5 − 2y ⇒ 2y = (3s + 6t)/5 − s = (-2s + 6t)/5 ⇒ y = (3t − s)/5.
Therefore, we have **x = (s + 2t)/5** and **y = (3t − s)/5**.
Since **y = f(x) = ax**, we get **(3t − s)/5 = f{(s + 2t)/5} = a{(s + 2t)/5}**.
Then, **(3t − s)/5 = a(s + 2t)/5 ⇒ 3t − s = a(s + 2t) ⇒ 2at − 3t = -as − s = -(a + 1)s**
⇒ (2a − 3)t = -(a + 1)s ⇒ t = (a + 1)s/(3 − 2a). So, **t = g(s) = (a + 1)s/(3 − 2a)**.
Then, **g(x) = (a + 1)x/(3 − 2a) = {(a + 1)/(3 − 2a)}x**.
Now, since **f(x) = ax** and **f(x) = g(x)**, we get **(a + 1)/(3 − 2a) = a** where **a ≠ 3/2**.
So, **a + 1 = a(3 − 2a) = $-2a^2$ + 3a ⇒ $2a^2$ − 2a + 1 = 0**.
Now, **$2a^2 - 2a + 1 = 2(a^2 - a + 1/4 - 1/4) + 1 = 2(a - 1/2)^2 + 1/2$**, which however, can't be 0 since the **a** is a real number. Therefore, there is no value of the **a**.

Suggestions or Solutions
To the Problem in the Example X

Suppose that
T: (x, y) ⟶ (2x + y + 3, 3x + 2y − 2).
y = f(x) = ax + b where the a and the b are constants.
T•f = f.

Then, find the values of the a and the b.

We need to determine the values of the **a** and the **b** so that the function **f** having the curve of an equation **y = ax + b** does not get changed even if we transform the **f** by the transformation **T**.
Suppose that **g = T•f**, and that **t = g(s)**. Then, we get **g = f**, and also, we have:
T: (x, y) ⟶ (s, t) = (2x + y + 3, 3x + 2y − 2).
So, we get **s = 2x + y + 3** and **t = 3x + 2y − 2**.

Then, we take them for the simultaneous equations for the **x** and the **y** to put each of the **x** and the **y** in terms of the **s** and the **t**.

First, $2s - t = 4x + 2y + 6 - (3x + 2y - 2) = x + 8 \Rightarrow 2s - t = x + 8 \Rightarrow x = 2s - t - 8$.
Next, $x = 2s - t - 8 \Rightarrow s = 2(2s - t - 8) + y + 3 = 4s - 2t - 16 + y + 3 \Rightarrow y = -3s + 2t + 13$.
Now, we have $x = 2s - t - 8$ and $y = -3s + 2t + 13$.

We have $y = f(x) = ax + b$. Then, $-3s + 2t + 13 = f(2s - t - 8) = a(2s - t - 8) + b$.
Then, $-3s + 2t + 13 = 2as - at - 8a + b \Rightarrow (2 + a)t = (2a + 3)s - 8a + b - 13$
$\Rightarrow t = (2a + 3)s/(a + 2) + (b - 8a - 13)/(a + 2)$.
Therefore, $t = g(s) = (2a + 3)s/(a + 2) + (b - 8a - 13)/(a + 2)$.

Now, since the **g** is in the x-y system, we need to change the letters **s** and **t** to **x** and **y** respectively.
Then, we get $y = g(x) = (2a + 3)x/(a + 2) + (b - 8a - 13)/(a + 2)$.
Now, since $f(x) = ax$ and $f(x) = g(x)$:
We get $(2a + 3)/(a + 2) = a$ and $(b - 8a - 13)/(a + 2) = b$ where $a \neq -2$.
Let's first, find the value of the **a**. Then, $2a + 3 = a(a + 2) = a^2 + 2a \Rightarrow a^2 = 3$. So, $a = \pm 3^{1/2}$.
Now, we need to find the value of the **b**.
$(b - 8a - 13)/(a + 2) = b \Rightarrow b - 8a - 13 = b(a + 2) = ab + 2b \Rightarrow b(a + 1) = -8a - 13$
$\Rightarrow b = -(8a + 13)/(a + 1)$. Then:

$a = \sqrt{3} \Rightarrow b = -(8\sqrt{3} + 13)/(\sqrt{3} + 1) = -(8\sqrt{3} + 13)(\sqrt{3} - 1)/(\sqrt{3} + 1)(\sqrt{3} - 1)$
$= -(8\sqrt{3} + 13)(\sqrt{3} - 1)/2 = -(24 + 5\sqrt{3} - 13)/2 = -(11 + 5\sqrt{3})/2$.
$a = -\sqrt{3} \Rightarrow b = -(-8\sqrt{3} + 13)/(1 - \sqrt{3}) = (8\sqrt{3} - 13)/(1 - \sqrt{3}) = (13 - 8\sqrt{3})/(\sqrt{3} - 1)$
$= (13 - 8\sqrt{3})(\sqrt{3} + 1)/2 = (5\sqrt{3} - 24 + 13)/2 = (5\sqrt{3} - 11)/2$.
Therefore, $a = \sqrt{3}$ and $b = -(11 + 5\sqrt{3})/2$, or $a = -\sqrt{3}$ and $b = (5\sqrt{3} - 11)/2$.

Now, even if we transform by the **T** the line $y = \sqrt{3}x - (11 + 5\sqrt{3})/2$, we get the same line, and the same is true for the line $y = -\sqrt{3}x + (5\sqrt{3} - 11)/2$.

0.1. Several Lines

We normally put one line in one equation. However, we can also put many lines in one equation as stated earlier. When we put in an equation more than one line, we usually use the general form of the equation: $\mathbf{ax + by + c = 0}$ where **a**, **b**, and **c** are constants.

Now, how do we specify two lines in one equation?
To begin with, let's put two lines in one equation.
We can put one line in a general equation as follows: $\mathbf{ax + by + c = 0}$ where **a**, **b**, and **c** are constants.
Let's put another line in the general form as follows: $\mathbf{px + qy + r = 0}$ where **p**, **q**, and **r** are constants.
Now, we can put the two lines in one equation in such a way as follows:

$\mathbf{(ax + by + c)(px + qy + r) = 0}$.

How come?
Suppose that we have an equation as follows: $\mathbf{AB = 0}$.
Then, we get $\mathbf{A = 0}$ or $\mathbf{B = 0}$.
Likewise, we get $\mathbf{(ax + by + c)(px + qy + r) = 0 \Rightarrow ax + by + c = 0}$ or $\mathbf{px + qy + r = 0}$.
So, $\mathbf{(ax + by + c)(px + qy + r) = 0}$ indicates two lines one of which is $\mathbf{ax + by + c = 0}$ and the other of which is $\mathbf{px + qy + r = 0}$.

The same idea applies to the equation of more than two lines.

Suppose now, that $\mathbf{L_1L_2L_3 \ldots L_n = 0}$.
Suppose also, that $\mathbf{L_k = a_kx + b_ky + c_k}$ where $\mathbf{a_k}$, $\mathbf{b_k}$, and $\mathbf{c_k}$ are constants and $\mathbf{k = 1, 2, \ldots n}$.
That is, $\mathbf{L_1 = a_1x + b_1y + c_1}$, $\mathbf{L_2 = a_2x + b_2y + c_2}$, . . . $\mathbf{L_n = a_nx + b_ny + c_n}$

Then, $\mathbf{L_1L_2L_3 \ldots L_n = 0 \Rightarrow (a_1x + b_1y + c_1)(a_2x + b_2y + c_2) \ldots (a_nx + b_ny + c_n) = 0}$.

Then, we get $\mathbf{L_1 = 0}$, $\mathbf{L_2 = 0}$, . . . or $\mathbf{L_n = 0}$.
That is, we get $\mathbf{L_k = 0}$ where $\mathbf{k = 1, 2, \ldots n}$.
In other words, we get $\mathbf{a_1x + b_1y + c_1 = 0}$, $\mathbf{a_2x + b_2y + c_2 = 0}$ . . . or $\mathbf{a_nx + b_ny + c_n = 0}$.

Therefore, $\mathbf{L_1L_2L_3 \ldots L_n = 0}$ is an equation that indicates **n** lines.

We know that a line is a set of points.
Then, we can put two lines in such a set **A** of points as follows:

$\mathbf{A = \{(x, y) | (ax + by + c)(px + qy + r) = 0}$ where **a**, **b**, **c**, **p**, **q**, and **r** are constants**}**
Of course, the **a** and the **b** are not 0 at the same time, neither are the **p** and **q**.

Then, we can put more than three lines in such a set **B** of points as follows:

$\mathbf{B = \{(x, y) | (a_1x + b_1y + c_1)(a_2x + b_2y + c_2) \ldots (a_nx + b_ny + c_n) = 0}$ where $\mathbf{a_k}$, $\mathbf{b_k}$, and $\mathbf{c_k}$ are constants**}**
Of course, $\mathbf{k = 1, 2, \ldots n}$, and the $\mathbf{a_k}$ and the $\mathbf{b_k}$ are not 0 at the same time.

Usually, we put a line in an equation or a set of points.
Making a line, we need to have at least two points. Then, we can find the equation of the line.
Suppose now, that the two points are **(s, t)** and **(u, v)**.
Then, the equation of the line is $y - t = \{(t - v)/(s - u)\}(x - s)$ or $y - v = \{(t - v)/(s - u)\}(x - u)$ where $\{(t - v)/(s - u)\}$ is the slope. Of course, the two equations above are the same. Make an example and give it a check, though.
Also, we can find the equation of a line if we have the slope and a point in the line, of course.
Suppose that **a** is the slope and the point is **(m, n)**. Then, the equation is $y - n = a(x - m)$.

Now, we have another method where we can make a line. We make a line by means of lines.
We use two different lines to make another line.
Let's suppose that the two lines are original lines.
Then, we make a new line by means of the two original lines.

Now, if the original lines are parallel to each other, the new line is parallel, too.
Otherwise, that is, if the original lines meet at a point, the new line includes the point, too.
That is, all the three lines are either parallel to each other or meeting at a point.

How do we make the new line, then?
If we find the average of the two original lines, the average is the new line.

Now, what do we mean by the average of the lines?
The average is the quotient from the division of the sum of the two equations of lines by 2.

Suppose for instance, that we have equations of line as follows: $y = a_1x + b_1$ and $y = a_2x + b_2$. Then:
$y + y = a_1x + b_1 + (a_2x + b_2) \Rightarrow 2y = (a_1 + a_2)x + (b_1 + b_2) \Rightarrow y = (a_1 + a_2)x/2 + (b_1 + b_2)/2$.

Now, the equation of the new line is $y = (a_1 + a_2)x/2 + (b_1 + b_2)/2$.

In fact, the new line doesn't have to be the average. Let's now work with the general form.
Suppose that $ax + by + c = 0$ and $ux + vy + w = 0$.
Then, we get $(a + u)x + (b + v)y + c + w = 0$, which is an equation of a new line.
Also, we can have $2(ax + by + c) = 0$. Then, we can get $(2a + u)x + (2b + v)y + 2c + w = 0$, which, too, is an equation of a new line.

Suppose now, that we have two lines $y = ax + b$ and $y = ax + c$.
Then, the two lines are parallel to each other. Then:
$\{y = ax + b\} + \{y = ax + c\} \Rightarrow 2y = 2ax + b + c \Rightarrow y = ax + (b + c)/2$, which is parallel to the original two lines.

Let's this time, make a simple example where two original lines meet each other at a point.
Suppose that we have two lines $y = x + 2$ and $y = -3x + 2$.
Let's first, find the point where the original lines meet each other.
Then, $x + 2 = -3x + 2 \Rightarrow 4x = 0 \Rightarrow x = 0 \Rightarrow y = x + 2 = 2 \Rightarrow (0, 2)$.
Therefore, the original lines meet at the point **(0, 2)**. Let's now, find the new line.
$\{y = x + 2\} + \{y = -3x + 2\} \Rightarrow 2y = -2x + 4 \Rightarrow y = -x + 2$.
Now, let's check to see if the new line includes the point **(0, 2)**.
$y = -x + 2 \Rightarrow 2 = 0 + 2$. It does.

Fig. 0.1.0

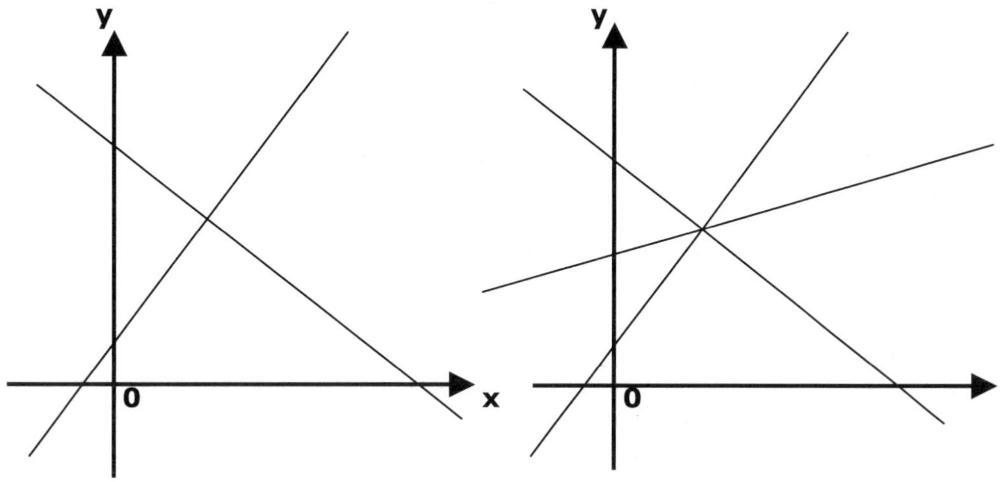

Also, with any two of the three lines, we make a new line, the new line, too, includes the point **(0, 2)**. Let's for instance, make a new one with **y = -3x + 2** and **y = -x + 2**. Then:
{y = -3x + 2} + {y = -x + 2} ⇒ **2y = -4x + 4** ⇒ **y = -2x + 2**.
Now, let's check to see if the new line includes the point **(0, 2)**.
y = -2x + 2 ⇒ **2 = 0 + 2**. It does. Therefore, having two lines meeting each other at a point, we can make infinitely many new line that include the point.

Fig. 0.1.1

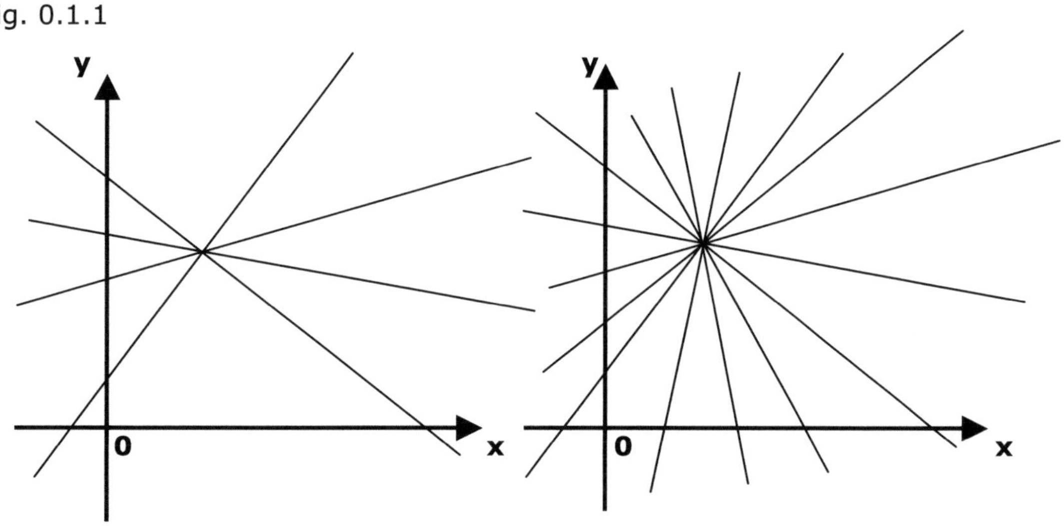

The same is true for the lines parallel to each other.
We can use two lines parallel to each other to make another parallel line, and also, with any two of the three lines, we make a new line, the new line is parallel, too.
What if we add together all the three lines, then?
The new line is parallel, too.
Taking the previous example, we have **y = ax + b**, **y = ax + c**, and **y = ax + (b + c)/2**.

Then, **{y = ax + b} + {y = ax + c} + {y = ax + (b + c)/2}** ⇒ **3y = 3ax + 3(b + c)/2**
⇒ **y = ax + (b + c)/2**, which the third line found previously.

For this time, let's make a new line with such three lines parallel to each other as follows:
y = ax + b, **y = ax + c**, and **y = ax + d**.
Then:
{y = ax + b} + {y = ax + c} + {y = ax + d} ⇒ **3y = 3a + (b + c + d)** ⇒ **y = a + (b + c + d)/3**, which is a parallel line, too.

The same is true for the lines meeting each other at a point.

What is the point of the stories above, then?
The point is a point and a slope.
First of all, the point is that we have infinitely many lines that include a particular point, and each of the lines is made up of two lines, which include the particular point, too.
Next, the point is that we have infinitely many lines parallel to each other, and each of the lines is made up of two lines, which are parallel, too.

Now, in fact, if we add together the original lines, the sum indicates the new line.
More precisely, if we multiply by a nonzero number each of the equations of the original lines and add together all the products, the sum indicates a new line, which includes a particular point where the original points meet each other. In addition, if the original lines are parallel to each other, the new line, too, is parallel to the original lines.

Now, let's take some examples.

Examples
in Several Lines

0. Suppose we have two lines $\mathbf{y = x}$ and $\mathbf{y = -x}$, and then, put them in one equation.

1. Suppose we have two lines $\mathbf{y = -x + 1}$ and $\mathbf{y = x}$, and then, put them in one equation.

2. The equation $\mathbf{2x^2 - 2y^2 - 3xy - x + 2y = 0}$ represents two lines. Then, find the two lines.

3. Suppose that the equation below represents two lines, and then, find the value of the **t**:
$\mathbf{2x^2 + 4xy - ty^2 + 4x + 2y - 2 = 0}$ where **t** is a constant.

4. Suppose that we have two lines as follows: $\mathbf{2x - y - 1 = 0}$ and $\mathbf{3x + 2y - 2 = 0}$.
Then, find a point **P(s, t)** that is the same distance away from each of the lines specified above.

5. Suppose that we have equations of lines as follows: $\mathbf{ax + by + c = 0}$ and $\mathbf{ux + vy + w = 0}$ where **a**, **b**, **c**, **u**, **v**, and **w** are constants, $\mathbf{abc \neq 0}$, and $\mathbf{uvw \neq 0}$.

5.0. Suppose also, that the equations in fact, indicate the same line, and then, find the connective expression between the constants.

5.1. Suppose also, that the two lines that are parallel to each other, and then, find the connective expression between the constants.

5.2. Suppose also, that the two lines meet each other at one point, and then, find the connective expression between the constants.

5.3. Suppose also, that the two lines are perpendicular to each other, and then, find the connective expression between the constants.

5.4. Then, find the equation of a line that includes the point where the two lines meet each other.

6. Suppose that we have an equation that represents a group of lines, and the equation is as follows:
$\mathbf{3(k - 1)x + (2k + 3)y + 4k + 5 = 0}$ where **k** is a constant.
Suppose also, that all the lines in the group include a particular point **P(s, t)**, and find the values of the **s** and the **t**.

7. Suppose that we have an equation that represents a group of lines, and the equation is as follows:
$\mathbf{y = ax - b}$ such that $\mathbf{3a + b = 2}$ where **a** and **b** are constant.
Suppose also, that all the lines in the group include a particular point **P(s, t)**, and find the values of the **s** and the **t**.

8. Suppose that we have an equation that represents a group of lines, and the equation is as follows: $2x/a + y/2b = 1$ where **a** and **b** are constants such that $1/2a + 1/b = 1/5$.
Suppose also, that all the lines include a particular point **P(u, v)**, and then, find the values of the **u** and the **v**.

9. Suppose that we have an equation that represents a group of lines, and the equation is as follows: $5ax - (3b - 2)y + 2b = 8$ where **a** and **b** are constant such that $2a - b = 2$ and $ab \neq 0$.

9.0. Suppose also, the lines in the group include a particular point **P(s, t)**, and then, find the point **P**.

9.1. Then, find the line in the group that is parallel to a line $2x + 5y = 10$.

A. Suppose that two lines meet at a point **P(s, t)** where $s > 0$ and $t > 0$, and that the equations of the lines are as follows: $2x + y - 2 = 0$ and $2kx - y + k - 1 = 0$ where **k** is a constant.

Then, find the extent of the **k**.

B. Suppose that:
$(2k - 1)x - (3k + 1)y + 5 + 2k = 0$ where **k** is a constant.
$y = 2x$ for $|x| \leq 1$.

Then, determine the value of the **k** so that the two lines above can meet each other.

Suggestions or Solutions
To the Problem in the Example 0

Suppose we have two lines y = x and y = -x, and put them in one equation.

(y = -x or **y = x)** ⇒ **(x + y = 0** or **x − y = 0)** ⇒ **(x + y)(x − y) = 0**.

Then, we can put the two lines in one equation **(x + y)(x − y) = 0**.

Now, let's work in reverse manner.

$(x + y)(x - y) = 0 \Rightarrow x^2 - xy + xy - y^2 = 0 \Rightarrow x^2 - y^2 = 0 \Rightarrow y^2 = x^2 \Rightarrow y = \pm x$.

⇒ **y = x** or **y = -x**.

Fig. 0.0

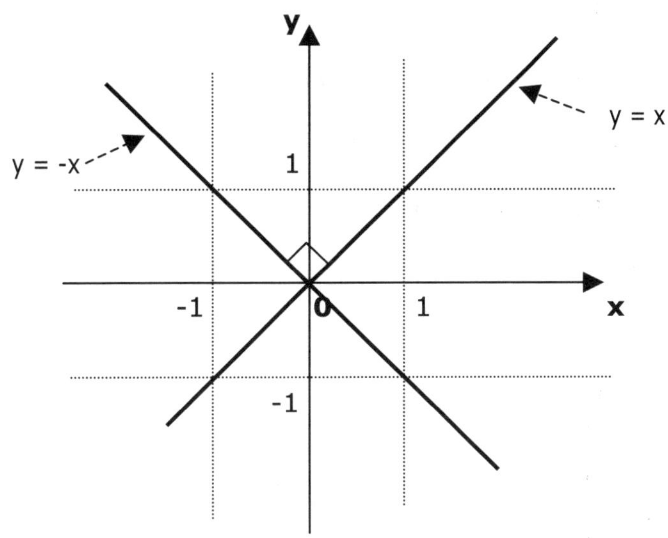

Note:
If now sure of factorization of polynomials, we get much limitation on solving equations. If not sure of it, refer to the books on factorizations of polynomials or **Run Factorizations** in the series of **Runmath**.

Suggestions or Solutions
To the Problem in the Example 1

Suppose we have two lines y = -x + 1 and y = x, and put them in one equation.

(y = -x + 1 or **y = x)** $\Rightarrow$ **(x + y − 1 = 0** or **x − y = 0)** $\Rightarrow$ **(x + y − 1)(x − y) = 0.**

Then, we can put the two lines in one equation **(x + y − 1)(x − y) = 0**.

Now, let's work in reverse manner.

$(x + y - 1)(x - y) = 0 \Rightarrow x^2 - xy + xy - y^2 - x + y = 0$

$\Rightarrow x^2 - y^2 - x + y = 0 \Rightarrow y^2 - y = x^2 - x$

$\Rightarrow y^2 - y + 1/4 - 1/4 = x^2 - x + 1/4 - 1/4 \Rightarrow (y - 1/2)^2 - 1/4 = (x - 1/2)^2 - 1/4$
$\Rightarrow (y - 1/2)^2 = (x - 1/2)^2$

$\Rightarrow y - 1/2 = \pm(x - 1/2) \Rightarrow y = \pm(x - 1/2) + 1/2$

$\Rightarrow y = (x - 1/2) + 1/2$ **or** $y = -(x - 1/2) + 1/2$

$\Rightarrow y = x$ **or** $y = -x + 1/2 + 1/2 = -x + 1$

$\Rightarrow$ **y = x** or **y = -x + 1**.

Fig. 1.0

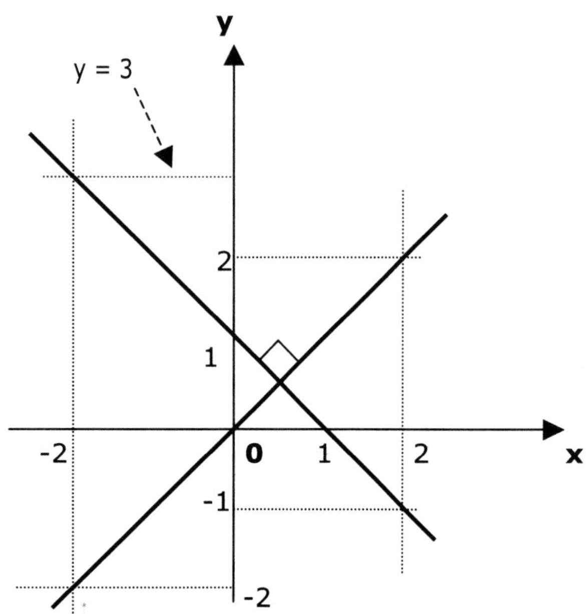

Suggestions or Solutions
To the Problem in the Example 2

The equation $2x^2 - 2y^2 - 3xy - x + 2y = 0$ represents two lines. Then, find the two lines.

The solution is a matter of factorization. In fact, solutions to most equations are matters of factorization. In this case, the factorization is much easier than ordinary factorization. That's because we've already known the form of the factors. Each of the factors is an equation of line. Therefore, we can assume the factors to be **$ax + by + c$** and **$dx + ey + f$**, for instance.

Now, let's factorize the equation given.

$$2x^2 - 2y^2 - 3xy - x + 2y = 2x^2 - 3xy - x - 2y^2 + 2y = x(2x - 3y - 1) - 2y(y - 1)$$
$$= x(2x - 4y + y - 1) - 2y(y - 1) = x(2x - 4y) + x(y - 1) - 2y(y - 1)$$
$$= 2x(x - 2y) + (x - 2y)(y - 1) = (x - 2y)(2x + y - 1).$$

Therefore, the two lines are $2x + y - 1 = 0$ and $x - 2y = 0$.

Now, what if we can't factorize the polynomial in the equation?
Of course, factorization itself is not easy in general, and many equations of lines can have the coefficients or constants in fractional numbers or even irrational numbers. Now, looking at the exponents in the polynomial in the equation, we can notice that the polynomial is of 2nd degree. Then, we may want to use the complete square method. Now, let's work it out with the complete square method.

$$2x^2 - 2y^2 - 3xy - x + 2y = 2x^2 - (1 + 3y)x - 2y^2 + 2y = 2\{x^2 - (1 + 3y)x/2\} - 2y^2 + 2y$$
$$= 2\{x^2 - (1 + 3y)x/2 + (1 + 3y)^2/16 - (1 + 3y)^2/16\} - 2y^2 + 2y$$
$$= 2\{x - (1 + 3y)/4\}^2 - (1 + 3y)^2/8 - 2y^2 + 2y$$

$$-(1 + 3y)^2/8 - 2y^2 + 2y = -(1 + 6y + 9y^2)/8 - 2y^2 + 2y = -1/8 - 3y/4 - 9y^2/8 - 2y^2 + 2y$$
$$= -25y^2/8 + 5y/4 - 1/8 = (-1/8)(25y^2 - 10y + 1) = (-1/8)(5y - 1)^2$$

Then, $2x^2 - 2y^2 - 3xy - x + 2y = 2\{x - (1 + 3y)/4\}^2 - 1/8(5y - 1)^2$
$= (1/8)[16\{x - (1 + 3y)/4\}^2 - (5y - 1)^2]$.

We know that $A^2 - B^2 = (A + B)(A - B)$. Then:
$$16\{x - (1 + 3y)/4\}^2 - (5y - 1)^2$$
$$= \{4(x - (1 + 3y)/4) + (5y - 1)\}\{4(x - (1 + 3y)/4) - (5y - 1)\}$$
$$= (4x - 1 - 3y + 5y - 1)(4x - 1 - 3y - 5y + 1)$$
$$= (4x + 2y - 2)(4x - 8y) = 2(2x + y - 1)4(x - 2y) = 8(2x + y - 1)(x - 2y)$$

Then, $(1/8)[16\{x - (1 + 3y)/4\}^2 - (5y - 1)^2] = (1/8)8(2x + y - 1)(x - 2y)$.
Therefore, $2x^2 - 2y^2 - 3xy - x + 2y = (2x + y - 1)(x - 2y)$.

In short:

$$2x^2 - 2y^2 - 3xy - x + 2y = 2x^2 - 3xy - x - 2y^2 + 2y = x(2x - 3y - 1) - 2y(y - 1)$$
$$= x(2x - 4y + y - 1) - 2y(y - 1) = x(2x - 4y) + x(y - 1) - 2y(y - 1)$$
$$= 2x(x - 2y) + (x - 2y)(y - 1) = (x - 2y)(2x + y - 1) \Rightarrow 2x + y - 1 = 0 \text{ and } x - 2y = 0.$$

Suggestions or Solutions

To the Problem in the Example 3

Suppose that the equation below represents two lines, and find the value of the t:
$2x^2 + 4xy - ty^2 + 4x + 2y - 2 = 0$ where t is a constant.

We can put two lines in an equation $(ax + by + c)(ux + vy + w) = 0$.
In other words, we factorize the polynomial and separate the individual lines apart.
Now, we can see that each of the two factors is a simple polynomial of 1st degree in the **x** and the **y**.
How do we do the factorization, then?
We can try the complete square method.
Let's now, begin the factorization.
Suppose first, that $P = 2x^2 + 4xy - ty^2 + 4x + 2y - 2$. Then:

$$P = 2x^2 + (4y + 4)x - ty^2 + 2y - 2$$
$$= 2\{x^2 + 2(y + 2)x\} - ty^2 + 2y - 2$$
$$= 2\{x^2 + 2(y + 2)x + (y + 2)^2 - (y + 2)^2\} - ty^2 + 2y - 2$$
$$= 2\{x + (y + 2)\}^2 - 2(y + 2)^2 - ty^2 + 2y - 2$$
$$= 2\{x + (y + 2)\}^2 - \{2(y + 2)^2 + ty^2 - 2y + 2\}.$$

Now, we need to convert $2(y + 2)^2 + ty^2 - 2y + 2$ into the complete square so that we can put the entire polynomial **P** in such a form as $A^2 - B^2$ which can be easily factorized into $(A + B)(A - B)$.

$$2(y + 2)^2 + ty^2 - 2y + 2 = 2(y^2 + 4y + 4) + ty^2 - 2y + 2 = (t + 2)y^2 + 6y + 10.$$

Now, if the polynomial $(t + 2)y^2 + 6y + 10$ can be converted into the complete square, then the determinant has to be 0. The determinant is what's inside the square root sign in the quadratic formula.

Suppose for instance, that $ax^2 + bx + c = 0$.

Then, the quadratic formula is $x = (-b \pm (b^2 - 4ac)^{1/2})/2a = \dfrac{-b \pm \sqrt{b^2 - 4ac}}{2a}$.

Now, the determinant is $b^2 - 4ac$.
If the determinant is 0, the solution is $-b/2a$ only, and in turn, the polynomial $ax^2 + bx + c$ can be in the complete square. In fact, $ax^2 + bx + c = (x + b/2a)^2$ if $b^2 - 4ac = 0$.

Now, the determinant of the polynomial $(t + 2)y^2 + 6y + 10$ has to be 0.

The determinant $= (-6)^2 - (4)(-(t + 2))(-10) = 36 - 40(t + 2) = 36 - 40t - 80 = -40t - 44$.
Then, $-40t - 44 = 0 \Rightarrow t = -44/40 = -11/10$.

Now, if $t = -11/10$, the polynomial in the equation of two lines, $2x^2 + 4xy + 11y^2/10 + 4x + 2y - 2$ can be factorized into polynomials in the two separate equations for the two lines.

Therefore, $2x^2 + 4xy + 11y^2/10 + 4x + 2y - 2 = 0$ is an equation of two lines.

Let's see now, what the two separate equations look like. Then:

$t = -11/10 \Rightarrow (t + 2)y^2 + 6y + 10 = 9y^2/10 + 6y + 10 = (1/10)(9y^2 + 60 + 100)$
$= (1/10)(3y + 10)^2$.

Now, **$2x^2 + 4xy + 11y^2/10 + 4x + 2y - 2 = 2\{x + (y + 2)\}^2 - (1/10)(3y + 10)^2$**.

We know that **$A^2 - B^2 = (A + B)(A - B)$**. Then:

$2\{x + (y + 2)\}^2 - (1/10)(3y + 10)^2$
$= \{2^{1/2}(x + y + 2) + (1/10)^{1/2}(3y + 10)\}\{ 2^{1/2}(x + y + 2) - (1/10)^{1/2}(3y + 10)\}$

$2^{1/2}(x + y + 2) + (1/10^{1/2})(3y + 10)$
$= 2^{1/2}x + 2^{1/2}y + 2^{3/2} + (3/10^{1/2})y + 10^{1/2}$
$= 2^{1/2}x + \{2^{1/2} + (3/10^{1/2})\}y + 2^{3/2} + 10^{1/2}$
$= 2^{1/2}x + \{2^{1/2} + 3(10^{1/2})/10\}y + 2^{3/2} + 10^{1/2}$.
$= \sqrt{2}x + (\sqrt{2} + 3\sqrt{10}/10)y + \sqrt{2^3} + \sqrt{10}$
$= \sqrt{2}x + (\sqrt{2} + 3\sqrt{10}/10)y + 2\sqrt{2} + \sqrt{10}$

$2^{1/2}(x + y + 2) - (1/10)^{1/2}(3y + 10)$
$= 2^{1/2}x + 2^{1/2}y + 2^{3/2} - (3/10^{1/2})y - 10^{1/2}$
$= 2^{1/2}x + \{2^{1/2} - (3/10^{1/2})\}y + 2^{3/2} - 10^{1/2}$
$= 2^{1/2}x + \{2^{1/2} - 3(10^{1/2})/10\}y + 2^{3/2} - 10^{1/2}$.
$= \sqrt{2}x + (\sqrt{2} - 3\sqrt{10}/10)y + \sqrt{2^3} - \sqrt{10}$
$= \sqrt{2}x + (\sqrt{2} - 3\sqrt{10}/10)y + 2\sqrt{2} - \sqrt{10}$

Therefore:

One of the two lines is $\sqrt{2}x + (\sqrt{2} + 3\sqrt{10}/10)y + 2\sqrt{2} + \sqrt{10} = 0$.

The other is $\sqrt{2}x + (\sqrt{2} - 3\sqrt{10}/10)y + 2\sqrt{2} - \sqrt{10} = 0$.

In short:

Suppose that **$P = 2x^2 + 4xy - ty^2 + 4x + 2y - 2$**. Then:
$P = 2x^2 + (4y + 4)x - ty^2 + 2y - 2 = 2\{x^2 + 2(y + 2)x\} - ty^2 + 2y - 2$
$= 2\{x^2 + 2(y + 2)x + (y + 2)^2 - (y + 2)^2\} - ty^2 + 2y - 2$
$= 2\{x + (y + 2)\}^2 - 2(y + 2)^2 - ty^2 + 2y - 2$
$= 2\{x + (y + 2)\}^2 - \{2(y + 2)^2 + ty^2 - 2y + 2\}$.

Now, we need to convert **$2(y + 2)^2 + ty^2 - 2y + 2$** into the complete square so that we can put the entire polynomial **P** in such a form as **$A^2 - B^2$** which can be easily factorized into **$(A + B)(A - B)$**.

$2(y + 2)^2 + ty^2 - 2y + 2 = 2(y^2 + 4y + 4) + ty^2 - 2y + 2 = (t + 2)y^2 + 6y + 10$.

Then, the determinant of the polynomial **$(t + 2)y^2 + 6y + 10$** has to be 0.
The determinant **$= 6^2 - 4(t + 2)10 = 36 - 40(t + 2) = 36 - 40t - 80 = -40t - 44$**.

Therefore, **$-40t - 44 = 0 \Rightarrow t = -44/40 = -11/10$**.

Suggestions or Solutions

To the **Problem** in the Example **4**

Suppose that we have two lines as follows: $2x - y - 1 = 0$ and $3x + 2y - 2 = 0$.
Then, find a point $P(s, t)$ that is the same distance away from each of the lines specified above.

When we take a measurement of a distance between two objects in math, we take the length of the line segment connecting the two objects, and the line segment is perpendicular to both of the objects.
Suppose that **A** is the line $2x - y - 1 = 0$, and that **B** is the line $3x + 2y - 2 = 0$.
Then, if we want to take the distance between the point **P** and the line **A**, we first, need to find the equation of a line including the point **P** and perpendicular to the line **A**. Also, the same true for the distance between the point **P** and the line **B**. Suppose now, that **C** is the line perpendicular to the line **A**, and that **D** is the line perpendicular to the line **B**. Then, we can notice that the point **P** is included in both of the perpendicular lines. Then, where do the lines **C** and **D** meet each other? It is the point **P**.

Fig. 4.0

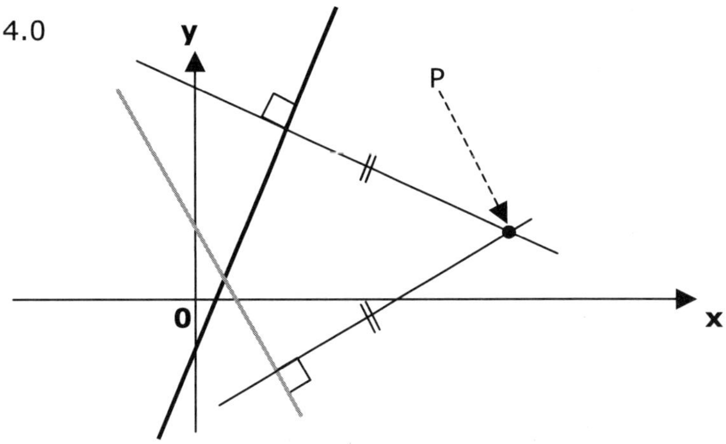

However, we can see another point that is equal distance away from both of the lines.

Fig. 4.1

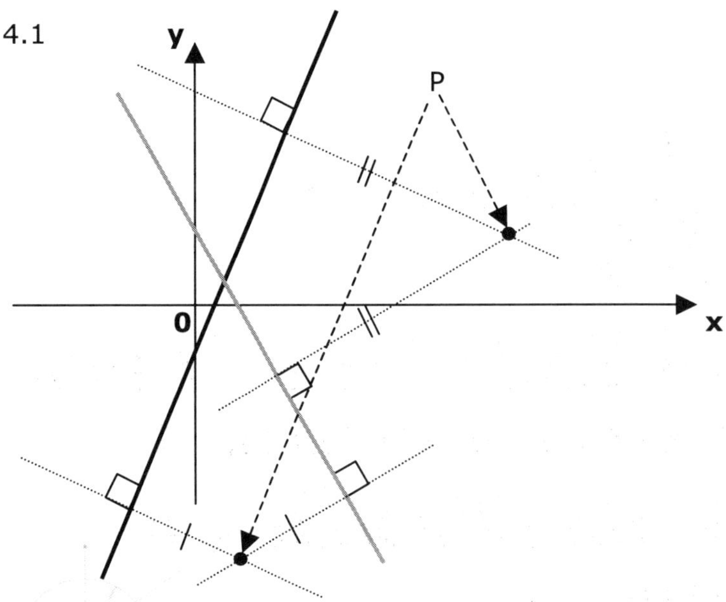

In fact, we can see infinitely many points each of which is equal distance away from both of the lines. All the points constitute two dotted lines as shown below.

Fig. 4.2

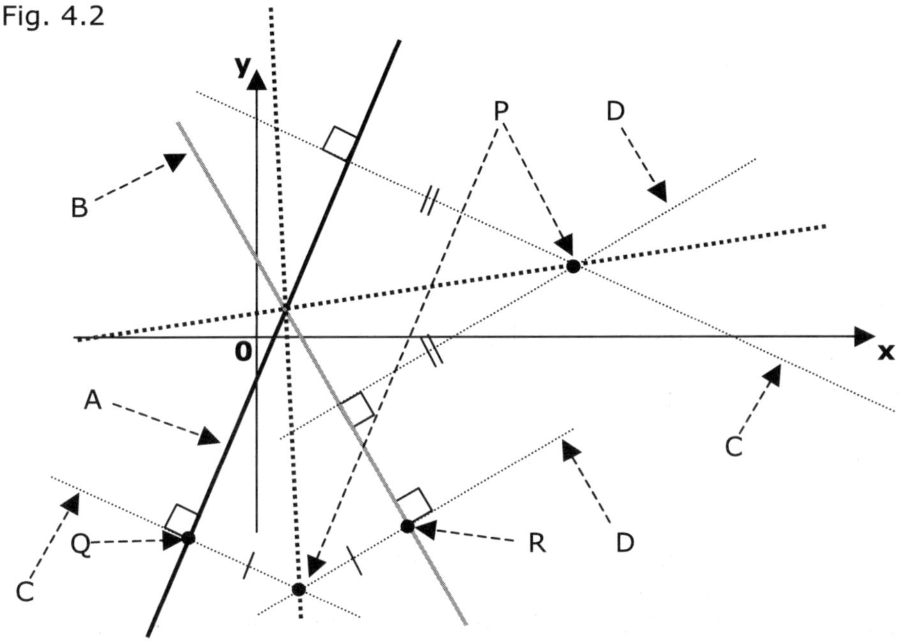

Then, we can see that that the point **P(s, t)** is not just a particular point but a general point representing all the points in each of the two dotted lines. Now, let's find each of the two dotted lines. That is, we use the **P** twice.

First, we need to begin with the two lines **C** and **D**. Then, we need to find the point where the two lines **A** and **C** meet each other and the point where the two lines **B** and **D** meet each other. Now, we can find the two dotted lines.

The line **C** is perpendicular to the line **A**, and the line **D** is perpendicular to the line **B**.
We know that if two lines are perpendicular to each other, the product of the slopes is -1.
Then, let's first, convert the given general equations to the standard ones. We have:
The line **A**: $\mathbf{2x - y - 1 = 0 \Rightarrow y = 2x - 1}$.
The line **B**: $\mathbf{3x + 2y - 2 = 0 \Rightarrow y = -3x/2 + 1}$.

Suppose now, that $\mathbf{y = a_1x + b_1}$ is the equation of the line **C**.
Then, because the slope of the line **A** is 2, the $\mathbf{a_1 = -1/2}$ since the product of the slopes is -1.
Then, $\mathbf{y = -x/2 + b_1}$.
Also, sine the **C** includes the point **P(s, t)**, $\mathbf{t = -s/2 + b_1 \Rightarrow b_1 = t + s/2}$.
Therefore, the equation of the **C** is $\mathbf{y = -x/2 + t + s/2}$.

Suppose this time, that $\mathbf{y = a_2x + b_2}$ is the equation of the line **D**.
Then, because the slope of the line **B** is -3/2, the $\mathbf{a_2 = 2/3}$ since the product of the slopes is -1.
Then, $\mathbf{y = 2x/3 + b_2}$.
In addition, since the **D** includes the point **P(s, t)**, too, $\mathbf{t = 2s/3 + b_2 \Rightarrow b_2 = t - 2s/3}$.
Therefore, the equation of the **D** is $\mathbf{y = 2x/3 + t - 2s/3}$.

Next, suppose that **Q** is the point where the two lines **A** and **C** meet each other, and let's find the **Q**.
The line **A** is $y = 2x - 1$, and the line **C** is $y = -x/2 + t + s/2$.
Then, we begin with the x-coordinate of the **Q**.
$2x - 1 = -x/2 + t + s/2.$
Then, $2x + x/2 = t + s/2 + 1 \Rightarrow 5x/2 = t + s/2 + 1 \Rightarrow x = (2t + s + 2)/5$. So:
$y = 2x - 1 = 2(2t + s + 2)/5 - 1 = (4t + 2s + 4)/5 - 1 = (4t + 2s + 4 - 5)/5 = (4t + 2s - 1)/5.$
Therefore, the **Q** is at $((2t + s + 2)/5, (4t + 2s - 1)/5)$.

Next, suppose that **R** is the point where the two lines **B** and **D** meet each other, and let's find the **R**.
The line **B** is $y = -3x/2 + 1$, and the line **D** is $y = 2x/3 + t - 2s/3$.
Then, we begin with the x-coordinate of the **R**.
$-3x/2 + 1 = 2x/3 + t - 2s/3.$
Then, $(2/3 + 3/2)x = 1 - t + 2s/3 \Rightarrow 13x/6 = 1 - t + 2s/3 \Rightarrow x = (6 - 6t + 4s)/13$.
So, $y = -3x/2 + 1 = (-3/2)(6 - 6t + 4s)/13 + 1 = (-3)(3 - 3t + 2s)/13 + 1$
$= (9t - 6s - 9)/13 + 1 = (9t - 6s + 4)/13$.
Therefore, the **R** is at $((6 - 6t + 4s)/13, (9t - 6s + 4)/13)$.

Now, let's use your favorite distance formula (usually called Pythagorean Theorem) to find the two dotted lines.

We know that the distance between the **P** and the **Q** is the same as the one between the **P** and the **R**, and that the formula for the distance is $d^2 = (\Delta x)^2 + (\Delta y)^2$.
Suppose first, that the distance between the **P** and the **Q** is d_1, that the one between the **P** and the **R** is d_2, and let's begin with the d_1.

$\Delta x = s - (2t + s + 2)/5 = (4s - 2t - 2)/5.$
$\Delta y = t - (4t + 2s - 1)/5 = (t - 2s + 1)/5.$
Then, $d_1^2 = (4s - 2t - 2)^2/5^2 + (t - 2s + 1)^2/5^2 = \{(4s - 2t - 2)^2 + (t - 2s + 1)^2\}/5^2$.

Next, let's move on to the d_2.

$\Delta x = s - (6 - 6t + 4s)/13 = (6t + 9s - 6)/13.$
$\Delta y = t - (9t - 6s + 4)/13 = (4t + 6s - 4)/13.$
Then, $d_2^2 = (6t + 9s - 6)^2/13^2 + (4t + 6s - 4)^2/13^2 = \{(6t + 9s - 6)^2 + (4t + 6s - 4)^2\}/13^2$.

Now, $\{(4s - 2t - 2)^2 + (t - 2s + 1)^2\}/5^2 = \{(6t + 9s - 6)^2 + (4t + 6s - 4)^2\}/13^2$.

The equation above is in fact, an equation of two lines which are the two dotted lines.
Then, we should be able to separate the lines apart so that we can see what they look like. We can try factorization to do the work. In other words, we factorize the equation into the form as follows:
$(ax + by + c)(dx + ey + f) = 0$ where **a**, **b**, **c**, **d**, **e**, and **f** are constants.
It looks however, quite messy. Let's proceed a little bit further, though.

$\{(4s - 2t - 2)^2 + (t - 2s + 1)^2\}/5^2 = \{(6t + 9s - 6)^2 + (4t + 6s - 4)^2\}/13^2$
$\Rightarrow 13^2\{(4s - 2t - 2)^2 + (t - 2s + 1)^2\} = 5^2\{(6t + 9s - 6)^2 + (4t + 6s - 4)^2\}$
$\Rightarrow 13^2(4s - 2t - 2)^2 + 13^2(t - 2s + 1)^2 = 5^2(6t + 9s - 6)^2 + 5^2(4t + 6s - 4)^2$
$\Rightarrow 13^2(4s - 2t - 2)^2 - 5^2(6t + 9s - 6)^2 + 13^2(t - 2s + 1)^2 - 5^2(4t + 6s - 4)^2 = 0.$

$13^2(4s - 2t - 2)^2 - 5^2(6t + 9s - 6)^2$
$= \{13(4s - 2t - 2) + 5(6t + 9s - 6)\}\{13(4s - 2t - 2) - 5(6t + 9s - 6)\}$

$= (52s - 26t - 26 + 30t + 45s - 30)(52s - 26t - 26 - 30t - 45s + 30)$
$= (97s + 4t - 56)(7s - 56t + 4)$
$= 679s^2 - 5432st + 388s + 28st - 224t^2 + 16t - 392s + 3136t - 224$
$= 679s^2 - 224t^2 - 5404st - 4s + 3152t - 224$

$13^2(t - 2s + 1)^2 - 5^2(4t + 6s - 4)^2$
$= \{13(t - 2s + 1) + 5(4t + 6s - 4)\}\{13(t - 2s + 1) - 5(4t + 6s - 4)\}$
$= (13t - 26s + 13 + 20t + 30s - 20)(13t - 26s + 13 - 20t - 30s + 20)$
$= (33t + 4s - 7)(-7t - 56s + 33)$
$= -231t^2 - 1848st + 1089t - 28st - 224s^2 + 132s + 49t + 392s - 231$
$= -224s^2 - 231t^2 - 1876st + 524s + 1138t - 231$

$13^2(4s - 2t - 2)^2 - 5^2(6t + 9s - 6)^2 + 13^2(t - 2s + 1)^2 - 5^2(4t + 6s - 4)^2$
$= 679s^2 - 224t^2 - 5404st - 4s + 3152t - 224 - 224s^2 - 231t^2 - 1876st + 524s + 1138t - 231$
$= 455s^2 - 455t^2 - 7280st + 520s + 4290t - 455$
$= 0$

Now, dividing by 65 both sides of the last equality, we get:
$\mathbf{7s^2 - 7t^2 - 112st + 8s + 66t - 7 = 0.}$

It is a real mess!
Now, we should work out this example in some other way around.

We often have to find a distance from a point to a line. Therefore, we have a formula for it. Now, let's use the formula. However, we are going to find out first, how the formula gets constructed.

The general equation of line is as follows: $ax + by + c = 0$ where **a**, **b**, and **c** are constants.
Suppose now, that **L** is the line $ax + by + c = 0$, and that we find the distance from a point **P(s, t)** to the line **L**.
Then, as usual, we find the perpendicular line and the point where the two lines meet each other, and then, use the famous distance formula.

Then, let's suppose that **T** is the line perpendicular to the line **L**, and find the **T** first.
We know that the **T** includes the point **P(s, t)**. Suppose that the slope of the **T** is **m**.
Then, the equation of the **T** is as follows: $y - t = m(x - s)$. Let's now, get the **m** specified.
We first, convert the equation of the **L** into the standard form and then, find the slope.
$ax + by + c = 0 \Rightarrow y = -ax/b - c/b \Rightarrow$ The slope of the **L** is $-a/b$.
Then, we can see that b/a is the slope of the **T** because the product of the slopes is -1.

Then, we can see that $y - t = m(x - s) \Rightarrow y - t = (b/a)(x - s)$.
So, we get $b(x - s) - a(y - t) = 0$. Then:

The line **L**: $ax + by + c = 0$. The line **T**: $b(x - s) - a(y - t) = 0$.

Then, we need to find where the two lines meet. In other words, we take the equations above for simultaneous equations for the **x** and the **y** and get the solution.
Let's multiply the equation of the **L** by the **a** and the equation of the **T** by the **b**. Then:
$a(ax + by + c) = a^2x + aby + ac = 0$.
$b\{b(x - s) - a(y - t)\} = b^2(x - s) - ab(y - t) = 0$.

Now, if we add both together, then the **y** will be removed.
$$(a^2x + aby + ac) + \{b^2(x - s) - ab(y - t)\} = a^2x + aby + ac + b^2x - b^2s - aby + abt$$
$$= a^2x + ac + b^2x - b^2s + abt = (a^2 + b^2)x - b^2s + abt + ac = 0.$$
Then, $(a^2 + b^2)x = b^2s - abt - ac$.
Therefore, $x = (b^2s - abt - ac)/(a^2 + b^2)$.

Let's now, multiply the equation of the **L** by the **b** and the equation of the **T** by the **a**. Then:
$$b(ax + by + c) = abx + b^2y + bc = 0.$$
$$a\{b(x - s) - a(y - t)\} = ab(x - s) - a^2(y - t) = 0.$$

Now, if we subtract the second equation from the first, then the **x** will be eliminated.
$$(abx + b^2y + bc) - \{ab(x - s) - a^2(y - t)\} = abx + b^2y + bc - abx + abs + a^2y - a^2t$$
$$= b^2y + bc + abs + a^2y - a^2t = (a^2 + b^2)y - a^2t + abs + bc = 0.$$
Then, $(a^2 + b^2)y = a^2t - abs - bc$.
Therefore, $y = (a^2t - abs - bc)/(a^2 + b^2)$.

So, the two lines **L** and **T** meet at $((b^2s - abt - ac)/(a^2 + b^2), (a^2t - abs - bc)/(a^2 + b^2))$.

Suppose now, that the two lines **L** and **T** meet at a point $Q(u, v)$, and that **d** is the distance between the point $P(s, t)$ and the point $Q(u, v)$.

Then, we have $u = (b^2s - abt - ac)/(a^2 + b^2)$, and $v = (a^2t - abs - bc)/(a^2 + b^2)$.

So, we get $(a^2 + b^2)u = b^2s - abt - ac$, and $(a^2 + b^2)v = a^2t - abs - bc$.

We know that $d^2 = (u - s)^2 + (v - t)^2$.
Then, we find **u − v** and **v − t** out of the expressions as follows:
$(a^2 + b^2)u = b^2s - abt - ac$, and $(a^2 + b^2)v = a^2t - abs - bc$.

$$(a^2 + b^2)u = b^2s - abt - ac = a^2s - a^2s + b^2s - abt - ac = (a^2 + b^2)s - a^2s - abt - ac.$$
$$\Rightarrow (a^2 + b^2)u = (a^2 + b^2)s - a^2s - abt - ac$$
$$\Rightarrow (a^2 + b^2)u - (a^2 + b^2)s = -(a^2s + abt + ac)$$
$$\Rightarrow (a^2 + b^2)(u - s) = -(a^2s + abt + ac)$$
$$\Rightarrow u - s = -(a^2s + abt + ac)/(a^2 + b^2) = -a(as + bt + c)/(a^2 + b^2).$$

$$(a^2 + b^2)v = a^2t - abs - bc = a^2t + b^2t - b^2t - abs - bc = (a^2 + b^2)t - (b^2t + abs + bc)$$
$$\Rightarrow (a^2 + b^2)(v - t) = -(b^2t + abs + bc).$$
$$\Rightarrow v - t = -(b^2t + abs + bc)/(a^2 + b^2) = -b(b^2t + abs + bc)/(a^2 + b^2).$$

Now, we can see that:

$$d^2 = (u - s)^2 + (v - t)^2$$
$$= a^2(as + bt + c)^2/(a^2 + b^2)^2 + b^2(bt + as + c)^2/(a^2 + b^2)^2$$
$$= (a^2 + b^2)(as + bt + c)^2/(a^2 + b^2)^2$$
$$= (as + bt + c)^2/(a^2 + b^2).$$

Then, $d = \pm(as + bt + c)/(a^2 + b^2)^{1/2}$.
We know however, that the $d > 0$ since **d** is the distance.
Therefore, $d = |as + bt + c|/(a^2 + b^2)^{1/2}$.

Now, we have found the formula that produces the distance from a point **P(s, t)** to a line **ax + by + c = 0**. Then, with this formula, let's get back to the problem in this example.
At the moment, we are working on the point **P(s, t)** that is an equal distance away from both of the lines **2x − y − 1 = 0** and **3x + 2y − 2 = 0**.

The line **A**: **2x − y − 1 = 0**. The line **B**: **3x + 2y − 2 = 0**.

Then, by means of the formula:
The distance between the line **A** and the point **P** is $|2s - t - 1|/(2^2 + (-1)^2)^{1/2}$.
The distance between the line **B** and the point **P** is $|3s + 2t - 2|/(3^2 + 2^2)^{1/2}$.

Then, we get $|2s - t - 1|/(2^2 + (-1)^2)^{1/2} = |3s + 2t - 2|/(3^2 + 2^2)^{1/2}$.
So, $(2s - t - 1)/5^{1/2} = \pm(3s + 2t - 2)/13^{1/2}$.

Now, why ±?
That's because what's inside the absolute signs can be positive or negative. That is, it means two lines.
What lines?
They are two dotted lines stated earlier.
We have found earlier that each of all the points constituting the two dotted lines can be the **P(s, t)**.
That is, the **P** is the general point in each of the dotted lines.
Since the lines are in the x-y system, let's now, change the **s** and the **t** to **x** and **y** respectively.

Now, one of the two dotted lines is as follows:

$$(2x - y - 1)/5^{1/2} = (3x + 2y - 2)/13^{1/2}$$
$$\Rightarrow (2/5^{1/2} - 3/13^{1/2})x - (1/5^{1/2} + 2/13^{1/2})y + 2/13^{1/2} - 1/5^{1/2} = 0$$

$$\Rightarrow (2X13^{1/2} - 3X5^{1/2})x/65^{1/2} - (13^{1/2} + 2X5^{1/2})y/65^{1/2} + (2X5^{1/2} - 13^{1/2})/65^{1/2} = 0$$

$$\Rightarrow (2X13^{1/2} - 3X5^{1/2})x - (13^{1/2} + 2X5^{1/2})y + 2X5^{1/2} - 13^{1/2}$$
$$= (2\sqrt{13} - 3\sqrt{5})x - (\sqrt{13} + 2\sqrt{5})y + 2\sqrt{5} - \sqrt{13} = 0.$$

Next, the other line is as follows:

$$2x - y - 1)/5^{1/2} = -(3x + 2y - 2)/13^{1/2}$$
$$\Rightarrow (2/5^{1/2} + 3/13^{1/2})x + (2/13^{1/2} - 1/5^{1/2})y - 1/5^{1/2} - 2/13^{1/2} = 0.$$

$$\Rightarrow (2X13^{1/2} + 3X5^{1/2})x/65^{1/2} + (2X5^{1/2} - 13^{1/2})y/65^{1/2} - (13^{1/2} + 2X5^{1/2})/65^{1/2} = 0$$

$$\Rightarrow (2X13^{1/2} + 3X5^{1/2})x + (2X5^{1/2} - 13^{1/2})y - 13^{1/2} - 2X5^{1/2}$$
$$= (2\sqrt{13} + 3\sqrt{5})x + (2\sqrt{5} - \sqrt{13})y - \sqrt{13} - 2\sqrt{5} = 0.$$

Now, each of all the points constituting the two lines above is the same distance away from both of the two lines **A** and **B**.

In short:

Suppose that:
Line **A**: $\mathbf{2x - y - 1 = 0}$. Line **B**: $\mathbf{3x + 2y - 2 = 0}$.

Then, by the distance formula:
The distance between the line **A** and the point **P** is $|2s - t - 1|/(2^2 + (-1)^2)^{1/2}$.
The distance between the line **B** and the point **P** is $|3s + 2t - 2|/(3^2 + 2^2)^{1/2}$.

Then, we get $|2s - t - 1|/(2^2 + (-1)^2)^{1/2} = |3s + 2t - 2|/(3^2 + 2^2)^{1/2}$.
So, $(2s - t - 1)/5^{1/2} = \pm(3s + 2t - 2)/13^{1/2}$.

Now, the expressions above are equations indicating two lines, and each of all the points constituting the lines is an equal distance away from both of the lines **A** and **B**. Therefore, the **P(s, t)** is the general point in each of the two lines.

Since the lines are in the x-y system, let's now, change the **s** and the **t** to **x** and **y** respectively.

Now, one of the two lines is as follows:

$$(2x - y - 1)/5^{1/2} = (3x + 2y - 2)/13^{1/2}$$
$$\Rightarrow (2/5^{1/2} - 3/13^{1/2})x - (1/5^{1/2} + 2/13^{1/2})y + 2/13^{1/2} - 1/5^{1/2} = 0$$

$$\Rightarrow (2X13^{1/2} - 3X5^{1/2})x/65^{1/2} - (13^{1/2} + 2X5^{1/2})y/65^{1/2} + (2X5^{1/2} - 13^{1/2})/65^{1/2} = 0$$

$$\Rightarrow (2X13^{1/2} - 3X5^{1/2})x - (13^{1/2} + 2X5^{1/2})y + 2X5^{1/2} - 13^{1/2}$$
$$= (2\sqrt{13} - 3\sqrt{5})x - (\sqrt{13} + 2\sqrt{5})y + 2\sqrt{5} - \sqrt{13} = 0.$$

Next, the other line is as follows:

$$(2x - y - 1)/5^{1/2} = -(3x + 2y - 2)/13^{1/2}$$
$$\Rightarrow (2/5^{1/2} + 3/13^{1/2})x + (2/13^{1/2} - 1/5^{1/2})y - 1/5^{1/2} - 2/13^{1/2} = 0.$$

$$\Rightarrow (2X13^{1/2} + 3X5^{1/2})x/65^{1/2} + (2X5^{1/2} - 13^{1/2})y/65^{1/2} - (13^{1/2} + 2X5^{1/2})/65^{1/2} = 0$$

$$\Rightarrow (2X13^{1/2} + 3X5^{1/2})x + (2X5^{1/2} - 13^{1/2})y - 13^{1/2} - 2X5^{1/2}$$
$$= (2\sqrt{13} + 3\sqrt{5})x + (2\sqrt{5} - \sqrt{13})y - \sqrt{13} - 2\sqrt{5} = 0.$$

By the way, over shoes, over boots, let's construct the other formula.
We often need to find an area of a triangle. Then, we usually use the simple formula where the area is half the product of the base and the height of the triangle. However, we sometimes face trouble finding the height. Now, there are several other formulas we can use to find the area of a triangle without the specification on the height. Among those, the one we are about to construct can be quite useful when we work on a triangle specified in a perpendicular (or Cartesian) coordinate plane such as the x-y plane.

Suppose now, that we are given three points $\mathbf{P_1(x_1, y_1)}$, $\mathbf{P_2(x_2, y_2)}$, and $\mathbf{P_3(x_3, y_3)}$, and that the points make a triangle. Then, let's put the points in the x-y plane and make a triangle.

Fig. 4.3

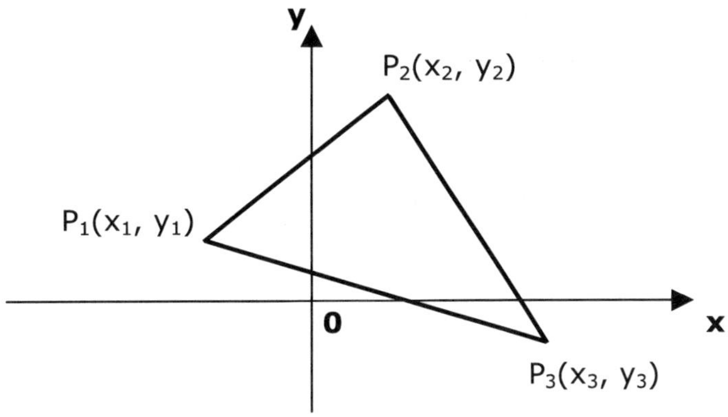

However, we still need to consult the basic formula where the area = (the base X the height)/2. Suppose now, that the line segment $\mathbf{P_1P_2}$ is the base. Then, the perpendicular distance from the $\mathbf{P_3}$ to the base is the height. Let's now, begin with the equation of the line that includes the two points $\mathbf{P_1}$ and $\mathbf{P_2}$.

To begin with, the slope of the line is $(y_2 - y_1)/(x_2 - x_1)$. Also, since the line includes $\mathbf{P_1}$, the equation of the line is as follows: $y - y_1 = \{(y_2 - y_1)/(x_2 - x_1)\}(x - x_1)$.
$\Rightarrow (y - y_1)(x_2 - x_1) = (y_2 - y_1)(x - x_1)$.
Now, let's convert the equation into the equation in general form.

Then, we get $y - y_1 = (y_2 - y_1)(x - x_1)/(x_2 - x_1)$. **So,** $(y - y_1)(x_2 - x_1) = (y_2 - y_1)(x - x_1)$.
Meanwhile:
$(y - y_1)(x_2 - x_1) = y(x_2 - x_1) - y_1(x_2 - x_1)$. $(y_2 - y_1)(x - x_1) = x(y_2 - y_1) - x_1(y_2 - y_1)$.

Then, $(y - y_1)(x_2 - x_1) = (y_2 - y_1)(x - x_1) \Rightarrow (y - y_1)(x_2 - x_1) - (y_2 - y_1)(x - x_1) = 0$
$\Rightarrow y(x_2 - x_1) - y_1(x_2 - x_1) - \{x(y_2 - y_1) - x_1(y_2 - y_1)\} = 0$
$\Rightarrow y(x_2 - x_1) - y_1(x_2 - x_1) - x(y_2 - y_1) + x_1(y_2 - y_1)\} = 0$
$\Rightarrow y(x_2 - x_1) - x(y_2 - y_1) - y_1x_2 + y_1x_1 + x_1y_2 - x_1y_1 = y(x_2 - x_1) - x(y_2 - y_1) - y_1x_2 + x_1y_2 = 0$
$\Rightarrow (y_2 - y_1)x - (x_2 - x_1)y - x_1y_2 + y_1x_2 = 0$, which is the general equation of line.

Now, we can use the distance formula just constructed earlier to find the distance between the line and the point $\mathbf{P_3}$. Then, the distance will be the height of the triangle. Therefore:
The height = $|(y_2 - y_1)x_3 - (x_2 - x_1)y_3 - x_1y_2 + y_1x_2|/\{(y_2 - y_1)^2 + (x_2 - x_1)^2\}^{1/2}$.
Also, the distance between the $\mathbf{P_1}$ and the $\mathbf{P_2}$, that is, the length of the base is:
$\{(y_2 - y_1)^2 + (x_2 - x_1)^2\}^{1/2}$.

Then, the base X the height
$= \{(y_2 - y_1)^2 + (x_2 - x_1)^2\}^{1/2}|(y_2 - y_1)x_3 - (x_2 - x_1)y_3 - x_1y_2 + y_1x_2|/\{(y_2 - y_1)^2 + (x_2 - x_1)^2\}^{1/2}$
$= |(y_2 - y_1)x_3 - (x_2 - x_1)y_3 - x_1y_2 + y_1x_2|$.

Therefore, the area of the triangle is $|(y_2 - y_1)x_3 - (x_2 - x_1)y_3 - x_1y_2 + y_1x_2|/2$.
Let's make look nicer what's inside the absolute signs.
$(y_2 - y_1)x_3 - (x_2 - x_1)y_3 - x_1y_2 + y_1x_2 = (x_1 - x_2)y_3 + y_2x_3 - y_1x_3 - x_1y_2 + y_1x_2$
$= (x_1 - x_2)y_3 + (x_2 - x_3)y_1 + (x_3 - x_1)y_2$.

Now, the area of the triangle $\mathbf{P_1P_2P_3}$ = $|(x_1 - x_2)y_3 + (x_2 - x_3)y_1 + (x_3 - x_1)y_2|/2$.

Suggestions or Solutions

To the Problem 0 in the Example 5

Suppose that we have equations of lines as follows: $ax + by + c = 0$ and $ux + vy + w = 0$ where a, b, c, u, v, and w are constants, $abc \neq 0$, and $uvw \neq 0$. Suppose also, that the equations in fact, indicate the same line, and find the connective expression between the constants.

Some equations looking quite different can actually indicate the same line.
Then, how can we find out if they are actually the same?
We should give some test to them.

The constants in the given equations are often called coefficients.
Rigorously though, the **c** and the **w** are not called coefficients.
Examining the relationship between the coefficients, we can check to see if the equations in question are the same, indeed.

First, let's see how the same equations can look different.
For instance, all of $2x + 3y - 5 = 0$, $4x + 6y - 10 = 0$, $-2x - 3y + 5 = 0$, and $x + 3y/2 - 5/2 = 0$ indicate the same line. That's because they are multiples of each other.
Multiplying the first equation by 2, we get the second one, and however, multiplying the first one by −1, we get the third one.
If we multiply the first one by 2, each of the constants gets multiplied by 2 and the products are respectively the constants in the second one.
Also, multiplying the second by −1/2, we get the third, and multiplying by −1/2 again each of the constants in the third, we get the forth.
Therefore, if the ratios between corresponding constants are the same, the equations indicate the same line. Of the examples above, let's take the first two and compare them.
They are $2x + 3y - 5 = 0$ and $4x + 6y - 10 = 0$.
Then, the ratios between the corresponding constants are as follows: $2/4 = 3/6 = -5/-10$.
All the ratios are 1/2 and are the same as each other.

Let's now, put the equations of the general form in the standard form. Then:
$ax + by + c = 0 \Rightarrow y = (-a/b)x - c/b$
$ux + vy + w = 0 \Rightarrow y = (-u/v)x - w/v$

Now, if the two equations are the same, the slopes need to be the same and so do the y-intercepts.
Therefore, we need to get $-a/b = -u/v$ and $-c/b = -w/v$.

Then:
$-a/b = -u/v \Rightarrow a/b = u/v \Rightarrow av = bu \Rightarrow a/u = b/v$.
$-c/b = -w/v \Rightarrow c/b = w/v \Rightarrow cv = bw \Rightarrow b/v = c/w$.
So, if $a/u = b/v$ and $b/v = c/w$, then $a/u = b/v = c/w$.

Therefore, if $ax + by + c = 0$ and $ux + vy + w = 0$ indicate the same line, we get $a/u = b/v = c/w$.

That is, comparing general equations of line, if we want to find out if they indicate the same line, we need to check to see if the ratios between the corresponding constants are the same.

Suggestions or Solutions

To the Problem 1 in the Example 5

Suppose that we have equations of lines as follows: $ax + by + c = 0$ and $ux + vy + w = 0$ where a, b, c, u, v, and w are constants, $abc \neq 0$, and $uvw \neq 0$. Suppose also, that the two lines that are parallel to each other, and find the connective expression between the constants.

Two lines parallel to each other have the same slope.
Therefore, if two equations indicate two lines parallel to each other, the slopes need to be the same.
What has to be different, then?
The y-intercepts or the x-intercepts need to be different.
Therefore, in the two equations, if the slopes are the same but the y-intercepts (or the x-intercepts) are different, the two equations represent two other lines parallel to each other.

Let's now, put the equations of the general form in the standard form. Then:
$ax + by + c = 0 \Rightarrow y = (-a/b)x - c/b$
$ux + vy + w = 0 \Rightarrow y = (-u/v)x - w/v$

Then, if $-a/b = -u/v$ but $-c/b \neq -w/v$, then the two equations indicate two different lines parallel to each other. Now:
$-a/b = -u/v \Rightarrow a/b = u/v \Rightarrow av = bu \Rightarrow a/u = b/v.$
$-c/b \neq -w/v \Rightarrow c/b \neq w/v \Rightarrow cv \neq bw \Rightarrow b/v \neq c/w.$

So, if $a/u = b/v$ but $b/v \neq c/w$, then $a/u = b/v \neq c/w$.

Therefore, if the two equations $ax + by + c = 0$ and $ux + vy + w = 0$ indicate two different lines parallel to each other, we get $a/u = b/v \neq c/w$.

That is, comparing two general equations of line, if we want to find out if they indicate two different lines parallel to each other, we need to check to see if the ratio between the coefficients of the **x** is the same as the one between those of the **y**, and then, if the same ratio, we need to check to see if it is different from the ratio between the constant terms.

However, if all the ratios are the same, the equations indicate the same line.

Suggestions or Solutions

To the Problem 2 in the Example 5

Suppose that we have equations of lines as follows: $ax + by + c = 0$ and $ux + vy + w = 0$ where a, b, c, u, v, and w are constants, $abc \neq 0$, and $uvw \neq 0$. Suppose also, that the two lines meet at one point, and find the connective expression between the constants.

If two lines are parallel to each other, the slopes are the same and the lines do not cross each other.
If the slopes of the two lines are different, the lines cross each other.
Therefore, if two lines meet each other at a point, the slopes need to be different from each other.

Let's now, put the equations of the general form in the standard form. Then:
$ax + by + c = 0 \Rightarrow y = (-a/b)x - c/b$
$ux + vy + w = 0 \Rightarrow y = (-u/v)x - w/v$

Then, if $-a/b \neq -u/v$, the two equations indicate two different lines that meet each other at a point.

Now, $-a/b \neq -u/v \Rightarrow a/b \neq u/v \Rightarrow av \neq bu \Rightarrow a/u \neq b/v$.

Therefore, if two equations $ax + by + c = 0$ and $ux + vy + w = 0$ indicate two different lines meeting each other at a point, we get $a/u \neq b/v$.

That is, comparing two general equations of line, if we want to see if they indicate two lines meeting each other at a point, we need to check to see if the ratio between the coefficients of the **x** is different from the one between those of the **y**.

Now, if the slopes of two lines are different, do the y-intercepts have to be different, too?
No. The y-intercepts can be the same as each other.
How come?
The lines can meet at a point in the y-axis.
Also, since lines with different slopes can meet at a point in the x-axis, the x-intercepts can be the same as each other, too.

Suggestions or Solutions
To the Problem 3 in the Example 5

Suppose that we have equations of lines as follows: $ax + by + c = 0$ and $ux + vy + w = 0$ where a, b, c, u, v, and w are constants, $abc \neq 0$, and $uvw \neq 0$. Suppose also, that the two lines are perpendicular to each other, and find the connective expression between the constants.

What is the condition under which two lines are perpendicular to each other?
The product of the slopes is -1.

Let's now, put the equations of the general form in the standard form. Then:
$ax + by + c = 0 \Rightarrow y = (-a/b)x - c/b$
$ux + vy + w = 0 \Rightarrow y = (-u/v)x - w/v$

Then, if $(-a/b)(-u/v) = -1$, the two equations indicate two lines perpendicular to each other.

Now, $(-a/b)(-u/v) = -1 \Rightarrow au/bv = -1 \Rightarrow au = -bv \Rightarrow au + bv = 0$.

Therefore, if two equations $ax + by + c = 0$ and $ux + vy + w = 0$ indicate two lines perpendicular to each other, we get $au + bv = 0$.

That is, comparing two general equations of line, if we want to see if they indicate two lines perpendicular to each other, we can check to see if the sum of the respective products of coefficients of the **x** and the **y** is 0 instead of checking the product of the slopes.

The example 4 in the previous section explains how the product of the slopes is -1 if the two lines are perpendicular to each other.

Suggestions or Solutions

To the Problem 4 in the Example 5

Suppose that we have equations of lines as follows: $ax + by + c = 0$ and $ux + vy + w = 0$ where a, b, c, u, v, and w are constants, $abc \neq 0$, and $uvw \neq 0$. Then, find the equation of a line that includes the point where the two lines meet.

How many lines are there to include a point where two lines meet each other?
There are infinitely many.
Not any line can do, though.

Suppose that the two lines above meet at a point (s, t).
Then, since the two lines include the point (s, t), we get $as + bt + c = 0$ and $us + vt + w = 0$.
Then, we can see that $m(as + bt + c) + n(us + vt + w) = 0$ where m and n are constants.

Now, we can see that $m(ax + by + c) + n(ux + vy + w) = 0$ is an equation of line, too.

The reason is as follows:

First, we have $m(ax + by + c) + n(ux + vy + w) = (am + nu)x + (bm + nv)y + cm + nw = 0$.
Next, in $(am + nu)x + (bm + nv)y + cm + nw = 0$, $am + nu$, $bm + nv$, and $cm + nw$ are constants.
Therefore, $(am + nu)x + (bm + nv)y + cm + nw = 0$ is an equation of line.

Now, if $m(as + bt + c) + n(us + vt + w) = 0$, the line $m(ax + by + c) + n(ux + vy + w) = 0$ includes the point (s, t).

Therefore, $m(ax + by + c) + n(ux + vy + w) = 0$ is the equation of the line including the point (s, t) where the two lines $ax + by + c = 0$ and $ux + vy + w = 0$ meet each other.

Particularly in this case, we may say that the two equations $ax + by + c = 0$ and $ux + vy + w = 0$ are basic equations.
That is, they are like the bases.
Then, the two basic equations indicate two basic lines.
The two basic lines meet at a particular point, and all the lines that can be indicated by the equation $m(ax + by + c) + n(ux + vy + w) = 0$ include the particular point, too.
In fact, we can arbitrarily choose any two lines from all the lines that can be indicated by the equation $m(ax + by + c) + n(ux + vy + w) = 0$, and then, can take them for the basic lines, too.

Now, if we multiply the basic lines respectively by nonzero numbers and add the products together, the sum is to be another equation of a line, which includes the point (s, t).

Besides, the same is true for the lines parallel to each other.
Suppose now, that two lines $ax + by + c = 0$ and $ux + vy + w = 0$ are parallel to each other.
Then, $m(ax + by + c) + n(ux + vy + w) = 0$ is an equation of a line that is parallel to both of the two lines $ax + by + c = 0$ and $ux + vy + w = 0$.

Suggestions or Solutions

To the Problem in the Example 6

Suppose that we have an equation that represents a group of lines, and the equation is as follows: $3(k - 1)x + (2k + 3)y + 4k + 5 = 0$ where k is a constant. Suppose also, that all the lines in the group include a particular point P(s, t), and find the values of the s and the t.

Suppose that:
A is an equation $m(ax + by + c) + n(ux + vy + w) = 0$.
B is an equation $ax + by + c = 0$. **C** is an equation $ux + vy + w = 0$.

Then, **A** is an equation of any line that includes the point where the two lines **B** and **C** meet each other. That is, the **A** is an equation of line, which represents all the lines that include the point where the two lines **B** and **C** meet each other. Therefore, the **B** and the **C** are the basic equations.

Now, the equation $3(k - 1)x + (2k + 3)y + 4k + 5 = 0$ is made up of equations of the two basic lines. Then, if we extract the two basic lines and find the point where the basic lines meet each other, the point is the particular point **P(s, t)**.
How do we extract them, then?
We convert the given equation $3(k - 1)x + (2k + 3)y + 4k + 5 = 0$ into the form as follows:
$m(ax + by + c) + n(ux + vy + w) = 0$. How?

$3(k - 1)x + (2k + 3)y + 4k + 5 = 0 \Rightarrow 3kx - 3x + 2ky + 3y + 4k + 5 = 0$
$\Rightarrow 3kx + 2ky + 4k - 3x + 3y + 5 = 0 \Rightarrow k(3x + 2y + 4) - (3x - 3y - 5) = 0$, which is in the form of $m(ax + by + c) + n(ux + vy + w) = 0$.

Therefore, we have $m = k$ and $n = -1$; also, $a = 3$, $b = 2$, $c = 4$, $u = 3$, $v = -3$, and $w = -5$.

Then, the equation $3(k - 1)x + (2k + 3)y + 4k + 5 = 0$ indicates each of all the lines that include the point where the two lines $3x + 2y + 4 = 0$ and $3x - 3y - 5 = 0$ meet each other.

Now, we can take the two equations of the lines above for the system of equations for the **x** and **y**, and take the solution for the values of the **s** and the **t**.
($3x + 2y + 4 = 0$ and $3x - 3y - 5 = 0$)
$\Rightarrow 3x + 2y + 4 - (3x - 3y - 5) = 0 \Rightarrow 5y + 9 = 0 \Rightarrow y = -9/5$.
Then, $y = -9/5 \Rightarrow 3x + 2y + 4 = 3x - 18/5 + 4 = 0 \Rightarrow 3x = -2/5 \Rightarrow x = -2/15$.
Then, not only the two basic lines but all the other lines in the group of the lines include **(-2/15, -9/5)**.
Therefore, we get $s = -2/15$ and $t = -9/5$.

In short:

$3(k - 1)x + (2k + 3)y + 4k + 5 = 0 \Rightarrow 3kx - 3x + 2ky + 3y + 4k + 5 = 0$
$\Rightarrow 3kx + 2ky + 4k - 3x + 3y + 5 = 0 \Rightarrow k(3x + 2y + 4) - (3x - 3y - 5) = 0$.

($3x + 2y + 4 = 0$ and $3x - 3y - 5 = 0$)
$\Rightarrow 3x + 2y + 4 - (3x - 3y - 5) = 0 \Rightarrow 5y + 9 = 0 \Rightarrow y = -9/5$.
Then, $y = -9/5 \Rightarrow 3x + 2y + 4 = 3x - 18/5 + 4 = 0 \Rightarrow 3x = -2/5 \Rightarrow x = -2/15$.
Therefore, $s = -2/15$ and $t = -9/5$.

Suggestions or Solutions

To the Problem in the Example 7

Suppose that we have an equation that represents a group of lines, and the equation is as follows: $y = ax - b$ such that $3a + b = 2$ where a and b are constant. Suppose also, that all the lines in the group include a particular point P(s, t), and find the values of the s and the t.

Let's first, listen to the problem.
"If we choose many pairs of numbers for the **a** and **b** in the connective expression $3a + b = 2$, then we get as many lines as the number of the choices, and the equation $y = ax - b$ indicates each of all the lines we get. Then, each of all the lines includes the particular point **P(s, t)**. Now, find the values of the **s** and the **t**."

Now, let's reason on what to do and how to do it.
Each of all the lines that meet each other at a particular point gets made up of two basic lines.
Then, the two basic lines meet each other at the particular point.
Now, we need to find the two basic lines and find the point where the basic lines meet each other.
We know that $y = ax - b$ has the equations indicating the two basic lines.

Now, we convert the equation $y = ax - b$ into the form of $m(px + qy + r) + n(ux + vy + w) = 0$.
Then, $y = ax - b \Rightarrow ax - y - b = 0$.
We now, need to eliminate either of **a** and **b** by means of the connective expression. Let's get rid of the **a**.

Then, $3a + b = 2 \Rightarrow a = (2 - b)/3 \Rightarrow y = ax - b = (2 - b)x/3 - b = 2x/3 - bx/3 - b$
$\Rightarrow y - 2x/3 + bx/3 + b = 0 \Rightarrow y - 2x/3 + b(x/3 + 1) = 0$. Then, in the form above, we can say that $m = 1$ and $n = b$; also, $p = -2/3$, $q = 1$, $r = 0$, $u = 1/3$, $v = 0$, and $w = 1$.

Therefore, the two equations $y - 2x/3 = 0$ and $x/3 + 1 = 0$ indicate the two basic lines.
Then, we need to find the point where the basic lines meet each other.

$y - 2x/3 = 0 \Rightarrow y = 2x/3.$
$x/3 + 1 = 0 \Rightarrow x = -3 \Rightarrow y = 2x/3 = -2.$

Then, we can see that the two basic lines meet each other at **(-3, -2)**.
Therefore, we get $s = -3$ and $t = -2$.

Now, what if we eliminate the **b**? We should get the same result. Let's see if it is the case.

$3a + b = 2 \Rightarrow b = 2 - 3a \Rightarrow y = ax - b = ax - (2 - 3a) = ax - 2 + 3a = a(x + 3) - 2$
$\Rightarrow a(x + 3) - 2 - y = 0 \Rightarrow a(x + 3) - (2 + y) = 0.$

Therefore, the two equations $x + 3 = 0$ and $2 + y = 0$ indicate the two basic lines.
Then, we need to find the point where the basic lines meet each other.

$x + 3 = 0 \Rightarrow x = -3$. $2 + y = 0 \Rightarrow y = -2$.

Then, the two basic lines meet each other at **(-3, -2)**. Therefore, we get $s = -3$ and $t = -2$.

Fig. 7.0

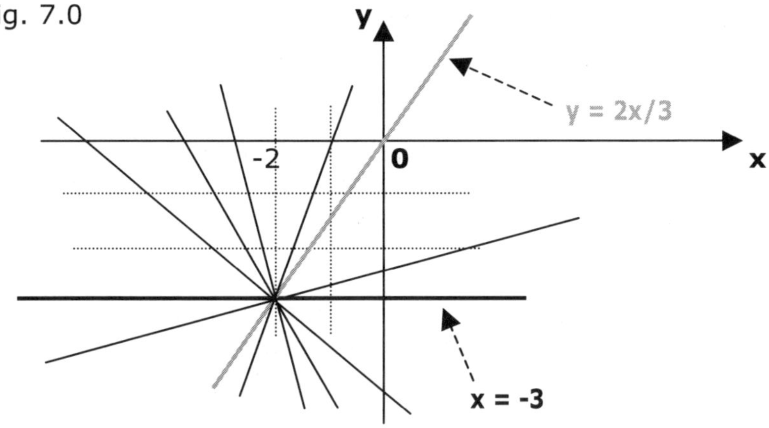

Fig. 7.1

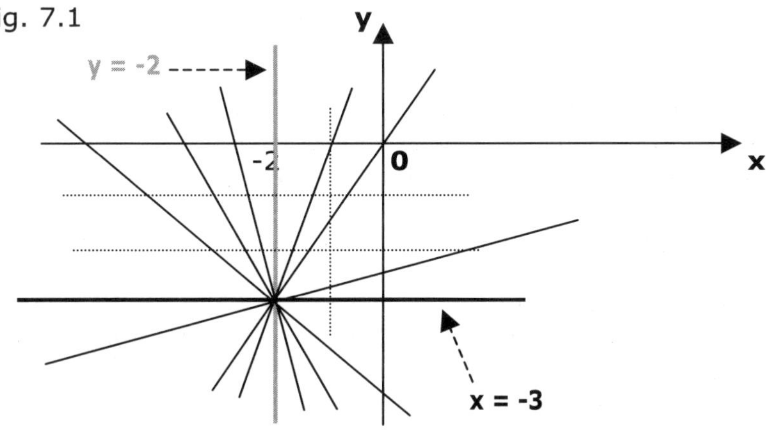

In short:

First, we convert the equation $\mathbf{y = ax - b}$ into the form of $\mathbf{m(ax + by + c) + n(ux + vy + w) = 0}$.
$\mathbf{y = ax - b \Rightarrow ax - y - b = 0}$. By means of the connective expression given, let's eliminate the **a**.

Then, $\mathbf{3a + b = 2 \Rightarrow a = (2 - b)/3 \Rightarrow y = ax - b = (2 - b)x/3 - b = 2x/3 - bx/3 - b}$
$\mathbf{\Rightarrow y - 2x/3 + bx/3 + b = 0 \Rightarrow y - 2x/3 + b(x/3 + 1) = 0}$.

Therefore, the point where the two lines $\mathbf{y - 2x/3 = 0}$ and $\mathbf{x/3 + 1 = 0}$ meet each other is the point where all the lines meet each other altogether.
Then, we need to find the point where the two lines meet each other.

$\mathbf{y - 2x/3 = 0 \Rightarrow y = 2x/3}$.
$\mathbf{x/3 + 1 = 0 \Rightarrow x = -3 \Rightarrow y = 2x/3 = -2}$.

Then, the lines meet each other at the point **(-3, -2)**.
Therefore, $\mathbf{s = -3}$ and $\mathbf{t = -2}$.

Suggestions or Solutions

To the **Problem** in the Example 8

Suppose that we have an equation that represents a group of lines, and the equation is as follows: $2x/a + y/2b = 1$ where a and b are constants such that $1/2a + 1/b = 1/5$. Suppose also, that all the lines include a particular point P(u, v), and find the values of the u and the v.

The problem says:
"Choosing many pairs of numbers for the **a** and **b** in the connective expression $1/2a + 1/b = 1/5$, we get as many lines as the number of the choices, and the equation $2x/a + y/2b = 1$ indicates each of all the lines we get. Then, each of all the lines includes the particular point **P(u, v)**. Now, find the values of the **u** and the **v**."

Each of all the lines that meet each other at a particular point gets made up of two basic lines.
Then, the two basic lines meet each other at the particular point.
Now, we need to find the two basic lines and find the point where the basic lines meet each other.
We know that $2x/a + y/2b = 1$ has the equations indicating the two basic lines.

Now, we convert $2x/a + y/2b = 1$ into the form of $m(ax + by + c) + n(ux + vy + w) = 0$.
Let's first, set $t = 1/a$ and $s = 1/b$. Then, the **t** and the **s** are constants, and also:
$2x/a + y/2b = 1 \Rightarrow 2tx + sy/2 = 1 \Rightarrow 2tx + sy/2 - 1 = 0 \Rightarrow 4tx + sy - 2 = 0$.

We now, need to eliminate either of **s** and **t** by means of the connective expression. Let's get rid of the **s**. Then:
$1/2a + 1/b = 1/5 \Rightarrow t/2 + s = 1/5 \Rightarrow s = 1/5 - t/2$
$\Rightarrow 4tx + sy - 2 = 4tx + (1/5 - t/2)y - 2 = 4tx + y/5 - ty/2 - 2 = 0 \Rightarrow t(4x - y/2) + y/5 - 2 = 0$.
Then, in the form above, we can say $m = t$ and $n = 1$.
Therefore, the two equations $4x - y/2 = 0$ and $y/5 - 2 = 0$ indicate the two basic lines.

Then, we need to find the point where the basic lines meet.
$y/5 - 2 = 0 \Rightarrow y = 10$. $4x - y/2 = 0 \Rightarrow x = y/8 = 10/8 = 5/4$.
Then, the two basic lines meet each other at the point **(5/4, 10)**.
Therefore, we get $u = 5/4$ and $v = 10$.

In short:

We first, convert $2x/a + y/2b = 1$ into the form of $m(ax + by + c) + n(ux + vy + w) = 0$.
Suppose that $t = 1/a$ and $s = 1/b$. Then:
$2x/a + y/2b = 1 \Rightarrow 2tx + sy/2 = 1 \Rightarrow 2tx + sy/2 - 1 = 0 \Rightarrow 4tx + sy - 2 = 0$.
$2tx + sy/2 = 1 \Rightarrow 2tx + sy/2 - 1 = 0 \Rightarrow 4tx + sy - 2 = 0$.
Let's eliminate the **t**.
$1/2a + 1/b = 1/5 \Rightarrow t/2 + s = 1/5 \Rightarrow t = 2/5 - 2s$
$\Rightarrow 4tx + sy - 2 = 4(2/5 - 2s)x + sy - 2 = 8x/5 - 8sx + sy - 2 = 8x/5 - 2 + s(y - 8x) = 0$.
Then, we need to find the point where $8x/5 - 2 = 0$ and $y - 8x = 0$ meet each other.
$8x/5 - 2 = 0 \Rightarrow x = 10/8 = 5/4 \Rightarrow x = 5/4$.
$y - 8x = 0 \Rightarrow y = 8x = 8(5/4) = 10$.
Therefore, $u = 5/4$ and $v = 10$.

Suggestions or Solutions

To the Problem 0 in the Example 9

Suppose that we have an equation that represents a group of lines, and the equation is as follows: $5ax - (3b - 2)y + 2b = 8$ where a and b are constant such that $2a - b = 2$ and $ab \neq 0$. Suppose also, the lines in the group include a particular point P(s, t), and find the point P.

The equation $m(ax + by + c) + n(dx + ey + f) = 0$ is called a linear combination of $ax + by + c = 0$ and $dx + ey + f = 0$. Then, if two lines meet each other at a particular point, a linear combination of the two lines includes the particular point. In other words, the sum of any multiples of the two lines includes the particular point. That is, no matter what numbers we may choose for the two constants **m** and **n**, the line indicated by the sum includes the particular point.

Suppose first, that **A** is the given equation $5ax - (3b - 2)y + 2b = 8$.
Then, the **A** is the equation indicating each of all the lines that include the particular point **P(s, t)**.
Now, we need to extract the two basic lines from the **A** to find the point where the two lines meet.
The, that point is the particular point **P(s, t)** where each of all the lines that the **A** indicates includes.

Now, wee need to put the **A** in the form of $m(ax + by + c) + n(ux + vy + w) = 0$.

Then, $5ax - (3b - 2)y + 2b = 8 \Rightarrow 5ax - 3by + 2y + 2b - 8 = 0$.
We now, need to eliminate either **a** or **b** by means of the connective expression given. Let's remove the **a**.

Then, $2a - b = 2 \Rightarrow a = (2 + b)/2 \Rightarrow 5ax - 3by + 2y + 2b - 8 = 5(2 + b)x/2 - 3by + 2y + 2b - 8$ $= 5x + 5bx/2 - 3by + 2y + 2b - 8 = b(5x/2 - 3y + 2) + 5x + 2y - 8 = 0$.

Then, in the form above, we can say that $m = b$ and $n = 1$.
Therefore, the two equations $5x/2 - 3y + 2 = 0$ and $5x + 2y - 8 = 0$ indicate the two basic lines.
Then, we need to find the point where the two lines meet each other.

$5x/2 - 3y + 2 = 0 \Rightarrow 5x - 6y + 4 = 0 \Rightarrow 5x = 6y - 4$.
$5x + 2y - 8 = 0 \Rightarrow 6y - 4 + 2y - 8 = 8y - 12 = 0 \Rightarrow y = 3/2$.
Then, $5x = 6y - 4 = 6(3/2) - 4 = 5 \Rightarrow x = 1$.

Then, the two basic lines meet each other at the point **(1, 3/2)**.
Consequently, the point **(1, 3/2)** is the particular point **P(s, t)** where each of all the lines that the **A** indicates includes.
Therefore, we get $s = 1$ and $t = 3/2$.

In short:

Suppose that **A** is the equation $5ax - (3b - 2)y + 2b = 8$.
Then, we put **A** in the form of $m(ax + by + c) + n(ux + vy + w) = 0$.
$5ax - (3b - 2)y + 2b = 8 \Rightarrow 5ax - 3by + 2y + 2b - 8 = 0$.
By means of the connective expression given, let's eliminate the **b**. Then:
$2a - b = 2 \Rightarrow b = 2a - 2 \Rightarrow 5ax - 3by + 2y + 2b - 8 = 5ax - 3(2a - 2)y + 2y + 2(2a - 2) - 8$
$= 5ax - 6ay + 6y + 2y + 4a - 4 - 8 = a(5x - 6y + 4) + 8y - 12 = 0$.

Then, we need to find the point where **$5x - 6y + 4 = 0$** and **$8y - 12 = 0$** meet each other.
$8y - 12 = 0 \Rightarrow y = 3/2$
$5x - 6y + 4 = 0 \Rightarrow 5x = 6y - 4 = 6(3/2) - 4 = 5 \Rightarrow x = 1.$
Then, the two lines meet each other at the point **(1, 3/2)**.
Therefore, **s = 1** and **t = 3/2**.

Fig. 9.0.0

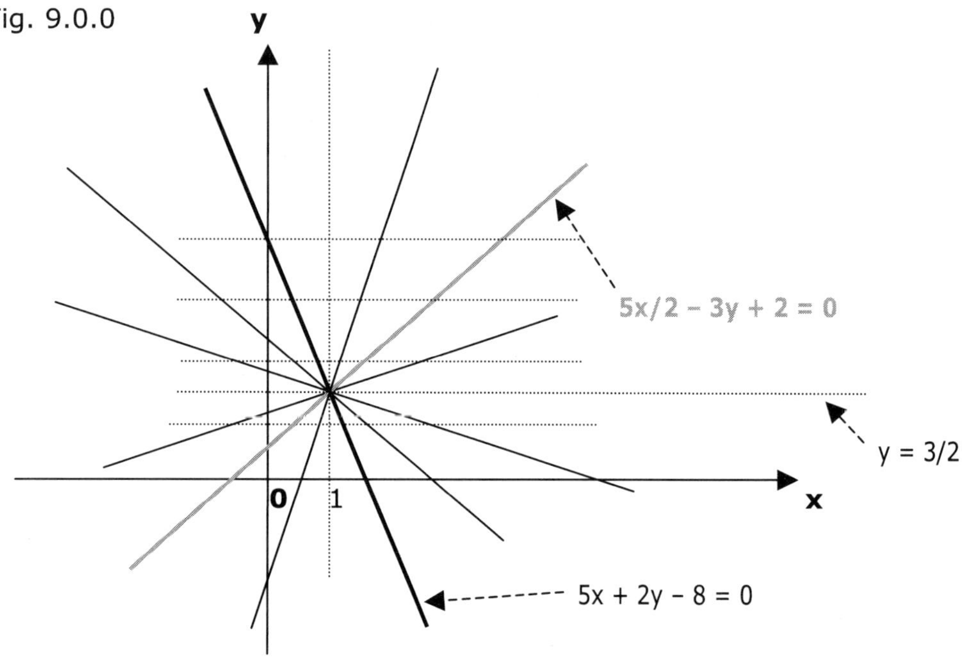

Fig. 9.0.1

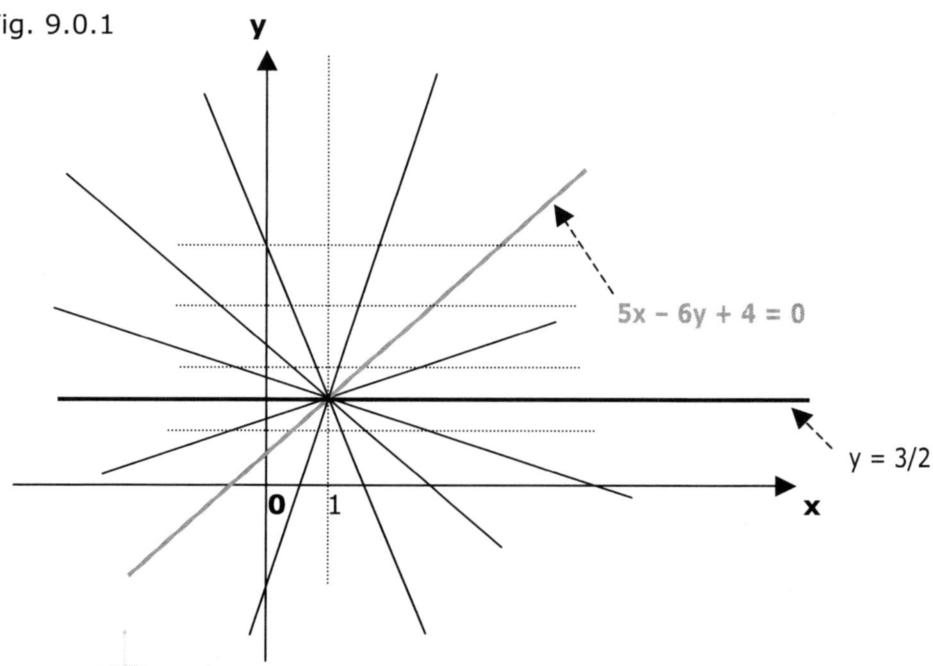

Suggestions or Solutions

To the Problem 1 in the Example 9

Suppose that we have an equation that represents a group of lines, and the equation is as follows: $5ax - (3b - 2)y + 2b = 8$ where a and b are constant such that $2a - b = 2$ and $ab \neq 0$. Then, find the line in the group that is parallel to a line $2x + 5y = 10$.

The statement, "Suppose that we have an equation indicating a group of lines, and the equation is as follows: $5ax - (3b - 2)y + 2b = 8$ where **a** and **b** are constants such that $2a - b = 2$ and $ab \neq 0$." means "If we choose many pairs of numbers for **a** and **b** such that $2a - b = 2$ and $ab \neq 0$, we get as many lines as the number of the choices, and the equation $5ax - (3b - 2)y + 2b = 8$ indicates each of the lines we get."

Therefore, among the lines stated above, we need find a line parallel to the line $2x + 5y = 10$. That is, we need to find the values of the **a** and **b** that determine such a line. Then, putting the values into the given equation $5ax - (3b - 2)y + 2b = 8$, we get such an equation of line. Now, we need to find the values and plug them in.

If two lines are parallel to each other, the slopes are the same. So, the slope of the equation we are after is the same as that of the line $2x + 5y = 10$. Then, let's put the given equations in the standard form to find the slopes.
$2x + 5y = 10 \Rightarrow y = -2x/5 + 2 \Rightarrow$ The slope $= -2/5$.
$5ax - (3b - 2)y + 2b = 8 \Rightarrow y = 5ax/(3b - 2) + (8 - 2b)/(3b - 2) \Rightarrow$ The slope $= 5a/(3b - 2)$.

In fact, in the general equation of a line, the negative value of the ratio of the coefficient of the **x** to that of the **y** is the slope of the line.
Then, the values of the **a** and **b** we are after need to satisfy the connective expressions as follows:
$5a/(3b - 2) = -2/5$ and $2a - b = 2$.

Now, we need to take the expressions above for the system of equations for the **a** and **b**, solve them, and find the values of the **a** and **b**. $2a - b = 2 \Rightarrow a = (2 + b)/2 \Rightarrow 5a = 5(2 + b)/2$.
Meanwhile, $5a/(3b - 2) = -2/5 \Rightarrow 5a = 2(2 - 3b)/5$. Then, $5(2 + b)/2 = 2(2 - 3b)/5$
$\Rightarrow 25(2 + b) = 4(2 - 3b) \Rightarrow 50 + 25b = 8 - 12b \Rightarrow 37b = -42 \Rightarrow b = -42/37$.
Then, $a = (2 + b)/2 = \{2 - (42/37)\}/2 = 1 - 42/37/2 = 1 - 21/37 = 16/37$.

Now, $5ax - (3b - 2)y + 2b = 8 \Rightarrow 80x/37 - (-126/37 - 2)y - 84/37 = 8$
$\Rightarrow 80x/37 + 200y/37 - 380/37 = 0 \Rightarrow 80x + 200y - 380 = 0 \Rightarrow 4x + 10y - 19 = 0$.

Therefore, the line parallel to the line $2x + 5y = 10$ is $4x + 10y - 19 = 0$.

In short:

$2x + 5y = 10 \Rightarrow$ The slope $= -2/5$. $5ax - (3b - 2)y + 2b = 8 \Rightarrow$ The slope $= 5a/(3b - 2)$.
Now, we need to take the following expressions for the system of equations for the **a** and **b**, solve them, and find the values of the **a** and **b**: $5a/(3b - 2) = -2/5$ and $2a - b = 2$.
$2a - b = 2 \Rightarrow a = (2 + b)/2 \Rightarrow 5a = 5(2 + b)/2$.
Meanwhile, $5a/(3b - 2) = -2/5 \Rightarrow 5a = 2(2 - 3b)/5$. Then, $5(2 + b)/2 = 2(2 - 3b)/5$
$\Rightarrow 25(2 + b) = 4(2 - 3b) \Rightarrow 50 + 25b = 8 - 12b \Rightarrow 37b = -42 \Rightarrow b = -42/37$.
Then, $a = (2 + b)/2 = \{2 - (42/37)\}/2 = 1 - 42/37/2 = 1 - 21/37 = 16/37$.

Now, $5ax - (3b - 2)y + 2b = 8 \Rightarrow 80x/37 - (-126/37 - 2)y - 84/37 = 8$
$\Rightarrow 80x/37 + 200y/37 - 380/37 = 0 \Rightarrow 80x + 200y - 380 = 0 \Rightarrow 4x + 10y - 19 = 0$.

Suggestions or Solutions
To the **Problem** in the Example **A**

Suppose that two lines meet at a point P(s, t) where s > 0 and t > 0, and that the equations of the lines are as follows: $2x + y - 2 = 0$ and $2kx - y + k - 1 = 0$ where k is a constant. Then, find the extent of the k.

Let's construct the graph of $2x + y - 2 = 0$ so that we can see better where the solution will be around.
$x = 0 \Rightarrow y = 2$. $y = 0 \Rightarrow x = 1$.

Fig. A.0.

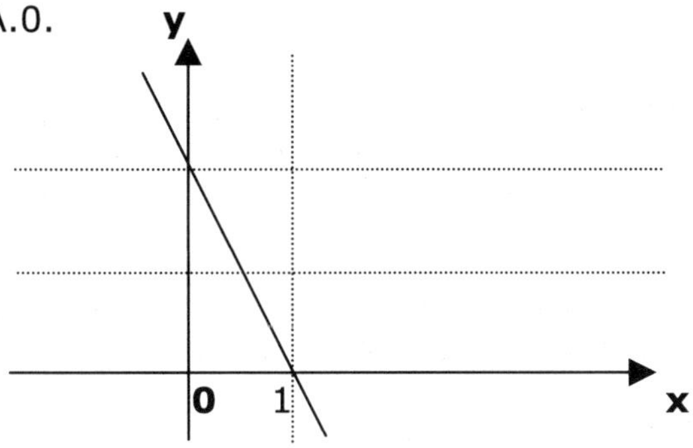

Only the line segment in the first quadrant is applicable since in **P(s, t)**, $s > 0$ and $t > 0$.

Now, since the second equation indicates a group of lines, let's find the two basic lines so that we can see where the group is.
$2kx - y + k - 1 = 0 \Rightarrow k(2x + 1) - y - 1 = 0$.

Then, the two basic lines are $2x + 1 = 0$ and $-y - 1 = 0$.
$-y - 1 = 0 \Rightarrow y = -1$. $2x + 1 = 0 \Rightarrow x = -1/2$.
Therefore, all the lines in the group meet each other at **(-1/2, -1)**. Let's now, put those in the graph, too.

Fig. A.1

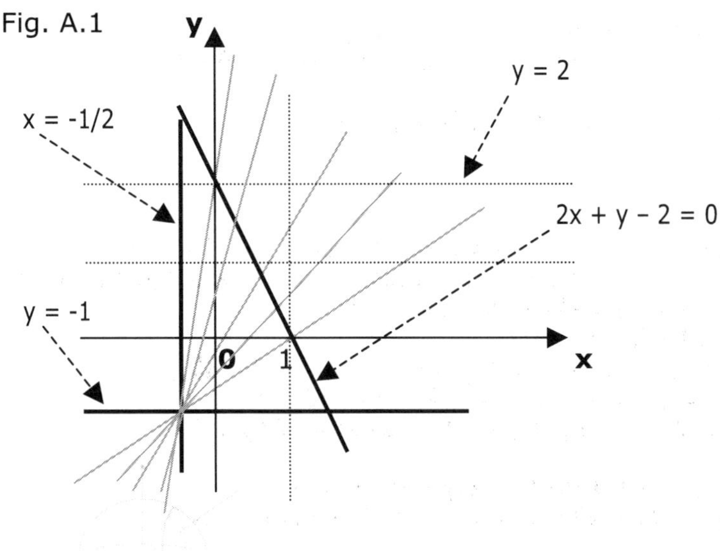

Now, we can see how the lines in gray should meet the line $\mathbf{2x + y - 2 = 0}$.
The y-intercept of the topmost gray line is 2, and the x-intercept of the bottommost gray line is 1.
Also, the slope **2k** is the largest when the y-intercept is 2, and keeps decreasing until the x-intercept becomes 1 to be the smallest.

$\mathbf{2kx - y + k - 1 = 0} \Rightarrow$ The slope $= \mathbf{2k}$.

The slope of the topmost line in gray = $\{2 - (-1)\}/\{0 - (-1/2)\} = 3/(1/2) = 6$.
The slope of the bottommost line in gray = $\{0 - (-1)\}/\{1 - (-1/2)\} = 1/(3/2) = 2/3$.

Therefore, $\mathbf{2/3 < 2k < 6 \Rightarrow 1/3 < k < 3}$. The boundary values get excluded since $\mathbf{s > 0}$ and $\mathbf{t > 0}$ at the point $\mathbf{P(s, t)}$.

In short:

$\mathbf{2kx - y + k - 1 = 0}$ indicates each of all the lines that include a particular point.
$\mathbf{2kx - y + k - 1 = 0 \Rightarrow k(2x + 1) - (y + 1) = 0.}$
Then, all the lines meet each other altogether at the point where the lines $\mathbf{2x + 1 = 0}$ and $\mathbf{y + 1 = 0}$ meet each other.

$\mathbf{y + 1 = 0 \Rightarrow y = -1}$. $\mathbf{2x + 1 = 0 \Rightarrow x = -1/2}$.
Therefore, the particular point is $\mathbf{(-1/2, -1)}$.

Fig. A.1

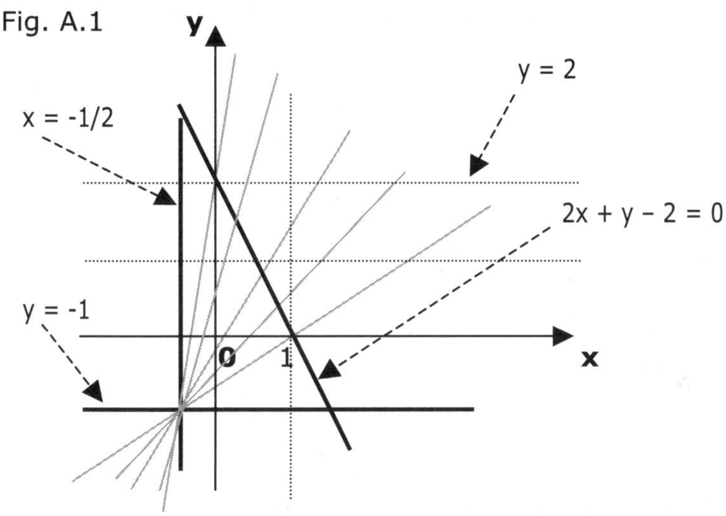

The y-intercept of the topmost gray line is 2, and the x-intercept of the bottommost gray line is 1.
Also, the slope **2k** is the greatest when the y-intercept is 2, and keeps decreasing until the x-intercept becomes 1 to be the smallest. $\mathbf{2kx - y + k - 1 = 0} \Rightarrow$ The slope $= \mathbf{2k}$.

The slope of the topmost line in gray = $\{2 - (-1)\}/\{0 - (-1/2)\} = 3/(1/2) = 6$.
The slope of the bottommost line in gray = $\{0 - (-1)\}/\{1 - (-1/2)\} = 1/(3/2) = 2/3$.

Therefore, $\mathbf{2/3 < 2k < 6 \Rightarrow 1/3 < k < 3}$. The boundary values are excluded since $\mathbf{s > 0}$ and $\mathbf{t > 0}$ at the point $\mathbf{P(s, t)}$.

Suggestions or Solutions

To the Problem in the Example B

Suppose that:
$(2k - 1)x - (3k + 1)y + 5 + 2k = 0$ where k is a constant.
$y = 2x$ for $|x| \le 1$.
Then, determine the value of the k so that the two lines above can meet each other.

The following equation indicates a group of lines: $(2k - 1)x - (3k + 1)y + 5 + 2k = 0$ where **k** is a constant. The following equation indicates a line segment: $y = 2x$ for $|x| \le 1$.
Therefore, we need to find the extent of the **k** so that all the lines in the group can meet the line segment.
Let's first, find the basic lines.
$(2k - 1)x - (3k + 1)y + 5 + 2k = 0 \Rightarrow k(2x - 3y + 2) - (x + y - 5) = 0$
Then, the basic lines are $2x - 3y + 2 = 0$ and $x + y - 5 = 0$.

We now, need to find the point where the basic lines meet each other.
We multiply the second equation by −2, add the product to the first equation to eliminate the **x**, and then, find the value of the **y**.

$-2(x + y - 5) = 0 \Rightarrow -2x - 2y + 10 = 0$.
$2x - 3y + 2 + (-2x - 2y + 10) = 0 \Rightarrow -5y + 12 = 0 \Rightarrow y = 12/5$.
Then, $y = 12/5 \Rightarrow x + y - 5 = x + 12/5 - 5 = 0 \Rightarrow x = 13/5$.
Therefore, the two basic lines along with the other lines meet each other at **(13/5, 12/5)**.
Putting the basic lines and the line segment in a graph, we can see better where the solution is.

Fig. B.0

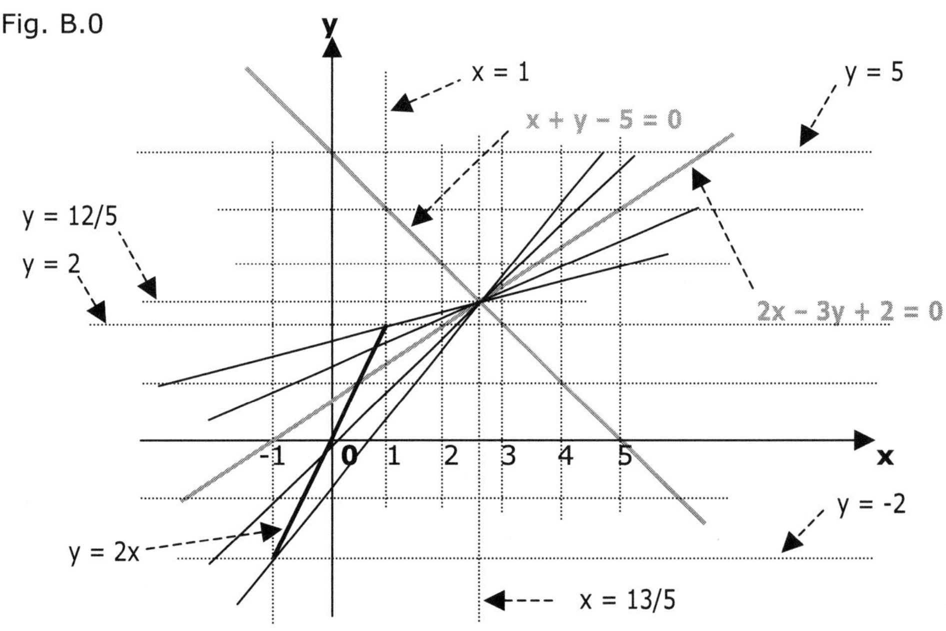

In the graph above, we can see the black lines meeting the line segment.
The topmost line in black includes the two points **(1, 2)** and **(13/5, 12/5)**.

The bottommost line in black includes the two points **$(-1, -2)$** and **$(13/5, 12/5)$**.
We can see that the slope of each of the black lines above can be indicated by **$(2k - 1)/(3k + 1)$** from the equation **$(2k - 1)x - (3k + 1)y + 5 + 2k = 0$**.

The slope of the topmost line = **$\Delta y/\Delta x$** = $(12/5 - 2)/(13/5 - 1) = (2/5)/(8/5) = 1/4$.
Then, **$(2k - 1)/(3k + 1) = 1/4 \Rightarrow 2k - 1 = 4(3k + 1) = 12k + 4 \Rightarrow 10k = -5 \Rightarrow k = -1/2$**.

The slope of the bottommost line = **$\Delta y/\Delta x$** = $(12/5 - (-2))/(13/5 - (-1)) = (22/5)/(18/5) = 11/9$.
Then, **$(2k - 1)/(3k + 1) = 11/9 \Rightarrow 9(2k - 1) = 11(3k + 1) \Rightarrow 18k - 9 = 33k + 11$**
$\Rightarrow 15k = -20 \Rightarrow k = -4/3$.

Therefore, **$-4/3 \le k \le -1/2$**.

In short:

The following equation indicates a group of lines: **$(2k - 1)x - (3k + 1)y + 5 + 2k = 0$** where **$k$** is a constant.
$(2k - 1)x - (3k + 1)y + 5 + 2k = 0 \Rightarrow k(2x - 3y + 2) - (x + y - 5) = 0$
Then, the basic lines are **$2x - 3y + 2 = 0$** and **$x + y - 5 = 0$**.
$-2(x + y - 5) = 0 \Rightarrow -2x - 2y + 10 = 0$.
$2x - 3y + 2 + (-2x - 2y + 10) = 0 \Rightarrow -5y + 12 = 0 \Rightarrow y = 12/5$.
$y = 12/5 \Rightarrow x + y - 5 = x + 12/5 - 5 = 0 \Rightarrow x = 13/5$.
Therefore, the two basic lines meet each other at **$(13/5, 12/5)$**.

Fig. B.0

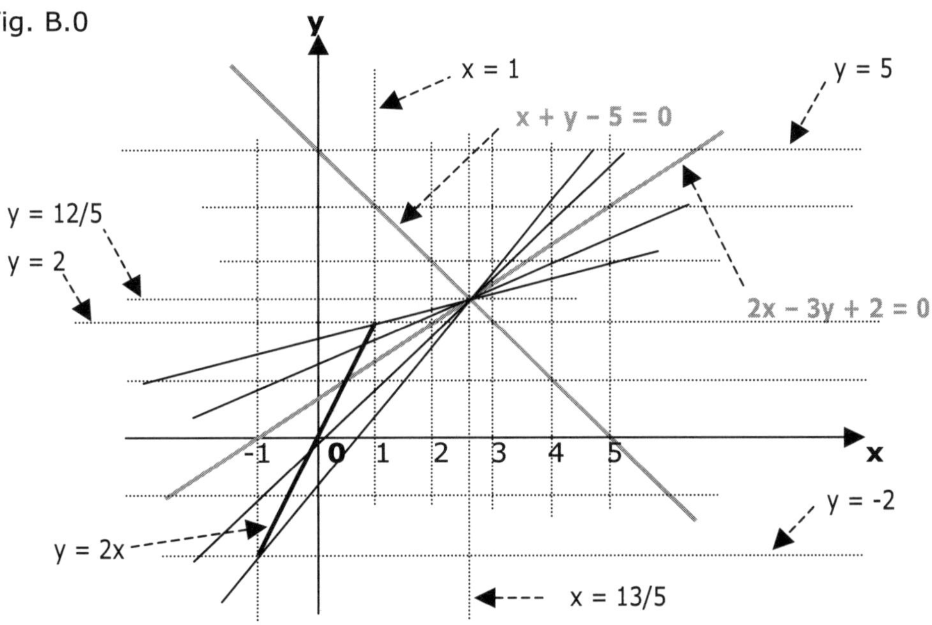

The slope of the topmost line = **$\Delta y/\Delta x$** = $(12/5 - 2)/(13/5 - 1) = (2/5)/(8/5) = 1/4$.
Then, **$(2k - 1)/(3k + 1) = 1/4 \Rightarrow 2k - 1 = 4(3k + 1) = 12k + 4 \Rightarrow 10k = -5 \Rightarrow k = -1/2$**.
The slope of the bottommost line = **$\Delta y/\Delta x$** = $(12/5 - (-2))/(13/5 - (-1)) = (22/5)/(18/5) = 11/9$.
Then, **$(2k - 1)/(3k + 1) = 11/9 \Rightarrow 9(2k - 1) = 11(3k + 1) \Rightarrow 18k - 9 = 33k + 11$**
$\Rightarrow 15k = -20 \Rightarrow k = -4/3$. Therefore, **$-4/3 \le k \le -1/2$**.

1. What is a parabola?

A parabola is a line but is a curved line. It is a curve which is formed by the intersection of a cone with a plane parallel to its side. Suppose that we cut a cone with a plane parallel to the cone's side. Then, the cross section is a parabola. A line is a collection of points. In a line, all the points have the same direction. A parabola is a collection of points, too, but all the points have different directions respectively. However, the directions are not random but are in such a way that the curve is symmetric. Symmetric about what? We will get to find out about it as we cover the next sections.

A parabola is an idea just as in the case of a line, and does not have a thickness or width. What about a length? In general math, we hardly talk about a length of a parabola. More precisely, we say it is a magnitude of a trajectory in a parabola. A trajectory in a parabola is a parabola segment. A magnitude of a parabola segment can be calculated by a tool in Calculus. In math, we usually mean a distance by a length, which is a magnitude of a line segment. In fact, a trajectory is a sum of tiny line segments when we take a measurement of a length of a trajectory.

We use a parabola a lot. It does have much to do with a trajectory. A parabola segment is often used for the path of a simple projectile. For instance, if we throw a ball, the ball travels in the air along a path that is a parabola. If we hit baseballs, they usually travel in such a manner. That's not all where we can find a parabola. For instance, we use a parabola dish when we watch satellite TV. Actually, the cross section of the dish is a parabola. The same is true for a reflector in a flashlight.

Just as in the case of a line, we put a parabola in an equation when work on it in math. Also, we put it in a graph, of course. We can put two or more parabolas in an equation if we want to. Now, let's see what the equation is and how it works.

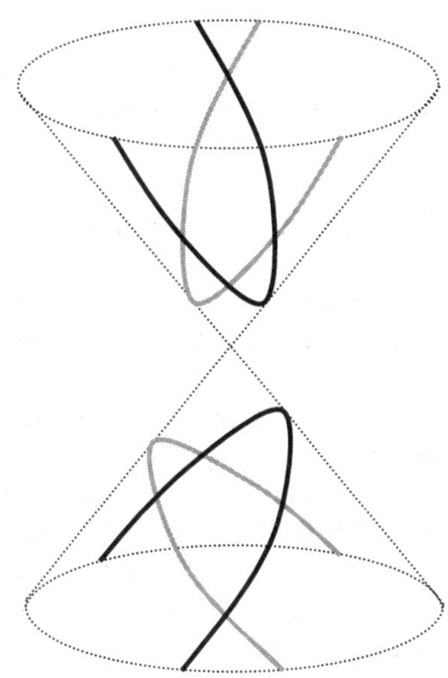

A parabola is a set of points. A set of what points, then? Let's take for an example a point in a parabola. Then, the point is the same distance away from both a particular point and a particular line. That is, the distance between the point and the particular point is the same as the distance between the point and the particular line. The same is true for each of all the other points in the parabola. Now, a set of such points is a parabola. Suppose now, that a parabola is in the x-y system. Then, each of the points has a pair of coordinates, and we can put the parabola in a set of points as follows.. To begin with, let's see the standard form.

A = {(x, y)| y^2 = 4px where **p** is a constant.**}**
B = {(x, y)| x^2 = 4py where **p** is a constant.**}**
C = {(x, y)| $(y - v)^2$ = 4p(x - u) where **p**, **u**, and **v** are constants.**}**
D = {(x, y)| $(x - u)^2$ = 4p(y - v) where **p**, **u**, and **v** are constants.**}**

The statement that **A = {(x, y)| y^2 = 4px** where **p** is a constant.**}** is saying that the set **A** is composed of points in the x-y system and the connective expression between the coordinates of each of the points is in general, **y^2 = 4px** where **p** is a constant. Therefore, a set of such points is called a parabola, and **A** is the name of the parabola. If we are told to put down the equation of the parabola **A**, we can put down **A(x, y) = y^2 - 4px = 0**, or just, **y^2 = 4px**. Therefore, in this case, **A(x, y)** is not a function but an equation.

The next is the general form.

P = {(x, y)| $ax^2 + by^2 + cxy + dx + ey + f = 0$ where **a**, **b**, **c**, **d**, **e**, and **f** are constants, and also, **ab > 0**; otherwise, **c** is 0, and also, either **a** or **b** is 0.**}**

The set **P** above is a comprehensive presentation of a general parabola. That is, any parabola is applicable. However, we often use such forms as follows:

Q = {(x, y)| $ax^2 + bx + c + dy = 0$ where **a**, **b**, **c**, and **d** are constants, and also, **ad ≠ 0}**
R = {(x, y)| $ay^2 + by + c + dx = 0$ where **a**, **b**, **c**, and **d** are constants, and also, **ad ≠ 0}**

The set **S** below indicates the **Q** and the **R** in a comprehensive manner:
S = {(x, y)| $ax^2 + by^2 + cx + dy + e = 0$ where **a**, **b**, **c**, **d**, and **e** are constants, and also, either **a** or **b** is 0.**}**

In general math, we usually put down general equations of parabola in such ways follows:

Q(x, y) = $ax^2 + bx + c + dy = 0$, or just, **$ax^2 + bx + c + dy = 0$**.
R(x, y) = $ay^2 + by + c + dx = 0$, or just, **$ay^2 + by + c + dx = 0$**.

As in the case of a line, when we work on a parabola in math, we put the parabola in an equation. That is, we construct an equation of parabola for use. Now, as in the case of lines, we have two forms in the equations, one of which is a standard form, and the other of which is a general form. The purposes of the two forms are the same as in the case of lines. That is, the general form can present any parabola, and the standard form can be said to be mainly for the presentation of the geometric elements of a parabola. That's because if we use the standard form, it is convenient to see the vertex, focus, directrix, axis of symmetry, and the points where the curve meets the axes (roots or intercepts). In addition, we can readily use the standard form for a function. On the other hand, the general form is appropriate to present all parabolas in general. Then, let's see how to construct such equations. Now, we are going to begin with the standard form.

1.0. Standard Equations of Parabola

Suppose now, that we have a line **D** in the x-y plane, and the equation of the **D** is $\mathbf{x = d}$. Then, we are making a of points, and the distance from each of the points to the origin is the same as the distance from the each to the line **D**. Then, let's specify what the points are. However, the set is composed of infinitely many points. Then, since we can't explain such many points individually, let's make a point that can represent all the points and work with it. The point is called a general point. Such a point is also called a variable point or point variable. Suppose now, the general point is **P(s, t)**. Then, the general point **P** is the same distance from the origin and the line **D**. Now, the general point **P(s, t)** is moving in the x-y plane covering all the points in the set. Then, if we capture a moment while the **P** is moving, we get the snap shot as follows.

Fig. 1.0.0

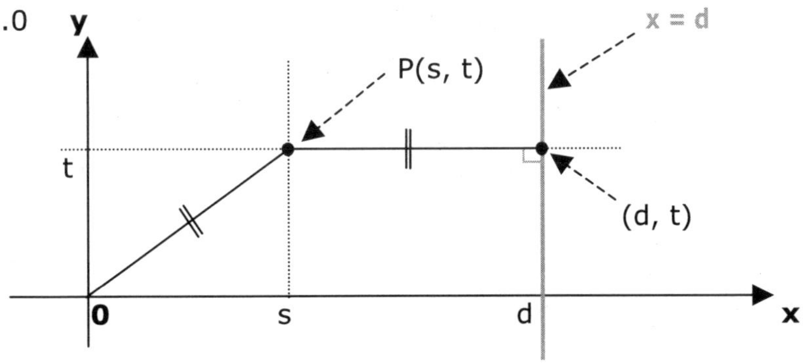

Let's now, keep track of the trace of the **P**. The **P** keeps the same distance away from the origin and the line **D**. Assuming that the distance is Δ, we get $\Delta^2 = (\Delta \mathbf{x})^2 + (\Delta \mathbf{y})^2$ from the distance formula. Suppose now, that the distance between **P(s, t)** and the origin is **U**, and the distance between **P(s, t)** and the line **D** is **V**. Then, we get $\mathbf{U^2 = (s - 0)^2 + (t - 0)^2}$ and $\mathbf{V^2 = (d - s)^2 + (t - t)^2}$. Therefore, we have $\mathbf{U^2 = s^2 + t^2}$ and $\mathbf{V^2 = (d - s)^2}$. Now, since the **P** has to keep the same distance away from the origin and the line **D** respectively, we need to have $\mathbf{U^2 = V^2}$. Therefore, $\mathbf{U^2 = V^2 \Rightarrow s^2 + t^2 = (d - s)^2 = d^2 - 2ds + s^2 \Rightarrow t^2 = d^2 - 2ds}$. Now, let's put in a graph the trace the **P** is making in the x-y plane.

Fig. 1.0.1

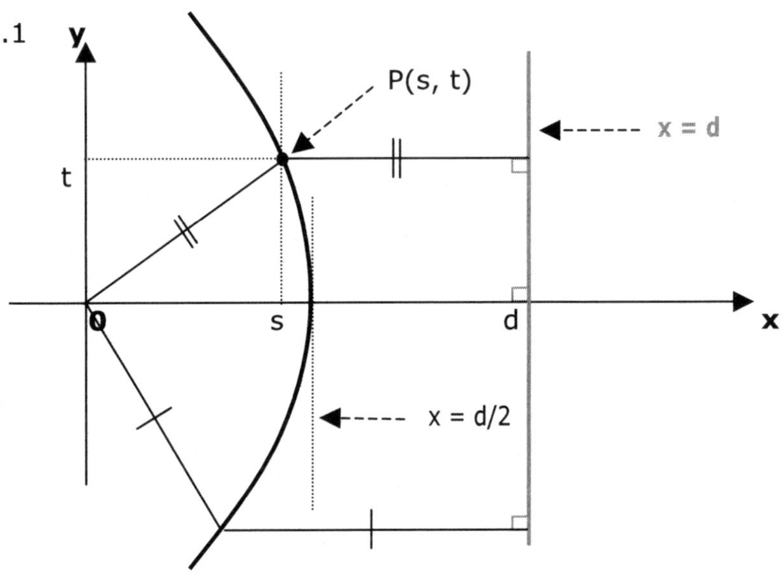

Then, the trace the point **P(s, t)** makes is a parabola. In other words, the **P** moves along a parabola. In the graph above, we can see that the trace, that is, the parabola is symmetric about the x-axis.
Now, the **P(s, t)** is the general point in the x-y plane.
Then, the **s** and the **t** gets respectively the values of the **x** and the **y** in the x-y system.
When we specify the points in the x-y system, we use values in the x-axis and y-axis for the coordinate values. Therefore, the **s** and **t** are the variables. Also, we use **x** and **y** for the variables in the x-y system. In addition, as presented above, the connective expression between the **s** and **t** is $\mathbf{t^2 = d^2 - 2ds}$.
Then, in the connective expression $\mathbf{t^2 = d^2 - 2ds}$, if we replace the **s** and the **t** with **x** and **y** respectively, the connective expression is no other than the connective expression between the **x** and **y**.
Then, the connective expression between the coordinates of each of all the points constituting a parabola is $\mathbf{y^2 = d^2 - 2dx}$.
Now, the connective expression is called an equation of parabola.

What is the line **D**, then?
It is called a directrix. In general, it is used for constructions of not only parabolas but ellipses and hyperbolas. In such a curve, the ratio between the distance from each point to the directrix and the distance from the point to a particular point is constant, and the ratio is 1 in cases of a parabola. Besides, the particular point is called a focus, which is the origin in this case we are working on now. In case of a parabola, the directrix indicates the orientation of the parabola, which tells us the direction where the parabola is open. A focus is, for instance, a position where a light bulb is put in a flashlight. Those will be covered shortly.

Now, let's change the directrix **D**. This time, let's put in a graph a horizontal line $\mathbf{y = c}$ instead of the vertical line $\mathbf{x = d}$ as shown below, and see what happens.

Fig. 1.0.2

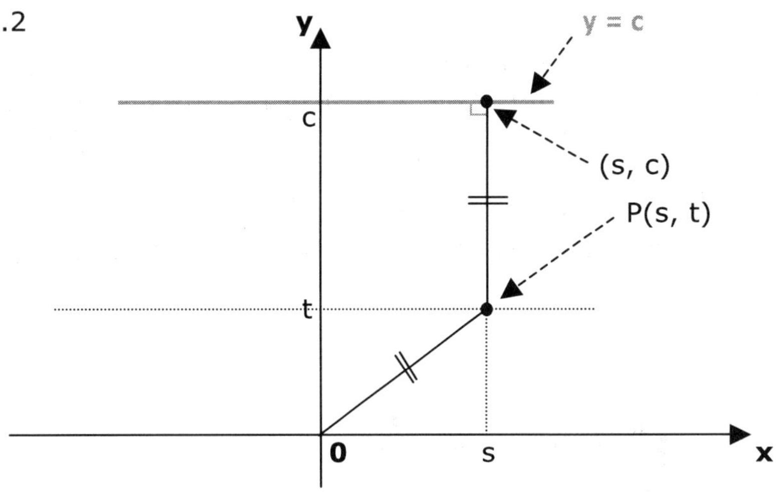

For this time also, suppose that **U** is the distance between **P(s, t)** and the origin, and **V** is the distance between **P(s, t)** and the line **D**. Then, we get $\mathbf{U^2 = (s - 0)^2 + (t - 0)^2}$ and $\mathbf{V^2 = (s - s)^2 + (c - t)^2}$. Therefore, we have $\mathbf{U^2 = s^2 + t^2}$ and $\mathbf{V^2 = (c - t)^2}$.
Now, since the **P** needs to keep the same distance away from both the origin and the line **D**, we need to get $\mathbf{U^2 = V^2}$. Therefore, $\mathbf{U^2 = V^2 \Rightarrow s^2 + t^2 = (c - t)^2 = c^2 - 2ct + t^2 \Rightarrow s^2 = c^2 - 2ct}$.
Now, the connective expression between the **s** and **t** is $\mathbf{s^2 = c^2 - 2ct}$.
Also, the **P(s, t)** is the general point in the x-y plane. Then, the **s** and **t** get respectively the values in the x-axis and the y-axis in the x-y system. Therefore, those are variables. In addition, we use **x** and **y** for the variables in the x-y system. To specify the points in the x-y system, we use for the coordinate values the values in the x-axis and the y-axis.. Then, in the connective expression $\mathbf{s^2 = c^2 - 2ct}$, if we replace the **s** and the **t** with **x** and **y** respectively, the connective expression is no other than a connective expression between the **x** and the **y**.

Then, the connective expression between the coordinates of each of all the points constituting the parabola is $\mathbf{x^2 = c^2 - 2cy}$. Now, let's put in a graph the trace the **P** makes in the x-y plane.

Fig. 1.0.3

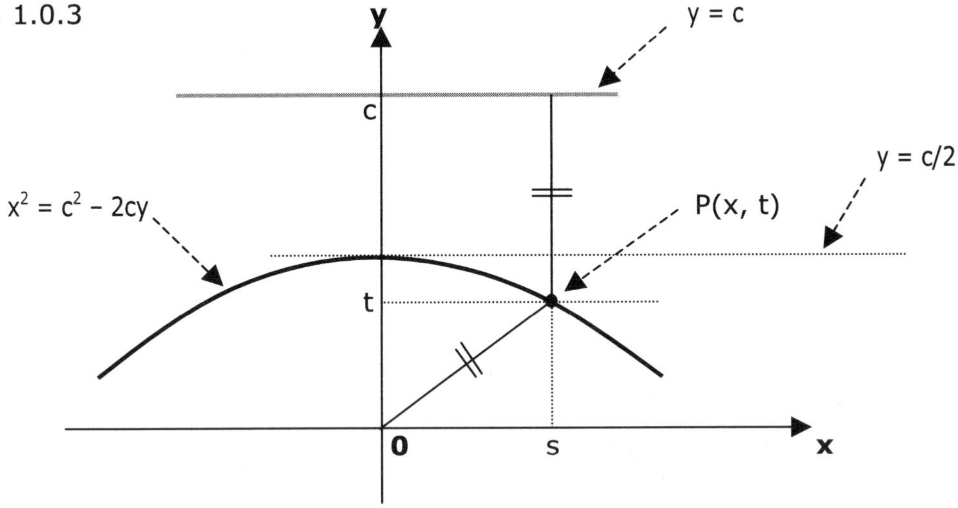

In the graph above, we can see that the trace, that is, the parabola is symmetric about the y-axis.
Now, we can move the focus to a different position, and change the position of the directrix, too. The reason that we take a horizontal or vertical line for the directrix is that the calculation gets simple. We can see that a parabola is bent toward the focus. Suppose now, that light rays are coming into a parabola. Then, all the light rays concentrate at the focus. In reverse, if we put a light source (a light bulb, for instance) at the focus, all the light rays get perpendicular to the directrix. If we move the directrix in a parallel manner, it will be like the dotted vertical line in the figure below.

Fig. 1.0.4

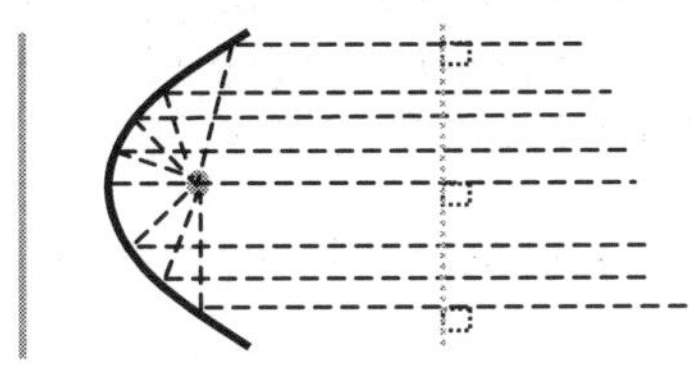

Now, let's suppose that the focus is at **(0, f)** and that the directrix **D** is a line $\mathbf{y = -f}$.

Fig. 1.0.5

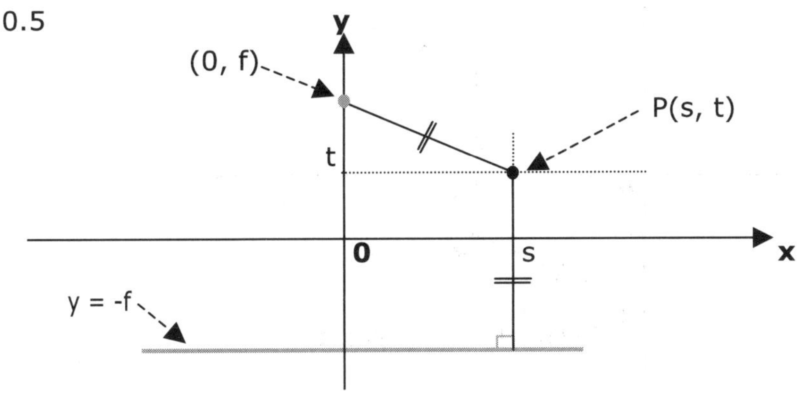

Suppose that **U** is the distance between **P(s, t)** and the focus, and **V** is the distance between the **P(s, t)** and the directrix **D**. Then, we get $\mathbf{U^2 = (s - 0)^2 + (t - f)^2}$ and $\mathbf{V^2 = (s - s)^2 + \{t - (-f)\}^2}$.
Therefore, we have $\mathbf{U^2 = s^2 + (t - f)^2}$ and $\mathbf{V^2 = (t + f)^2}$.

Now, since the **P** needs to keep the same distance away from the focus and the directrix **D**, we need to get $\mathbf{U^2 = V^2}$. Therefore, $\mathbf{U^2 = V^2 \Rightarrow s^2 + (t - f)^2 = (t + f)^2 \Rightarrow s^2 + t^2 - 2ft + f^2 = t^2 + 2ft + f^2}$
$\Rightarrow \mathbf{s^2 = 4ft}$. Now, the connective expression between the **s** and the **t** is $\mathbf{s^2 = 4ft}$.
Also, since the **P(s, t)** is the general point in the x-y plane, the **s** and the **t** are the variables that get respectively the values in the x-axis and the y-axis in the x-y system, and, also, we use **x** and **y** for the variables in the x-y system. Besides, when we specify points in the x-y system, we use for the coordinate values the values in the x-axis and the y-axis. Then, in the connective expression $\mathbf{s^2 = 4ft}$, if we replace the **s** and the **t** with **x** and **y** respectively, the connective expression is no other than a connective expression between the **x** and the **y**. Then, the connective expression between the coordinates at each of all the points constituting the parabola is $\mathbf{x^2 = 4fy}$.

Fig. 1.0.6

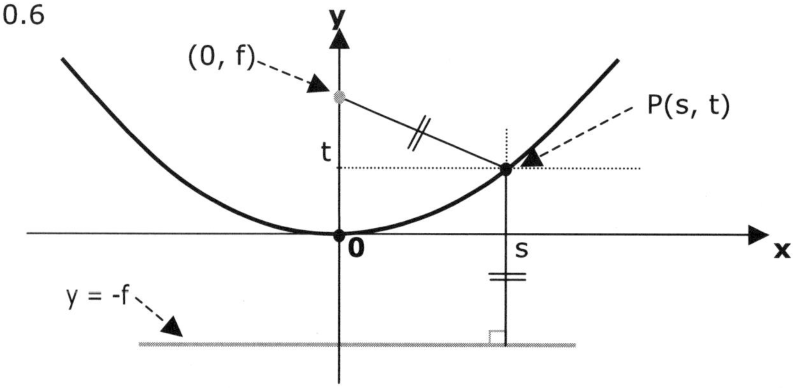

Let $\mathbf{f = 1/4}$, now. Then, we get $\mathbf{y = x^2}$.
Now, in such a curve, we have another special point, which is called the vertex of the parabola. In the case above, the vertex is at the origin. The vertex is the midpoint between the focus and the directrix. Also, the parabola is symmetric about the y-axis. Therefore, the axis of symmetry is perpendicular to the directrix and includes the vertex. So, in the case above, the axis of symmetry is the y-axis, that is, a line $\mathbf{x = 0}$. Besides, the entrance of the parabola is above the parabola.

For this time, let's suppose that the directrix is a line $\mathbf{y = f}$, and that the focus is at **(0, -f)**.

Fig. 1.0.7

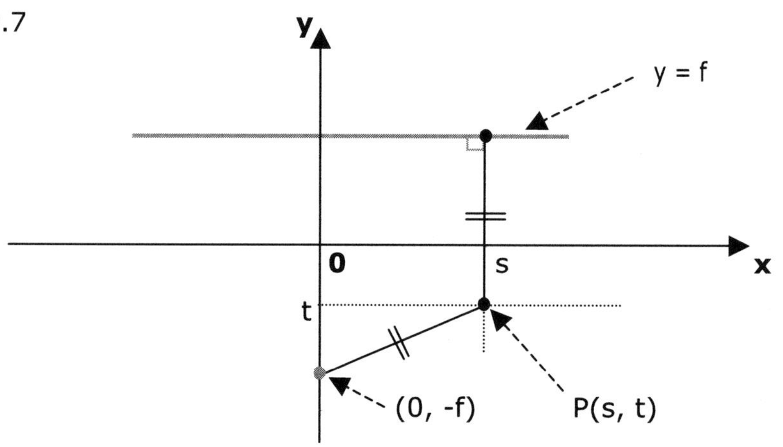

Let **U** be the distance between the **P(s, t)** and the focus, and **V** be the distance between the **P(s, t)** and the directrix. Then, we get $\mathbf{U^2 = (s - 0)^2 + \{t - (-f)\}^2}$ and $\mathbf{V^2 = (s - s)^2 + (t - f)^2}$.
Therefore, we have $\mathbf{U^2 = s^2 + (t + f)^2}$ and $\mathbf{V^2 = (t - f)^2}$.

Now, since the **P** needs to keep the same distance away from the focus and the directrix, we need to get $\mathbf{U^2 = V^2}$. Therefore, $\mathbf{U^2 = V^2 \Rightarrow s^2 + (t + f)^2 = (t - f)^2 \Rightarrow s^2 + t^2 + 2ft + f^2 = t^2 - 2ft + f^2 \Rightarrow s^2 = -4ft}$.
Now, the connective expression between the **s** and the **t** is $\mathbf{s^2 = -4ft}$.
Also, since the **P(s, t)** is the general point in the x-y plane, the **s** and the **t** are the variables that get respectively the values in the x-axis and the y-axis in the x-y system, and we use **x** and **y** for the variables in the x-y system. Besides, when we specify points in the x-y system, we use for the coordinate values the values in the x-axis and the y-axis. Then, in the connective expression $\mathbf{s^2 = -4ft}$, if we replace the **s** and the **t** respectively with **x** and **y**, the connective expression between the coordinates at each of all the points constituting the parabola is $\mathbf{x^2 = -4fy}$.
Of course, we can put it in $\mathbf{y = -x^2/4f}$, too.

Fig. 1.0.8

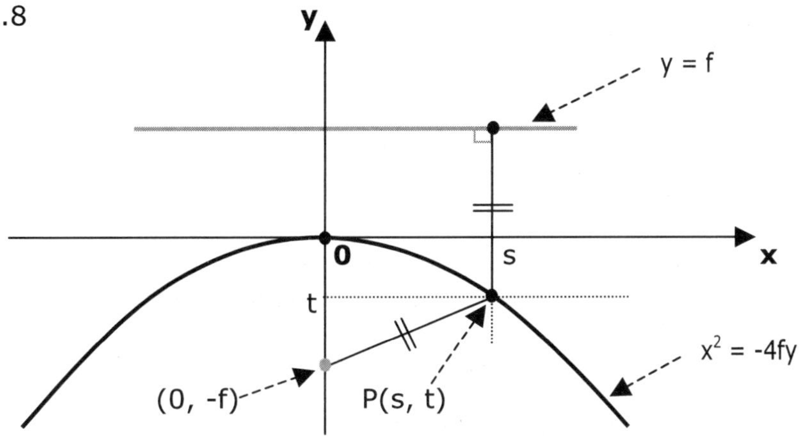

Also, let's set $\mathbf{f = 1/4}$. Then, we get $\mathbf{y = -x^2}$.
In the parabola, too, above, the vertex is at the origin, and the axis of symmetry is the y-axis, that is, the line $\mathbf{x = 0}$, too. The entrance of the parabola is below the parabola, however.

For this time, let's suppose that the focus is at **(f, 0)**, and that the directrix is a line $\mathbf{x = -f}$.

Fig. 1.0.9

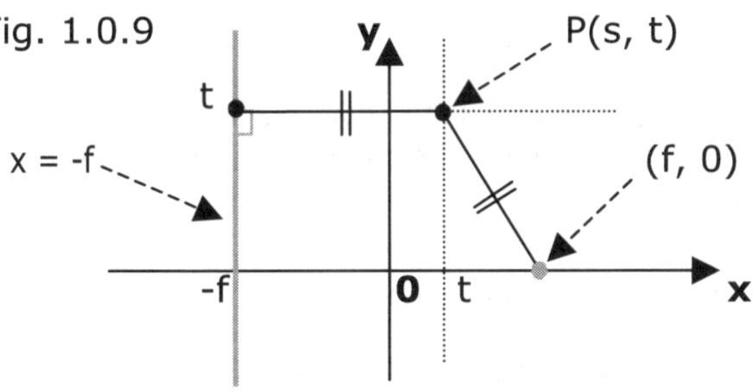

Let **U** be the distance between **P(s, t)** and the focus, and **V** be the distance between **P(s, t)** and the directrix **D**. Then, we get $\mathbf{U^2 = (s - f)^2 + (t - 0)^2}$ and $\mathbf{V^2 = \{s - (-f)\}^2 + (t - t)^2}$.
Therefore, we have $\mathbf{U^2 = (s - f)^2 + t^2}$ and $\mathbf{V^2 = (s + f)^2}$.
Now, since **P** needs to keep the same distance from the focus and the directrix, we need to get $\mathbf{U^2 = V^2}$.
Therefore, $\mathbf{U^2 = V^2 \Rightarrow (s - f)^2 + t^2 = (s + f)^2 \Rightarrow s^2 - 2fs + f^2 + t^2 = s^2 + 2fs + f^2 \Rightarrow t^2 = 4fs}$.
Now, the connective expression between the **s** and the **t** is $\mathbf{t^2 = 4fs}$. Also, since **P(s, t)** is the general point in the x-y plane, the **s** and the **t** are variables that get respectively the values in the x-axis and the y-axis in the x-y system, and we use **x** and **y** for the variables in the x-y system.

Besides, to specify points in the x-y system, we use for the coordinate values the values in the x-axis and the y-axis. Then, in the connective expression $\mathbf{t^2 = 4fs}$, if we replace the **s** and the **t** with **x** and **y** respectively, the connective expression between the coordinates at each of all the points constituting the parabola is $\mathbf{y^2 = 4fx}$. Of course, we can put it in $\mathbf{x = y^2/4f}$, too. Then, if $\mathbf{f = 1/4}$, the equation of the parabola is $\mathbf{x = y^2}$.

Fig. 1.0.A

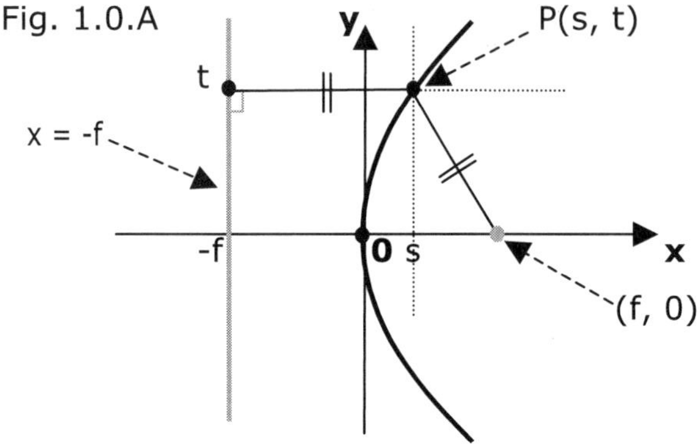

In the parabola, too, above, the vertex is at the origin, but the axis of symmetry is not the y-axis but the x-axis, that is, a line $\mathbf{y = 0}$. For this time, the entrance of the parabola is on the right.

For this time, let's suppose that the focus is at **(0, v)**.

When do we put the directrix, then? Is the equation of the directrix $\mathbf{y = -v}$?
Not necessarily That is, we can put the directrix anywhere we want to put. Nevertheless, for simplicity, we merely put the directrix to be parallel to one of the coordinate axes in general math.

Of course, the midpoint between the focus and the directrix doesn't have to be the origin, either.
What midpoint?
It is the vertex of a parabola, which is a very significant point in a parabola. In fact, it is often a point we have to find working on many problems.

This book covers right parabolas. In a right parabola, the directrix is parallel to one of the coordinate axes. Therefore, the axis of symmetry of the parabola is perpendicular to that coordinate axis. For a parabola where the axis of symmetry or the directrix is slanted to one of the coordinate axes, refer to the **Book 2: Graph Operations**.

Now, the components in a parabola we need to take note of are the vertex and the focus.
What about the directrix?
The directrix is important, too, of course. However, we can see it if we know the vertex and the focus. That's because the vertex is the midpoint between the focus and the directrix. Therefore, the vertex and the focus are also in the line perpendicular to the directrix. What is the line? It is the axis of symmetry.

Suppose that the vertex **V** is at **(u, v)**, the axis of symmetry is a line $\mathbf{x = u}$, the focus **F** is at **(@, f)**, and we don't know the @ yet. Then, what can be the @?
The @ needs to be the **u** since the vertex and the focus are in the axis of symmetry, which is a line $\mathbf{x = u}$.
Suppose now, that the focus **F** is at **(u, f)** and we find the equation of the parabola.
Let's first, put in a graph the vertex and the focus along with the general point **P(s, t)**. Then, we can see better what we can do to come up with the equation.

Fig. 1.0.B

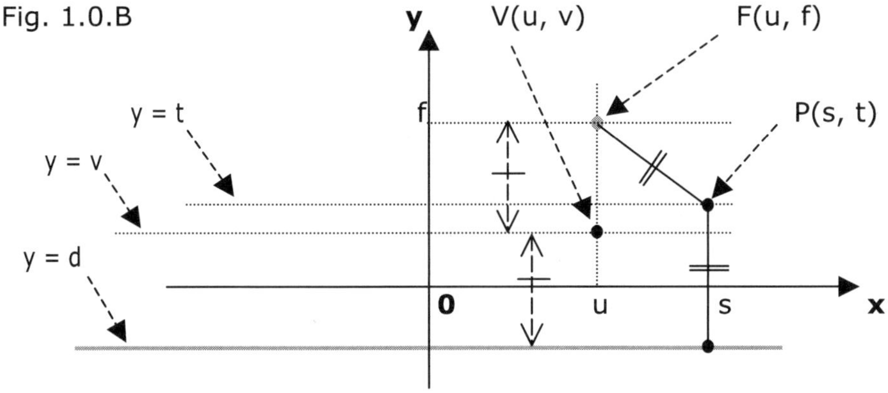

Suppose now, that the directrix is a line **y = d**.
Then, since the **V** is a midpoint between the directrix and focus, the **v** is an average of the **f** and **d**.
Then, **v = (f + d)/2** ⇒ **d = 2v − f**. So, the directrix is a line **y = 2v − f**.

Suppose that **Q** is the distance between **P(s, t)** and focus, and **R** is the distance between **P(s, t)** and the directrix **D**.
Then, we get $Q^2 = (s - u)^2 + (t - f)^2$ and $R^2 = (s - s)^2 + \{t - (2v - f)\}^2 = \{t - (2v - f)\}^2$.
Now, since the **P** needs to keep the same distance away from the focus and the directrix, we need to have $Q^2 = R^2$. Therefore, $Q^2 = R^2 \Rightarrow (s - u)^2 + (t - f)^2 = \{t - (2v - f)\}^2$
$\Rightarrow (s - u)^2 = \{t - (2v - f)\}^2 - (t - f)^2 = \{t - (2v - f) + (t - f)\}\{t - (2v - f) - (t - f)\}$
$= (t - 2v + f + t - f)(t - 2v + f - t + f) = 2(t - v)2(f - v)$.

$\Rightarrow (s - u)^2 = 4(t - v)(f - v)$.

Now, the connective expression between the **s** and the **t** is $(s - u)^2 = 4(f - v)(t - v)$.
Also, since the **P(s, t)** is the general point in the x-y plane, the **s** and **t** are variables that get respectively the values in the x-axis and the y-axis in the x-y system, and we use **x** and **y** for the variables in the x-y system. Besides, specifying points in the x-y system, we use for the coordinate values the values in the x-axis and the y-axis. Then, in the connective expression $(s - u)^2 = 4(f - v)(t - v)$, if we replace the **s** and the **t** respectively with **x** and **y**, the connective expression between the coordinates at each of all the points constituting the parabola is $(x - u)^2 = 4(f - v)(y - v)$.

Fig. 1.0.C

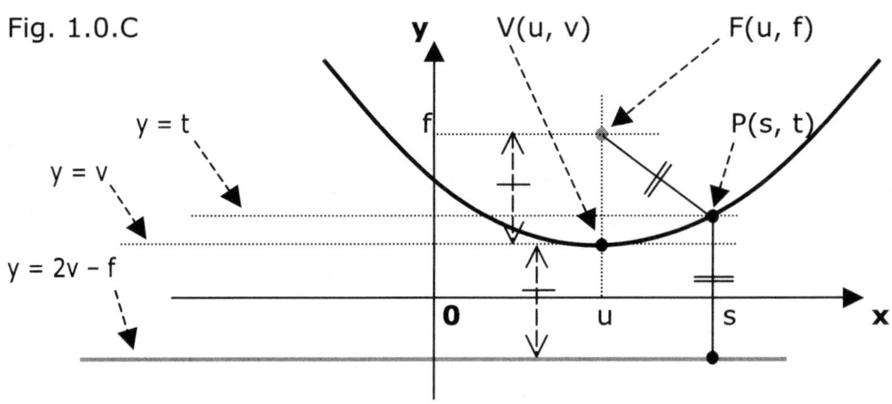

The curve above can also be taken for the result of the translation of the dotted curve in gray by **u** in the direction of the x-axis and by **v** in the direction of the y-axis as shown below.

Fig. 1.0.D

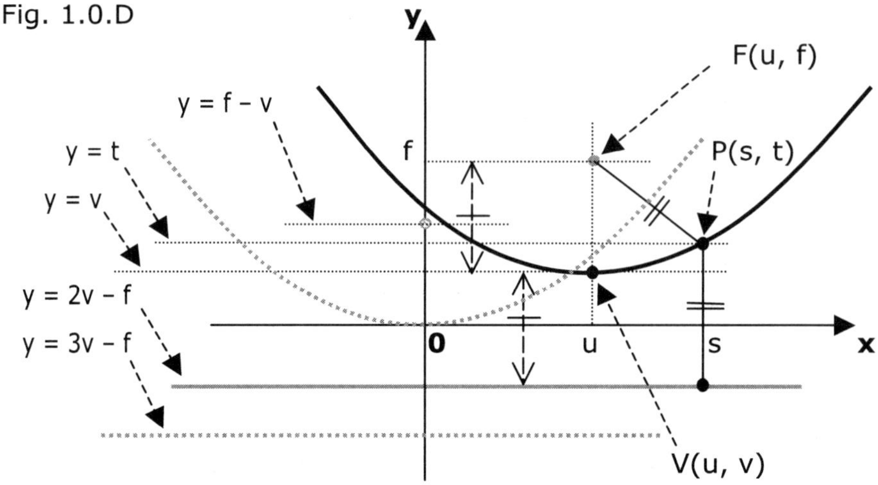

The equation of the dotted parabola in gray is $\mathbf{x^2 = 4(f - v)y}$, the vertex is at the origin, and the focus is at $\mathbf{(0, f - v)}$. The axis of symmetry of the parabola in black is a line $\mathbf{x = u}$.

Suppose this time, that the vertex **V** is at $\mathbf{(u, v)}$, and that the focus **F** is at $\mathbf{(u, -f)}$.
The n, we can immediately see that the axis of symmetry is still the line $\mathbf{x = u}$.

Fig. 1.0.E

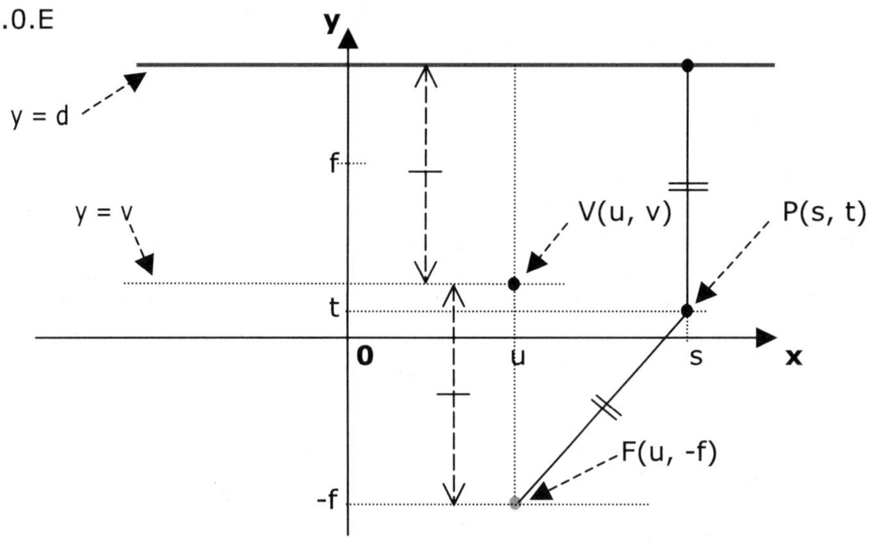

Suppose next, that the directrix is a line $\mathbf{y = d}$. Then, since the **V** is a midpoint between the directrix and focus, the **v** is an average of the **–f** and the **d**. Then, $\mathbf{v = (-f + d)/2 \Rightarrow d = 2v + f}$.
So, the directrix is a line $\mathbf{y = 2v + f}$. Suppose now, that **Q** is the distance between **P(s, t)** and focus, and **R** is the distance between **P(s, t)** and the directrix **D**.
Then, we get $\mathbf{Q^2 = (s - u)^2 + \{t - (-f)\}^2}$ and $\mathbf{R^2 = (s - s)^2 + \{t - (2v + f)\}^2 = \{t - (2v + f)\}^2}$.
Now, since the **P** needs to keep the same distance away from the focus and the directrix:

$$U^2 = V^2 \Rightarrow (s - u)^2 + (t + f)^2 = \{t - (2v + f)\}^2$$

$$\Rightarrow (s - u)^2 = \{t - (2v + f)\}^2 - (t + f)^2 = \{t - (2v + f) + (t + f)\}\{t - (2v + f) - (t + f)\}$$

$$= (t - 2v - f + t + f)(t - 2v - f - t - f) = 2(t - v)(-2)(f + v) \Rightarrow (s - u)^2 = -4(t - v)(f + v).$$

Now, the connective expression between the **s** and the **t** is $\mathbf{(s - u)^2 = -4(f + v)(t - v)}$.

Also, since the **P(s, t)** is the general point in the x-y plane, the **s** and **t** are the variables that get respectively the values in the x-axis and the y-axis in the x-y system, we use **x** and **y** for the variables in the x-y system. Besides, specifying points in the x-y system, we use for the coordinate values the values in the x-axis and the y-axis. Then, in the connective expression $\mathbf{(s - u)^2 = -4(f + v)(t - v)}$, if we replace the **s** and the **t** respectively with **x** and **y**, the connective expression between the coordinates of each of all the points constituting the parabola is $\mathbf{(x - u)^2 = -4(f + v)(y - v)}$.

Fig. 1.0.F

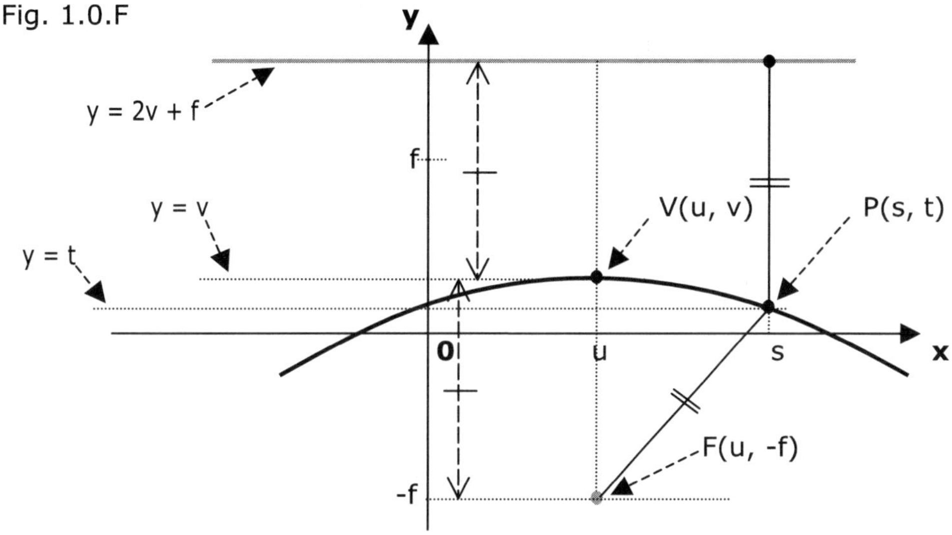

The curve above can also be taken for the result of the translation of the dotted curve in gray by **u** in the direction of the x-axis and by **v** in the direction of the y-axis as shown below.

Fig. 1.0.G

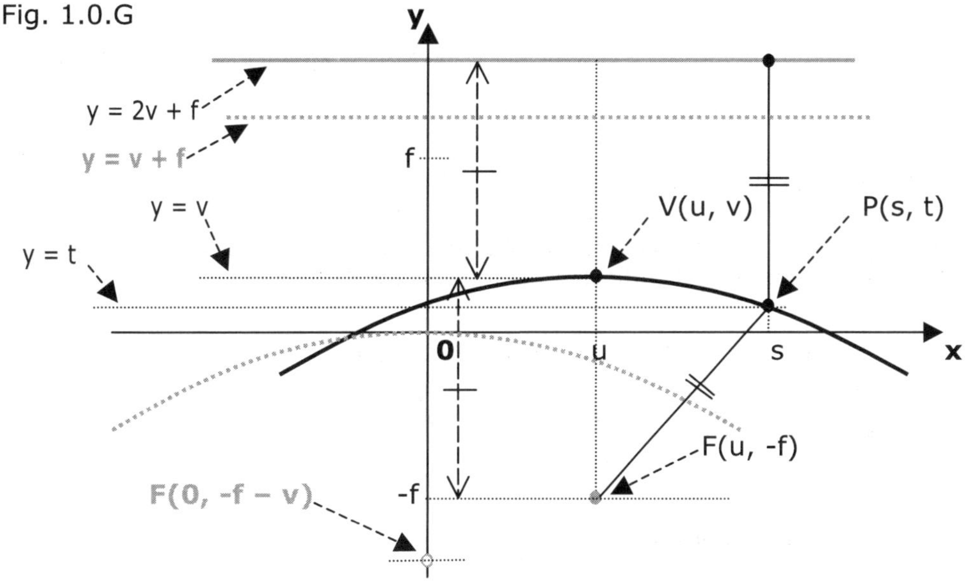

The equation of the dotted parabola in gray is $\mathbf{x^2 = 4(f + v)y}$, the focus is at **(0, -f – v)**, and the axis of symmetry is a line **x = 0**. The equation of the parabola in black is $\mathbf{(x - u)^2 = -4(f + v)(y - v)}$, the focus is at **(u, –f)**, and the axis of symmetry is a line **x = u**.
Now, let's compare the graphs of the two parabolas as follows:

$(x - u)^2 = -4(f + v)(y - v)$ and $(x - u)^2 = 4(f - v)(y - v)$.

Fig. 1.0.C

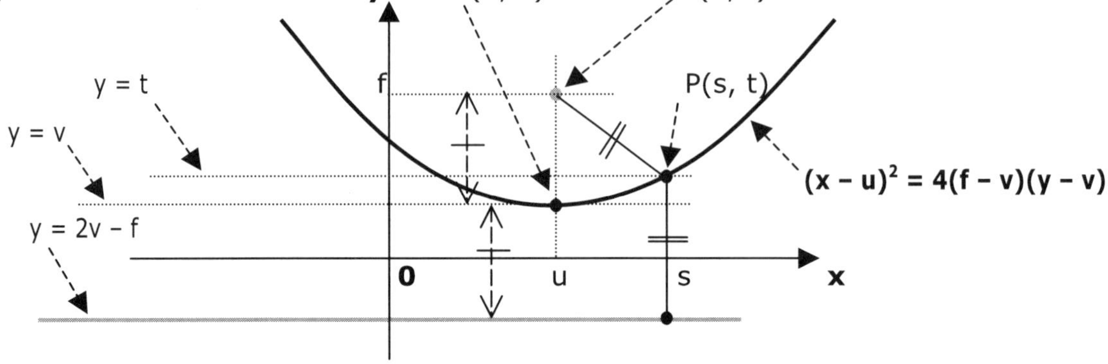

Fig. 1.0.F

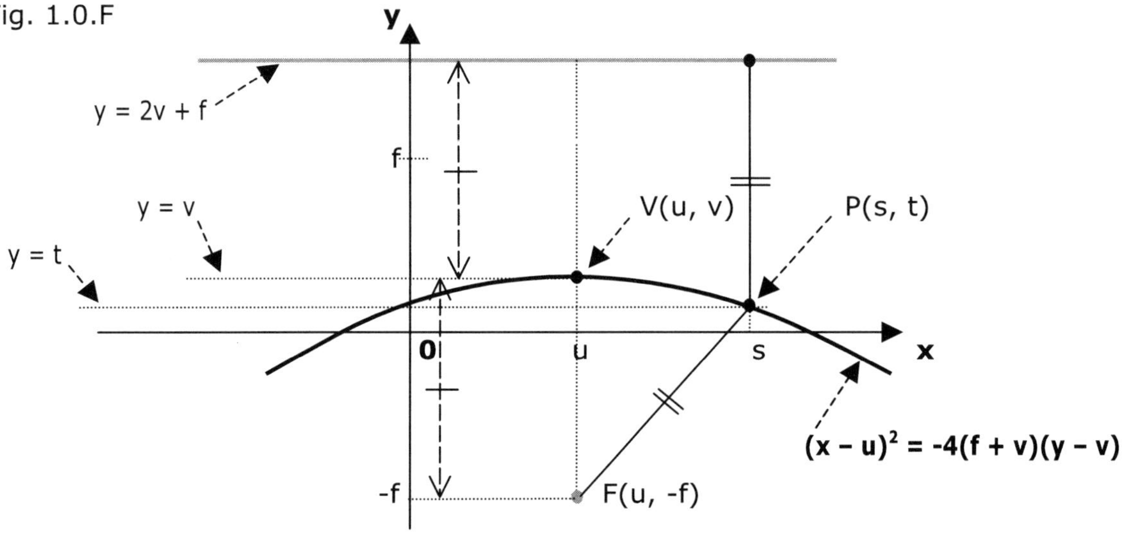

Suppose this time, that the vertex **V** is at **(u, v)**, and the focus **F** is at **(f, v)**.
Then, we can immediately see that the axis of symmetry is a line **y = v**. Let's first, put in a graph the vertex and the focus along with the general point **P(s, t)**. We know that the vertex is a midpoint between the focus and the directrix. Suppose now, that the directrix is a line **x = d**. Then:

Fig. 1.0.H

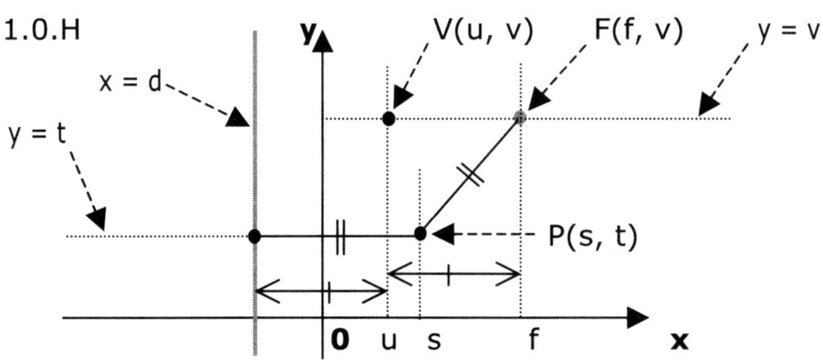

Then, the **u** is an average of the **f** and the **d** as we can see it above.

Then, $\mathbf{u = (f + d)/2 \Rightarrow d = 2u - f}$. Therefore, the directrix is a line $\mathbf{x = 2u - f}$.
Suppose that **Q** is the distance between **P(s, t)** and the focus, and **R** is the distance between **P(s, t)** and the directrix **D**.
Then, we get $\mathbf{Q^2 = (s - f)^2 + (t - v)^2}$ and $\mathbf{R^2 = \{s - (2u - f)\}^2 + (t - t)^2 = \{s - (2u - f)\}^2}$.

Now, since the **P** keeps the same distance away from the focus and the directrix:
$\mathbf{Q^2 = R^2 \Rightarrow (s - f)^2 + (t - v)^2 = \{s - (2u - f)\}^2}$
$\mathbf{\Rightarrow (t - v)^2 = \{s - (2u - f)\}^2 - (s - f)^2 = \{s - (2u - f) + (s - f)\}\{s - (2u - f) - (s - f)\}}$
$\mathbf{= (s - 2u + f + s - f)(s - 2u + f - s + f) = 2(s - u)2(f - u) \Rightarrow (t - v)^2 = 4(f - u)(s - u)}$.

Now, the connective expression between the **s** and the **t** is $\mathbf{(t - v)^2 = 4(f - u)(s - u)}$.
Also, since the **P(s, t)** is the general point in the x-y plane, the **s** and the **t** are the variables that get respectively the values in the x-axis and the y-axis in the x-y system, and we use **x** and **y** for the variables in the x-y system. Besides, specifying points in the x-y system, we use for the coordinate values the values in the x-axis and the y-axis. Then, in the connective expression above, if we replace the **s** and the **t** respectively with **x** and **y**, the connective expression between the coordinates at each of all the points constituting the parabola is $\mathbf{(y - v)^2 = 4(f - u)(x - u)}$.

Fig. 1.0.I

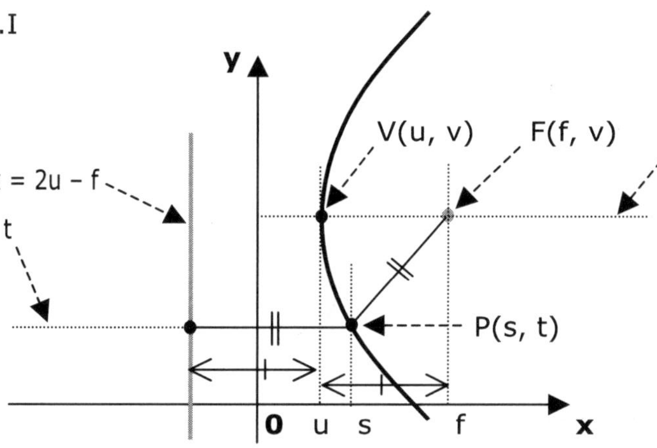

Suppose next, that the vertex **V** is at **(u, v)**, and the focus **F** is at **(-f, v)**. Then, the axis of symmetry is still the line **y = v**. Suppose now that the directrix is a line **x = d**. Then, since the vertex is a midpoint between the focus and directrix, putting in a graph the vertex and focus along with the directrix, we get:

Fig. 1.0.J

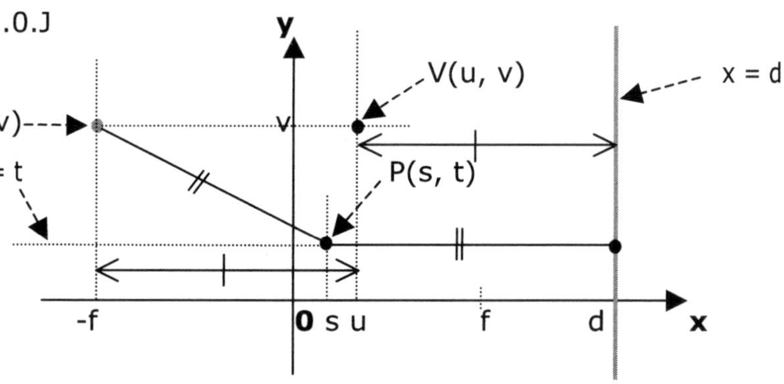

Then, since the **u** is an average of the **–f** and the **d**, $\mathbf{u = (-f + d)/2 \Rightarrow d = 2u + f}$.
So, the directrix is a line $\mathbf{x = 2u + f}$.

Let **Q** be the distance between **P(s, t)** and the focus, and **R** be the distance between **P(s, t)** and the directrix **D**. Then, we get $\mathbf{Q^2 = (s + f)^2 + (t - v)^2}$ and $\mathbf{R^2 = \{s - (2u + f)\}^2 + (t - t)^2}$
$\mathbf{= \{s - (2u + f)\}^2}$. Now, $\mathbf{Q^2 = R^2 \Rightarrow (s + f)^2 + (t - v)^2 = \{s - (2u + f)\}^2}$
$\mathbf{\Rightarrow (t - v)^2 = \{s - (2u + f)\}^2 - (s + f)^2 = \{s - (2u + f) + (s + f)\}\{s - (2u + f) - (s + f)\}}$
$\mathbf{= (s - 2u - f + s + f)(s - 2u - f - s - f) = 2(s - u)(-2)(f + u)}$
$\mathbf{\Rightarrow (t - v)^2 = -4(s - u)(f + u)}$.

Now, the connective expression between the **s** and the **t** is $\mathbf{(t - v)^2 = -4(f + u)(s - u)}$.
Also, since the **P(s, t)** is the general point in the x-y plane, the **s** and the **t** are the variables that get respectively the values in the x-axis and the y-axis in the x-y system, and we use **x** and **y** for the variables in the x-y system. Besides, specifying points in the x-y system, we use for the coordinate values the values in the x-axis and the y-axis. Then, in the connective expression above, if we replace the **s** and the **t** respectively with **x** and **y**, the connective expression between the coordinates at each of all the points constituting the parabola is $\mathbf{(y - v)^2 = -4(f + u)(x - u)}$.

Fig. 1.0.K

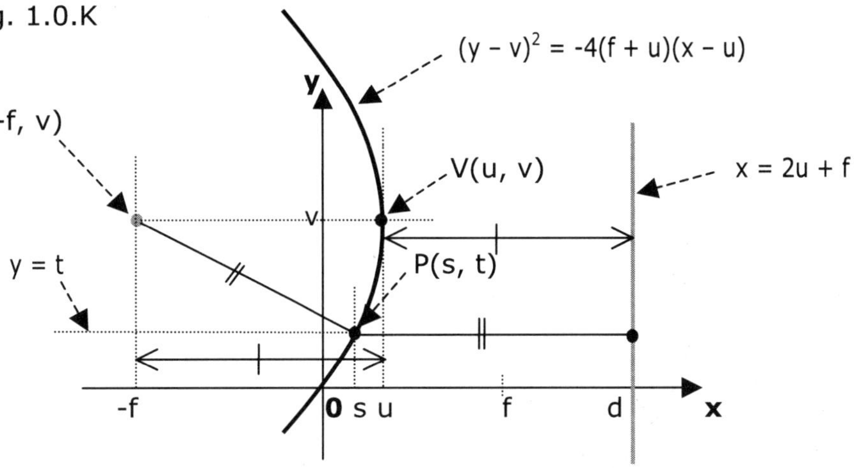

The parabola above is symmetric about a line **y = v**.
Then, let's compare each other the graph above and the previous graph presented again below.

Fig. 1.0.I

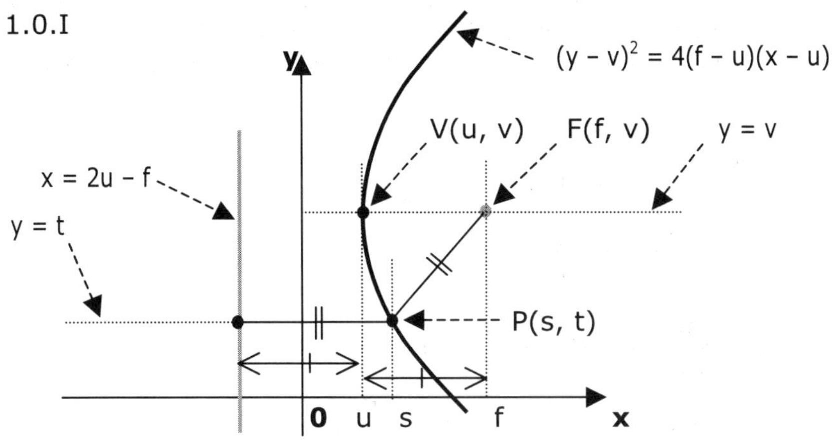

Now, let's talk about the standard equation of parabola.
To begin with, those ones we've constructed above can be said to be the examples of standard equation of parabola. Let **V** be the vertex, **F** be the focus, and **D** be the directrix. Then:

$y^2 = d^2 - 2dx$ (V: (d/2, 0), F: the origin, D: x = d)
$x^2 = c^2 - 2cy$ (V: (0, c/2), F: the origin, D: y = c)
$x^2 = 4fy$ (V: the origin, F: (0, f), D: y = -f)
$x^2 = -4fy$ (V: the origin, F: (0, -f), D: y = f)
$y^2 = 4fx$ (V: the origin, F: (f, 0), D: x = -f)
$(x - u)^2 = 4(f - v)(y - v)$ (V: (u, v), F: (u, f)), D: y = 2v − f)
$x^2 = 4(f - v)y$ (V: the origin, F: (0, f − v), D: y = 3v − f)
$(x - u)^2 = -4(f + v)(y - v)$ (V: (u, v), F: (u, -f), D: y = 2v + f)
$x^2 = 4(f + v)y$ (V: the origin, F: (0, -f − v), D: y = v + f)
$(y - v)^2 = 4(f - u)(x - u)$ (V: (u, v), F: (f, v), D: x = 2u − f)
$(y - v)^2 = -4(f + u)(x - u)$ (V: (u, v), F: (-f, v), D: x = 2u + f)

Of the equations above, the ones in bold face can be said to be the basic forms in the standard form. Of the basic forms, in particular, **$y^2 = 4fx$** and **$x^2 = 4fy$** are simple basic forms, which we actually call the standard equations of parabola. Now, when we present the standard form, we conventionally use **p** instead of the letter **f**. Therefore, the standard equations of parabola are as follows:

$x^2 = 4py$ where the vertex is the origin, the focus is **(0, p)**, and the directrix is a line **y = -p**.
$y^2 = 4px$ where the vertex is the origin, the focus is **(p, 0)**, and the directrix is a line **x = -p**.

Now, we have equations as follows:
$(x - u)^2 = 4(f - v)(y - v)$ (vertex: **(u, v)**, focus: **(u, f)**, directrix: **y = 2v − f**)
$(y - v)^2 = 4(f - u)(x - u)$ (vertex: **(u, v)**, focus: **(f, v)**, directrix: **x = 2u − f**)

In the first equation right above, set **f − v = p**.
Then, **$f - v = p \Rightarrow f = p + v$**.
So, in the case of the directrix, **$y = 2v - f = 2v - (p + v) = 2v - p - v = v - p$**. Therefore:
$(x - u)^2 = 4p(y - v)$ where the vertex is **(u, v)**, the focus is **(u, v + p)**, and the directrix is **y = v − p**.

In the second equation, set **f − u = p**.
Then, **$f - u = p \Rightarrow f = p + u$**.
So, in the case of the directrix, **$x = 2u - f = 2u - (p + u) = 2u - p - u = u - p$**. Therefore:
$(y - v)^2 = 4p(x - u)$ where the vertex is **(u, v)**, the focus is **(u + p, v)**, and the directrix is **x = u − p**.

Consequently, we may say that the following equations are the standard forms generalized a bit further:
$(x - u)^2 = 4p(y - v)$ where the vertex is **(u, v)**, the focus is **(u, v + p)**, and the directrix is **y = v − p**.
$(y - v)^2 = 4p(x - u)$ where the vertex is **(u, v)**, the focus is **(u + p, v)**, and the directrix is **x = u − p**.

Now, let's look at the concavity of the curve in accordance with each of the standard forms.

Suppose first, that **p > 0**. Then, the following curves are concave up:
$x^2 = 4py$ where the vertex is the origin, the focus is **(0, p)**, and the directrix is a line **y = -p**.
$(x - u)^2 = 4p(y - v)$ where the vertex is **(u, v)**, the focus is **(u, v + p)**, and the directrix is **y = v − p**.

Fig. 1.0.L

Next, the curves as follows are concave down:
$\mathbf{x^2 = -4py}$ where the vertex is the origin, the focus is **(0, -p)**, and the directrix is a line **y = p**.
$\mathbf{(x - u)^2 = -4p(y - v)}$ where the vertex is **(u, v)**, the focus is **(u, v − p)**, and the directrix is **y = v + p**.

Fig. 1.0.M

Next, the curves as follows are concave right:
$\mathbf{y^2 = 4px}$ where the vertex is the origin, the focus is **(p, 0)**, and the directrix is a line **x = -p**.
$\mathbf{(y - v)^2 = 4p(x - u)}$ where the vertex is **(u, v)**, the focus is **(u + p, v)**, and the directrix is **x = u − p**.

Fig. 1.0.N

Next, the curves as follows are concave left:
$\mathbf{y^2 = -4px}$ where the vertex is the origin, the focus is **(-p, 0)**, and the directrix is a line **x = p**.
$\mathbf{(y - v)^2 = -4p(x - u)}$ where the vertex is **(u, v)**, the focus is **(u − p, v)**, and the directrix is **x = u + p**.

Fig. 1.0.O

Fig. 1.0.P The standard form when **p = 1/4**

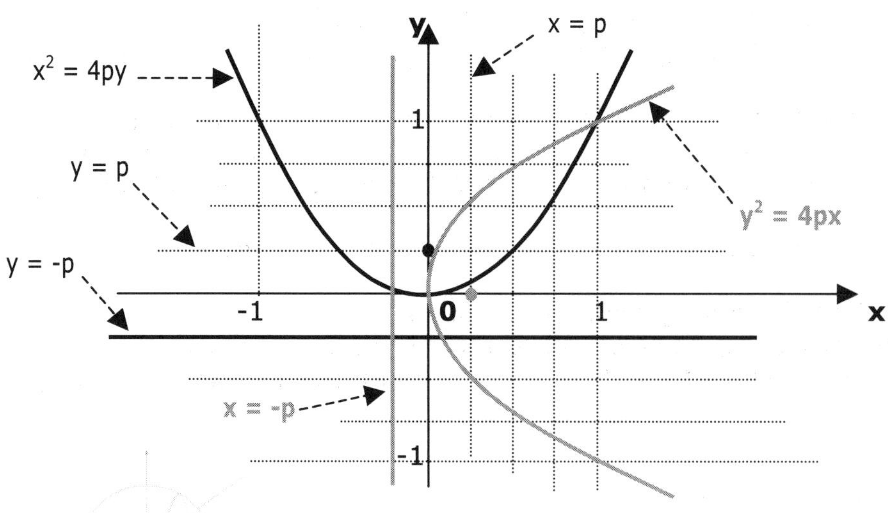

Fig. 1.0.Q $x^2 = 4py$ and $(x - u)^2 = 4p(y - v)$

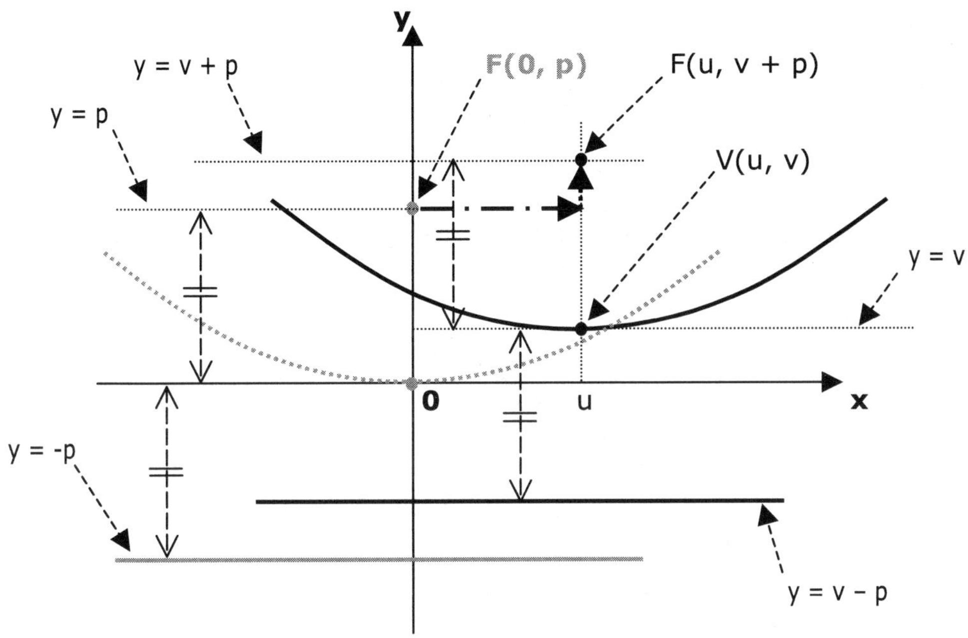

Fig. 1.0.R $y^2 = 4px$ and $(y - v)^2 = 4p(x - u)$

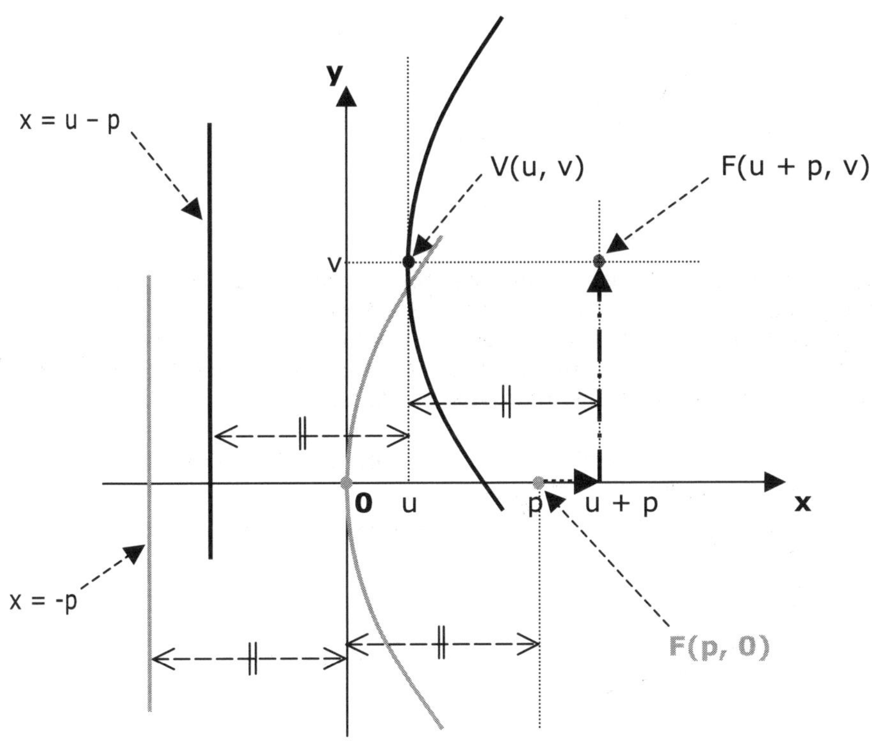

A parabola is a se of points. Suppose now, that a parabola is in the x-y system. Then, each of the points has a pair of coordinates, and we can put a parabola in a set as follows by means of an equation of parabola:

$\mathbf{A = \{(x, y) | \ y^2 = 4px}$ where $\mathbf{p}$ is a constant.$\}$
$\mathbf{B = \{(x, y) | \ x^2 = 4py}$ where $\mathbf{p}$ is a constant.$\}$
$\mathbf{C = \{(x, y) | \ (y - v)^2 = 4p(x - u)}$ where $\mathbf{p}$, $\mathbf{u}$, and $\mathbf{v}$ are constants.$\}$
$\mathbf{D = \{(x, y) | \ (x - u)^2 = 4p(y - v)}$ where $\mathbf{p}$, $\mathbf{u}$, and $\mathbf{v}$ are constants.$\}$

The parabola **A** is a set of points in the x-y system, the connective expression between the coordinates at each of the points is $\mathbf{y^2 = 4px}$ where **p** is a constant.
We can put the equation of the parabola **A** in $\mathbf{A(x, y) = y^2 - 4px = 0}$ or just, $\mathbf{y^2 = 4px}$.
The same is true for all the other parabolas above.

Some properties of a parabola is as follows:

Fig. 1.0.S

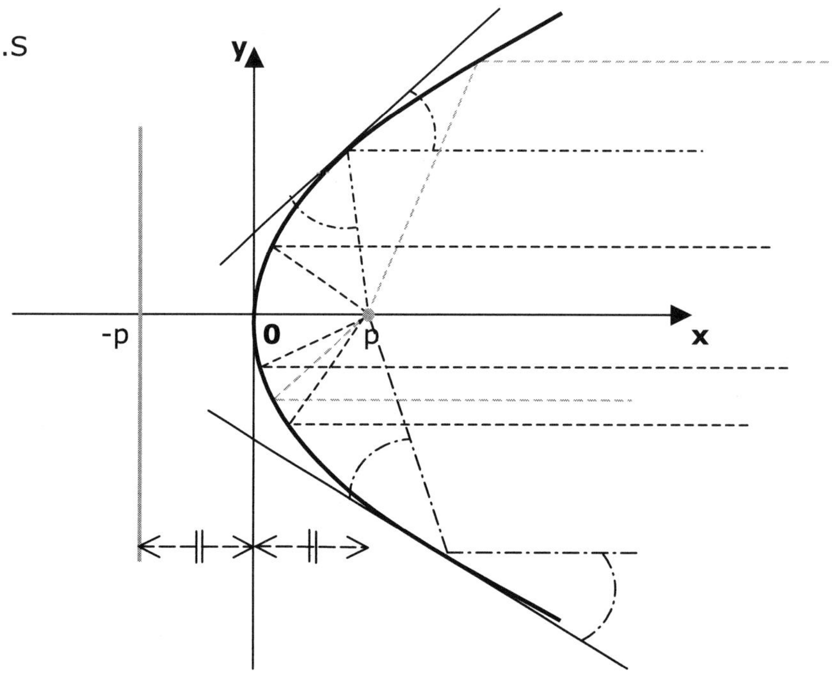

Rays coming into a parabola and perpendicular to the directrix concentrate at the focus, and rays emitting from the focus are perpendicular to the directrix, too. That's because when a ray reflects from a parabola, the incidence angle is the same as the angle of reflection. A ray can be such a wave as light or sound. Suppose now, that **A** is the length of a ray from the entrance of a parabola to the parabola, and **B** is the length of a ray from the parabola to the focus. Then, **A + B** is constant.

Fig. 1.0.T

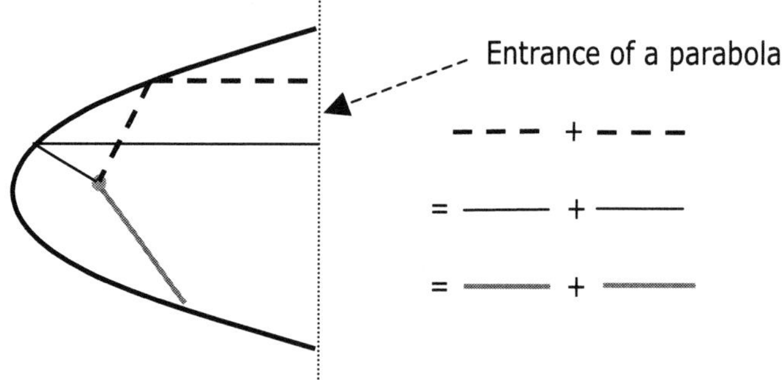

The Kepler's method for construction of a parabola presented below uses the property above.
First, pick a position of the focus as shown below, and then, put a vertical bar on a horizontal bar that serves as the parabola's directrix. Also, we need a string of a length same as the length of the vertical bar. Next, pin one end of the string to the focus and the other end to the end of the vertical bar. The string is in gray. Hold the string taught against the bar with a pencil. Then, slide the vertical rod along the horizontal rod. Then, the parabola gets made.

Fig. 1.0.U The Kepler's Method for Construction of a Parabola

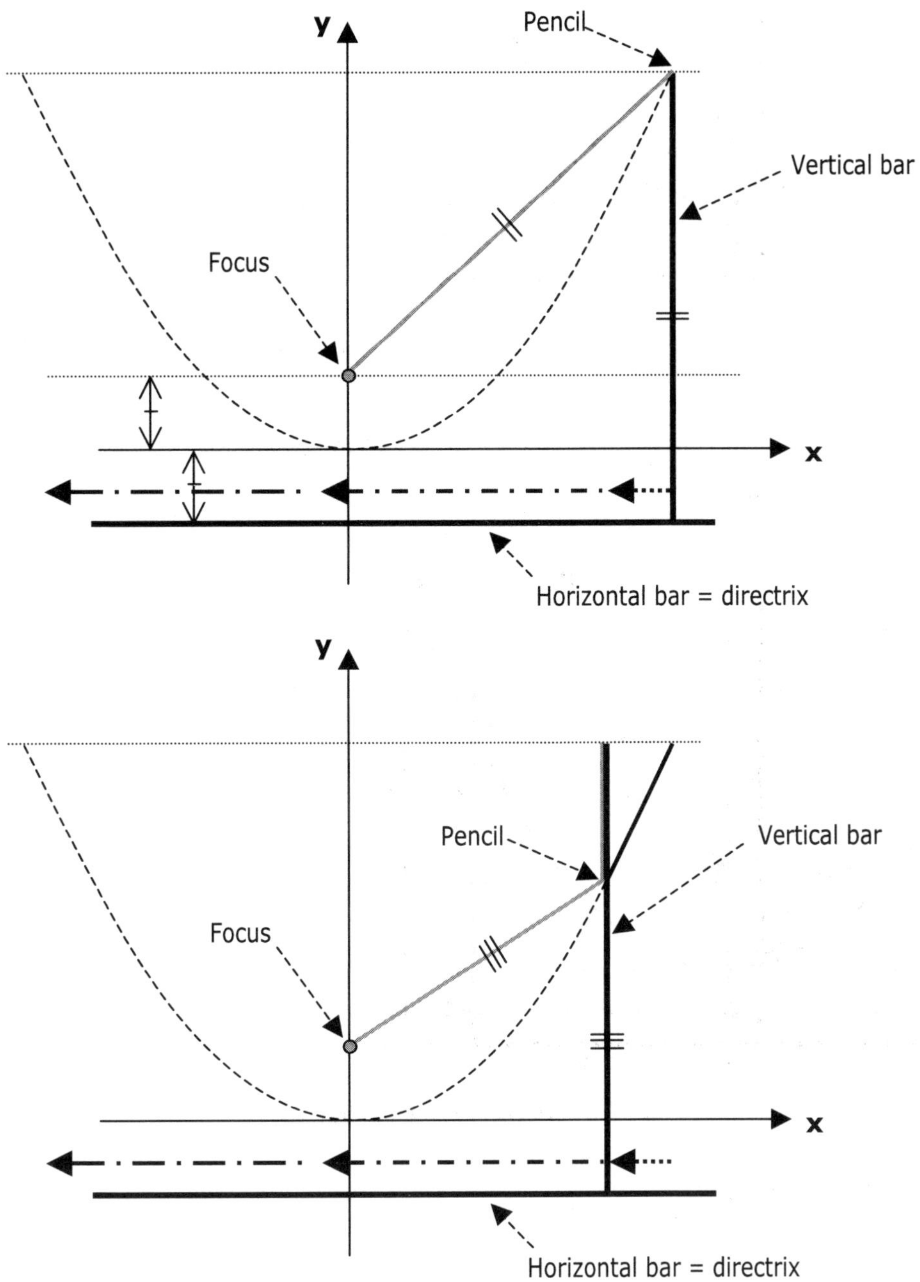

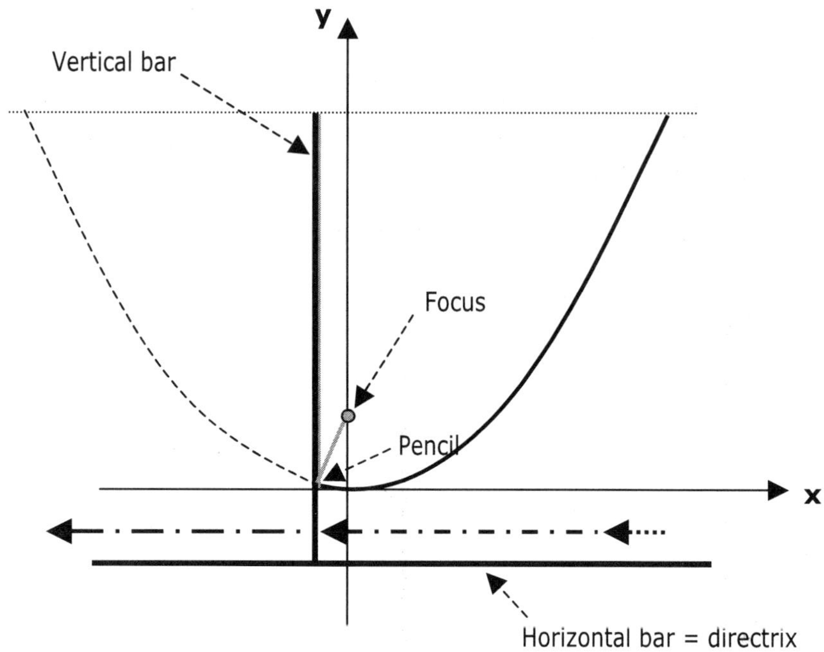

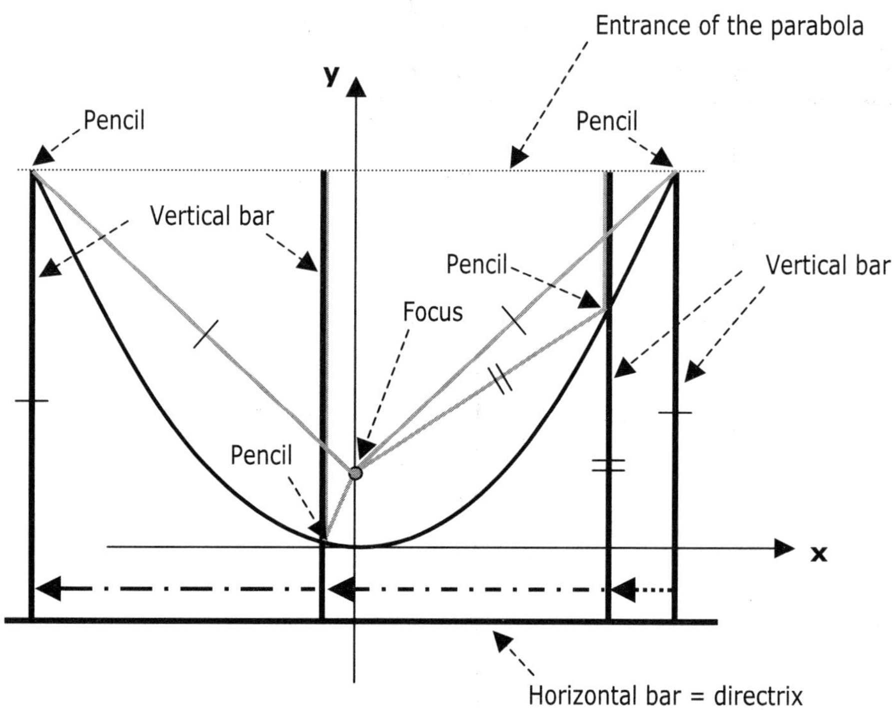

Now, let's take care of some examples.

Examples
in Standard Equations of Parabola

Suppose that **V** is the vertex, **F** is the focus, and **D** is the directrix, and then, find the equations of parabola as follows:

0. V(3, 0), F(0, 0)

1. V(0, -1), F(3, -1)

2. V(1, 0), F(3, 0)

3. V(1, -1), F(3, -1)

4. V(0, 0), F(0, 3)

5. V(-1, 0), F(-1, 3)

6. V(0, 1), F(0, 3)

7. V(-1, 1), F(-1, 3)

8. V(3, 1), D: x = -2

Suggestions or Solutions

To the **Problem** in the Example **0**

The vertex is the origin, and the focus is (3, 0).

We know that each point in a parabola is the same distance away from the focus and the directrix. Besides, the vertex is the midpoint between the directrix and the focus, and the vertex and focus are in the axis of symmetry.
Since the vertex is **(0, 0)** and the focus is **(3, 0)**, the y-coordinates are the same to be 0, and therefore, the axis of symmetry is a line **y = 0**, that is, the x-axis. So, since the directrix is perpendicular to the axis of symmetry, it needs to be perpendicular to the x-axis or parallel to the y-axis.
Now, suppose that the directrix is a line **x = d**. Then, the graph of the parabola will be similar to the one shown below.

Fig. 0.0

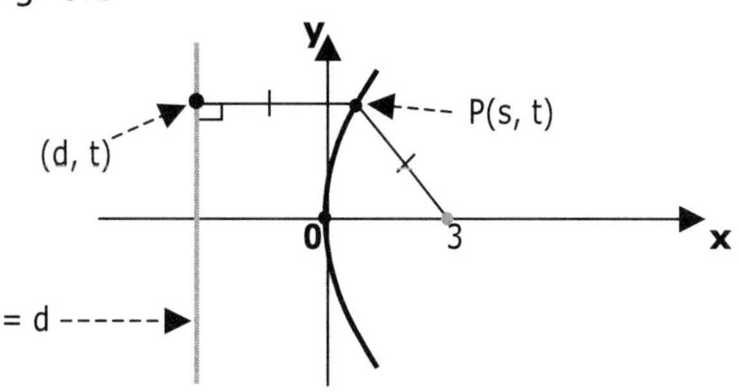

Then, since the vertex is the midpoint between the directrix and the focus:
$0 = (d + 3)/2 \Rightarrow d = -3$.
Therefore, the directrix is a line **$x = -3$**.

Suppose now, that the general point of the parabola is **P(s, t)**. Then:
(the distance between the point **P** and the directrix)^2 = **$\{s - (-3)\}^2 + (t - t)^2 = (s + 3)^2$.**
(the distance between the point **P** and the focus)^2 = **$(s - 3)^2 + (t - 0)^2 = (s - 3)^2 + t^2$.**

Since the two distances are the same, **$(s + 3)^2 = (s - 3)^2 + t^2$**
$\Rightarrow t^2 = (s + 3)^2 - (s - 3)^2 = (s + 3 + s - 3)(s + 3 - s + 3) = 2s(6) = 12s$
$\Rightarrow t^2 = 12s$.

Now, since **P(s, t)** is in the x-y system, the equation of the parabola is **$y^2 = 12x$**.

Since the vertex is the origin and the focus is **(3, 0)**, **p = 3** in the standard equation **$y^2 = 4px$**. If this is in our memory, we can find the solution immediately. We need to do memorizations, but keeping things in our memories is not an easy task.

In short:

Suppose that the directrix is a line **x = d**.
Then, **$0 = (d + 3)/2 \Rightarrow d = -3$.**
Let the general point of the parabola be **P(s, t)**. Then:
$\{s - (-3)\}^2 + (t - t)^2 = (s + 3)^2 = (s - 3)^2 + (t - 0)^2 = (s - 3)^2 + t^2 \Rightarrow (s + 3)^2 = (s - 3)^2 + t^2$
$\Rightarrow t^2 = (s + 3)^2 - (s - 3)^2 = (s + 3 + s - 3)(s + 3 - s + 3) = 2s(6) = 12s \Rightarrow t^2 = 12s$.
Now, since **P(s, t)** is in the x-y system, the equation of the parabola is **$y^2 = 12x$**.

Suggestions or Solutions
To the Problem in the Example 1

The vertex is (0, -1), and the focus is (3, -1).

We know that a parabola is a set of points, each of which is the same distance away from the focus and the directrix. Besides, the vertex is the midpoint between the directrix and the focus, and the vertex and focus are in the axis of symmetry.
Since the vertex is **(0, -1)** and the focus is **(3, -1)**, the y-coordinates are the same to be -1, and therefore, the axis of symmetry is a line **y = -1**. So, since the directrix is perpendicular to the axis of symmetry, it needs to be perpendicular to the x-axis or parallel to the y-axis.
Now, suppose that the directrix is a line **x = d**. Then, the graph of the parabola will be similar to the one shown below.

Fig. 1.0

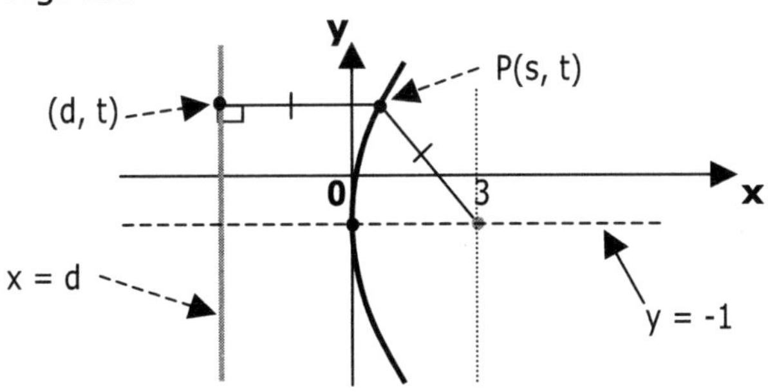

Then, since the vertex is the midpoint between the directrix and the focus:
$\mathbf{0 = (d + 3)/2 \Rightarrow d = -3}$.
Therefore, the directrix is a line **x = -3**.

Suppose now, that the general point of the parabola is **P(s, t)**. Then:
(the distance between the point **P** and the directrix)^2 = $\{s - (-3)\}^2 + (t - t)^2 = (s + 3)^2$.
((the distance between the point **P** and the focus)^2 = $(s - 3)^2 + \{t - (-1)\}^2 = (s - 3)^2 + (t + 1)^2$.

Since the two distances are the same, $(s + 3)^2 = (s - 3)^2 + (t + 1)^2$
$\Rightarrow (t + 1)^2 = (s + 3)^2 - (s - 3)^2 = (s + 3 + s - 3)(s + 3 - s + 3) = 2s(6) = 12s$.

Now, since **P(s, t)** is in the x-y system, the equation of the parabola is $(y + 1)^2 = 12x$.

In short:

Suppose that the directrix is a line **x = d**.
Then, $0 = (d + 3)/2 \Rightarrow d = -3$.
Let the general point of the parabola be **P(s, t)**. Then:
$\{s - (-3)\}^2 + (t - t)^2 = (s + 3)^2 = (s - 3)^2 + \{t - (-1)\}^2 = (s - 3)^2 + (t + 1)^2$
$\Rightarrow (t + 1)^2 = (s + 3)^2 - (s - 3)^2 = (s + 3 + s - 3)(s + 3 - s + 3) = 2s(6) = 12s$
$\Rightarrow (t + 1)^2 = 12s$.

Now, since **P(s, t)** is in the x-y system, the equation of the parabola is $(y + 1)^2 = 12x$.

Suggestions or Solutions
To the Problem in the Example 2

The vertex is (1, 0), and the focus is (3, 0).

We know that a parabola is a set of points, each of which is the same distance away from the focus and the directrix. Besides, the vertex is the midpoint between the directrix and the focus, and the vertex and focus are in the axis of symmetry.
Since the vertex is **(1, 0)** and the focus is **(3, 0)**, the y-coordinates are the same to be 0, and therefore, the axis of symmetry is a line $y = 0$, that is, the x-axis. So, since the directrix is perpendicular to the axis of symmetry, it needs to be perpendicular to the x-axis or parallel to the y-axis.
Now, suppose that the directrix is a line $x = d$. Then, the graph of the parabola will be similar to the one shown below.

Fig. 2.0

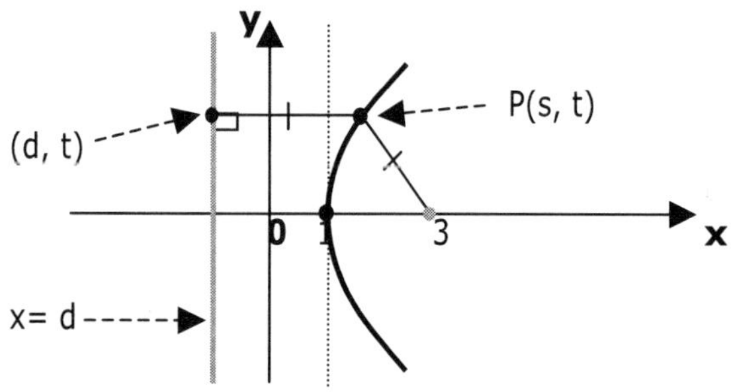

Then, since the vertex is the midpoint between the directrix and the focus:
$1 = (d + 3)/2 \Rightarrow d = -1$.
Therefore, the directrix is a line $x = -1$.

Suppose now, that the general point of the parabola is **P(s, t)**. Then:
(the distance between the point **P** and the directrix)^2 = $\{s - (-1)\}^2 + (t - t)^2 = (s + 1)^2$.
((the distance between the point **P** and the focus)^2 = $(s - 3)^2 + (t - 0)^2 = (s - 3)^2 + t^2$.

Since the two distances are the same, $(s + 1)^2 = (s - 3)^2 + t^2$
$\Rightarrow t^2 = (s + 1)^2 - (s - 3)^2 = (s + 1 + s - 3)(s + 1 - s + 3) = 2(s - 1)(4) = 8(s - 1)$.

Now, since **P(s, t)** is in the x-y system, the equation of the parabola is $y^2 = 8(x - 1)$.

In short:

Suppose that the directrix is a line $x = d$.
Then, $1 = (d + 3)/2 \Rightarrow d = -1$.
Let the general point of the parabola be **P(s, t)**. Then:
$\{s - (-1)\}^2 + (t - t)^2 = (s + 1)^2 = (s - 3)^2 + (t - 0)^2 = (s - 3)^2 + t^2$
$\Rightarrow t^2 = (s + 1)^2 - (s - 3)^2 = (s + 1 + s - 3)(s + 1 - s + 3) = 2(s - 1)(4) = 8(s - 1)$
$\Rightarrow t^2 = 8(s - 1)$.

Now, since **P(s, t)** is in the x-y system, the equation of the parabola is $y^2 = 8(x - 1)$.

Suggestions or Solutions
To the Problem in the Example 3

The vertex is (1, -1), and the focus is (3, -1).

We know that a parabola is a set of points, each of which is the same distance away from the focus and the directrix. Besides, the vertex is the midpoint between the directrix and the focus, and the vertex and focus are in the axis of symmetry.
Since the vertex is **(1, -1)** and the focus is **(3, -1)**, the y-coordinates are the same to be -1, and therefore, the axis of symmetry is a line **y = -1**. So, since the directrix is perpendicular to the axis of symmetry, it needs to be perpendicular to the x-axis or parallel to the y-axis.
Now, suppose that the directrix is a line **x = d**. Then, the graph of the parabola will be similar to the one shown below.

Fig. 3.0

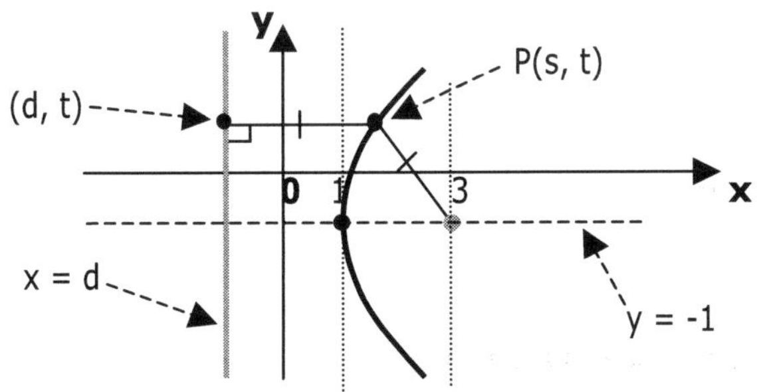

Then, since the vertex is the midpoint between the directrix and the focus:
1 = (d + 3)/2 ⇒ d = -1.
Therefore, the directrix is a line **x = -1**.

Suppose now, that the general point of the parabola is **P(s, t)**. Then:
(the distance between the point **P** and the directrix)^2 = **{s – (-1)}^{2} + (t – t)^2 = (s + 1)^2**.
((the distance between the point **P** and the focus)^2 = **(s – 3)^2 + {t – (-1)}^{2} = (s – 3)^2 + (t + 1)^2**.

Since the two distances are the same, **(s + 1)^2 = (s – 3)^2 + (t + 1)^2**
⇒ (t + 1)^2 = (s + 1)^2 – (s – 3)^2 = (s + 1 + s – 3)(s + 1 – s + 3) = 2(s – 1)(4) = 8(s – 1)
⇒ (t + 1)^2 = 8(s – 1).

Now, since **P(s, t)** is in the x-y system, the equation of the parabola is **(y + 1)^2 = 8(x – 1)**.

In short:
Suppose that the directrix is a line **x = d**.
Then, **1 = (d + 3)/2 ⇒ d = -1.**
Let the general point of the parabola be **P(s, t)**. Then:
{s – (-1)}^{2} + (t – t)^2 = (s + 1)^2 = (s – 3)^2 + {t – (-1)}^{2} = (s – 3)^2 + (t + 1)^2
⇒ (t + 1)^2 = (s + 1)^2 – (s – 3)^2 = (s + 1 + s – 3)(s + 1 – s + 3) = 2(s – 1)(4) = 8(s – 1)
⇒ (t + 1)^2 = 8(s – 1).

Now, since **P(s, t)** is in the x-y system, the equation of the parabola is **(y + 1)^2 = 8(x – 1)**.

Suggestions or Solutions
To the Problem in the Example 4

The vertex is the origin, and the focus is (0, 3).

We know that each point in a parabola is the same distance away from the focus and the directrix. Besides, the vertex is the midpoint between the directrix and the focus, and the vertex and focus are in the axis of symmetry.
Since the vertex is **(0, 0)** and the focus is **(0, 3)**, the x-coordinates are the same to be 0, and therefore, the axis of symmetry is a line **x = 0**, that is, the y-axis. So, since the directrix is perpendicular to the axis of symmetry, it needs to be perpendicular to the y-axis or parallel to the x-axis.
Now, suppose that the directrix is a line **y = d**. Then, the graph of the parabola will be similar to the one shown below.

Fig. 4.0

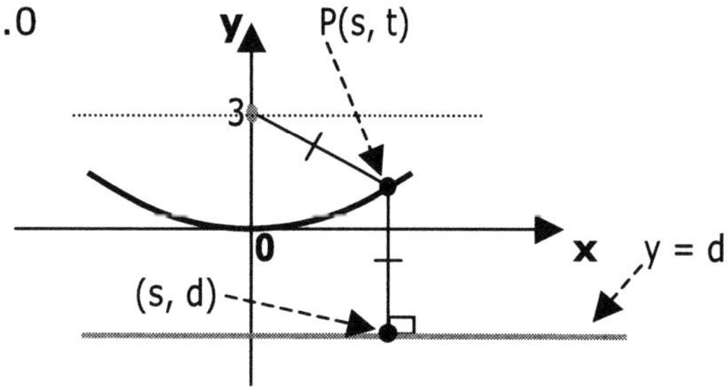

Then, since the vertex is the midpoint between the directrix and the focus:
$\mathbf{0 = (d + 3)/2 \Rightarrow d = -3}$.
Therefore, the directrix is a line **y = -3**.

Suppose now, that the general point of the parabola is **P(s, t)**. Then:
(the distance between the point **P** and the directrix)^2 = $\mathbf{(s - s)^2 + \{t - (-3)\}^2 = (t + 3)^2}$.
(the distance between the point **P** and the focus)^2 = $\mathbf{(s - 0)^2 + (t - 3)^2 = s^2 + (t - 3)^2}$.

Since the two distances are the same, $\mathbf{(t + 3)^2 = (t - 3)^2 + s^2}$
$\Rightarrow \mathbf{s^2 = (t + 3)^2 - (t - 3)^2 = (t + 3 + t - 3)(t + 3 - t + 3) = 2t(6) = 12t}$
$\Rightarrow \mathbf{s^2 = 12t}$.

Now, since **P(s, t)** is in the x-y system, the equation of the parabola is $\mathbf{x^2 = 12y}$.

Since the vertex is the origin and the focus is **(3, 0)**, **p = 3** in the standard equation $\mathbf{x^2 = 4py}$. If this is in our memory, we can find the solution immediately. We need to do memorizations, but keeping things in our memories is not an easy task.

In short:

Suppose that the directrix is a line **y = d**.
Then, $\mathbf{0 = (d + 3)/2 \Rightarrow d = -3}$.
Let the general point of the parabola be **P(s, t)**. Then:
$\mathbf{(s - s)^2 + \{t - (-3)\}^2 = (t + 3)^2 = (s - 0)^2 + (t - 3)^2 = s^2 + (t - 3)^2 \Rightarrow (t + 3)^2 = (t - 3)^2 + s^2}$
$\Rightarrow \mathbf{s^2 = (t + 3)^2 - (t - 3)^2 = (t + 3 + t - 3)(t + 3 - t + 3) = 2t(6) = 12t}$
$\Rightarrow \mathbf{s^2 = 12t}$.
Now, since **P(s, t)** is in the x-y system, the equation of the parabola is $\mathbf{x^2 = 12y}$.

Suggestions or Solutions
To the Problem in the Example 5

The vertex is (-1, 0), and the focus is (-1, 3).

We know that each point in a parabola is the same distance away from the focus and the directrix.
Besides, the vertex is the midpoint between the directrix and the focus, and the vertex and focus are in the axis of symmetry.
Since the vertex is **(-1, 0)** and the focus is **(-1, 3)**, the x-coordinates are the same to be -1, and therefore, the axis of symmetry is **x = -1**. So, since the directrix is perpendicular to the axis of symmetry, it needs to be perpendicular to the y-axis or parallel to the x-axis.
Now, suppose that the directrix is **y = d**. Then, the graph of the parabola will be similar to the one shown below.

Fig. 5.0

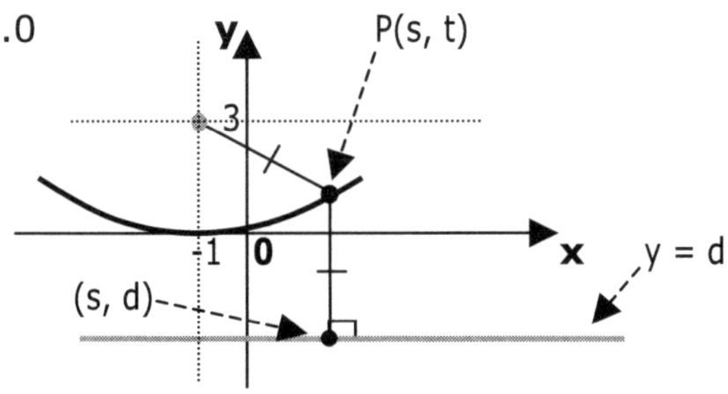

Then, since the vertex is the midpoint between the directrix and the focus:
$\mathbf{0 = (d + 3)/2 \Rightarrow d = -3}$.
Therefore, the directrix is $\mathbf{y = -3}$.

Suppose now, that the general point of the parabola is **P(s, t)**. Then:
(the distance between the point **P** and the directrix)^2 $= \mathbf{(s - s)^2 + \{t - (-3)\}^2 = (t + 3)^2}$.
(the distance between the point **P** and the focus)^2 $= \mathbf{\{s - (-1)\}^2 + (t - 3)^2 = (s + 1)^2 + (t - 3)^2}$.

Since the two distances are the same, $\mathbf{(t + 3)^2 = (t - 3)^2 + (s + 1)^2}$
$\mathbf{\Rightarrow (s + 1)^2 = (t + 3)^2 - (t - 3)^2 = (t + 3 + t - 3)(t + 3 - t + 3) = 2t(6) = 12t}$
$\mathbf{\Rightarrow (s + 1)^2 = 12t}$.

Now, since **P(s, t)** is in the x-y system, the equation of the parabola is $\mathbf{(x + 1)^2 = 12y}$.

In short:

Suppose that the directrix is **y = d**.
Then, $\mathbf{0 = (d + 3)/2 \Rightarrow d = -3}$.
Let the general point of the parabola be **P(s, t)**. Then:
$\mathbf{(s - s)^2 + \{t - (-3)\}^2 = (t + 3)^2 = \{s - (-1)\}^2 + (t - 3)^2 = (s + 1)^2 + (t - 3)^2}$
$\mathbf{\Rightarrow (t + 3)^2 = (t - 3)^2 + (s + 1)^2}$
$\mathbf{\Rightarrow (s + 1)^2 = (t + 3)^2 - (t - 3)^2 = (t + 3 + t - 3)(t + 3 - t + 3) = 2t(6) = 12t}$
$\mathbf{\Rightarrow (s + 1)^2 = 12t}$.

Now, since **P(s, t)** is in the x-y system, the equation of the parabola is $\mathbf{(x + 1)^2 = 12y}$.

Suggestions or Solutions

To the Problem in the Example 6

The vertex is (0, 1), and the focus is (0, 3).

We know that each point in a parabola is the same distance away from the focus and the directrix. Besides, the vertex is the midpoint between the directrix and the focus, and the vertex and focus are in the axis of symmetry.
Since the vertex is **(0, 1)** and the focus is **(0, 3)**, the x-coordinates are the same to be 0, and therefore, the axis of symmetry is $x = 0$, that is, the y-axis. So, since the directrix is perpendicular to the axis of symmetry, it needs to be perpendicular to the y-axis or parallel to the x-axis.
Now, suppose that the directrix is $y = d$. Then, the graph of the parabola will be similar to the one shown below.

Fig. 6.0

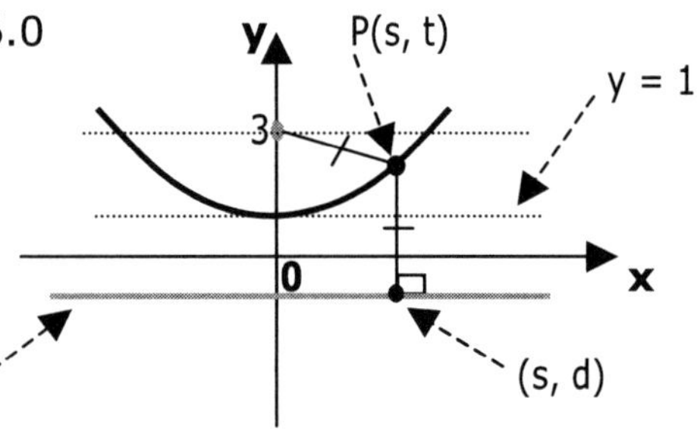

Then, since the vertex is the midpoint between the directrix and the focus:
$1 = (d + 3)/2 \Rightarrow d = -1$.
Therefore, the directrix is $y = -1$.

Suppose now, that the general point of the parabola is **P(s, t)**. Then:
(the distance between the point **P** and the directrix)^2 = $(s - s)^2 + \{t - (-1)\}^2 = (t + 1)^2$.
(the distance between the point **P** and the focus)^2 = $(s - 0)^2 + (t - 3)^2 = s^2 + (t - 3)^2$.

Since the two distances are the same, $(t + 1)^2 = (t - 3)^2 + s^2$
$\Rightarrow s^2 = (t + 1)^2 - (t - 3)^2 = (t + 1 + t - 3)(t + 1 - t + 3) = 2(t - 1)(4) = 8(t - 1)$
$\Rightarrow s^2 = 8(t - 1)$.

Now, since **P(s, t)** is in the x-y system, the equation of the parabola is $x^2 = 8(y - 1)$.

In short:

Suppose that the directrix is $y = d$.
Then, $1 = (d + 3)/2 \Rightarrow d = -1$.
Let the general point of the parabola be **P(s, t)**. Then:
$(s - s)^2 + \{t - (-1)\}^2 = (t + 1)^2 = (s - 0)^2 + (t - 3)^2 = s^2 + (t - 3)^2 \Rightarrow (t + 1)^2 = (t - 3)^2 + s^2$
$\Rightarrow s^2 = (t + 1)^2 - (t - 3)^2 = (t + 1 + t - 3)(t + 1 - t + 3) = 2(t - 1)(4) = 8(t - 1)$
$\Rightarrow s^2 = 8(t - 1)$.

Now, since **P(s, t)** is in the x-y system, the equation of the parabola is $x^2 = 8(y - 1)$.

Suggestions or Solutions
To the Problem in the Example 7

The vertex is (-1, 1), and the focus is (-1, 3).

We know that each point in a parabola is the same distance away from the focus and the directrix. Besides, the vertex is the midpoint between the directrix and the focus, and the vertex and focus are in the axis of symmetry.
Since the vertex is **(-1, 1)** and the focus is **(-1, 3)**, the x-coordinates are the same to be -1, and therefore, the axis of symmetry is $x = -1$. So, since the directrix is perpendicular to the axis of symmetry, it needs to be perpendicular to the y-axis or parallel to the x-axis.
Now, suppose that the directrix is $y = d$. Then, the graph of the parabola will be similar to the one shown below.

Fig. 7.0

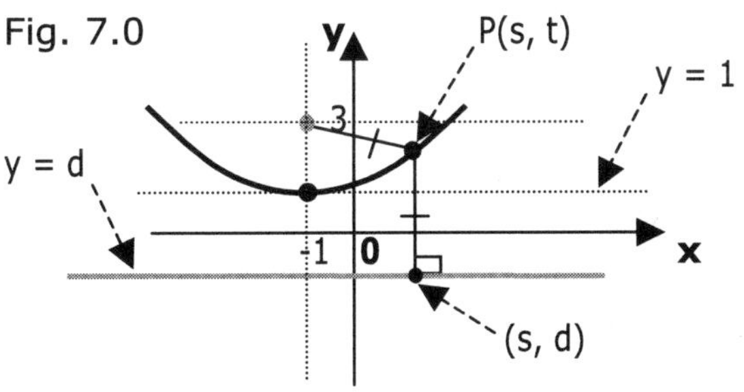

Then, since the vertex is the midpoint between the directrix and the focus:
$1 = (d + 3)/2 \Rightarrow d = -1$.
Therefore, the directrix is $y = -1$.

Suppose now, that the general point of the parabola is **P(s, t)**. Then:
(the distance between the point **P** and the directrix)^2 = $(s - s)^2 + \{t - (-1)\}^2 = (t + 1)^2$.
(the distance between the point **P** and the focus)^2 = $\{s - (-1)\}^2 + (t - 3)^2 = (s + 1)^2 + (t - 3)^2$.

Since the two distances are the same, $(t + 1)^2 = (t - 3)^2 + (s + 1)^2$
$\Rightarrow (s + 1)^2 = (t + 1)^2 - (t - 3)^2 = (t + 1 + t - 3)(t + 1 - t + 3) = 2(t - 1)(4) = 8(t - 1)$
$\Rightarrow (s + 1)^2 = 8(t - 1)$.

Now, since **P(s, t)** is in the x-y system, the equation of the parabola is $(x + 1)^2 = 8(y - 1)$.

In short:

Suppose that the directrix is $y = d$.
Then, $1 = (d + 3)/2 \Rightarrow d = -1$.
Let the general point of the parabola be **P(s, t)**. Then:
$(s - s)^2 + \{t - (-1)\}^2 = (t + 1)^2 = \{s - (-1)\}^2 + (t - 3)^2 = (s + 1)^2 + (t - 3)^2$
$\Rightarrow (t + 1)^2 = (t - 3)^2 + (s + 1)^2$
$\Rightarrow (s + 1)^2 = (t + 1)^2 - (t - 3)^2 = (t + 1 + t - 3)(t + 1 - t + 3) = 2(t - 1)(4) = 8(t - 1)$
$\Rightarrow (s + 1)^2 = 8(t - 1)$.

Now, since **P(s, t)** is in the x-y system, the equation of the parabola is $(x + 1)^2 = 8(y - 1)$.

Suggestions or Solutions
To the Problem in the Example 8

The vertex is (3, 1), and the directrix is x = -2.

We know that a parabola is a set of points, each of which is the same distance away from the focus and the directrix. Besides, the vertex is the midpoint between the directrix and the focus, and the vertex and focus are in the axis of symmetry.
Since the directrix is $\mathbf{x = -2}$, the axis of symmetry is parallel to the x-axis.
Since the axis of symmetry includes **(3, 1)**, it is a line $\mathbf{y = 1}$.
Since the vertex and focus are in the axis of symmetry, the y-coordinate of the focus is 1.
Now, suppose that the focus is **(f, 1)**. Then, the graph of the parabola will be similar to the one shown below.

Fig. 8.0

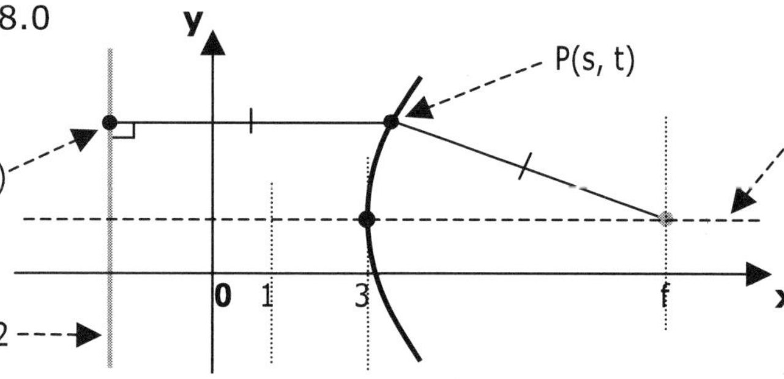

Then, since the vertex is the midpoint between the directrix and the focus:
$\mathbf{3 = (-2 + f)/2 \Rightarrow f = 8}$.
Therefore, the focus is **(8, 1)**.

Suppose now, that the general point of the parabola is **P(s, t)**. Then:
(the distance between the point **P** and the directrix)^2 = $\{s - (-2)\}^2 + (t - t)^2 = (s + 2)^2$.
(the distance between the point **P** and the focus)^2 = $(s - 8)^2 + (t - 1)^2 = (s - 8)^2 + (t - 1)^2$.

Since the two distances are the same, $(s + 2)^2 = (s - 8)^2 + (t - 1)^2$
$\Rightarrow (t - 1)^2 = (s + 2)^2 - (s - 8)^2 = (s + 2 + s - 8)(s + 2 - s + 8) = 2(s - 3)(10) = 20(s - 3)$
$\Rightarrow (t - 1)^2 = 20(s - 3)$.

Now, since **P(s, t)** is in the x-y system, the equation of the parabola is $(y - 1)^2 = 20(x - 3)$.

In short:

Suppose that the focus is **(f, 1)**.
Then, $\mathbf{3 = (-2 + f)/2 \Rightarrow f = 8}$.
Let the general point of the parabola be **P(s, t)**. Then:
$\{s - (-2)\}^2 + (t - t)^2 = (s + 2)^2 = (s - 8)^2 + (t - 1)^2 = (s - 8)^2 + (t - 1)^2$
$\Rightarrow (t - 1)^2 = (s + 2)^2 - (s - 8)^2 = (s + 2 + s - 8)(s + 2 - s + 8) = 2(s - 3)(10) = 20(s - 3)$
$\Rightarrow (t - 1)^2 = 20(s - 3)$.
Now, since **P(s, t)** is in the x-y system, the equation of the parabola is $(y - 1)^2 = 20(x - 3)$.

1.1. General Equations of Parabola

A general equation of parabola is an equation of parabola that is in the general form, which can indicate every parabola. Therefore, such an equation can indicate any parabola. Now, we put a right parabola in such a standard equation as $\mathbf{y^2 = 4px}$. In a right parabola, the axis of symmetry is parallel to one of the axes in the coordinate system. For instance, in $\mathbf{(y - 1)^2 = 4x}$, the axis of symmetry is $\mathbf{y = 1}$, which is parallel to the x-axis. The standard equation has the advantage that we can readily see the vertex, the focus, the directrix, and the roots. Of course, we can put a right parabola in a general equation of parabola, too. In addition, we can convert an equation in the general form to the one in the standard form, which is called the complete square method. Now, since the general equation can indicate not only a right parabola but a parabola where the axis of symmetry is slanted to a coordinate axis, we can present any parabola in such an equation.

A parabola is a set of points. What points, then? Suppose that a parabola is in the x-y system. Then, each of the points has a pair of coordinates, and when we express a parabola in general, we can put it in such a way as follows:
$\mathbf{P = \{(x, y) | \ ax^2 + by^2 + cxy + dx + ey + f = 0}$ where **a**, **b**, **c**, **d**, **e**, and **f** are constants, and $\mathbf{ab > 0}$; otherwise, **c** is 0, and also, ether of **a** and **b** is 0.**}**

Note: "$\mathbf{ab > 0}$" also means that the sings of the **a** and the **b** are the same.

The set **P** above is a comprehensive presentation of all parabolas. That is, any parabola is applicable. Therefore, we can say that a general equation of parabola is as follows:
$\mathbf{P(x, y) = ax^2 + by^2 + cxy + dx + ey + f = 0}$ or simply, $\mathbf{ax^2 + by^2 + cxy + dx + ey + f = 0}$.

However, such an equation merely indicates all parabolas in a comprehensive manner, and we don't actually use it frequently. In general math, we usually work with right parabolas. Then, let's have a look at such a set of points as follows: $\mathbf{A = \{(x, y) | \ ax^2 + by^2 + cx + dy + e = 0}$ where **a**, **b**, **c**, **d**, and **e** are constants, and either **a** or **b** is 0.**}**

The **A** indicates all right parabolas in a comprehensive manner. Now, the **A** can be said to indicate such two kinds of parabolas at the same time:
$\mathbf{B = \{(x, y) | \ ax^2 + bx + c + dy = 0}$ where **a**, **b**, **c**, and **d** are constants, and also, $\mathbf{ad \neq 0\}}$.
$\mathbf{C = \{(x, y) | \ ay^2 + by + c + dx = 0}$ where **a**, **b**, **c**, and **d** are constants, and also, $\mathbf{ad \neq 0\}}$.

Note: $\mathbf{ad \neq 0}$ also means that neither **a** nor **d** is 0.

Both the **B** and **C** are right parabolas, the axis of symmetry of the **B** is parallel to the y-axis, and that of the **C** is parallel to the x-axis. Therefore, in general math, we mainly use right parabolas, we can put in writing the general equations of right parabola in such a way as follows:
$\mathbf{B(x, y) = ax^2 + bx + c + dy = 0}$ or just, $\mathbf{ax^2 + bx + c + dy = 0}$.
$\mathbf{C(x, y) = ay^2 + by + c + dx = 0}$ or just, $\mathbf{ay^2 + by + c + dx = 0}$.

Now, we often use such a semi general form as follows: $\mathbf{y = ax^2 + bx + c}$. Such a form is convenient when we take a parabola for a curve of a function. For instance, $\mathbf{y = f(x) = 2x^2 + 3x + 5}$ and $\mathbf{y = g(x) = (x - 1)^{1/2}} = \sqrt{x-1}$. The curve of the function **f** is the same as that of the equation $\mathbf{y = 2x^2 + 3x + 5}$, and that of the function **g** is the same as that of the equation of a half parabola $\mathbf{y = (x - 1)^{1/2}} = \sqrt{x-1}$. This curve is half the curve of $\mathbf{y^2 = x - 1}$, which is the same as the part above the x-axis in the parabola shown below.

Fig. 1.1.0

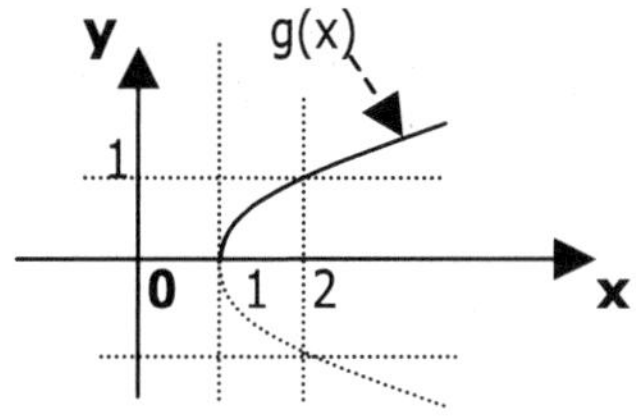

Also, if we set **d = 0** in the equation **B(x, y)** or **C(x, y)**, the resulting equality is a quadratic equation. That is, $ax^2 + bx + c = 0$ or $ay^2 + by + c = 0$. Such an equation is called an equation for an unknown. That is, we find the value of the **x** or **y** that makes 0 the value of the expression $ax^2 + bx + c$ or the expression $ay^2 + by + c$, and the **x** or **y** is called the unknown.

In general math, we often use such a semi general form, which is usually just called the general form. Then, let's find out about such a general form. It is used for the presentation of a right parabola. A right parabola is a parabola where the axis of symmetry is parallel to one of the two axes in the x-y system, that is, either the x-axis or the y-axis. Now, let's begin with the case where the axis of symmetry is parallel to the y-axis. That is, we are going to find out about the equation $y = ax^2 + bx + c$.

First of all, the left hand side of the equation $y = ax^2 + bx + c$ is composed of one term, and the right hand side is composed of three terms. The term with a variable is called a variable term, and a term with a constant only is called a constant term. Individually, the ax^2 is called the **x^2-term**, and the **bx** is called the **x-term**. Now, the constant with a variable is also often called a coefficient. Therefore, the coefficient of the x^2 is the **a**, and that of the **x** is the **b**.

As in the case of equations of line, an equation of a parabola, too, shows the slope and the intercepts. Of course, in case of lines, the standard equations are the case. Now, looking at the general equation of parabola $y = ax^2 + bx + c$, what we can readily see are the slope and the y-intercept. In the equation, the **a** which is the coefficient of the x^2 is the slope, and the **c** is the y-intercept. As in case of a line, the y-intercept **c** is the value of the y-coordinate at the point where the parabola meets the y-axis. However, the slope **a** is somewhat different from the one in case of a line. First, the slope **a** tells us the concavity of the parabola. That is, it says where the parabola is bent. For instance, in case where the axis of symmetry is parallel to the y-axis, if the a is positive, the parabola is concave up, and if negative, it is concave down. In addition, if the magnitude, that is, the absolute value of the **a** is big, the parabola is bent a lot.

Now, a slope in math says the direction of a particular point in a curve, which indicates the slope of a line including the particular point. That is, it says how the line is slanted to the x-axis, for instance. Since the directions of all the points in a line are the same, the direction of a point of them can be just called the slope of the line. However, all the points in such a bent curve as a parabola do not have the same direction. In fact, all the points in a parabola have different directions. What in the world is the slope, then? The precise definition of a slope is covered when we work on differential Calculus. Therefore, let's discuss the rigorous definition at that time. However, the property that a slope itself has was explained when we worked on lines. First of all, a slope says a slope at a point, which is the direction the point has. Next, suppose that a line is tangent to a curve. That is, without crossing the curve, the line meets the curve. Then, the line usually is tangent to the curve at a point. Now, the slope at the point in the curve is the same as that of the tangent line. Let's stop at this point, for now.

In a general equation of parabola, the coefficient of the x^2-term, that is, the **a** is called the slope of the parabola, but it can be said to indicate the concavity. Besides, if the magnitude is big, the parabola is bent a lot. The y-intercept is the value of the y-coordinate of the point where the parabola meets the y-axis, which is the value of the constant term, that is, the **c**. What about the x-intercept, then? Since the x-intercept is the value of the x-coordinate at the point where the parabola meets the x-axis, we need to set $\mathbf{y = 0}$ and solve $\mathbf{ax^2 + bx + c = 0}$ to find it. Then, is the **b** which is the coefficient of the x-term not much useful? It is not the case. It will turn out to be significant when we cover a determinant later.

We need to set to 0 the **y** in the general equation and solve the equation for unknown. The unknown is of course, the **x**, and the equation is called a quadratic equation. Then, we can find the roots by means of the quadratic formula. If we forget the formula, we can use the complete square method instead. That is, we convert the general equation to the standard equivalent: $\mathbf{ax^2 + bx + c = 0 \rightarrow a(x + p)^2 + q = 0}$. Then, solving the standard equation for the **x**, we can find the roots. It is in fact, no other than the derivation of the quadratic formula. Now, only if the parabola meets the x-axis, the x-intercept exists. If the parabola meets the x-axis, the meeting happens at either one or two points. By means of a determinant, we can see if the parabola has the x-intercept, that is, the root of the quadratic equation, and if it has, we can see how many it has. The determinant is $\mathbf{b^2 - 4ac}$, if it is negative, there is no root. If the determinant is 0, there is one root, which is called a double root, and the x-axis is tangent to the parabola. If the determinant is positive, there are two roots, that is, two x-intercepts exist.

Now, let's find out more specifically about the general equation. Then, we begin with the simplest parabola $\mathbf{y = x^2}$. It is the most basic parabola, which is called the prototype in the series of **Runmath**. We know the standard equation $\mathbf{x^2 = 4py}$. In the prototype $\mathbf{y = x^2}$, we have $\mathbf{p = 1/4}$. In this equation, we can readily see that the axis of symmetry is the y-axis, the focus is **(0, 1/4)**, and the directrix is a line $\mathbf{y = -1/4}$. Besides, since the slope is 1 which is positive, it is concave up. In this prototype, the **c** is 0, and so, the y-intercept is 0. Also, if we set $\mathbf{y = 0}$, we get $\mathbf{x^2 = 0}$, and so, the x-intercept is 0, too. Now, since both the x-intercept and y-intercept are 0, we can immediately see the vertex is the origin **(0, 0)**.

Fig.1.1.1 $y = x^2 \Leftrightarrow x^2 = 4(1/4)y$

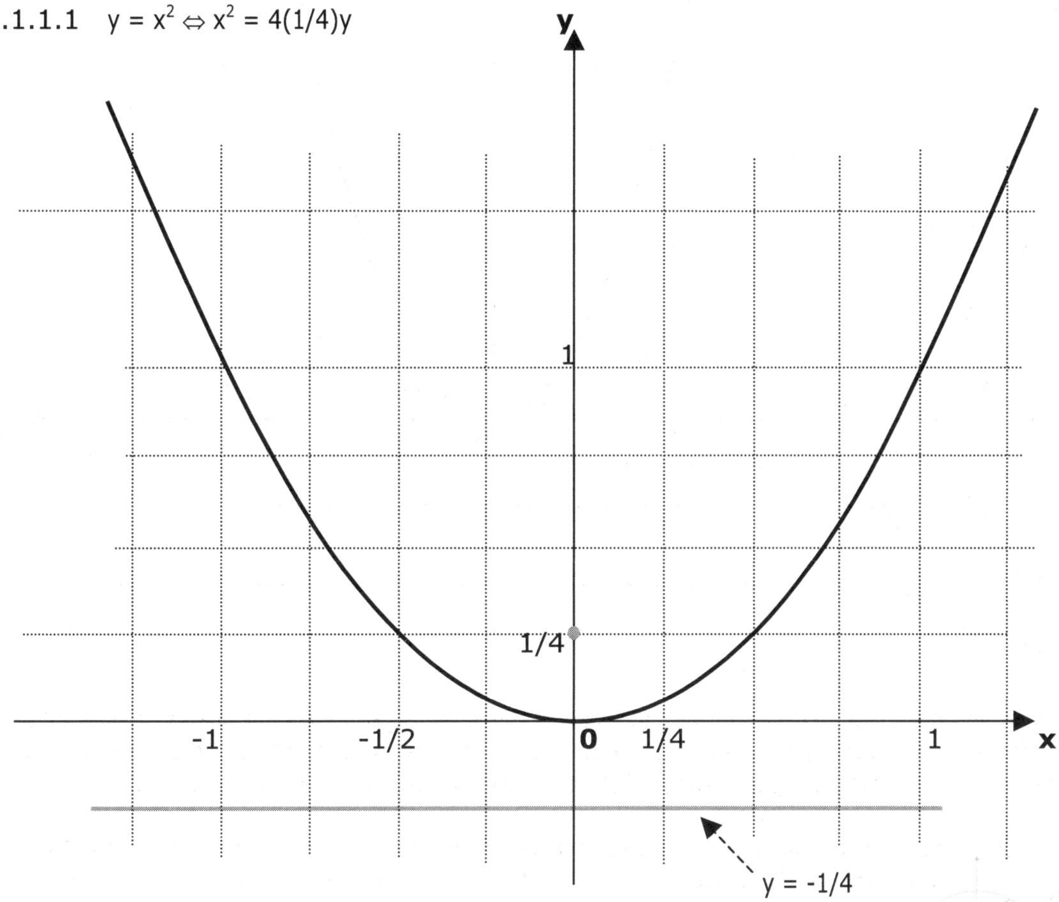

Now, let's see some parabolas in the form of $\mathbf{y = ax^2}$.
For instance, let's consider the parabolas as follows. Then:

$\mathbf{y = (1/2)x^2} \Rightarrow \mathbf{x^2 = 2y = 4(1/2)y} \Rightarrow$ focus = **(0, 1/2)**, directrix: **y = -1/2**, vertex = **(0, 0)**.
$\mathbf{y = 2x^2} \Rightarrow \mathbf{x^2 = (1/2)y = 4(1/8)y} \Rightarrow$ focus = **(0, 1/8)**, directrix: **y = -1/8**, vertex = **(0, 0)**.
$\mathbf{y = 4x^2} \Rightarrow \mathbf{x^2 = y/4 = 4(1/16)y} \Rightarrow$ focus = **(0, 1/16)**, directrix: **y = -1/16**, vertex = **(0, 0)**.

Now, we can see that as the slope **a** gets bigger, the focus and the directrix get closer to the vertex. Also, the parabola gets bent more.

Fig. 1.1.2

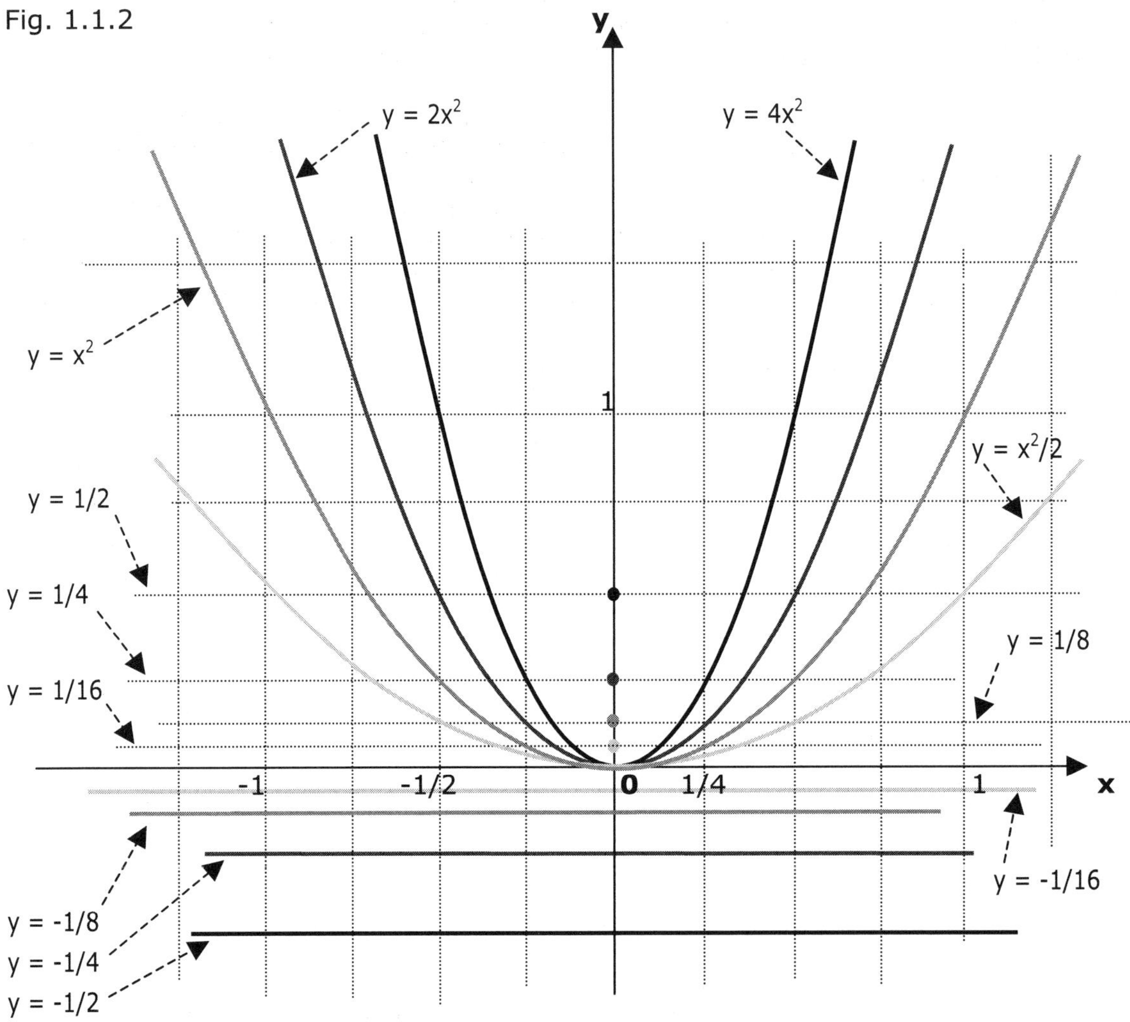

In the graph above, we can see that as the value of the slope a gets larger or smaller, how the positions of the focus and directrix get changed, and how much the parabola gets bent. If the slope **a** is positive, the parabola is concave up (open above). If the **a** is negative, the parabola needs to be concave down.

Suppose now, that we move the parabola $\mathbf{y = ax^2}$ by **m** along the x-axis and by **n** along the y-axis. That is, we translate the parabola. Then, the vertex, focus, etc. get moved in the same manner. Then, we get a new parabola, which is the same as the original parabola, and the position only has changed. The equation of the new parabola is as follows: $\mathbf{y - n = a(x - m)^2}$.

Fig. 1.1.3

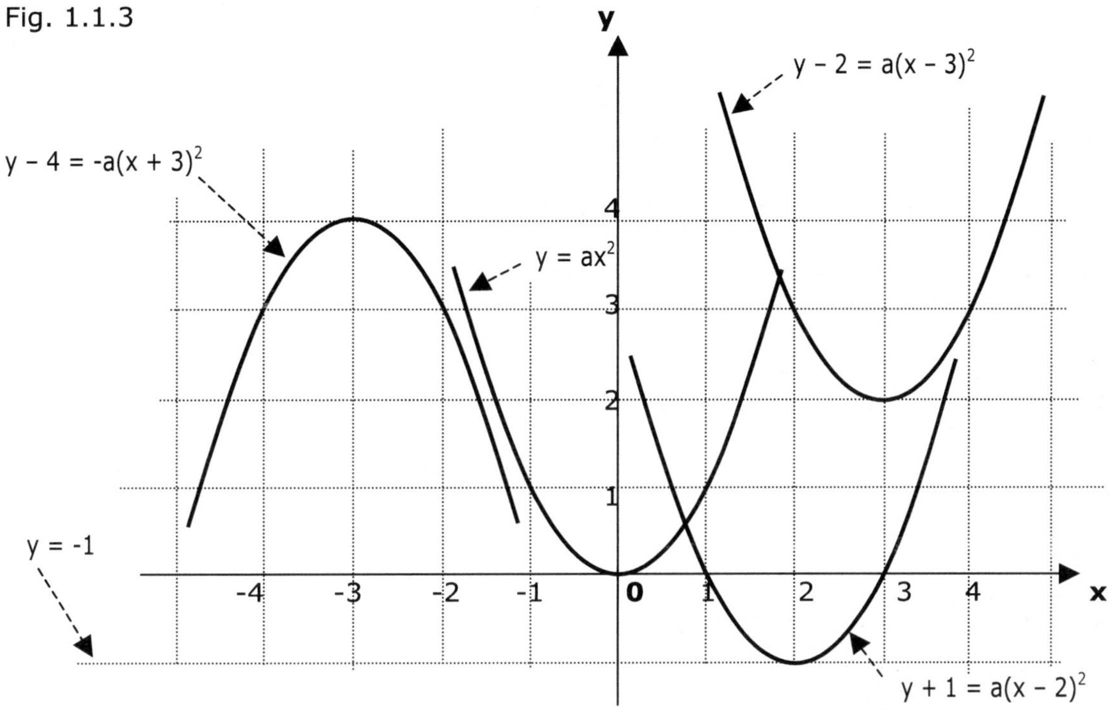

The same idea applies to the parabolas $\mathbf{x = ay^2}$ and $\mathbf{x - m = a(y - n)^2}$, too. The parabolas are concave sideways (left or right). The directrices are parallel to the y-axis.

$\mathbf{y = ax^2 \Rightarrow x^2 = (1/a)y = 4(1/4a)y \Rightarrow x^2 = 4(1/4a)y}$. Therefore:

$\mathbf{y = ax^2 \Rightarrow x^2 = 4(1/4a)y}$.

Then, in the parabola $\mathbf{y = ax^2}$, we have:

Vertex = $\mathbf{(0, 0)}$.
Focus = $\mathbf{(0, 1/4a)}$.
Directrix: $\mathbf{y = -1/4a}$.

Now, we have $\mathbf{y - n = a(x - m)^2 \Rightarrow (x - m)^2 = (1/a)(y - n) = 4(1/4a)(y - n)}$.
Therefore, we get $\mathbf{y - n = a(x - m)^2 \Rightarrow (x - m)^2 = 4(1/4a)(y - n)}$.

Then, since the entire parabola $\mathbf{y = ax^2}$ gets moved by **m** along the x-axis and by **n** along the y-axis, we get:

Vertex = $\mathbf{(0 + m, 0 + n) = (m, n)}$.
Focus = $\mathbf{(0 + m, n + 1/4a) = (m, 1/4a + n)}$.
Directrix: $\mathbf{y = -1/4a + n = n - 1/4a}$.

Now, if the equations are different, the shapes or the concavities are different, too?
Not necessarily! If the slopes are the same in the equations of parabola, the shapes and concavities are all the same.
For instance, the ones such as $\mathbf{y = x^2}$, $\mathbf{y = x^2 - x + 1}$, $\mathbf{y = x^2 + 2x + 5}$, $\mathbf{y = x^2 + x/3} + \sqrt{5}$, etc. have the same concavities and the shapes themselves are the same, too. In other words, the same parabolas are merely put in different positions.
However, we should not just take for the slope the coefficient of the x^2- term. We need to first, put the given equation in the general or standard form, and then, take for the slope the coefficient of the x^2-term.

Now, let's expand the equation $\mathbf{y - n = a(x - m)^2}$ and see the details.

$\mathbf{y - n = a(x - m)^2 = a(x^2 - 2mx + m^2) = ax^2 - 2amx + am^2}$.
$\Rightarrow \mathbf{y = ax^2 - 2amx + am^2 + n}$.

Set $\mathbf{-2am = b}$ and $\mathbf{am^2 + n = c}$, now.
The **a**, **m**, and **n** are constants. Therefore, **b** and **c** are constants, too.
Then, we get $\mathbf{y = ax^2 + bx + c}$.
Now, what we want to do is that converting an equation in the form of $\mathbf{y = ax^2 + bx + c}$ so that it can have the form of $\mathbf{y - n = a(x - m)^2}$.

$\mathbf{ax^2 + bx + c}$
$\mathbf{= a\{x^2 + (b/a)x\} + c}$
$\mathbf{= a\{x^2 + (b/a)x + (b/2a)^2 - (b/2a)^2\} + c}$
$\mathbf{= a\{x + (b/2a)\}^2 - a(b/2a)^2 + c}$
$\mathbf{= a\{x + (b/2a)\}^2 - b^2/4a + c}$
$\mathbf{= a\{x + (b/2a)\}^2 - (b^2 - 4ac)/4a}$

Then, we get $\mathbf{y + (b^2 - 4ac)/4a = a\{x + (b/2a)\}^2}$.

Let's now, set $\mathbf{n = -(b^2 - 4ac)/4a}$ and $\mathbf{m = -b/2a}$.
Then, we get $\mathbf{y - n = a(x - m)^2}$.

Then, we have the vertex = **(m, n)**. The axis of symmetry is parallel to the y-axis and includes the vertex.
Then, the axis of symmetry is a vertical line $\mathbf{x = m}$. Besides, we have $\mathbf{m = -b/2a}$.
Then, the axis of symmetry is the line $\mathbf{x = -b/2a}$.

Let's take some examples, now.

Examples
in General Equations of Parabola

Find the vertex, the focus and the directrix of each of the parabolas as follows:

0. $\mathbf{y = x^2 - x + 1}$.

1. $\mathbf{y = 2x^2 - 4x + 2}$ and $\mathbf{2x^2 - 3x + 2y - 1 = 0}$.

2. $\mathbf{x = 3y^2 - 6y + 3}$ and $\mathbf{2y^2 - 3x - 2y - 1 = 0}$.

Suggestions or Solutions
To the Problem in the Example 0

Find the vertex, the focus, and the directrix of the parabola as follows: $y = x^2 - x + 1$.

The equation given is in the general form. If we put the equation in the standard form, we can readily see the vertex, the focus, and the directrix. Now, let's try the conversion.

$x^2 - x + 1 = x^2 - x + (1/2)^2 - (1/2)^2 + 1 = (x - 1/2)^2 - 1/4 + 1 = (x - 1/2)^2 - 3/4.$

Then, we can see that:
$y = x^2 - x + 1 \Rightarrow y = (x - 1/2)^2 - 3/4 \Rightarrow y + 3/4 = (x - 1/2)^2 \Rightarrow y - (-3/4) = (x - 1/2)^2$
$\Rightarrow (x - 1/2)^2 = 4(1/4)\{y - (-3/4)\}.$

We know that the standard equations are as follows:
$x^2 = 4py$ and $(x - m)^2 = 4p(y - n)$.

In the parabola $x^2 = 4py$, the vertex $= (0, 0)$, the focus $= (0, p)$, and the directrix is $y = -p$.
Moving the parabola by **m** in the direction of the x-axis and by **n** in the direction of the y-axis, we get a parabola $(x - m)^2 = 4p(y - n)$.
So, the vertex $= (0 + m, 0 + n) = (m, n)$, the focus $= (0 + m, p + n) = (m, p + n)$, and the directrix is $y = -p + n = n - p$.

Now, we have $y = x^2 - x + 1 \Rightarrow (x - 1/2)^2 = 4(1/4)\{y - (-3/4)\}$.

Then, $p = 1/4$, $m = 1/2$, and $n = -3/4$. So:
The vertex $= (m, n) = (1/2, -3/4)$, the focus $= (m, p + n) = (1/2, 1/4 - 3/4) = (1/2, -1/2)$, and the directrix is $y = n - p = -3/4 - 1/4 = -1$.

Therefore, the vertex $= (1/2, -3/4)$, the focus $= (1/2, -1/2)$, and the directrix: $y = -1$.

In short:

$(x - m)^2 = 4p(y - n) \Rightarrow$ the vertex $= (m, n)$, the focus $= (m, p + n)$, and the directrix is $y = n - p$.

$y = x^2 - x + 1 = x^2 - x + (1/2)^2 - (1/2)^2 + 1 = (x - 1/2)^2 - 1/4 + 1 = (x - 1/2)^2 - 3/4.$
$\Rightarrow y = (x - 1/2)^2 - 3/4 \Rightarrow y + 3/4 = (x - 1/2)^2 \Rightarrow y - (-3/4) = (x - 1/2)^2$
$\Rightarrow (x - 1/2)^2 = 4(1/4)\{y - (-3/4)\}.$

Therefore, $y = x^2 - x + 1 \Rightarrow (x - 1/2)^2 = 4(1/4)\{y - (-3/4)\}$.
So, $p = 1/4$, $m = 1/2$, $n = -3/4$. Then:

The vertex $= (m, n) = (1/2, -3/4)$, the focus $= (0 + m, p + n) = (1/2, 1/4 - 3/4) = (1/2, -1/2)$, and the directrix is $y = n - p = -3/4 - 1/4 = -1$.

Therefore, the vertex $= (1/2, -3/4)$, the focus $= (1/2, -1/2)$, and the directrix: $y = -1$.

Fig. 0.0 $y = x^2 - x + 1 \Leftrightarrow (x - 1/2)^2 = 4(1/4)\{y - (-3/4)\}$

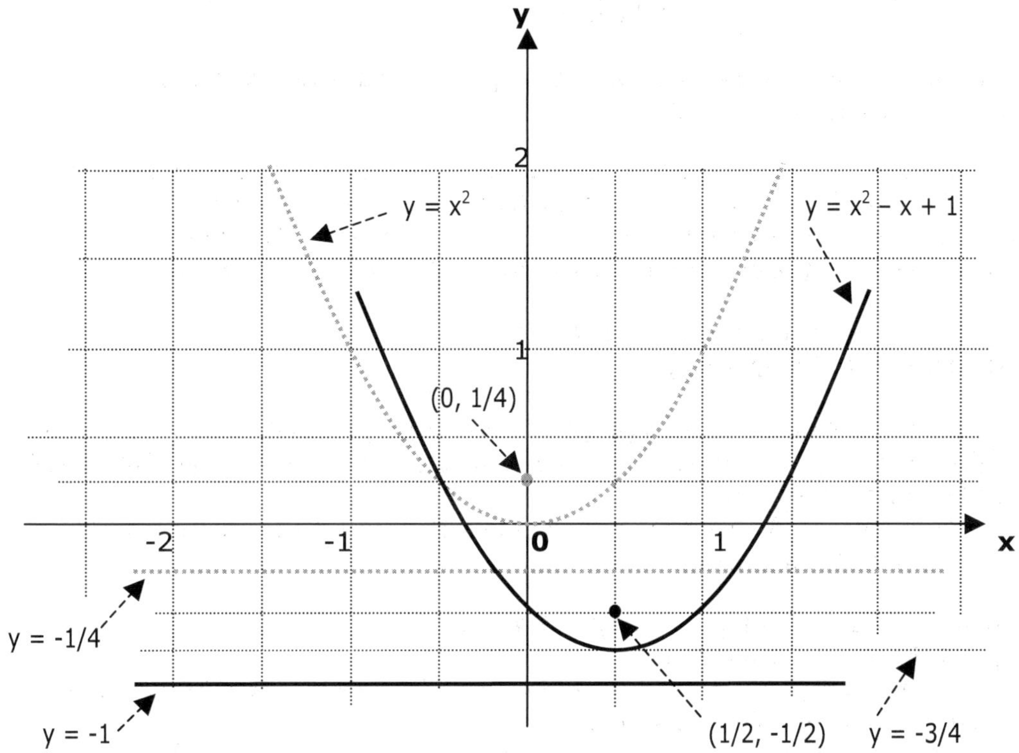

Suggestions or Solutions
To the Problem in the Example 1

Find the vertices, the focuses, and the directrices of the parabolas as follows:
$y = 2x^2 - 4x + 2$ and $2x^2 - 3x + 2y - 1 = 0$.

The equations given are in the general form. If we put the equations in the standard form, we can readily see the vertices, the focuses, and the directrices. Now, let's try the conversion.
$y = 2x^2 - 4x + 2 = 2(x^2 - 2x + 1) = 2(x - 1)^2 \Rightarrow (x - 1)^2 = y/2 = 4(1/8)y$.
Then, we can see that: $y = 2x^2 - 4x + 2 \Rightarrow y = 2(x - 1)^2 \Rightarrow (x - 1)^2 = 4(1/8)y$.

We know that the standard equations are as follows:
$x^2 = 4py$ and $(x - m)^2 = 4p(y - n)$.

In the parabola $x^2 = 4py$, the vertex = $(0, 0)$, the focus = $(0, p)$, and the directrix is $y = -p$.
Moving the parabola by **m** in the direction of the x-axis and by **n** in the direction of the y-axis, we get a parabola $(x - m)^2 = 4p(y - n)$.
So, the vertex = $(0 + m, 0 + n) = (m, n)$, the focus = $(0 + m, p + n) = (m, p + n)$, and the directrix is $y = -p + n = n - p$.

Now, we have $y = 2x^2 - 4x + 2 \Rightarrow (x - 1)^2 = 4(1/8)y$.
Then, $p = 1/8$, $m = 1$, and $n = 0$. So:
The vertex = $(m, n) = (1, 0)$, the focus = $(0 + m, p + n) = (0 + 1, 1/8 + 0) = (1, 1/8)$, and the directrix is $y = n - p = 0 - 1/8 = -1/8$.
Therefore, the vertex = $(1, 0)$, the focus = $(1, 1/8)$, and the directrix: $y = -1/8$.

Fig. 1.0

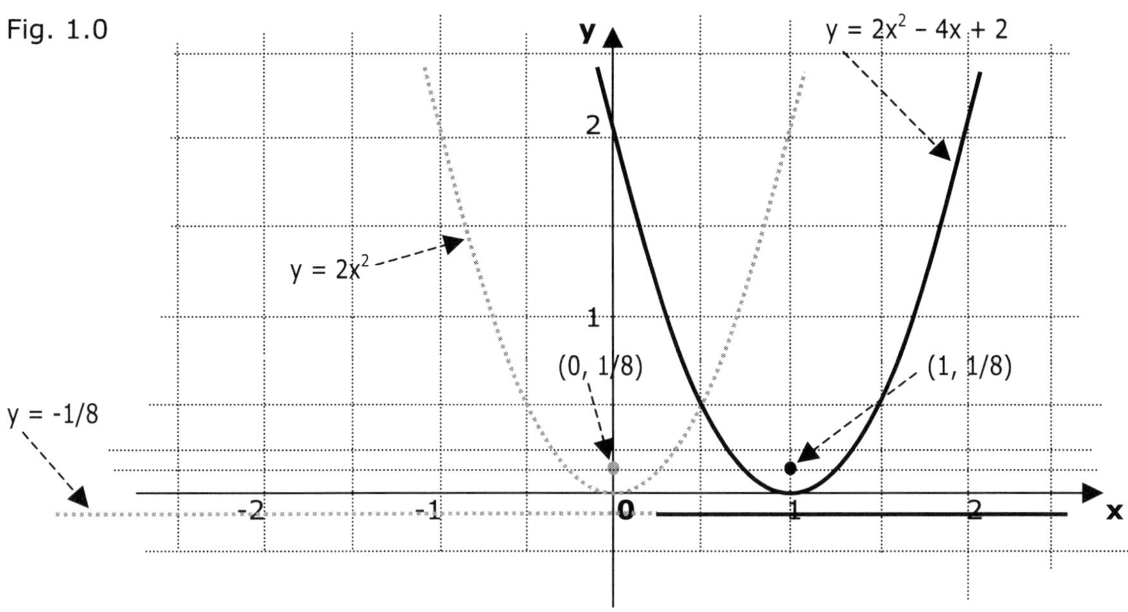

Now, let's move on to the next parabola: $2x^2 - 3x + 2y - 1 = 0$.
Then, $2x^2 - 3x + 2y - 1 = 0 \Rightarrow 2y = -2x^2 + 3x + 1 \Rightarrow y = -x^2 + 3x/2 + 1/2$.
$-x^2 + 3x/2 + 1/2 = -(x^2 - 3x/2) + 1/2 = -\{x^2 - 3x/2 + (3/4)^2 - (3/4)^2\} + 1/2$
$= -(x - 3/4)^2 + (3/4)^2 + 1/2 = -(x - 3/4)^2 + 17/16$.

So, $\mathbf{2x^2 - 3x + 2y - 1 = 0 \Rightarrow y = -(x - 3/4)^2 + 17/16 \Rightarrow y - 17/16 = -(x - 3/4)^2}$
$\mathbf{\Rightarrow (x - 3/4)^2 = -(y - 17/16) = 4(-1/4)(y - 17/16)}$.

Then, we have: $\mathbf{2x^2 - 3x + 2y - 1 = 0 \Rightarrow (x - 3/4)^2 = 4(-1/4)(y - 17/16)}$.

We know that the standard equations are as follows: $\mathbf{x^2 = 4py}$ and $\mathbf{(x - m)^2 = 4p(y - n)}$.
In the parabola $\mathbf{x^2 = 4py}$, the vertex = **(0, 0)**, the focus = **(0, p)**, and the directrix is **y = -p**.
Moving the parabola by **m** in the direction of the x-axis and by **n** in the direction of the y-axis, we get a parabola $\mathbf{(x - m)^2 = 4p(y - n)}$.
So, the vertex = **(0 + m, 0 + n) = (m, n)**, the focus = **(0 + m, p + n) = (m, p + n)**, and the directrix is **y = -p + n = n - p**.

Now, we have $\mathbf{2x^2 - 3x + 2y - 1 = 0 \Rightarrow (x - 3/4)^2 = 4(-1/4)(y - 17/16)}$.
Then, **p = -1/4**, **m = 3/4**, and **n = 17/16**. So, we get:
The vertex = **(m, n) = (3/4, 17/16)**, the focus = **(0 + m, p + n) = (0 + 3/4, -1/4 + 17/16) = (3/4, 13/16)**, and the directrix is **y = n - p = 17/16 + 1/4 = 21/16**.
Therefore, the vertex = **(3/4, 17/16)**, the focus = **(3/4, 13/16)**, and the directrix: **y = 21/16**.

Fig. 1.1 $2x^2 - 3x + 2y - 1 = 0 \Leftrightarrow (x - 3/4)^2 = 4(-1/4)(y - 17/16)$

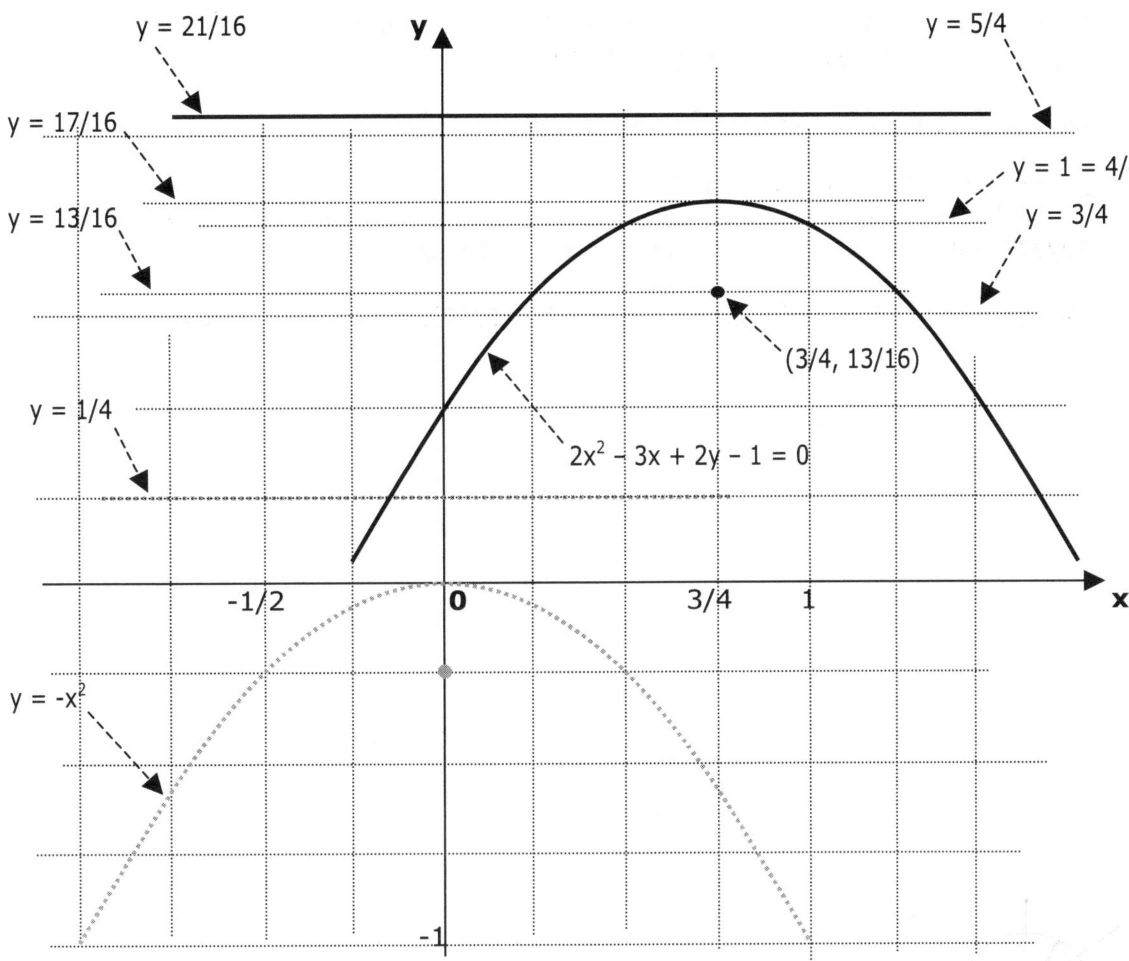

In short:

$\mathbf{(x - m)^2 = 4p(y - n)} \Rightarrow$ the vertex = **(m, n)**, the focus = **(m, p + n)**, and the directrix is **y = n − p**.

$\mathbf{y = 2x^2 - 4x + 2 = 2(x^2 - 2x + 1) = 2(x - 1)^2 \Rightarrow (x - 1)^2 = y/2 = 4(1/8)y.}$

Then, $\mathbf{y = 2x^2 - 4x + 2 \Rightarrow (x - 1)^2 = 4(1/8)y}$.

Then, we have **p = 1/8**, **m = 1**, and **n = 0**.
So, the vertex = **(m, n) = (1, 0)**, the focus = **(0 + m, p + n) = (0 + 1, 1/8 + 0) = (1, 1/8)**, and the directrix is **y = n − p = 0 − 1/8 = -1/8**.

Therefore, the vertex = **(1, 0)**, the focus = **(1, 1/8)**, and the directrix: **y = -1/8**.

Next, we have $\mathbf{2x^2 - 3x + 2y - 1 = 0}$.
Then, $\mathbf{2x^2 - 3x + 2y - 1 = 0 \Rightarrow 2y = -2x^2 + 3x + 1 \Rightarrow y = -x^2 + 3x/2 + 1/2}$.

$\mathbf{-x^2 + 3x/2 + 1/2 = -(x^2 - 3x/2) + 1/2 = -\{x^2 - 3x/2 + (3/4)^2 - (3/4)^2\} + 1/2}$
$\mathbf{= -(x - 3/4)^2 + (3/4)^2 + 1/2 = -(x - 3/4)^2 + 17/16}$.

So, $\mathbf{2x^2 - 3x + 2y - 1 = 0 \Rightarrow y = -(x - 3/4)^2 + 17/16 \Rightarrow y - 17/16 = -(x - 3/4)^2}$
$\mathbf{\Rightarrow (x - 3/4)^2 = -(y - 17/16) = 4(-1/4)(y - 17/16)}$.

Then, $\mathbf{2x^2 - 3x + 2y - 1 = 0 \Rightarrow (x - 3/4)^2 = 4(-1/4)(y - 17/16)}$.

Then, **p = -1/4**, **m = 3/4**, and **n = 17/16**. So, we get:
The vertex = **(m, n) = (3/4, 17/16)**, the focus = **(0 + m, p + n) = (0 + 3/4, -1/4 + 17/16) = (3/4, 13/16)**, and the directrix is **y = n − p = 17/16 + 1/4 = 21/16**.

Therefore, the vertex = **(3/4, 17/16)**, the focus = **(3/4, 13/16)**, and the directrix: **y = 21/16**.

Suggestions or Solutions
To the **Problem** in the Example **2**

Find the vertices, the focuses, and the directrices of the parabolas as follows:
$x = 3y^2 - 6y + 3$ and $2y^2 + x - y - 1 = 0$.

Putting the equations in the standard form, we can readily see the vertices, the focuses, and the directrices. Now, let's try the conversion.

$x = 3y^2 - 6y + 3 = 3(y^2 - 2y + 1) = 3(y - 1)^2 \Rightarrow (y - 1)^2 = x/3$.

Then, we have: **$x = 3y^2 - 6y + 3 \Rightarrow (y - 1)^2 = 4(1/12)x$.**

We know that the standard equations are as follows: **$y^2 = 4px$** and **$(y - n)^2 = 4p(x - m)$.**

In the parabola **$y^2 = 4px$**, the vertex **= (0, 0)**, the focus **= (p, 0)**, and the directrix is **$x = -p$**.
Moving the parabola by **m** in the direction of the x-axis and by **n** in the direction of the y-axis, we get a parabola **$(y - n)^2 = 4p(x - m)$**.
So, the vertex **= (0 + m, 0 + n) = (m, n)**, the focus **= (p + m, n) = (p + m, n)**, and the directrix is **$x = -p + m = m - p$**.

Now, we have **$x = 3y^2 - 6y + 3 \Rightarrow (y - 1)^2 = 4(1/12)x$.**

Then, **$p = 1/12$, $m = 0$**, and **$n = 1$**.
So, we get the vertex **= (m, n) = (0, 1)**, the focus **= (p + m, n) = (1/12 + 0, 1) = (1/12, 1)**, and the directrix is **$x = m - p = 0 - 1/12 = -1/12$**.

Therefore, the vertex **= (0, 1)**, the focus **= (1/12, 1)**, and the directrix: **$x = -1/12$**.

Since the directrix is parallel to the y-axis and the slope is positive, the parabola is concave right.

In short:

$(y - n)^2 = 4p(x - m)$ $\Rightarrow$ the vertex **= (m, n)**, the focus **= (p + m, n)**, and the directrix is **$x = m - p$**.

$x = 3y^2 - 6y + 3 = 3(y^2 - 2y + 1) = 3(y - 1)^2 \Rightarrow (y - 1)^2 = x/3$.

Then, **$x = 3y^2 - 6y + 3 \Rightarrow (y - 1)^2 = 4(1/12)x$.**

Then, **$p = 1/12$, $m = 0$**, and **$n = 1$**.
So, we get the vertex **= (m, n) = (0, 1)**, the focus **= (p + m, n) = (1/12 + 0, 1) = (1/12, 1)**, and the directrix is **$x = m - p = 0 - 1/12 = -1/12$**.

Therefore, the vertex **= (0, 1)**, the focus **= (1/12, 1)**, and the directrix: **$x = -1/12$**.

Fig. 2.0

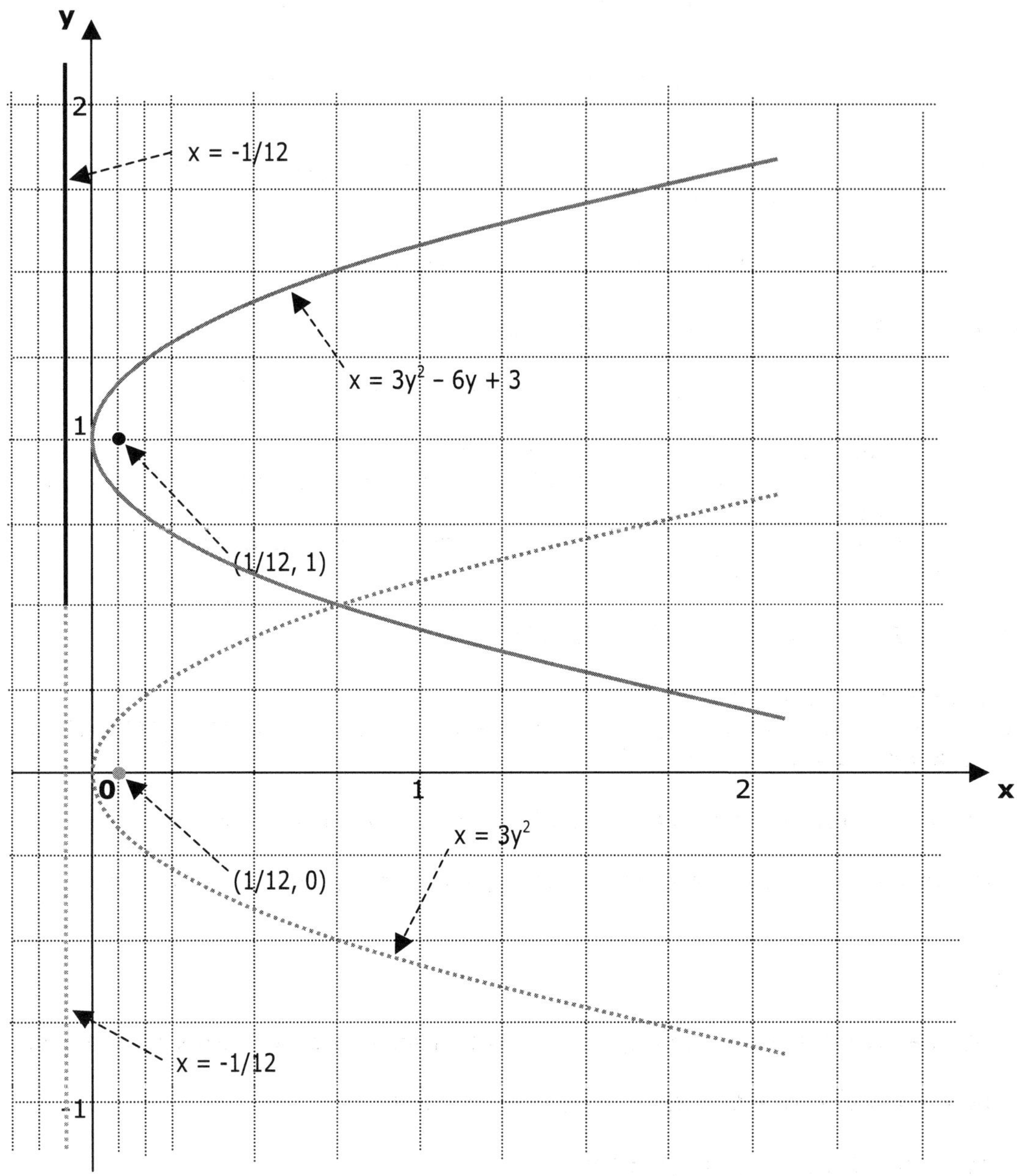

Next, we have $2y^2 + x - y - 1 = 0$.

Then, $2y^2 + x - y - 1 = 0 \Rightarrow x = -2y^2 + y + 1$.

$-2y^2 + y + 1 = -2(y^2 - y/2) + 1 = -2\{y^2 - y/2 + (1/4)^2 - (1/4)^2\} + 1$
$= -2(y - 1/4)^2 + 2(1/16) + 1 = -2(y - 1/4)^2 + 9/8$.

Then, $2y^2 + x - y - 1 = 0 \Rightarrow x = -2(y - 1/4)^2 + 9/8 \Rightarrow x - 9/8 = -2(y - 1/4)^2$
$\Rightarrow (y - 1/4)^2 = (-1/2)(x - 9/8) = 4(-1/8)(x - 9/8)$.

Then, we have: $2y^2 + x - y - 1 = 0 \Rightarrow (y - 1/4)^2 = 4(-1/8)(x - 9/8)$.

We know that the standard equations are as follows: $y^2 = 4px$ and $(y - n)^2 = 4p(x - m)$.
In the parabola $y^2 = 4px$, the vertex $= (0, 0)$, the focus $= (p, 0)$, and the directrix is $x = -p$.
Moving the parabola by **m** in the direction of the x-axis and by **n** in the direction of the y-axis, we get a parabola $(y - n)^2 = 4p(x - m)$.
So, the vertex $= (0 + m, 0 + n) = (m, n)$, the focus $= (p + m, n) = (p + m, n)$, and the directrix is $x = -p + m = m - p$.

Now, we have $2y^2 + x - y - 1 = 0 \Rightarrow (y - 1/4)^2 = 4(-1/8)(x - 9/8)$.
Then, we have $p = -1/8$, $m = 9/8$, and $n = 1/4$. So:
The vertex $= (m, n) = (9/8, 1/4)$, the focus $= (p + m, n) = (-1/8 + 9/8, 1/4) = (1, 1/4)$, and the directrix is $x = m - p = 9/8 + 1/8 = 5/4$.
Therefore, the vertex $= (9/8, 1/4)$, the focus $= (1, 1/4)$, and the directrix: $x = 5/4$.

Now, since the directrix is parallel to the y-axis and the slope is –2 which is negative, the parabola is concave left.

In short:

$(y - n)^2 = 4p(x - m) \Rightarrow$ the vertex $= (m, n)$, the focus $= (p + m, n)$, and the directrix is $x = m - p$.

Then, $2y^2 + x - y - 1 = 0 \Rightarrow x = -2y^2 + y + 1$.

$-2y^2 + y + 1 = -2(y^2 - y/2) + 1 = -2\{y^2 - y/2 + (1/4)^2 - (1/4)^2\} + 1$
$= -2(y - 1/4)^2 + 2(1/16) + 1 = -2(y - 1/4)^2 + 9/8$.

Then, $2y^2 + x - y - 1 = 0 \Rightarrow x = -2(y - 1/4)^2 + 9/8 \Rightarrow x - 9/8 = -2(y - 1/4)^2$
$\Rightarrow (y - 1/4)^2 = (-1/2)(x - 9/8) = 4(-1/8)(x - 9/8)$.

Then, $2y^2 + x - y - 1 = 0 \Rightarrow (y - 1/4)^2 = 4(-1/8)(x - 9/8)$.

Then, we have $p = -1/8$, $m = 9/8$, and $n = 1/4$. So:
The vertex $= (m, n) = (9/8, 1/4)$, the focus $= (p + m, n) = (-1/8 + 9/8, 1/4) = (1, 1/4)$, and the directrix is $x = m - p = 9/8 + 1/8 = 5/4$.

Therefore, the vertex $= (9/8, 1/4)$, the focus $= (1, 1/4)$, and the directrix: $x = 5/4$.

Fig. 2.1

$x = -2y^2 \Leftrightarrow y^2 = 4(-1/8)x$

$2y^2 + x - y - 1 = 0 \Leftrightarrow x = -2y^2 + y + 1 \Leftrightarrow (y - 1/4)^2 = 4(-1/8)(x - 9/8)$

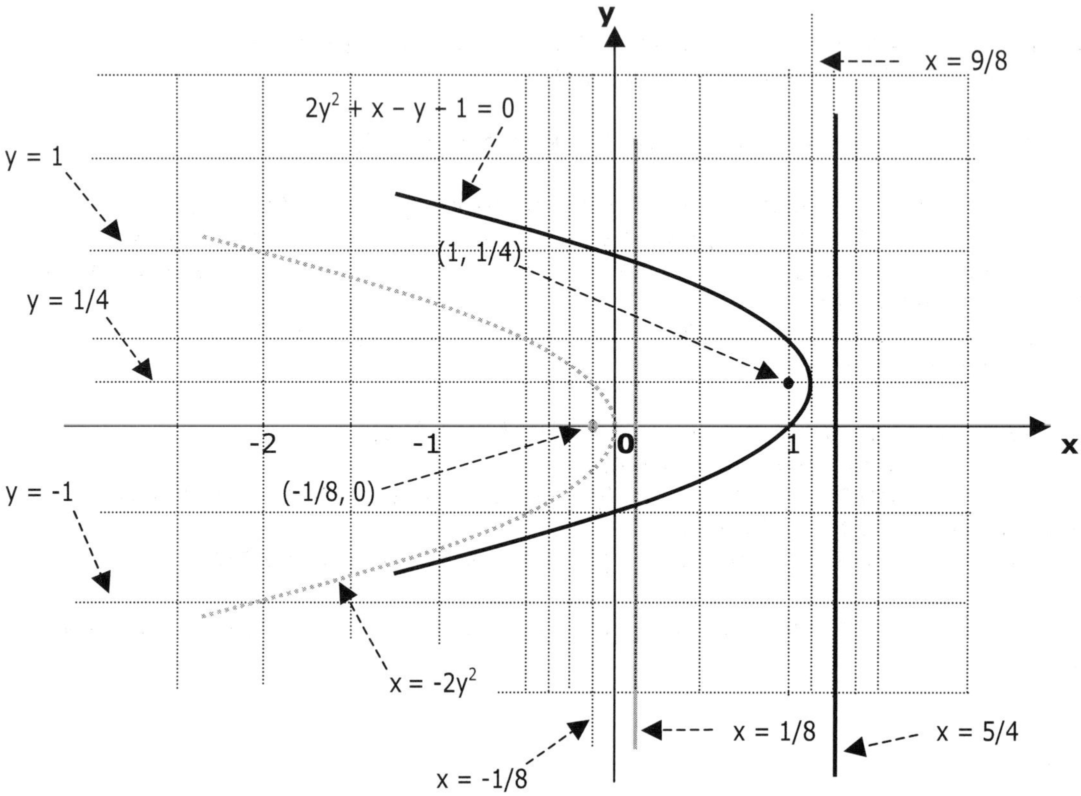

1.2. What do we need to find an equation of a parabola?

Working on problems, we often have a case where we need to find an equation of a curve such as a line or parabola. In other words, we need to construct the equation of such a curve. Of course, we can't just construct it. Then, what do we construct it with? The solution to a problem is inside the problem. Therefore, we need to extract the necessary fact from the description of the problem. So, we need to first, understand the problem properly. Sometime, such a fact is explicit in the definition of the problem, but usually, a part or the whole of such is hidden. Some other times, such a fact is implicit. If such is not explicit, we need to reason to find it. Then, let's find out specifically what fact we need to sort out. First, let's begin with necessary fact to construct an equation of a line.

Constructing an equation of a line, we need either of the two kinds of fact as follows:

Two points in the line
One point in the line and the slope of the line

With either kind of fact above, we may use such a standard form as follows: $\mathbf{y = ax + b}$.
Then, we need to find the **a** and **b**. That is, the **a** and **b** are the unknowns. Then, since we have two unknowns, we need two pieces of information. Now, each of the two kinds of fact above is composed of two pieces of information. One is composed of two points, that is, two pairs of coordinates, and the other is composed of a point and the slope. Therefore, we can set up two equations. They are of course, the equations for the **a** and **b**.

Now, what do we need to construct an equation of a parabola?
We need to have either of the three kinds of fact as follows:

Three points in the parabola
Two points in the parabola and another piece of information such as the slope, the y-intercept, etc.
The vertex and the slope of the parabola

In case where we have either of the first two kinds, we may use either of the two forms as follows:

$\mathbf{y = ax^2 + bx + c}$
$\mathbf{x = ay^2 + by + c}$

In fact, to choose either of the two forms above, we need another piece of information, which tells us the orientation of the axis of symmetry or the directrix. For instance, we need to know if the directrix is parallel to the x-axis.

Now, we need to find the **a**, **b**, and **c**. That is, the **a**, **b**, and **c** are the unknowns. Then, since we have three unknowns, we need three pieces of information. Now, each of the two kinds of fact is composed of three pieces of information. One kind is composed of three points, that is, three pairs of coordinates, and the other is composed of two points and the slope. Therefore, we can set up three equations. They are of course, the equations for the **a**, **b**, and **c**.

Next, in cases where we know the slope and the vertex, we can use either of the two forms as follows:
$\mathbf{y - n = a(x - m)^2}$ and $\mathbf{x - m = a(y - n)^2}$.

However, we have three unknowns, don't we? Of course, we do.
Then, we need to have another piece of information, don't we? No, we don't.
How come? The vertex is **(m, n)**, and therefore, we need to find in fact, one unknown only, which is the slope **a**.

Besides, to choose either of the two forms, we need another piece of information, which tells us the orientation of the axis of symmetry or the directrix. For instance, we need to know if the axis of symmetry is perpendicular to the x-axis.

Now, when we work on problems with finding equations of lines or parabolas, we should be able to somehow extract such information specified as above from the problem descriptions.

Let's do some examples now.

Examples
in "What do we need to find an equation of a parabola?"

0. Suppose that a parabola includes three points P_1**(1, 6)**, P_2**(-1, 4)**, and P_3**(-2, 9)**, and the directrix is parallel to the x-axis. Then, find the equation of the parabola.

1. Suppose that a parabola includes three points P_1**(1, 6)**, P_2**(-1, 4)**, and P_3**(-2, 9)**, and the axis of symmetry is parallel to the x-axis. Then, find the equation of the parabola.

2. Suppose that we have an equation of a parabola $y = 2x^2 + bx + c$, and that the parabola includes two points where another parabola $y = -2x^2 + 3x/2 + 9/4$ meets a line $y = x/2 - 3/4$. Then, find the values of the **b** and **c**.

3. Suppose that we have an equation of a parabola $y = ax^2 - x/2 + c$, and that the parabola includes two points where another parabola $y = -2x^2 + 3x/2 + 9/4$ meets a line $y = x/2 - 3/4$. Then, find the values of the **a** and **c**.

4. Suppose that we have an equation of a parabola $x = ay^2 - y/2 + c$, and that the parabola includes two points where another parabola $y = 2x^2 + x - 3$ meets a line $y = 3x + 1$. Then, find the values of the **a** and **c**.

5. Suppose a parabola $y/2 = ax^2 - 2x + c/2$ includes a point at **(1/2, -1/2)**, and a line $y = x - 1$ includes the vertex. Then, find the values of the **a** and **c**.

6. Suppose that a parabola $y = 6x^2 + bx + c$ includes a point at **(2, 2)**, and that a line $y = x - 1$ includes the vertex. Then, find the values of the **b** and **c**.

7. Suppose that a parabola $y = ax^2 + bx + c$ includes two points **(1, 1)** and **(2, 3)**, and that the vertex is in a line $y = 2x - 1$. Then, find the values of the **a**, **b**, and **c**.

8. Suppose that a parabola $y = ax^2 + bx + c$ includes two points **(1, 1)** and **(2, 4)**, and has the vertex in a line $y = 2x - 1$. Then, find the values of the **a**, **b**, and **c**.

9. Suppose that the equation of a parabola **A** is $y = -4x^2 + 8mx - n(n + 2)$, and that the equation of a parabola **B** is $y = 8x^2 - 2x + 5/4$. Suppose also, that the vertex of the **A** is in the **B**. Then, find the values of the **m** and **n**.

A. Suppose that the equation of a parabola **A** is $y = x^2 + bx + c$, and that the equation of a parabola **B** is $y = -2x^2 - x + 5$. Suppose also, that the vertex of the **A** is in the **B**, and the vertex of the **B**, too, is in the **A**. Then, find the values of the **b** and **c**.

B. Suppose that the equation of a parabola **A** is $y = 2x^2 + bx + c$, and that the equation of a parabola **B** is $y = -2x^2 + 8x - 5$. Suppose also, that the vertex of the **A** is in the **B**, and the vertex of the **B**, too, is in the **A**. Then, find the values of the **b** and **c**.

C. Suppose that the equation of a parabola **A** is $y = x^2 + bx + c$, and that the equation of a parabola **B** is $y = -2x^2 + 8x - 5$. Suppose also, that the vertex of the **A** is in the **B**, and the vertex of the **B**, too, is in the **A**. Then, find the values of the **b** and **c**.

D. In the next several examples, we are going to do some practices on graphing equations of parabola.

Note:
The pair of vertical bars in the expressions below is called an absolute sign. Taking care of expressions where absolute signs exist, we need to remove the sign, and then, proceed. Now, before we remove the sign, we need to consider two cases. One is the case where what's inside the absolute sign is greater than or equal to 0. The other is the case where what's inside the absolute sign is less than 0. Then, in the case where it is less than 0, we need to remove the absolute sign, and then, add a minus sign (-) to it. For instance:

Suppose that $t = |s|$ for **s** that is a real number. Then:
If $s \geq 0$, then $t = s$; that is, $|s| = s$. If $s < 0$, then $t = -s$; that is, $|s| = -s$ since $|s|$ is positive.

Now, construct the graphs of the equations as follows.

0. $y = 2(x - 1)^2 - 2|x - 2| + 1$

1. $y = (|2x| - 1)(x + 1)$

2. $y = |x^2 + 2x - 3|$

3. $y = |x|^2 + 2|x| - 3$

4. $|y| = x^2 + 2x - 3$

5. $|y| = |x|^2 + 2|x| - 3$

6. $|y| = x^2 + |2x - 3|$

7. $y = (x^2 - 2x + 1)^{1/2} = \sqrt{x^2 - 2x + 1}$

8. $y = \{(x^2 - 4x + 4)^2\}^{1/2} = \sqrt{(x^2 - 4x + 4)^2}$

9. $y = \{(x^2 - 2x + 2)^2\}^{1/2} = \sqrt{(x^2 - 2x + 2)^2}$
A. $y = \{(x^2 - 2x)^2\}^{1/2} = \sqrt{(x^2 - 2x)^2}$

B. $\mathbf{y = \{(-x^2 + 2x)^2\}^{1/2}} = \sqrt{(-x^2+2x)^2}$

C. $\mathbf{y = \{(-x^2 + 2x - 1)^2\}^{1/2}} = \sqrt{(-x^2+2x-1)^2}$

D. $\mathbf{y = \{(-x^2 + 2x - 2)^2\}^{1/2}} = \sqrt{(-x^2+2x-2)^2}$

E. Suppose that $\mathbf{y = f(x) = 2x^2 + bx + c}$ for $\mathbf{0 \le x \le 1}$ and $\mathbf{0 \le y \le 2}$, and then, find the values of the **b** and **c**.

F. Suppose that $\mathbf{y = f(x) = ax^2 + bx + 1}$ for $\mathbf{0 \le x \le 1}$ and $\mathbf{0 \le y \le 2}$, and then, find the values of the **a** and **b**.

G. Suppose $\mathbf{y = f(x) = ax^2 + bx + c}$ for $\mathbf{0 \le x \le 1}$ and $\mathbf{0 \le y \le 2}$, and then, find the values of the **a**, **b**, and **c**.

H. Suppose that $\mathbf{y = f(x) = (2x^2 + 3)/7}$ for $\mathbf{m \le x \le n}$ and $\mathbf{m \le y \le n}$ where **m** and **n** are constants, and then, find the values of the **m** and **n**.

I. Suppose that $\mathbf{y = f(x) = (2x^2 - 3)/5}$ for $\mathbf{m \le x \le n}$ and $\mathbf{m \le y \le n}$ where **m** and **n** are constants, and then, find the values of the **m** and **n**.

J. Suppose that $\mathbf{y = f(x) = (2x^2 + c)/5}$ for $\mathbf{m \le x \le n}$ and $\mathbf{m \le y \le n}$ where **c**, **m** and **n** are constant, and then, find the values of the **c**, **m**, and **n**.

K. Suppose we have an equation of a parabola as follows: $\mathbf{2x^2 + (2c + 1)x - (c + 1)y + c + 1 = 0}$ where **c** is constant. Then, assuming that the parabola includes two particular points for any value of **c**, find the two points.

L. Suppose an equation of a parabola is as follows: $\mathbf{(c + 2)x^2 + (2c + 1)x - (c + 1)y + 3c + 1 = 0}$ where **c** is constant. Then, assuming that the parabola includes two particular points for any value of **c**, find the two points.

M. Suppose we have an equation of a parabola as follows: $\mathbf{cx^2 + 2cx - (c + 1)y + 3c + 3 = 0}$ where **c** is constant. Then, assuming that the parabola includes two particular points for any value of **c**, find the two points.

N. Suppose we have an equation of a parabola as follows: $\mathbf{cx^2 + (2c + 1)x - cy + 3c + 1 = 0}$ where **c** is constant. Then, assuming that the parabola includes a particular point for any value of **c**, find the particular point.

O. Suppose we have an equation of a parabola as follows: $\mathbf{y = cx^2 - (a - 2c)x - 2b(c - 1)}$ where **c** is constant. Then, assuming that the parabola includes two points **(1, 2)** and **(s, t)** for any value of **c**, find the two points.

In the next several examples, let's do some practices on function transformations. The curves are parabolas, of course. For the details on transformations, refer to the **Book 2: Graph Operations**.

P. Suppose that: $y = f(x) = 2(x + 1)^2$ and $f(x) = T_3 \bullet T_2 \bullet T_1 \bullet f$.
T_1: $(x, y) \longrightarrow (3 - x, -2 - y)$.
T_2: $(x, y) \longrightarrow (-x, -y)$.
T_3: $(x, y) \longrightarrow (x + u, y + v)$ where **u** and **v** are constants.

Then, find the values of the **u** and **v**.

Q. Suppose that: $y = f(x) = 2x^2 - 2x + 3$ and $g = T_2 \bullet T_1 \bullet f$.
T_1: $(x, y) \longrightarrow (x - 3, y - 2)$.
T_2: $(x, y) \longrightarrow (y, x)$.

Then, find the **g**.

R. Suppose that: $y = f(x) = -2x^2 + 8x - 5$, $y = g(x) = 2x^2 - 6x + 5/2$, and $g = T \bullet f$.
T: $(x, y) \longrightarrow (u - x, v - y)$

Then, find the values of the **u** and **v**.

S. Suppose that: $x = f(y) = 2(y - 1)^2 + 2$, $y = g(x) = (x - 3/2)^2 + 2$, and $g = T_2 \bullet T_1 \bullet f$.
T_1: $(x, y) \longrightarrow (cy, 2x)$ where **c** is constant.
T_2: $(x, y) \longrightarrow (x + u, y + v)$ where **u** and **v** are constant.

Then, find the values of the **c**, **u**, and **v**.

T. Suppose that: $f(x) = 2x^2 - 2cx + c$ where **c** is constant and ≥ -1.
The curve of $y = f(x)$ is a parabola, and the vertex is (s, t).
T: $c \longrightarrow (s, t)$.
$t = g(s)$.

Then, specify the **g**.

U. Suppose that:
The vertex of a parabola $x = y^2 - 2(c + 1)y + 3c^2$ is (s, t) where **c** is constant and ≥ -1.
T: $c \longrightarrow (s, t)$.
$t = g(s)$.

Then, find the **g**.

Suggestions or Solutions
To the Problem in the Example 0

Suppose that a parabola includes three points $P_1(1, 6)$, $P_2(-1, 4)$, and $P_3(-2, 9)$, and the directrix is parallel to the x-axis. Then, find the equation of the parabola.

Since the directrix is parallel to the x-axis, we may want to use the form of the equation as follows:
$y = ax^2 + bx + c$ where **a**, **b**, and **c** are constant.

Now, we need to construct three equations for the **a**, **b**, and **c** by means of the three points given.
They are to be simultaneous equations.

We have $y = ax^2 + bx + c$. Then:

$P_1(1, 6) \Rightarrow 6 = a + b + c$
$P_2(-1, 4) \Rightarrow 4 = a - b + c$
$P_3(-2, 9) \Rightarrow 9 = 4a - 2b + c$

Now, we've got to do some algebra. We are going to eliminate the **b** first.
$(a + b + c) + (a - b + c) = 6 + 4 \Rightarrow 2a + 2c = 10 \Rightarrow a + c = 5.$
$2(a + b + c) + (4a - 2b + c) = 12 + 9 \Rightarrow 6a + 3c = 21 \Rightarrow 2a + c = 7.$

Next, let's get rid of the **c** and find the **a**.
$(2a + c) - (a + c) = 7 - 5 \Rightarrow a = 2.$

We can find the **c**, now.
$(a + c = 5$ and $a = 2) \Rightarrow 2 + c = 5 \Rightarrow c = 3.$

Then, we need to find the **b**. Since $a = 2$ and $c = 3$:
$6 = a + b + c \Rightarrow 6 = 2 + b + 3 \Rightarrow b = 1.$

Therefore, the equation is $y = 2x^2 + x + 3$.

Let's now, put the parabola in a graph.
Then, we may want to convert it into the complete square to expose the vertex.

$y = 2x^2 + x + 3 = 2(x^2 + x/2) + 3$
$= 2\{x^2 + x/2 + (1/4)^2 - (1/4)^2\} + 3 = 2(x + 1/4)^2 - 1/8 + 3 = 2(x + 1/4)^2 + 23/8.$
$\Rightarrow y - 23/8 = 2(x + 1/4)^2.$

Now, we can see that the vertex is at $(-1/4\ ,\ 23/8)$.

Fig. 0.0

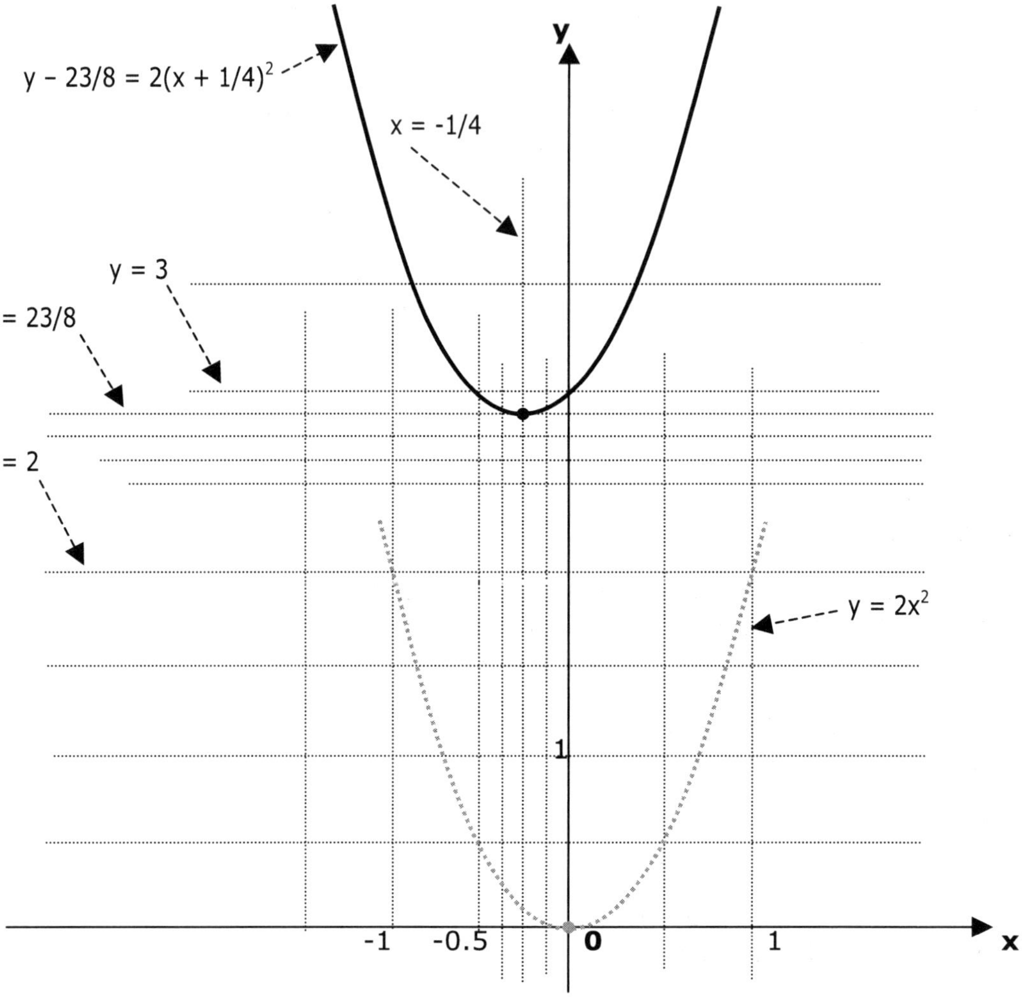

In short:

Since the directrix is parallel to the x-axis, we can use the form as follows:
$\mathbf{y = ax^2 + bx + c}$ where **a**, **b**, and **c** are constants. Then:
$\mathbf{P_1(1, 6) \Rightarrow 6 = a + b + c}$
$\mathbf{P_2(-1, 4) \Rightarrow 4 = a - b + c}$
$\mathbf{P_3(-2, 9) \Rightarrow 9 = 4a - 2b + c}$
Eliminate the **b** first.
$\mathbf{(a + b + c) + (a - b + c) = 6 + 4 \Rightarrow 2a + 2c = 10 \Rightarrow a + c = 5.}$
$\mathbf{2(a + b + c) + (4a - 2b + c) = 12 + 9 \Rightarrow 6a + 3c = 21 \Rightarrow 2a + c = 7.}$
Next, eliminate the **c** and find the **a**.
$\mathbf{(2a + c) - (a + c) = 7 - 5 \Rightarrow a = 2.}$
We can now, find the **c**.
$\mathbf{a = 2 \Rightarrow a + c = 2 + c = 5 \Rightarrow c = 3.}$
Then, we need to find the **b**.
($\mathbf{6 = a + b + c}$, $\mathbf{a = 2}$, and $\mathbf{c = 3}$**)** $\mathbf{\Rightarrow 6 = 2 + b + 3 \Rightarrow b = 1.}$
Therefore, the equation is $\mathbf{y = 2x^2 + x + 3}$.

Suggestions or Solutions

To the **Problem** in the Example **1**

Suppose that a parabola includes three points $P_1(1, 6)$, $P_2(-1, 4)$, and $P_3(-2, 9)$, and the axis of symmetry is parallel to the x-axis. Then, find the equation of the parabola.

Since the axis of symmetry is parallel to the x-axis, we may want to use the form of the equation as follows: $\mathbf{x = ay^2 + by + c}$ where **a, b**, and **c** are constant. Now, we need to use the three points to set up three simultaneous equations for the **a**, **b**, and **c**, and then, solve the system. Then:

$P_1(1, 6) \Rightarrow 1 = 36a + 6b + c$
$P_2(-1, 4) \Rightarrow -1 = 16a + 4b + c$
$P_3(-2, 9) \Rightarrow -2 = 81a + 9b + c$

$(36a + 6b + c) - (16a + 4b + c) = 1 - (-1) \Rightarrow 20a + 2b = 2 \Rightarrow 10a + b = 1.$
$(81a + 9b + c) - (16a + 4b + c) = -2 - (-1) \Rightarrow 65a + 5b = -1 \Rightarrow 13a + b = -1/5.$
$(13a + b) - (10a + b) = -1/5 - 1 \Rightarrow 3a = -6/5 \Rightarrow a = -2/5.$

$a = -2/5 \Rightarrow 10a + b = -20/5 + b = 1 \Rightarrow b = 1 + 4 = 5.$

Since $\mathbf{a = -2/5}$ and $\mathbf{b = 5}$:
$-1 = 16a + 4b + c = -32/5 + 20 + c \Rightarrow c = (32 - 105)/5 = -73/5.$

Therefore, the equation is $\mathbf{x = -2y^2/5 + 5y - 73/5}$.

In short:

Since the axis of symmetry is parallel to the x-axis: $\mathbf{x = ay^2 + by + c}$ where **a**, **b**, and **c** are constant. Then:
$P_1(1, 6) \Rightarrow 1 = 36a + 6b + c$, $P_2(-1, 4) \Rightarrow -1 = 16a + 4b + c$, and $P_3(-2, 9) \Rightarrow -2 = 81a + 9b + c$.

$(36a + 6b + c) - (16a + 4b + c) = 1 - (-1) \Rightarrow 20a + 2b = 2 \Rightarrow 10a + b = 1.$
$(81a + 9b + c) - (16a + 4b + c) = -2 - (-1) \Rightarrow 65a + 5b = -1 \Rightarrow 13a + b = -1/5.$
$(13a + b) - (10a + b) = -1/5 - 1 \Rightarrow 3a = -6/5 \Rightarrow a = -2/5.$

$a = -2/5 \Rightarrow 10a + b = -20/5 + b = 1 \Rightarrow b = 1 + 4 = 5.$
Since $\mathbf{a = -2/5}$ and $\mathbf{b = 5}$, $-1 = 16a + 4b + c = -32/5 + 20 + c \Rightarrow c = (32 - 105)/5 = -73/5.$
Therefore, the equation is $\mathbf{x = -2y^2/5 + 5y - 73/5}$.

Let's now, put the parabola in a graph. Let's first, expose the vertex **V** by the complete square method.
$x = -2y^2/5 + 5y - 73/5 \Rightarrow 5x = -2y^2 + 25y - 73 = -2(y^2 - 25y/2) - 73$
$= -2\{y^2 - 25y/2 + (25/4)^2 - (25/4)^2\} - 73 = -2(y - 25/4)^2 + 2(25/4)^2 - 73$
$= -2(y - 25/4)^2 + 25^2/8 - 73 = -2(y - 25/4)^2 + (625 - 584)/8 = -2(y - 25/4)^2 + 41/8$
$\Rightarrow 5x = -2(y - 25/4)^2 + 41/8 \Rightarrow x = (-2/5)(y - 25/4)^2 + 41/40$
$\Rightarrow x - 41/40 = (-2/5)(y - 25/4)^2.$

Now, the vertex **V** is at $\mathbf{(41/40\ ,\ 25/4)}$.

Fig. 1.0

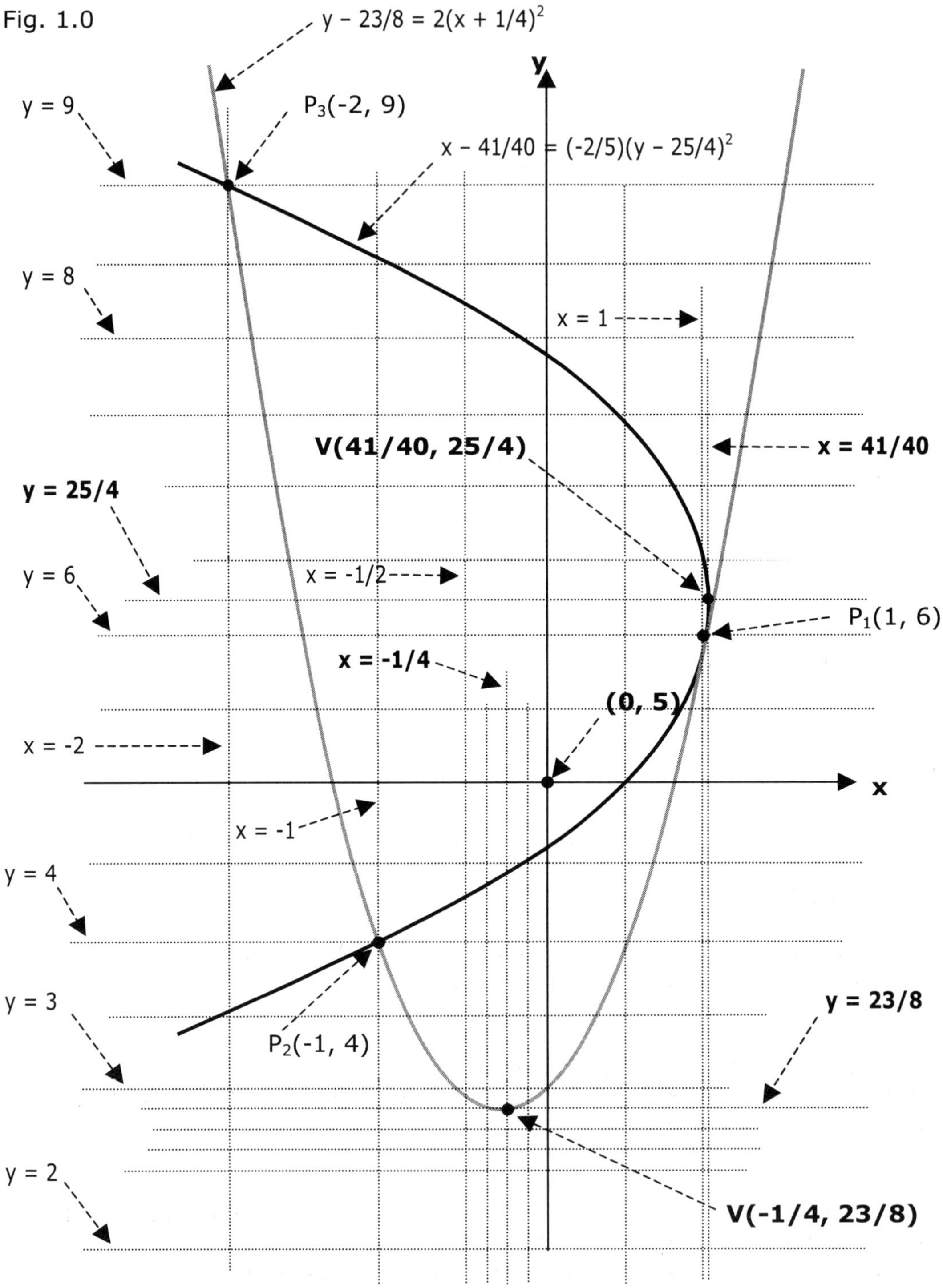

Note:
The origin is not shown in the graph above, but is somewhere below the line **y = 2**, of course.
The parabola in black is the solution to the problem, and the parabola in gray was found in the example 0, and also, includes the three given points.

Suggestions or Solutions
To the Problem in the Example 2

Suppose that we have an equation of a parabola $y = 2x^2 + bx + c$, and that the parabola includes two points where another parabola $y = -2x^2 + 3x/2 + 9/4$ meets a line $y = x/2 - 3/4$. Then, find the values of the b and the c.

Suppose that **A** is the parabola $y = 2x^2 + bx + c$. Then, for the **A**, the axis of symmetry is parallel to the y-axis, and the parabola includes two points, at which the other parabola and the line described in the problem meet each other. Then, we need to find the two points. At the two points where the two curves $y = -2x^2 + 3x/2 + 9/4$ and $y = x/2 - 3/4$ meet each other, the values of the **y** need to be the same. Then, we get to have an equation for the **x**, $-2x^2 + 3x/2 + 9/4 = x/2 - 3/4$.
Now, $-2x^2 + 3x/2 + 9/4 = x/2 - 3/4 \Rightarrow 2x^2 - x - 3 = 0$
$\Rightarrow 2x^2 - x - 3 = (2x - 3)(x + 1) = 0 \Rightarrow x = 3/2$ or -1.
Then, $x = 3/2 \Rightarrow y = x/2 - 3/4 = 3/4 - 3/4 = 0$. $x = -1 \Rightarrow y = x/2 - 3/4 = -1/2 - 3/4 = -5/4$.
Therefore, the other parabola and the line meet each other at $(3/2, 0)$ and $(-1, -5/4)$.

Now, the parabola $y = 2x^2 + bx + c$ includes the two points above. Then:
$0 = 2(3/2)^2 + (3/2)b + c = 9/2 + 3b/2 + c \Rightarrow 3b + 2c + 9 = 0 \Rightarrow 3b + 2c = -9$.
$-5/4 = 2(-1)^2 + (-1)b + c = 2 - b + c \Rightarrow b - c = 2 + 5/4 = 13/4 \Rightarrow b - c = 13/4$.

$(3b + 2c) + 2(b - c) = -9 + 26/4 = -10/4 \Rightarrow 5b = -10/4 \Rightarrow b = -1/2$.

Then, $b = -1/2 \Rightarrow b - c = -1/2 - c = 13/4 \Rightarrow c = -1/2 - 13/4 = -15/4$.
So, the parabola **A** is $y = 2x^2 - x/2 - 15/4$, and therefore, $b = -1/2$ and $c = -15/4$.

In short:
We need to find the two points where the parabola $y = -2x^2 + 3x/2 + 9/4$ and the line $y = x/2 - 3/4$ meet each other. $-2x^2 + 3x/2 + 9/4 = x/2 - 3/4 \Rightarrow 2x^2 - x - 3 = 0$
$\Rightarrow 2x^2 - x - 3 = (2x - 3)(x + 1) = 0 \Rightarrow x = 3/2$ or -1.
Then, $x = 3/2 \Rightarrow y = x/2 - 3/4 = 3/4 - 3/4 = 0$. $x = -1 \Rightarrow y = x/2 - 3/4 = -1/2 - 3/4 = -5/4$.
Therefore, the other parabola and the line meet each other at $(3/2, 0)$ and $(-1, -5/4)$.
Now, the parabola $y = 2x^2 + bx + c$ includes the two points above. Then:
$0 = 2(3/2)^2 + (3/2)b + c = 9/2 + 3b/2 + c \Rightarrow 3b + 2c + 9 = 0 \Rightarrow 3b + 2c = -9$.
$-5/4 = 2(-1)^2 + (-1)b + c = 2 - b + c \Rightarrow b - c = 2 + 5/4 = 13/4 \Rightarrow b - c = 13/4$.
Therefore, first, we get $(3b + 2c) + 2(b - c) = -9 + 26/4 = -10/4 \Rightarrow 5b = -10/4 \Rightarrow b = -1/2$.
Next, $b = -1/2 \Rightarrow b - c = -1/2 - c = 13/4 \Rightarrow c = -1/2 - 13/4 = -15/4$.

Let's now, put in a graph the parabolas and the line.
$y = 2x^2 - x/2 - 15/4$
$\Rightarrow 4y = 8x^2 - 2x - 15 = 8(x^2 - x/4) - 15 = 8\{x^2 - x/4 + (1/8)^2 - (1/8)^2\} - 15$
$= 8(x - 1/8)^2 - 8(1/8)^2 - 15 = 8(x - 1/8)^2 - 1/8 - 15 = 8(x - 1/8)^2 - 121/8$.
$\Rightarrow y = 2(x - 1/8)^2 - 121/32 \Rightarrow y + 121/32 = 2(x - 1/8)^2$.
$\Rightarrow$ The vertex $= (1/8, -121/32) = (0.125, -3.78125)$.

$y = -2x^2 + 3x/2 + 9/4 = -2\{x^2 - 3x/4 + (3/8)^2 - (3/8)^2\} + 9/4$
$= -2(x - 3/8)^2 + 2(3/8)^2 + 9/4 = -2(x - 3/8)^2 + 9/32 + 9/4 = -2(x - 3/8)^2 + 81/32$
$\Rightarrow y - 81/32 = -2(x - 3/8)^2$. $\Rightarrow$ The vertex $= (3/8, 81/32) = (0.375, 2.53125)$.

Since the magnitude of the slopes are the same, so are the shapes of the parabolas. However, the concavities are directly opposite of each other since the sings are opposite.
Now, is the parabola $\mathbf{y = 2x^2 - x/2 - 15/4}$ the only parabola that includes the two points?
Of course, not! There can be infinitely many parabolas that include the two points.
However, among the many parabolas, there is only one that has the slope of 2 and the axis of symmetry parallel to the y-axis and includes the two points, and the only one is the very parabola **A**.

Fig. 2.0

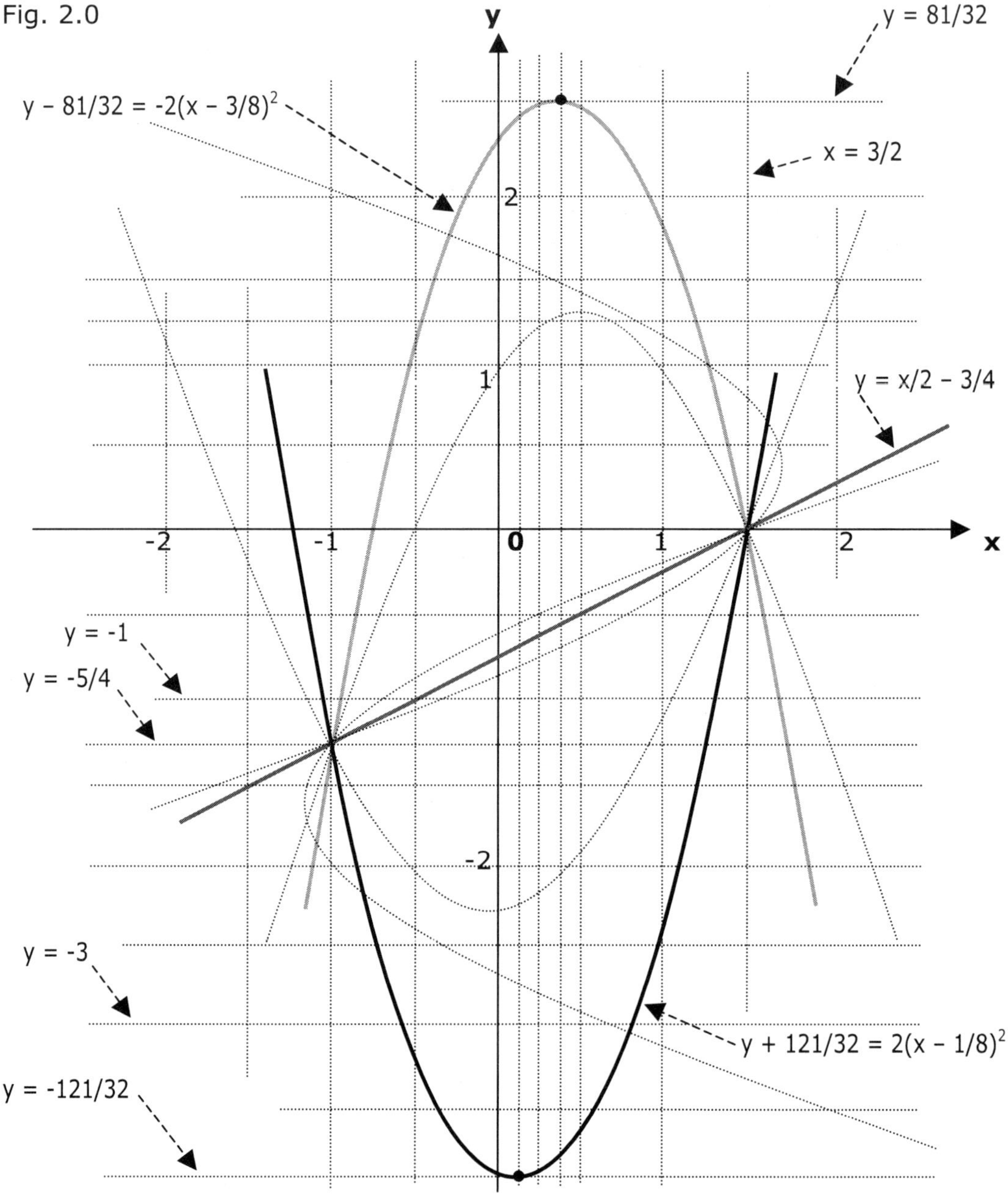

Now, let's try a similar one in the next example.

Suggestions or Solutions
To the Problem in the Example 3

Suppose that we have an equation of a parabola $y = ax^2 - x/2 + c$, and that the parabola includes two points where another parabola $y = -2x^2 + 3x/2 + 9/4$ meets a line $y = x/2 - 3/4$. Then, find the values of the a and the c.

Suppose that **B** is the parabola $y = ax^2 - x/2 + c$. Then, for the **B**, the axis of symmetry is parallel to the y-axis, and the **B** includes the two points where the other parabola and the line described in the problem meet each other.
Now, we've already found in the previous example, that the two points are **(3/2, 0)** and **(-1, -5/4)**.

Also, we can see that the coefficient of the x-term is -1/2, which is the same as that of the x-term in the equation of the parabola **A** in the previous example.
Then, is the parabola **B** the same as the parabola **A** in the previous example?
In other words, is the value of the **a** equal to 2, and is the value of the **c** equal to -15/4?
Probably the same! Let's see if the **A** and the **B** are the same as each other.

We know that the equation of the parabola **B** given is $y = ax^2 - x/2 + c$, and that the parabola includes the two points **(3/2, 0)** and **(-1, -5/4)**. Then:

$0 = a(3/2)^2 - (3/2)/2 + c = 9a/4 - 3/4 + c \Rightarrow 9a + 4c - 3 = 0 \Rightarrow 9a + 4c = 3.$
$-5/4 = a(-1)^2 - (-1)/2 + c = a + 1/2 + c \Rightarrow a + c = -1/2 - 5/4 = -7/4 \Rightarrow a + c = -7/4.$
$(9a + 4c) - 4(a + c) = 3 - (-7) = 10 \Rightarrow 5a = 10 \Rightarrow a = 2.$
Then, $a = 2 \Rightarrow a + c = 2 + c = -7/4 \Rightarrow c = -7/4 - 2 = -15/4.$

Now, the **A** and the **B** are the same as each other.

What if the given equation is $y = ax^2 + bx - 15/4$ instead of the **B**?
It will be the same equation, too.
That's because there is only one particular equation for one particular parabola.
Some same equations can look different.
For instance, $y = 2x^2 + 3x + 5/2$ is the same as $4x^2 + 6x - 2y + 5 = 0$.

Now, let's take another similar one in the next example.

Suggestions or Solutions
To the Problem in the Example 4

Suppose that we have an equation of a parabola $x = ay^2 - y/2 + c$, and that the parabola includes two points where another parabola $y = 2x^2 + x - 3$ meets a line $y = 3x + 1$. Then, find the values of the a and the c.

Suppose that **C** is the parabola $x = ay^2 - y/2 + c$. Then, for the **C**, the axis of symmetry is parallel to the x-axis, and the **C** includes two points, at which the other parabola and the line described in the problem meet each other. Now, we need to find the two points.
At the two points in the two curves $y = 2x^2 + x - 3$ and $y = 3x + 1$, the values of the **y** need to be the same. Then, we get to have an equation for the **x**, $2x^2 + x - 3 = 3x + 1$. Now:

$2x^2 + x - 3 = 3x + 1 \Rightarrow 2x^2 - 2x - 4 = 0 \Rightarrow 2(x + 1)(x - 2) = 0 \Rightarrow x = -1$ or 2.

Then, $x = -1 \Rightarrow y = 3x + 1 = -3 + 1 = -2$. $x = 2 \Rightarrow y = 3x + 1 = 6 + 1 = 7$.

Therefore, the other parabola and the line meet each other at **(-1, -2)** and **(2, 7)**.
Now, the parabola $x = ay^2 - y/2 + c$ includes the two points above. Then:
$-1 = a(-2)^2 - (-2)/2 + c = 4a + 1 + c \Rightarrow 4a + c + 1 = -1 \Rightarrow 4a + c = -2$.
$2 = 7^2a - 7/2 + c = 49a + c - 7/2 \Rightarrow 49a + c = 2 + 7/2 = 11/2 \Rightarrow 49a + c = 11/2$.

$(49a + c) - (4a + c) = 11/2 - (-2) = 15/2 \Rightarrow 45a = 15/2 \Rightarrow a = 1/6$.

Then, $a = 1/6 \Rightarrow 4a + c = 4/6 + c = -2 \Rightarrow c = -16/6 = -8/3$.
Consequently, the parabola **C** is $x = y^2/6 - y/2 - 8/3$ and therefore, $a = 1/6$ and $c = -8/3$.

In short:
We need to find the two points where the parabola $y = 2x^2 + x - 3$ and the line $y = 3x + 1$ meet each other.
$2x^2 + x - 3 = 3x + 1 \Rightarrow 2x^2 - 2x - 4 = 0 \Rightarrow 2(x + 1)(x - 2) = 0 \Rightarrow x = -1$ or 2.
Then, $x = -1 \Rightarrow y = 3x + 1 = -3 + 1 = -2$. $x = 2 \Rightarrow y = 3x + 1 = 6 + 1 = 7$.
Therefore, the other parabola and the line meet each other at **(-1, -2)** and **(2, 7)**.

Now, the parabola $x = ay^2 - y/2 + c$ includes the two points above. Then:
$-1 = a(-2)^2 - (-2)/2 + c = 4a + 1 + c \Rightarrow 4a + c + 1 = -1 \Rightarrow 4a + c = -2$.
$2 = 7^2a - 7/2 + c = 49a + c - 7/2 \Rightarrow 49a + c = 2 + 7/2 = 11/2 \Rightarrow 49a + c = 11/2$.
$(49a + c) - (4a + c) = 11/2 - (-2) = 15/2 \Rightarrow 45a = 15/2 \Rightarrow a = 1/6$.
Then, $a = 1/6 \Rightarrow 4a + c = 4/6 + c = -2 \Rightarrow c = -16/6 = -8/3$.

Let's now, put in a graph the parabolas and the line.

$x = y^2/6 - y/2 - 8/3 \Rightarrow 6x = y^2 - 3y - 16 = y^2 - 3y + (3/2)^2 - (3/2)^2 - 16$
$= (y - 3/2)^2 - 9/4 - 16 = (y - 3/2)^2 - 73/4 \Rightarrow 6x = (y - 3/2)^2 - 73/4$
$\Rightarrow x = (1/6)(y - 3/2)^2 - 73/24 \Rightarrow x + 73/24 = (y - 3/2)^2/6$.
$\Rightarrow$ The vertex $= (-73/24, 3/2) \approx (-3.04, 1.5)$.

$$\mathbf{y = 2x^2 + x - 3 = 2\{x^2 + x/2 + (1/4)^2 - (1/4)^2\} - 3 = 2(x + 1/4)^2 - 2(1/4)^2 - 3}$$
$$\mathbf{= 2(x + 1/4)^2 - 1/8 - 3 = 2(x + 1/4)^2 - 25/8 \Rightarrow y + 25/8 = 2(x + 1/4)^2}$$
$\Rightarrow$ The vertex = **(-1/4, -25/8) = (-0.25, -3.125)**.

Now, is the parabola $\mathbf{x = y^2/6 - y/2 - 8/3}$ the only parabola that includes the two points?
Of course, not! There can be infinitely many parabolas that include the two points.
However, among the many parabolas, there is only one that has the slope of 2 and the axis of symmetry parallel to the x-axis and includes the two points, and the only one is the very parabola **C**.

Fig. 4.0.

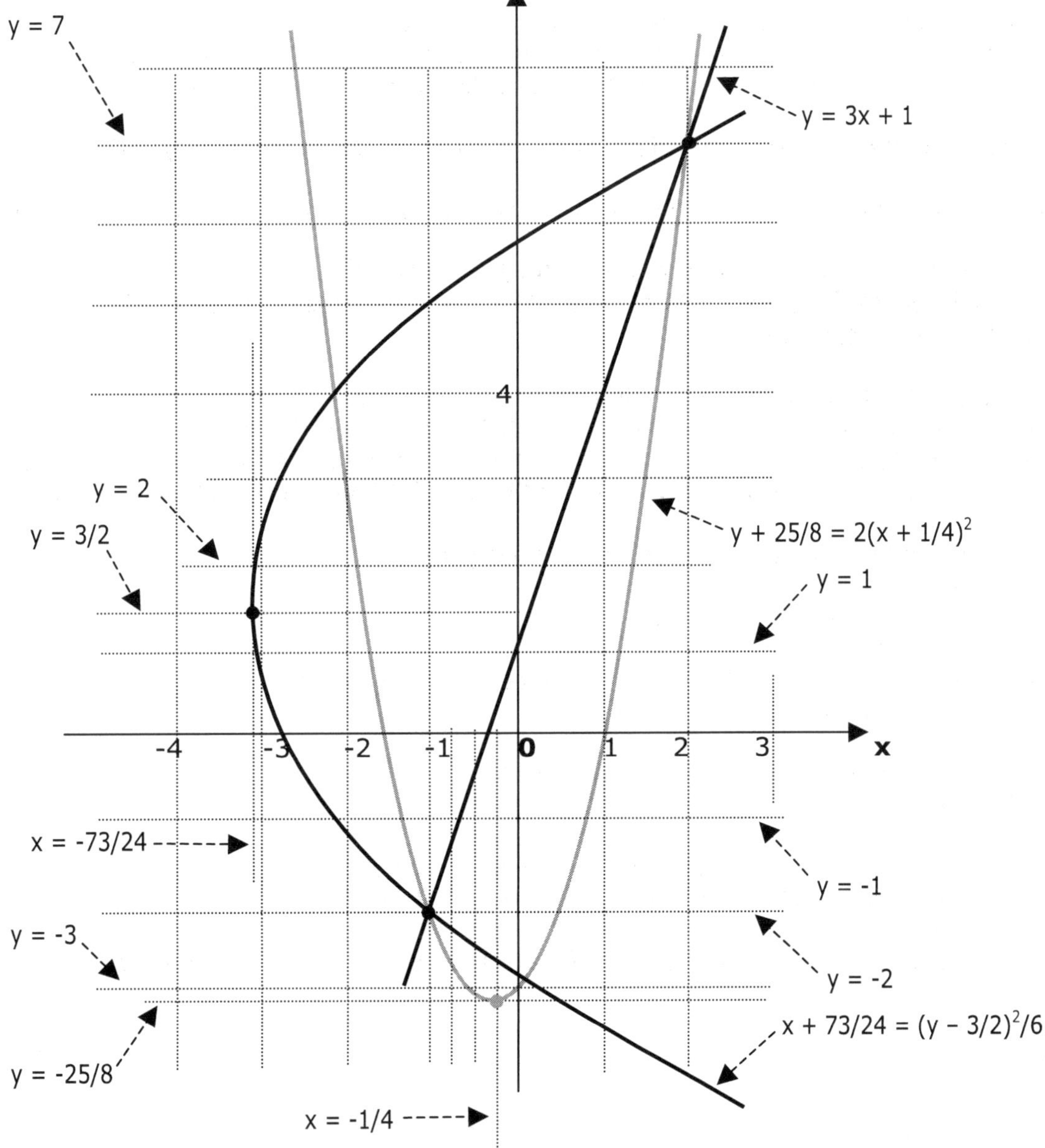

Now, let's take the next example.

Suggestions or Solutions

To the **Problem** in the Example **5**

Suppose that a parabola $y/2 = ax^2 - 2x + c/2$ includes a point at $(1/2, -1/2)$, and that a line $y = x - 1$ includes the vertex. Then, find the values of the a and the c.

Since the parabola $y/2 = ax^2 - 2x + c/2$ includes the point at $(1/2, -1/2)$:
$(-1/2)/2 = -1/4 = a(1/2)^2 - 2(1/2) + c/2 = a/4 - 1 + c/2 \Rightarrow -1 = a - 4 + 2c \Rightarrow a + 2c = 3$.

Since we have two unknowns a and c, we need to get one more equation.
Now, the vertex is in the line. Therefore, putting the coordinates of the vertex in terms of the **a** and **c**, we get to have one more equation for them. The vertex of a parabola $y = ax^2 + bx + c$ is as follows:

$y = ax^2 + bx + c = a(x + b/2a)^2 - (b^2 - 4ac)/4a \Rightarrow$ The vertex is $\{-b/2a, -(b^2 - 4ac)/4a\}$.

It is often said to be the formula for the vertex, but is nothing but the result of the complete square method as follows:
$ax^2 + bx + c = a(x^2 + bx/a) + c = a\{x^2 + bx/a + (b/2a)^2 - (b/2a)^2\} + c$
$= a(x + b/2a)^2 - ab^2/4a^2 + c = a(x + b/2a)^2 - b^2/4a + c = a(x + b/2a)^2 - (b^2/4a - c)$
$= a(x + b/2a)^2 - (b^2 - 4ac)/4a$.

Now, let's directly apply to the given equation the complete square method:
$y/2 = ax^2 - 2x + c/2 \Rightarrow y = 2ax^2 - 4x + c = 2a(x^2 - 2x/a + 1/a^2 - 1/a^2) + c$
$= 2a(x^2 - 2x/a + 1/a^2) - 2a/a^2 + c = 2a(x^2 - 2x/a + 1/a^2) - 2/a + c = 2a(x - 1/a)^2 - 2/a + c$
$\Rightarrow y + 2/a - c = 2a(x - 1/a)^2 \Rightarrow y - (-2/a + c) = 2a(x - 1/a)^2 \Rightarrow$ **The vertex** $= (1/a, c - 2/a)$.

We know that the vertex $(1/a, -2/a + c)$ is to be in the line $y = x - 1$.
Now, let's put the vertex in the line.
How do we put it in the line? Plug it in the equation.
Then, since the vertex $(1/a, -2/a + c)$ is in the line $y = x - 1$:
$y = x - 1 \Rightarrow -2/a + c = 1/a - 1 \Rightarrow 3/a - c = 1 \Rightarrow 3 - ac = a \Rightarrow a + ac = 3$
$\Rightarrow a(1 + c) = 3 \Rightarrow a = 3/(1 + c)$.

Now, we have two equations for the two unknowns, **a** and **c**:
$a + 2c = 3$ and $a = 3/(1 + c)$.

Then, $a + 2c = 3/(1 + c) + 2c = 3 \Rightarrow 3 + 2c(1 + c) = 3(1 + c) = 3 + 3c$
$\Rightarrow 3 + 3c = 3 + 2c + 2c^2 \Rightarrow 2c^2 - c = 0 \Rightarrow c(2c - 1) = 0 \Rightarrow c = 0$ or $1/2$.

Consequently, $c = 0 \Rightarrow a + 2c = a + 0 = 3 \Rightarrow a = 3$. $c = 1/2 \Rightarrow a + 2c = a + 1 = 3 \Rightarrow a = 2$.

In short:

Since the parabola $y/2 = ax^2 - 2x + c/2$ includes the point at $(1/2, -1/2)$:
$(-1/2)/2 = -1/4 = a(1/2)^2 - 2(1/2) + c/2 = a/4 - 1 + c/2 \Rightarrow -1 = a - 4 + 2c \Rightarrow a + 2c = 3$.
$y/2 = ax^2 - 2x + c/2 \Rightarrow y = 2ax^2 - 4x + c = 2a(x^2 - 2x/a + 1/a^2 - 1/a^2) + c$
$= 2a(x^2 - 2x/a + 1/a^2) - 2a/a^2 + c = 2a(x^2 - 2x/a + 1/a^2) - 2/a + c = 2a(x - 1/a)^2 - 2/a + c$

$\Rightarrow$ **y + 2/a − c = 2a(x − 1/a)²** $\Rightarrow$ **y − (−2/a + c) = 2a(x − 1/a)²** $\Rightarrow$ The vertex **= (1/a, c − 2/a).**

The vertex **(1/a, −2/a + c)** is to be in the line **y = x − 1**. Therefore, **−2/a + c = 1/a − 1**
$\Rightarrow$ **3/a − c = 1** $\Rightarrow$ **3 − ac = a** $\Rightarrow$ **a + ac = 3** $\Rightarrow$ **a(1 + c) = 3** $\Rightarrow$ **a = 3/(1 + c).**
Now, we have **a + 2c = 3** and **a = 3/(1 + c).**
Then, **a + 2c = 3/(1 + c) + 2c = 3** $\Rightarrow$ **3 + 2c(1 + c) = 3(1 + c) = 3 + 3c**
$\Rightarrow$ **3 + 3c = 3 + 2c + 2c²** $\Rightarrow$ **2c² − c = 0** $\Rightarrow$ **c(2c − 1) = 0** $\Rightarrow$ **c = 0** or **1/2.**
So, **c = 0** $\Rightarrow$ **a + 2c = a + 0 = 3** $\Rightarrow$ **a = 3. c = 1/2**$\Rightarrow$ **a + 2c = a + 1 = 3** $\Rightarrow$ **a = 2.**
Therefore, **a = 3** and **c = 0**, or **a = 2** and **c = 1/2.**

Now, we can see that two parabolas can be the solution.
We have **y/2 = ax² − 2x + c/2** $\Rightarrow$ **y = 2ax² − 4x + c.**

First, if **a = 3** and **c = 0**, then **y = 2ax² − 4x + c** $\Rightarrow$ **y = 6x² − 4x = 6{x² − (2/3)x + 1/9 − 1/9}**
= 6(x − 1/3)² − 2/3. $\Rightarrow$ **y + 2/3 = 6(x − 1/3)²** $\Rightarrow$ **y − (−2/3) = 6(x − 1/3)².**

Next, if **a = 2** and **c = 1/2**, then **y = 2ax² − 4x + c** $\Rightarrow$ **y = 4x² − 4x + 1/2 = 4(x² − x) + 1/2**
= 4(x² − x + 1/4 − 1/4) + 1/2 = 4(x − 1/2)² − 1 + 1/2 = 4(x − 1/2)² − 1/2.
$\Rightarrow$ **y + 1/2 = 4(x − 1/2)²** $\Rightarrow$ **y − (−1/2) = 4(x − 1/2)².**

Fig. 5.0

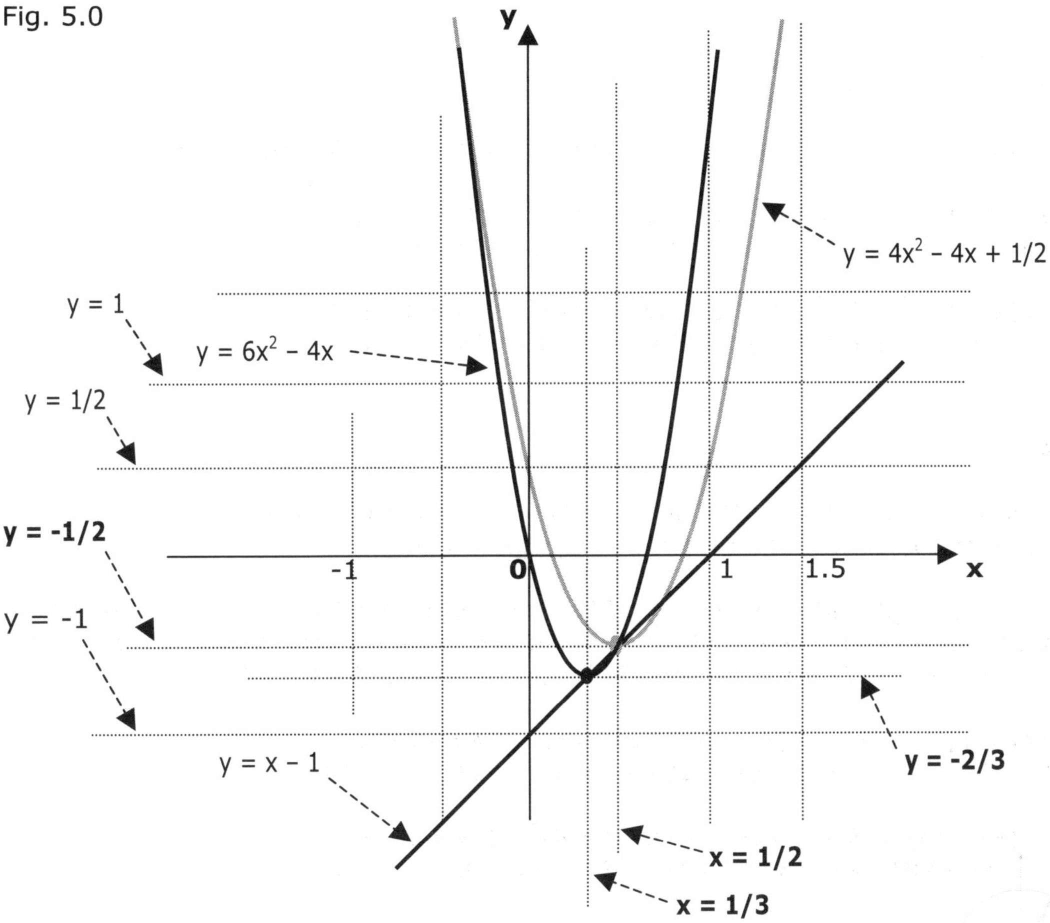

Suggestions or Solutions
To the Problem in the Example 6

Suppose that a parabola $y = 6x^2 + bx + c$ includes a point at (2, 2), and that a line $y = x - 1$ includes the vertex. Then, find the values of the b and the c.

In this example, we may want to use the form as follows:
$y - n = a(x - m)^2$ where the **a** is the slope, and **(m, n)** is the vertex.

We can see that the slope is 6, and know that the point **(2, 2)** is in the parabola.
Then, we get $y - n = 6(x - m)^2$, which is now, an equation for the **m** and the **n**.

Besides, we have the information on the vertex.
The vertex **(m, n)** is in the line $y = x - 1$, and therefore, we get $n = m - 1$.

Now, we are all set to go and get the equation of the parabola.
We have $2 - n = 6(2 - m)^2$ and $n = m - 1$.

Then, $2 - (m - 1) = 6(2 - m)^2 \Rightarrow 2 - m + 1 = 6(2 - m)^2 = 6(4 - 4m + m^2) = 24 - 24m + 6m^2$
$\Rightarrow 3 - m = 24 - 24m + 6m^2 \Rightarrow 6m^2 - 23m + 21 = 0 \Rightarrow (2m - 3)(3m - 7) = 0 \Rightarrow m = 3/2$ or $7/3$.

$m = 3/2 \Rightarrow n = m - 1 = 3/2 - 1 = 1/2.$
$m = 7/3 \Rightarrow n = m - 1 = 7/3 - 1 = 4/3.$

So, we have two parabolas as follows: $y - 1/2 = 6(x - 3/2)^2$ and $y - 4/3 = 6(x - 7/3)^2$.

Now, $y - 1/2 = 6(x - 3/2)^2 = 6(x^2 - 3x + 9/4) = 6x^2 - 18x + 27/2$
$\Rightarrow y = 6x^2 - 18x + 27/2 + 1/2 = 6x^2 - 18x + 14 \Rightarrow b = -18$ and $c = 14$.

$y - 4/3 = 6(x - 7/3)^2 = 6(x^2 - 14x/3 + 49/9) = 6x^2 - 28x + 98/3$
$\Rightarrow y = 6x^2 - 28x + 98/3 + 4/3 = 6x^2 - 28x + 34 \Rightarrow b = -28$ and $c = 34$.

In short:

Since the slope is 6 and the point **(2, 2)** is in the parabola, set $y - n = 6(x - m)^2$.
since the vertex **(m, n)** is to be in the line $y = x - 1$, we get $n = m - 1$.

Then, $2 - (m - 1) = 6(2 - m)^2 \Rightarrow 2 - m + 1 = 6(2 - m)^2 = 6(4 - 4m + m^2) = 24 - 24m + 6m^2$
$\Rightarrow 3 - m = 24 - 24m + 6m^2 \Rightarrow 6m^2 - 23m + 21 = 0 \Rightarrow (2m - 3)(3m - 7) = 0 \Rightarrow m = 3/2$ or $7/3$.
$m = 3/2 \Rightarrow n = m - 1 = 3/2 - 1 = 1/2$. $m = 7/3 \Rightarrow n = m - 1 = 7/3 - 1 = 4/3$.

Now, $y - 1/2 = 6(x - 3/2)^2 = 6(x^2 - 3x + 9/4) = 6x^2 - 18x + 27/2$
$\Rightarrow y = 6x^2 - 18x + 27/2 + 1/2 = 6x^2 - 18x + 14 \Rightarrow b = -18$ and $c = 14$.

$y - 4/3 = 6(x - 7/3)^2 = 6(x^2 - 14x/3 + 49/9) = 6x^2 - 28x + 98/3$
$\Rightarrow y = 6x^2 - 28x + 98/3 + 4/3 = 6x^2 - 28x + 34 \Rightarrow b = -28$ and $c = 34$.

Fig. 6.0

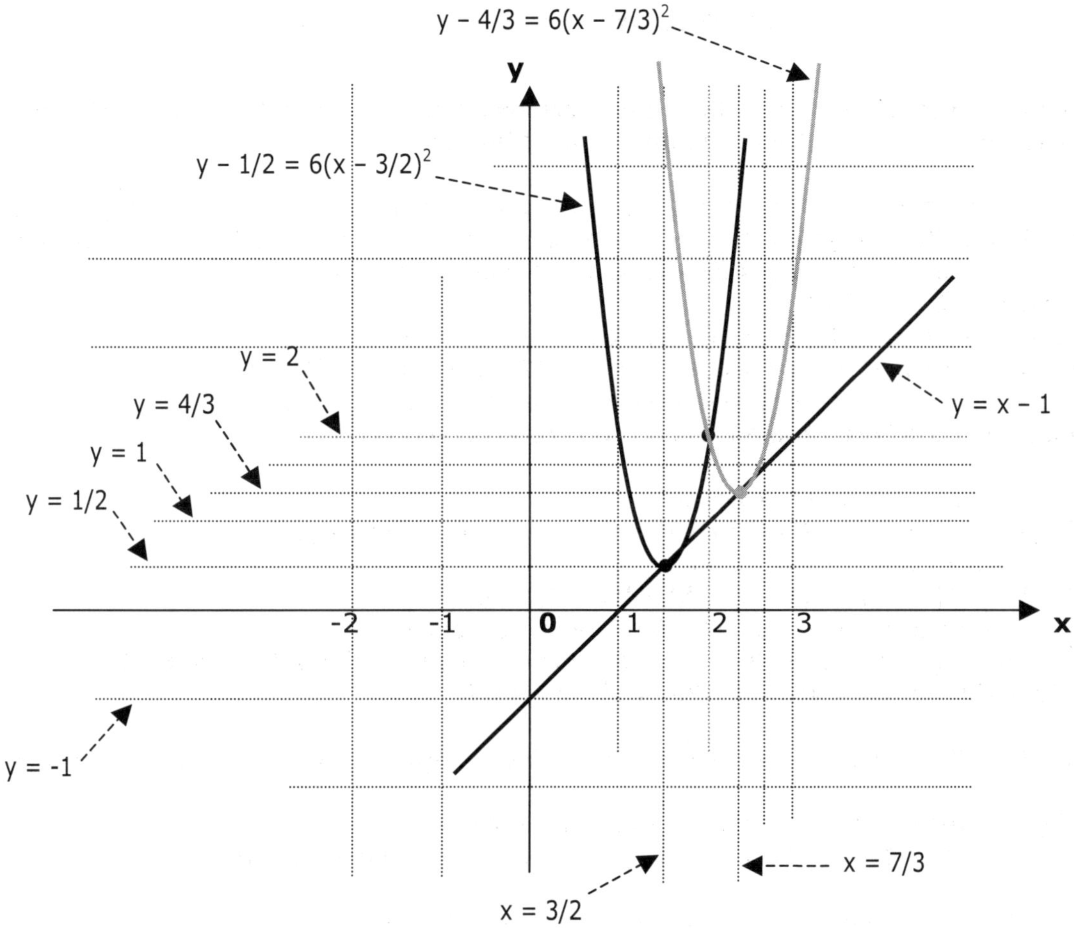

Let's take a similar one in the next example, now.

Suggestions or Solutions
To the Problem in the Example 7

Suppose that a parabola $y = ax^2 + bx + c$ includes two points (1, 1) and (2, 3), and that the vertex is in a line $y = 2x - 1$. Then, find the values of the a, the b, and the c.

If we remember the formula for the vertex of a parabola $y = ax^2 + bx + c$, then we can readily put the vertex in $\{-b/2a, -(b^2 - 4ac)/4a\}$. If we don't, then we can derive it by means of the complete square.
Now, since the vertex is in the line $y = 2x - 1$, we get an equation as follows:
$-(b^2 - 4ac)/4a = 2(-b/2a) - 1$.
Besides, since the two points (1, 1) and (2, 3) are in the parabola:
$(1, 1) \Rightarrow 1 = a(1^2) + b(1) + c = a + b + c$, and also, $(2, 3) \Rightarrow 3 = 4a + 2b + c$.
Now, all we have to do is the algebra. We need to solve the system of equations as follows:
$-(b^2 - 4ac)/4a = 2(-b/2a) - 1$, $a + b + c = 1$, and $4a + 2b + c = 3$.

To begin with, let's simplify the first equation:
$-(b^2 - 4ac)/4a = 2(-b/2a) - 1 = -b/a - 1 = -(b + a)/a \Rightarrow b^2 - 4ac = 4(b + a)$.
Meanwhile, $a + b + c = 1 \Rightarrow a + b = 1 - c$.
Then, $b^2 - 4ac = 4(b + a) = 4(1 - c) \Rightarrow b^2 - 4ac = 4(1 - c)$.
Now, $4a + 2b + c = 3 \Rightarrow 4ac + 2bc + c^2 = 3c \Rightarrow 4ac = 3c - 2bc - c^2$.
Then, $b^2 - 4ac = 4(1 - c) \Rightarrow b^2 - (3c - 2bc - c^2) = b^2 - 3c + 2bc + c^2 = 4(1 - c) = 4 - 4c$.
So, $b^2 - 3c + 2bc + c^2 = 4 - 4c \Rightarrow b^2 + 2bc + c^2 + c - 4 = 0$.

Now, since $4a + 2b + c = 3$ and $a + b + c = 1$, $4a + 2b + c - 4(a + b + c) = 3 - 4 \Rightarrow -2b - 3c = -1$.

Meanwhile, $b^2 + 2bc + c^2 + c - 4 = 0 \Rightarrow (b + c)^2 + c - 4 = 0$.
Also, $-2b - 3c = -1 \Rightarrow -2b - 2c - c = -2(b + c) - c = -1 \Rightarrow 2(b + c) = 1 - c \Rightarrow b + c = (1 - c)/2$.

Then, since $(b + c)^2 + c - 4 = 0$ and $b + c = (1 - c)/2$, we get $(1 - c)^2/4 + c - 4 = 0$.
So, $1 - 2c + c^2 + 4c - 16 = 0 \Rightarrow c^2 + 2c - 15 = 0 \Rightarrow (c + 5)(c - 3) = 0 \Rightarrow c = -5$ or 3.

Let's now, put the values of c into the system as follows: $a + b + c = 1$ and $4a + 2b + c = 3$.
Let's however, put the system in such a way as follows: $a + b = 1 - c$ and $4a + 2b = 3 - c$.
Now, if $c = -5$, we get $a + b = 1 + 5 = 6 \Rightarrow a + b = 6$, and also, $4a + 2b = 3 + 5 = 8 \Rightarrow 2a + b = 4$.
Then, $2a + b - (a + b) = 4 - 6 = -2 \Rightarrow a = -2 \Rightarrow a + b = -2 + b = 6 \Rightarrow b = 8$.
Next, if $c = 3$, we get $a + b = 1 - 3 = -2 \Rightarrow a + b = -2$, and also, $4a + 2b = 3 - 3 = 0 \Rightarrow 2a + b = 0$.
Then, $2a + b - (a + b) = 0 - (-2) = 2 \Rightarrow a = 2 \Rightarrow a + b = 2 + b = -2 \Rightarrow b = -4$.
Therefore, we get to have such two parabolas as follows:
$y = -2x^2 + 8x - 5 = -2(x^2 - 4x) - 5 = -2(x^2 - 4x + 4 - 4) - 5 = -2(x - 2)^2 + 3$
$\Rightarrow y - 3 = -2(x - 2)^2$.
$y = 2x^2 - 4x + 3 = 2(x^2 - 2x) + 3 = 2(x - 1)^2 + 1 \Rightarrow y - 1 = 2(x - 1)^2$.

In short:
Because the vertex of the parabola $y = ax^2 + bx + c$ is $\{-b/2a, -(b^2 - 4ac)/4a\}$ and is in the line $y = 2x - 1$: $-(b^2 - 4ac)/4a = 2(-b/2a) - 1$.
Also, since the two points (1, 1) and (2, 3) are in the parabola:
$(1, 1) \Rightarrow 1 = a(1^2) + b(1) + c = a + b + c$, and $(2, 3) \Rightarrow 3 = 4a + 2b + c$.

Now: $-(b^2 - 4ac)/4a = 2(-b/2a) - 1$, $a + b + c = 1$, and $4a + 2b + c = 3$.
To begin with:
$-(b^2 - 4ac)/4a = 2(-b/2a) - 1 = -b/a - 1 = -(b + a)/a \Rightarrow b^2 - 4ac = 4(b + a)$.
$a + b + c = 1 \Rightarrow a + b = 1 - c$.
Then, $b^2 - 4ac = 4(b + a) = 4(1 - c) \Rightarrow b^2 - 4ac = 4(1 - c)$.
Now, $4a + 2b + c = 3 \Rightarrow 4ac + 2bc + c^2 = 3c \Rightarrow 4ac = 3c - 2bc - c^2$.
Then, $b^2 - 4ac = 4(1 - c) \Rightarrow b^2 - (3c - 2bc - c^2) = b^2 - 3c + 2bc + c^2 = 4(1 - c) = 4 - 4c$.
So, $b^2 - 3c + 2bc + c^2 = 4 - 4c \Rightarrow b^2 + 2bc + c^2 + c - 4 = 0$.
Now, $4a + 2b + c - 4(a + b + c) = 3 - 4 \Rightarrow -2b - 3c = -1$.
Meanwhile, $b^2 + 2bc + c^2 + c - 4 = 0 \Rightarrow (b + c)^2 + c - 4 = 0$.
Also, $-2b - 3c = -1 \Rightarrow -2b - 2c - c = -2(b + c) - c = -1 \Rightarrow 2(b + c) = 1 - c \Rightarrow b + c = (1 - c)/2$.
Then, since $(b + c)^2 + c - 4 = 0$ and $b + c = (1 - c)/2$, we get $(1 - c)^2/4 + c - 4 = 0$.
So, $1 - 2c + c^2 + 4c - 16 = 0 \Rightarrow c^2 + 2c - 15 = 0 \Rightarrow (c + 5)(c - 3) = 0 \Rightarrow c = -5$ or 3.
Now, since $a + b + c = 1$ and $4a + 2b + c = 3$: $a + b = 1 - c$ and $4a + 2b = 3 - c$.

First, if $c = -5$, we get $a + b = 1 + 5 = 6 \Rightarrow a + b = 6$, and also, $4a + 2b = 3 + 5 = 8 \Rightarrow 2a + b = 4$.
Then, $2a + b - (a + b) = 4 - 6 = -2 \Rightarrow a = -2 \Rightarrow a + b = -2 + b = 6 \Rightarrow b = 8$.
Therefore, $a = -2$, $b = 8$, and $c = -5$.
Next, if $c = 3$, we get $a + b = 1 - 3 = -2 \Rightarrow a + b = -2$, and also, $4a + 2b = 3 - 3 = 0 \Rightarrow 2a + b = 0$.
Then, $2a + b - (a + b) = 0 - (-2) = 2 \Rightarrow a = 2 \Rightarrow a + b = 2 + b = -2 \Rightarrow b = -4$.
Therefore, $a = 2$, $b = -4$, and $c = 3$.

Fig. 7.0

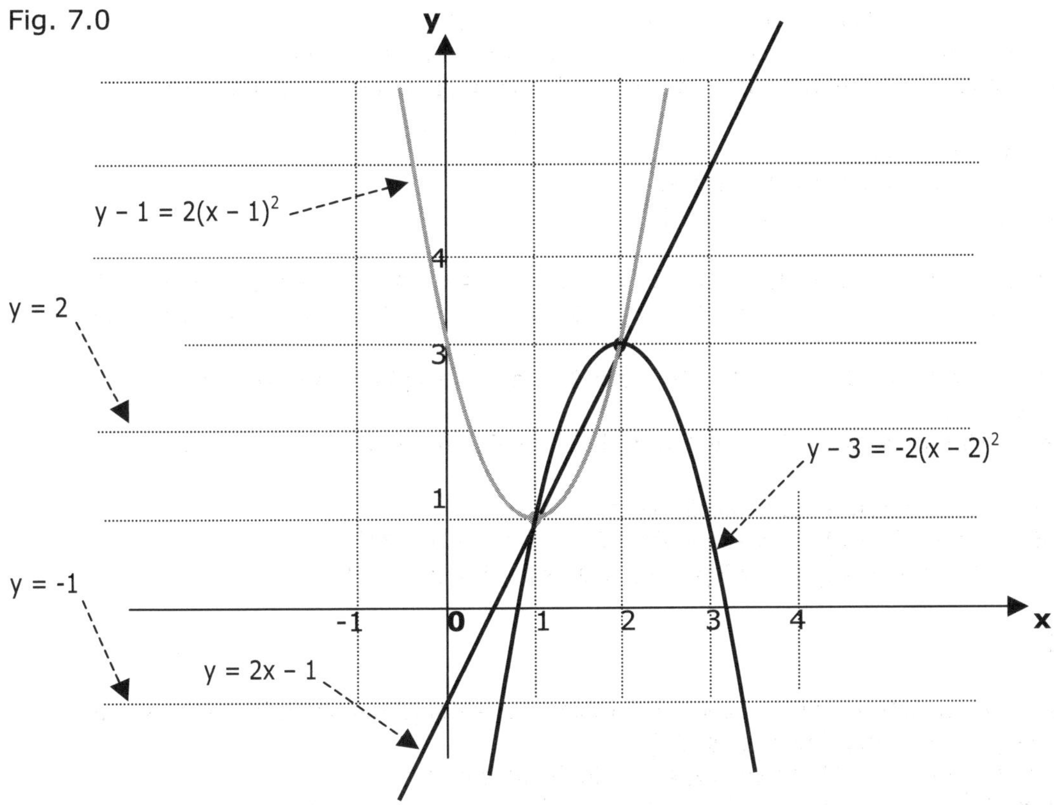

Suggestions or Solutions

To the Problem in the Example 8

Suppose that a parabola $y = ax^2 + bx + c$ includes two points (1, 1) and (2, 4), and has the vertex in a line $y = 2x - 1$. Then, find the values of the a, the b, and the c.

In this example, the difference from the previous one is only the position of one of the two points. In the previous one, we had **(2, 3)** instead of **(2, 4)**.
The vertex of the parabola $y = ax^2 + bx + c$ is $\{-b/2a, -(b^2 - 4ac)/4a\}$. If not sure, then we may want to derive it.
Now, since the vertex is in the line $y = 2x - 1$, we get an equation as follows:
$-(b^2 - 4ac)/4a = 2(-b/2a) - 1$.
Besides, since the two points **(1, 1)** and **(2, 4)** are in the parabola:
$(1, 1) \Rightarrow 1 = a(1^2) + b(1) + c = a + b + c$, and also, $(2, 4) \Rightarrow 4 = 4a + 2b + c$.
Now, all we have to do is the algebra. We need to solve the system of equations as follows:
$-(b^2 - 4ac)/4a = 2(-b/2a) - 1$, $a + b + c = 1$, and $4a + 2b + c = 4$.

To begin with, let's simplify the first equation:
$-(b^2 - 4ac)/4a = 2(-b/2a) - 1 = -b/a - 1 = -(b + a)/a \Rightarrow b^2 - 4ac = 4(b + a)$.
Meanwhile, $a + b + c = 1 \Rightarrow a + b = 1 - c$.
Then, $b^2 - 4ac = 4(b + a) = 4(1 - c) \Rightarrow b^2 - 4ac = 4(1 - c)$.
Now, $4a + 2b + c = 4 \Rightarrow 4ac + 2bc + c^2 = 4c \Rightarrow 4ac = 4c - 2bc - c^2$.
Then, $b^2 - 4ac = 4(1 - c) \Rightarrow b^2 - (4c - 2bc - c^2) = b^2 - 4c + 2bc + c^2 = 4(1 - c) = 4 - 4c$.
So, $b^2 - 4c + 2bc + c^2 = 4 - 4c \Rightarrow b^2 + 2bc + c^2 = 4 \Rightarrow (b + c)^2 = 4 \Rightarrow b + c = 2$ or -2.

Now, since $a + b + c = 1$:
$b + c = 2 \Rightarrow a + b + c = a + 2 = 1 \Rightarrow a = -1$.
$b + c = -2 \Rightarrow a + b + c = a - 2 = 1 \Rightarrow a = 3$.

Then first, if $a = -1$, we get $4a + 2b + c = 4 \Rightarrow 2b + c = 4 - 4a = 4 - 4(-1) = 8 \Rightarrow 2b + c = 8$.
Also, we have $b + c = 2$. Then, $2b + c - (b + c) = 8 - 2 = 6 \Rightarrow b = 6 \Rightarrow b + c = 6 + c = 2 \Rightarrow c = -4$.

Next, if $a = 3$, we get $4a + 2b + c = 4 \Rightarrow 2b + c = 4 - 4a = 4 - 4(3) = -8 \Rightarrow 2b + c = -8$.
Also, we have $b + c = -2$.
Then, $2b + c - (b + c) = -8 - (-2) = -8 + 2 = -6 \Rightarrow b = -6 \Rightarrow b + c = -6 + c = -2 \Rightarrow c = 4$.

Therefore, we get to have such two parabolas as follows:
$y = -x^2 + 6x - 4 = -(x^2 - 6x) - 4 = -(x - 3)^2 + 5 \Rightarrow y - 5 = -(x - 3)^2$.
$y = 3x^2 - 6x + 4 = 3(x^2 - 2x) + 4 = 3(x - 1)^2 + 1 \Rightarrow y - 1 = 3(x - 1)^2$.

In short:

The vertex of the parabola $y = ax^2 + bx + c$ is $\{-b/2a, -(b^2 - 4ac)/4a\}$ and is in the line $y = 2x - 1$.
Therefore, we get $-(b^2 - 4ac)/4a = 2(-b/2a) - 1$.
Besides, since the two points **(1, 1)** and **(2, 4)** are in the parabola:
$(1, 1) \Rightarrow 1 = a(1^2) + b(1) + c = a + b + c$, and also, $(2, 4) \Rightarrow 4 = 4a + 2b + c$.
Now, we need to solve a system of equations as follows:
$-(b^2 - 4ac)/4a = 2(-b/2a) - 1$, $a + b + c = 1$, and $4a + 2b + c = 4$.

$-(b^2 - 4ac)/4a = 2(-b/2a) - 1 = -b/a - 1 = -(b + a)/a \Rightarrow b^2 - 4ac = 4(b + a)$.
$a + b + c = 1 \Rightarrow a + b = 1 - c$.

Then, $b^2 - 4ac = 4(b + a) = 4(1 - c) \Rightarrow b^2 - 4ac = 4(1 - c)$.
Now, $4a + 2b + c = 4 \Rightarrow 4ac + 2bc + c^2 = 4c \Rightarrow 4ac = 4c - 2bc - c^2$.
Then, $b^2 - 4ac = 4(1 - c) \Rightarrow b^2 - (4c - 2bc - c^2) = b^2 - 4c + 2bc + c^2 = 4(1 - c) = 4 - 4c$.
So, $b^2 - 4c + 2bc + c^2 = 4 - 4c \Rightarrow b^2 + 2bc + c^2 = 4 \Rightarrow (b + c)^2 = 4 \Rightarrow b + c = 2$ or -2.

Now, since $a + b + c = 1$:
$b + c = 2 \Rightarrow a + b + c = a + 2 = 1 \Rightarrow a = -1$.
$b + c = -2 \Rightarrow a + b + c = a - 2 = 1 \Rightarrow a = 3$.

First, when $a = -1$ and $b + c = 2$:
$4a + 2b + c = 4 \Rightarrow 2b + c = 4 - 4a = 4 - 4(-1) = 8 \Rightarrow 2b + c = 8$.
Then, $2b + c - (b + c) = 8 - 2 = 6 \Rightarrow b = 6 \Rightarrow b + c = 6 + c = 2 \Rightarrow c = -4$.
Therefore, $a = -1$, $b = 6$, and $c = -4$.

Next, when $a = 3$ and $b + c = -2$:
$4a + 2b + c = 4 \Rightarrow 2b + c = 4 - 4a = 4 - 4(3) = -8 \Rightarrow 2b + c = -8$.
Then, $2b + c - (b + c) = -8 - (-2) = -8 + 2 = -6 \Rightarrow b = -6 \Rightarrow b + c = -6 + c = -2 \Rightarrow c = 4$.
Therefore, $a = 3$, $b = -6$, and $c = 4$.

Fig. 8.0

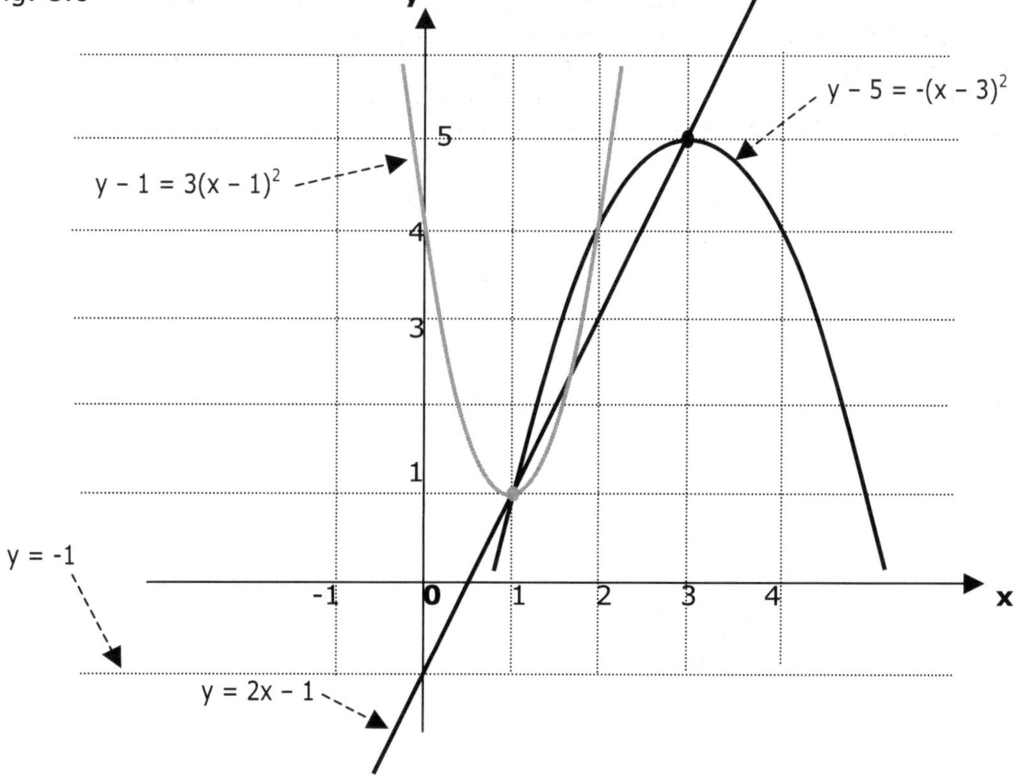

Suggestions or Solutions
To the Problem in the Example 9

Suppose that the equation of a parabola A is $y = -4x^2 + 8mx - n(n + 2)$, and that the equation of a parabola B is $y = 8x^2 - 2x + 5/4$. Suppose also, that the vertex of the A is in the B. Then, find the values of the m and the n.

To begin with, let's find the vertex of the parabola **A** and put it into the parabola **B**.
Then, $y = -4x^2 + 8mx - n(n + 2) = -4(x^2 - 2mx) - n(n + 2) = -4(x - m)^2 + 4m^2 - n(n + 2)$.
So, we can see that the vertex = $\{m, 4m^2 - n(n + 2)\}$. Now, we need to put it into the parabola **B**.
Then, $4m^2 - n(n + 2) = 8m^2 - 2m + 5/4 \Rightarrow 4m^2 - 2m + n(n + 2) + 5/4 = 0$.

Now, examining the expression above, we can see that it has an n^2-term. Then, it could be the sum of the complete squares for the **m** and the **n** respectively. Let's try the complete square, then.
$4m^2 - 2m + n(n + 2) + 5/4 = 4m^2 - 2m + 1/4 + n^2 + 2n + 1 = (2m - 1/2)^2 + (n + 1)^2 = 0$.

It is the case. Now, we need to consider the fact as follows : **if $A^2 + B^2 = 0$**, both **A** and **B** are 0.
So, we need to have $2m - 1/2 = 0$, and also, $n + 1 = 0$. Therefore, $m = 1/4$ and $n = -1$.

In short:

$y = -4x^2 + 8mx - n(n + 2) = -4(x^2 - 2mx) - n(n + 2) = -4(x - m)^2 + 4m^2 - n(n + 2)$.
So, the vertex = $\{m, 4m^2 - n(n + 2)\}$. The vertex is in the parabola **B**.
Therefore, $4m^2 - n(n + 2) = 8m^2 - 2m + 5/4 \Rightarrow 4m^2 - 2m + n(n + 2) + 5/4 = 0$. Now:
$4m^2 - 2m + n(n + 2) + 5/4 = 4m^2 - 2m + 1/4 + n^2 + 2n + 1 = (2m - 1/2)^2 + (n + 1)^2 = 0$.
So, we need to have $2m - 1/2 = 0$, and also, $n + 1 = 0$. Therefore, $m = 1/4$ and $n = -1$.

Fig. 9.0

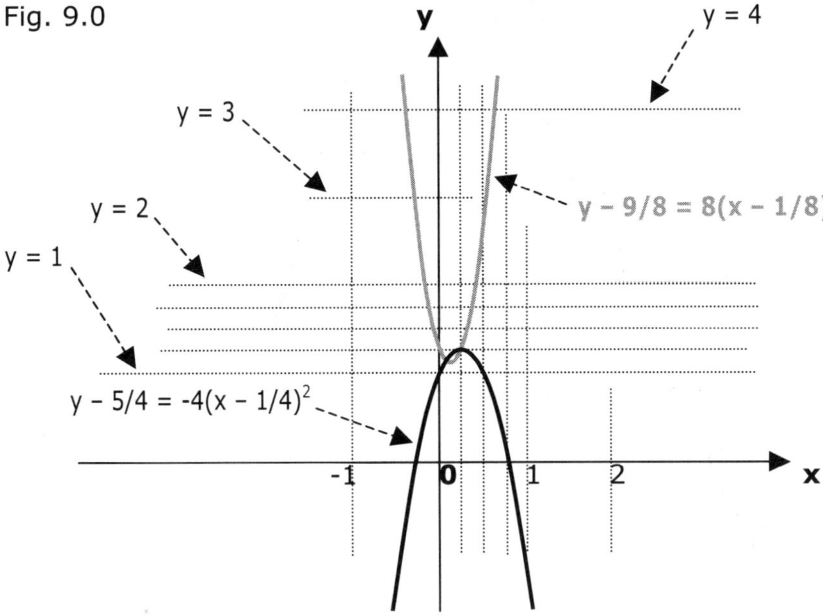

Suggestions or Solutions

To the Problem in the Example A

Suppose that the equation of a parabola A is $y = x^2 + bx + c$, and that the equation of a parabola B is $y = -2x^2 - x + 5$. Suppose also, that the vertex of the A is in the B, and the vertex of the B, too, is in the A. Then, find the values of the b and the c.

Since the signs of the slopes of the **A** and the **B** are opposite of each other, so are their concavities. Suppose in the figure below, that **A** is one of the parabolas in gray, and that **B** is the parabola in black. Then, we can see that there can be three ways where the two parabolas include each other's vertex. Therefore, there needs to be either two solutions or one solution. If the **A** and **B** share one point as the vertex, there is one solution and the **A** is the dotted gray parabola; otherwise, there are two solutions and the **A** needs to be one of the two solid gray parabolas.

Fig. A.0

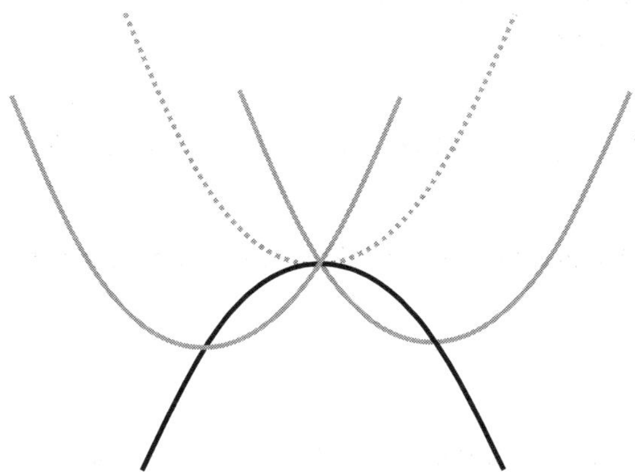

Now, finding the values of the **b** and **c** in the equation of the **A** is no other than finding the parabola **A**. Then, let's find the parabola **A**. However, we are first going to find the vertex of the parabola **A**, and then, put the vertex into the parabola **B**.

$y = x^2 + bx + c = x^2 + bx + (b/2)^2 - (b/2)^2 + c = (x + b/2)^2 - (b/2)^2 + c$
$= (x + b/2)^2 - (b^2 - 4c)/4.$

Therefore, we can see that the vertex of the **A** is $\{-b/2, -(b^2 - 4c)/4\}$.
Let's now, put the vertex into the **B**. The **B** is $y = -2x^2 - x + 5$. Then:
$\{-b/2, -(b^2 - 4c)/4\} \Rightarrow -(b^2 - 4c)/4 = -2(-b/2)^2 - (-b/2) + 5 = -b^2/2 + b/2 + 5$
$\Rightarrow b^2 - 4c = 2b^2 - 2b - 20 \Rightarrow b^2 - 2b + 4c - 20 = 0.$

Also, since the vertex of the parabola **B** needs to be in the parabola **A**, we need to find the vertex, too, and put it into the **A**. Then, $y = -2x^2 - x + 5 = -2(x^2 + x/2) + 5 = -2(x + 1/4)^2 + (-2)(-1/16) + 5$
$= -2(x + 1/4)^2 + 1/8 + 5 = -2(x + 1/4)^2 + 41/8$.

Therefore, the vertex of the **B** is $(-1/4, 41/8)$. We now, put it into the **A**, which is $y = x^2 + bx + c$.
$41/8 = (-1/4)^2 + (-1/4)b + c = 1/16 - b/4 + c$
$\Rightarrow 41 = 1/2 - 2b + 8c \Rightarrow 8c - 2b = 81/2 \Rightarrow 4c - b = 81/4$.

Now, we need to solve a system of equations as follows: $b^2 - 2b + 4c - 20 = 0$ and $4c - b = 81/4$.

$4c - b = 81/4 \Rightarrow 4c = b + 81/4$. Then:
$b^2 - 2b + 4c - 20 = 0 \Rightarrow b^2 - 2b + b + 81/4 - 20 = b^2 - b + 1/4 = 0 \Rightarrow (b - 1/2)^2 = 0 \Rightarrow b = 1/2.$

Then, $\mathbf{b = 1/2 \Rightarrow 4c - b = 4c - 1/2 = 81/4 \Rightarrow 4c = 81/4 + 1/2 = 83/4 \Rightarrow c = 83/16}$.
Therefore, the equation of the parabola **A** is $\mathbf{y = x^2 + x/2 + 83/16}$.

Now, since we have only one solution, the **A** needs to meet the **B** in such a way as below.

Fig. A.1

In other words, the vertices are the same as each other.
Let's now, check to see if the vertices are the same as each other indeed.
Since the vertex of the **A** is $\mathbf{\{-b/2, -(b^2 - 4c)/4\}}$:

$\mathbf{b = 1/2 \Rightarrow -b/2 = -1/4}$.
Since $\mathbf{b = 1/2}$ and $\mathbf{c = 83/16}$, we get $\mathbf{-(b^2 - 4c)/4 = -\{1/4 - 4(83/16)\}/4 = -(1/4 - 83/4)/4}$ $\mathbf{= 82/4/4 = 41/8}$.

Therefore, the vertex of the **A** is $\mathbf{(-1/4, 41/8)}$, and the vertices of the **A** and **B** are the same.

In short:

We first find the vertex of the parabola **A**, and then, put the vertex into the parabola **B**.
$\mathbf{y = x^2 + bx + c = x^2 + bx + (b/2)^2 - (b/2)^2 + c = (x + b/2)^2 - (b/2)^2 + c}$
$\mathbf{= (x + b/2)^2 - (b^2 - 4c)/4}$.

Therefore, the vertex is $\mathbf{\{-b/2, -(b^2 - 4c)/4\}}$.
We now, put the vertex into the **B**. The **B** is $\mathbf{y = -2x^2 - x + 5}$. Then:
$\mathbf{\{-b/2, -(b^2 - 4c)/4\} \Rightarrow -(b^2 - 4c)/4 = -2(-b/2)^2 - (-b/2) + 5 = -b^2/2 + b/2 + 5}$
$\mathbf{\Rightarrow b^2 - 4c = 2b^2 - 2b - 20 \Rightarrow b^2 - 2b + 4c - 20 = 0}$.

Also, since the vertex of the **B** is to be in the **A**, we need to find the vertex, too, and put it into the **A**.
Then, $\mathbf{y = -2x^2 - x + 5 = -2(x^2 + x/2) + 5 = -2(x + 1/4)^2 + (-2)(-1/16) + 5}$
$\mathbf{= -2(x + 1/4)^2 + 1/8 + 5 = -2(x + 1/4)^2 + 41/8}$.
Therefore, the vertex of the **B** is $\mathbf{(-1/4, 41/8)}$. We now, put it into the **A**, which is $\mathbf{y = x^2 + bx + c}$.
$\mathbf{41/8 = (-1/4)^2 + (-1/4)b + c = 1/16 - b/4 + c}$
$\mathbf{\Rightarrow 41 = 1/2 - 2b + 8c \Rightarrow 8c - 2b = 81/2 \Rightarrow 4c - b = 81/4}$.

Now, we need to solve a system of equations as follows: $\mathbf{b^2 - 2b + 4c - 20 = 0}$ and $\mathbf{4c - b = 81/4}$.

$\mathbf{4c - b = 81/4 \Rightarrow 4c = b + 81/4}$. Then:
$\mathbf{b^2 - 2b + 4c - 20 = 0 \Rightarrow b^2 - 2b + b + 81/4 - 20 = b^2 - b + 1/4 = 0 \Rightarrow (b - 1/2)^2 = 0 \Rightarrow b = 1/2}$.

Then, $\mathbf{b = 1/2 \Rightarrow 4c - b = 4c - 1/2 = 81/4 \Rightarrow 4c = 81/4 + 1/2 = 83/4 \Rightarrow c = 83/16}$.

Suggestions or Solutions

To the Problem in the Example B

Suppose that the equation of a parabola A is $y = 2x^2 + bx + c$, and that the equation of a parabola B is $y = -2x^2 + 8x - 5$. Suppose also, that the vertex of the A is in the B, and the vertex of the B, too, is in the A. Then, find the values of the b and the c.

This example is almost the same as the one presented earlier.
We are first going to find the vertex of the parabola **A**, and then, put the vertex into the parabola **B**.
$y = 2x^2 + bx + c = 2\{x^2 + bx/2 + (b/4)^2 - (b/4)^2\} + c = 2(x + b/4)^2 - 2(b/4)^2 + c$
$= 2(x + b/4)^2 - (b^2 - 8c)/8.$
Then, we can see that the vertex is $\{-b/4, -(b^2 - 8c)/8\}$. Let's now, put the vertex into the **B**.
The **B** is $y = -2x^2 + 8x - 5$. Then:
$-(b^2 - 8c)/8 = -2(-b/4)^2 + 8(-b/4) - 5 = -b^2/8 - 2b - 5$
$\Rightarrow b^2 - 8c = b^2 + 16b + 40 \Rightarrow 16b + 8c + 40 = 0 \Rightarrow 2b + c + 5 = 0.$

Also, since the vertex of the parabola **B** needs to be in the parabola **A**, we need to find the vertex and put it into the **A**. Then:
$y = -2x^2 + 8x - 5 = -2(x^2 - 4x) - 5 = -2(x - 2)^2 + 3.$
Then, we can see that the vertex of the **B** is **(2, 3)**. We now, put it into the **A**, which is $y = 2x^2 + bx + c$.
$3 = 2(2^2) + 2b + c = 8 + 2b + c \Rightarrow 2b + c = -5.$

Now, we have two unknowns, but have one equation only: $2b + c = -5$ is the same as $2b + c + 5 = 0$.
Now, what else can we do?
Nothing else, and there are infinitely many solutions. In other words, all the values of the **b** and **c** that satisfy the connective expression $2b + c + 5 = 0$ can be the solutions. That is, there are infinitely many parabolas that can be the **A** where the values of the **b** and **c** satisfy $2b + c = -5$.

Fig. B.0

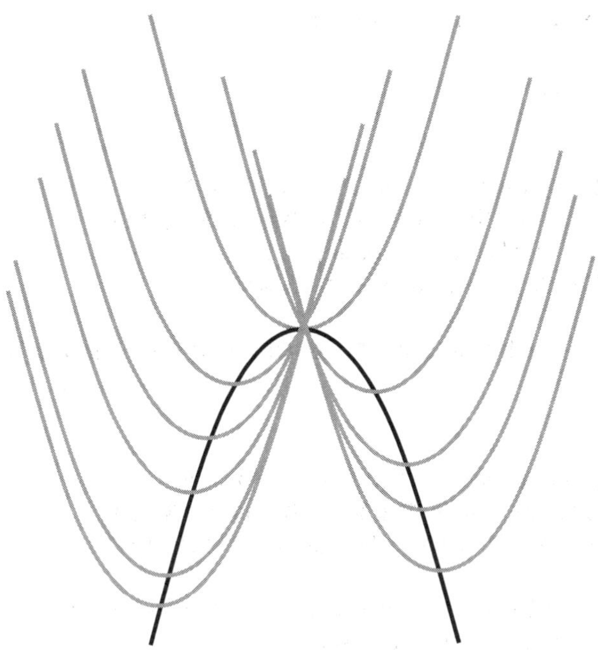

Suggestions or Solutions

To the Problem in the Example C

Suppose that the equation of a parabola A is $y = x^2 + bx + c$, and that the equation of a parabola B is $y = -2x^2 + 8x - 5$. Suppose also, that the vertex of the A is in the B, and the vertex of the B, too, is in the A. Then, find the values of the b and the c.

This example, too, is almost the same as the previous ones.
We are first going to find the vertex of the parabola **A**, and then, put the vertex into the parabola **B**.

$y = x^2 + bx + c = (x + b/2)^2 - (b^2 - 4c)/4$. Then, the vertex is $\{-b/2, -(b^2 - 4c)/4\}$.
Let's put the vertex into the **B**. The **B** is $y = -2x^2 + 8x - 5$. Then:
$-(b^2 - 4c)/4 = -2(-b/2)^2 + 8(-b/2) - 5 = -b^2/2 - 4b - 5 \Rightarrow b^2 - 4c = 2b^2 + 16b + 20$
$\Rightarrow b^2 + 16b + 4c + 20 = 0$.

Also, since the vertex of the parabola **B** needs to be in the parabola **A**, we need to find the vertex and put it into the **A**.

Then, $y = -2x^2 + 8x - 5 = -2(x - 2)^2 + 3$. Then, the vertex of the **B** is $(2, 3)$.
We now, put it into the **A**, which is $y = x^2 + bx + c$. Then:
$3 = (2^2) + 2b + c = 4 + 2b + c \Rightarrow 2b + c = -1$.

Now, we need to solve a system of equations as follows: $b^2 + 16b + 4c + 20 = 0$ and $2b + c = -1$.

$2b + c = -1 \Rightarrow c = -1 - 2b \Rightarrow 4c = -4 - 8b$. Then:
$b^2 + 16b + 4c + 20 = 0 \Rightarrow b^2 + 16b - 4 - 8b + 20 = b^2 + 8b + 16 = (b + 4)^2 = 0 \Rightarrow b = -4$.
Then, $b = -4 \Rightarrow 2b + c = -8 + c = -1 \Rightarrow c = 7$. Therefore, the equation of the **A** is $y = x^2 - 4x + 7$.

Now, $y = x^2 - 4x + 7 = (x - 2)^2 + 3 \Rightarrow$ The vertex $= (2, 3)$, which is the same as the vertex of the **B**.

Now, we can make a summary as follows.

Suppose that:
We have a particular parabola **B**, and want to find another parabola **A**.
The vertex of the **A** is in the **B**, and also, the vertex of the **B** is in the **A**, too.
The axes of symmetry of those are parallel to each other.
The magnitudes of the slopes of those are the same as each other.

Then, infinitely many parabolas can be the parabola **A**, but each of those parabolas doesn't have to share the vertex with the parabola **B**.

Suppose now, that:
We have a particular parabola **B**, and want to find another parabola **A**.
The vertex of the **A** is in the **B**, and also, the vertex of the **B** is in the **A**, too.
However, the magnitude of the slopes of those are different from each other.

Then, infinitely many parabolas can be the parabola **A**, and each of those parabolas shares the vertex with the parabola **B**; that is, the vertices of the **A** and **B** are the same.

Suggestions or Solutions

To the Problem 0 in the Example D

The equation is as follows: $y = 2(x - 1)^2 - 2|x - 2| + 1$.

We know that $|x - 2| \geq 0$ by definition of the absolute sign.
We also know that there are two cases of $x - 2 \geq 0$ and $x - 2 < 0$.

If $x - 2 \geq 0$, then $|x - 2| = x - 2$.
If $x - 2 < 0$, then $|x - 2| = -(x - 2) = -x + 2$.

$x - 2 \geq 0 \Rightarrow x \geq 2 \Rightarrow y = 2(x - 1)^2 - 2(x - 2) + 1 = 2x^2 - 4x + 2 - 2x + 4 + 1 = 2x^2 - 6x + 7$
$= 2(x^2 - 3x) + 7 = 2(x - 3/2)^2 - 2(3/2)^2 + 7 = 2(x - 3/2)^2 + 5/2$
$\Rightarrow y - 5/2 = 2(x - 3/2)^2$ for $x \geq 2$.

$x - 2 < 0 \Rightarrow x < 2 \Rightarrow y = 2(x - 1)^2 - \{-2(x - 2)\} + 1 = 2x^2 - 4x + 2 + 2x - 4 + 1 = 2x^2 - 2x - 1$
$= 2(x^2 - x) - 1 = 2(x - 1/2)^2 - 2(1/2)^2 - 1 = 2(x - 1/2)^2 - 3/2$
$\Rightarrow y + 3/2 = 2(x - 1/2)^2$ for $x < 2$.

Fig. D.0.0

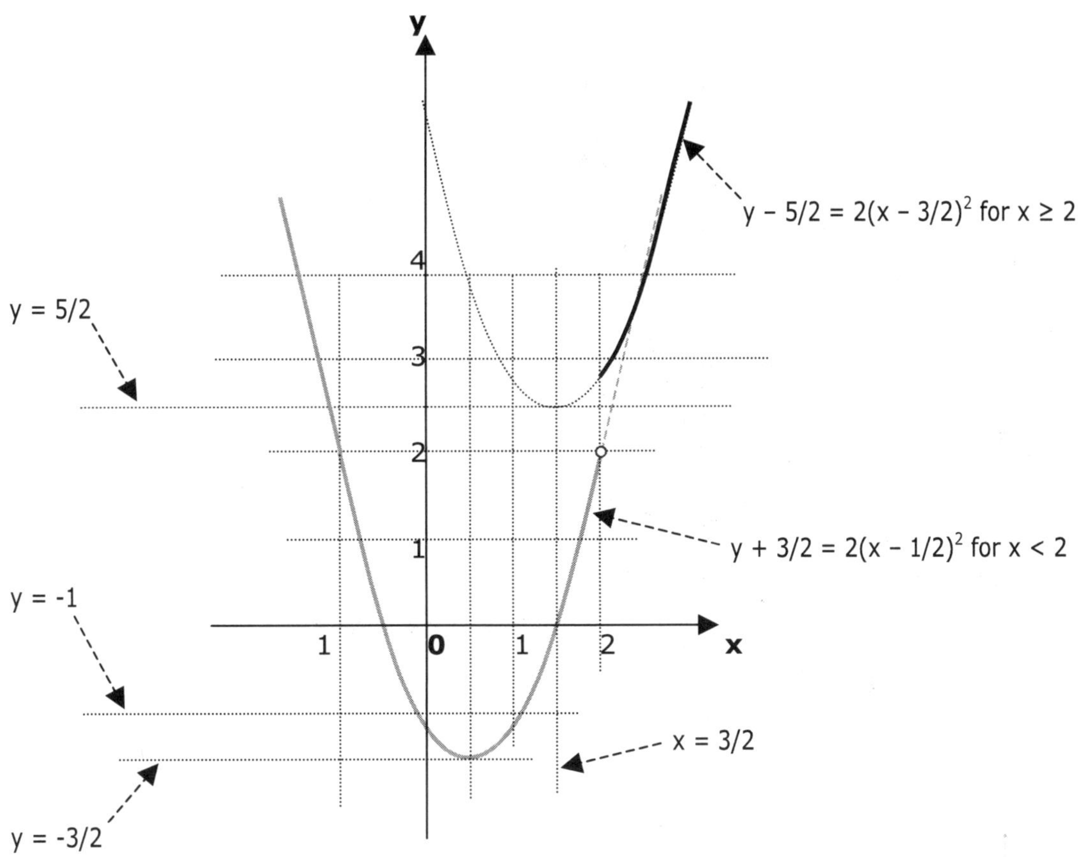

Suggestions or Solutions

To the Problem 1 in the Example D

The equation is as follows: $y = (|2x| - 1)(x + 1)$.

We know that $|2x| \geq 0$ by definition of the absolute sign.
We also, know that there are two cases of $x \geq 0$ and $x < 0$.
Why not 2 in there? That's because 2 is positive and does not affect the negative or positive sign.

If $x \geq 0$, then $|2x| = 2x$. If $x < 0$, then $|2x| = -2x$.

$x \geq 0 \Rightarrow y = (|2x| - 1)(x + 1) = (2x - 1)(x + 1)$ for $x \geq 0$ where we can see the x-intercepts, which are the roots, which are 1/2 and -1.

$x < 0 \Rightarrow y = (|2x| - 1)(x + 1) = (-2x - 1)(x + 1) = -(2x + 1)(x + 1)$ for $x < 0$ where we can see the x-intercepts are -1/2 and -1.

Fig. D.1.0

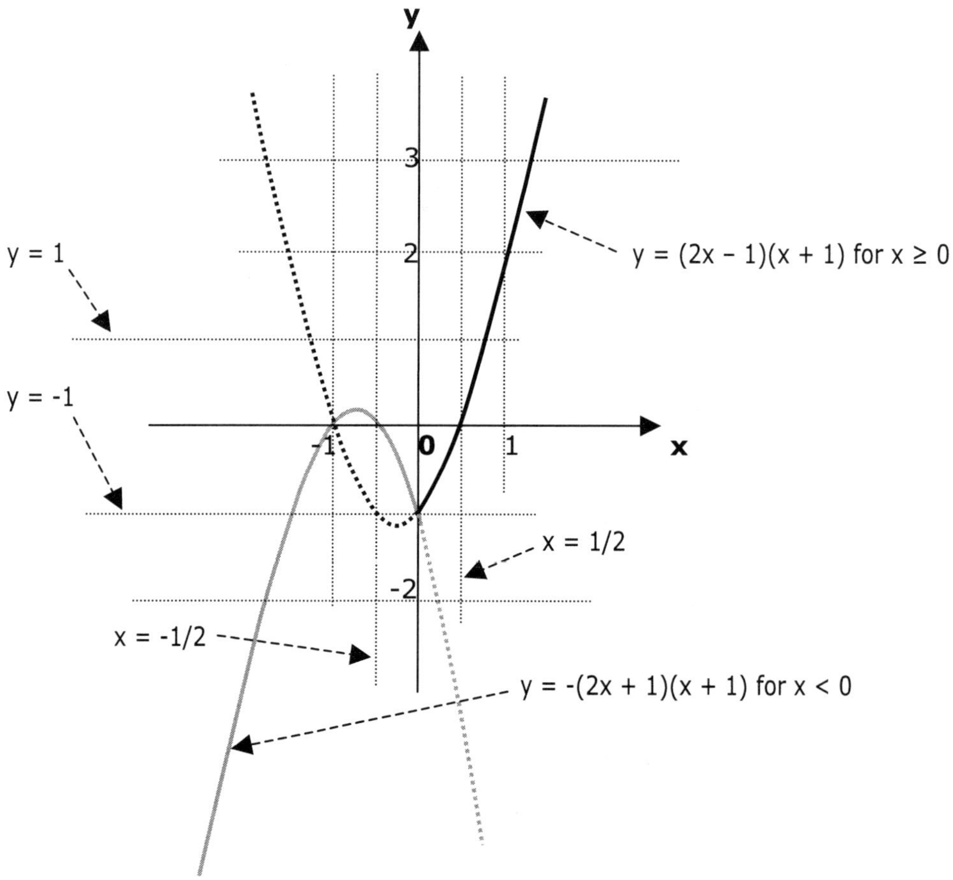

Suggestions or Solutions
To the Problem 2 in the Example D

The equation is as follows: $\mathbf{y = |x^2 + 2x - 3|}$.

We know that $\mathbf{|x^2 + 3x + 2| \geq 0}$ by definition of the absolute sign.

$\mathbf{x^2 + 3x + 2 = (x + 1)(x + 2) \geq 0 \Rightarrow x \geq -1}$ or $\mathbf{x \leq -2}$
$\Rightarrow \mathbf{y = |x^2 + 3x + 2| = x^2 + 3x + 2 = (x + 1)(x + 2)}$ **for** $\mathbf{x \geq -1}$ or $\mathbf{x \leq -2}$ where we can see the x-intercepts are -1 and -2.

$\mathbf{x^2 + 3x + 2 = (x + 1)(x + 2) < 0 \Rightarrow -2 < x < -1}$
$\Rightarrow \mathbf{y = |x^2 + 3x + 2| = -(x^2 + 3x + 2) = -(x + 1)(x + 2)}$ for $\mathbf{-2 < x < -1}$ where we can see the x-intercepts are -1 and -2.

Fig. D.2.0

y = (x + 1)(x + 2) for x ≥ -1 or x ≤ -2

y = -(x + 1)(x + 2) for -2 < x < -1

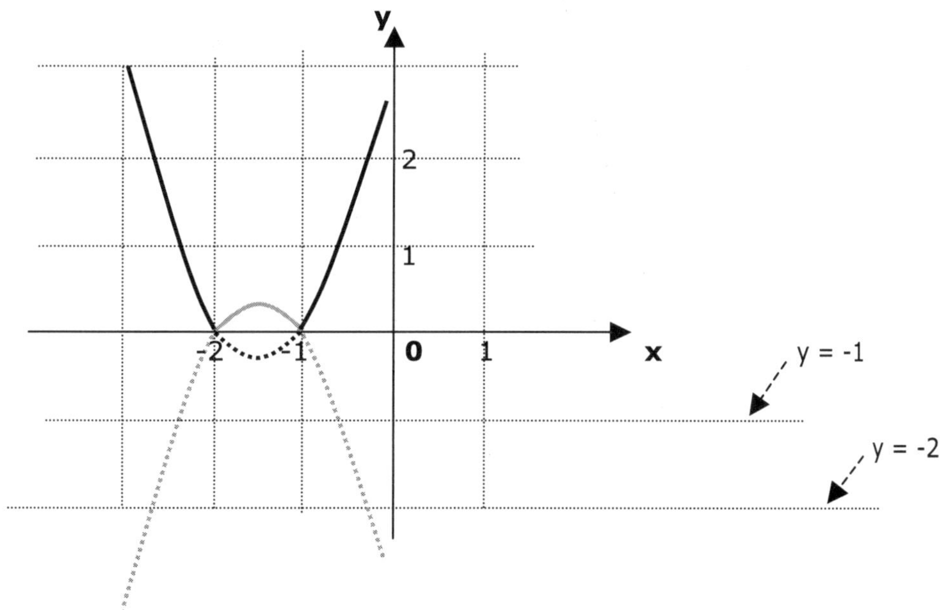

Suggestions or Solutions

To the Problem 3 in the Example D

The equation is as follows: $y = |x|^2 + 2|x| - 3$.

$x \ge 0 \Rightarrow y = |x|^2 + 2|x| - 3 = x^2 + 2x - 3 = (x - 1)(x + 3)$ for $x \ge 0$ where we can see the x-intercepts are 1 and -3.

$x < 0 \Rightarrow y = |x|^2 + 2|x| - 3 = x^2 - 2x - 3 = (x + 1)(x - 3)$ for $x < 0$ where we can see the x-intercepts are -1 and 3.

Fig. D.3.0

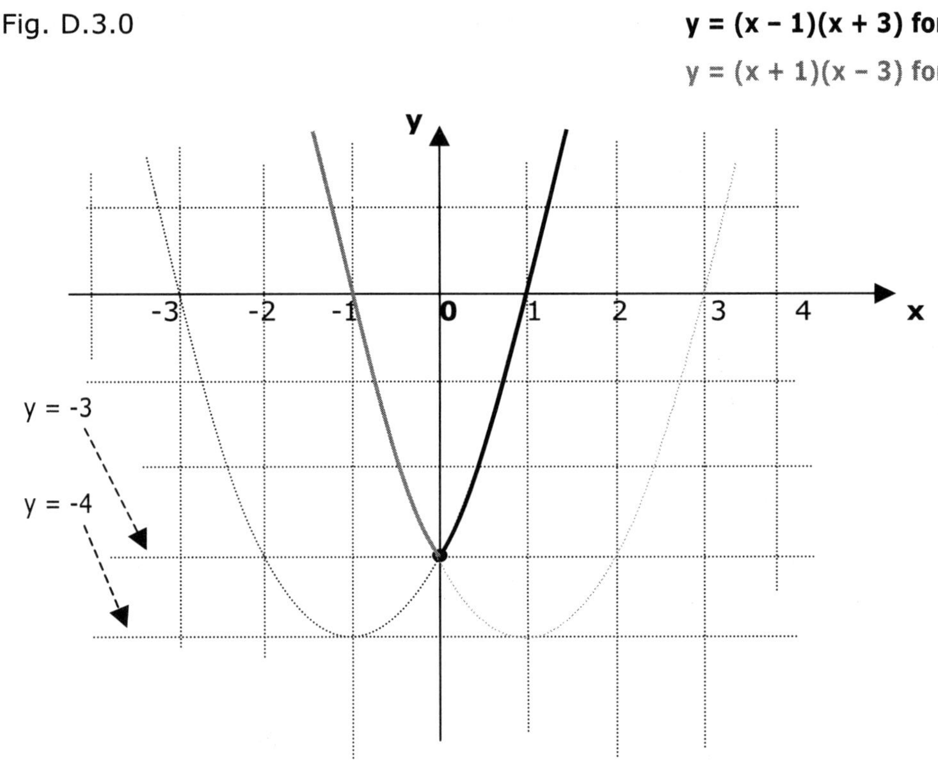

Suggestions or Solutions

To the **Problem 4** in the Example **D**

The equation is as follows: $|y| = x^2 + 2x - 3$.

We know that $|y| \geq 0$ by definition of the absolute sign.
We also know that we need to consider two cases of $y \geq 0$ and $y < 0$.

If $y \geq 0$, then $|y| = y$, and if $y < 0$, then $|y| = -y$.

$y \geq 0$
$\Rightarrow y = x^2 + 2x - 3 = (x + 3)(x - 1)$ for $y \geq 0$ where we can see the x-intercepts are 1 and -3.
$y < 0$
$\Rightarrow -y = x^2 + 2x - 3 \Rightarrow y = -(x^2 + 2x - 3) = -(x + 3)(x - 1)$ for $y < 0$ where we can see that x-intercepts are 1 and -3.

Fig. D.4.0 $y = (x + 3)(x - 1)$ for $y \geq 0$ $y = -(x + 3)(x - 1)$ for $y < 0$

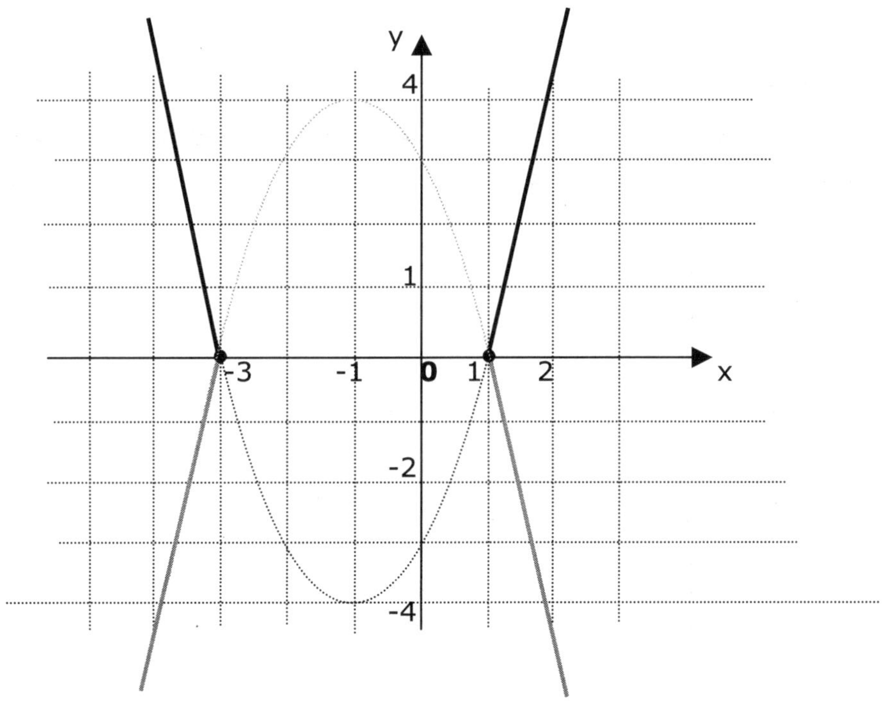

Suggestions or Solutions
To the Problem 5 in the Example D

The equation is as follows: $|y| = |x|^2 + 2|x| - 3$.

Since both the **x** and the **y** are covered by the absolute signs, we need to consider 4 different cases.

$x \geq 0$ and $y \geq 0 \Rightarrow y = x^2 + 2x - 3 = (x - 1)(x + 3)$ for $x \geq 0$ and $y \geq 0$.

$x \geq 0$ and $y < 0 \Rightarrow -y = x^2 + 2x - 3 \Rightarrow y = -(x^2 + 2x - 3) = -(x - 1)(x + 3)$ for $x \geq 0$ and $y < 0$.

$x < 0$ and $y \geq 0 \Rightarrow y = x^2 - 2x - 3 = (x + 1)(x - 3)$ for $x < 0$ and $y \geq 0$.

$x < 0$ and $y < 0 \Rightarrow -y = x^2 - 2x - 3 \Rightarrow y = -(x^2 - 2x - 3) = -(x + 1)(x - 3)$ for $x < 0$ and $y < 0$.

Now, the equation $|y| = |x|^2 + 2|x| - 3$ can be expanded to be put in such a way as follows:

$y = (x - 1)(x + 3)$ for $x \geq 0$ and $y \geq 0$. $y = -(x - 1)(x + 3)$ for $x \geq 0$ and $y < 0$.
$y = (x + 1)(x - 3)$ for $x < 0$ and $y \geq 0$. $y = -(x + 1)(x - 3)$ for $x < 0$ and $y < 0$.

Fig. D.5.0

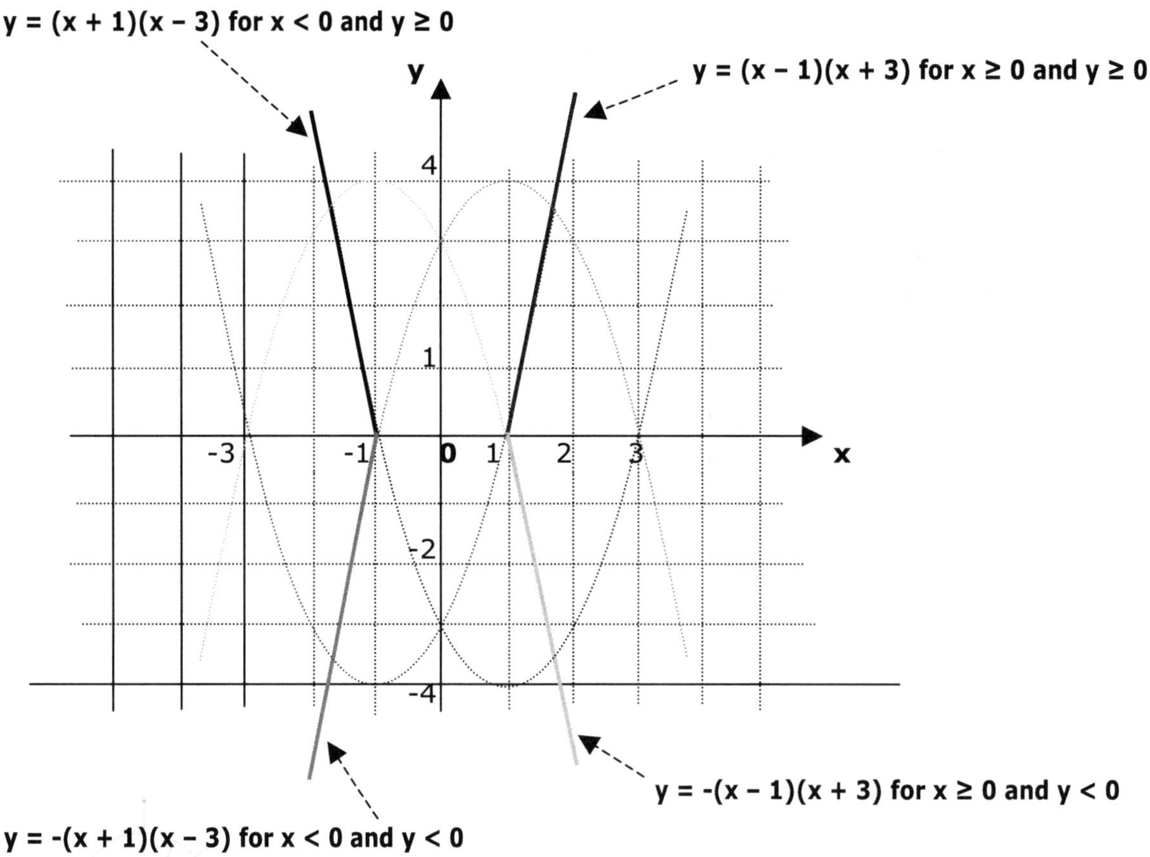

Suggestions or Solutions
To the Problem 6 in the Example D

The equation is as follows: $|y| = x^2 + |2x - 3|$.

Since we have '**y**' and '**2x − 3**' covered by the absolute sign, we need to consider two cases each, that is, 4 cases in total.

$2x - 3 \geq 0$ and $y \geq 0 \Rightarrow x \geq 3/2$ and $y \geq 0$
$\Rightarrow y = x^2 + 2x - 3 = (x - 1)(x + 3)$ for $x \geq 3/2$ and $y \geq 0$.

$2x - 3 \geq 0$ and $y < 0 \Rightarrow x \geq 3/2$ and $y < 0$
$\Rightarrow -y = x^2 + 2x - 3 \Rightarrow y = -(x^2 + 2x - 3) = -(x - 1)(x + 3)$ for $x \geq 3/2$ and $y < 0$.

$2x - 3 < 0$ and $y \geq 0 \Rightarrow x < 3/2$ and $y \geq 0$
$\Rightarrow y = x^2 - (2x - 3) = x^2 - 2x + 3 = (x - 1)^2 - 1 + 3 = (x - 1) + 2$ for $x < 3/2$ and $y \geq 0$.

$2x - 3 < 0$ and $y < 0 \Rightarrow x < 3/2$ and $y < 0 \Rightarrow -y = x^2 - (2x - 3) = x^2 - 2x + 3$
$\Rightarrow y = -(x^2 - 2x + 3) = -\{(x - 1) + 2\} = -(x - 1)^2 - 2$ for $x < 0$ and $y < 0$.

Now, the equation $|y| = x^2 + |2x - 3|$ can be expanded to be put in such a way as follows:

$y = (x - 1)(x + 3)$ for $x \geq 3/2$ and $y \geq 0$.
$y = -(x - 1)(x + 3)$ for $x \geq 3/2$ and $y < 0$.
$y = (x - 1)^2 + 2$ for $x < 3/2$ and $y \geq 0$.
$y = -(x - 1)^2 - 2$ for $x < 3/2$ and $y < 0$.

Let's put the equation in a graph now.

Fig. D.6.0

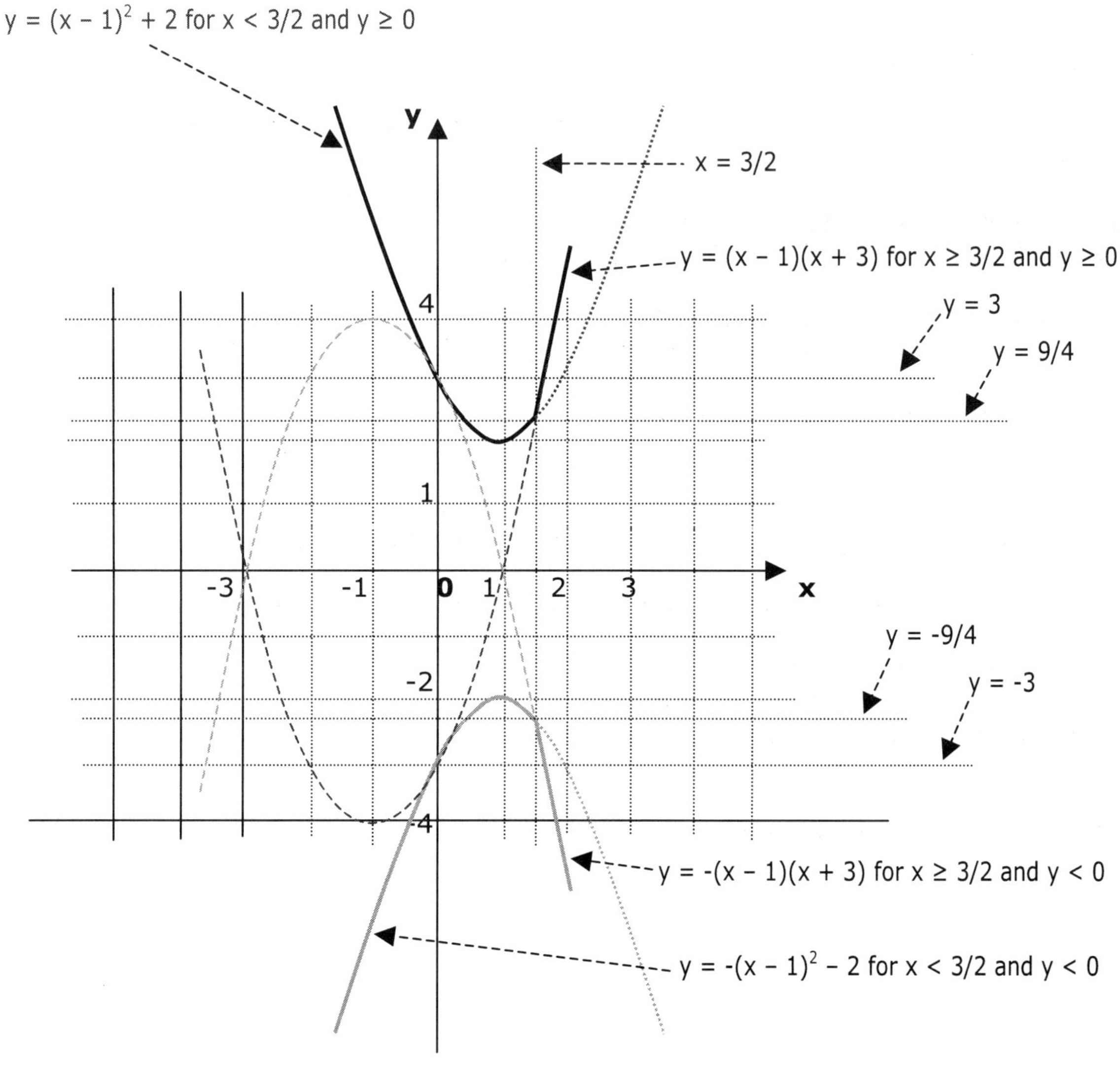

y = (x − 1)² + 2 for x < 3/2 and y ≥ 0
y
x = 3/2
y = (x − 1)(x + 3) for x ≥ 3/2 and y ≥ 0
y = 3
y = 9/4
4
1
-3
-1
0
1
2
3
x
y = -9/4
y = -3
-2
-4
y = -(x − 1)(x + 3) for x ≥ 3/2 and y < 0
y = -(x − 1)² − 2 for x < 3/2 and y < 0

Suggestions or Solutions

To the Problem 7 in the Example D

The equation is as follows: $y = (x^2 - 2x + 1)^{1/2} = \sqrt{x^2 - 2x + 1}$.

Although it is self-evident that a square root of a legitimate real number is greater than or equal to zero, we tend to make some occasional errors. That's because if a thing is too close to us, then we can hardly see it. Therefore, we should check to see if the **y** is greater than or equal to 0. What do we mean by the legitimate real number, then? What's inside the square root is supposed to be greater than or equal to a zero if it is a real number.
Suppose now, that $g(x) = x^2 - 2x + 1$. Then, the curve of the **g** is as follows:

Fig. D.7.0

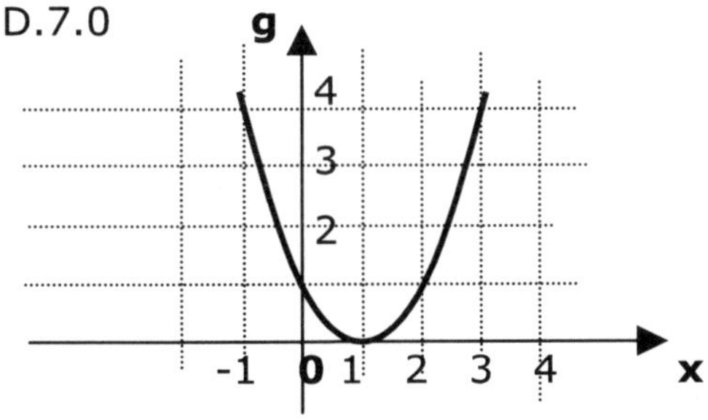

Then, we can see that the **g(x)** produces values greater than or equal to 0 for all the values of the **x**. Therefore, what's inside the square root is legitimate for all **x**. Now, we can see that $g(x) = x^2 - 2x + 1 = (x - 1)^2$, which is a complete square. Then, we can see that $y = (x^2 - 2x + 1)^{1/2} = \{(x - 1)^2\}^{1/2}$. We need to note however, that $\{(x - 1)^2\}^{1/2}$ is not $x - 1$ but $|x - 1|$. That's because a square root of a real number, that is, $\sqrt{x^2 - 2x + 1}$ is greater than or equal to 0. Now, we need to consider two cases as follows:
$x - 1 \geq 0 \Rightarrow x \geq 1 \Rightarrow y = x - 1$ for $x \geq 1$.
$x - 1 < 0 \Rightarrow x < 1 \Rightarrow y = -(x - 1) = -x + 1$ for $x < 1$.

Let's now put the equation $y = (x^2 - 2x + 1)^{1/2}$ in a graph.

Fig. D.7.1

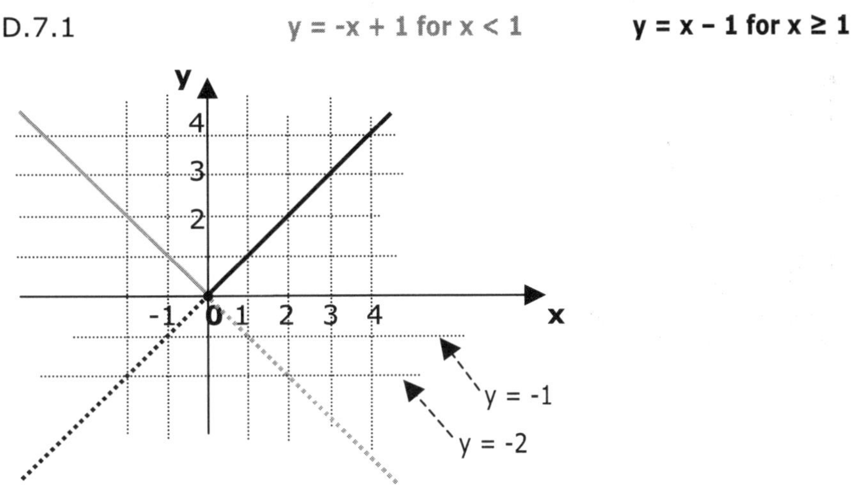

Suggestions or Solutions
To the Problem 8 in the Example D

The equation is as follows $y = \{(x^2 - 4x + 4)^2\}^{1/2} = \sqrt{(x^2 - 4x + 4)^2}$.

We know that what's inside the square root is supposed to be greater than or equal to a zero.
Let's see what's inside the square root, then.

Suppose now, that $g(x) = x^2 - 4x + 4$. Then, the curve of the **g** is as follows:

Fig. D.8.0

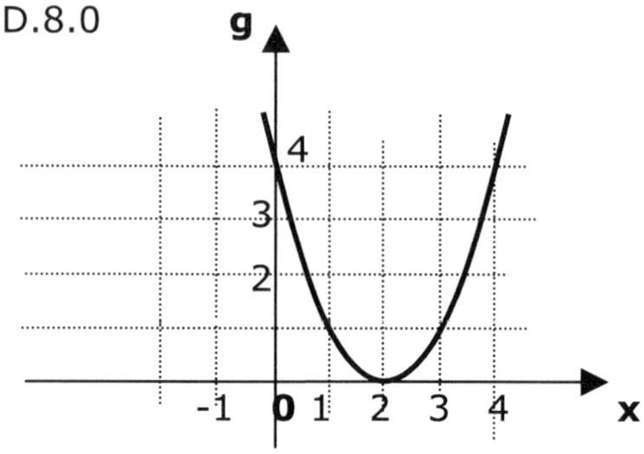

Then, we can see in the graph that the **g(x)** produces all the values greater than or equal to 0 for any **x** is real. Therefore, what's inside the square root is legitimate for all **x**.
Now, we can see that $\mathbf{g(x) = x^2 - 4x + 4 = (x - 2)^2}$, which is a complete square.
Then, we can see that:

$y = \{(x^2 - 4x + 4)^2\}^{1/2} = \{(x - 2)^4\}^{1/2} = (x - 2)^2 = g(x)$ since the $g \geq 0$.

Therefore, we don't need to consider different cases regarding the extent of the **x**.
Now, the graph of the equation given is as follows:

Fig. D.8.1 $y = \{(x - 4x + 4)^2\}^{1/2} = \sqrt{(x^2 - 4x + 4)^2}$

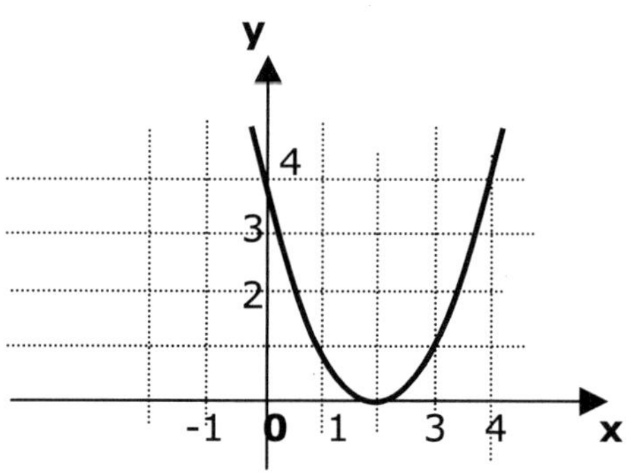

Suggestions or Solutions

To the Problem 9 in the Example D

The equation is as follows: $y = \{(x^2 - 2x + 2)^2\}^{1/2} = \sqrt{(x^2 - 2x + 2)^2}$.

We know that a square root of a legitimate real number is greater than or equal to 0.
We also know that what's inside the square root is supposed to be greater than or equal to a zero.

Let's see what's inside the square root first, then.
It is $(x^2 - 2x + 2)^2$, which is greater than or equal to 0, and therefore, it seems to be OK for **x** that has any real number.

Now, let's see what we should do after the removal of the square root sign.

Suppose now, that $g(x) = x^2 - 2x + 2$.
Then, $g(x) = (x - 1)^2 + 1$, and the curve of the **g** is as follows:

Fig. D.9.0

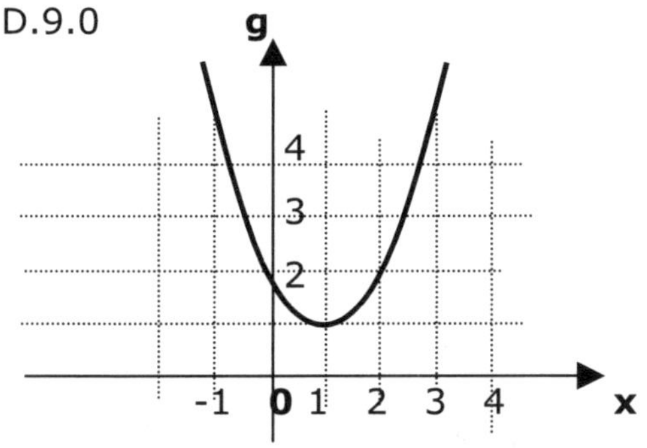

Then, we can see that the **g(x)** produces all the values greater than 0 for every **x** that is real. Therefore, what's inside the square root is legitimate for all **x**. Then, we can see that:

$y = \{(x^2 - 2x + 2)^2\}^{1/2} = x^2 - 2x + 2 = g(x)$ since the $g > 0$.

Therefore, we don't need to consider different cases regarding the extent of the **x**
Now, the graph of the equation given is as follows:

Fig. D.9.1

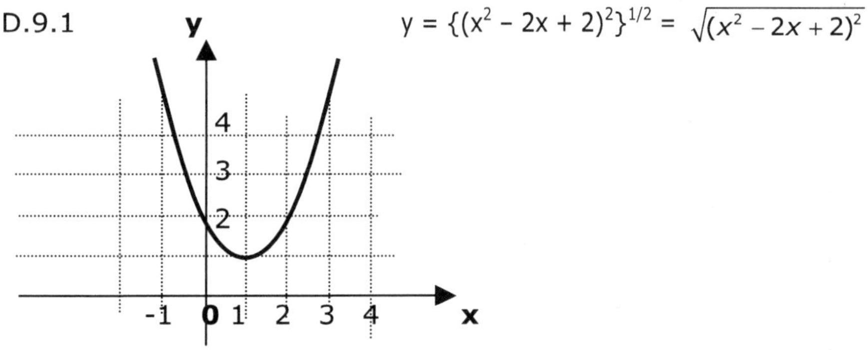

Suggestions or Solutions
To the Problem A in the Example D

The equation is as follows: $y = \{(x^2 - 2x)^2\}^{1/2} = \sqrt{(x^2 - 2x)^2}$

Though it is obvious that square root of a legitimate real number is greater than or equal to zero, we tend to make some occasional errors. That's because if it is too natural, we can hardly notice it. Therefore, when we remove the square root sign, we should check to see if the result, too, is greater than or equal to 0. Now, the legitimate real number inside the square root means that what's inside the square root sign has to be greater than or equal to 0 if it is a real number. Let's see what's inside the square root, then.

It is $(x^2 - 2x)^2$, which is greater than or equal to 0 for any x that is real, and therefore, it is fine with the square root sign. Now, let see what we should do after the removal of the square root sign.
Suppose now, that $g(x) = x^2 - 2x$.
Then, $g(x) = x^2 - 2x + 1 - 1 = (x - 1)^2 - 1$, and the curve of the g is as follows:

Fig. D.A.0

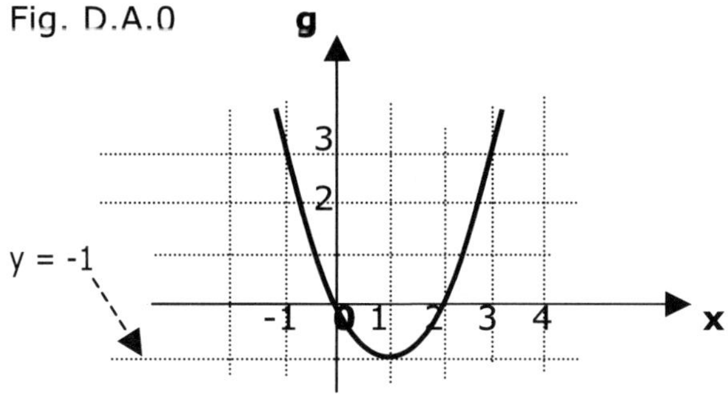

Then, we can see that the $g(x)$ does not produce all the values greater than or equal to 0 for all the real numbers that the x takes. In other words, we cannot just say that $y = g(x) = x^2 - 2x$. Then, in this case, when the stuff inside the square root is outside, it can matter. In other words, we can't just remove the square root sign along with the exponent 2 in this case. We know that the $y \geq 0$ since square root of a legitimate real number is greater than or equal to 0. Then, we need to consider two different cases regarding the extent of the x. We can see in the graph above that:

$y = x^2 - 2x$ for $x \leq 0$ or $x \geq 2$ and $y = -(x^2 - 2x) = -x^2 + 2x$ for $0 < x < 2$.

Now, let's see the graph of the given equation.

Fig. D.A.1

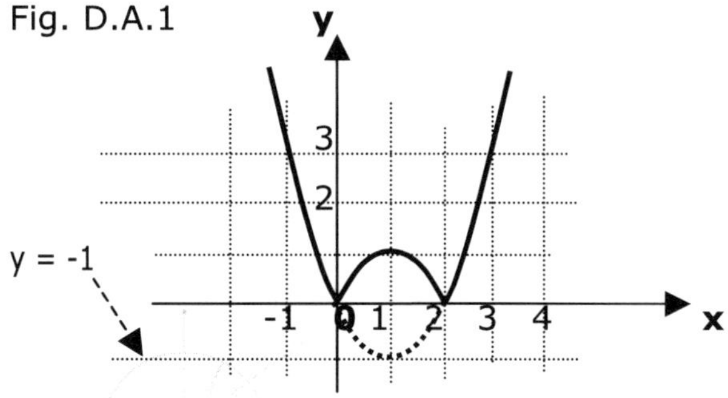

Suggestions or Solutions

To the Problem B in the Example D

The equation is as follows: $y = \{(-x^2 + 2x)^2\}^{1/2} = \sqrt{(-x^2+2x)^2}$.

Let's begin with what's inside the square root.
It is $(-x^2 + 2x)^2$, which is greater than or equal to 0 for any x that is real, and therefore, it is OK with the square root sign.

We know that when the stuff inside the square root gets outside, it can matter.
Now, let's see what we should do after the removal of the square root sign.

Suppose now, that $g(x) = -x^2 + 2x$.
Then, $g(x) = -x^2 + 2x = -(x^2 - 2x + 1 - 1) = -(x - 1)^2 + 1 = (2 - x)x$, and the curve of the **g** is as follows:

Fig. D.B.0

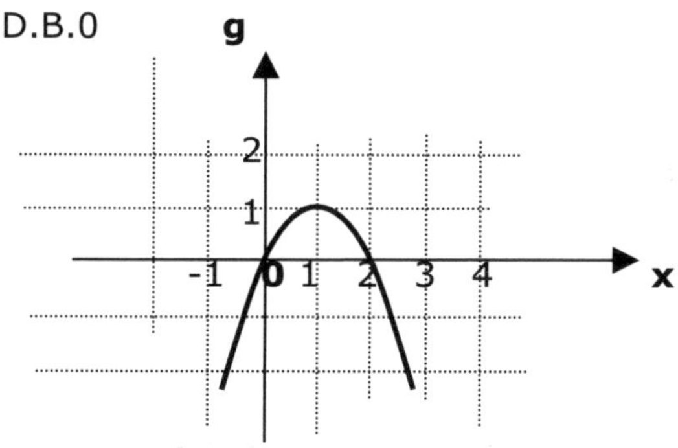

Then, we can see that the **g(x)** does not always produce a value greater than or equal to 0 for any real number that the **x** takes. That is, we can just put $y = g(x) = -x^2 + 2x$.

We know that the $y \geq 0$ since a square root of a legitimate real number is greater than or equal to 0.

Then, from the graph above, we can see that we need to have:
$y = -x^2 + 2x$ for $0 \leq x \leq 2$; but, $y = -(-x^2 + 2x) = x^2 - 2x$ for $x < 0$ or $x > 2$.

Therefore, we can put the given equation in a graph in such a way as follows:

Fig. D.B.1

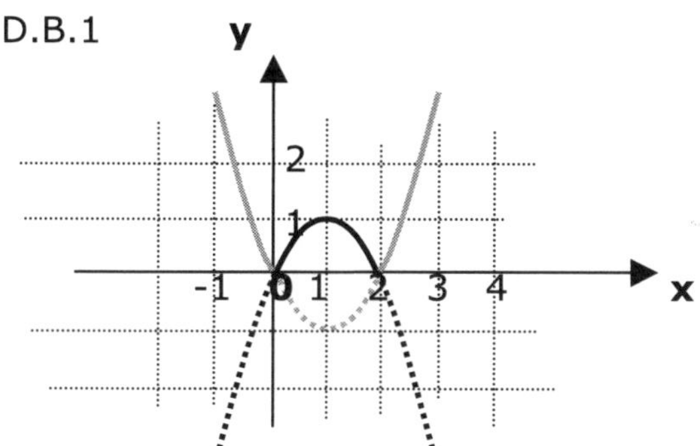

Suggestions or Solutions

To the Problem C in the Example D

The equation is as follows: $y = \{(-x^2 + 2x - 1)^2\}^{1/2} = \sqrt{(-x^2+2x-1)^2}$.

Let's begin with what's inside the square root.
It is $(-x^2 + 2x - 1)^2$, which is greater than or equal to 0 for any x that is real, and therefore, it is legal with the square root sign.

We know that when what's inside the square root is outside, it can matter.
Now, let's see what we should do after the removal of the square root sign.

Suppose now, that $g(x) = -x^2 + 2x - 1$.
Then, $g(x) = -x^2 + 2x - 1 = -(x^2 - 2x + 1) = -(x - 1)^2$, and the curve of the **g** is as follows:

Fig. D.C.0

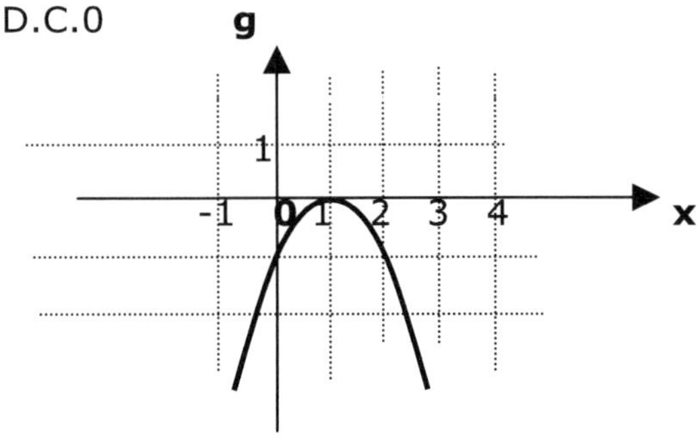

Then, we can see that the **g(x)** does not produce any positive value for any real number that the **x** takes; however, it does produce 0 when $x = 1$. In other words, $y = -x^2 + 2x - 1 \le 0$ for **x** that is real.
The $y \ge 0$ since a square root of a legitimate real number is greater than or equal to 0.
Then, we need to change the sign of every value the **y** takes in $y = g(x)$.
Then, let's change the sign of the expression on the left hand side of the equal sign. We can change the sign of either side, though.
Then, we get $-y = -x^2 + 2x - 1 \Rightarrow y = -(-x^2 + 2x - 1) = x^2 - 2x + 1 = (x - 1)^2 \ge 0$ for **x** that is real.

Now, the graph of the equation given is as follows:

Fig. D.C.1

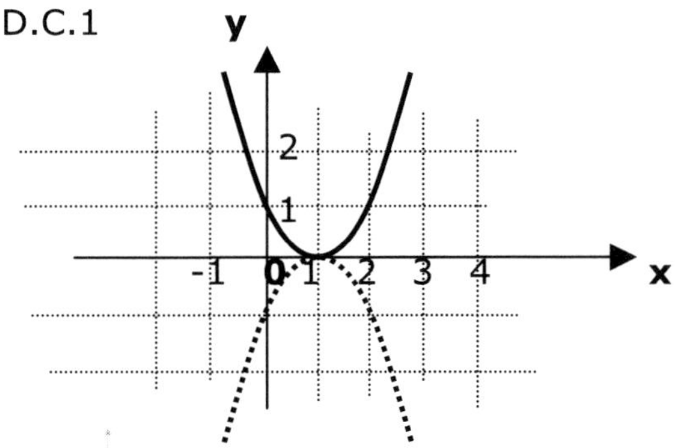

Suggestions or Solutions
To the **Problem D** in the Example **D**

We have an equation that $\mathbf{y = \{(-x^2 + 2x - 2)^2\}^{1/2}} = \sqrt{(-x^2+2x-2)^2}$.

Let's begin with what's inside the square root. It is $\mathbf{(-x^2 + 2x - 2)^2}$, which is greater than or equal to 0 for any **x** that is real, and therefore, it does get along with the square root sign. Now, when what's inside the square root gets outside, it can matter. Now, we need to see what we should do after the removal of the square root sign. Suppose now, that $\mathbf{g(x) = -x^2 + 2x - 2}$.
Then, $\mathbf{g(x) = -x^2 + 2x - 2 = -(x^2 - 2x + 2) = -(x^2 - 2x + 1 + 1) = -(x - 1)^2 - 1}$, and the curve of the **g** is as follows:

Fig. D.D.0

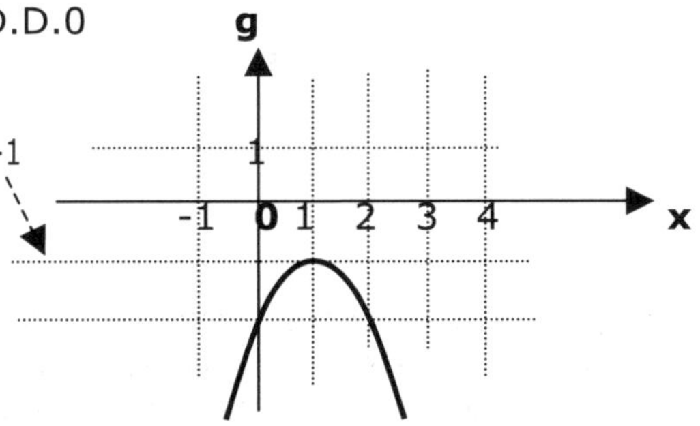

Then, we can see that the g(x) produces negative values only for all real numbers that the **x** takes.
That is, $\mathbf{y = -x^2 + 2x - 2 < 0}$ for **x** that is real. The $\mathbf{y \geq 0}$ since square root of a legitimate real number is greater than or equal to 0. Therefore, we need to change the sign of every value the **y** takes in **y = g(x)**. Let's change the sign of the expression on the right hand side of the equal sign. We can change the sign of either side, of course.
Then, we get $\mathbf{y = -(-x^2 + 2x - 2) = x^2 - 2x + 2 = (x - 1)^2 + 1 > 0}$ for **x** that is real.
Now, the graph of the equation given is as follows:

Fig. D.D.1

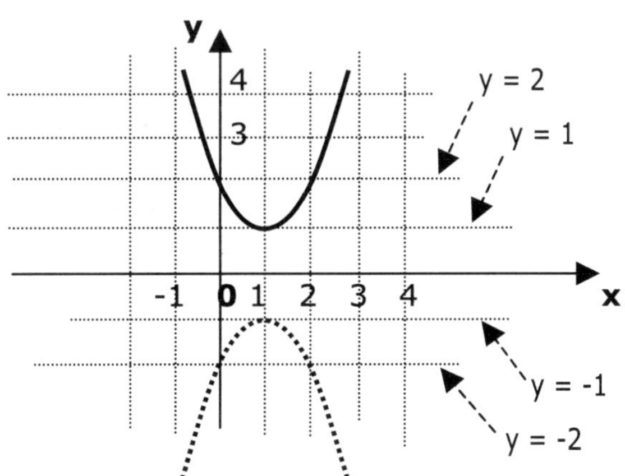

Suggestions or Solutions

To the **Problem** in the Example **E**

Suppose that $y = f(x) = 2x^2 + bx + c$ for $0 \le x \le 1$ and $0 \le y \le 2$, and find then, the values of the b and c.

We want the function **f** to produce the values from 0 to 2 for $\mathbf{0 \le x \le 1}$. Then, we need to see how the function should behave in its playground specified by $\mathbf{0 \le x \le 1}$ and $\mathbf{0 \le y \le 2}$. We can see the behavior by means of the curve, and we know that the curve of the **f** is a parabola. Then, let's put the curve in a graph in the x-y system, and have a look at the curve in the playground. The playground is the light gray area in the x-y plane below.

Fig. E.0

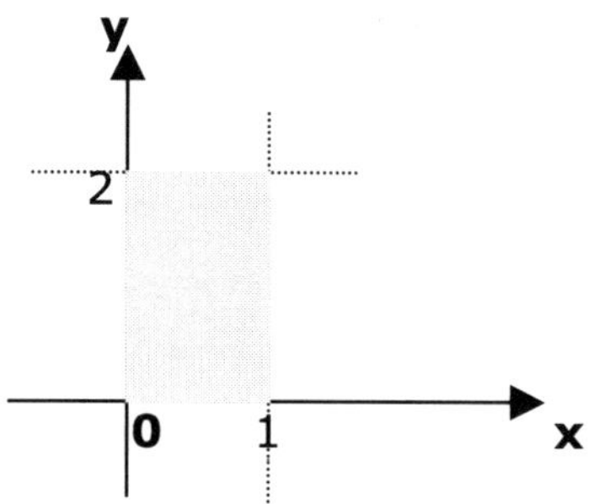

However, we don't just put the curve in the graph. We put the curve so that a segment of the curve (parabolic line segment) is placed in the area. More specifically, we want the segment to fulfill the area spanning from 0 to 1 along the x-axis and from 0 to 2 along the y-axis. Let's put some curves in the graph now.

Fig. E.1

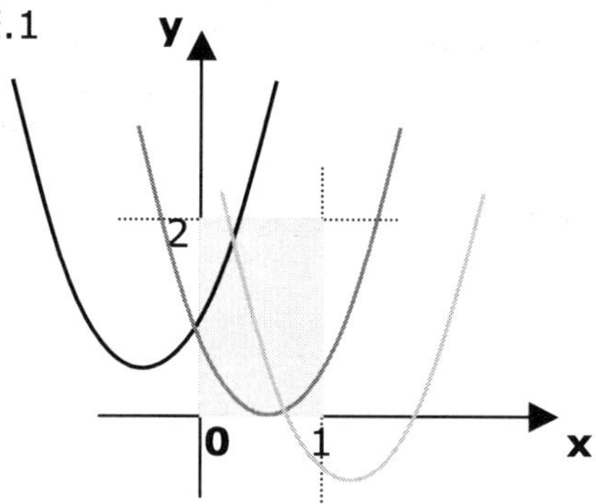

None of the curves put as above will do. That's because neither of those fulfills the playground spanning in the amount specified as above. Let's try again now.

Fig. E.2

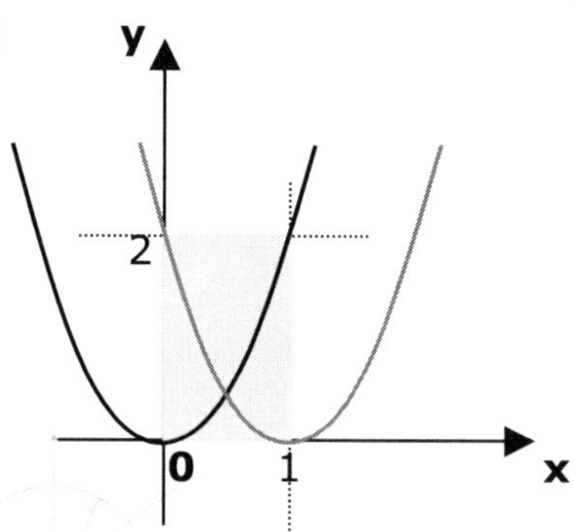

This time, we've done it right. Both of the curves are put in the right place so that each of the segments fulfills the playground spanning in the right amount. Now that the two parabolas can do, we can begin construction of the equations of the parabolas.
Wait a minute! What about the parabolas shown below?

Fig. E.3

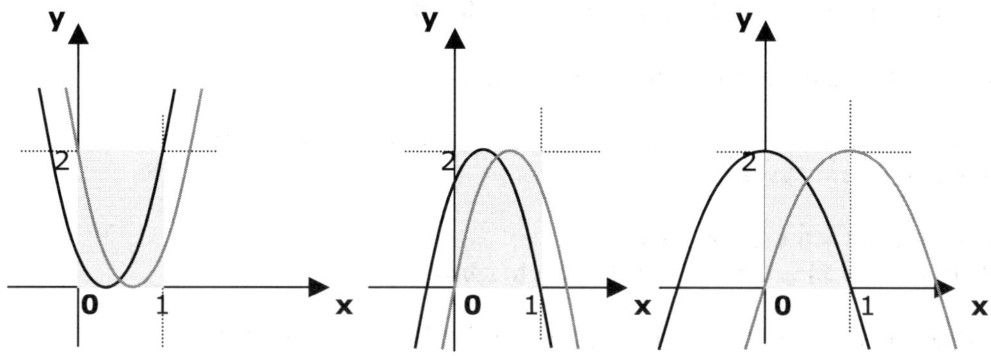

Those parabolas as above can do also, of course; however, the slope of each of those matters. In other words, they are not eligible. In the equation, we can see that the slope of the parabola is 2, but the parabolas shown above do not have the slope of 2. However, if the slope is a constant, and therefore, is in question, then we should be able to consider the parabolas that can be placed as shown above. If that is the case, we should also consider such parabolas as shown below.

Fig. E.4

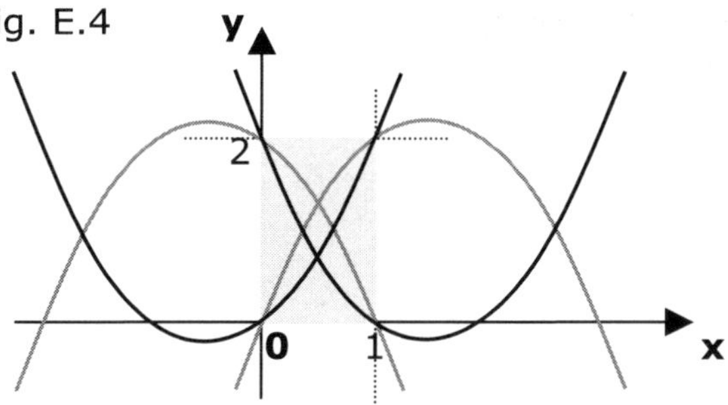

Let's get back to the right parabolas now. Those are as follows:

Fig. E.2

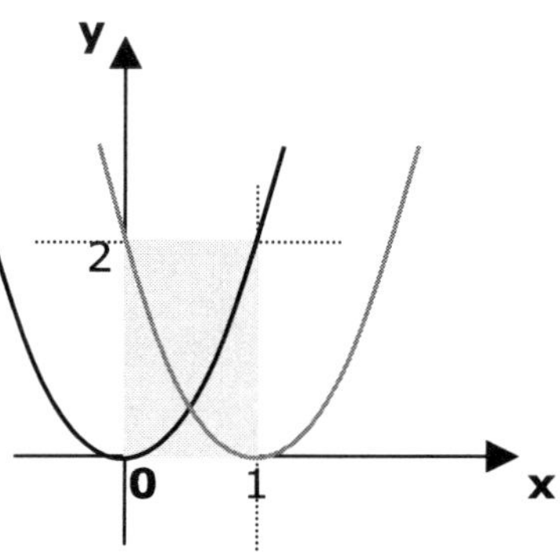

We have the equation $\mathbf{y = 2x^2 + bx + c}$ for $\mathbf{0 \le x \le 1}$ and $\mathbf{0 \le y \le 2}$.
We want to find the values of the **b** and the **c**. That is, we want to find the equations of the two parabolas. In the graph above, we can see two points we can use for the construction of each of the two equations we want.

For the parabola on the left, the points to be used are the origin and **(1, 2)**.
We have $\mathbf{y = 2x^2 + bx + c}$. Then:

$\mathbf{(0, 0) \Rightarrow 0 = 0 + 0 + c = c \Rightarrow c = 0}$.
$\mathbf{((1, 2)}$ and $\mathbf{c = 0) \Rightarrow 2 = 2 + b + 0 = 2 + b \Rightarrow b = 0}$.

Therefore, the equation of the parabola is $\mathbf{y = 2x^2}$.

For the parabola on the right, the points to be used are **(0, 2)** and **(1, 0)**.
We have $\mathbf{y = 2x^2 + bx + c}$. Then:

$\mathbf{(0, 2) \Rightarrow 2 = 0 + 0 + c = c \Rightarrow c = 2}$.
$\mathbf{((1, 0)}$ and $\mathbf{c = 2) \Rightarrow 0 = 2 + b + 2 = 4 + b \Rightarrow b = -4}$.

Therefore, the equation of the parabola is $\mathbf{y = 2x^2 - 2x + 2 = 2(x - 1)^2}$.

By the way, some students can find the equations by looking at the graph only.

In short:

We have the equation $\mathbf{y = 2x^2 + bx + c}$ for $\mathbf{0 \le x \le 1}$ and $\mathbf{0 \le y \le 2}$.
Then, the parabola should be either of the two as follows.

Fig. E.2

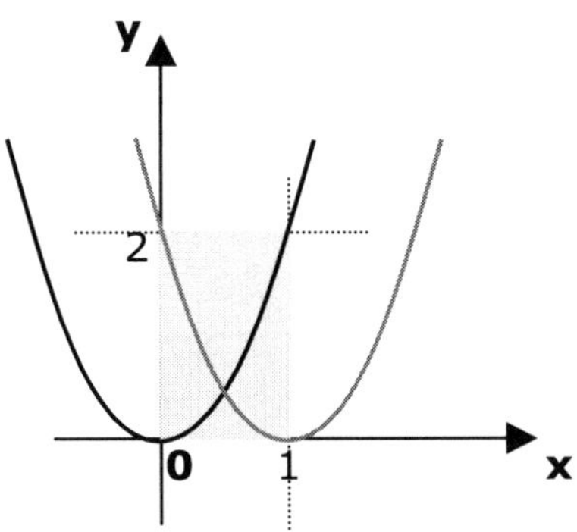

The parabola on the left needs to include the origin and **(1, 2)**.
We have $\mathbf{y = 2x^2 + bx + c}$. Then:
$\mathbf{(0, 0) \Rightarrow 0 = 0 + 0 + c = c \Rightarrow c = 0}$.
$\mathbf{((1, 2)}$ and $\mathbf{c = 0) \Rightarrow 2 = 2 + b + 0 = 2 + b \Rightarrow b = 0}$.

The parabola on the right needs to include **(0, 2)** and **(1, 0)**.
We have $\mathbf{y = 2x^2 + bx + c}$. Then:
$\mathbf{(0, 2) \Rightarrow 2 = 0 + 0 + c = c \Rightarrow c = 2}$.
$\mathbf{((1, 0)}$ and $\mathbf{c = 2) \Rightarrow 0 = 2 + b + 2 = 4 + b \Rightarrow b = -4}$.

Therefore, $\mathbf{b = 0}$ and $\mathbf{c = 0}$, or $\mathbf{b = -4}$ and $\mathbf{c = 2}$.

Suggestions or Solutions
To the Problem in the Example F

Suppose that $y = f(x) = ax^2 + bx + 1$ for $0 \le x \le 1$ and $0 \le y \le 2$, and then, find the values of the a and b.

In this example, too, we want the function f to produce values from 0 to 2 for $0 \le x \le 1$. Then, we need to see how the function should behave in its playground specified by $0 \le x \le 1$ and $0 \le y \le 2$. We can see the behavior of a function by means of the curve, and we know the curve of the **f** is a parabola. Then, let's put the curve in a graph in the x-y system, and have a look at the curve in the playground. The light gray area in the x-y plane below is the playground.

Fig. F.0

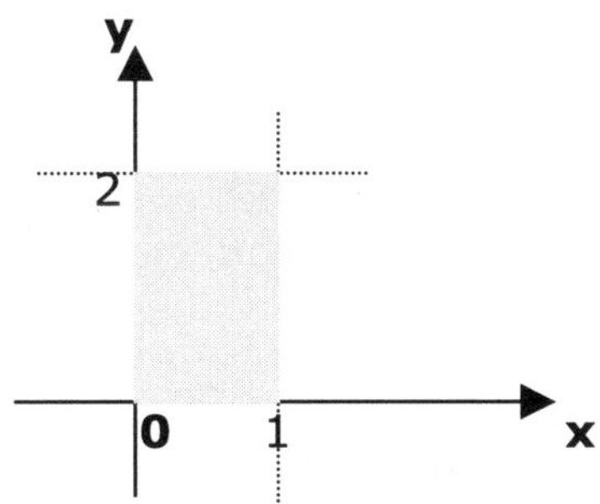

Now, we are going to put the curve so that a segment of the curve (parabolic line segment) is placed in the area. More specifically, we want the segment to fulfill the area spanning from 0 to 1 along the x-axis and from 0 to 2 along the y-axis. Let's put some curves in the graph now.

Fig. F.1

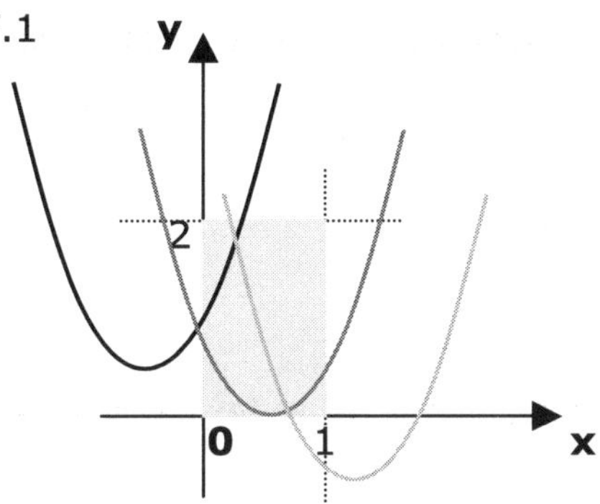

None of the curves put as above will do. That's because neither of those fulfills the playground spanning in the amount specified as above. Let's try again now.

Fig. F.2

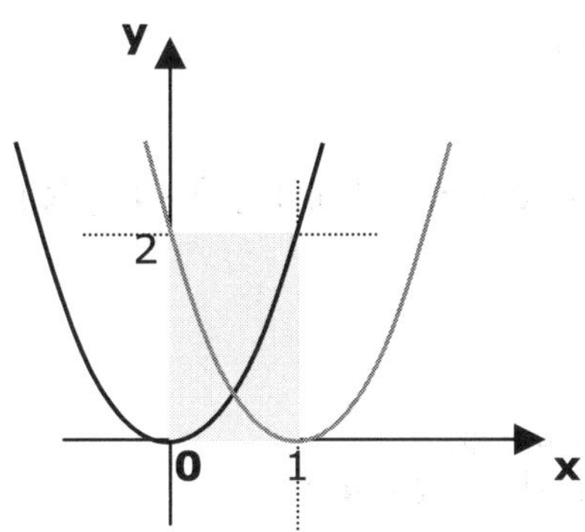

Sorry but wrong again! Both of the curves respectively fulfill the playground spanning in the right amount; however, the y-intercepts are not correct. In the given equation $\mathbf{y = ax^2 + bx + 1}$, we can see that the y-intercept is 1. Now, let' try again.

Fig. F.3

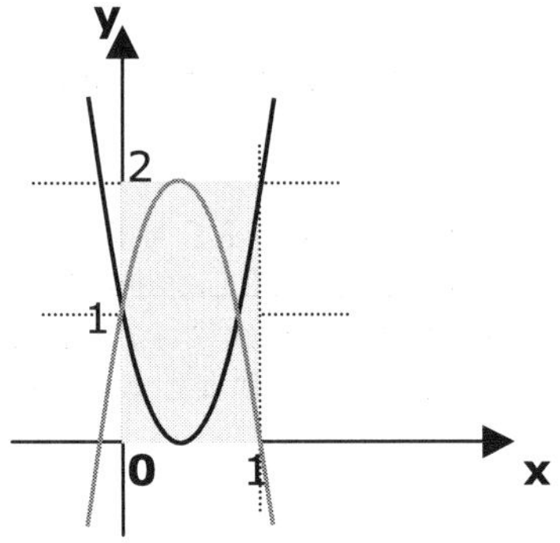

This time, we've done it right.
Both of the curves put themselves respectively in the right places so that each of the parts fulfills the playground spanning in the right amount and have the correct y-intercept. Now, the two parabolas can do, and we can begin construction of the equations of the parabolas.

We have the equation $\mathbf{y = ax^2 + bx + 1}$ for $\mathbf{0 \le x \le 1}$ and $\mathbf{0 \le y \le 2}$.
Now, we want to find the values of the **a** and the **b**. That is, we need to find the equations of the two parabolas. Looking at the graph above, it seems that we can readily use two points for the construction of each of the two equations we want. However, one of the two points is not useful for the construction of each of the equations. The point useful is at **(1, 0)** only, but the point at **(0, 1)** is not. That's because, for instance, for the parabola with the slope negative:
Since the equation is $\mathbf{y = ax^2 + bx + 1}$, $\mathbf{(0, 1) \Rightarrow 1 = 0 + 0 + 1}$, which can leave us nothing significant.

Let's use the point at **(1, 0)** now:

Since $\mathbf{y = ax^2 + bx + 1}$, $\mathbf{(1, 0) \Rightarrow 0 = a + b + 1 \Rightarrow a + b = -1}$.

Now, what else can we use?
We can use the limitation on the vertex of the parabola.
Specifically, the x-coordinate needs to be positive, and the y-coordinate of the vertex needs to be 2.
Then, let's find the vertex of the parabola now.

$$\mathbf{y = ax^2 + bx + 1 = a\{x^2 + bx/a + (b/2a)^2 - (b/2a)^2\} + 1 = a(x + b/2a)^2 - a(b/2a)^2 + 1}$$
$$\mathbf{= a(x + b/2a)^2 - b^2/4a + 1 = a(x + b/2a)^2 - (b^2 - 4a)/4a.}$$

Then, we can see that the vertex is at $\mathbf{\{-b/2a, -(b^2 - 4a)/4a\}}$.
Therefore, we get $\mathbf{-(b^2 - 4a)/4a = 2}$.

Now, we have two simultaneous equations for the **a** and the **b**: $\mathbf{a + b = -1}$ and $\mathbf{-(b^2 - 4a)/4a = 2}$.
Then:
$\mathbf{a + b = -1 \Rightarrow a = -(1 + b).}$
$\mathbf{-(b^2 - 4a)/4a = 2 \Rightarrow b^2 - 4a = -8a \Rightarrow b^2 + 4a = 0.}$

Then, $\mathbf{a = -(1 + b) \Rightarrow b^2 + 4a = b^2 - 4(1 + b) = b^2 - 4b - 4 = 0.}$
$\mathbf{b^2 - 4b - 4 = (b - 2)^2 - 8 = 0 \Rightarrow b - 2 = \pm 8^{1/2} \Rightarrow b = 2 + 8^{1/2}}$ or $\mathbf{2 - 8^{1/2}}$.

Then:
$\mathbf{b = 2 + 8^{1/2} \Rightarrow a = -(1 + b) = -(3 + 8^{1/2})}$.
$\mathbf{b = 2 - 8^{1/2} \Rightarrow a = -(1 + b) = -(3 - 8^{1/2})}$.

Now, we need to choose the value of the **a** from the two values above, and we know that the slope is negative. However, we don't know which one of the two values is for the **a** since both values are negative. How do we know if $-(3 - 8^{1/2})$ is negative, then?
We know that $3^2 = 9$ and $(8^{1/2})^2 = 8$. Then, $3 > 8^{1/2} \Rightarrow 3 - 8^{1/2} > 0 \Rightarrow -(3 - 8^{1/2}) < 0$.

Now, we are going to use the limitation on the x-coordinate at the vertex.
We know that **the x-coordinate = $-b/2a > 0$**.
Then, $\mathbf{-b/2a > 0 \Rightarrow b/2a < 0 \Rightarrow b/a < 0}$. Now:

$\mathbf{b = 2 + 8^{1/2} > 0}$ and $\mathbf{a = -(3 + 8^{1/2}) < 0}$. Then, $\mathbf{b/a < 0} \Rightarrow$ **appropriate**.
$\mathbf{b = 2 - 8^{1/2} < 0}$ and $\mathbf{a = -(3 - 8^{1/2}) < 0}$. Then, $\mathbf{b/a > 0} \Rightarrow$ **inappropriate**.

Therefore, the equation of the parabola is $\mathbf{y = -(3 + 8^{1/2})x^2 + (2 + 8^{1/2})x + 1}$.

Now, let's move on to the parabola with the slope positive.

Fig. F.4

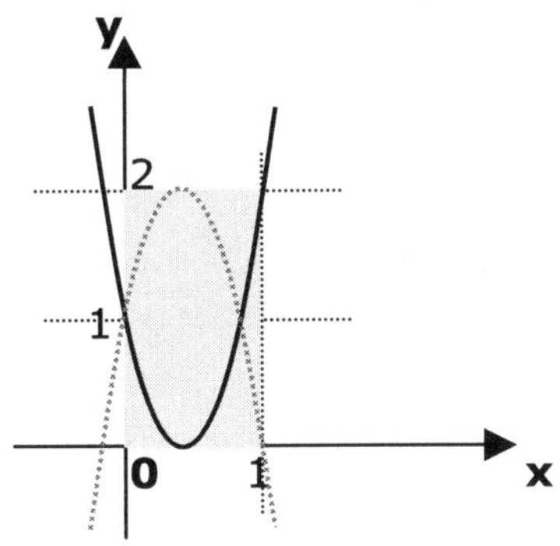

Examining closely the curves above, we can notice that they are symmetric each other about a line **y = 1**. Then, we can readily see that the magnitudes of the slopes are the same as each other. In other words, changing the sign of the slope of the parabola we just have found, we get the slope of the parabola we are after now. Therefore, the $\mathbf{a = 3 + 8^{1/2}}$.

Now, what about the value of the **b**?
We can find the value of the **b** by the employment of the point at **(1, 2)**. Let's put the point at **(1, 2)** in use now.
Since $\mathbf{y = ax^2 + bx + 1}$, $\mathbf{(1, 2) \Rightarrow 2 = a + b + 1 \Rightarrow a + b = 1 \Rightarrow b = 1 - a}$.
Then, $\mathbf{a = 3 + 8^{1/2} \Rightarrow b = 1 - a = 1 - (3 + 8^{1/2}) = -2 - 8^{1/2} = -(2 + 8^{1/2})}$.
Therefore, the equation is $\mathbf{y = (3 + 8^{1/2})x^2 - (2 + 8^{1/2})x + 1}$.

However, for practice, let's find the a first by the straight method to find the parabola again.
Then, we are going to employ the point **(1, 2)** and the vertex.

To begin with, since $\mathbf{y = ax^2 + bx + 1}$, $\mathbf{(1, 2) \Rightarrow 2 = a + b + 1 \Rightarrow a + b = 1 \Rightarrow a = 1 - b}$.
Next, we are going to employ such specifics on the vertex as follows:
The x-coordinate of the vertex is positive, and the y-coordinate is 0.

We know that the vertex is at **$\{-b/2a, -(b^2 - 4a)/4a\}$**. Therefore, **$-(b^2 - 4a)/4a = 0$**.

Now, we have two simultaneous equations for the **a** and the **b**: **$a = 1 - b$** and **$-(b^2 - 4a)/4a = 0$**. Then:

$-(b^2 - 4a)/4a = 0 \Rightarrow b^2 - 4a = 0$.
$a = 1 - b \Rightarrow b^2 - 4a = b^2 - 4(1 - b) = b^2 + 4b - 4 = (b + 2)^2 - 8 = 0 \Rightarrow b + 2 = \pm 8^{1/2}$
$\Rightarrow b = -2 + 8^{1/2}$ or $-2 - 8^{1/2}$.

Then:
$b = -2 + 8^{1/2} \Rightarrow a = 1 - b = 1 - (-2 + 8^{1/2}) \Rightarrow a = 3 - 8^{1/2}$.
$b = -2 - 8^{1/2} \Rightarrow a = 1 - b = 1 - (-2 - 8^{1/2}) \Rightarrow a = 3 + 8^{1/2}$.

Now, we need to choose the value of the **a** from the two values above, and we know that the slope is positive. However, we don't know which one to choose for the **a** since both values are positive.

Now, we are going to use the limitation on the x-coordinate at the vertex.
We know that the **x-coordinate = $-b/2a > 0$**. Then, **$-b/2a > 0 \Rightarrow b/2a < 0 \Rightarrow b/a < 0$**. Now:
$b = -2 + 8^{1/2} > 0$ and **$a = 3 - 8^{1/2} > 0$**. Then, **$b/a > 0 \Rightarrow$ inappropriate**.
$b = -2 - 8^{1/2} < 0$ and **$a = 3 + 8^{1/2} > 0$**. Then, **$b/a < 0 \Rightarrow$ appropriate**.

Therefore, the equation of the parabola is **$y = (3 + 8^{1/2})x^2 - (2 + 8^{1/2})x + 1$**.

In short:

The equation given is **$y = ax^2 + bx + 1$** for **$0 \le x \le 1$** and **$0 \le y \le 2$**.
Then, the parabola needs to be either of the two as follows.

Fig. F.3

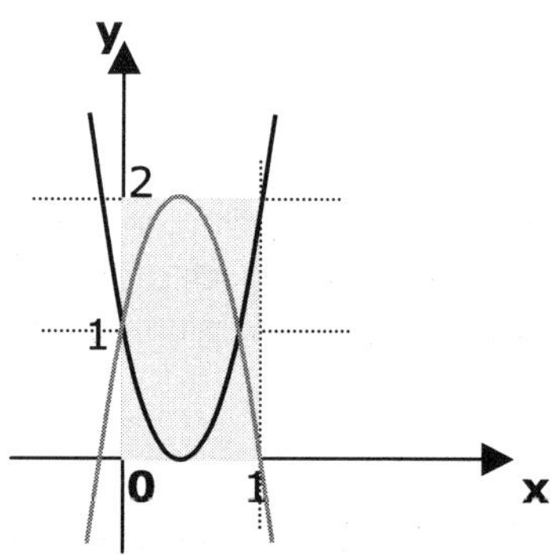

We are going to begin with the parabola with the slope positive.
Then, it includes the point **(1, 0)**, and the x-coordinate of the vertex is positive and the y-coordinate is 2.

To begin with, since **$y = ax^2 + bx + 1$**, **$(1, 0) \Rightarrow 0 = a + b + 1 \Rightarrow a + b = -1$**.
Next, we are going to find the vertex.
$y = ax^2 + bx + 1 = a\{x^2 + bx/a + (b/2a)^2 - (b/2a)^2\} + 1 = a(x + b/2a)^2 - a(b/2a)^2 + 1$
$= a(x + b/2a)^2 - b^2/4a + 1 = a(x + b/2a)^2 - (b^2 - 4a)/4a$.

Then, the vertex is **$\{-b/2a, -(b^2 - 4a)/4a\}$**. Therefore, **$-(b^2 - 4a)/4a = 2$**.

Now, we get a system of two equations for the **a** and the **b**: $a + b = -1$ and $-(b^2 - 4a)/4a = 2$. Then:

$a + b = -1 \Rightarrow a = -(1 + b)$.
$-(b^2 - 4a)/4a = 2 \Rightarrow b^2 - 4a = -8a \Rightarrow b^2 + 4a = 0$.

Then, $a = -(1 + b) \Rightarrow b^2 + 4a = b^2 - 4(1 + b) = b^2 - 4b - 4 = 0$.
$b^2 - 4b - 4 = (b - 2)^2 - 8 = 0 \Rightarrow b - 2 = \pm 8^{1/2} \Rightarrow b = 2 + 8^{1/2}$ or $2 - 8^{1/2}$.

Then, $b = 2 + 8^{1/2} \Rightarrow a = -(1 + b) = -(3 + 8^{1/2})$, and also, $b = 2 - 8^{1/2} \Rightarrow a = -(1 + b) = -(3 - 8^{1/2})$.

We know that **the x-coordinate of the vertex** $= -b/2a > 0$.
Then, $-b/2a > 0 \Rightarrow b/2a < 0 \Rightarrow b/a < 0$.
Now:
$b = 2 + 8^{1/2} > 0$ and $a = -(3 + 8^{1/2}) < 0$. Then, $b/a < 0 \Rightarrow$ **appropriate**.
$b = 2 - 8^{1/2} < 0$ and $a = -(3 - 8^{1/2}) < 0$. Then, $b/a > 0 \Rightarrow$ **inappropriate**.

So, $a = -(3 + 8^{1/2})$ and $b = 2 + 8^{1/2}$.

Now, let's move on to the parabola with the slope positive.
Then, it includes **(1, 2)**, and the x-coordinate of the vertex is positive and the y-coordinate is 0.
To begin with, since $y = ax^2 + bx + 1$, $(1, 2) \Rightarrow 2 = a + b + 1 \Rightarrow a + b = 1 \Rightarrow a = 1 - b$.
Next, we know that the vertex is at $\{-b/2a, -(b^2 - 4a)/4a\}$. Therefore, $-(b^2 - 4a)/4a = 0$.

Now, we get a system of two equations for the **a** and the **b**: $a = 1 - b$ and $-(b^2 - 4a)/4a = 0$.

Then:
$-(b^2 - 4a)/4a = 0 \Rightarrow b^2 - 4a = 0$.
$a = 1 - b \Rightarrow b^2 - 4a = b^2 - 4(1 - b) = b^2 + 4b - 4 = (b + 2)^2 - 8 = 0 \Rightarrow b + 2 = \pm 8^{1/2}$
$\Rightarrow b = -2 + 8^{1/2}$ or $-2 - 8^{1/2}$.

Then:
$b = -2 + 8^{1/2} \Rightarrow a = 1 - b = 1 - (-2 + 8^{1/2}) \Rightarrow a = 3 - 8^{1/2}$.
$b = -2 - 8^{1/2} \Rightarrow a = 1 - b = 1 - (-2 - 8^{1/2}) \Rightarrow a = 3 + 8^{1/2}$.

We know that **the x-coordinate** $= -b/2a > 0$. Then, $-b/2a > 0 \Rightarrow b/2a < 0 \Rightarrow b/a < 0$. Now:
$b = -2 + 8^{1/2} > 0$ and $a = 3 - 8^{1/2} > 0$. Then, $b/a > 0 \Rightarrow$ **inappropriate**.
$b = -2 - 8^{1/2} < 0$ and $a = 3 + 8^{1/2} > 0$. Then, $b/a < 0 \Rightarrow$ **appropriate**.

So, $a = 3 + 8^{1/2}$ and $b = -(2 + 8^{1/2})$.

Therefore, $a = -(3 + 8^{1/2})$ and $b = 2 + 8^{1/2}$, or $a = 3 + 8^{1/2}$ and $b = -(2 + 8^{1/2})$.

Suggestions or Solutions
To the Problem in the Example G

Suppose that $y = f(x) = ax^2 + bx + c$ for $0 \le x \le 1$ and $0 \le y \le 2$, and then, find the values of the a, b, and c.

In this example also, we want the function **f** to produce values from 0 to 2 for $\mathbf{0 \le x \le 1}$, and therefore, we need to see how the function has to behave in its playground specified by $\mathbf{0 \le x \le 1}$ and $\mathbf{0 \le y \le 2}$. We can see the behavior of a function by the curve, and we know that the curve of the **f** is a parabola. Then, let's put the curve in a graph in the x-y system, and have a look at the curve in the playground. The area in light gray in the x-y plane below is the playground:

Fig. G.0

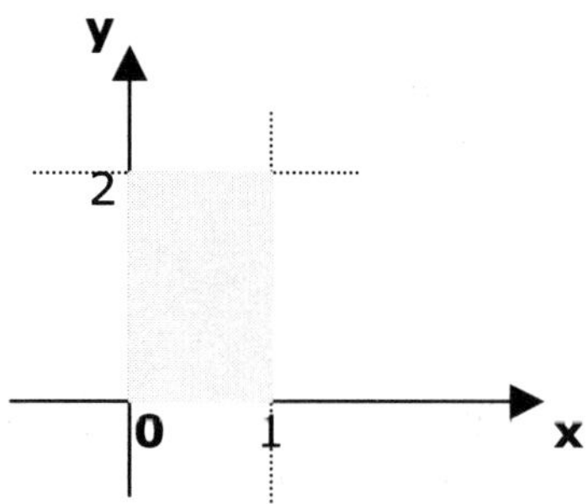

Now, we are going to put the curve in the graph so that a segment of the curve (parabolic line segment) is in the are. However, the segment is not just put in the area. We need to make the segment fulfill the area spanning from 0 to 1 along the x-axis and from 0 to 2 along the y-axis. Let's put some curves in the graph now.

Fig. G.1

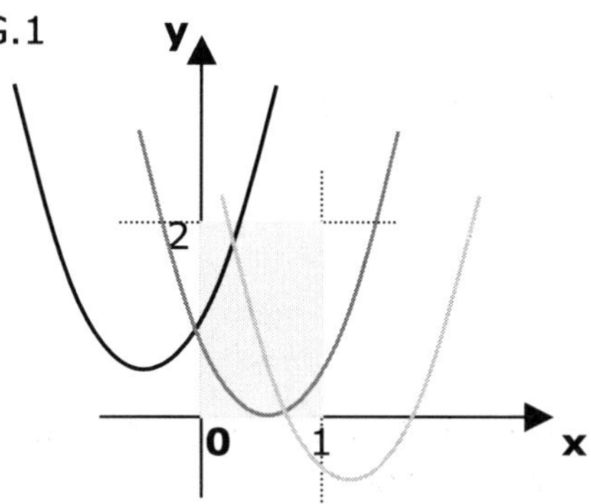

None of the curves put as above will do. That's because neither of those fulfills the playground spanning in the amount specified as above. Let's try again now.

Fig. G.2

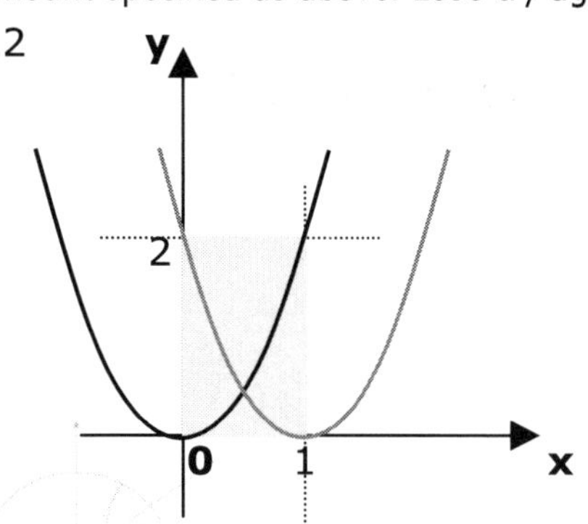

They will do but are not the only ones. Therefore, let's try more.

Fig. G.3

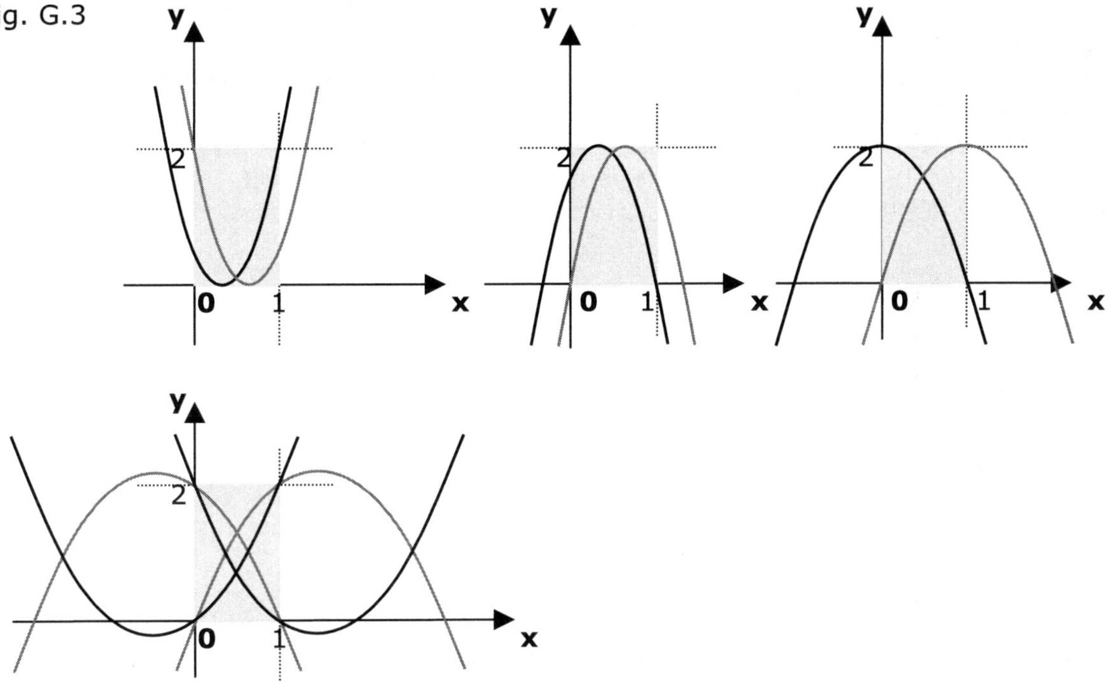

There are quite a few cases we have to consider. Therefore, we are going to put all the cases in two different categories. In one of the two, the parabolas have positive slopes, and in the other, the parabolas have negative slopes. Now, in each of the two categories, we will see that there are many such parabolas that can do. How many? Indefinitely many! Therefore, we will get to find the not particular values of the constants **a**, **b**, and **c** in $\mathbf{y = ax^2 + bx + c}$ but the extent of each of those constants.

Now, let's begin with the category where the parabolas have positive slopes.

Fig. G.4

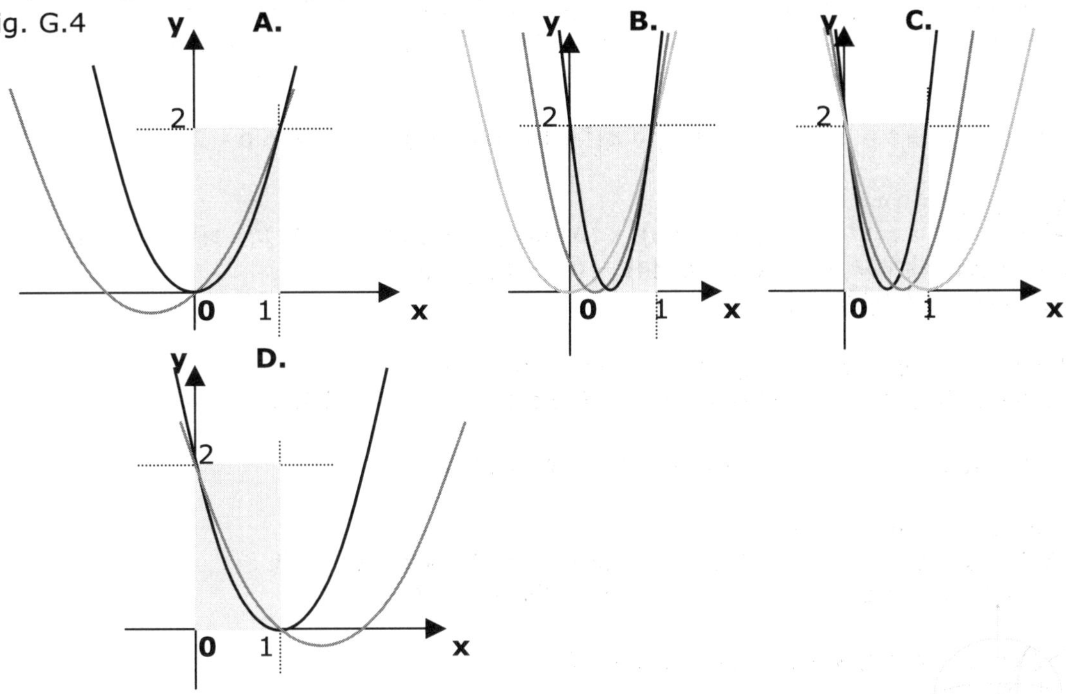

In case of the **A**, $\mathbf{f(0) = 0}$, and $\mathbf{f(1) = 2}$; also at the vertex, **the x coordinate ≤ 0**.
In case of the **B**, $\mathbf{0 \le f(0) \le 2}$, and $\mathbf{f(1) = 2}$; also at the vertex, **0 ≤ the x-coordinate ≤ 1/2**, and **the y-coordinate = 0**.
In case of the **C**, $\mathbf{f(0) = 2}$, and $\mathbf{0 \le f(1) \le 2}$; also at the vertex, **1/2 ≤ the x-coordinate ≤ 1**, and **the y-coordinate = 0**.
In case of the **D**, $\mathbf{f(0) = 2}$, and $\mathbf{f(1) = 0}$; also at the vertex, **the x-coordinate ≥ 1**.

Now, let's begin with the case **A**.

Fig. G.4

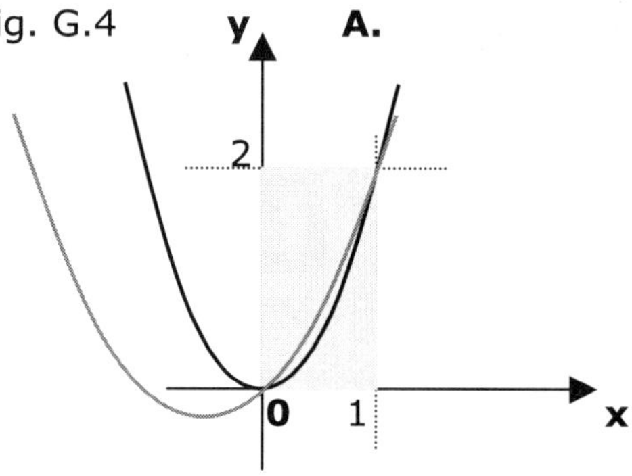

We need to find the values of the constants **a**, **b**, and **c** such that:
$\mathbf{y = f(x) = ax^2 + bx + c}$ for $\mathbf{0 \le x \le 1}$ and $\mathbf{0 \le y \le 2}$.

Then, we need to meet the conditions that $\mathbf{f(0) = 0}$, and $\mathbf{f(1) = 2}$; also, **the x-coordinate ≤ 0** at the vertex.

Why not consider the y-coordinate at the vertex, then? Shouldn't it be less than or equal to zero?
It is to be less than or equal to zero, of course. However, we don't have to. If we do, it will be redundant. The conditions specified above is exhaustive now, and will force the y-coordinate to be so. In other words, if all the conditions are met, then the y-coordinate at the vertex cannot be positive. Now, let's satisfy the conditions.

$\mathbf{f(0) = 0 \Rightarrow 0 + 0 + c = 0 \Rightarrow c = 0 \Rightarrow f(x) = ax^2 + bx}$. Then, $\mathbf{f(1) = 2 \Rightarrow a + b = 2}$.

Let's get the vertex now.
$\mathbf{y = ax^2 + bx + c = a\{x^2 + bx/a + (b/2a)^2 - (b/2a)^2\} + c = a(x + b/2a)^2 - a(b/2a)^2 + c}$
$\mathbf{= a(x + b/2a)^2 - b^2/4a + c = a(x + b/2a)^2 - (b^2 - 4ac)/4a}$.
Then, the vertex = $\mathbf{\{-b/2a, -(b^2 - 4ac)/4a\}}$.

Now, the **a** is the slope, and the slope **a** is positive at the moment.
Then, at the vertex, **the x-coordinate** $\mathbf{\le 0 \Rightarrow -b/2a \le 0 \Rightarrow b/a \ge 0 \Rightarrow b \ge 0}$ since $\mathbf{a > 0}$.

Now, as an interim result, we have: $\mathbf{c = 0}$, $\mathbf{a + b = 2}$, and $\mathbf{b \ge 0}$.

Let's find the extent of the **a** now.
$\mathbf{a + b = 2 \Rightarrow b = 2 - a}$. Then, since $\mathbf{b = 2 - a}$ and $\mathbf{b \ge 0}$, $\mathbf{2 - a \ge 0 \Rightarrow a \le 2}$.
And yet, we know that $\mathbf{a > 0}$. Therefore, we can see that $\mathbf{0 < a \le 2}$.

So, we now have: $\mathbf{c = 0}$, $\mathbf{a + b = 2}$, and $\mathbf{0 < a \le 2}$.

Therefore in the case **A**, the equation is $\mathbf{y = ax^2 + bx}$ such that:
$\mathbf{a + b = 2}$ for $\mathbf{0 < a \leq 2}$.

Now, let's begin the case **B**.

Fig. G.4

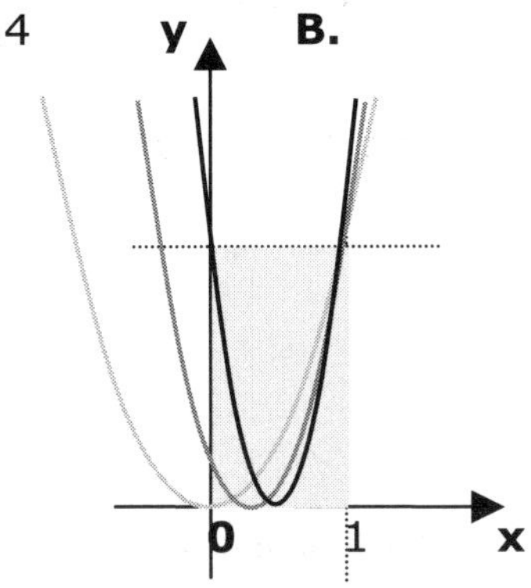

We want to find the values of the constants **a**, **b**, and **c** such that:
$\mathbf{y = f(x) = ax^2 + bx + c}$ for $\mathbf{0 \leq x \leq 1}$ and $\mathbf{0 \leq y \leq 2}$.

Then, we need to meet the conditions that:
$\mathbf{f(1) = 2}$, and also, at the vertex, **0 ≤ the x-coordinate ≤ 1/2**, and **the y-coordinate = 0**.

Why not consider the value of **f(0)**, then? That is, shouldn't we have $\mathbf{0 \leq f(0) \leq 2}$?
Of course, we should. However, we don't have to consider the case. If we do, it will be redundant. The conditions specified above is exhaustive now, and will force **f(0)** to be so. In other words, if all the conditions are met, then we will get $\mathbf{0 \leq f(0) \leq 2}$. Now, let's satisfy the conditions.

First, we begin with **f(1)**: $\mathbf{f(1) = 2 \Rightarrow 2 = a + b + c}$.

We know that the vertex is at $\mathbf{\{-b/2a, -(b^2 - 4ac)/4a\}}$.

Next, we know that the vertex is at $\mathbf{\{-b/2a, -(b^2 - 4ac)/4a\}}$. Then:

The y-coordinate = 0 $\mathbf{\Rightarrow -(b^2 - 4ac)/4a = 0 \Rightarrow b^2 - 4ac = 0}$.
0 ≤ the x-coordinate ≤ 1/2 $\mathbf{\Rightarrow 0 \leq -b/2a \leq 1/2 \Rightarrow -1/2 \leq b/2a \leq 0 \Rightarrow -1 \leq b/a \leq 0}$.

As an interim result, we now have: $\mathbf{a + b + c = 2}$, $\mathbf{b^2 - 4ac = 0}$, and $\mathbf{-1 \leq b/a \leq 0}$.
Now, we need to find the extents of the **a**, **b**, and **c**, but we don't have much to work with.
However, looking at closely the graph above, we can see that $\mathbf{2 \leq a \leq 8}$. How come?

We can see that the parabola in light gray has the slope of 2. How come?
Suppose that the equation is $\mathbf{y = ax^2}$ since the vertex is at the origin. Then:
Since we have $\mathbf{y = ax^2}$ and the parabola includes **(1, 2)**, $\mathbf{(1, 2) \Rightarrow 2 = a}$.

We can also, see that the parabola in black has the slope of 8. How come?

Suppose that the equation is $\mathbf{y = a(x - 1/2)^2}$ since the vertex is at **(1/2, 0)**.
Then, since we have $\mathbf{y = a(x - 1/2)^2}$, which includes **(1, 2)**:
$\mathbf{(1, 2) \Rightarrow 2 = a(1 - 1/2)^2 = a(1/2)^2 = a/4 \Rightarrow a = 8}$.

Therefore, as another interim result, we have:
$\mathbf{2 \le a \le 8}$, $\mathbf{a + b + c = 2}$, $\mathbf{b^2 - 4ac = 0}$, and $\mathbf{-1 \le b/a \le 0}$.

Now, let's find the possible values of the **b**.
$\mathbf{-1 \le b/a \le 0 \Rightarrow -a \le b \le 0}$. Then, since $\mathbf{2 \le a \le 8}$, we get $\mathbf{-a \le b \le 0 \Rightarrow -8 \le b \le 0}$. How come?

$\mathbf{2 \le a \le 8}$ means that the **a** can have any value from 2 to 8. Then:
First, if $\mathbf{a = 2}$, then $\mathbf{-a \le b \le 0 \Rightarrow -2 \le b \le 0}$. Next, if $\mathbf{a = 8}$, then $\mathbf{-a \le b \le 0 \Rightarrow -8 \le b \le 0}$.
Therefore, we can see that $\mathbf{-8 \le b \le 0}$.

Now, we have: $\mathbf{a + b + c = 2}$, $\mathbf{b^2 - 4ac = 0}$, $\mathbf{2 \le a \le 8}$, and $\mathbf{-8 \le b \le 0}$.

What about the **c**, then?
Let's get rid of the **c**. Then, $\mathbf{a + b + c = 2 \Rightarrow c = 2 - (a + b)}$.

Also, we can combine the two connective expressions $\mathbf{a + b + c = 2}$ and $\mathbf{b^2 - 4ac = 0}$.

First, $\mathbf{a + b + c = 2 \Rightarrow c = 2 - (a + b) \Rightarrow -4ac = -8a + 4a(a + b) = -8a + 4a^2 + 4ab}$
$\mathbf{= 4a^2 + 4ab + b^2 - b^2 - 8a = (2a + b)^2 - b^2 - 8a}$.

Next, $\mathbf{b^2 - 4ac = b^2 + (2a + b)^2 - b^2 - 8a = (2a + b)^2 - 8a = 0}$.

Then, we get $\mathbf{(2a + b)^2 = 8a \Rightarrow 2a + b = \pm(8a)^{1/2} = \pm\sqrt{8a} \Rightarrow b = \pm\sqrt{8a} - 2a = \pm(8a) - 2a}$.

Now, when the **a** changes from 2 to 8:
$\mathbf{\{(8a)^{1/2} - 2a\}}$ changes from 0 to −8, but $\mathbf{\{-(8a)^{1/2} - 2a\}}$ changes from −8 to −24.

We know that $\mathbf{-8 \le b \le 0}$.
Therefore, we get $\mathbf{b = (8a)^{1/2} - 2a = \sqrt{8a} - 2a}$.

Now, we have: $\mathbf{b = (8a)^{1/2} - 2a}$, $\mathbf{2 \le a \le} 8$, and $\mathbf{c = 2 - (a + b)}$.

Therefore, we can put the equation we are after in such a way as follows:

$\mathbf{y = ax^2 + bx + 2 - (a + b)}$ such that $\mathbf{b = (8a)^{1/2} - 2a}$ and $\mathbf{2 \le a \le 8}$.

Now, examining the graph again, we can see that $\mathbf{0 \le c \le 2}$.
That's because we need to have $\mathbf{0 \le}$ **y-intercept** $\mathbf{\le 2}$ and in $\mathbf{y = ax^2 + bx + c}$, the **c** is the y-intercept.

Also, let's check the extent of the **b**.

($\mathbf{0 \le c \le 2}$ and $\mathbf{2 \le a \le 8}$) $\mathbf{\Rightarrow 2 \le a + c \le 10}$.
$\mathbf{a + b + c = 2 \Rightarrow a + c = 2 - b \Rightarrow 2 \le 2 - b \le 10 \Rightarrow 0 \le -b \le 8 \Rightarrow -8 \le b \le 0}$.

Now, let's continue with the case of the **C**.

Fig. G.4

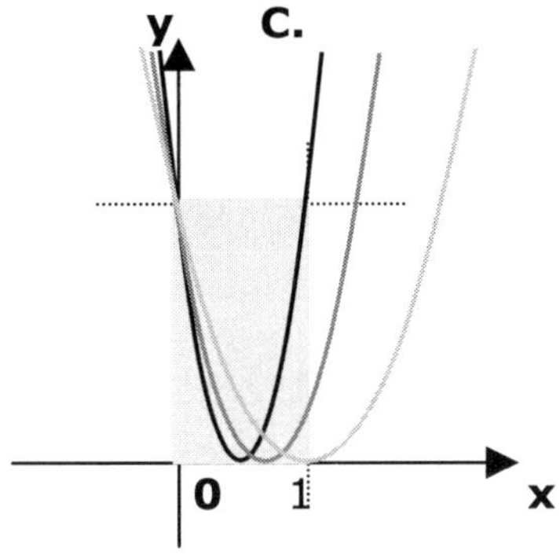

Now, we need to find the values of the constants **a**, **b**, and **c** such that:
$\mathbf{y = f(x) = ax^2 + bx + c}$ for $\mathbf{0 \le x \le 1}$ and $\mathbf{0 \le y \le 2}$.

Then, we need to meet the conditions that:
$\mathbf{f(0) = 2}$, and also, at the vertex, **1/2 ≤ the x-coordinate ≤ 1**, and **the y-coordinate = 0**.

To begin with, $\mathbf{f(0) = 2 \Rightarrow 2 = c \Rightarrow f(x) = ax^2 + bx + 2}$.

Next, let's move on to the vertex. We know that the vertex is at $\mathbf{\{-b/2a, -(b^2 - 4ac)/4a\}}$. Then:

First, **the y-coordinate = 0** $\mathbf{\Rightarrow -(b^2 - 4ac)/4a = 0 \Rightarrow b^2 - 4ac = 0 \Rightarrow b^2 - 8a = 0}$ **since c = 2**.
Next, **1/2 ≤ the x-coordinate ≤ 1** $\mathbf{\Rightarrow 1/2 \le -b/2a \le 1 \Rightarrow -1 \le b/2a \le -1/2 \Rightarrow -2 \le b/a \le -1}$.

Then, as an interim result, we have: $\mathbf{c = 2}$, $\mathbf{b^2 - 8a = 0}$, and $\mathbf{-2 \le b/a \le -1}$.

Now, since we have $\mathbf{b^2 - 8a = 0}$ the connective expression between the **a** and **b**, we are going to get the extent of the **a** or **b**. We have b^2-term in the expression but we don't know the sign of the **b**.
Then, let's find the extent of the **b** by means of the inequality $\mathbf{-2 \le b/a \le -1}$.
$\mathbf{b^2 - 8a = 0 \Rightarrow b^2 = 8a \Rightarrow b/a = 8/b}$.
Then, since $\mathbf{-2 \le b/a \le -1}$, $\mathbf{-2 \le 8/b \le -1}$. So, $\mathbf{-1 \le b/8 \le -1/2 \Rightarrow -8 \le b \le -4}$.

Now, we have: $\mathbf{c = 2}$, $\mathbf{-8 \le b \le -4}$, and $\mathbf{b^2 - 8a = 0}$.
Therefore, in the case **C**, the equation is $\mathbf{y = ax^2 + bx + 2}$ such that:
$\mathbf{b^2 - 8a = 0}$(or, $\mathbf{a = b^2/8}$) for $\mathbf{-8 \le b \le -4}$.

Now, as in the case of the **B** above, looking at only the graph above, we can see that $\mathbf{2 \le a \le 8}$.
First of all, we can see that the slope of the parabola in light gray is 2.
Suppose that the equation is $\mathbf{y = a(x - 1)^2}$ since the vertex is at **(1, 0)**. Then:
Since we have $\mathbf{y = a(x - 1)^2}$ and the parabola includes **(0, 2)**, $\mathbf{(0, 2) \Rightarrow 2 = a}$.
Next, we can see that the slope of the parabola in black is 8.
Suppose that the equation is $\mathbf{y = a(x - 1/2)^2}$ since the vertex is at **(1/2, 0)**. Then:
Since we have $\mathbf{y = a(x - 1/2)^2}$ and the parabola includes **(1, 2)**, $\mathbf{(1, 2) \Rightarrow 2 = a(1 - 1/2)^2 = a(1/2)^2 = a/4 \Rightarrow a = 8}$.
Let's actually find the extent of the **a** to check to see if the above is correct.
$\mathbf{-8 \le b \le -4 \Rightarrow 16 \le b^2 \le 64 \Rightarrow 2 \le b^2/8 \le 8}$. Then, since $\mathbf{a = b^2/8}$, $\mathbf{2 \le a \le 8}$.

Now, let's move on to the case of the **D**.

Fig. G.4

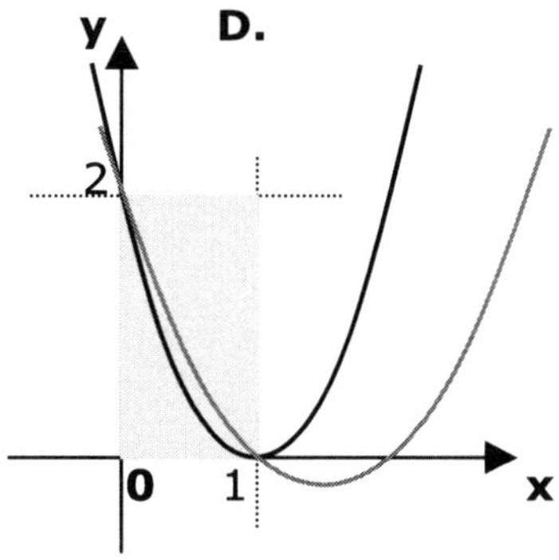

Now, we want to find the values of the constants **a**, **b**, and **c** such that:
$y = f(x) = ax^2 + bx + c$ for $0 \le x \le 1$ and $0 \le y \le 2$.
Then, we need to meet the conditions that $f(0) = 2$, and $f(1) = 0$; also, **the x-coordinate** ≥ 1 at the vertex. Then, $f(0) = 2 \Rightarrow 2 = c \Rightarrow f(x) = ax^2 + bx + 2$.
So, we get $f(1) = 0 \Rightarrow 0 = a + b + 2$. Therefore, $a + b = -2$.
Let's now move on to the vertex. We know that the vertex is at $\{-b/2a, -(b^2 - 4ac)/4a\}$.
Also, we know that the slope **a** is positive.
Then, **the x-coordinate** $\ge 1 \Rightarrow -b/2a \ge 1 \Rightarrow b/a \le -2 \Rightarrow b \le -2a$ since $a > 0$.
Then, as an interim result, we have: $c = 2$, $a + b = -2$, and $b \le -2a$.
Let's get the extent of the **a** now.
$a + b = -2 \Rightarrow b = -2 - a$. Then, $b \le -2a \Rightarrow -2 - a \le -2a \Rightarrow a \le 2$.
And yet, the $a > 0$ at the moment. Therefore, we can see that $0 < a \le 2$.
Therefore, the equation is $y = ax^2 + bx + 2$ such that $a + b = -2$ for $0 < a \le 2$.

Now, let's move on to the category where the parabolas have negative slopes.

Fig. G.5

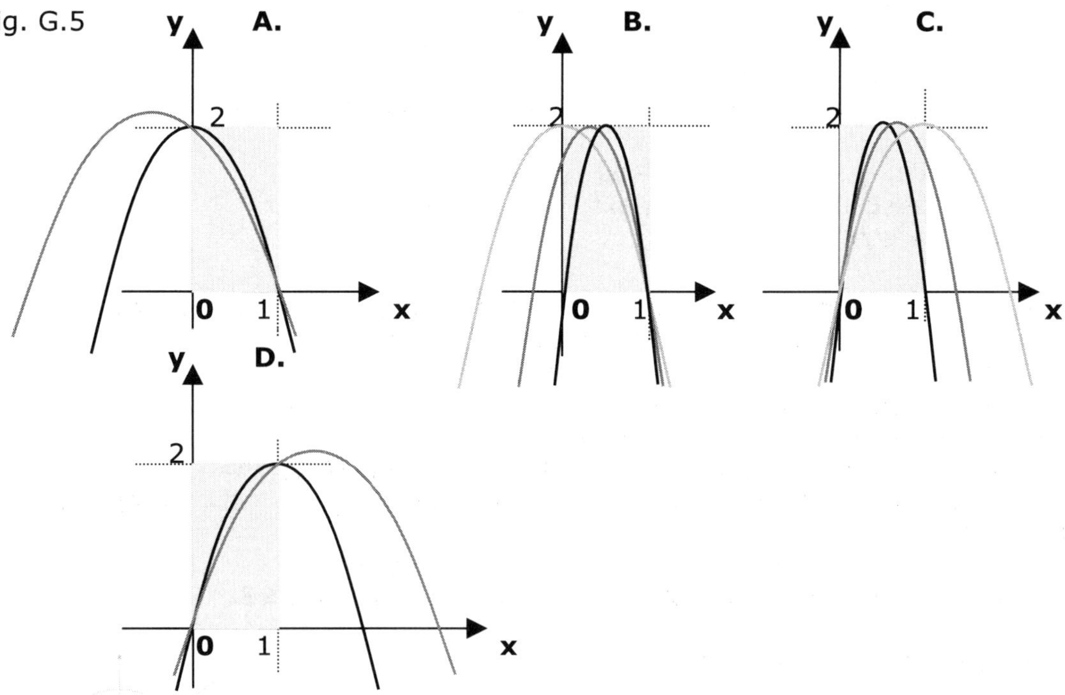

In case of the **A**, $f(0) = 2$, and $f(1) = 0$; also at the vertex, **the x-coordinate** ≤ 0.
In case of the **B**, $0 \le f(0) \le 2$, and $f(1) = 0$; also at the vertex, $0 \le$ **the x-coordinate** $\le 1/2$, and **the y-coordinate** $= 2$.
In case of the **C**, $f(0) = 0$, and $0 \le f(1) \le 2$; also at the vertex, $1/2 \le$ **the x-coordinate** ≤ 1, and **the y-coordinate** $= 2$.
In case of the **D**, $f(0) = 0$, and $f(1) = 2$; also at the vertex, **the x-coordinate** ≥ 1.

Now, let's begin with the case **A**.

Fig. G.5

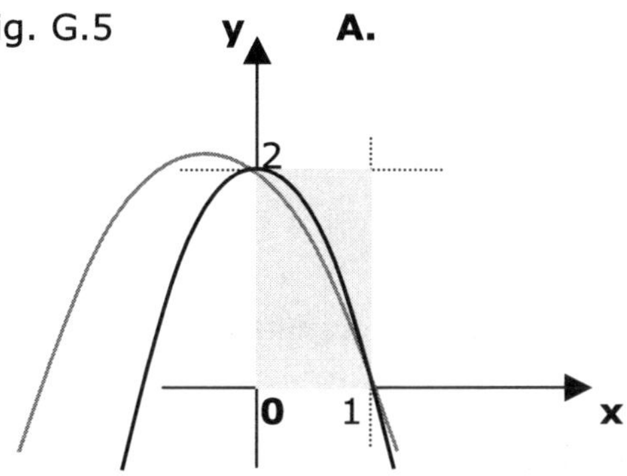

We want to find the values of the constants **a**, **b**, and **c** such that:
$y = f(x) = ax^2 + bx + c$ for $0 \le x \le 1$ and $0 \le y \le 2$.

Then, we need to satisfy the conditions that $f(0) = 2$, and $f(1) = 0$; also, **the x-coordinate** ≤ 0 at the vertex.

To begin with, $f(0) = 2 \Rightarrow 2 = 0 + 0 + c \Rightarrow c = 2 \Rightarrow f(x) = ax^2 + bx + 2$.
Then, $f(1) = 0 \Rightarrow 0 = a + b + 2 \Rightarrow a + b = -2$.

Next, we know that the vertex is $\{-b/2a, -(b^2 - 4ac)/4a\}$.
Now, the **a** is the slope, and we know that the slope **a** is negative now.

Then, at the vertex, **the x-coordinate** $\le 0 \Rightarrow -b/2a \le 0 \Rightarrow b/a \ge 0 \Rightarrow b \le 0$ since $a < 0$.

Now, as an interim result, we have: $c = 2$, $a + b = -2$, and $b \le 0$.

Let's find the extent of the **a** now.

$a + b = -2 \Rightarrow b = -2 - a$. Then, since $b = -2 - a$ and $b \le 0$, $-2 - a \le 0 \Rightarrow a \ge -2$.
Now, we also know that $a < 0$. Therefore, we can see that $-2 \le a < 0$.

So, we have: $c = 2$, $a + b = -2$, and $-2 \le a < 0$.

Therefore, in the case of the **A**, the equation is $y = ax^2 + bx + 2$ such that:
$a + b = 2$ for $-2 \le a < 0$.

Now, let's begin the case of the **B**.

Fig. G.5

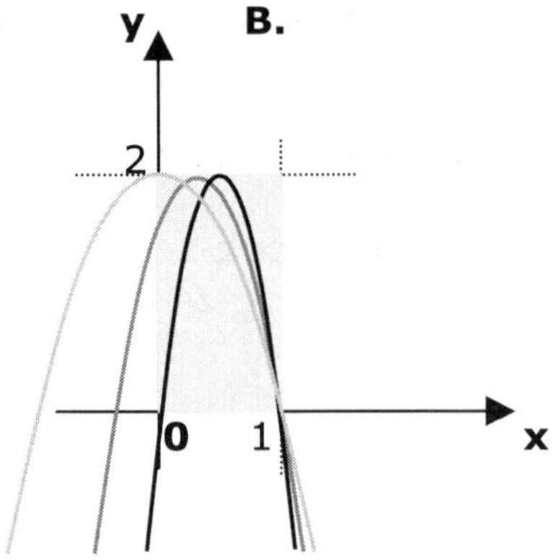

Now, we want to find the values of the constants **a**, **b**, and **c** such that:
$y = f(x) = ax^2 + bx + c$ for **$0 \le x \le 1$** and **$0 \le y \le 2$**.

Then, we need to satisfy the conditions that:
$f(1) = 0$, and also, at the vertex, **$0 \le$ the x-coordinate $\le 1/2$**, and **the y-coordinate = 2**.

First, we begin with **f(1)**: **$f(1) = 0 \Rightarrow 0 = a + b + c$**.
Next, we know that the vertex is at **$\{-b/2a, -(b^2 - 4ac)/4a\}$**. Then:
The y-coordinate = $2 \Rightarrow -(b^2 - 4ac)/4a = 2 \Rightarrow b^2 - 4ac = -8a \Rightarrow b^2 - 4ac + 8a = 0$.
$0 \le$ x-coordinate $\le 1/2 \Rightarrow 0 \le -b/2a \le 1/2 \Rightarrow -1/2 \le b/2a \le 0 \Rightarrow -1 \le b/a \le 0$.

Now, as an interim result, we have: **$a + b + c = 0$, $b^2 - 4ac + 8a = 0$**, and **$-1 \le b/a \le 0$**.

Now, we need to find the extents of the **a**, **b**, and **c**, but we don't have much to work with.
However, looking at closely the graph above, we can see that **$-8 \le a \le -2$**. How come?

In the graph above, we can see that the slope of the parabola in light gray is -2. How come?
Since the vertex is at **(0, 2)**, suppose that the equation is **$y = ax^2 + 2$**. Then:
Since we have **$y = ax^2 + 2$** and the parabola includes **(1, 0)**, **$(1, 0) \Rightarrow 0 = a + 2 \Rightarrow a = -2$**.

Also, w can see that the slope of the parabola in black is -8. How come?
Since the vertex is at **(1/2, 2)**, suppose that the equation is **$y = a(x - 1/2)^2 + 2$**.
Then, since we have **$y = a(x - 1/2)^2 + 2$** and the parabola includes **(1, 0)**:
$(1, 0) \Rightarrow 0 = a(1 - 1/2)^2 + 2 = a(1/2)^2 + 2 = a/4 + 2 \Rightarrow a = -8$.

Therefore, as another interim result, we have:
$-8 \le a \le -2$, $a + b + c = 0$, $b^2 - 4ac + 8a = 0$, and **$-1 \le b/a \le 0$**.

Then, let's find the extent of the **b**.
$1 \le b/a \le 0 \Rightarrow 0 \le b \le -a$ since **$a < 0$**. Then, since **$-8 \le a \le -2$, $0 \le b \le -a \Rightarrow 0 \le b \le 8$**.
How come?
$-8 \le a \le -2$ means that the **a** can have any value from −8 to -2. Then:
First, if **$a = -8$**, then **$0 \le b \le -a \Rightarrow 0 \le b \le -(-8) = 8 \Rightarrow 0 \le b \le 8$**.
Next, if **$a = -2$**, then **$0 \le b \le -a \Rightarrow 0 \le b \le -(-2) = 2 \Rightarrow 0 \le b \le 2$**.

Therefore, we can see that **$0 \le b \le 8$**.

Now, we have: $\mathbf{a + b + c = 0}$, $\mathbf{b^2 - 4ac + 8a = 0}$, $\mathbf{-8 \le a \le -2}$, and $\mathbf{0 \le b \le 8}$.

Let's eliminate the **c**. Then, $\mathbf{a + b + c = 0 \Rightarrow c = -(a + b)}$.
Also, we can combine the two connective expressions $\mathbf{a + b + c = 0}$ and $\mathbf{b^2 - 4ac + 8a = 0}$.

To begin with, $\mathbf{a + b + c = 0 \Rightarrow c = -(a + b)}$
$\mathbf{\Rightarrow -4ac = 4a(a + b) = 4a^2 + 4ab = 4a^2 + 4ab + b^2 - b^2 = (2a + b)^2 - b^2}$.

Next, $\mathbf{b^2 - 4ac + 8a = b^2 + (2a + b)^2 - b^2 + 8a = (2a + b)^2 + 8a = 0 \Rightarrow (2a + b)^2 = -8a}$, which is positive since the **a** is negative.
So, $\mathbf{2a + b = \pm(-8a)^{1/2} \Rightarrow b = -2a \pm (-8a)^{1/2} \Rightarrow b = -2a + \sqrt{-8a}}$ or $\mathbf{b = -2a - \sqrt{-8a}}$.

Now, when the **a** changes from −8 to -2:
$\mathbf{\{-2a + \sqrt{-8a}\}}$ changes from 24 to 8, but $\mathbf{\{-2a - \sqrt{-8a}\}}$ changes from 8 to 0.

We know that $\mathbf{0 \le b \le 8}$.
Therefore, we get $\mathbf{b = -2a - \sqrt{-8a}}$.

Now, we have: $\mathbf{b = -2a - \sqrt{-8a}}$, $\mathbf{-8 \le a \le -2}$, and $\mathbf{c = -(a + b)}$.

Therefore, we can put the equation we are after in such a way as follows:

$\mathbf{y = ax^2 + bx - (a + b)}$ such that $\mathbf{b = -2a - \sqrt{-8a}}$ for $\mathbf{-8 \le a \le -2}$.

Now, examining the graph again, we can see that $\mathbf{0 \le c \le 2}$.
That's because we need to have **0 ≤ y-intercept ≤ 2** and in $\mathbf{y = ax^2 + bx + c}$, the **c** is the y-intercept.

In addition, let's check the extent of the **b**.

$\mathbf{(0 \le c \le 2}$ and $\mathbf{-8 \le a \le -2) \Rightarrow -8 \le a + c \le 0}$.
$\mathbf{a + b + c = 0 \Rightarrow a + c = -b \Rightarrow -8 \le -b \le 0 \Rightarrow 0 \le b \le 8}$.

Now, let's continue with the case of the **C**.

Fig. G.5

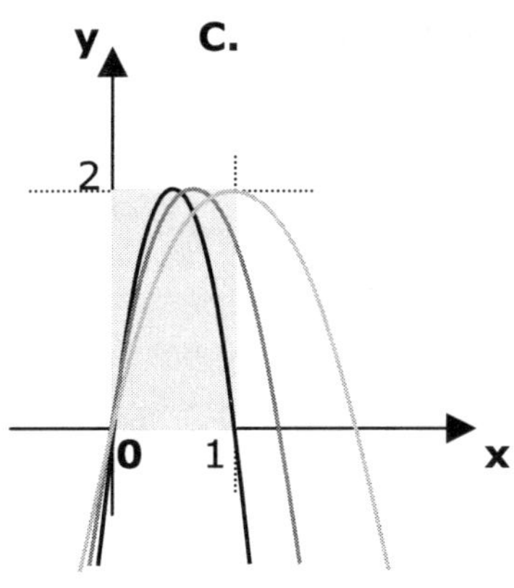

Now, we want to find the values of the constants **a**, **b**, and **c** such that:
$y = f(x) = ax^2 + bx + c$ for $0 \le x \le 1$ and $0 \le y \le 2$.

Then, we need to meet the conditions that:
$f(0) = 0$, and also, at the vertex, **1/2 ≤ the x-coordinate ≤ 1**, and **the y-coordinate = 2**.

To begin with, $f(0) = 0 \Rightarrow 0 = c \Rightarrow f(x) = ax^2 + bx$.

Next, let's move on to the vertex. We know that the vertex is at $\{-b/2a, -(b^2 - 4ac)/4a\}$. Then:
First, **the y-coordinate** $= 2 \Rightarrow -(b^2 - 4ac)/4a = 2 \Rightarrow b^2 - 4ac = -8a \Rightarrow b^2 + 8a = 0$ since $c = 0$.
Next, **1/2 ≤ the x-coordinate ≤ 1** $\Rightarrow 1/2 \le -b/2a \le 1 \Rightarrow -1 \le b/2a \le -1/2 \Rightarrow -2 \le b/a \le -1$.

Then, as an interim result, we have: $c = 0$, $b^2 + 8a = 0$, and $-2 \le b/a \le -1$.
Let's use the two expressions, $b^2 + 8a = 0$ and $-2 \le b/a \le -1$ to get the extent of the **b**.

Then, $b^2 + 8a = 0 \Rightarrow b^2 = -8a \Rightarrow a = -b^2/8$. Then, since $a = -b^2/8$:
$-2 \le b/a \le -1 \Rightarrow -2 \le -8/b \le -1 \Rightarrow 1 \le 8/b \le 2 \Rightarrow 1/2 \le b/8 \le 1 \Rightarrow 4 \le b \le 8$.

Now, we have: $c = 0$, $b^2 + 8a = 0$, and $4 \le b \le 8$.

Therefore, the equation we are after can be put in such a way as follows:

$y = ax^2 + bx$ such that $b^2 + 8a = 0$ (or, $a = -b^2/8$) for $4 \le b \le 8$.

Now, as in the case of the **B** above, looking at the graph only, we can see that $-8 \le a \le -2$.

First of all, we can see that the slope of the parabola in light gray is -2.
Since the vertex is at $(1, 2)$, suppose that the equation is $y = a(x - 1)^2 + 2$.

Then, since we have $y = a(x - 1)^2 + 2$ and the parabola includes $(0, 0)$:
$(0, 0) \Rightarrow 0 = a(0 - 1)^2 + 2 \Rightarrow 0 = a + 2 \Rightarrow a = -2$.

Next, we can see that the slope of the parabola in black is -8.
Since the vertex is at $(1/2, 2)$, suppose that the equation is $y = a(x - 1/2)^2 + 2$.

Then, since we have $y = a(x - 1/2)^2 + 2$ and the parabola includes $(0, 0)$:
$(0, 0) \Rightarrow 0 = a(0 - 1/2)^2 + 2 = a(1/2)^2 + 2 = a/4 + 2 \Rightarrow a/4 = -2 \Rightarrow a = -8$.

Let's actually find the extent of the **a** to check to see if the above is correct.

$4 \le b \le 8 \Rightarrow 16 \le b^2 \le 64 \Rightarrow 2 \le b^2/8 \le 8 \Rightarrow -8 \le -b^2/8 \le -2$.
Then, since $a = -b^2/8$, $-8 \le a \le -2$.

Now, let's move on to the case of the **D**.

Fig. G.5

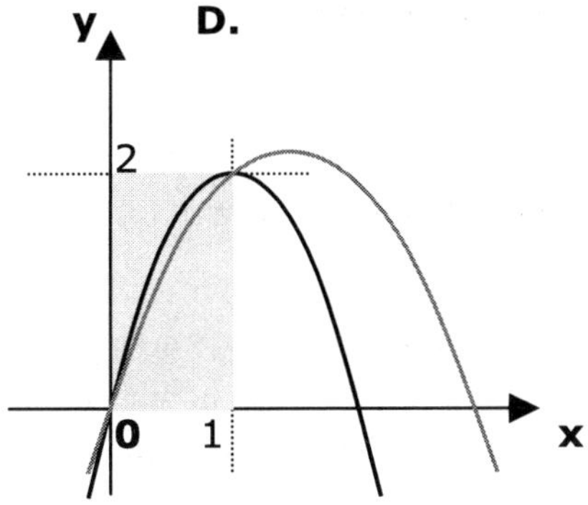

Now, we want to find the values of the constants **a**, **b**, and **c** such that:
$y = f(x) = ax^2 + bx + c$ for **$0 \le x \le 1$** and **$0 \le y \le 2$**.

Then, we need to meet the conditions that **$f(0) = 0$**, and **$f(1) = 2$**; also, the **x-coordinate $\ge$ 1** at the vertex.

Then, **$f(0) = 0 \Rightarrow 0 = c \Rightarrow f(x) = ax^2 + bx$**.
So, we get **$f(1) = 2 \Rightarrow 2 = a + b$**. Therefore, **$a + b = 2$**.

Let's now move on to the vertex. we know that the vertex is at **$\{-b/2a, -(b^2 - 4ac)/4a\}$**.
Also, we know that the slope **a** is negative.

Then, **the x-coordinate $\ge 1 \Rightarrow -b/2a \ge 1 \Rightarrow b/a \le -2 \Rightarrow b \ge -2a$** since **$a < 0$**.

Then, as an interim result, we have: **$c = 0$**, **$a + b = 2$**, and **$b \ge -2a$**.

Let's get the extent of the **a** now.
$a + b = 2 \Rightarrow b = 2 - a$. Then, **$b \ge -2a \Rightarrow 2 - a \ge -2a \Rightarrow a \ge -2$**.
And yet, the **$a < 0$** now. Therefore, **$-2 \le a < 0$**.

Therefore, in the case of the **D**, the equation is **$y = ax^2 + bx$** such that:
$a + b = 2$ for **$-2 \le a < 0$**.

Suggestions or Solutions
To the Problem in the Example H

Suppose that $y = f(x) = (2x^2 + 3)/7$ for $m \le x \le n$ and $m \le y \le n$ where m and n are constants, and then, find the values of the m and n.

In this example, we get the extent of the output of the function **f** to be the same as that of the input. Then, we need to first find two particular input values for each of which the output is the same as the input itself. Therefore, when we input the particular values to the function, the outputs respectively have to be the same as the particular values.
Now, we have a function where inputs and outputs are the same respectively, and what is the function? It is called an identity function. The curve of the function is a line $\mathbf{y = x}$ in the x-y system.
Now, the curve of the function **f** is a parabola. Then, if we find the points where the two curves, that is, the parabola and the line meet each other, we can find the extent. The coordinate values of the points are to be the values of the **m** and the **n**.
In fact, the same is true for the curves other than parabolas. Now, let's find the points.
The equation of the parabola is $\mathbf{y = (2x^2 + 3)/7}$ and the equation of the line is $\mathbf{y = x}$. Then:

$$x = (2x^2 + 3)/7 \Rightarrow 7x = 2x^2 + 3 \Rightarrow 2x^2 - 7x + 3 = 0 \Rightarrow (2x - 1)(x - 3) = 0 \Rightarrow x = 1/2 \text{ or } 3.$$

Then, the curves meet each other when the **x** is 1/2 and 3. Therefore, we can see that the $\mathbf{m = 1/2}$ and the $\mathbf{n = 3}$. However, that's not the way it is all the time. We get to see why not if we look at the next example.

Fig. H.0

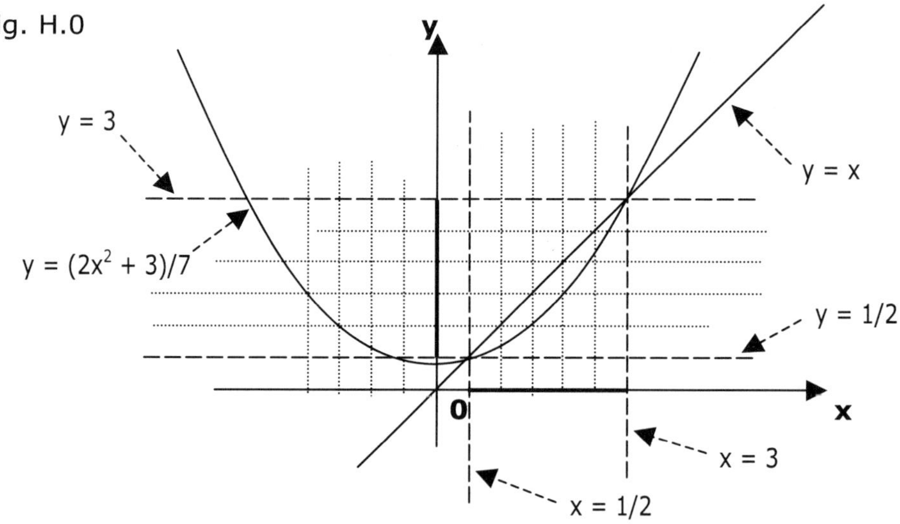

In short:

The values of the x-coordinates of the points where the curve of the **f** and the line, that is, the parabola $\mathbf{y = (2x^2 + 3)/7}$ and the line $\mathbf{y = x}$ meet each other are the **m** and **n**.

$\mathbf{x = (2x^2 + 3)/7 \Rightarrow 7x = 2x^2 + 3 \Rightarrow 2x^2 - 7x + 3 = 0 \Rightarrow (2x - 1)(x - 3) = 0 \Rightarrow x = 1/2}$ or **3**.
Therefore, $\mathbf{m = 1/2}$ and $\mathbf{n = 3}$.

Suggestions or Solutions

To the Problem in the Example I

Suppose that $y = f(x) = (2x^2 - 3)/5$ for $m \le x \le n$ and $m \le y \le n$ where m and n are constants, and then, find the values of the m and n.

As in the previous example, we get the extent of the output of a function to be the same as the extent of the input. Then, we need to first find two particular input values for each of which the output is the same as the input itself. Therefore, when we input the particular values to the function, the outputs respectively have to be the same as the particular values.
Now, we have a function where inputs and outputs are the same respectively, the function is called an identity function, and the curve of the function is a line **$y = x$** in the x-y system. Then, finding the points where the two curves, that is, the parabola and the line meet each other, we can find the extent. Now, let's find the points.

The equation of the parabola is **$y = (2x^2 - 3)/5$**, and the equation of the line is **$y = x$**. Then:

$x = (2x^2 - 3)/5 \Rightarrow 5x = 2x^2 - 3 \Rightarrow 2x^2 - 5x - 3 = 0 \Rightarrow (2x + 1)(x - 3) = 0 \Rightarrow x = -1/2$ or **3**.

Then, it looks like that **$m = -1/2$** and **$n = 3$**. Therefore, if **$-1/2 \le x \le 3$**, we need to have **$-1/2 \le y \le 3$**. However, there is something wrong with the extent of the **y**. That is, it is not case where **$-1/2 \le y \le 3$**. Why not?

We know that the curve of the function **f** is a parabola, the slope is positive, and the axis of symmetry is a line **$x = 0$**; in other words, it is the y-axis. Then, it is concave down. Therefore, the y-coordinate at the vertex is less than -1/2. So, it is not the case where **$-1/2 \le y \le 3$**. In fact, we can see that the value of the y-coordinate is -3/5: **$y = (2x^2 - 3)/5 = 2x^2/5 - 3/5$**. Therefore, if **$-1/2 \le x \le 3$**, then we get to have **$-3/5 \le y \le 3$**, which can't be the solution. That is, there is no solution to this problem.

We should always consider the shape of a parabola. The same is true for the other curves.

Fig. I.0

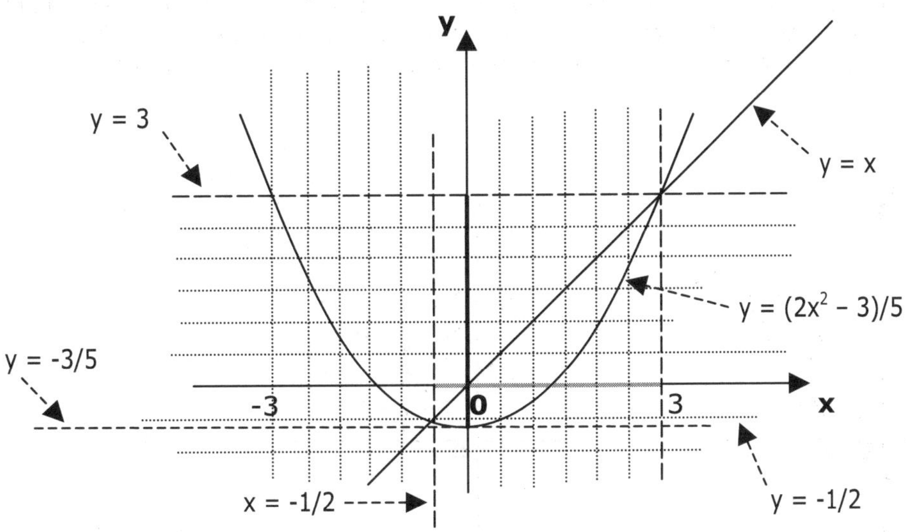

Let's take another similar example now.

Suggestions or Solutions
To the Problem in the Example J

Suppose that y = f(x) = $(2x^2 + c)/5$ for m ≤ x ≤ n and m ≤ y ≤ n where c, m, and n are constants, and then, find the values of the c , m, and n.

As in the examples presented earlier, we get the extent of the output of a function to be the same as the extent of the input. Then, we need to first find two particular input values for each of which the output is the same as the input itself. Therefore, when we input the particular values to the function, the outputs respectively have to be the same as the particular values. Now, we have a function where inputs and outputs are the same respectively, the function is called an identity function, and the curve of the function is a line **y = x** in the x-y system. The curve of the **f** is a parabola. Then, finding the points where the two curves, that is, the parabola and the line meet each other, we can find the extent. However, finding the points only doesn't complete the solution. That's because the existence of the solution depends on the location of the vertex. Then, let's construct a schematic graph to see in what circumstances the solution can exist. Let's first check the concavity of the parabola. The slope of the parabola **y = $(2x^2 + c)/5$** is positive, and the axis of symmetry is the y-axis. Then, it is concave down. Then, let's put some parabolas so that the vertices are in the y-axis.

Fig J.0

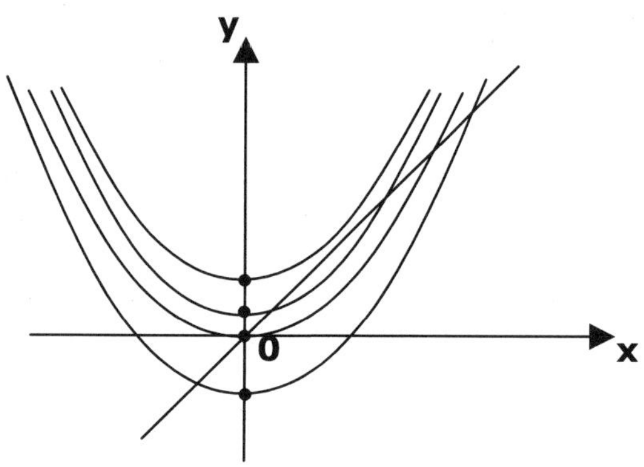

Now, the parabola and the line need to meet each other first at two points. Next, although they meet, the vertex shouldn't be below the origin or the line. That is, the y-coordinate of the vertex is greater than or equal to the x-coordinate, and the parabola and line need to meet each other at two points. Then, let's see some other cases where the axes of symmetry are not the y-axis but are parallel to the y-axis.

Fig. J.1

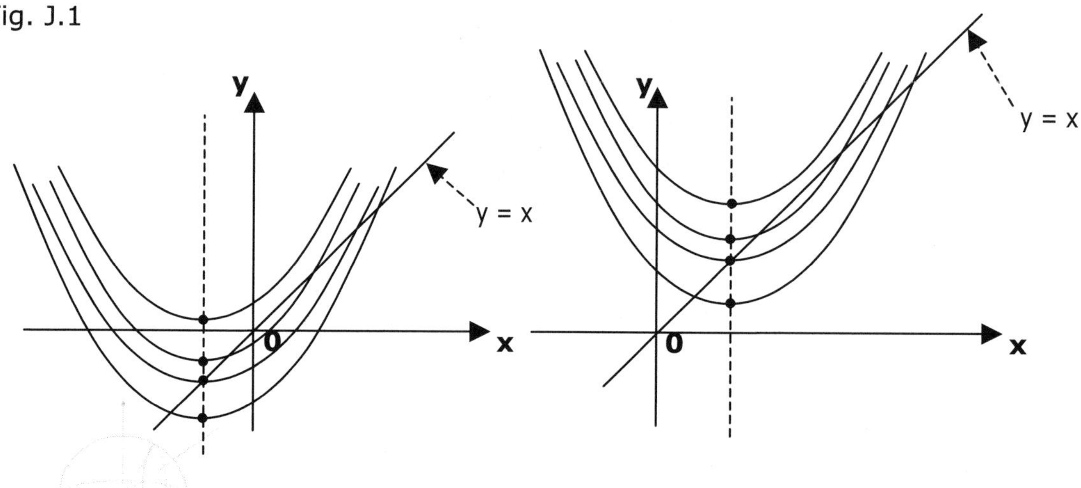

Now, first if the parabola is concave down, the y-coordinate at the vertex cannot be less than the x-coordinate regardless of the position of the axis of symmetry. That is, the y-coordinate has to be greater than or equal to the x-coordinate. In addition, the parabola and the line need to meet each other at two points.

Therefore, if there exists such an extent that the problem is asking, the y-coordinate of the vertex has to be greater than or equal to the x-coordinate, and also, the two curves, that is, the parabola and the line need to meet each other at two points.

Then, is there any other way we can see if there is such an extent?
There is. Examining the graphs above, we can also see the fact as follows.

First, the two curves, that is, the parabola and the line need to meet each other at two points.
The equation of the parabola is $\mathbf{y = (2x^2 + c)/5}$ and that of the line is $\mathbf{y = x}$.

Then, $\mathbf{x = (2x^2 + c)/5 \Rightarrow 5x = 2x^2 + c \Rightarrow 2x^2 - 5x + c = 0}$.

Now, suppose that the roots of the equation are $\mathbf{s_1}$ and $\mathbf{s_2}$; that is, $\mathbf{2x^2 - 5x + c = (x - s_1)(x - s_2) = 0}$.
Let's suppose also, that $\mathbf{s_1 < s_2}$.

Then, the x-coordinate at the vertex has to be less than or equal to the $\mathbf{s_1}$.

Now, in both of the methods, the parabola and the lines need to meet each other first at two points.
We can check it by means of the determinant of the equation where we find the two points. If the determinant is greater than 0, the equation has two different real roots. Then, the determinant of the equation $\mathbf{2x^2 - 5x + c = 0}$ needs to be greater than 0.

Suppose now, that the equation is $\mathbf{ax^2 + bx + c = 0}$ and the determinant is $\mathbf{D}$.
Then, the $\mathbf{D}$ is as follows: $\mathbf{D = b^2 - 4ac}$.

Now, let's find the determinant of $\mathbf{2x^2 - 5x + c = 0}$.
We can see that the determinant $\mathbf{D = (-5)^2 - 4(2)c = 25 - 8c}$.
Then, we need to have $\mathbf{D > 0}$. That is, we need to have $\mathbf{25 - 8c > 0 \Rightarrow c < 25/8}$.

Now, let's find the vertex and the roots:

$\mathbf{y = (2x^2 + c)/5 = 2x^2/5 + c/5 \Rightarrow}$ **The vertex =** $\mathbf{(0, c/5)}$.

$\mathbf{2x^2 - 5x + c = 2(x^2 - 5x/2) + c = 2(x^2 - 5x/2 + 25/16 - 25/16) + c = 2(x - 5/4)^2 - 25/8 + c}$
$\mathbf{= 0 \Rightarrow 2(x - 5/4)^2 = 25/8 - c \Rightarrow (x - 5/4)^2 = 25/16 - c/2 \Rightarrow x - 5/4 = \pm(25/16 - c/2)^{1/2}}$
$\mathbf{\Rightarrow x = 5/4 + (25/16 - c/2)^{1/2}}$ **or** $\mathbf{5/4 - (25/16 - c/2)^{1/2}}$.
Then, since $\mathbf{5/4 + (25/16 - c/2)^{1/2} > 5/4 - (25/16 - c/2)^{1/2}}$:
$\mathbf{m = 5/4 - (25/16 - c/2)^{1/2}}$ and $\mathbf{n = 5/4 + (25/16 - c/2)^{1/2}}$.

Now, let's first use the first method.
Then, the y-coordinate of the vertex is greater than or equal to the x-coordinate, and we need to have **the determinant > 0**.
To begin with, since **the vertex =** $\mathbf{(0, c/5)}$, we get $\mathbf{c \geq 0}$ because **the y-coordinate =** $\mathbf{c/5 \geq 0}$ **= the x-coordinate**. Next, we get $\mathbf{c < 25/8}$ since $\mathbf{D > 0}$.
So, we need to have $\mathbf{0 \leq c < 25/8}$.
Therefore, $\mathbf{m = 5/4 - (25/16 - c/2)^{1/2}}$ and $\mathbf{n = 5/4 + (25/16 - c/2)^{1/2}}$ where $\mathbf{0 \leq c < 25/8}$.

Now, let's give a check to the second method.
Then, the x-coordinate of the vertex has to be less than or equal to the smaller of the two roots.

Since the vertex = **(0, c/5)**, the x-coordinate at the vertex is 0.
Then, we can see that **the x-coordinate = 0 ≤ m = 5/4 – (25/16 – c/2)$^{1/2}$**. Therefore:

$5/4 - (25/16 - c/2)^{1/2} \geq 0 \Rightarrow (25/16 - c/2)^{1/2} \leq 5/4 \Rightarrow 25/16 - c/2 \leq 25/16 \Rightarrow -c/2 \leq 0$
$\Rightarrow c/2 \geq 0 \Rightarrow c \geq 0$.

Also, we have **$c < 25/8$** since the determinant **$D > 0$**.

So, we get **$0 \leq c < 25/8$**.

Therefore, **$m = 5/4 - (25/16 - c/2)^{1/2}$** and **$n = 5/4 + (25/16 - c/2)^{1/2}$** where **$0 \leq c < 25/8$**.

In short:

The curve of the function **f** is a parabola, and needs to meet a line **y = x** at two points, and the y-coordinate of the vertex has to be greater than or equal to the x-coordinate.

To begin with, we find the vertex.
$y = (2x^2 + c)/5 = 2x^2/5 + c/5 \Rightarrow$ The vertex = (0, c/5).

Next, we find the x-coordinates of the points where the curve of the **f** and the line **y = x** meet each other.
$2x^2 - 5x + c = 2(x^2 - 5x/2) + c = 2(x^2 - 5x/2 + 25/16 - 25/16) + c = 2(x - 5/4)^2 - 25/8 + c$
$= 0 \Rightarrow 2(x - 5/4)^2 = 25/8 - c \Rightarrow (x - 5/4)^2 = 25/16 - c/2 \Rightarrow x - 5/4 = \pm(25/16 - c/2)^{1/2}$
$\Rightarrow x = 5/4 + (25/16 - c/2)^{1/2}$ or $5/4 - (25/16 - c/2)^{1/2}$.

Then, since **$5/4 + (25/16 - c/2)^{1/2} > 5/4 - (25/16 - c/2)^{1/2}$**:
$m = 5/4 - (25/16 - c/2)^{1/2}$ and **$n = 5/4 + (25/16 - c/2)^{1/2}$**.

The determinant of the equation above is **$D = (-5)^2 - 4(2)c = 25 - 8c$**.

Then, the y-coordinate of the vertex has to be greater than or equal to the x-coordinate, and we need to have **the determinant > 0**.
To begin with, **the vertex = (0, c/5) ⇒ the y-coordinate = c/5 and the x-coordinate = 0**.
Then, **the y-coordinate = $c/5 \geq 0 \Rightarrow c \geq 0$**. Next, **$D > 0 \Rightarrow c < 25/8$**.
So, **$0 \leq c < 25/8$**.

Therefore, **$m = 5/4 - (25/16 - c/2)^{1/2}$** and **$n = 5/4 + (25/16 - c/2)^{1/2}$** where **$0 \leq c < 25/8$**.

Suggestions or Solutions

To the Problem in the Example K

Suppose we have an equation of a parabola as follows: $2x^2 + (2c + 1)x - (c + 1)y + c + 1 = 0$ where c is constant. Then, assuming that the parabola includes two particular points for any value of c, find the two points.

In fact, the given equation of parabola is not an equation of a particular parabola, but indicates many parabolas forming a group. When we work on the equations of line, we covered the fact that if two lines meet at a particular point, then any linear combination (or sum of any multiples) of the two lines includes the particular point.
Suppose for instance, that **a**, **b**, **c**, **d**, **e**, **f**, **m**, and **n** are constant, and that two lines $ax + by + c = 0$ and $dx + ey + f = 0$ meet at a particular point. Then, a line $m(ax + by + c) + n(dx + ey + f)$ includes the particular point, too. Then, we say that $m(ax + by + c) + n(dx + ey + f) = 0$ is a linear combination of $ax + by + c = 0$ and $dx + ey + f = 0$. In other words, the sum of any multiples of the two lines includes the particular point. That is, for any choices for the constants **m** and **n**, the line indicated by the sum includes the particular point, too.

Now, the similar idea applies to two parabolas. In addition, the same is true for a parabola and a line, too. That is, if two parabolas meet each other at a particular point, all the linear combinations meet each other at the particular point, too, and if a parabola and a line meet each other at a particular point, then all the linear combinations meet each other at the particular point, too.
Suppose for instance, that we add together two equations **A** and **B** indicating parabolas and make an equation of a new parabola **C**. Suppose also, that the parabolas **A** and **B** meet each other at two points. Then, the new parabola **C** includes the two points, and therefore, all the three parabolas meet each other altogether at the two points.

Suppose for another instance, that any letter except the **x** and the **y** is constant in the equations below:
$A(x, y) = ax^2 + bx + c + dy = 0$.
$B(x, y) = ex + fy + g = 0$.
$C(x, y) = m(ax^2 + bx + c + dy) + n(ex + fy + g) = 0$.
Suppose next, that the parabola **A** and the line **B** meet each other at particular points.
Then, the parabola **C**, too, meets the **A** and **B** at the particular points.

Suppose for one more instance, that any letter except the **x** and **y** is constant in the equations below.
$A(x, y) = ax^2 + bx + c + dy = 0$.
$B(x, y) = ex^2 + fx + g + hy = 0$.
$C(x, y) = m(ax^2 + bx + c + dy) + n(ex^2 + fx + g + hy) = 0$.
Suppose also, that the parabolas **A** and **B** meet each other at particular points.
Then, the parabola **C**, too, meet the parabolas **A** and **B** at the particular points, too.

In fact, the similar idea applies to equations of curves in other kinds.
Suppose for instance, that **m** and **n** are constants, and that two curves $f(x, y) = 0$ and $g(x, y) = 0$ meet each other at particular points. Then, all the curves represented by $mf(x, y) + ng(x, y) = 0$ include the particular points, too.

When we say curves in math in general, we mean not only curved lines but straight lines. In addition, an equation of line can indicate any line. For instance, $x = 1$ and $x = -1/2$ can be equations of line. Of course, $y = 2$ can be an equation of line, too. An equation $x = 0$, too, can be an equation of line, which indicates the y-axis, and an equation $y = 0$, too, can be an equation of line, which indicates the x-axis. Also, a line $x = 1$ includes all the points $(1, y)$ for all the values of the **y**. That is, the line $x = 1$ includes $(1, 0)$, $(1, -1/2)$, $(1, 12)$, etc.

Suppose now, that **f(x, y) = 0** and **g(x, y) = 0** are basic curves. Then, if the basic curves meet each other at particular points, all the curves represented by **mf(x, y) + ng(x, y) = 0** meet each other altogether at the particular points.

Now, in this example, we have $2x^2 + (2c + 1)x - (c + 1)y + c + 1 = 0$, which indicates a group of parabolas.
Then, it is composed of two basic curves, and the two basic curves meet each other at the two points.
Therefore, we need to extract the two basic curves, and find out where the basic curves meet each other.
That is, we need to find the two points where the curves meet each other.

Then, let's first begin with the two basic curves.

$$2x^2 + (2c + 1)x - (c + 1)y + c + 1 = 2cx - cy + c + 2x^2 + x - y + 1$$
$$= c(2x - y + 1) + 2x^2 + x - y + 1 = 0.$$

Then, the two basic curves are as follows:
A line $2x - y + 1 = 0$ and a parabola $2x^2 + x - y + 1 = 0$.

Next, we need to find the points where the curves meet each others.
Let's begin with the x-coordinate.
$2x - y + 1 = 0 \Rightarrow y = 2x + 1$, and also, $2x^2 + x - y + 1 = 0 \Rightarrow y = 2x^2 + x + 1$.
Then, we need to set $2x + 1 = 2x^2 + x + 1$, and then, solve the equation.
$2x + 1 = 2x^2 + x + 1 \Rightarrow 2x^2 - x = 0 \Rightarrow x(2x - 1) = 0 \Rightarrow x = 0$ or $1/2$.
Then: $x = 0 \Rightarrow y = 2x + 1 = 0 + 1 \Rightarrow y = 1$, and also, $x = 1/2 \Rightarrow y = 2x + 1 = 1 + 1 = 2 \Rightarrow y = 2$.
Therefore, every parabola $2x^2 + (2c + 1)x - (c + 1)y + c + 1 = 0$ includes **(0, 1)** and **(1/2, 2)**.

In short:

$2x^2 + (2c + 1)x - (c + 1)y + c + 1 = 0$ indicates a group of parabolas, and the equation is composed of two equations of two basic curves, and the two basic curves meet each other at the two points.
Then, we first find the two basic curves.

$$2x^2 + (2c + 1)x - (c + 1)y + c + 1 = 2cx - cy + c + 2x^2 + x - y + 1$$
$$= c(2x - y + 1) + 2x^2 + x - y + 1 = 0.$$

Then, the two basic curves are as follows: a line $2x - y + 1 = 0$ and a parabola $2x^2 + x - y + 1 = 0$.

Next, we find the points where the curves meet each other.
$2x - y + 1 = 0 \Rightarrow y = 2x + 1$, and also, $2x^2 + x - y + 1 = 0 \Rightarrow y = 2x^2 + x + 1$.
$2x + 1 = 2x^2 + x + 1 \Rightarrow 2x^2 - x = 0 \Rightarrow x(2x - 1) = 0 \Rightarrow x = 0$ or $1/2$.
Then: $x = 0 \Rightarrow y = 2x + 1 = 0 + 1 \Rightarrow y = 1$, and also, $x = 1/2 \Rightarrow y = 2x + 1 = 1 + 1 = 2 \Rightarrow y = 2$.

Therefore, the points are **(0, 1)** and **(1/2, 2)**.

Suggestions or Solutions

To the Problem in the Example L

Suppose an equation of a parabola is as follows: $(c + 2)x^2 + (2c + 1)x - (c + 1)y + 3c + 1 = 0$ where c is constant. Then, assuming that the parabola includes two particular points for any value of c, find the two points.

As in the previous example, the equation $(c + 2)x^2 + (2c + 1)x - (c + 1)y + 3c + 1 = 0$, too, seems to indicate a group of parabolas. If so, the equation will be a linear combination of some basic curves. Then, let's rearrange the equation in terms of the **c** in such a way as follows:

$(c + 2)x^2 + (2c + 1)x - (c + 1)y + 3c + 1 = cx^2 + 2cx - cy + 3c + 2x^2 + x - y + 1$
$= c(x^2 + 2x - y + 3) + 2x^2 + x - y + 1 = 0.$

Then, the basic curves are two parabolas as follows:
$x^2 + 2x - y + 3 = 0$ and $2x^2 + x - y + 1 = 0$.

Now, we need to find the points where the parabolas meet each other.
Then, let's begin with the x-coordinates:
$x^2 + 2x - y + 3 = 0 \Rightarrow y = x^2 + 2x + 3$, and also, $2x^2 + x - y + 1 = 0 \Rightarrow y = 2x^2 + x + 1$.
Then, we need to solve the equation $x^2 + 2x + 3 = 2x^2 + x + 1$.
$x^2 + 2x + 3 = 2x^2 + x + 1 \Rightarrow x^2 - x - 2 = 0 \Rightarrow (x + 1)(x - 2) = 0 \Rightarrow x = -1$ or 2.
Then, we find the y-coordinates:
$x = -1 \Rightarrow y = x^2 + 2x + 3 = 1 - 2 + 3 = 2.$
$x = 2 \Rightarrow y = x^2 + 2x + 3 = 4 + 4 + 3 = 11.$

Then, the basic parabolas meet each other at **(-1, 2)** and **(2, 11)**.
Therefore, for each value of the **c**, every parabola $(c + 2)x^2 + (2c + 1)x - (c + 1)y + 3c + 1 = 0$, too, includes the two points.

In short:

$(c + 2)x^2 + (2c + 1)x - (c + 1)y + 3c + 1 = cx^2 + 2cx - cy + 3c + 2x^2 + x - y + 1$
$= c(x^2 + 2x - y + 3) + 2x^2 + x - y + 1 = 0.$

Then, the parabola $(c + 2)x^2 + (2c + 1)x - (c + 1)y + 3c + 1 = 0$ includes the points where the following parabolas meet each other: $x^2 + 2x - y + 3 = 0$ and $2x^2 + x - y + 1 = 0$.

Then, let's begin with the x-coordinates:
$x^2 + 2x - y + 3 = 0 \Rightarrow y = x^2 + 2x + 3$, and also, $2x^2 + x - y + 1 = 0 \Rightarrow y = 2x^2 + x + 1$.
Then, $x^2 + 2x + 3 = 2x^2 + x + 1 \Rightarrow x^2 - x - 2 = 0 \Rightarrow (x + 1)(x - 2) = 0 \Rightarrow x = -1$ or 2.
Then, we find the y-coordinates:
$x = -1 \Rightarrow y = x^2 + 2x + 3 = 1 - 2 + 3 = 2.$
$x = 2 \Rightarrow y = x^2 + 2x + 3 = 4 + 4 + 3 = 11.$

Therefore, the two points are **(-1, 2)** and **(2, 11)**.

Fig. L.0

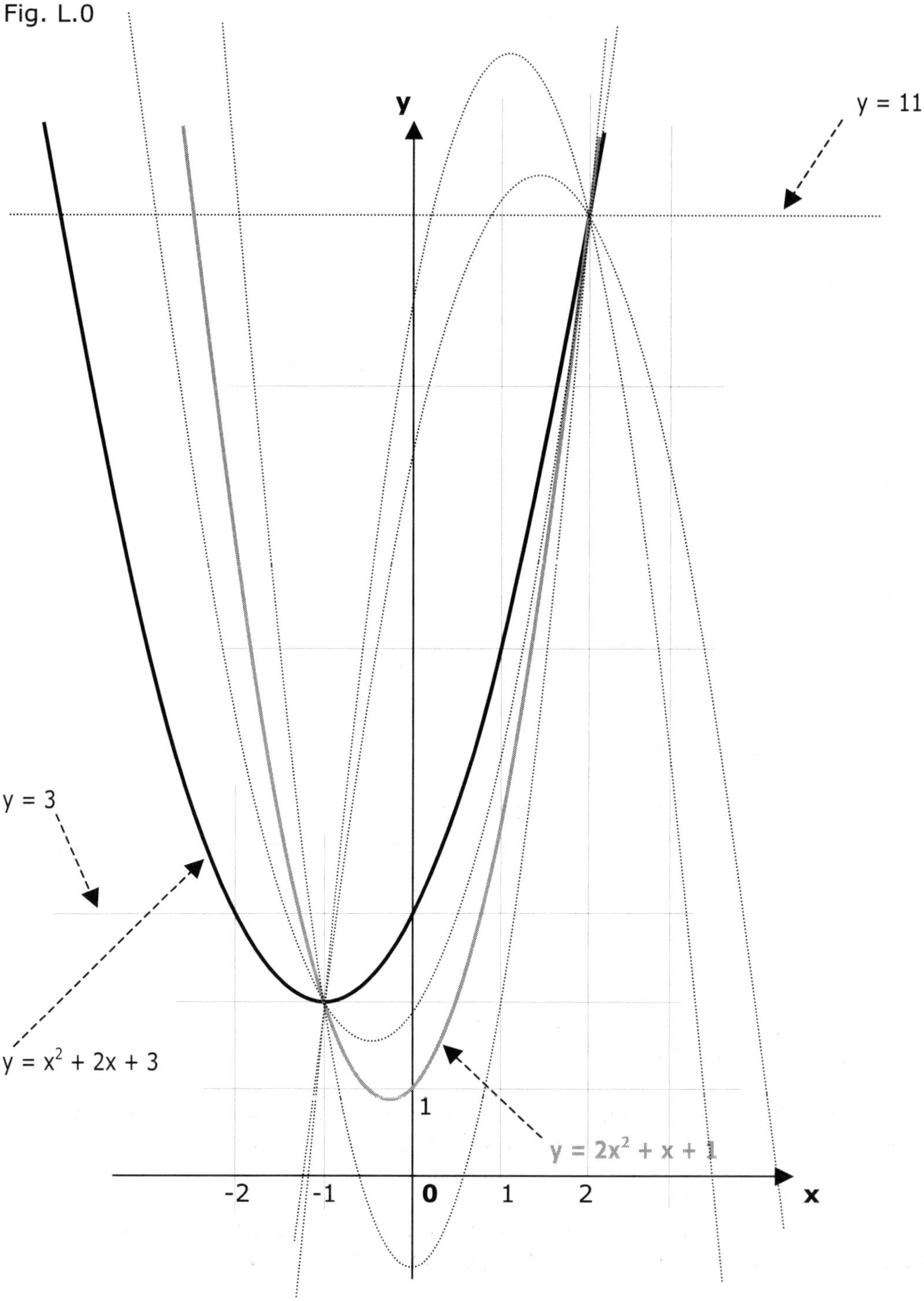

Suggestions or Solutions
To the Problem in the Example M

Suppose we have an equation of a parabola as follows: $cx^2 + 2cx - (c + 1)y + 3c + 3 = 0$ where c is constant. Then, assuming that the parabola includes two particular points for any value of c, find the two points.

As in the previous example, the equation $cx^2 + 2cx - (c + 1)y + 3c + 3 = 0$, too, seems to indicate a group of parabolas. Then, let's rearrange the equation in terms of the **c** in such a way as follows:
$cx^2 + 2cx - (c + 1)y + 3c + 3 = cx^2 + 2cx - cy + 3c - y + 3 = c(x^2 + 2x - y + 3) - y + 3 = 0$.
Then, the basic equations are as follows: $x^2 + 2x - y + 3 = 0$ and $-y + 3 = 0$.
Therefore, the basic curves are a parabola $x^2 + 2x - y + 3 = 0$ and a line $y = 3$. The line is parallel to the x-axis. Now, we need to find the points where the curves meet each other. Since the line $y = 3$ meets the parabola, the y-coordinates at the points are 3. Then, let's find the x-coordinates.
$y = 3 \Rightarrow y = x^2 + 2x + 3 \Rightarrow 3 = x^2 + 2x + 3 \Rightarrow x^2 + 2x = 0 \Rightarrow x(x + 2) = 0 \Rightarrow x = 0$ or -2.
Therefore, the curves meet each other at **(0, 3)** and **(-2, 3)**.
Therefore, for each value of the **c**, every parabola $cx^2 + 2cx - (c + 1)y + 3c + 3 = 0$ includes the two points **(0, 3)** and **(-2, 3)**.

In short:
$cx^2 + 2cx - (c + 1)y + 3c + 3 = cx^2 + 2cx - cy + 3c - y + 3 = c(x^2 + 2x - y + 3) - y + 3 = 0.$
Then, the basic curves are a parabola $x^2 + 2x - y + 3 = 0$ and a line $-y + 3 = 0$, and since the line $y = 3$ meets the parabola, the y-coordinates at the points are 3. Then:
$y = 3 \Rightarrow y = x^2 + 2x + 3 \Rightarrow 3 = x^2 + 2x + 3 \Rightarrow x^2 + 2x = 0 \Rightarrow x(x + 2) = 0 \Rightarrow x = 0$ or -2.
Therefore, the points are **(0, 3)** and **(-2, 3)**.

Fig. M.0

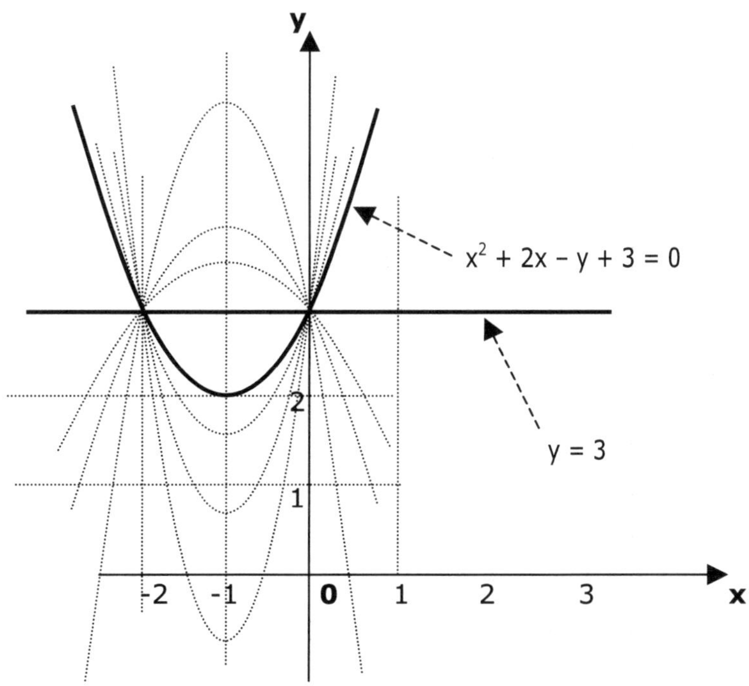

Suggestions or Solutions

To the Problem in the Example N

Suppose we have an equation of a parabola as follows: $cx^2 + (2c + 1)x - cy + 3c + 1 = 0$ where c is constant. Then, assuming that the parabola includes a particular point for any value of c, find the particular point.

As in the previous example, the equation $cx^2 + (2c + 1)x - cy + 3c + 1 = 0$, too, seems to indicate a group of parabolas. Then, let's rearrange the equation in terms of the **c** in such a way as follows:
$cx^2 + (2c + 1)x - cy + 3c + 1 = cx^2 + 2cx - cy + 3c + x + 1 = c(x^2 + 2x - y + 3) + x + 1 = 0$.
Then, the basic equations are as follows: $x^2 + 2x - y + 3 = 0$ and $x + 1 = 0$.
Therefore, the basic curves are a parabola $x^2 + 2x - y + 3 = 0$ and a line $x = -1$. The line is parallel to the y-axis. Now, we need to find the point where the curves meet each other, and since the line $x = -1$ meets the parabola, the x-coordinate at the particular point is -1. Then, let's find the y-coordinate.
$x = -1 \Rightarrow y = x^2 + 2x + 3 = 1 - 2 + 3 \Rightarrow y = 2$. So, the curves meet each other at $(-1, 2)$. Therefore, for each value of the **c**, every parabola $cx^2 + (2c + 1)x - cy + 3c + 1 = 0$ includes the point $(-1, 2)$.

In short:

$cx^2 + (2c + 1)x - cy + 3c + 1 = cx^2 + 2cx - cy + 3c + x + 1 = c(x^2 + 2x - y + 3) + x + 1 = 0$.
Then, the basic curves are a parabola $x^2 + 2x - y + 3 = 0$ and a line $x = -1$, and since the line $x = -1$ meets the parabola, the x-coordinate at the particular point is -1.
Then, $x = -1 \Rightarrow y = x^2 + 2x + 3 = 1 - 2 + 3 \Rightarrow y = 2$.
Therefore, the particular point is $(-1, 2)$.

Fig. N.0

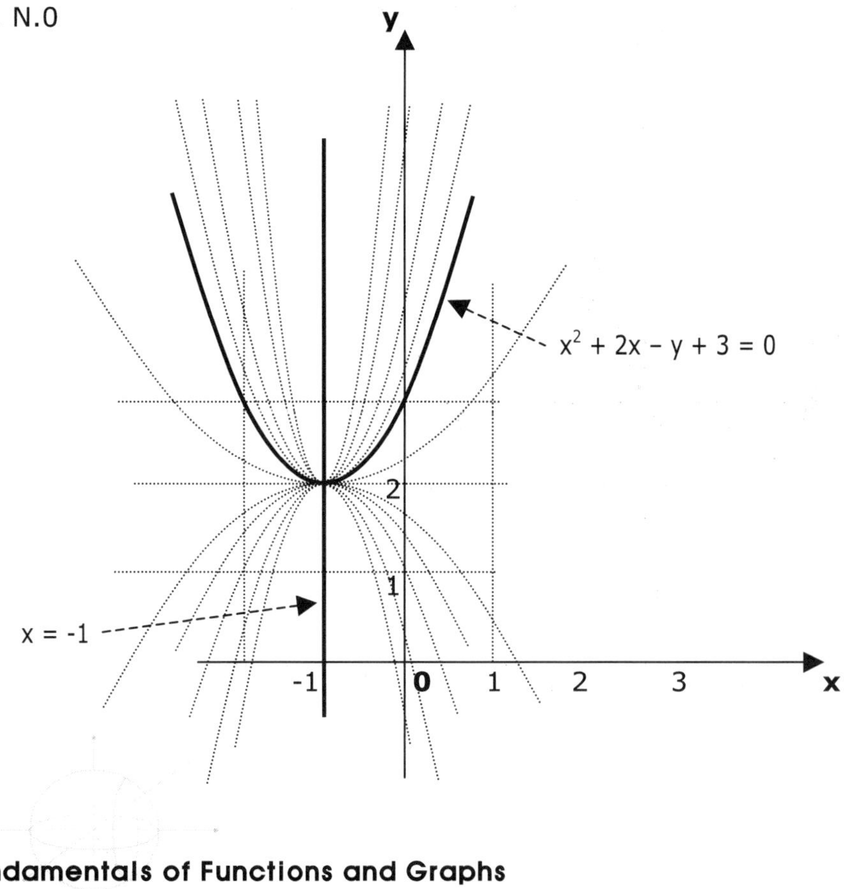

Suggestions or Solutions
To the Problem in the Example O

Suppose we have an equation of a parabola as follows: $y = cx^2 - (a - 2c)x - 2b(c - 1)$ where c is constant. Then, assuming that the parabola includes two points (1, 2) and (s, t) for any value of c, find the two points.

As in the previous example, the equation $y = cx^2 - (a - 2c)x - 2b(c - 1)$, too, seems to indicate a group of parabolas. If that is the case, the equation should be a linear combination of basic curves. Then, we need check to see if it is the case. Then, since the highest degree term in the equation is the x^2-term, let's rearrange the equation in terms of the **c** in such a way as follows:

$y = cx^2 - (a - 2c)x - 2b(c - 1) \Rightarrow cx^2 + 2cx - 2cb - ax + 2b - y = 0.$
$\Rightarrow c(x^2 + 2x - 2b) - (ax - 2b + y) = 0.$ That is, $c(x^2 + 2x - 2b) + (-1)(ax - 2b + y) = 0.$

Then, the basic curves are $x^2 + 2x - 2b = 0$ and $ax - 2b + y = 0$.
However, the curve of the equation $x^2 + 2x - 2b = 0$ is not a parabola. Is there anything wrong with it, then? Nothing wrong with it The equation indicates two lines. How come?
The basic curves need to include the point **(1, 2)**. The x-coordinate of the point is 1, and the y-coordinate is 2. Then, if $x = 1$, then we get $x^2 + 2x - 2b = 1 + 2 - 2b = 0 \Rightarrow b = 3/2$.
So, we get $x^2 + 2x - 3 = 0$.
Therefore, $x^2 + 2x - 3 = 0 \Rightarrow (x - 1)(x + 3) = 0 \Rightarrow x = 1$ or $x = -3$.
So, the two lines are $x = 1$ and $x = -3$ which are parallel to the y-axis.

Now, the basic curves include the two points **(1, 2)** and **(s, t)**.
Then, the x-coordinates at the two points are 1 and −3.
Therefore, at the point **(s, t)**, the **s** is −3.

Now, let's move on to the other basic line $ax - 2b + y = 0$.
First, we have $b = 3/2 \Rightarrow ax - 2b + y = ax - 3 + y = 0$.
Next, since the line, too, includes **(1, 2)**:
We get $(1, 2) \Rightarrow ax - 3 + y = a - 3 + 2 = 0 \Rightarrow a = 1$.
Therefore, the other basic line is $x - 3 + y = 0$.
Now, the basic line, too, includes the point **(s, t)**, that is, **(-3, t)**.
Therefore, when $x = -3$ in the equation of the basic line $x - 3 + y = 0$, we get $y = t$.
Then, we can see that $x = -3 \Rightarrow x - 3 + y = -3 - 3 + t = 0 \Rightarrow t = 6$.

Now, the basic curves are two lines $x = 1$ and $x = -3$ from $x^2 + 2x - 3 = (x - 1)(x + 3) = 0$, and another line $x - 3 + y = 0$, and the two points are **(1, 2)** and **(-3, 6)**.
Now, $y = cx^2 - (a - 2c)x - 2b(c - 1)$ is an equation of parabola, and why there is no parabola among the basic curves?
In a figurative sense , there is no parabola, but there are roots of parabolas. What root?
The equation $x^2 + 2x - 3 = 0$ is from a parabola $y = x^2 + 2x - 3$. The roots are the x-intercepts of the parabola $y = x^2 + 2x - 3$. Then, the solution to the equation $x^2 + 2x - 3 = 0$ can be called the roots of the curve of $y = x^2 + 2x - 3$ called a parabola. The solution looks as if they were roots of a tree. If we take the x-axis for the surface of the ground, and the curve for a tree, then the solution can be taken for the roots. Now, the two lines which are the roots of a parabola get combined with a particular line, and then, becomes another tree, which is another parabola. Other parabolas can have the same roots, and the equation of each of such parabolas is $y = c(x^2 + 2x - 3)$.

In short:

$y = cx^2 - (a - 2c)x - 2b(c - 1) \Rightarrow cx^2 + 2cx - 2cb - ax + 2b - y = 0$.
$\Rightarrow c(x^2 + 2x - 2b) - (ax - 2b + y) = 0$. That is, **$c(x^2 + 2x - 2b) + (-1)(ax - 2b + y) = 0$.**

Then, the basic curves are **$x^2 + 2x - 2b = 0$** and **$ax - 2b + y = 0$**.
The equation **$x^2 + 2x - 2b = 0$** indicates two lines.
Since the basic curves include the point **(1, 2)**, if **$x = 1$**, then **$x^2 + 2x - 2b = 1 + 2 - 2b = 0$**
$\Rightarrow b = 3/2$. So, **$x^2 + 2x - 2b = x^2 + 2x - 3 = 0 \Rightarrow (x - 1)(x + 3) = 0 \Rightarrow x = 1$** or **$x = -3$**.
Therefore, the two lines are **$x = 1$** and **$x = -3$**, which are parallel to the y-axis.

Since the basic curves includes the two points **(1, 2)** and **(s, t)** and the x-coordinates at the two points are 1 and –3, the **s** at the point **(s, t)** is –3.
$b = 3/2 \Rightarrow ax - 2b + y = ax - 3 + y = 0$.
Since the line, too, includes **(1, 2)**, **$(1, 2) \Rightarrow ax - 3 + y = a - 3 + 2 = 0 \Rightarrow a = 1 \Rightarrow x - 3 + y = 0$.**
Since the basic line, too, includes the point **(s, t)**, that is, **(-3, t)**:
When **$x = -3$** in the equation of the basic line **$x - 3 + y = 0$**, we get **$y = t$**.
Then, **$x = -3 \Rightarrow x - 3 + y = -3 - 3 + t = 0 \Rightarrow t = 6$.**
Therefore, **$a = 1$**, **$b = 3/2$**, **$s = -3$**, and **$t = 6$**.

Fig. O.0

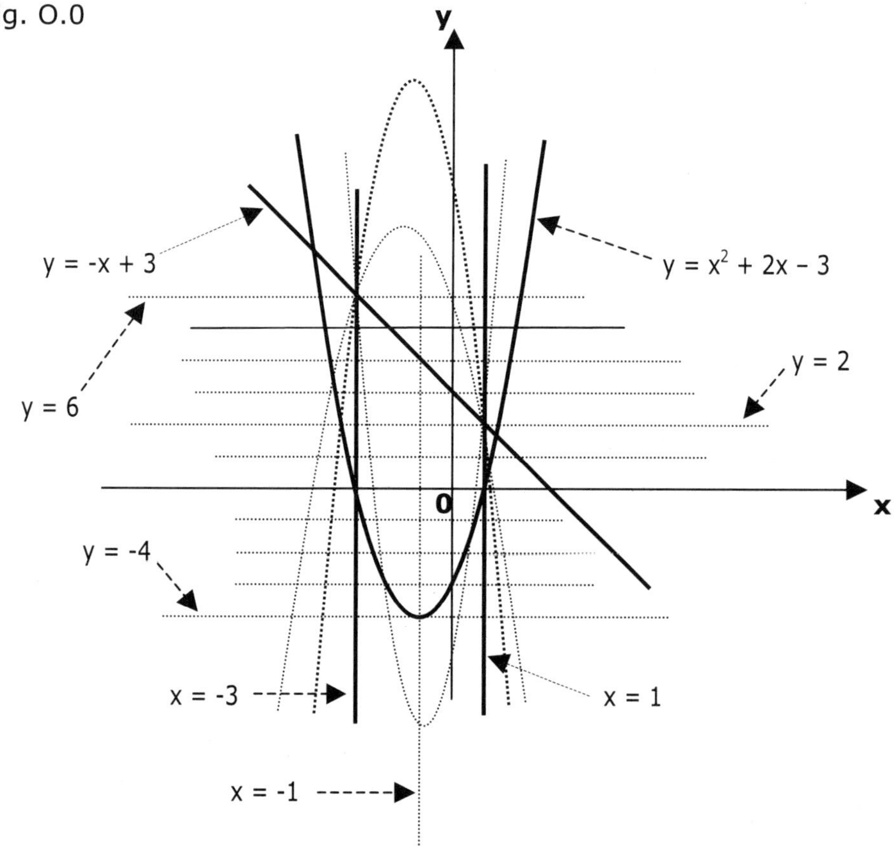

Suggestions or Solutions
To the Problem in the Example P

Suppose that: $y = f(x) = 2(x + 1)^2$ and $f(x) = T_3 \bullet T_2 \bullet T_1 \bullet f$.
T_1: $(x, y) \longrightarrow (3 - x, -2 - y)$.
T_2: $(x, y) \longrightarrow (-x, -y)$.
T_3: $(x, y) \longrightarrow (x + u, y + v)$ where u and v are constants.

Then, find the values of the u and the v.

We can transform the function **f** in analytic or algebraic manner. When we make a new function through a function transformation, we actually transform each and every point in the curve that the original function has. Therefore, we make a general point of the curve, and apply the transformation to the general point. However, we are going to try the transformation by a graph method. So to speak, it can be said to be a rule of thumb, but is useful when we want to get a quick access to only the final solution. Now, let's move a point **P** in the curve of the **f**. Then, the rest of the points, too, will move in the same manner as the point **P** moves. Suppose that the point **P** is in the x-y plane as follows:

Fig. P.0

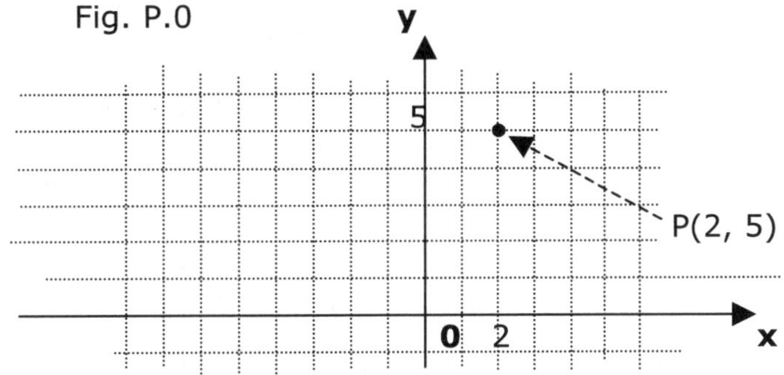

Then, let's begin with the application of the transformation $\mathbf{T_1}$ to the point **P**.
T_1: $(x, y) \longrightarrow (3 - x, 2 - y)$.

The $\mathbf{T_1}$ is in fact, a composite transformation of the two transformations as follows:
A: $(x, y) \longrightarrow (-x, -y)$
B: $(x, y) \longrightarrow (x + 3, y + 2)$

Fig. P.1

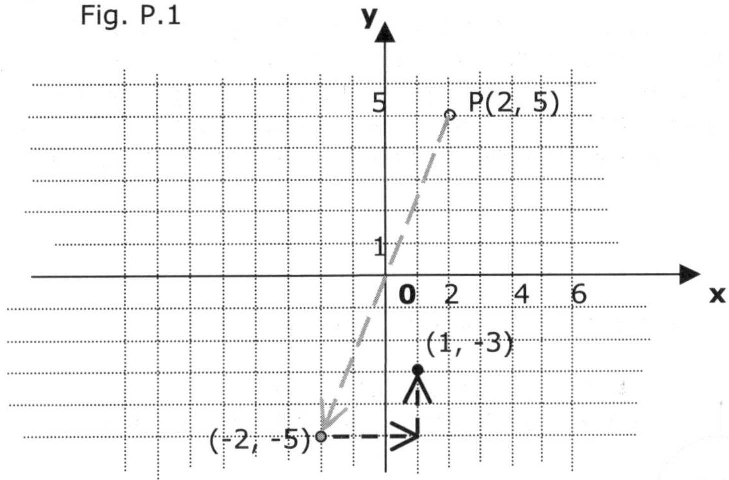

The composite transformation T_1 = **B•A** can be said to begin with the transformation **A** and get ended up with the transformation **B**.
Therefore, T_1 = **B•A**: (x, y) ⟶ (3 - x, 2 - y).
The composite transformation T_1 works as follows. As shown in the figure below, in the transformation **A**, the point **P** at **(2, 5)** gets transformed symmetrically about the origin to be the point at **(-2, -5)**. In the transformation **B**, the point at **(-2, -5)** gets translated by 3 in the direction of the x-axis and by 2 in the direction of the y-axis to be the point at **(1, -3)**.

Next, the T_2 is as follows. T_2: (x, y) ⟶ (-x, -y).

In the transformation T_2, the point at **(1, -3)** gets transformed symmetrically about the origin to be the point at **(-1, 3)**.

Fig. P.2

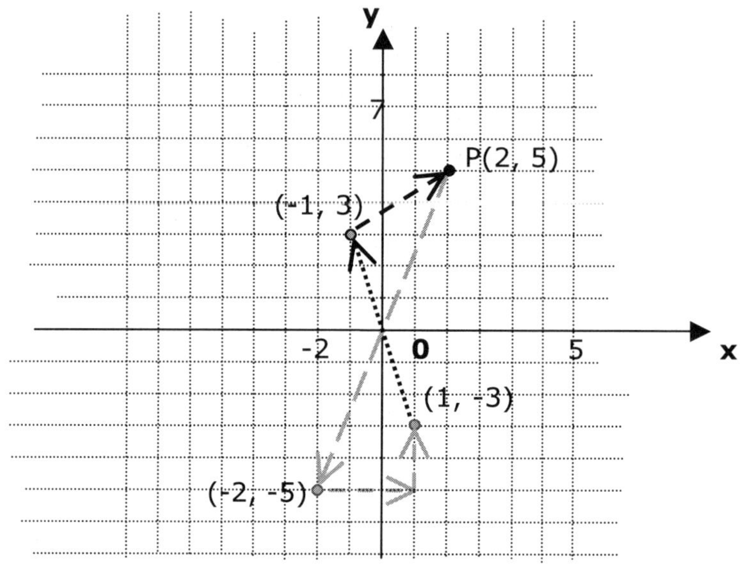

Therefore, in short, we get $T_2 \bullet T_1$:(x, y) ⟶ {-(3 - x), -(2 - y)} = (x - 3, y - 2).

Now, the T_3 is as follows. T_3: (x, y) ⟶ (x + u, y + v) where **u** and **v** are constants.

Then, in the T_3, a parallel transformation is applied to a point at **(-1, 3)**.
Now, if we want to move a point at **(-1, 3)** back to the original position **(2, 5)**, what should the values of the **u** and **v**? As we can see in the graph, the point at **(-1, 3)** needs to move by 3 in the direction of the x-axis and by 2 in the direction of the y-axis. Therefore, we get **u = 3** and **v = 2**.
The same is true for each of all the other points in the curve of the function **f**.

In short:

We can do it analytically in such a way as follows:

$T_3 \bullet T_2 \bullet T_1$:(x, y) ⟶ (x - 3 + u, y - 2 + v) = (x, y).
Therefore, **x − 3 + u = x ⇒ u = 3**, and also, **y − 2 + v = y ⇒ v = 2.**

Suggestions or Solutions
To the Problem in the Example Q

Suppose that: $y = f(x) = 2x^2 - 2x + 3$ and $g = T_2 \cdot T_1 \cdot f$.
T_1: (x, y) ⟶ (x − 3, y − 2).
T_2: (x, y) ⟶ (y, x).

Then, find the g.

We can transform the function **f** analytically, of course. However, this time, too, let's transform it by a graph method. In the composite transformation **$T_2 \cdot T_1$**, we begin with the transformation **T_1** and end up with the **T_2**. We know that the **T_1** is a parallel transformation. In the **T_1**, every point in the curve of the **f** will move by -3 in the direction of the x-axis and by -2 in the direction of the y-axis. Then, a new curve will be made, and a new function will have the new curve. The new function is **$T_1 \cdot f$**.
Now, in the **T_2**, the curve of the new function **$T_1 \cdot f$** will move symmetrically about the line **y = x**. Then, another new curve will be made, and another new function will have the new curve. The new function is **$T_2 \cdot T_1 \cdot f$**, which is the **g**.
Now, let's move one particular point in the curve of the **f**. Then, we can move all the other points in the same manner as we move the particular point. Then, let's take the vertex for the particular point. Then, we can move all the other points in the same way as the way we move the vertex.
Let's first, see the particulars of the curve of the **f**. The slope is 2, and the curve is concave-up.
Next is the vertex: **$2x^2 - 2x + 3 = 2(x^2 - x) + 3 = 2(x - 1/2)^2 - 1/2 + 3 = 2(x - 1/2)^{1/2} + 5/2$**
⇒ The vertex = (1/2, 5/2).
Now, we begin the transformations with the vertex.

Fig. Q.0

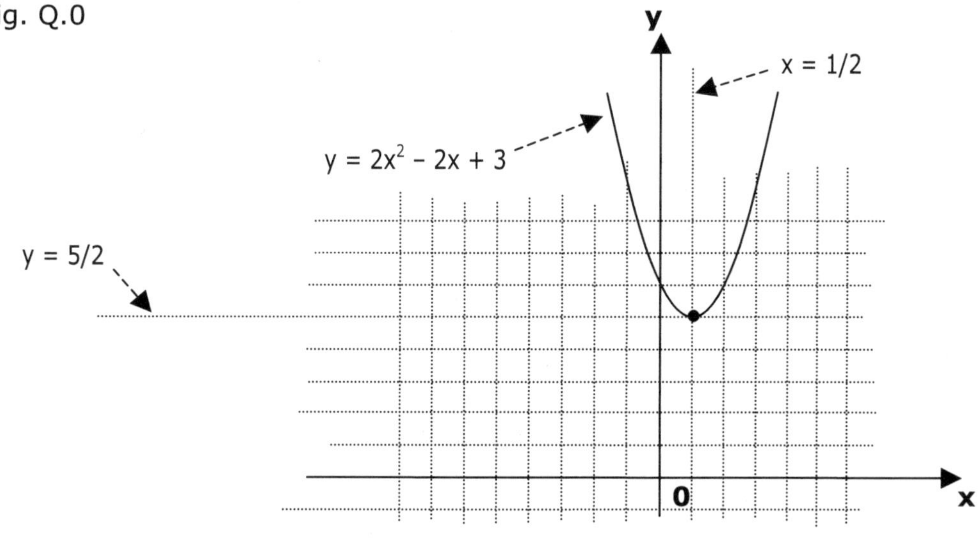

T_1: (x, y) ⟶ (x − 3, y − 2).

In the **T_1**, the vertex will move by -3 in the direction of the x-axis and by -2 in the direction of the y-axis.
Now, as soon as we move the vertex, we move the entire curve.
The concavity and the slope remain the same since the transformation is done in parallel manner.

Fig. Q.1

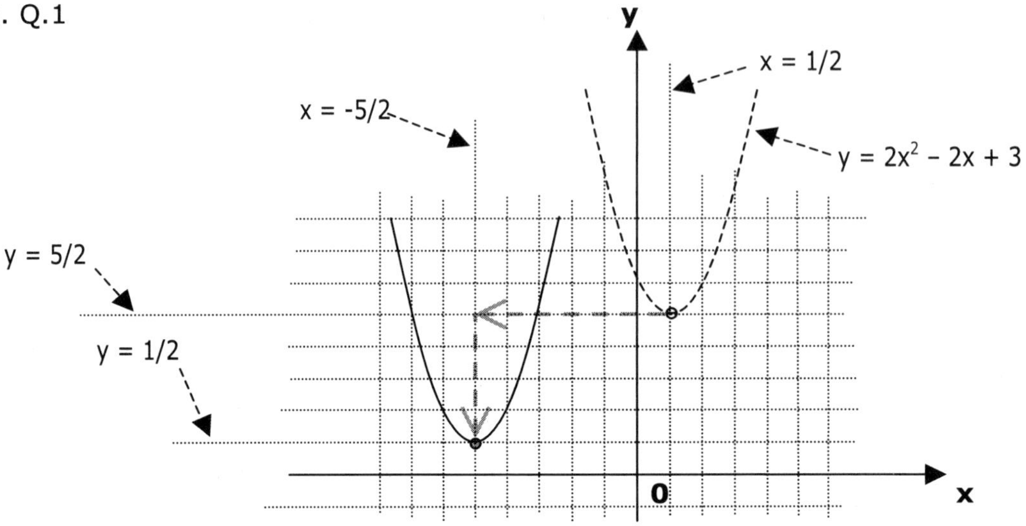

T_2: (x, y) ⟶ (y, x).

In the $\mathbf{T_2}$, the vertex gets moved symmetrically about the line **y = x**.
Then, the ,concavity of the curve will be changed through the transformation, and it will be concave-right. In other words, the parabola will be open to the right. However, the symmetric transformation will not affect the magnitude of the slope of the curve. Therefore, the magnitude of the slope will remain 2.

Fig. Q.2

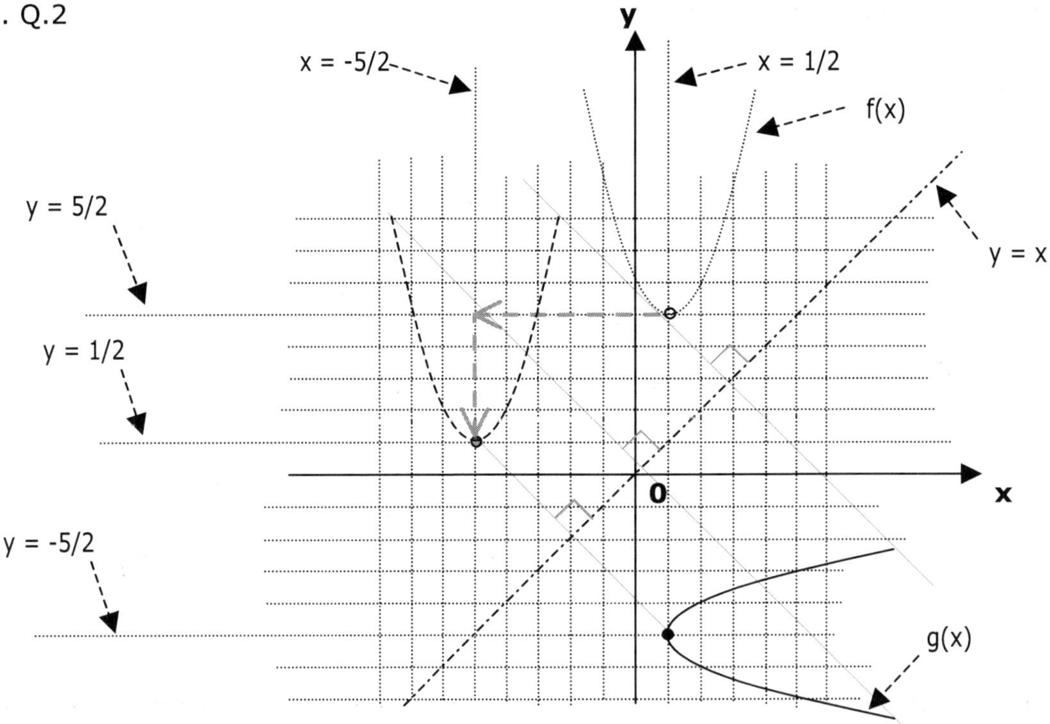

Then, the parabola concave right is the result of the composite transformation of the $\mathbf{T_1}$ and the $\mathbf{T_2}$.

Now, let's find the equation of the curve of the **g**.

The standard form of such an equation is $\mathbf{x - m = a(y - n)^2}$ where **(m, n)** is the vertex and the **a** is the slope. We can see in the graph that the new vertex is now at **(1/2, -5/2)**.
We know that the magnitude of the slope remains the same. Then, the slope is 2 since the curve is concave right. Therefore, the equation is as follows:

$$x - 1/2 = 2\{y - (-5/2)\}^2 = 2(y + 5/2)^2 \Rightarrow x - 1/2 = 2(y + 5/2)^2.$$

Therefore, the **g** is as follows: $g(x, y) = x - 2(y + 5/2)^2 - 1/2 = 0$.

In short:

We can do it analytically in such a way as follows:

We have: $y = f(x) = 2x^2 - 2x + 3$ and $g = T_2 \bullet T_1 \bullet f$.
T_1: $(x, y) \longrightarrow (x - 3, y - 2)$.
T_2: $(x, y) \longrightarrow (y, x)$.

Suppose now, that **(u, v)** is the general point of the **g**. Then:
$T_2 \bullet T_1$: $(x, y) \longrightarrow (u, v) = (y - 2, x - 3)$.

Then, $u = y - 2 \Rightarrow y = u + 2$, and also, $v = x - 3 \Rightarrow x = v + 3$.

Now, $y = 2x^2 - 2x + 3 = 2(x^2 - x) + 3 = 2(x - 1/2)^2 - 1/2 + 3 = 2(x - 1/2)^2 + 5/2$
$\Rightarrow y = 2(x - 1/2)^2 + 5/2$.

Then, since $y = u + 2$ and $x = v + 3$:
$y = 2(x - 1/2)^2 + 5/2 \Rightarrow u + 2 = 2\{(v + 3) - 1/2\}^2 + 5/2 = 2(v + 3 - 1/2)^2 + 5/2$
$= 2(v + 5/2)^2 \Rightarrow u + 2 - 5/2 = u - 1/2 = 2(v + 5/2)^2$.

Now, the general point **(u, v)** is in the x-y plane.
Therefore, replacing the **u** and **v** with **x** and **y** respectively, we get the **g**.
Then, $u - 1/2 = 2(v + 5/2)^2 \Rightarrow x - 1/2 = 2(y + 5/2)^2$.
Therefore, $g(x, y) = x - 2(y + 5/2)^2 - 1/2 = 0$.

Suggestions or Solutions
To the Problem in the Example R

Suppose that: $y = f(x) = -2x^2 + 8x - 5$, $y = g(x) = 2x^2 - 6x + 5/2$, and $g = T \cdot f$.
$T: (x, y) \longrightarrow (u - x, v - y)$

Then, find the values of the u and the v.

We have $\mathbf{u - x = -x + u}$ and $\mathbf{v - y = -y + v}$.
Therefore, the transformation **T** is a composite transformation of the two as follows:
$A: (x, y) \longrightarrow (-x, -y)$
$B: (x, y) \longrightarrow (x + u, y + v)$.

In the composite transformation $\mathbf{T_1 = B \cdot A}$, the transformation begins with the **A** and gets ended up with the **B**. Therefore, $\mathbf{T_1 = B \cdot A}$: $(x, y) \longrightarrow (u - x, v - y)$.

This time also, let's use the graph method. Now, in the **A**, every point in the curve of the **f** gets moved symmetrically about the origin, and becomes a point in the new curve. In the **B**, every point in the new curve gets translated by **u** in the direction of the x-axis and by **v** in the direction of the y-axis, and becomes a point in the curve of a new function, which is the **g(x)**.
Now, we need to find the **u** and **v**, that is, the amounts of parallel movements in the respective directions in the x-y plane. Then, let's put in a graph the curves of the **f** and the **g**.

$$y = f(x) = -2x^2 + 8x - 5 = -2(x^2 - 4x + 4 - 4) - 5 = -2(x - 2)^2 + 8 - 5 = -2(x - 2)^2 + 3.$$

$$y = g(x) = 2x^2 - 6x + 5/2 = 2(x^2 - 3x + 9/4 - 9/4) + 5/2 = 2(x - 3/2)^2 - 9/2 + 5/2$$
$$= 2(x - 3/2)^2 - 2.$$

Then, we can see that:
First, the vertex of the curve of the **f** is **(2, 3)**, the slope is −2, and therefore, the curve is concave down.
Next, the vertex of the curve of the **g** is **(3/2, -2)**, the slope is 1, and therefore, the curve is concave up.

Fig. R.0

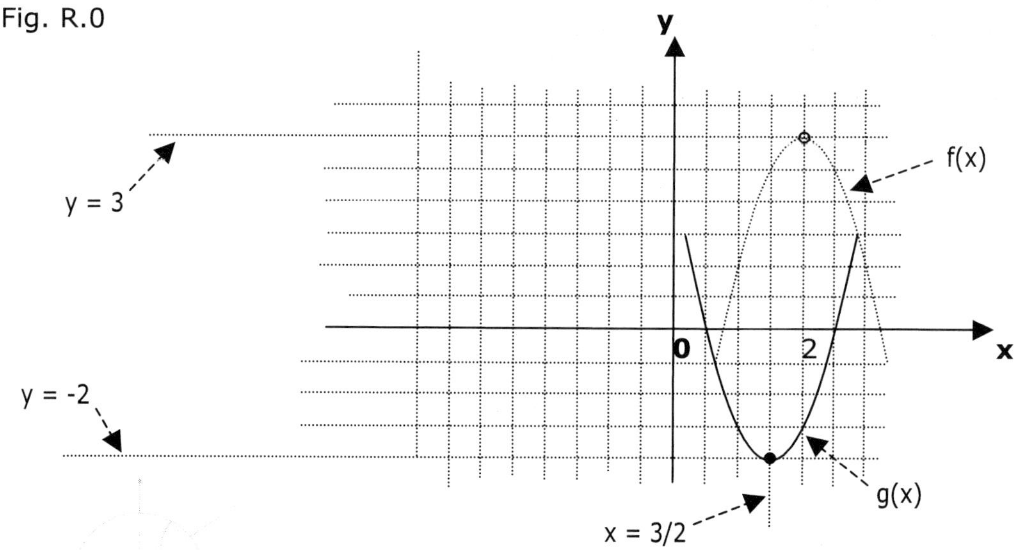

In the transformation **T**, the two transformations happen as follows:
A: $(x, y) \longrightarrow (-x, -y)$.
B: $(x, y) \longrightarrow (x + u, y + v)$.

The **A** is a symmetric transformation about the origin, and the **B** is a parallel transformation. Then, the vertex of the f gets first, transformed symmetrically about the origin by the **A** to be moved to the position **(-2, -3)**. Then, the curve of the **f** gets transformed to the dotted parabola in gray.

Fig. R.1

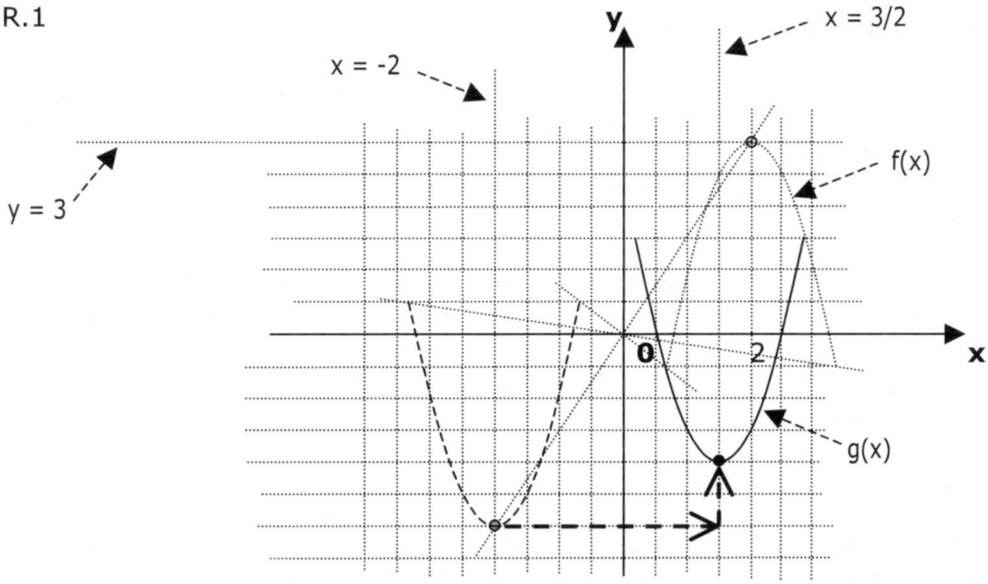

Then, the dotted parabola in gray will be moved by the **B** in the amount of the **u** along the x-axis and in the amount of the **v** along the y-axis to be the parabola in gray, which is the curve of the **g**.
In the graph above, we can see that the **u = 7/2** and the **v = 1**.

In short:

We can do it analytically in such a way as follows:

We have: $\mathbf{y = f(x) = -2x^2 + 8x - 5}$, $\mathbf{y = g(x) = 2x^2 - 6x + 5/2}$, and $\mathbf{g = T \circ f}$.
T: $(x, y) \longrightarrow (u - x, v - y)$.

Then, the magnitudes of the slopes of the f and the g are the same, and the vertices are as follows:
$\mathbf{y = f(x) = -2x^2 + 8x - 5 = -2(x^2 - 4x + 4 - 4) - 5 = -2(x - 2)^2 + 8 - 5 = -2(x - 2)^2 + 3}$
$\Rightarrow$ **The vertex = (2, 3).**
$\mathbf{y = g(x) = 2x^2 - 6x + 5/2 = 2(x^2 - 3x + 9/4 - 9/4) + 5/2 = 2(x - 3/2)^2 - 9/2 + 5/2}$
$\mathbf{= 2(x - 3/2)^2 - 2}$ $\Rightarrow$ **The vertex = (3/2, -2)**.

Then, the vertex **(2, 3)** gets transformed to be the vertex **(3/2, -2)**.
Now, every point in the curve to be transformed gets transformed in the same manner.
In the transformation **T**, the point **(x, y)** gets transformed to be the point **(u – x, v – y)**.
So, since **(2, 3)** in the curve gets transformed to be **(3/2, -2)**, we get **u – 2 = 3/2** and **v – 3 = -2**.

Therefore, we get **u = 2 + 3/2 = 7/2** and **v = -2 + 3 = 1**.

Suggestions or Solutions
To the Problem in the Example S

Suppose that: $x = f(y) = 2(y - 1)^2 + 2$, $y = g(x) = (x - 3/2)^2 + 2$, and $g = T_2 \bullet T_1 \bullet f$.
T_1: $(x, y) \longmapsto (cy, 2x)$ where c is constant.
T_2: $(x, y) \longmapsto (x + u, y + v)$ where u and v are constant.

Then, find the values of the c, the u, and the v.

The T_2 is a parallel transformation, but the T_1 is not a transformation symmetric about the line $y = x$. For this time, because of the transformation T_1, the graph method seems to be inappropriate, and therefore, let's use the analytic method.
Now, suppose that a point (s, t) is the general point of the new function $T_1 \bullet f$. Then:
$T_1:(x, y) \longmapsto (s, t) = (cy, 2x)$.

Then, we have $s = cy$ and $t = 2x$. Therefore, we get $y = s/c$ and $x = t/2$. So:
$x = 2(y - 1)^2 + 2 \Rightarrow t/2 = 2(s/c - 1)^2 + 2 = 2s^2/c^2 - 4s/c + 4 \Rightarrow t = 4s^2/c^2 - 8s/c + 8$.

Since the general point is in the x-y plane, let's replace the **s** and the **t** with **x** and **y** respectively.
Then, we get $T_1 \bullet f = y = 4x^2/c^2 - 8x/c + 8$.

Now, let's move on to the T_2.
Then, $T_1 \bullet f$ gets input to the T_2, and the output, that is, $T_2 \bullet T_1 \bullet f$ is the **g**.
Now, in a parallel transformation of a parabola, the slope of the parabola doesn't get changed.
Then, since the slope of the **g** is 1, the slope of the $T_1 \bullet f$ needs to be 1, too.

Therefore, we need to have $4/c^2 = 1$ in the equation of the parabola $y = 4x^2/c^2 - 8x/c + 8$.
Then, we get $4/c^2 = 1 \Rightarrow c^2 = 4 \Rightarrow c = 2$ or -2.

Then, first, we can see that $c = 2 \Rightarrow y = 4x^2/c^2 - 8x/c + 8 = x^2 - 4x + 8 = (x - 2)^2 + 4$.
Therefore, we can see that $y = T_1 \bullet f = (x - 2)^2 + 4$, that is, $y - 4 = (x - 2)^2$.
Now, comparing the positions of the vertices of the $T_1 \bullet f$ and the **g**, we can find the values of the **u** and **v**.

However, we can try another method as follows. Suppose now, that the general point if the **g** is (s, t).
The T_2 is as follows:
T_2: $(x, y) \longmapsto (x + u, y + v)$ where **u** and **v** are constants.
Then, we can see that $(s, t) = (x + u, y + v) \Rightarrow s = x + u$ and $t = y + v$.
So, we get $x = s - u$ and $y = t - v$.
Therefore, we get $y - 4 = (x - 2)^2 \Rightarrow (t - v) - 4 = \{(s - u) - 2\}^2$.
The general point (s, t) is the general point of the **g**, the **g** is in the x-y system, and therefore, we need to replace the **s** and the **t** with **x** and **y** respectively.
Therefore, we get $(y - v) - 4 = \{(x - u) - 2\}^2$, which is for the **g**.
Now, we have $y = g(x) = (x - 3/2)^2 + 2 \Rightarrow y - 2 = (x - 3/2)^2$.
Then, we can see that $(y - v) - 4 = y - 2$, and also, $(x - u) - 2 = x - 3/2$.
Therefore, we have $c = 2$, $v = -2$, and $u = -1/2$.

Next, we can see that $c = -2 \Rightarrow y = 4x^2/c^2 - 8x/c + 8 \Rightarrow y = x^2 + 4x + 8 = (x + 2)^2 + 4$.
Therefore, we can see that $y = T_1 \bullet f = (x + 2)^2 + 4$, that is, $y - 4 = (x + 2)^2$.

Now, if **(s, t)** is the general point of the **g** and **(x, y)** is that of the **$T_1 \cdot f$**, we get **x = s − u** and **y = t − v**.
Therefore, we can see that $\mathbf{y - 4 = (x + 2)^2 \Rightarrow (t - v) - 4 = \{(s - u) + 2\}^2}$.
The general point **(s, t)** is the general point of the **g**, the **g** is in the x-y system, and therefore, we need to replace the **s** and the **t** with **x** and **y** respectively.
Therefore, we can see that $\mathbf{(y - v) - 4 = \{(x - u) + 2\}^2}$, which is for the **g**.
Now, we have $\mathbf{y = g(x) = (x - 3/2)^2 + 2 \Rightarrow y - 2 = (x - 3/2)^2}$.
Then, we can see that **(y − v) − 4 = y − 2**, and also, **(x − u) + 2 = x − 3/2**.
Therefore, we have **c = -2**, **v = -2**, and **u = 7/2**.

In short:

Suppose that a point **(s, t)** is the general point of the function **$T_1 \cdot f$**.
Then, T_1:(x, y) ⟶ (s, t) = (cy, 2x).
Then, we have **s = cy** and **t = 2x**. Therefore, we get **y = s/c** and **x = t/2**. So:
$\mathbf{x = 2(y - 1)^2 + 2 \Rightarrow t/2 = 2(s/c - 1)^2 + 2 = 2s^2/c^2 - 4s/c + 4 \Rightarrow t = 4s^2/c^2 - 8s/c + 8}$.
Then, $\mathbf{T_1 \cdot f = y = 4x^2/c^2 - 8x/c + 8}$.

Now, we have $\mathbf{g = T_2 \cdot T_1 \cdot f}$, and $\mathbf{y = g(x) = (x - 3/2)^2 + 2 \Rightarrow y - 2 = (x - 3/2)^2}$.
The $\mathbf{T_1}$ is a parallel transformation, in a parallel transformation of a parabola, the slope doesn't get changed, the slope of the **g** is 1, and therefore, the slope of the **$T_1 \cdot f$** is 1, too.
Therefore, we need to have $\mathbf{4/c^2 = 1}$ in the equation of the parabola $\mathbf{y = 4x^2/c^2 - 8x/c + 8}$.
Then, we get $\mathbf{4/c^2 = 1 \Rightarrow c^2 = 4 \Rightarrow c = 2}$ or **−2**.

Suppose now, that the general point of the **g** is **(s, t)**.
Then, we get $\mathbf{(s, t) = (x + u, y + v) \Rightarrow s = x + u}$ and **t = y + v**.
So, we get **x = s − u** and **y = t − v**.

Then, first, we can see that $\mathbf{c = 2 \Rightarrow y = 4x^2/c^2 - 8x/c + 8 = x^2 - 4x + 8 = (x - 2)^2 + 4}$.
Then, we get $\mathbf{y = T_1 \cdot f = (x - 2)^2 + 4}$, that is, $\mathbf{y - 4 = (x - 2)^2}$.
Therefore, we can see that $\mathbf{y - 4 = (x - 2)^2 \Rightarrow (t - v) - 4 = \{(s - u) - 2\}^2}$.
Now, we get $\mathbf{(y - v) - 4 = \{(x - u) - 2\}^2}$, which is for the $\mathbf{g \Rightarrow y - 2 = (x - 3/2)^2}$.
Then, we get **(y − v) − 4 = y − 2**, and also, **(x − u) − 2 = x − 3/2**.
Therefore, we have **c = 2**, **v = -2**, and **u = -1/2**.

Next, we can see that $\mathbf{c = -2 \Rightarrow y = 4x^2/c^2 - 8x/c + 8 \Rightarrow y = x^2 + 4x + 8 = (x + 2)^2 + 4}$.
Then, we get $\mathbf{y = T_1 \cdot f = (x + 2)^2 + 4}$, that is, $\mathbf{y - 4 = (x + 2)^2}$.
Therefore, we can see that $\mathbf{y - 4 = (x + 2)^2 \Rightarrow (t - v) - 4 = \{(s - u) + 2\}^2}$.
Now, we get $\mathbf{(y - v) - 4 = \{(x - u) + 2\}^2}$, which is for the $\mathbf{g \Rightarrow y - 2 = (x - 3/2)^2}$.
Then, we get **(y − v) − 4 = y − 2**, and also, **(x − u) + 2 = x − 3/2**.
Therefore, we have **c = -2**, **v = -2**, and **u = 7/2**.

Note:

There are two paths where the function **f(y)** changes to the function **g(x)**. One of the two is as follows:
T_1: (x, y) ⟶ (2y, 2x).
T_2: (x, y) ⟶ (x - 1/2, y - 2).

In short:
$T_2 \bullet T_1$:(x, y) ⟶ (2y - 1/2, 2x - 2).

The other of the two is as follows:
T_1: (x, y) ⟶ (-2y, 2x).
T_2: (x, y) ⟶ (x + 7/2, y - 2).

In short:
$T_2 \bullet T_1$:(x, y) ⟶ (-2y + 7/2, 2x - 2).

Fig. S.0

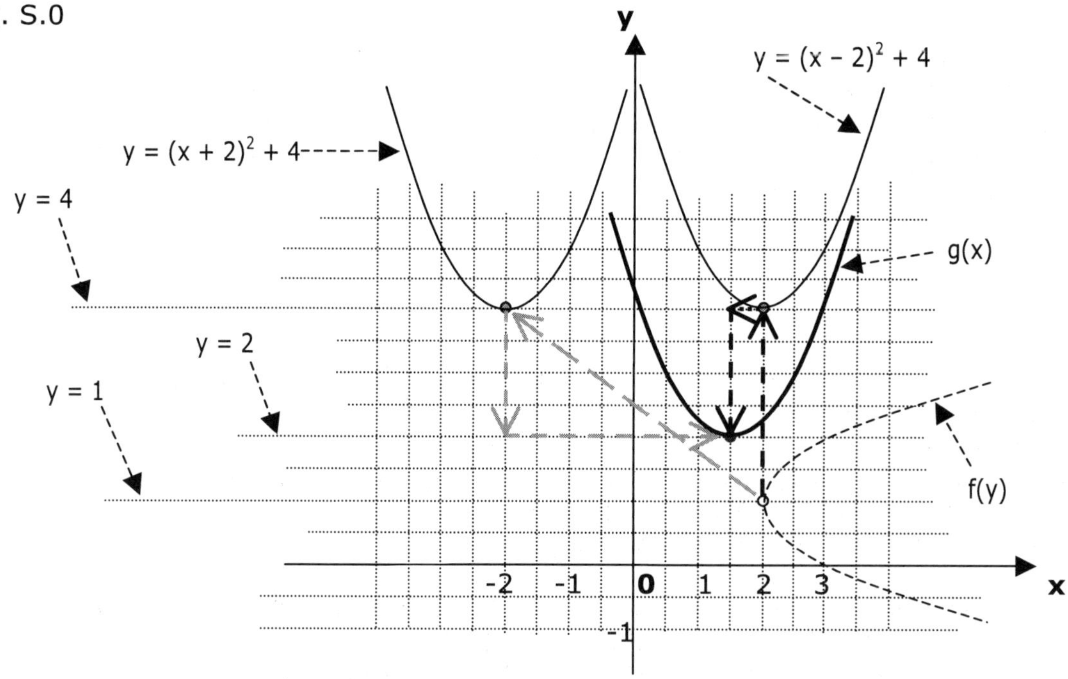

Suggestions or Solutions
To the Problem in the Example T

Suppose that: $f(x) = 2x^2 - 2cx + c$ where c is constant and ≥ -1.
The curve of $y = f(x)$ is a parabola, and the vertex is (s, t).
T: c ⟶ (s, t).
$t = g(s)$.

Then, specify the g.

The **f** is a function that indicates a group of parabolas. That is, if we choose a vale of the **c**, the **f** gets determined. The value needs to be greater than or equal to −1, of course.
The vertex of a parabola $y = 2x^2 - 2cx + c$ is **(s, t)** where **c** is constant and ≥ -1.
Now, the **T** is a parametric transformation where the parameter is the **c**.
T: c ⟶ (s, t), in which **(s, t)** is the vertex of the parabola. Then, the **c** connects the **s** and the **t**. That is, the **s** and the **t** get expressed in terms of the **c** respectively. Therefore, the **s** and the **t** are functions of the **c** respectively. For the details on parametric transformations, refer to the **Book 2: Graph Operations**.

Now, suppose that $s = u(c)$ and $t = v(c)$ for $c \geq -1$.
Let's now, begin with the vertex.
$y = 2x^2 - 2cx + c = 2(x^2 - cx + c^2/4 - c^2/4) + c = 2(x - c/2)^2 - c^2/2 + c$ where $c \geq -1$.

Therefore, the vertex is $(c/2, -c^2/2 + c)$ where $c \geq -1$.
Then, we get $s = u(c) = c/2$ and $t = v(c) = -c^2/2 + c$ where $c \geq -1$.

We now, have:
T: c ⟶ $(s, t) = (c/2, -c^2/2 + c)$ where $c \geq -1$.

Then, $s = c/2 \Rightarrow c = 2s$. So, $c \geq -1 \Rightarrow c/2 \geq -1/2 \Rightarrow s \geq -1/2$.
Besides, $c = 2s \Rightarrow t = -c^2/2 + c \Rightarrow t = -4s^2/2 + 2s = -2s^2 + 2s$ for $s \geq -1/2$.
Therefore, $t = g(s) = -2s^2 + 2s$ for $s \geq -1/2$.

Putting the function **g** in the x-y system, we need to replace the **s** and the **t** with **x** and **y** respectively.
Then, we get $y = g(x) = -2x^2 + 2x$ for $x \geq -1/2$.

In short:

$y = 2x^2 - 2cx + c = 2(x^2 - cx + c^2/4 - c^2/4) + c = 2(x - c/2)^2 - c^2/2 + c$ where $c \geq -1$.
Therefore, the vertex $(s, t) = (c/2, -c^2/2 + c)$ where $c \geq -1$.

Suppose now, that $s = u(c)$ and $t = v(c)$ where $c \geq -1$.
Then, we get $s = u(c) = c/2$ and $t = v(c) = -c^2/2 + c$ where $c \geq -1$.
Then, $s = c/2 \Rightarrow c = 2s$. So, $c \geq -1 \Rightarrow c/2 \geq -1/2 \Rightarrow s \geq -1/2$.
Besides, $c = 2s \Rightarrow t = -c^2/2 + c \Rightarrow t = -4s^2/2 + 2s = -2s^2 + 2s$ for $s \geq -1/2$.

Therefore, we get $t = g(s) = -2s^2 + 2s$ for $s \geq -1/2$.

Let's now, put in the x-y plane the curves of the **f** and the **g**.
$y = g(x) = -2x^2 + 2x = -2(x^2 - x) = -2(x^2 - x + 1/4 - 1/4) = -2(x - 1/2)^2 + 1/2$ for $x \geq -1/2$.

Fig. T.0

$y = g(x) = -2(x - 1/2)^2 + 1/2$ for $x \geq -1/2$. $y = f(x) = 2x^2 - 2cx + c$ for $c \geq -1$.

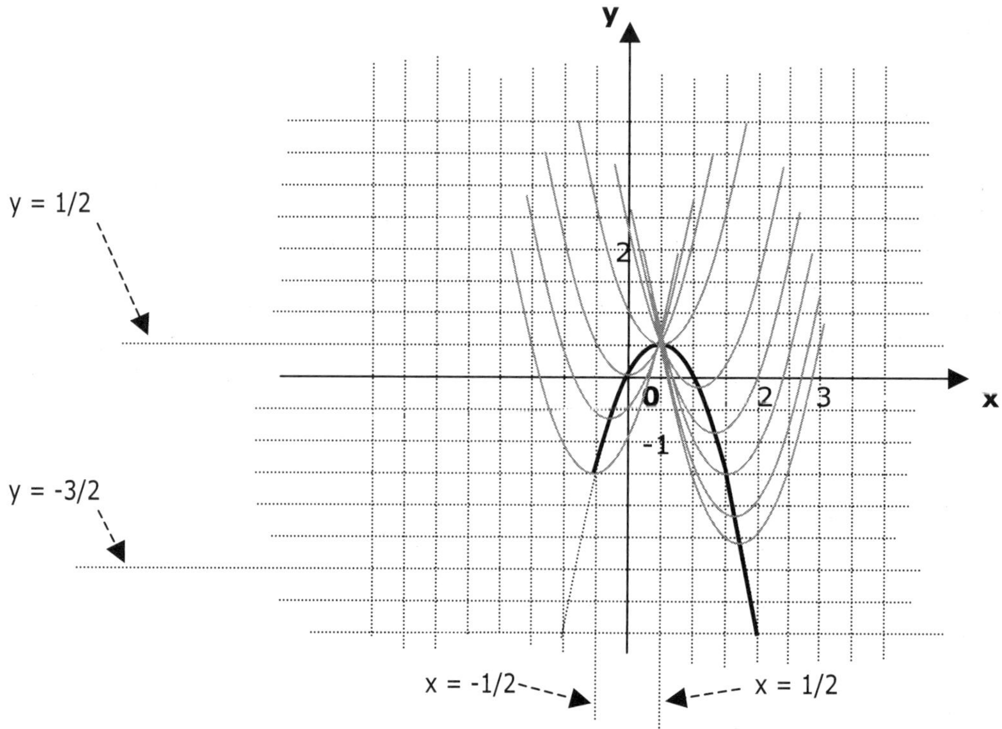

Now, we can say that the curve of the **g** indicates the trace that the vertex of the **f** makes as the **c** changes.

By the way, as in the cases of equations of line, we can put in one equation many parabolas or lines, or combinations of both. For instance:
Suppose first, that $k = 0, 1, 2, \ldots$ and a_k, b_k, etc. are constants. Then:

Two parabolas: $(a_0y + b_0x^2 + c_0x + d_0)(a_1y + b_1x^2 + c_1x + d_1) = 0$.
Many parabolas: $(a_0y + b_0x^2 + c_0x + d_0)(a_1y + b_1x^2 + c_1x + d_1)\ldots(a_ny + b_nx^2 + c_nx + d_n) = 0$.
A parabola and a line: $(ay + bx^2 + cx + d)(ey + fx + g) = 0$.
A parabola and two lines: $(ay + bx^2 + cx + d)(ey + fx + g)(hy + ix + j) = 0$.
Many parabolas and lines: $(a_0y + b_0x^2 + c_0x + d_0)(a_1y + b_1x^2 + c_1x + d_1)\ldots(a_ny + b_nx^2 + c_nx + d_n)$ $(a_1x + b_1y + c_1)(a_2x + b_2y + c_2)\ldots(a_nx + b_ny + c_n) = 0$.

How come?
For instance, $(ay + bx^2 + cx + d)(ey + fx + g) = 0 \Rightarrow ay + bx^2 + cx + d = 0$ or $ey + fx + g = 0$.

Besides, putting such curves in a set of points, we can take examples as follows.
A parabola: $\{(x, y) \mid ax + by^2 + cy + d = 0\}$ where the axis of symmetry is parallel to the x-axis.
Two parabolas and a line: $\{(x, y) \mid (ay + bx^2 + cx + d)(px + qy^2 + ry + s)(ty + ux + v) = 0\}$.

Suggestions or Solutions
To the Problem in the Example U

Suppose that:
The vertex of a parabola $x = y^2 - 2(c + 1)y + 3c^2$ is (s, t) where c is constant and ≥ -1.
T: c ⟶ (s, t).
t = g(s).

Then, find the g.

As in the previous example, the equation of the parabola given indicates a group of parabolas.

The vertex of the parabola $x = y^2 - 2(c + 1)y + 2c^2$ is **(s, t)** where **c** is constant and **c ≥ -1**.
Now, the **T** is a parametric transformation where the parameter is the **c**.
T: c ⟶ (s, t), in which **(s, t)** is the vertex of the parabola. Then, the **c** connects the **s** and the **t**. That is, the **s** and the **t** get expressed in terms of the **c** respectively. Therefore, the **s** and the **t** are functions of the **c** respectively.

Now, suppose that **s = u(c)** and **t = v(c)** where **c ≥ -1**.
Let's now, begin with the vertex.
$x = y^2 - 2(c + 1)y + 2c^2 = y^2 - 2(c + 1)y + (c + 1)^2 - (c + 1)^2 + 2c^2$
$= \{y - (c + 1)\}^2 - (c + 1)^2 + 2c^2 = \{y - (c + 1)\}^2 - c^2 - 2c - 1 + 2c^2$
$= \{y - (c + 1)\}^2 + c^2 - 2c - 1 = \{y - (c + 1)\}^2 + (c - 1)^2 - 2$ where $c \geq -1$.

Therefore, the vertex is $\{(c - 1)^2 - 2, c + 1\}$ where $c \geq -1$.
Then, we have $s = u(c) = (c - 1)^2 - 2$ and $t = v(c) = c + 1$ where $c \geq -1$.

We now, have:
T: c ⟶ $(s, t) = \{(c - 1)^2 - 2, c + 1\}$ where $c \geq -1$.

Therefore, we can see that $s = (c - 1)^2 - 2$ and $t = c + 1$.
Then, $s = (c - 1)^2 - 2 \Rightarrow (c - 1)^2 = s + 2 \Rightarrow c = \pm(s + 2)^{1/2} + 1$.
So, we can see that $c = (s + 2)^{1/2} + 1$ or $c = -(s + 2)^{1/2} + 1$.
Meanwhile, we can see that $t = c + 1 \Rightarrow c = t - 1$.
Now, since $c = t - 1$, we can see that $t - 1 = (s + 2)^{1/2} + 1$ or $t - 1 = -(s + 2)^{1/2} + 1$.
Therefore, we have $t = (s + 2)^{1/2} + 2$ or $-(s + 2)^{1/2} + 2$.

Now, we need to have $c \geq -1$.
First of all, $c = \pm(s + 2)^{1/2} + 1$ holds when the **s** is greater than or equal to −2.
Next, we have $c = (s + 2)^{1/2} + 1 \geq 1$. However, we also have $c = -(s + 2)^{1/2} + 1 \leq 1$.
Therefore, we need to find the extent of the **s** that can make the **c** always greater than or equal to −1.
Then, $c \geq -1 \Rightarrow -(s + 2)^{1/2} + 1 \geq -1 \Rightarrow (s + 2)^{1/2} \leq -2 \Rightarrow s + 2 \leq 4 \Rightarrow s \leq 2$.
However, the **s** needs to be greater than or equal to −2.
Consequently, we have $-2 \leq s \leq 2$.
Therefore, we have $t = g(s) = (s + 2)^{1/2} + 2$ for $s \geq -2$, or $t = g(s) = -(s + 2)^{1/2} + 2$ for $-2 \leq s \leq 2$.

Putting the function **g** in the x-y system, we need to replace the **s** and the **t** with **x** and **y** respectively.
Then, we get $y = g(x) = (x + 2)^{1/2} + 2$ for $x \geq -2$, or $y = g(x) = -(x + 2)^{1/2} + 2$ for $-2 \leq x \leq 2$.

Let's now, put in the x-y plane the parabolas given and the curves of the **g**.
The equation of the parabolas given is $x = y^2 - 2(c + 1)y + 2c^2$ where **c** is a constant and **c ≥ -1**.

Fig. U.0.

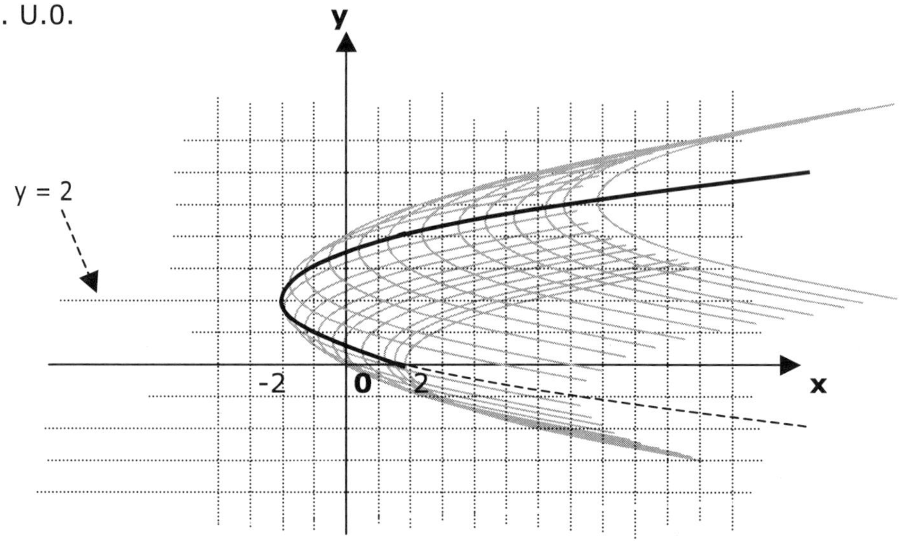

Now, we can see that the curves of the **g** are made up of the vertices of all the parabolas represented by the equation given as follows: $\mathbf{x = y^2 - 2(c + 1)y + 2c^2}$ for $\mathbf{c \geq -1}$. That is, the curve of the **g** is the trace that the vertex of $\mathbf{x = y^2 - 2(c + 1)y + 2c^2}$ makes as the **c** changes.

In short:

$\mathbf{x = y^2 - 2(c + 1)y + 2c^2 = y^2 - 2(c + 1)y + (c + 1)^2 - (c + 1)^2 + 2c^2}$
$\mathbf{= \{y - (c + 1)\}^2 - (c + 1)^2 + 2c^2 = \{y - (c + 1)\}^2 - c^2 - 2c - 1 + 2c^2}$
$\mathbf{= \{y - (c + 1)\}^2 + c^2 - 2c - 1 = \{y - (c + 1)\}^2 + (c - 1)^2 - 2}$ where $\mathbf{c \geq -1}$.
Therefore, the vertex $\mathbf{(s, t) = \{(c - 1)^2 - 2, c + 1\}}$ where $\mathbf{c \geq -1}$.
Suppose now, that $\mathbf{s = u(c)}$ and $\mathbf{t = v(c)}$ where $\mathbf{c \geq -1}$.
Then, we have $\mathbf{s = u(c) = (c - 1)^2 - 2}$ and $\mathbf{t = v(c) = c + 1}$ where $\mathbf{c \geq -1}$.
Then, $\mathbf{s = (c - 1)^2 - 2 \Rightarrow (c - 1)^2 = s + 2 \Rightarrow c = \pm(s + 2)^{1/2} + 1}$.
So, $\mathbf{c = (s + 2)^{1/2} + 1}$ or $\mathbf{c = -(s + 2)^{1/2} + 1}$.
Meanwhile, $\mathbf{t = c + 1 \Rightarrow c = t - 1}$.
Now, since $\mathbf{c = t - 1}$, $\mathbf{t - 1 = (s + 2)^{1/2} + 1}$ or $\mathbf{t - 1 = -(s + 2)^{1/2} + 1}$.
Therefore, $\mathbf{t = (s + 2)^{1/2} + 2}$ or $\mathbf{-(s + 2)^{1/2} + 2}$.

Now, we need to have $\mathbf{c \geq -1}$. First of all, $\mathbf{c = \pm(s + 2)^{1/2} + 1}$ holds when the **s** is greater than or equal to –2. Next, we have $\mathbf{c = (s + 2)^{1/2} + 1 \geq 1}$. However, we also have $\mathbf{c = -(s + 2)^{1/2} + 1 \leq 1}$.
Therefore, we need to find the extent of the **s** that can make the **c** always greater than or equal to –1.
Then, $\mathbf{c \geq -1 \Rightarrow -(s + 2)^{1/2} + 1 \geq -1 \Rightarrow (s + 2)^{1/2} \leq -2 \Rightarrow s + 2 \leq 4 \Rightarrow s \leq 2}$.
However, the **s** is greater than or equal to –2. So, we get $\mathbf{-2 \leq s \leq 2}$.

Therefore, we get $\mathbf{t = g(s) = (s + 2)^{1/2} + 2}$ for $\mathbf{s \geq -2}$, or $\mathbf{t = g(s) = -(s + 2)^{1/2} + 2}$ for $\mathbf{-2 \leq s \leq 2}$.

Putting the function **g** in the x-y system, we need to replace the **s** and the **t** with **x** and **y** respectively.
Then, we get $\mathbf{y = g(x) = (x + 2)^{1/2} + 2}$ for $\mathbf{x \geq -2}$, or $\mathbf{y = g(x) = -(x + 2)^{1/2} + 2}$ for $\mathbf{-2 \leq x \leq 2}$.

함수와 그래프의 기초
– 제 3권 : 원뿔곡선 1 (직선과 포물선) –

2007년 12월 15일 초판 인쇄
2008년 1월 1일 초판 발행

저 자 / 김성렬
발행자 / 오판근
발행처 / 교우사

주 소 / 서울 동대문구 제기2동 148-19
전 화 / 925-2861(대), 편집부 / 925-2825
팩 스 / 925-2860, 953-2860
E-mail / kw2861 @ chollian. net
http: // www. kyowoo. co. kr

등록 / 1994. 3. 24 제6-0256호

값(한글 · 영문판) **25,000원**

함수와 그래프의 기초

제 3 권: 원뿔곡선 1(직선과 포물선)

김성렬

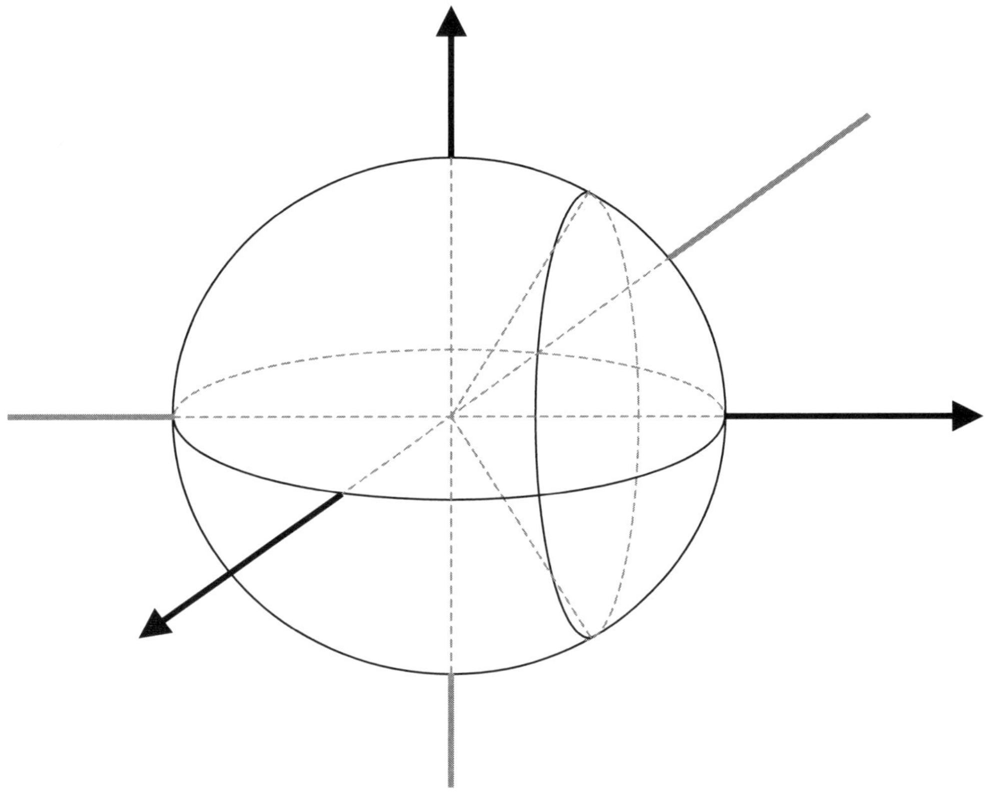

Runmath

Printed in Korea

함수와 그래프의 기초
Runmath

머리말

이 책은 **함수와 그래프의 기초**(영문판: **Fundamentals of Functions and Graphs**)시리즈(series)에서의 세 번째 책입니다. 먼저, 제 1권: **무슨 함수**?(영문판: **What Functions?**)에서는 기본(基本) 함수(函數)란 무엇이고 어떻게 작동하는지에 대하여 이야기하였죠. 다음으로, 제 2 권: **그래프 조작**(영문판: **Graph Operations**)에서는 기본 함수나 방정식(方程式)의 그래프(graph)를 기본적으로 어떻게 조작하는지에 대하여 이야기했어요. 함수가 무엇이고 어떻게 만들어지며 어떻게 사용하는지를 알고 또한, 그 그래프를 어떻게 만들고 사용하며 조작하는지를 이해한다면 그 때는 함수와 그래프의 기초의 대부분을 제대로 공부한 겁니다. 이 책에서는 원뿔곡선이라는 기본 커브(curve)들 중의 두 가지인 직선(直線)과 포물선(抛物線)에 대하여 중점적으로 이야기할거예요. 그 기본 커브들의 개별적인 그래프에 관해서 무엇을 구체적으로 어떻게 할 수 있는지를 이해하게 될 겁니다. 이 시리즈의 나머지 책에서 공부하는 것은 그 나머지 기본 커브들인 원과 타원 그리고 쌍곡선입니다.

위에 나타낸 바와 같이 원뿔곡선이란 꼭지점에서 맞닿아있는 두개의 원뿔들을 자른 단면(斷面)이에요. 직선이란 원뿔곡선이라고 하는 기본 커브들 중의 하나입니다. 그런데, math 에서 선(線)이라고 하면, 이는 직선인 겁니다. 두개의 직각 원뿔들이 그 꼭지점들에서 서로 만나는데 그 원형 바닥들이 서로 평행하다 하세요. 또한, 하나의 평면이 그 원뿔들을 자르고 지나가는데 이는 그 꼭지점들이 맞닿은 점을 포함한다고 하세요. 그러면, 그 단면은 서로 교차하는 두개의 직선들로 되어있죠. 그런데, 그 평면이 그 원뿔들의 옆면들과 맞닿아있고 그 점을 지나간다면 하나의 직선만이 만들어지는 겁니다. 이번에는, 그 평면이 그 원뿔의 옆면과 평행하지만 닿지는 않으며 그 꼭지점을 포함하지도 않는다고 하세요. 그러면, 그 단면은 포물선인 겁니다.

직선과 포물선은 기본 커브들 중에서도 기본적인 것들이죠. 그렇게 기본적이니 만큼 매우 중요한 커브들입니다. 기본적일수록 더욱 중요한 거죠. 사실, 이 두 가지의 커브들 중에서도 더욱 기본적이어서 더욱 중요한 것이 직선이에요. 물론, 가장 기본이 되는 것은 하나의 점(點)이긴 하지만요. 사실, 하나의 점도 원뿔곡선들 중의 하나라고 할 수도 있어요. 그 평면이 원뿔의 꼭지점만을 포함할 수도 있거든요. 그래서 엄밀히 말하면, 하나의 점도 원뿔곡선이다라고 해도 되겠죠. 하지만, 하나의 점을 하나의 커브라고 하기는 좀 그렇죠?

직선 없이는 일반(一般) math 에서 별로 할 수 있는 것이 없어요. 직선을 제대로 모르고서는 기본 기하(幾何)뿐만 아니라 기본적인 방정식을 다루는 일반 기본 대수학(代數學)도 해결하기 곤란한 경우가 많다는 거죠. Math 에서의 다른 모든 사물(事物)과 마찬가지로 직선도 하나의 개념(槪念)입니다. 그러니까, 직선을 가지고 무엇을 말하자는 건지 알아야 하는 거죠. 모르면 많은 문제풀이에서 맥없이 주저앉게 되지만 알면 엄청난 힘을 발휘할 수도 있는 겁니다.

다음으로, 포물선도 일반 math 에서 자주 등장하는 중요한 커브죠. 일상생활에서도 포물선이 실제로 많이 관측되고 있어요. 많은 자연 현상들이 포물선을 따르는 거죠. 예를 들면, 야구공이 날아가는 경로가 포물선이죠. 또한, 위성 방송 안테나(satellite dish)의 모양도 기본적으로 포물선입니다. 포물선은 빛이나 소리 등의 파동(波動)들을 모아서 특정 지점으로 집중시킬 때 또한, 특정 지점에서 나오는 그런 파동들을 효율적으로 보낸다거나 할 때에도 널리 사용되죠.

그런데, math 에서 하는 일은 개념에 대한 겁니다. Math 에서는 그 어느 사물이든 이는 하나의 개념이며 하나의 정의(定義)를 갖는 거죠. 현실에서의 물질적인 것은 그것이 제아무리 작다 하더라도 3 차원의 사물이에요. 즉, 종이에 제아무리 작은 점을 찍는다 해도 그것은 잉크(inc)나 흑연 덩어리죠. Math 에서 사물은 개념이기 때문에 현실에서의 그것과 상당히 다릅니다. 현실에서는 굵은 선이나 가는 선이 있는데 math 에서는 선이라는 건 개념이기 때문에 그 굵기가 없어요. 또한, 현실에는 완벽한 직선이란 없는데 math 에서는 직선은 완벽하게 직선이어야 하는 거죠. 이는 평면이나 삼각형, 사각형, 원, 등등 그 어느 것에 대해서도 마찬가집니다. 현실에 있는 것은 모두 그 개념에 근사(近似)한 것에 지나지 않는 거죠. 즉, 그 개념처럼 보이는 것 뿐이죠.

간단한 기하학적 사물을 수식으로 나타낼 때가 많죠. 예를 들어, 하나의 직선이나 포물선을 방정식으로 나타내는 겁니다. 그러한 방정식은 많은 문제를 풀 때 사용할 수 있는 하나의 편리한 도구이죠. 그런데, 그런 직선과 같은 사물은 하나의 개념이고 개념을 설명한다거나 개념들간의 관계를 이야기 할 때 방정식만으로는 좀 너무 추상적일 때가 많아요. 개념 자체는 눈으로 볼 수 있는 게 아니거든요. 그래서, 다른 이들에게 전달하려는 개념을 눈으로 볼 수 있게 나타낸다는 것은 매우 중요한 능력이에요. 그 능력이란 그래프를 작성하는 능력이고 그 그래프에 있는 커브가 그 사물인 거죠. 그러면, 그러한 방정식의 실제 내용물을 그래프라는 도구를 사용해서 눈으로 볼 있게 할 수 있는 겁니다.

Math 에서, 사물 자체나 그 사물이 어떻게 행동하는지 나타내고자 할 때 주로 방정식이나 그래프를 사용합니다. 즉, 개념이나 현상을 설명할 때 그 것을 말만으로 보다는 체계적이거나 물리적으로 나타내고자 하는 거죠. 그 사물이란 예를 들면, 생산이나 저축, 소비나 지출, 거리, 입력, 출력, 비율, 가격, 수요, 공급, 온도, 부피 등등의 양이 될 수 있어요. 또한, 그 현상이란 것도 예를 들면, 이런 사물이 어떻게 변하는 가를 말해주는 거죠.

그러니까, 이 두 가지 도구들 모두 제대로 작성해서 다룰 수 있어야 하겠어요. 방정식으로 개념이나 현상을 분석적으로 나타내고 설명할 수 있는 겁니다. 즉, 해당 원리나 법칙에 따라 그 개념이나 현상을 방정식을 가지고 설명하기가 편리한 거죠. 또한, 그 방정식의 커브로 그 개념이나 현상을 눈으로 볼 수 있게 표현할 수 있는 겁니다.

그러면, 함수는 어떻고요? 함수는 방정식하고는 다르죠. 경우에 따라, 함수와 방정식의 커브들이 동일할 수는 있어요. 기본적으로 함수는 데이터(data) 집단들 간의 관계를 설명해서 한 집단에 있는 개별적인 데이터들이 그 다른 집단의 개별적인 데이터들과 서로 어떻게 연결되어있는가를 나타내죠. 말하자면, 함수는 인과관계(因果關係)를 나타내거나 설명한다고 할 수 있어요. 그런데, 일반 math 나 미적분학(微積分學)(Calculus)에서는 하나의 특정 함수가 여러 개의 다른 원인(原因)들로 하나의 동일한 결과(結果)를 만들어낼 수는 있지만 하나의 원인만으로 두개 이상의 다른 결과들을 만들어낼 수는 없는 겁니다. 포물선이나 원(圓)과 같은 기하학적 사물의 방정식은 그 사물을 이루는 점들의 좌표(座標)들간의 관계를 설명하는 거예요. 다시 말해서, 그 방정식은 변수(變數)들간의 관계식인데 그 변수들이 커브로서의 그 사물이 위치한 좌표계(座標系)에서 사용되는 거죠. 하지만, 어떤 방정식들은 함수로 직접 사용될 수도 있어요. 예를 들면, 단일(單一) 직선의 방정식이나 어떤 포물선의 방정식을 함수로 사용할 수도 있는 거죠.

함수나 방정식의 그래프를 작성할 때 주로 직교(直交) 좌표계를 사용하는데 이를 흔히, 데카르트 좌표계(Cartesian coordinate system)라고도 하죠. 그 함수나 방정식에서의 변수들의 개수에 따라 그 좌표축(座標軸)들의 개수가 정해지고 그 축들은 서로 수직(垂直)입니다. 기본적인 일반 math 에서는 주로 x-y 좌표계를 사용해요. 그러면, 그 변수들이 x 와 y 인 것이고 x 축과 y 축이 서로 수직이죠. 또한, 이 축들로 이루어지는 평면을 x-y 평면이라고 합니다.

그러면 이제, 간단하고도 기본적인 기하학적 사물들을 어떻게 방정식으로 표현하는지 그리고 그런 방정식을 어떻게 그래프로 나타내는지 알아보기로 하죠. 그 기본적인 기하학적 사물들이 직선, 포물선, 원, 타원, 쌍곡선입니다. 이런 것들을 일반적으로 기본 커브들이라고 해요. 특히, 이것들을 원뿔곡선들이라고 하는 겁니다. 이는 왜냐하면 이것들이 평면으로 잘린 두개의 원뿔들의 단면들이기 때문이에요.

그러면 이제, 직선부터 시작하기로 하죠.

차례

0. 선(線)이란 무엇인가요?

선이란 하나의 기하학적 사물인데 이는 점들의 모임입니다. 물론, 점들의 모임이라는 것이 어떤 다른 것이 될 수도 있는 거죠. 보통, 선이라고 하면 직선을 의미합니다. 하지만, 선이 휘어져 있으면, 이를 곡선(曲線)이라고 하죠. 그런데, 그 선이 휘어져 있던 말든, 일반적으로는 커브라고 해요. 사실, 그래프에 있는 사물이 그 어떤 것이라 해도 이를 그저 커브라고 합니다. 즉, 커브라는 것은 그래프에서의 모든 기하학적 사물들의 통칭(通稱)인 거죠. 예를 들어, 그래프에서의 커브는 직선이나 포물선, 곡면(曲面) 또는 구(球)가 될 수도 있는 겁니다. Math 에서 선이라는 것은 하나의 직선이며 이는 그래프에서의 커브들 중의 하나입니다.

직선이란 원뿔곡선이라고 하는 기본 커브들 중의 하나이죠. 두개의 직각 원뿔들이 그 꼭지점들에서 서로 만나는데 그 원형 바닥들이 서로 평행하다 하세요. 또한, 하나의 평면이 그 원뿔들을 자르고 지나가는데 이는 그 꼭지점들이 맞닿은 점을 포함한다고 하세요. 그러면, 그 단면은 서로 교차하는 두개의 직선들로 되어있죠. 그런데, 그 평면이 그 원뿔들의 옆면들과 맞닿아있고 그 점을 지나간다면 하나의 직선만이 만들어지는 겁니다.

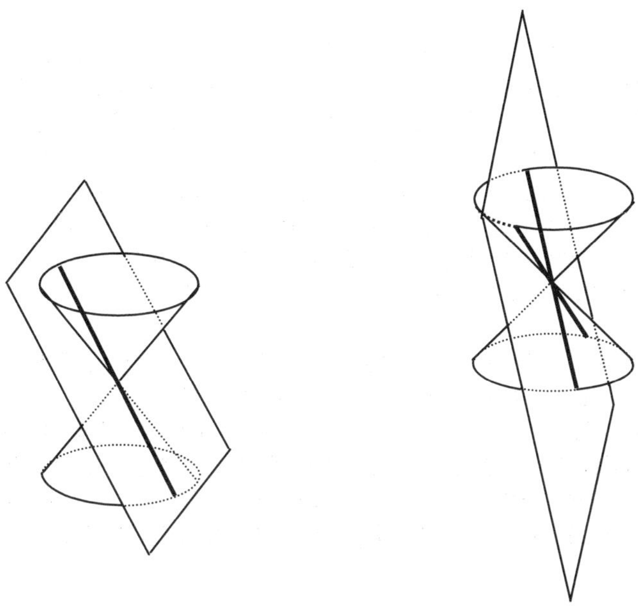

그런데, 직선은 두께나 폭을 갖지 않습니다. 이는 math 에서의 직선이란 하나의 개념이기 때문이죠. 기본적으로는 직선의 길이는 무한정이에요. 보통, 직선은 앞뒤로 즉, 서로 반대인 양방향으로 무한히 진행하죠. 다시 말해서, 양 끝이 무한정 진행하는 겁니다. 물론, 이는 한 방향으로만 진행할 수도 있어요. 이 경우에, 이를 반직선(半直線)이라고 해요. 그러면, 선분(線分)이란 뭐죠? 이는 직선의 일부이며 유한한 길이를 갖는 겁니다. 일반적으로, 선분이라 하면 직선분(直線分)을 의미하고 math 에서 거리(距離)라는 것은 이러한 선분의 길이를 의미합니다. 예를 들어, 두 지점들 간의 거리는 그 두 지점을 잇는 선분의 길이로 나타냅니다. 그럼, 휘어진 선분은 뭐라고 하나요? 이는 곡선분(曲線分)이라고 할 수 있죠. Math 에서 선이란 직선이고 이는 math 에서의 다른 모든 사물과 마찬가지로 하나의 개념입니다.

그러면 이제, 직선으로 뭘 말하자는 거죠? 길이를 나타내는데 직선의 일부 즉, 선분을 사용하긴 하지만 직선에서 흔히 중요한 것은 방향(方向)이라는 겁니다. 즉, 길이 보다는 방향을 나타내기 위해서 직선이 사용되는 거죠. 해석(解析)기하에서, 직선은 흔히 방향을 의미합니다. 해석기하는 좌표계에서의 그래프를 주로 이용하는 기하라고 할 수 있죠. 하나의 직선에서는 그 직선을 이루는 모든 점들 각각이 동일한 방향을 갖고 있어요. 그래서, 직선에는 굽힘이나 휨 또는 뒤틀림이 없죠. 직선이나 선분은 동일한 방향을 갖고있는 점들로 이루어진 사물입니다. 그 방향은 그 직선이 놓여있는 좌표계에서 그 직선이 그 좌표축들 중의 하나와 이루는 각도로 나타냅니다. 그 축은 보통, 수평 축이죠. 예를 들면, 이는 x-y 좌표계에서 x 축입니다. 그러니까, 직선은 방향을 갖고 있으며 하나의 평면에서 한 점의 방향이란 그 점을 포함하는 직선의 방향과 같은 겁니다. 그런데, 서로 평행인 직선들은 모두 동일한 방향을 갖는다는 것을 유의해야 합니다.

그런데, 그 각도라는 것은 어떻게 측정(測定)하나요? 하나의 직선이 x-y 좌표계에 있다고 하세요. 그러면, 그 각도는 그 직선의 기울기를 가지고 측정합니다. 기울기라니 요? 기울기란 하나의 직각삼각형에서 높이의 밑변에 대한 비(比)의 값이죠. 그러면 이제, 하나의 직각삼각형이 그 두개의 좌표축들과 그 직선으로 이루어집니다. 그러면, x 축의 일부가 그 삼각형의 밑변이고 y 축의 일부가 높이이며 그 직선의 일부가 빗변이 되는 거죠. 따라서, 그 높이의 그 밑변에 대한 비가 그 직선의 기울기죠. 그러니까, 그 기울기는 그 직선이 x 축에 대하여 어떻게 기울어져 있는지를 말해주는 거죠. 사실, 그 기울기란 비가 아니라 비율(比率)이에요. 그 자세한 내용은 다음 절(節)에서 알게 될 겁니다.

함수의 커브도 직선이 될 수 있어요. 즉, 어떤 함수의 그래프를 작성하면 그 커브가 직선인 거죠. 그런데, 모든 직선을 함수로 표현할 수는 없죠. 즉, 어떤 직선은 함수로 표현할 수 없다는 거죠. 예를 들어, x-y 좌표계에서의 y 축과 같은 출력 변수 축에 평행한 직선을 갖는 함수는 없어요. 즉, 그 x 축에 수직인 그 어떤 직선도 함수로는 표현할 수 없다는 거죠. 직선에 대하여 뭘 할 때, 그 직선을 주로 방정식으로 표현합니다. 이는 미지수(未知數) 방정식이 아니라 도형(圖形) 방정식이죠. 그 도형이 직선이면 그러한 방정식을 직선의 방정식이라고 해요. 보통은, 하나의 직선을 하나의 방정식으로 표현합니다. 하지만, 여러 개의 서로 다른 직선들을 하나의 방정식으로도 표현할 수도 있어요. 하나의 직선을 방정식으로 표현할 때 흔히 등호(等號)의 양변의 한쪽을 0 으로 놓죠. 그러나, 많은 경우에 그렇게 하지 않아도 되요. 그 방정식의 표현 방식은 필요에 따라 선택하면 되는 겁니다.

직선이란 점들의 집합이며 직선의 일부가 길이 또는 거리라는 값을 갖을 수도 있죠. 또한, 그 직선을 이루는 모든 점들이 동일한 방향을 갖고 있으며 그 방향은 그 직선의 기울기로 나타내어질 수 있는 거죠. 그리하여, 직선은 거리를 재거나 방향에 대하여 이야기할 때 사용합니다. 물론, 이는 다른 용도로도 사용되죠. 특히 해석기하에서는, 점의 방향에 대하여 설명하는데 직선을 사용합니다. 예를 들어, 곡선에서의 한 점의 방향을 구할 때, 그 점에서 그 곡선에 접하는 직선의 기울기를 구해서 사용하며 이러한 직선을 접선(接線)이라 하는데 이는 미적분학에서 다룹니다. 보통은, 직선의 방정식을 사용하여 직선을 표현해요. 그러면 이제, 단일한 직선을 나타내는 직선의 방정식부터 시작해보도록 하죠.

0.0. 단일(單一) 직선

단일 직선이란 그저 하나의 직선입니다. 어떻게 하나의 직선을 표현하고 이에 대하여 무엇을 어떻게 할 수 있을 지에 대하여 이야기 해보기로 하죠. 우선, Math 에서는 직선을 어디에 놓나요? 그걸 그래프에 놓죠. 그러면, 그래프에 있는 직선을 수식으로 어떻게 표현하나요? 이를 방정식으로 표현합니다. 그러면, 그 방정식은 직선의 방정식이죠. 즉, 그 방정식은 하나의 직선을 나타내고있는 겁니다. 그러면 이제, 먼저 하나의 직선을 방정식으로 어떻게 나타내는지 알아보기로 하죠.

잠깐만 이요! 함수를 사용하는 건 어떤가요?
함수로 대부분의 직선들을 표현할 수도 있지만 모든 직선을 함수로 표현할 수는 없어요. 예를 들어, 함수로는 x-y 좌표계에서 y 축에 평행한 직선을 표현할 수는 없는 겁니다.

직선은 무엇으로 이루어져있나요? 이는 점들로 되어있죠. 즉, 직선은 점들의 모임인 거죠. 그러니까, 기본적으로 직선이란 점들의 집합이라고 합니다. 그러나, 직선만이 점들의 집합은 아니죠. 사실, 그래프에서의 모든 도형은 점들로 이루어져있는 거죠. 이제, 하나의 직선이 하나의 평면에 있다고 해보세요. 그러면, 그 평면에서의 어떤 점들을 모아놓느냐에 따라서 그 도형의 모양이 결정되는 겁니다. 그러면, 어떤 점들이 모여서 직선을 이루는지 알아야 하겠네요. 그런데, 어떤 점들이 모여야 하는지는 이미 알고있죠. 그 점들을 모아 놓으면 그 점들이 모두 동일한 방향을 갖는 겁니다. 그러면 이제, 그 점들을 구체적으로 표현하는 것이 해야 하는 일이죠. 어떻게 요? 그것들을 직선의 방정식으로 표현하는 거죠. 어떻게 요? 지금 알아내려고 하는 것이 바로 그겁니다.

직선은 그래프에 놓여집니다. 그러면, 그 그래프는 어디에 있나요? 이는 특정 좌표계에 있는 거죠. 그럼, 직선을 이루는 점들 각각은 좌표들을 갖고있는 겁니다. 그러면 이제, 하나의 직선이 x-y 좌표계에 있는 그래프에 있다고 하세요. 그런데, x-y 좌표계에 있는 그래프를 x-y 평면이라고도 해요. 그러면, 이 x-y 평면에서의 직선에 있는 그 점들 각각은 한 쌍(雙)의 좌표들을 갖는 것이죠. 그러면, 그 점들을 표현해보도록 하죠. 그런데, 그 점들이 셀 수 없이 많아서 일일이 다 표현할 수는 없죠. 그럼, 어떻게 해야 하나요? 그 모든 점들을 대표하는 점을 만들어 사용하면 되죠. 그 점을 일반점(一般點)이라고 합니다. 이는 또한, 변점(變點)이나 점변수(點變數)라고 하기도 하죠. 일반점은 해당(該當) 변수들을 좌표로 사용할 수 있습니다. 그 직선이 x-y 좌표계에 있으므로 그 변수들은 x 와 y 네요. 그러면, 그 일반점은 (x, y)로 나타낼 수 있어요. 직선은 점들의 집합이라는 것을 알고있죠. 그러면 이제, L 이라 하는 직선을 x-y 평면에 다음과 같은 방식으로 놓을 수 있어요:

L = {(x, y)| ax + by + c = 0 여기에서 a, b, c 는 상수(常數)이지만 a 와 b 가 동시(同時)에 0 이 될 수 없다.}

위의 집합표현에서 **ax + by + c = 0** 를 변수들 간의 관계식이라고 할 수 있죠. 이 관계식이 바로 그 직선의 방정식입니다. 특히, 이런 직선의 방정식을 **일반형**(一般形)이라고 합니다. 그러면 이제, 일반형을 사용하면 x-y 좌표계에서의 직선 L 을 다음과 같이 나타냅니다:

L(x, y) = ax + by + c = 0 여기에서 a, b, c 는 상수이지만 a 와 b 가 동시에 0 이 될 수 없다.

그러면, 방정식 L(x, y) = 0 는 단일 직선을 나타내는 직선의 방정식인 겁니다.
그러면, 이 직선의 방정식의 **이름**은 **L(x, y)**이고 그 **방정식**은 **ax + by + c = 0** 이죠.
또한, L(x, y)에서 L 을 이 방정식의 호출명(呼出名)이라고 할 수도 있는 거죠. 호출명이 뭐죠? 이는 그 방정식을 간단히 부르는데 사용하는 이름이죠. 그러면, L(x, y)는 완전한 이름인 겁니다. 그러므로, 이 방정식은 '방정식 L', '방정식 L(x, y)' 아니면 '방정식 L(x, y) = 0'이라고 할 수 있어요.

또한, 흔히 다음과 같은 형태의 방정식으로 하나의 직선을 나타내기도 합니다: $y = ax + b$ 여기에서 a 와 b 는 상수. 하지만, 이 a 와 b 는 앞에서 사용된 a 와 b 와는 다르다는 것을 유의해야 합니다. 이러한 방정식을 **표준형**(標準形)이라 하는데 이를 가지고 거의 모든 직선을 나타낼 수 있어요. 어째서 **모두 다**가 아니고 **거의 모두**인 거죠? 이 표준형으로는 y 축에 평행한 직선을 나타낼 수가 없기 때문이죠. 하지만, 이것에는 이점도 있는 겁니다. 이 표준형은 직선을 방정식보다는 함수의 커브로 취급하는 데 사용하기 용이하죠. 그런데, 더욱 중요한 건 이러한 표준 방정식에서는 직선의 기울기 또는 그 방향을 즉각 알 수가 있다는 겁니다. 이 표준 방정식에서 a 가 그 기울기입니다. 또한, 이런 방정식에서는 직선이 어디에서 y 축과 만나는지도 바로 알 수 있죠. 직선이 어디에서 y 축과 만나는지 어떻게 알죠? 이 방정식에서 x 에 0 을 대입하면 $y = b$ 가 되죠. 따라서, 직선은 점(0, b)에서 y 축과 만나는 거죠. 즉, 직선이 (0, b)에서 y 축을 통과하거나 자르는 거죠. 그래서, 이 b 를 y 절편(截片)이라고 하지요. 그러면, x 절편은 어떤가요? 이 방정식에서 $y = 0$ 으로 놓으면 $x = -b/a$ 가 되는 거죠. 그러면, 직선은 점(-b/a, 0)에서 x 축과 교차하죠. 그러니, 이 $-b/a$ 가 그 x 절편인 겁니다.

그러면 이제, 위에서 언급한 두개의 방정식들이 어떻게 만들어지는지 알아내보기로 하죠. 먼저, 표준형부터 시작할 겁니다. 무엇인가를 작성하려면 쓸 재료가 있어야 하죠. 그러면, 이 방정식을 작성하는데 무엇이 필요한가요? 그 직선에 있는 적어도 두개의 점들이거나 아니면 그 직선에 있는 하나의 점과 그 직선의 기울기가 필요합니다. 직선의 방정식을 구할 때 보통 처음에 해야 하는 것이 그 직선의 기울기를 구하는 겁니다. 직선의 기울기(또는, 방향)는 그 직선에서 가장 의미 있는 요소이죠. 다시 말해서, 직선에 대하여 뭘 할 때 그 기울기가 가장 중요한 겁니다. 사실, 직선이 곧 기울기죠. 물론, 동일한 기울기를 갖는 직선은 무수히 많죠. 하지만, 동일한 기울기를 갖는 그 모든 직선들 중에 하나의 특정한 점을 포함할 수 있는 직선은 하나밖에는 없는 거죠.

직선의 방정식을 구한다는 것은 그 직선을 이루는 모든 점들을 구체적으로 나타내는 겁니다. 즉, 이는 "도대체 무슨 점들이 그 직선을 이루는가?" 라는 질문에 대한 답이죠. 하지만, 직선은 헤아릴 수 없이 많은 점들로 되어있으니 그 무수히 많은 점들을 일일이 명시할 수는 없는 거죠. 그래서, 어떻게 해야 하죠? 그 직선에 있는 모든 점들을 대표하는 일반점을 만들어 사용하는 거죠. 그 일반점은 그 좌표계의 변수들을 사용해서 만들 수 있어요. 즉, 그 좌표계의 변수들이 그 일반점의 좌표들인 거죠. 그러면 이제, 하나의 직선이 x-y 좌표계에 있다 하고 그 일반점의 이름이 P 라고 하세요. 그러면, 그 일반점은 P(x, y)이죠. 그러면, 그 좌표들 즉, 그 변수들 간의 관계를 도출(導出)해내어 제시해야 하는 겁니다. 즉, 그 변수들 간의 관계식이 직선의 방정식인 거죠. 자 이제, 그 직선이 특정한 점 $P_1(x_1, y_1)$을 포함하고 그 직선의 기울기가 a 라고 해보세요. 이제, 그 일반점 P(x, y)와 점 P_1 을 아래 그림에서와 같이 그래프에 놓도록 하죠.

그림. 0.0.0

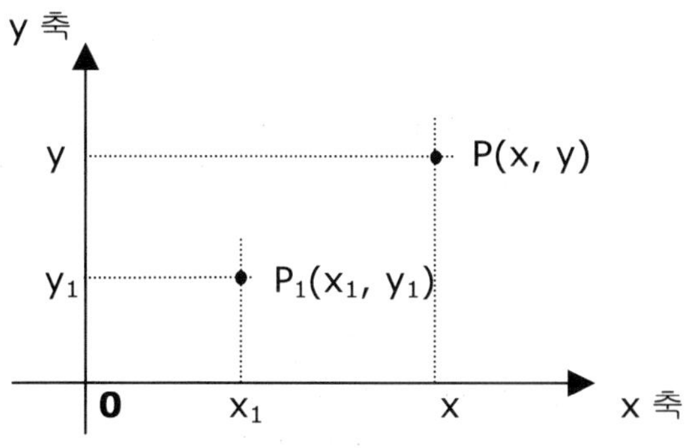

그럼 이제, 그래프에 위의 두 점들을 포함하는 직선을 놓아보도록 하죠. 보통, x-y 좌표계에서는, x 축을 따라 일어나는 변화는 Δx 로 나타내고 y 축을 따라 일어나는 변화는 Δy 로 나타냅니다.

먼저, 통상적으로 y 축에서의 값은 x 축에서의 값에 따릅니다.
그러면, x 축에서의 값은 x 축 방향으로 변화(증가)하므로 구하려는 직선은 점 $P_1(x_1, y_1)$에서 $P(x, y)$로 진행합니다.
그러면, x 축에서의 값이 x_1 에서 x 로 변하므로 y 축에서의 값은 y_1 에서 y 로 변하죠.
그러면, $\Delta x = x - x_1$ 이고 $\Delta y = y - y_1$ 입니다.

아래의 그림에서, $\Delta x(= x - x_1)$라는 양(量)은 x 축 방향으로의 변화이고 $\Delta y(= y - y_1)$라는 양은 y 축 방향으로의 변화라는 걸 알 수 있죠.

그러면 이제, 그 기울기에 대하여 이야기하도록 하죠.

직선의 기울기는 직선이 수평 축 즉, x 축에 대하여 어떻게 기울어져있는가를 말합니다. 좀더 구체적으로 이야기하면, 이는 그 직선이 x 축에 대하여 얼마만큼 어떤 식으로 기울어져있는가를 설명하는 거죠.

아래의 그림에서 그래프에 하나의 직각 삼각형이 있다는 것을 알 수 있죠.

그림. 0.0.1

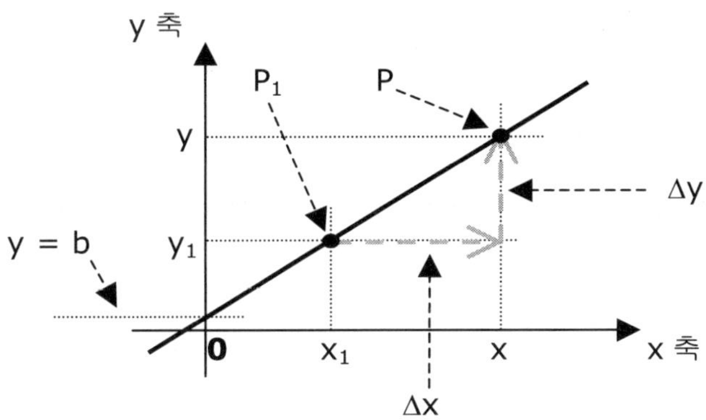

그 삼각형에서는 Δx 가 밑변이고 Δy 가 높이이며 P_1과 P 사이의 선분이 빗변입니다.
그러면, 직선의 기울기는 그 삼각형의 밑변에 대한 높이의 비(比)인 것이죠.

그러나 실제로는, 그 기울기는 단순히 비가 아니고 비율(比率)입니다.
어째서 요
그 이유는 이 절(節)에서 나중에 설명될 겁니다.

그리하여, 기울기 **a** 를 다음과 같이 정의(定義)할 수 있어요: $a = \Delta y/\Delta x = (y - y_1)/(x - x_1)$.

그런데, $a = (y_1 - y)/(x_1 - x)$ 라고 놓아도 되나요? 물론이죠. 이는 왜냐하면:

$$(y - y_1)/(x - x_1) = (-y_1 + y)/(-x_1 + x) = \{-(y_1 - y)\}/\{-(x_1 - x)\} = (y_1 - y)/(x_1 - x).$$

그러니, $a = \Delta y/\Delta x = (y - y_1)/(x - x_1) = (y_1 - y)/(x_1 - x)$ 인 것이죠.

한편, 아래의 그림에서와 같이 직선이 놓일 수도 있는 겁니다.
이 경우에는 기울기 **a** 가 음(陰)의 값을 갖게 되죠.

그림. 0.0.2

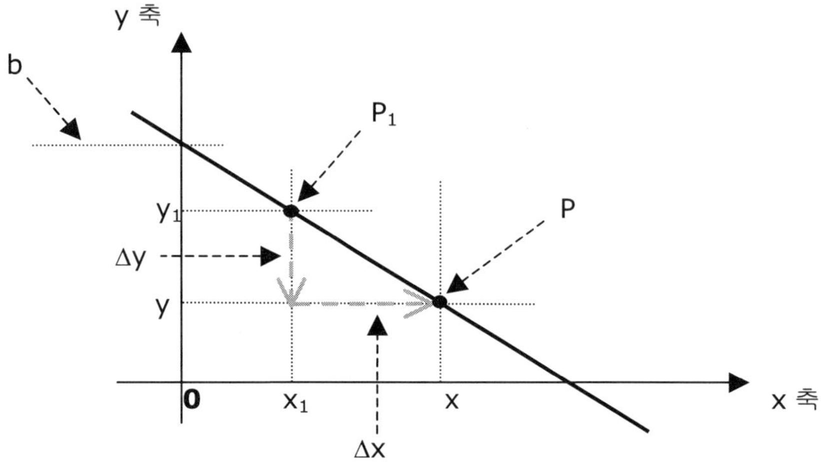

그러면 이제, 직선의 기울기가 직선의 방향을 어떻게 나타낼까요?
기울기의 값이 직선의 방향을 명시할 수 있는 거죠. 그 값은 직각삼각형의 밑변에 대한 높이의 비율이죠.
이는 비율의 값이므로 조절이 가능하죠. 예를 들어, 밑변의 길이를 1 로 잡으면 높이는 그 비율의 값이 될 겁니다.

높이/밑변 $= \Delta y/\Delta x = (\Delta y/\Delta x)/1 \Rightarrow$ 밑변 $= 1$ 이고 높이 $= \Delta y/\Delta x$.

그러면 이제, 기울기의 값이 양(陽)이면 그 삼각형은 아래의 그림에서 왼쪽에 있는 삼각형이고 그 값이 음이면 오른쪽에 있는 삼각형이 됩니다.

그림. 0.0.3

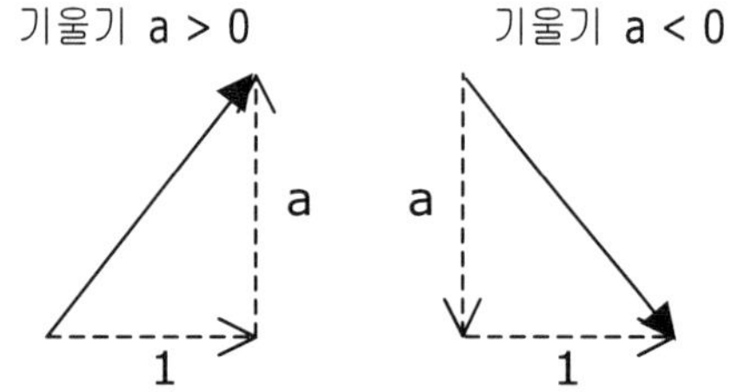

그러면, 밑변과 높이와 함께 빗변이 그 방향을 구체적으로 표현할 수 있는 것이죠.
서로 평행인 모든 직선은 동일한 기울기를 갖는 겁니다. 그러므로, 서로 평행인 모든 직선들은 동일한 방향을 갖는 것이죠.

그러면 이제, 직선의 표준형 방정식 $y = ax + b$ 에서 기울기 a 를 알아낸 겁니다.
그러면, y 절편 b 의 값을 알아내야 하죠.

이제, $a = (y - y_1)/(x - x_1)$을 다음과 같은 방식으로 놓아보죠: $y - y_1 = a(x - x_1)$.
그러면, $y - y_1 = a(x - x_1) \Rightarrow y = ax - ax_1 + y_1$.

이 a 는 상수입니다. 그러므로, $-ax_1 + y_1$ 도 상수이죠.
그러면, $y = ax + b$ 에서 b 가 $-ax_1 + y_1$ 인 것을 알 수 있죠.
수식 $y = ax - ax_1 + y_1$ 은 일반점의 좌표들 간의 관계식이죠.

그러면, 이 관계식은 직선에 있는 모든 점들 각각의 좌표들이 만족하는 수식이네요.
그러므로, x-y 평면에서 하나의 특정한 점 (x_1, y_1)을 포함하며 기울기가 a 인 직선의 방정식은 $y = ax + b$ 이며 여기에서 $b = -ax_1 + y_1$ 입니다.

그럼 이제, 직선의 일반형 방정식으로 넘어가보기로 하죠. 그 일반형은 $ax + by + c = 0$ 입니다.

표준형 방정식은 $y = ax + b$ 이죠.
그러면, $ax - y + b = 0$ 으로 놓을 수 있는데 이 형태에서는 y 의 계수가 특정한 수 즉, -1 이죠.
그러면, 이는 완전한 일반성을 갖지 않는 거죠.
그러면, 어떻게 그 일반성을 줄 수 있을까요?
그 y 의 계수를 상수로 바꾸면 되죠 상수는 수와는 달리 일반성을 갖으니까요.
예를 들어, 문자 q 를 그 상수로 사용할 수도 있죠.
그러나, 이 경우에는 기울기도 a 가 아니고 y 절편도 b 가 아닌 겁니다.
이 직선의 기울기는 $-a/q$ 이고 y 절편은 $-b/q$ 이죠.

$$ax + qy + b = 0 \Rightarrow y = (-a/q)x - b/q.$$

그런데, 통상적으로 그 상수들로 a, b, c 를 사용해요.
그러면 이제, 직선의 일반형 방정식은 다음과 같아요:

$ax + by + c = 0$ 여기에서 a, b, c 는 상수이지만 a 와 b 는 동시에 0 이 될 수 없다.

그러나, 위의 일반형에서 a 가 직선의 기울기도 아니고 b 가 y 절편도 아니라는 것을 유의해야 합니다.

그런데, 어째서 귀찮게 이런 일반형도 있는 겁니까?
직선의 표준형 방정식은 모든 직선을 표현할 수 없죠.
이는 y 축에 평행인(또는 x 축에 수직인) 직선을 나타낼 수 없어요.
예를 들면, y 축에 평행한 직선의 방정식은 다음과 같아요: $x = d$ 여기에서 d 는 상수.
그런데, y 축에 평행한 직선은 x 축에는 수직이라는 걸 알고있죠.
그러나, 그 일반형은 그러한 직선도 나타낼 수 있는 거죠.
직선의 일반형 방정식 $ax + by + c = 0$ 에서 $b = 0$ 이라 하세요.
그러면, $ax + c = 0 \Rightarrow x = -c/a$ 이며 여기에서 $-c/a$ 는 상수이죠.
그러므로, 이 방정식 $x = -c/a$ 는 직선을 나타내는데 이는 y 축에 평행(또는, x-축에 수직)입니다.

또한, 다음 절에서 일반형 방정식의 또 하나의 용도를 알게 될 겁니다.

그 표준형과 일반형 말고, 뭐든 다른 형식은 없나요?
있죠. 직선을 $x/p + y/q = 1$ 로 나타낼 수도 있는데 여기에서 p 와 q 는 0 이 아닌 상수입니다.
이 경우에는 p 가 x 절편이고 q 가 y 절편이죠. 하지만, 그런 직선은 원점을 포함할 수는 없는 거죠.

이번에는 직선을 함수의 커브로 해서 그 표준형 방정식을 구하는 과정을 복습하고 또한, 어째서 직선의 기울기가 하나의 비율인지를 알아보기로 하죠.
표준 방정식 $y = ax + b$ 에서는 a 가 기울기죠.
이제, 하나의 직선이 함수 $y = f(x) = ax + b$ 의 커브라고 해보세요 여기에서 x 는 물론 실수입니다. 또한, 아래의 그림에서와 같이 그 직선이 하나의 특정한 점 (x_1, y_1)을 포함한다 하고 이러한 직선에 있는 모든 점들을 대표하는 일반점을 (x, y)라 하세요.
그러면 이제, 이 변수들 x 와 y 사이의 관계식을 구하는 겁니다.

보통, x-y 좌표계에서는, 입력에서의 변화(증가)는 Δx 로 나타내고 출력에서의 변화는 Δy 로 나타내죠.
함수에서 출력은 입력에 따릅니다.
입력은 x 축의 방향으로 변화(증가)하므로 구하려는 직선은 점 (x_1, y_1)에서 (x, y)로 진행합니다.
그러면, 입력이 x_1에서 x로 변하므로 출력은 y_1에서 y로 변하죠.
그러면, $\Delta x = x - x_1$ 이고 $\Delta y = y - y_1$ 이죠.

그러면 이제, 그 기울기에 대하여 이야기하도록 하죠.
직선의 기울기는 직선이 수평 축 또는 입력변수 축 즉, x 축에 대하여 어떻게 기울어져있는가를 말합니다. 좀더 구체적으로 이야기하면, 이는 그 직선이 그 축에 대하여 얼마만큼 어떤 식으로 기울어져있는가를 설명하는 거죠.
아래의 그림에서 그래프에 하나의 직각 삼각형이 있다는 것을 알 수 있죠.

그림. 0.0.4

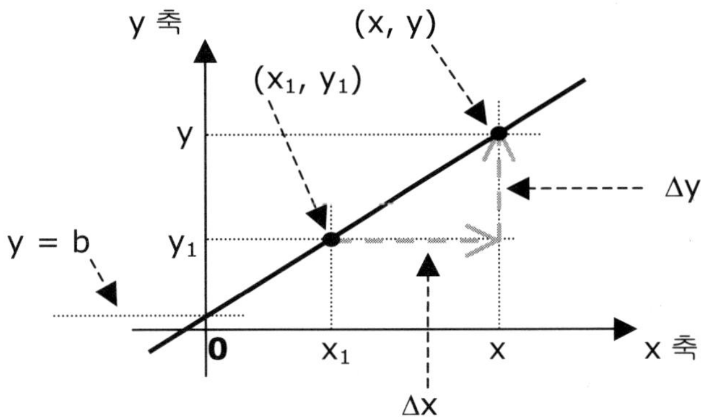

이 삼각형에서는 Δx가 밑변이고 Δy가 높이이며 직선의 일부가 빗변입니다.
그러면, 직선의 기울기는 이 삼각형의 밑변에 대한 높이의 비(比)인 것이죠.
그러나 실제로는, 그 기울기는 단순히 비가 아니고 비율(比率)입니다.
어째서 요?

우선, 입력과 출력은 주로 서로 본질이 다른 양들입니다. 예를 들면, 입력이 시간이고 그 출력은 거리인 경우이거나 또는 입력이 부피일 때 그 출력이 온도인 경우, 등등이 될 수 있죠.

다음으로, 이 삼각형의 밑변은 Δx 인데 이는 입력의 변화량(증가량)이고 그 높이는 Δy 인데 이는 출력의 변화량이죠. 즉, 이 두 가지의 양들은 변화하는 양들입니다.
다시 말해서, 기울기는 입력의 변화량 (증가량)에 대하여 출력이 변화하는 양을 명시하는 거죠.
그러므로, 기울기는 입력이 증가함에 따라 출력이 얼마나 증가하는지, 감소하는지, 아니면 그대로인지를 명시하므로, 그 기울기의 값은 양이거나 음 아니면 영이 될 수도 있는 겁니다.

그러면, 입력의 단위 증가 분에 대한 출력의 변화량을 명시하면 그 기울기를 이해하기가 용이하겠죠. 그러므로, **기울기는 단위 입력에 대한 출력 값**을 나타내는 것이며 **비율**입니다. 단위 입력이란 보통, 1 에 해당하는 입력의 양이죠. 예를 들면, 1 미터, 1 센티미터, 1 킬로그램, 1 그램, 1 리터, 등등

또한, 이러한 비율을 더욱 구체적으로 **변화율**이라고 해요.

그러면 이제, 이 기울기 **a** 는 출력이 입력이 변함에 따라 얼마나 빠르게 또는 천천히 (사실은, 크게 또는 작게) 변하는지를 말해주는 겁니다. 따라서, 이는 그래프에서 입력 축(수평 축)에 대하여 그 직선이 얼마나 기울어져

있는지를 말해줄 수 있는 거죠. 또한, 이 기울기 a 는 그 직선이 어느 쪽으로 기울어져있는지를 말해줍니다: a 가 양이면 그 직선은 왼쪽으로 기운 것이고 a 가 음이면 그 직선은 오른쪽으로 기운 것이죠.

그러면 이제, 이 기울기 a 자체는 양이거나 0 아니면 음이 될 수 있죠. 그러니까, **기울기 a** 는 **속도**로 취급될 수 있는 거죠. 그러면, 이 기울기 a 의 **크기**는 **속력**이라고 할 수도 있겠네요.

기울기가 양이면 입력이 증가함에 따라 출력이 늘어나는 거죠. 그럼, 그 속도도 양이죠. 한편, 기울기가 음이면 입력이 증가함에 따라 출력이 줄어드는 거죠. 그럼, 그 속도도 음이네요. 기울기가 0 이면 어떤가요? 그러면, 입력이 변한다 해도 출력의 변화는 없는 거죠; 다시 말해서, 출력이 일정한 겁니다. 그럼, 이는 무슨 직선이죠? 이는 수평 직선인데 이는 입력변수 축 즉, 예를 들어 x-y 좌표계에서 x 축에 평행한 거죠.

그럼 이제, 기울기는 비율이고 또한, 속도로 취급될 수 있죠.
또한, 이러한 비율을 더욱 구체적으로 변화율이라 하죠.
그러면, 변화율이 증가하면 속도도 증가하는 거죠. 따라서, 그 기울기도 증가하죠.
반면에, 변화율이 감소하면 속도도 감소하고 따라서, 그 기울기도 감소하는 거죠. 그러니까, 변화율, 속도, 기울기는 서로 같은 것들이 될 수 있는 겁니다.

이제, 표준 방정식 $y = ax + b$ 에서 $b = 0$ 이라 해보세요. 그러면, 점(x_1, y_1)은 원점에 있는 것이고 이 원점에서는 두 좌표들이 0 이죠. 그러면 $x_1 = 0$ 이고 $y_1 = 0$. 그러면, 그 직선은 원점을 포함하는 겁니다. 그러면, **제 2 권: 그래프 조작**을 공부한 학생은 $y - y_1 = a(x - x_1)$은 직선의 방정식 $y = ax$ 를 평행으로 이동(평행 변환)한 것이라는 걸 바로 알 수 있죠. 이는 또한, 점 (x_1, y_1)을 지나고 기울기가 a 인 직선의 방정식이라고도 할 수 있죠.

그림. 0.0.4

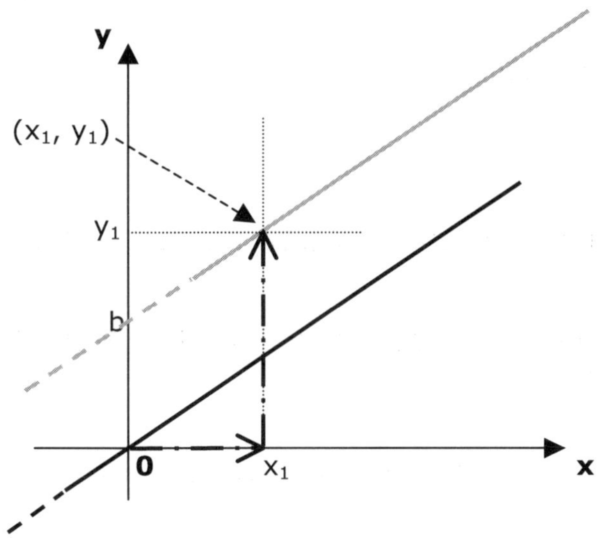

그러면 이제, 특정 직선의 방정식을 구할 때 무엇이 필요한가요? 그 직선에 있는 적어도 두개의 점들이거나 아니면 그 직선에 있는 하나의 점과 그 직선의 기울기가 필요합니다. 그런데, 왜 두개의 점들이 필요한 거죠? 기울기는 두개의 점들에 의해서 결정될 수 있기 때문이죠. 그러면, 직선의 방정식을 구하는 문제를 풀 때 그 문제의 내용으로부터 이러한 정보들을 어떻게 하든 추출해낼 수 있어야 하는 거죠. 어떤 때는 이 정보들이 명시적으로 표현되어있기도 하지만 어떤 다른 경우에는 그 정보의 일부 또는 모두가 암시적이거나 문제의 내용 속에 감추어져 있기도 합니다. 그럼 이제, 예제를 좀 다루어보기로 하죠.

예제

단일 직선

0. 하나의 직선에 세 개의 점들 $P_1(-2k - 1, 5)$, $P_2(1, k + 3)$, $P_3(k + 1, k - 1)$가 있다 하고 k의 값을 구하세요.

1. s와 t가 0이 아닌 상수라 하고 x-y 평면에서 점 (s, 0)와 점 (0, t)를 포함하는 직선의 방정식을 구하세요.

2. 다음과 같은 두개의 점들을 포함하는 직선의 방정식을 구하세요: (1, 2)와 (3, 5).

3. 직선이 한 점 (-2, 9)을 포함하며 그 기울기가 3이라 하고 그 직선의 방정식을 구하세요.

4. 직선 L은 두 개의 점들 (2, 1)과 (3, 5)를 포함하고 직선 N은 점 (1, 1)을 포함하며 직선 L에 수직이라 하고 이 직선 N의 방정식을 구하세요.

5. a와 b는 0이 아닌 상수라 하고 다음과 같이 가정하세요:
직선 A의 방정식은 $3x + y/a - 2 = 0$ 이고
직선 B의 방정식은 $5x - by + 3 = 0$ 이며
직선 C의 방정식은 $2x + y/(1 - 2b) - 5 = 0$ 이다.

또한, 직선 A는 직선 B에 수직이고 직선 C에 평행이라 하세요.

그러면, a와 b의 값을 구하세요.

6. 직선이 두 개의 점들 (1, 2)와 (3, 5)를 포함한다 하고 이 직선으로부터 점 (-5, 3)까지의 거리를 구하세요.

7. 하나의 직선이 점 (1, 3)을 포함하고 그 기울기는 2 라 하세요. 또한, 이 직선에 관하여 대칭인 두 점들 중의 하나가 (3, 2)라 하세요. 그러면, 그 나머지 한 점을 구하세요.

8. 직선이 한 점 (2, 5)를 포함하고 그 기울기는 2라 하세요. 그러면, 이 직선까지의 거리가 5인 점 (s, t)를 구하세요.

9. 점 (-2, 1)에서의 거리와 점 (3, 2)에서의 거리가 동일한 점 (s, t)를 구하세요.

A. 다음과 같이 가정하세요:
x-y 평면에 여섯 개의 점들 $P_1(1, 1)$, $P_2(1, 3)$, $P_3(2, 1)$, $P_4(4, 1)$, $P_5(s, t)$, $P_6(u, v)$가 있다.
P_5 와 P_6 는 P_3 와 P_4 의 위쪽에 있고 P_2 의 왼쪽에 있다.
삼각형 $P_5P_1P_2$ 와 또 하나의 삼각형 $P_5P_3P_4$ 를 만드는데 이들의 면적의 합이 3이다.

A.0. 점 P_5 를 명시하세요.

A.1. 또한, 다음과 같이 가정하세요:
D_1은 P_6 와 P_4 사이의 거리.
D_2는P_6 와 P_2 사이의 거리.
$D_1^2 - D_2^2 = 5$.

그러면 이제, 점 P_6 를 명시하세요.

B. x-y평면에 있는 두 점들 P_1 과 P_2 사이의 거리가 5라 하세요.
또한, 다음과 같이 가정하세요:
D_1 은 P_1 과 또 하나의 점 P_3 사이의 거리.
D_2 는 P_2 와 P_3 사이의 거리.
$D_1^2 - D_2^2 = 10$.

그러면, P_3를 명시하세요.

C. x-y평면에 4개의 점들 $P_1(-2, 3)$, $P_2(-2, -1)$, $P_3(2, 2)$, P_4 $(-2, 0)$가 있는데 세 개의 점들 P_1, P_2, P_3를 연결하는 삼각형 T를 만든다고 하세요. 또한, A는 P_1과 P_2를 잇는 직선이고 또 하나의 점 P_5가 A에 있다고 하세요.

C.0. P_5를 포함하며 직선 A에 수직인 직선을 그으면 그 수직선은 그 삼각형의 면적을 이등분합니다. 그러면 이제, P_5의 위치를 구하세요.

C.1. P_4를 포함하는 직선 B를 그으면 그 직선 B는 T의 면적을 이등분합니다. 그러면 이제, 직선 B의 방정식을 구하세요.

다음에 나오는 몇 개의 예제에서는 직선을 그래프에 넣는 연습을 할겁니다. 그래프는 방정식과 함수에 관한 문제를 풀 때 도움이 되고 많은 문제에서 필수적입니다.

D. 다음과 같은 방정식이 있다 하세요: $y = (t/4 + 1)x + t + 4$ 이며 여기에서 t는 상수.
그러면, $x > -3$ 일 때 $y > 0$ 가 되는 t의 값을 구하세요.

E. 다음과 같은 직선의 방정식이 있다 하세요: $y = (t/4 + 1)x + t + 5$ 이며 여기에서 t는 상수.
그러면, $x > -3$ 일 때 $y > 0$ 가 되는 t의 값을 구하세요.

F. 다음과 같은 직선의 방정식이 있다 하세요: $y = (t/4 + 1)x + t + 4$ 이며 여기에서 t는 상수.
그러면, $x > -3$ 일 때 $y \geq 0$ 가 되는 t의 값을 구하세요.

G. 다음과 같은 직선의 방정식이 있다 하세요: $y = (t/4 + 1)x + t + 5$ 이며 여기에서 t는 상수.
그러면, $x > -3$ 일 때 $y \geq 0$ 가 되는 t의 값을 구하세요.

H. 다음과 같은 직선의 방정식이 있다 하세요: $y = (t/4 + 1)x + t + 3$ 이며 여기에서 t는 상수.
그러면, $x > -3$ 일 때 $y \geq 0$ 가 되는 t의 값을 구하세요.

I. 다음과 같은 직선의 방정식이 있다 하세요: $y = (t/4 + 1)x + t + 3$ 이며 여기에서 t는 상수.
그러면, $-2 < x < 3$ 일 때 $y > 0$ 가 되는 t의 값을 구하세요.

J. 다음과 같은 직선의 방정식이 있다 하세요: $y = (t/4 + 1)x + t + 4$ 이며 여기에서 t는 상수.
그러면, $-2 < x < 3$ 일 때 $y > 0$ 가 되는 t의 값을 구하세요.

K. 다음과 같은 직선의 방정식이 있다 하세요: $y = (t/4 + 1)x + t + 4$ 이며 여기에서 t는 상수.
그러면, $-5 < x < -3$ 일 때 y가 양이 되는 t의 값을 구하세요.

L. 다음과 같은 직선의 방정식이 있다 하세요: $y = (t/4 + 1)x + t + 4$ 이며 여기에서 t는 상수.
그러면, $-5 \leq x \leq -3$ 일 때 y가 양이 되는 t의 값을 구하세요.

M. 다음과 같은 직선의 방정식이 있다 하세요: $y = (t/4 + 1)x + t + 4$ 이며 여기에서 t는 상수.
그러면, $-5 < x < -3$ 일 때 $y \geq 0$가 되는 t의 값을 구하세요.

N. 다음과 같은 직선의 방정식이 있다 하세요: $y = (t/4 + 1)x + t + 4$ 이며 여기에서 t는 상수.
그러면, $-5 \leq x \leq -3$ 일 때 $y \geq 0$가 되는 t의 값을 구하세요.

O. 다음과 같은 직선의 방정식이 있다 하세요: $y = (t/4 + 2)x + t + 3$ 이며 여기에서 t는 상수.
그러면, $-5 \leq x \leq -3$ 일 때 y가 음, 0, 또는 양이 되는 t의 값을 구하세요.

P. 다음과 같은 직선의 방정식이 있다 하세요: $y = (t/4 + 2)x + t + 3$ 이며 여기에서 t는 상수.
그러면, $1 < t < 5$ 일 때 $y > 0$가 되는 x의 범위를 구하세요.

Q. 다음과 같은 직선의 방정식이 있다 하세요: $y = 3(t - 1)x + 3 - 2t$ 이며 여기에서 t는 상수.
그러면, $0 < t < 2$ 일 때 $y > 0$가 되는 x의 범위를 구하세요.

문제를 풀 때 그래프를 작성하는 것이 좋을 때가 많아요. 그래프라는 건 보다 잘 보게끔 해주는 거죠. 뭘 더 잘 보게 해주는데요? 풀고있는 문제를 더 잘 볼 수 있죠 그래서 문제의 어디쯤에 와 있는지 답을 찾아서 어디로 가고 있는지 따라서, 그 답이 어디쯤에 있을지를 더욱 잘 알 수 있게 되는 겁니다. 그래프를 작성하는 도중에 그 답과 우연히 맞닥뜨리는 경우도 꽤 많아요. 물론, 더 정밀할수록 더 좋은 거죠! 하지만, 개략적으로 그려도 풀고있는 문제를 이해하고 이에 대하여 무엇을 할 수 있을지 알아내기에 충분할 때가 많아요.

그러면 이제, 그래프에 직선을 좀 넣어보기로 하죠.

R. $-1/2 \le x \le 2$ 일 때 $y = f(x) = 2x + 1$ 이고 $2 < x \le 9/2$ 일 때 $y = f(x) = 9 - 2x$ 이며 $g(x) = (f \bullet f)(x)$ 라 하세요. 그러면, g(x)의 그래프를 작성하세요.

유의사항:
절대값이라는 것은 크기이고 이는 0 이 아니면 양이죠. 값이라는 것은 양이나 0 아니면 음이 될 수 있죠. 값의 크기를 나타내고자 할 때에는 한 쌍의 짤막한 수직 막대들을 사용합니다. 그래서, $|v|$를 v 의 절대값이라 하고 그 두 개의 막대 $|\ |$를 절대 기호라고 하죠. 예를 들어, $|x|$는 x 의 값의 크기를 나타내는 거니까 $x = 1$ 일 때 뿐만 아니라 $x = -1$ 일 때도 $y = |x| = 1$ 이죠. 예를 또 하나 들자면, $|x - 3|$과 $|x - y|$는 $(x - 3)$과 $(x - y)$의 값들의 크기들을 나타내는 거니까 $x = 3$ 이고 $y = 5$ 이면 $|x - 3| = 0$ 이고 $|x - y| = 2$ 입니다. 그러니까, 절대값은 0 보다 크거나 같죠. 그러면 이제, $|x|$을 처리할 때는 $x \ge 0$ 인 경우와 $x < 0$ 인 경우 둘 다 고려해야 하는 거죠.

S. 다음과 같은 방정식들 각각의 그래프를 작성하세요:

S.0. $y = |3 - 2x| - 1$.

S.1. $y = |2x - 3| + 3x + 1$.

S.2. $y = x - |3x - 2|/2$.

S.3. $2x + |y/2| = 2$.

S.4. $|2x| + y/2 = 2$.

S.5. $|2x| + |y/2| = 2$.

S.6. $y = (x^2)^{1/2} = \sqrt{x^2}$.

S.7. $y = (x^2)^{1/2}/x = \sqrt{x^2}\ / x$.

S.8. $y = (x^2)^{1/2}/|x| = \sqrt{x^2}\ / \mid x \mid$.

S.9. $y = (x^2 - 5x + 25/4)^{1/2} = \sqrt{x^2 - 5x + 25/4}$.

S.A. $y = 2x\{|x^2 - 1|/(x^2 - 1)\}$.

S.B. $y = ((2x + 1)^2)^{1/2} + ((2x - 1)^2)^{1/2} = \sqrt{(2x+1)^2} + \sqrt{(2x-1)^2}$.

S.C. $||2x| - |3y|| = 1$.

S.D. $y = |2x - 1| + |3x + 2|$.

S.E. $y = ||2x - 1| - |3x - 1||$.

T. x-y평면에서의 세 개의 점들 $P_1(x_1, y_1)$, $P_2(x_2, y_2)$, $P_2(x_3, y_3)$가 하나의 직선에 있다고 하고 이들 좌표들간의 관계식을 구하세요.

U. 점 $P_1(2, 1)$에서 하나의 특정 직선까지의 거리가 또 하나의 점 $P_2(-2, -3)$에서 그 특정 직선까지의 거리와 같다고 하고 그 특정 직선의 방정식을 구하세요.

V. 다음과 같이 가정하세요:
$T: (x, y) \longrightarrow (3x - y, x - y)$.
$y = f(x) = ax$ 여기에서 a 는 상수.
$T \bullet f = f$.

그러면, a 의 값을 구하세요.

W. 다음과 같이 가정하세요:
$T: (x, y) \longrightarrow (3x - 2y, x + y)$.
$y = f(x) = ax$ 여기에서 a 는 상수.
$T \bullet f = f$.

그러면, a 의 값을 구하세요.

X. 다음과 같이 가정하세요:
$T: (x, y) \longrightarrow (2x + y + 3, 3x + 2y - 2)$.
$y = f(x) = ax + b$ 여기에서 a 와 b 는 상수.
$T \bullet f = f$.

그러면, a 와 b 의 값을 구하세요.

제안과 풀이
0 번 예제에서의 문제

하나의 직선에 세 개의 점들 **$P_1(-2k-1, 5)$, $P_2(1, k+3)$, $P_3(k+1, k-1)$**가 있다 하고 **k**의 값을 구하세요.

이 세 개의 점들이 모두 하나의 직선에 있으니까 이것들이 하나의 직선의 방정식을 만족해야 하는 거죠.
그러면, 직선의 방정식에 이 점들을 적용해야 하겠네요. 즉, 이 점들을 방정식에 대입하는 겁니다.
그럼, 결국은 k 에 대한 방정식을 갖게 되는 거죠.

그러면, 일반형과 표준형 중에서 어느것을 사용하는 것이 좋을까요?
일반형은 $ax + by + c = 0$ 이고 표준형은 $y = ax + b$ 이죠.

일반형을 사용하면 3 개의 방정식이 만들어지지만 4 개의 미지수를 갖게 되고 이것들은 a, b, c, k 이죠.
이 경우에는, 방정식의 개수가 미지수의 개수보다 적으므로 곤란합니다.

그러면, 표준형 방정식을 사용해보도록 하죠.
어떻게 사용하죠?
표준형 방정식을 수단으로 해서 주어진 3 개의 점들의 좌표들을 연결하여 하나의 관계식을 만드는 겁니다.
그러면, 그 관계식은 k 에 관한 방정식이 되는 거죠.

이제, 이 직선의 방정식이 $y = ax + b$ 라 하세요.
또한, $x_1 = -2k - 1$, $y_1 = 5$, $x_2 = 1$, $y_2 = k + 3$, $x_3 = k + 1$, $y_3 = k - 1$ 이라 하세요.

그러면, P_1이 직선에 있으므로 $y = ax + b \Rightarrow y_1 = ax_1 + b \Rightarrow b = y_1 - ax_1$.
따라서, $y = ax + b \Rightarrow y = ax + y_1 - ax_1 \Rightarrow y - y_1 = a(x - x_1)$.

또한, P_3도 직선에 있으므로 $y_3 - y_1 = a(x_3 - x_1)$ 라는 걸 알 수 있죠.
그럼 이제, P_1과 P_2를 사용하여 이 직선의 기울기 a를 구해보죠.

그러면, $a = (y_2 - y_1)/(x_2 - x_1)$.
그러면, $y_3 - y_1 = a(x_3 - x_1) = \{(y_2 - y_1)/(x_2 - x_1)\}(x_3 - x_1) = (y_2 - y_1)(x_3 - x_1)/(x_2 - x_1)$
$\Rightarrow (x_2 - x_1)(y_3 - y_1) = (y_2 - y_1)(x_3 - x_1)$.

자 이제, 좌표 값들을 위의 식에 대입하는 겁니다. 그러면, k 에 관한 방정식을 갖을 수 있어요.

$(x_2 - x_1)(y_3 - y_1) = \{1 - (-2k - 1)\}\{(k - 1) - 5\} = (1 + 2k + 1)(k - 6) = (2k + 2)(k - 6)$
$=2k^2 - 10k - 12$.

$(y_2 - y_1)(x_3 - x_1) = \{(k + 3) - 5\}\{(k + 1) - (-2k - 1)\} = (k - 2)(3k + 2) = 3k^2 - 4k - 4$.

그러면, $2k^2 - 10k - 12 = 3k^2 - 4k - 4 \Rightarrow k^2 + 6k + 8 = 0 \Rightarrow (k + 2)(k + 4) = 0$.
그러므로, $k = -2$ 또는 -4.

이 k 의 값을 각 점에 대입하면 다음과 같다는 것을 알 수 있죠:
(3, 5), (1, 1), (-1, -3)이 한 직선에 있다.
(7, 5), (1, -1), (-3, -5)가 한 직선에 있다.

요약:

이 직선의 방정식이 $y = ax + b$ 라 하세요.
또한, $x_1 = -2k - 1$, $y_1 = 5$, $x_2 = 1$, $y_2 = k + 3$, $x_3 = k + 1$, $y_3 = k - 1$ 이라 하세요.

그러면, P_1이 직선에 있으므로 $y = ax + b \Rightarrow y_1 = ax_1 + b \Rightarrow b = y_1 - ax_1$.
그러면, $y = ax + b \Rightarrow y = ax + y_1 - ax_1 \Rightarrow y - y_1 = a(x - x_1)$.

또한, P_3도 직선에 있으므로 $y_3 - y_1 = a(x_3 - x_1)$ 라는 걸 알 수 있죠.
그럼 이제, P_1과 P_2를 사용하여 이 직선의 기울기를 구해보죠.

그러면, $a = (y_2 - y_1)/(x_2 - x_1)$.
그러면, $y_3 - y_1 = a(x_3 - x_1) = \{(y_2 - y_1)/(x_2 - x_1)\}(x_3 - x_1) = (y_2 - y_1)(x_3 - x_1)/(x_2 - x_1)$
$\Rightarrow (x_2 - x_1)(y_3 - y_1) = (y_2 - y_1)(x_3 - x_1)$.

$(x_2 - x_1)(y_3 - y_1) = \{1 - (-2k - 1)\}\{(k - 1) - 5\} = (1 + 2k + 1)(k - 6) = (2k + 2)(k - 6)$
$= 2k^2 - 10k - 12$.

$(y_2 - y_1)(x_3 - x_1) = \{(k + 3) - 5\}\{(k + 1) - (-2k - 1)\} = (k - 2)(3k + 2) = 3k^2 - 4k - 4$.

따라서, $2k^2 - 10k - 12 = 3k^2 - 4k - 4 \Rightarrow k^2 + 6k + 8 = 0 \Rightarrow (k + 2)(k + 4) = 0$.
그러므로, $k = -2$ 또는 -4.

제안과 풀이
1 번 예제에서의 문제

s 와 **t** 가 **0** 이 아닌 상수라 하고 **x-y** 평면에서 점 **(s, 0)**와 점 **(0, t)**를 포함하는 직선의 방정식을 구하세요.

(s, 0)는 x 절편이 있는 점이고 (0, t)는 y 절편이 있는 점이죠.
이 두 점들을 그 표준형에 대입해서 그 기울기와 y 절편을 s 와 t 로 나타내보기로 하죠.
이제, 그 직선의 방정식이 y = ax + b 이며 여기에서 a 와 b 는 상수라 하세요.
그러면, 그 기울기 a = Δy/Δx = (t – 0)/(0 – s) = -t/s.
따라서, y = (-t/s)x + b 입니다.

그러면 이제, (s, 0)를 사용하여 b 를 구해보죠.

0 = (-t/s)s + b ⇒ b = t.

그러므로, 그 방정식은 y = (-t/s)x + t 입니다.

그런데, 이 방정식을 좀 보기 좋게 해보기로 하죠.
양변을 t 로 나누면, y/t = -x/s + 1 이죠.
그러면, 그 방정식의 새로운 모습은 x/s + y/t = 1.

그리하여, 방정식 x/s + y/t = 1 은 x 절편이 s 이고 y 절편이 t 인 직선의 방정식입니다.

요약:

표준형 방정식은 y = ax + b 이며 여기에서 a 와 b 는 상수.

그 기울기 a = Δy/Δx = (t – 0)/(0 – s) = -t/s.
그러면, y = (-t/s)x + b 입니다.

(s, 0)가 이 직선에 있으므로 0 = (-t/s)s + b ⇒ b = t.

그러면, y = (-t/s)x + t ⇒ y/t = -x/s + 1.
그리하여, 그 직선의 방정식은 x/s + y/t = 1.

제안과 풀이
2 번 예제에서의 문제

다음과 같은 두개의 점들을 포함하는 직선의 방정식을 구하세요**: (1, 2)**와 **(3, 5).**

점 (s, t)를 포함하고 기울기가 a 인 직선의 방정식은 $y - t = a(x - s)$입니다.
어떻게 요?
표준형 방정식을 $y = ax + b$ 라 하세요.
그러면, $t = as + b \Rightarrow b = t - as$.
따라서, $y = ax + t - as = a(x - s) + t \Rightarrow y - t = a(x - s)$.

점 (1, 2)를 포함하고 기울기가 a 인 직선의 방정식은 $y - 2 = a(x - 1)$ 입니다.
두개의 점들 (1, 2)와 (3, 5)를 사용해서 그 기울기 a 를 구할 수 있죠.

그러면, $a = \Delta y/\Delta x = (5 - 2)/(3 - 1) = 3/2$.

그리하여, 그 직선의 방정식은 $y - 2 = (3/2)(x - 1)$이죠.

직선 $y = (3/2)x$ 를 x 축 방향으로 1 만큼 y 축 방향으로 2 만큼 이동하면, 위의 직선을 갖게 되죠.
이를 평행 이동 (평행 변환)이라 하죠.

그림. 2.0

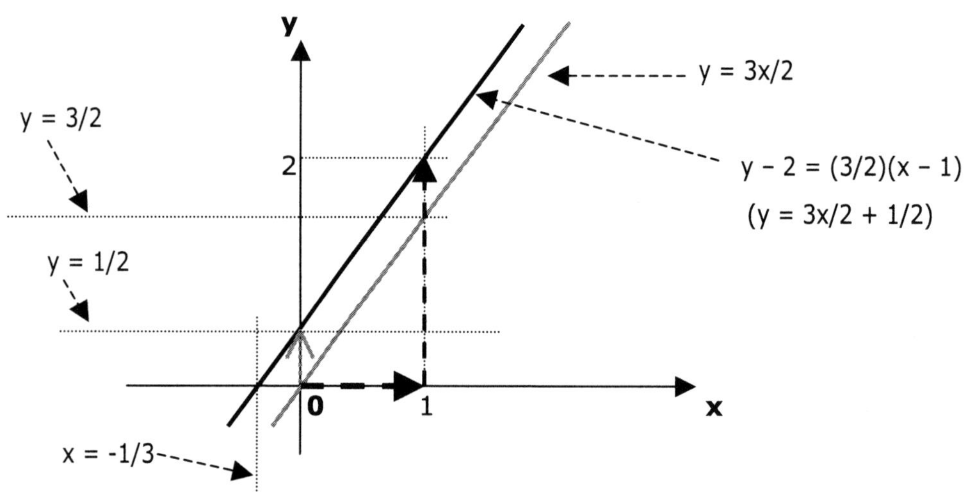

요약**:**

그 직선은 점 (1, 2)를 포함하고 그 기울기가 a 이므로 그 직선의 방정식은 $y - 2 = a(x - 1)$.

그 직선은 (1, 2)와 (3, 5)를 지나므로, $a = \Delta y/\Delta x = (5 - 2)/(3 - 1) = 3/2$.

그리하여, 그 직선의 방정식은 $y - 2 = (3/2)(x - 1)$.

제안과 풀이
3 번 예제에서의 문제

직선이 한 점 (-2, 9)을 포함하며 그 기울기가 3 이라 하고 그 직선의 방정식을 구하세요.

그 직선에 있는 점과 기울기가 명시되어있으니까 표준형을 사용하면 그 방정식을 바로 구할 수 있겠네요.
기울기가 3 이므로 방정식을 $y = 3x + b$ 라 하세요.
그러면, 이 직선은 (-2, 9)를 포함하므로 $9 =3(-2) + b = -6 + b$ 이죠.
그러면, $b = 9 + 6 = 15$ 이죠.
그러므로, 그 방정식은 $y = 3x + 15$ 이며 이는 직선 $y = 3x$ 를 y 축 방향으로 15 만큼 평행이동 한 결과로 해석될 수도 있죠.

그런데, 하나의 점 (s, t)를 포함하고 기울기가 a 인 직선의 방정식은 $\mathbf{y - t = a(x - s)}$이죠.
그러면, 이 직선은 (-2, 9)를 포함하고 그 기울기가 3 이니까, 그 방정식은 $y - 9 = 3(x - (-2)) = 3(x + 2)$이죠.
그러므로, 이는 $y - 9 = 3(x + 2)$로 나타낼 수도 있네요.
직선 $y = 3x$ 를 x 축 방향으로 –2 만큼 y 축 방향으로는 9 만큼 이동하면 위의 직선이 되는 거죠.
이 식을 풀어서 정리하면 $y = 3x + 15$ 와 같을 겁니다.

그림. 3.0.

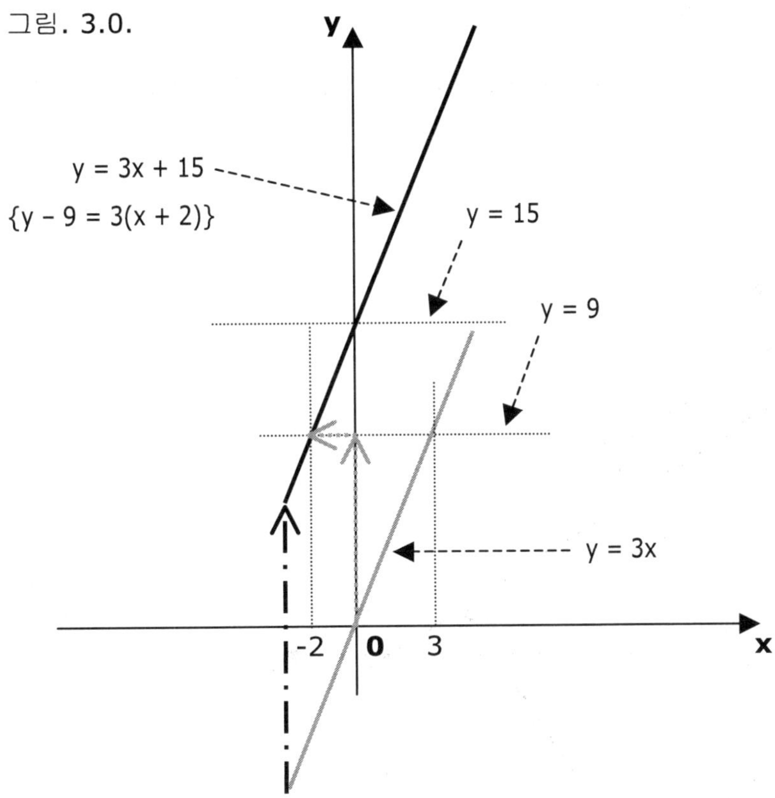

제안과 풀이
4 번 예제에서의 문제

직선 L 은 두 개의 점들 (2, 1)과 (3, 5)를 포함하고 직선 N 은 점 (1, 1)을 포함하며 직선 L 에 수직이라 하고 이 직선 N 의 방정식을 구하세요.

두개의 직선들 L 과 N 은 서로 수직입니다. 그러면, 두개의 직선들이 서로 수직이 되는 조건을 알아보아야 하겠네요. 아래 그래프에서와 같이 두개의 직선들이 서로 수직이라고 하세요.

그림. 4.0

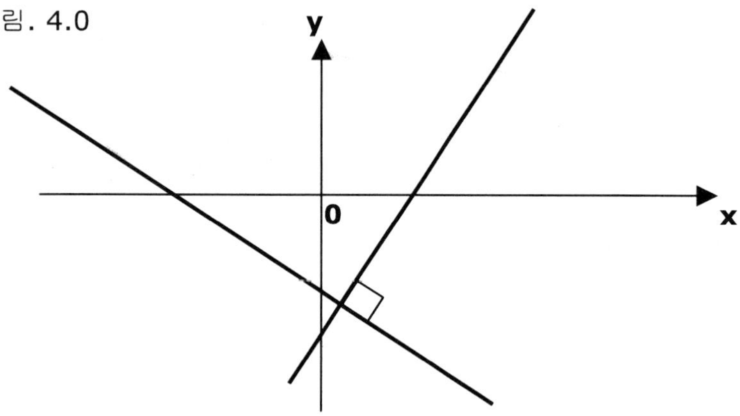

이 직선들을 평행이동하면 그들의 방향들은 그대로이고 여전히 서로 수직이죠. 그럼 이제, 아래에서와 같이 직선들을 옮기는 겁니다.

그림. 4.1

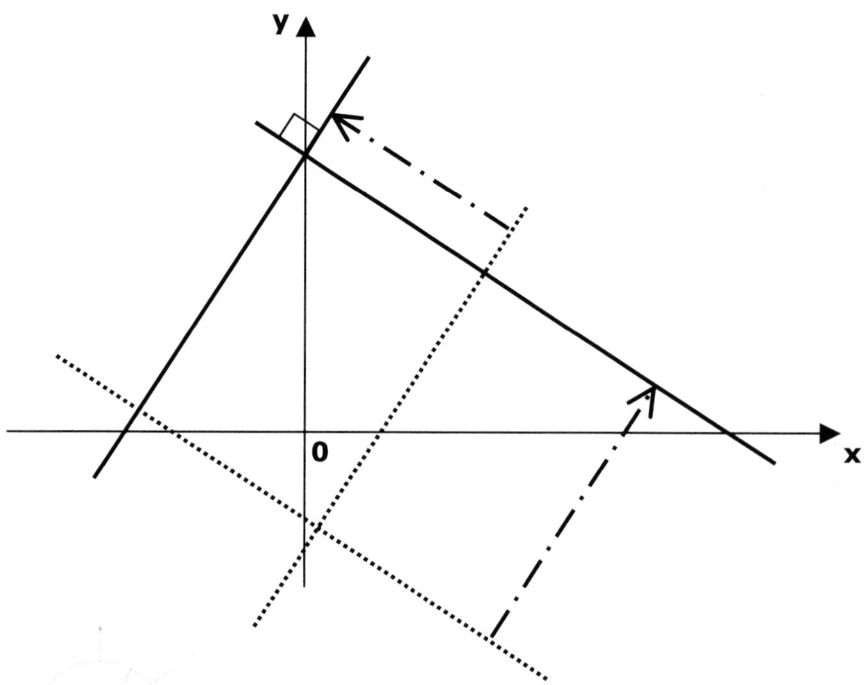

그림. 4.2

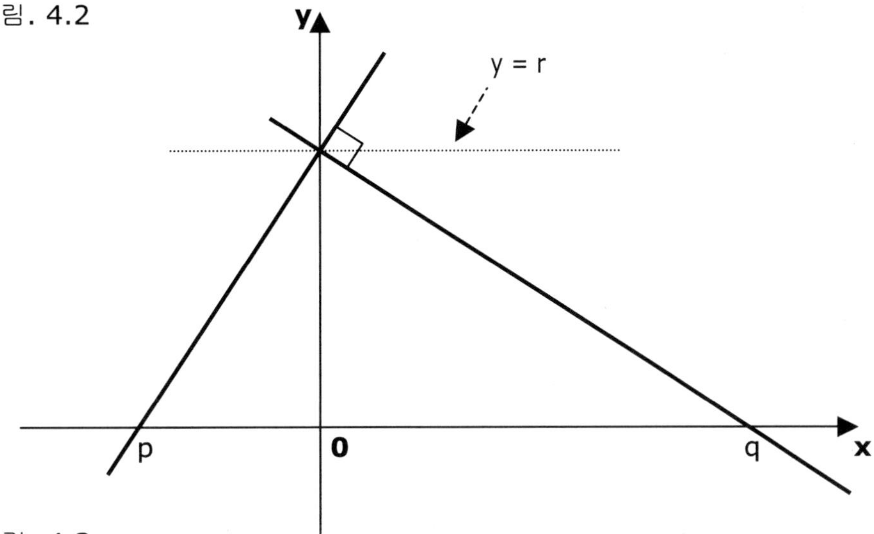

그림. 4.3

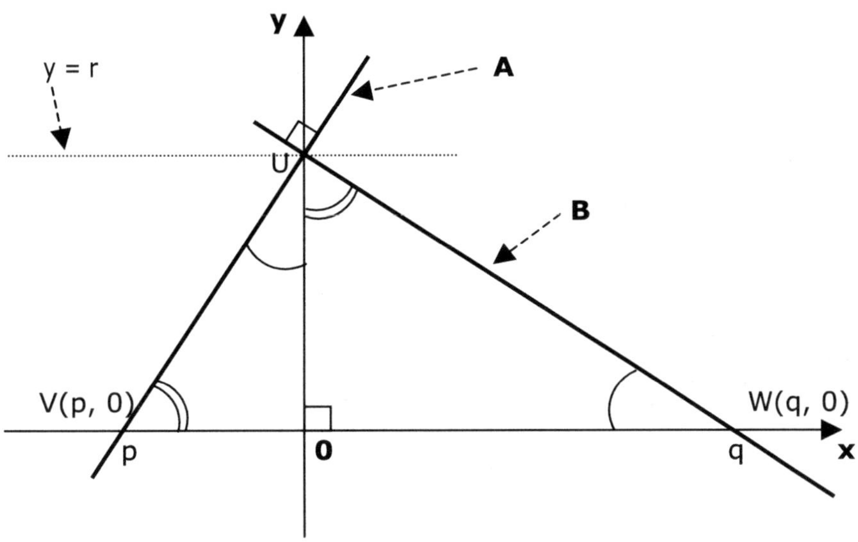

자 이제, 이 직선들 각각의 기울기를 구해보도록 하죠. 그 기울기는 $\Delta y/\Delta x$ 로 정의된다는 걸 알고있죠.
위의 그래프에서 다음과 같은 사실들을 알 수 있죠:
직선 A 는 V(p, 0)에서 x 축과 만나고 직선 B 는 W(q, 0)에서 x 축과 만난다.
두 직선들은 y 축에서의 U(0, r)에서 서로 만난다.

원점을 O 라고 하세요. 그러면, $OU = r$ 이므로 $UO = -r$ 이고 또한, $OV = p$ 이니까 $VO = -p$ 이죠. 따라서:
직선 A 의 기울기는 $\Delta y/\Delta x = OU/VO = r/(-p) = -r/p > 0$ 왜냐하면 $r > 0$ 이지만 $p < 0$ 이니까.
직선 B 의 기울기는 $\Delta y/\Delta x = UO/OW = (-r)/q = -r/q <0$ 왜냐하면 q 와 r 둘 다 양이니까.
그런데, 삼각형 UOV 와 삼각형 WOV 는 서로 닮은꼴이죠.
그러면, $|OW|/|OU| = |OU|/|OV| = |OU|/|VO|$. (주의: $|x|$는 x 의 크기를 의미합니다.)
따라서, $|OW|/|OU| = q/r$ 이니까 $|OU|/|VO| = q/r$ 이죠.
그런데, OU/VO 는 직선 A 의 기울기이고 이는 양이죠.
그러니까, $OU/VO = q/r$ 이죠 왜냐하면 $q/r > 0$ 이니까.

그러면, (OU/VO)(UO/OW) = (q/r)(-r/q) = -1.
다시 말해서, 직선 A의 기울기와 직선 B의 기울기의 곱이 −1이죠.
그리하여, 두 직선들이 서로 수직이면 이들의 기울기들의 곱은 −1입니다.

자 이제, 직선 N의 방정식을 구해보기로 하죠.
점 (1, 1)이 N에 있다는 걸 알고 있으니까 기울기를 구하면 N의 방정식을 구할 수 있죠.

N의 기울기와 L의 기울기의 곱이 −1이니까 N의 기울기는 L의 기울기로부터 알 수 있죠.

L이 포함하는 두 개의 점들을 알고있으므로 L의 기울기는 바로 알 수 있죠.
그 두 점들은 (2, 1) 과 (3, 5)입니다.
그러면, L의 기울기 = $\Delta y/\Delta x = (5 - 1)/(3 - 2) = 4/1 = 4$. 따라서, N의 기울기 = -1/4.

N은 (1, 1)을 포함한다는 걸 알고있죠.
그러면, N의 방정식은 $y - 1 = (-1/4)(x - 1)$이네요.
이를 표준형으로 나타내면, $y = (-1/4)x + 1/4 + 1 = (-1/4)x + 5/4 \Rightarrow y = (-1/4)x + 5/4$.
참고로 L의 방정식은 $y - 1 = 4(x - 2) \Rightarrow y = 4x - 8 + 1 = 4x - 7 \Rightarrow y = 4x - 7$.

그림. 4.4

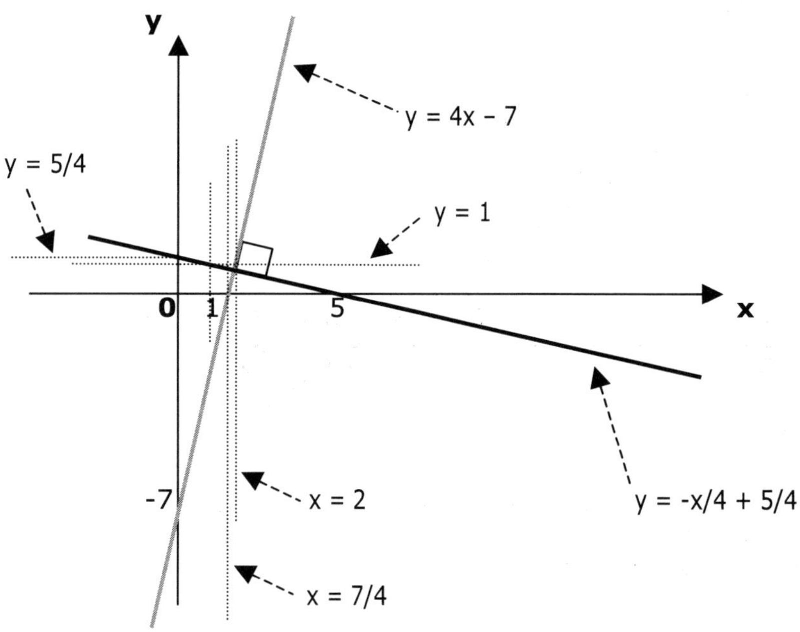

요약:

N과 L이 서로 수직이므로 이 N과 L의 기울기들의 곱이 −1.
L의 기울기 = $\Delta y/\Delta x = (5 - 1)/(3 - 2) = 4/1 = 4$. 따라서, N의 기울기 = -1/4.
N은 (1, 1)을 포함하므로, N의 방정식은 $y - 1 = (-1/4)(x - 1)$.

제안과 풀이
5 번 예제에서의 문제

a 와 b 는 0 이 아닌 상수라 하고 다음과 같이 가정하세요:
직선 **A** 의 방정식은 **$3x + y/a - 2 = 0$** 이고
직선 **B** 의 방정식은 **$5x - by + 3 = 0$** 이며
직선 **C** 의 방정식은 **$2x + y/(1 - 2b) - 5 = 0$** 이다.

또한, 직선 A 는 직선 B 에 수직이고 직선 C 에 평행이라 하세요.

그러면, a 와 b 의 값을 구하세요.

다음과 같은 사실들을 알고있죠:
서로 수직인 두 직선들의 기울기의 곱은 -1 이다.
서로 평행한 직선들의 기울기들은 동일하다.

그러면, 기울기를 살펴보도록 하죠.
먼저, 주어진 방정식들을 표준형으로 놓아보도록 하죠.

직선 A: $3x + y/a - 2 = 0 \Rightarrow y = -3ax + 2a$.
직선 B: $5x - by + 3 = 0 \Rightarrow y = 5x/b - 3/b$.
직선 C: $2x + y/(1 - 2b) - 5 = 0 \Rightarrow y = 2(2b - 1)x + 5(2b - 1)$.

그러면, A 의 기울기 $= -3a$ 이고 B 의 기울기 $= 5/b$
A 는 B 에 수직이므로 $(-3a)(5/b) = -1 \Rightarrow 15a = b$.

또한, C 의 기울기 $= 2(2b - 1)$이고 A 는 C 에 평행하므로 B 는 C 에 수직입니다.
그러므로, $(5/b)2(2b - 1) = -1$ 이네요.

$(5/b)2(2b - 1) = -1 \Rightarrow 10(2b - 1) = -b \Rightarrow 20b - 10 + b = 0 \Rightarrow 21b = 10 \Rightarrow b = 10/21$.

$15a = b$ 이죠. 그러면, $a = b/15 = (10/21)/15 = 10/(21X15) =2/(21X3) = 2/63$.
그러므로, $a = 2/63$ 이고 $b = 10/21$.

요약:

직선 A: $3x + y/a - 2 = 0 \Rightarrow y = -3ax + 2a$.
직선 B: $5x - by + 3 = 0 \Rightarrow y = 5x/b - 3/b$.
직선 C: $2x + y/(1 - 2b) - 5 = 0 \Rightarrow y = 2(2b - 1)x + 5(2b - 1)$.
그러면, A 의 기울기 $= -3a$, B 의 기울기 $= 5/b$, C 의 기울기 $= 2(2b - 1)$.
A 는 B 에 수직이므로 $(-3a)(5/b) = -1 \Rightarrow 15a = b$.
B 는 C 에 수직이므로 $(5/b)2(2b - 1) = -1 \Rightarrow 10(2b - 1) = -b \Rightarrow 21b = 10 \Rightarrow b = 10/21$.
$15a = b \Rightarrow a = b/15 = (10/21)/15 = 10/(21X15) =2/(21X3) = 2/63$.
그러므로, $a = 2/63$ 이고 $b = 10/21$.

제안과 풀이
6 번 예제에서의 문제

직선이 두 개의 점들 (1, 2)와 (3, 5)를 포함한다 하고 이 직선으로부터 점 (-5, 3)까지의 거리를 구하세요.

두 점들 사이의 거리라는 것은 그 두 점들은 잇는 선분의 길이죠.
그럼, 한 특정 점에서 직선까지의 거리는 어떻게 정의하나요?
Math 에서 그 거리는 그 직선과 그 특정 점 사이의 가장 짧은 거리입니다.
가장 짧은 거리라니 요?
그 특정 점을 지나고 그 직선에 수직인 직선을 A 라 하세요.
또한, 그 A 가 그 직선을 만나는 점을 B 라 하세요.
그러면, 그 최단 거리는 그 특정 점과 점 B 사이의 거리이죠.

그러면 이제, 두 점들 (1, 2)와 (3, 5)를 포함하는 직선을 L 이라 하고 점 (-5, 3)을 P 라 하세요.
그러면, L 에 수직이며 P 를 포함하는 직선을 구해야 하죠.
이제, 그 수직선을 N 이라 하세요.
또한, L 이 N 을 만나는 점을 Q 라 하세요.
그러면, L 과 P 사이의 거리는 P 와 Q 사이의 거리 즉, 이 두 점들 P 와 Q 를 잇는 선분의 길이죠.

그 길이는 그 유명한 거리 공식 즉, 피타고라스 정리로 구할 수 있어요.
이 정리에 의하면, 직각 삼각형에서 그 빗변의 제곱은 그 나머지 두 변들 각각의 제곱들의 합과 같죠.
즉, 두 점들 간의 거리의 제곱은 그 두 점들의 제각기 x 와 y 좌표들 간의 차들 각각의 제곱들의 합인 겁니다.
예를 들어, D가 두 점들 간의 거리라면 $D^2 = (\Delta x)^2 + (\Delta y)^2$ 인 겁니다.

그림. 6.0

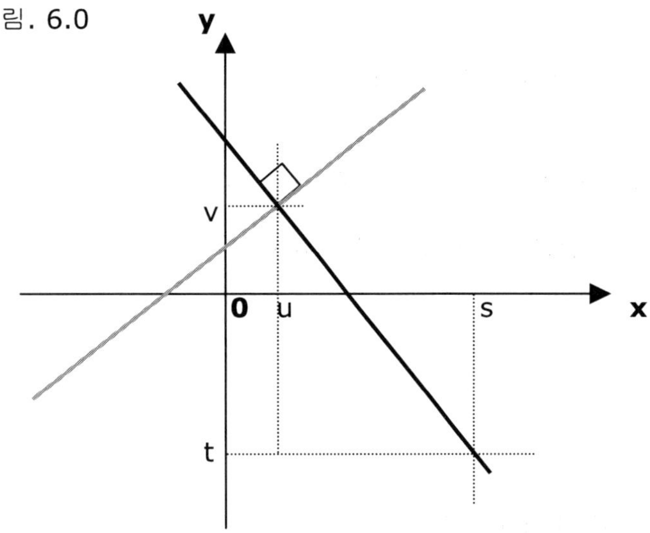

위의 그래프에서 D 가 (s, t)와 (u, v)사이의 거리라고 하세요.
그러면, $D^2 = (\Delta x)^2 + (\Delta y)^2 = (s - u)^2 + (t - v)^2 = (u - s)^2 + (v - t)^2 \Rightarrow D = \{(s - u)^2 + (t - v)^2\}^{1/2}$.
그러므로, $D = \sqrt{(s-u)^2 + (t-v)^2}$.

그러면 이제, P와 Q 사이의 거리를 구해보기로 하죠.
L의 방정식부터 구해야 하죠. L은 (1, 2)와 (3, 5)를 포함합니다.
그러면, 그 기울기부터 시작하죠.
L의 기울기 = $\Delta y/\Delta x = (5 - 2)/(3 - 1) = 3/2$.
그러면, L이 (1, 2)를 포함하므로, L 방정식은 $y - 2 = (3/2)(x - 1)$입니다.
서로 수직인 두 직선들의 기울기의 곱은 −1이라는 걸 알고있죠.
N의 기울기를 n이라 하세요. 그러면, $(3/2)n = -1$이죠.
그러면, $n = -2/3$입니다.
따라서, N이 P(-5, 3)을 포함하므로 N의 방정식은 $y - 3 = (-2/3)\{x - (-5)\} = (-2/3)(x + 5)$이죠.

그러면 이제, L이 N을 만나는 점 Q를 구해야 하죠.
이 두 직선들은 이 두 직선들이 동시에 포함하는 점에서 만나므로 그 점이 Q이죠.
그러면, 그 두 방정식들에서 그 두 y값들이 같을 때의 x값이 Q의 x좌표네요.
그리고 나서, 그 x값을 그 두 방정식들 중 어느 한 방정식에 대입하면 그 y값이 Q의 y좌표이죠.

L의 방정식은 $y - 2 = (3/2)(x - 1) \Rightarrow y = (3/2)(x - 1) + 2 = (3/2)x + 1/2$.

N의 방정식은 $y - 3 = (-2/3)(x + 5) \Rightarrow y = (-2/3)(x + 5) + 3 = (-2/3)x - 1/3$.

그러면, $(3/2)x + 1/2 = (-2/3)x - 1/3 \Rightarrow (3/2 + 2/3)x = -1/3 - 1/2$
$\Rightarrow (13/6)x = -5/6 \Rightarrow 13x = -5 \Rightarrow x = -5/13$.

$y = (3/2)x + 1/2$ 이죠.
그러면, $y = (3/2)(-5/13) + 1/2 = -15/26 + 1/2 = -2/26 = -1/13$.

그러므로, Q는 (-5/13, -1/13)에 있는 거죠. 또한, P는 (-5, 3)에 있죠.

$D^2 = (\Delta x)^2 + (\Delta y)^2$ 여기에서 D가 그 거리이죠.

$D^2 = \{-5/13 - (-5)\}^2 + (-1/13 - 3)^2 = (-5/13 + 5)^2 + (1/13 + 3)^2$
$= (60/13)^2 + (40/13)^2 = (60^2 + 40^2)/13^2 = 3600 + 1600 = 5200/13^2 = (13)(4)10^2/13^2 = 20^2(13)/13^2$
$\Rightarrow D = (20/13)(13)^{1/2} = 20\sqrt{13}/13 = \dfrac{20\sqrt{13}}{13}$.

그러므로, 그 거리는 $\dfrac{20\sqrt{13}}{13}$.

어째서 그저 $\dfrac{20}{\sqrt{13}}$라 하지 않나요? 보통, 분모에는 유리수를 선호하거든요. 하지만, 답을 $\dfrac{20}{\sqrt{13}}$로 두어도 되요.

물론, 직선과 점 사이의 거리 공식을 사용해도 됩니다.

점 (s, t)에서 직선 $ax + by + c = 0$까지의 거리를 D라 하세요.
그러면, $D = |as + bt + c|/(a^2 + b^2)^{1/2}$.

물론, 어떻게 이 공식이 만들어지는 지 알아야 하죠. 이는 다음 절에서 알아낼 겁니다.

요약:

다음과 같이 가정하세요.
그 두 점들 (1, 2)와 (5, 3)을 포함하는 그 직선을 L 이라 하고 점 (-5, 3)을 P.
L 에 수직이며 P 를 포함하는 직선을 N.
L 과 N 이 만나는 점을 Q.

그러면, L 의 기울기 = $\Delta y/\Delta x = (5 - 2)/(3 - 1) = 3/2$.
따라서, L 이 (1, 2)를 포함하므로, L 방정식은 $y - 2 = (3/2)(x - 1)$.
서로 수직인 두 직선들의 기울기의 곱은 −1.
N 의 기울기를 n 이라 하세요. 그러면, $(3/2)n = -1 \Rightarrow n = -2/3$.
N 이 P(-5, 3)을 포함하므로 N 의 방정식은 $y - 3 = (-2/3)\{x - (-5)\} = (-2/3)(x + 5)$.

$y - 2 = (3/2)(x - 1) \Rightarrow y = (3/2)(x - 1) + 2 = (3/2)x + 1/2$.
$y - 3 = (-2/3)(x + 5) \Rightarrow y = (-2/3)(x + 5) + 3 = (-2/3)x - 1/3$.

그러면, $(3/2)x + 1/2 = (-2/3)x - 1/3 \Rightarrow (3/2 + 2/3)x = -1/3 - 1/2$
$\Rightarrow (13/6)x = -5/6 \Rightarrow 13x = -5 \Rightarrow x = -5/13$.

$y = (3/2)x + 1/2 = (3/2)(-5/13) + 1/2 = -15/26 + 1/2 = -2/26 = -1/13$.

그러므로, Q 는 (-5/13, -1/13)이고 P 는 (-5, 3).

D 가 그 거리라고 하세요. 그러면:

$D^2 = \{-5/13 - (-5)\}^2 + (-1/13 - 3)^2 = (-5/13 + 5)^2 + (1/13 + 3)^2$
$= (60/13)^2 + (40/13)^2 = (60^2 + 40^2)/13^2 = 3600 + 1600 = 5200/13^2 = (13)(4)10^2/13^2 = 20^2(13)/13^2$
$\Rightarrow D = (20/13)(13)^{1/2} = 20\sqrt{13}/13 = \dfrac{20\sqrt{13}}{13}$.

그러므로, 그 거리는 $\dfrac{20\sqrt{13}}{13}$.

제안과 풀이
7 번 예제에서의 문제

하나의 직선이 점 (1, 3)을 포함하고 그 기울기는 2 라 하세요. 또한, 이 직선에 관하여 대칭인 두 점들 중의 하나가 (3, 2)라 하세요. 그러면, 그 나머지 한 점을 구하세요.

두 점들이 한 직선에 관하여 대칭이라면 그 두 점들 중의 한 점에서 그 직선까지의 거리가 그 나머지 한 점에서 그 직선까지의 거리와 같은 거죠. 이 경우, 그 직선을 대칭축이라고 해요. 또한, 그 두 점들은 그 대칭축에 수직인 직선에 있는 겁니다.

그러면, 위의 내용을 적합한 수식들로 표현해서 답을 구할 수 있을 겁니다.

또한, 다음과 같이 이 문제를 풀 수도 있는데 이 방법을 먼저 사용해보기로 하죠.

그 두 점들의 중점이 그 대칭축에 있는 겁니다.
그러면, 그 중점의 좌표들은 제각기 그 두 점들의 x 좌표들과 y 좌표들의 평균들이죠.

이제, 점(1, 3)을 P 라 하고 이 점 P 를 포함하며 기울기가 2 인 그 직선을 L 이라 하며 이 L 에 관하여 P 에 대칭인 점을 Q(s, t)라 하세요. 또한, P 와 Q 의 중점을 R(u, v)라 하세요.
그러면, 이 R 은 L 에 있는 겁니다. 또한, R 의 좌표들은 u 와 v 인데 이것들은 제각기 s 와 t 의 식이죠 왜냐하면 R 이 P 와 Q 의 중점이고 Q 가 (s, t)이거든요.
그래서, 이 R 을 L 의 방정식에 대입하면 s 와 t 에 관한 하나의 방정식을 갖게 되죠.

그런데, 구하고자 하는 미지수는 두 개이죠.
그러면, 방정식이 하나 더 필요하네요.
그 방정식은 무엇이 될 수 있을까요?
그 중점 R 은 L 에 수직인 그 직선에도 포함되는 거죠.
이제, 그 수직선을 N 이라 하세요. 그러면, 이 N 은 P 를 포함하죠.
또한, 이 N 은 L 에 수직이므로 그 기울기는 L 의 기울기로부터 구할 수 있죠.
서로 수직인 두 직선들의 기울기들의 곱은 −1 이라는 걸 알고있죠.
그러면 이제, 그 중점 R 을 N 의 방정식에 대입하면 u 와 v 에 관한 또 하나의 방정식을 갖게 되죠.

그러면, 그 두개의 방정식으로 된 연립방정식을 풀면 그 R 을 갖게 될 겁니다.

그러면 이제, L 의 방정식부터 시작하도록 하죠.
이 L 은 (1, 3)을 포함하고 그 기울기가 2 이므로 그 방정식은 다음과 같습니다: $y - 3 = 2(x - 1)$.
다음으로 N 의 방정식을 구해보도록 하죠.
N 의 기울기를 n 이라 하세요. 그러면, $2n = -1 \Rightarrow n = -1/2$.
그러면, 이 N 은 P(3, 2)를 포함하고 그 기울기가 −1/2 이므로 그 방정식은 다음과 같죠: $y - 2 = (-1/2)(x - 3)$.

이제, 그 중점 R 을 구해보도록 하죠.
이 R 의 좌표들은 제각기 P 와 Q 의 x 좌표들과 y 좌표들의 평균들이라는 걸 알고있죠.
P 는 (3, 2)에 있으며 Q 는 (s, t)에 있고 R 은 (u, v)에 있는 겁니다.

그러면, $u = (3 + s)/2$ 이고 $v = (2 + t)/2$ 이죠.

자 이제, 이 R은 L과 N 둘 다에 있어야 하죠.
L의 방정식은 $y - 3 = 2(x - 1)$이고 N의 방정식은 $y - 2 = (-1/2)(x - 3)$입니다.

먼저, R을 L에 대입해보죠. 그러면, $v - 3 = 2(u - 1) \Rightarrow (2 + t)/2 - 3 = 2\{(3 + s)/2 - 1\}$
$\Rightarrow 2 + t - 6 = t - 4 = 4\{(3 + s)/2 - 1\} = 2(3 + s) - 4 = 2s + 2 \Rightarrow t - 4 = 2s + 2 \Rightarrow 2s - t = -6$.

이제, R을 N에 대입해보죠. 그러면, $v - 2 = (-1/2)(u - 3) \Rightarrow (2 + t)/2 - 2 = (-1/2)\{(3 + s)/2 - 3\}$
$\Rightarrow 2 + t - 4 = t - 2 = -\{(3 + s)/2 - 3\} \Rightarrow 2t - 4 = -(3 + s) + 6 = 3 - s \Rightarrow 2t - 4 = 3 - s \Rightarrow s + 2t = 7$.

그러면 이제, 다음과 같은 연립방정식을 갖는 거죠: $2s - t = -6$ 와 $s + 2t = 7$.
먼저, t를 소거해보죠.
$2s - t = -6 \Rightarrow 4s - 2t = -12$.
그러면, $(4s - 2t) + (s + 2t) = -12 + 7 \Rightarrow 5s = -5 \Rightarrow s = -1$.
따라서, $s + 2t = 7 \Rightarrow -1 + 2t = 7 \Rightarrow 2t = 8 \Rightarrow t = 4$.

그러니까, Q는 (-1, 4)에 있네요.

이번에는 다음 내용을 수식들로 표현해서 답을 구해보기로 하죠.

두 점들이 한 직선에 관하여 대칭이라면 그 두 점들 중의 한 점에서 그 직선까지의 거리가 그 나머지 한 점에서 그 직선까지의 거리와 같은 거죠. 이 경우, 그 직선을 대칭축이라고 해요. 또한, 그 두 점들은 그 대칭축에 수직인 직선에 있는 겁니다.

그러면, 다음 두개의 사실들을 가지고 방정식을 세우면 되겠네요:
주어진 점 P에서 주어진 직선 L까지의 거리가 대칭점 Q에서 L까지의 거리와 같다는 사실
P와 Q가 L에 수직인 N에 있다는 사실

첫 번째 사실을 이용하자면 L과 N이 서로 만나는 점을 구해야 하죠. 사실 이 점은 위에서 구한 그 중점 R이죠.

그럼 이제, 위에서 이미 구하거나 정의한 방정식이나 점들은 그대로 사용하지만, 알고리즘 즉, 문제 해결 과정은 처음부터 작성해 보기로 하겠습니다.

구하는 점 Q 의 좌표들이 두개의 미지수들이므로 이 좌표들에 관한 두개의 방정식들로 된 연립방정식을 세워야 할 겁니다.

먼저, L의 방정식을 구하고 나서 N의 방정식을 구해야 하죠.
그 다음, L과 N이 만나는 점 즉, 그 중점 R을 구합니다.
다음으로, 주어진 점 P와 R사이의 거리를 구하죠. 이 거리를 D라 하세요..
그러면, 그 대칭점 Q와 R사이의 거리가 이 D와 같다는 방정식을 세웁니다. 그러면, 이것이 하나의 방정식이죠.
이제, Q가 N에 있어야 하는 점이므로, 이 Q를 N에 대입하여 그 나머지 방정식을 세우는 겁니다.

L의 방정식은 다음과 같죠: $y - 3 = 2(x - 1)$.
N의 방정식은 다음과 같죠: $y - 2 = (-1/2)(x - 3)$.

또한, 이 방정식들 자체에 제각기 이름을 붙이려면 다음과 같이해도 됩니다:
$L(x, y) = 2(x -1) - (y- 3) = 0$. $N(x, y) = (-1/2)(x - 3) - (y - 2) = 0$.

이제, L 이 N 을 만나는 그 중점 R 을 구해보기로 하죠.
그러면, 다음의 두개의 방정식의 연립방정식을 풀어야 하죠.
$L(x, y) = 0$ 과 $N(x, y) = 0$.

$L(x, y) = 0 \Rightarrow 2(x - 1) - (y - 3) = 0 \Rightarrow y - 3 = 2(x - 1) \Rightarrow y = 2x - 2 + 3 = 2x + 1 \Rightarrow y = 2x + 1$.

$N(x, y) = 0 \Rightarrow (-1/2)(x - 3) - (y - 2) = 0 \Rightarrow y - 2 = (-1/2)(x - 3) \Rightarrow y = -x/2 + 3/2 + 2$
$\Rightarrow y = -x/2 + 7/2$

그러므로, $2x + 1 = -x/2 + 7/2 \Rightarrow 4x + 2 = -x + 7 \Rightarrow 5x = 5 \Rightarrow x = 1$.
그러면, $y = 2x + 1 = 3$.
따라서, R 은 (1, 3)에 있는 거죠.

이제, P 와 R 사이의 거리 D 를 구해보기로 하죠. P 는 (3, 2)에 있죠.
$D^2 = (\Delta x)^2 + (\Delta y)^2$ 이라는 걸 알고있죠.

두 점들 사이의 거리 계산에서는 Δx 와 Δy 가 제곱 되니까 어느 점에서 어느 점을 빼도 상관없는 겁니다.
P 에서 R 을 뺀다고 하세요. 그러면:

$D^2 = (3 - 1)^2 + (2 - 3)^2 = 4 + 1 = 5 \Rightarrow D^2 = 5$.

이제, 그 대칭점 Q(s, t)와 R(1, 3)사이의 거리가 D 와 같다는 방정식을 세웁니다. 그러면, 이것이 그 연립 방정식에서의 하나의 방정식이죠. 그러면:

$D^2 = 5 = (s - 1)^2 + (t - 3)^2$.

이제, Q(s, t)가 N 에 있어야 하는 점이므로, 이 Q 를 N 에 대입하여 그 나머지 방정식을 세우는 겁니다:

$N(x, y) = (-1/2)(x - 3) - (y - 2) = 0 \Rightarrow y - 2 = (-1/2)(x - 3) \Rightarrow y = -x/2 + 7/2 \Rightarrow t = -s/2 + 7/2$.

그러면, 다음과 같은 연립방정식을 갖는 것이죠: $5 = (s - 1)^2 + (t - 3)^2$ 과 $t = -s/2 + 7/2$.

그러면, $5 = (s - 1)^2 + (-s/2 + 7/2 - 3)^2$ 이고 $-s/2 + 7/2 - 3 = -s/2 + 1/2 = (1 - s)/2$.
따라서, $5 = (s - 1)^2 + (1 - s)^2/4 = (s - 1)^2 + (s - 1)^2/4 = 5(s - 1)^2/4. \Rightarrow 5 = 5(s - 1)^2/4 \Rightarrow (s - 1)^2 = 4$.

그런데, 잠깐만 요! 어째서 2 차 방정식이죠? 2 차 방정식의 답은 보통 두개의 근들로 되 있죠. 그럼, L 에 관하여 (3, 2)에 대칭인 점이 두개가 있나요?
아니죠, L에 관하여 (3, 2)에 대칭인 점은 단 하나뿐이죠. 그 2 차 방정식이 $(cs + d)^2 = 0$ (여기에서 c와 d는 상수)과 같은 완전 제곱식이 아니면 이는 서로 다른 두개의 근들(답들)을 갖는 건 맞아요.
그런데, 현재 풀고있는 방정식은 완전 제곱식이 되어서는 안되죠.
어째서 요?
그 서로 다른 두 근들 중에 하나가 바로 P 의 것이어야 하거든요.
따라서, 그 나머지 하나가 그 대칭점 Q 의 것인 겁니다.
그러니까 그 연립방정식은 직선 N에 있는 두개의 점들을 구하는 연립방정식 인데 그 점들은 (1, 3)을 중심으로 하고 반지름이 $5^{1/2}$인 원을 이루는 점들 중에 있는 겁니다. 원의 방정식은 제 4 권에서 다룹니다.

자 이제, 계속진행하기로 하죠.

$(s-1)^2 = 4 \Rightarrow (s-1 = 2$ 또는 $-2) \Rightarrow (s = 3$ 또는 $-1)$.
이제, t를 구해야 하죠. $t = -s/2 + 7/2$ 이죠.
그러면, $s = 3$일 때 $t = -3/2 + 7/2 = 2$이죠.
따라서, $(s, t) = (3, 2)$이며 이는 이미 주어진 점 P이죠.
그러면 이제, $s = -1$일 때 $t = -(-1)/2 + 7/1 = 1/2 + 7/2 = 4$입니다.
그리하여, $Q(s, t) = Q(-1, 4)$이네요.

그러면 이제, 직선 $L(x, y) = 2x - y + 1 = 0$에 관하여 대칭인 두 점들 (3, 2)와 (-1, 4)를 아래의 그래프에서 보기로 하죠.

그림. 7.0

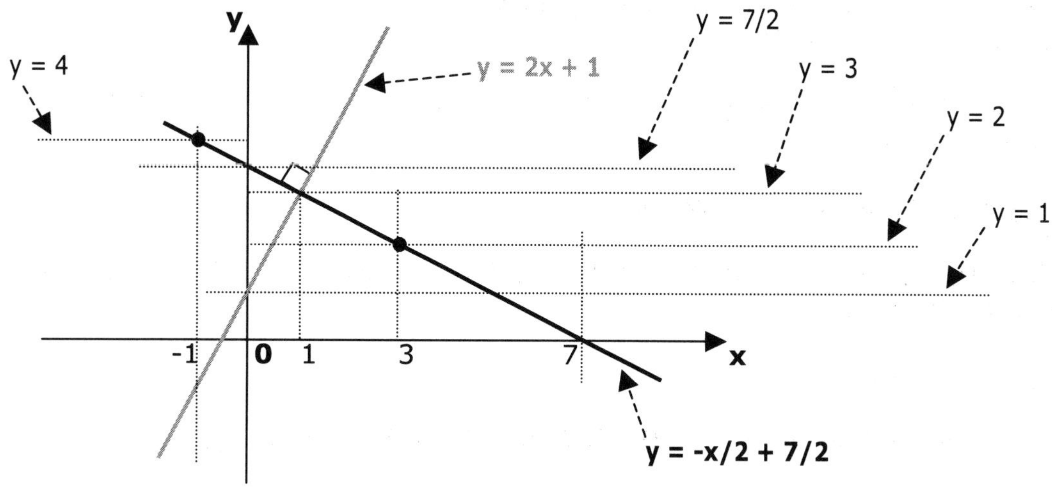

요약:

첫 번째 방법

주어진 직선을 L이라 하고 P를 (3, 2)라 하며 L에 관하여 P에 대칭인 점을 Q(s, t)라 하세요.
또한, P와 Q의 중점을 R(u, v)라 하고 L에 수직인 직선을 N이라 하세요.
L은 (1, 3)을 포함하고 그 기울기가 2이므로 L의 방정식은 $y - 3 = 2(x - 1)$이죠.
N의 기울기를 n이라 하세요. 그러면, $2n = -1 \Rightarrow n = -1/2$.
이 N은 P(3, 2)를 포함하고 그 기울기가 $-1/2$이므로 N의 방정식은 $y - 2 = (-1/2)(x - 3)$이죠.
이제, 그 중점 R(u, v)를 구해보도록 하죠.
이 R의 좌표들은 제각기 P와 Q의 x좌표들과 y좌표들의 평균들이라는 걸 알고있죠.
그러면, $u = (3 + s)/2$ 이고 $v = (2 + t)/2$.

먼저, R을 L에 대입하면, $v - 3 = 2(u - 1) \Rightarrow (2 + t)/2 - 3 = 2\{(3 + s)/2 - 1\}$
$\Rightarrow 2 + t - 6 = t - 4 = 4\{(3 + s)/2 - 1\} = 2(3 + s) - 4 = 2s + 2 \Rightarrow t - 4 = 2s + 2 \Rightarrow 2s - t = -6$.

다음으로, R을 N에 대입하면, $v - 2 = (-1/2)(u - 3) \Rightarrow (2 + t)/2 - 2 = (-1/2)\{(3 + s)/2 - 3\}$
$\Rightarrow 2 + t - 4 = t - 2 = -\{(3 + s)/2 - 3\} \Rightarrow 2t - 4 = -(3 + s) + 6 = 3 - s \Rightarrow 2t - 4 = 3 - s \Rightarrow s + 2t = 7$.

그러면 이제, 다음과 같은 연립방정식을 갖는 거죠: $2s - t = -6$ 와 $s + 2t = 7$.

먼저, t 를 소거해보죠.
$2s - t = -6 \Rightarrow 4s - 2t = -12$. 그러면, $(4s - 2t) + (s + 2t) = -12 + 7 \Rightarrow 5s = -5 \Rightarrow s = -1$.
그러면, $s + 2t = 7 \Rightarrow -1 + 2t = 7 \Rightarrow 2t = 8 \Rightarrow t = 4$.

그러니까, Q 는 (-1, 4).

두 번째 방법

주어진 직선을 L 이라 하고 P 를 (3, 2)라 하며 L 에 관하여 P 에 대칭인 점을 Q(s, t)라 하세요.
또한, L 에 수직인 직선을 N 이라 하고 L 이 N 을 만나는 점을 R(u, v)라 하세요.

L 은 (1, 3)을 포함하고 그 기울기가 2 이므로 그 방정식은 다음과 같습니다: $y - 3 = 2(x - 1)$.
N 의 기울기를 n 이라 하세요. 그러면, $2n = -1 \Rightarrow n = -1/2$.
이 N 은 P(3, 2)를 포함하고 그 기울기가 $-1/2$ 이므로 그 방정식은 다음과 같죠: $y - 2 = (-1/2)(x - 3)$.

이제, L 과 N 이 서로 만나는 점 R 을 구해보기로 하죠.

$y - 3 = 2(x - 1) \Rightarrow y = 2x - 2 + 3 = 2x + 1 \Rightarrow y = 2x + 1$.
$y - 2 = (-1/2)(x - 3) \Rightarrow y = -x/2 + 3/2 + 2 \Rightarrow y = -x/2 + 7/2$.

그러므로, $2x + 1 = -x/2 + 7/2 \Rightarrow 4x + 2 = -x + 7 \Rightarrow 5x = 5 \Rightarrow x = 1$.
그러면, $y = 2x + 1 = 3$. 따라서, R 은 (1, 3).

이제, P 와 R 사이의 거리 D 를 구해보기로 하죠. P 는 (3, 2).
$D^2 = (3 - 1)^2 + (2 - 3)^2 = 4 + 1 = 5 \Rightarrow D^2 = 5$.

Q(s, t)와 R(1, 3)사이의 거리가 D 와 같죠:
$D^2 = 5 = (s - 1)^2 + (t - 3)^2$.

Q(s, t)가 N 에 있어야 하는 점이죠:
$N(x, y) = (-1/2)(x - 3) - (y - 2) = 0 \Rightarrow y - 2 = (-1/2)(x - 3) \Rightarrow y = -x/2 + 7/2 \Rightarrow t = -s/2 + 7/2$.

그러면, 다음과 같은 연립방정식을 갖는 것이죠: $5 = (s - 1)^2 + (t - 3)^2$과 $t = -s/2 + 7/2$.

그러면, $5 = (s - 1)^2 + (-s/2 + 7/2 - 3)^2$ 이고 $-s/2 + 7/2 - 3 = -s/2 + 1/2 = (1 - s)/2$.
그러면, $5 = (s - 1)^2 + (1 - s)^2/4 = (s - 1)^2 + (s - 1)^2/4 = 5(s - 1)^2/4$.
$\Rightarrow 5 = 5(s - 1)^2/4 \Rightarrow (s - 1)^2 = 4 \Rightarrow s - 1 = 2$ 또는 $-2 \Rightarrow s = 3$ 또는 -1.
$t = -s/2 + 7/2$ 이죠.
그러면, $s = 3$ 일 때 $t = -3/2 + 7/2 = 2$.
따라서, (s, t) = (3, 2)이며 이는 이미 주어진 점 P 이죠.
그러면 이제, $s = -1$ 일 때 $t = -(-1)/2 + 7/1 = 1/2 + 7/2 = 4$.
그리하여, Q(s, t) = Q(-1, 4).

이제, 다음 예제를 다루어보기로 하죠.

제안과 풀이
8 번 예제에서의 문제

직선이 한 점 **(2, 5)**를 포함하고 그 기울기는 **2** 라 하세요. 그러면, 이 직선까지의 거리가 **5** 인 점 **(s, t)**를 구하세요.

하나의 직선으로부터의 특정한 거리만큼 떨어진 점은 하나뿐이 아니죠. 그럼, 몇 개나 되나요? 무수히 많을 수 있죠. 그렇다고 해서 이 점들이 아무데나 다 있는 건 물론 아니죠. 이 점들은 특정한 직선들을 이루고 있어요. 이제, 주어진 직선을 L 이라고 하세요. 그러면, 그 특정 직선들은 L 로부터 5 만큼 떨어진 것이고 이들은 L 에 평행인 겁니다. 즉, 그 특정 직선들과 L 은 서로 평행한 거죠. 그러면, 그 특정 직선들을 이루는 모든 점 하나 하나는 L 로부터 5 만큼 떨어져있다고 할 수 있죠. 그럼, 이 특정 직선들은 몇 개? 두개죠. L 의 양쪽으로 하나씩 입니다.

이제, 그 두개의 평행선들을 P 와 Q 라고 하세요.
그러면, P 와 Q 를 어떻게 구할 수 있나요?
일단, P 만 구해보기로 하죠. 결국, 이는 Q 에 대하여도 마찬가지거든요.
우선, 이 P 는 L 에 평행하니까 그 기울기는 L 의 것과 같은 거죠.
그러면, 이 P 가 포함하는 하나의 특정한 점이 있어야 하죠.
그 특정 점은 L 에서 5 만큼 떨어져있는 겁니다.
그럼, 이 특정 점은 어떻게 구하나요?
주어진 직선 L 에서의 한 점을 알고 있죠. 이점은 (2, 5)입니다.
그러면, 점 (2, 5)에서 5 만큼 떨어져있는 점을 구해야 하네요.
그런데, 그런 점이 단 하나만 있나요?
아니죠, 그런 게 무수히 많아요. 그 점 (2, 5)를 중심으로 하고 반지름이 5 인 원을 이루는 모든 점들이 그런 거죠.
그 모든 점들 중에서 하나가 바로 그 특정 점이죠.
그럼, 그 특정 점은 그 원에서 어디에 있나요?
그 특정 점은 점 (2, 5)를 포함하고 직선 L 에 수직인 직선에도 있죠.
사실, 그 특정 점은 두 개가 있는 거죠, 하나는 직선 P 에 있는 것이고 나머지 하나는 직선 Q 에 있는 겁니다.
이는 왜냐하면 그 수직선이 그 원의 지름을 포함하기 때문이고 그래서, 그 P 와 Q 가 제각기 그 지름 양끝에 있어야 하기 때문이죠.
이제, 이 수직선을 N 이라 하세요.
그러면, 그 원은 N 을 두개의 점들에서 만나는데 이 점들이 바로 그 특정 점들이죠.
그러면, 그 원의 방정식과 N 의 방정식으로 이루어진 연립방정식을 풀어야 하는 겁니다.
원의 방정식이라니 요?
원의 방정식이란 그 원의 중심에서 특정한 거리만큼 떨어져 있는 모든 점들을 나타내는 건데 그 특정한 거리가 반지름이죠. 원에 대한 자세한 내용은 제 4 권에서 다루고 있어요.
그 원의 중심은 점 (2, 5)이고 반지름은 5 인 거죠.
그런데, 직선과 같은 기하학적 사물을 이루는 모든 점들을 대표하는 하나의 점을 무엇이라고 하죠?
이를 일반점이라고 하죠.
이제, 그 원에서의 일반점이 (x, y)라고 하세요.
그러면, 점 (2, 5)와 (x, y)사이의 거리가 5 라고 하는 수식이 그 원의 방정식인 겁니다.
그러면, 그 원이 그 직선 N 을 만나는 점을 어떻게 구하나요?
그 원의 방정식과 N 의 방정식으로 이루어진 연립방정식을 풀어야 하는 거죠.
그러면, 그 답이 그 두 개의 특정 점들의 좌표들을 말하는 겁니다.
그러면, 그 두개의 평행선들 P 와 Q 를 구할 수 있는 거죠.

그럼 이제, L 에 수직인 직선 N 부터 시작하기로 하죠.
N 의 기울기를 n 이라 하세요. 그러면, $2n = -1$ 이죠 왜냐하면 L 의 기울기가 2 이고 또한, 서로 수직인 직선들의 기울기들의 곱은 –1 이니까.
그러면, $n = -1/2$ 이고 또한, N 도 (2, 5)를 포함하므로 그 방정식은 $y - 5 = (-1/2)(x - 2)$이죠.

다음으로, 그 원의 방정식을 구하기로 하죠.
점 (2, 5)가 중심이고 이 중심과 그 일반점 (x, y) 사이의 거리가 5 즉, 그 반지름이 5 이죠.
그러면, 그 원의 방정식은 다음과 같은 겁니다: $5^2 = (x - 2)^2 + (y - 5)^2$.

그러면 이제, 그 연립방정식은 다음과 같네요:
$5^2 = (x - 2)^2 + (y - 5)^2$과 $y - 5 = (-1/2)(x - 2)$.

먼저, y 를 소거하고 x 를 구하기로 하죠.
그러면 우선, 두 번째 방정식 즉, N 의 방정식을 표준형으로 나타내기로 하죠. 그러면:

$y - 5 = (-1/2)(x - 2) \Rightarrow y = -x/2 + 1 + 5 = -x/2 + 6$.

다음으로, $y = -x/2 + 6$ 를 첫번째 방정식 즉, 그 원의 방정식에 대입하는 겁니다. 그러면:

$5^2 = (x - 2)^2 + (6 - x/2 - 5)^2 = (x - 2)^2 + (1 - x/2)^2$
$= (x - 2)^2 + \{(1/2)(2 - x)\}^2 = (x - 2)^2 + (2 - x)^2/4 = (x - 2)^2 + (x - 2)^2/4 = 5(x - 2)^2/4$.

$\Rightarrow 25 = 5(x - 2)^2/4 \Rightarrow (x - 2)^2 = 20 \Rightarrow x - 2 = 20^{1/2}$ 또는 $-(20^{1/2}) \Rightarrow x = 2 + 20^{1/2}$ 또는 $2 - 20^{1/2}$.

$20^{1/2} = 2 \times 5^{1/2} = \sqrt{20} = 2\sqrt{5}$

그러면, $x = 2(1 + 5^{1/2})$ 또는 $2(1 - 5^{1/2})$.

그러면:
$x = 2(1 + 5^{1/2})$일 때 $t = -x/2 + 6 \Rightarrow t = -(1 + 5^{1/2}) + 6 = 5 - 5^{1/2}$이죠.
$x = 2(1 - 5^{1/2})$일 때 $t = -x/2 + 6 \Rightarrow t = -(1 - 5^{1/2}) + 6 = 5 + 5^{1/2}$입니다.

이제, $s_1 = 2(1 + 5^{1/2})$, $t_1 = 5 - 5^{1/2}$, $s_2 = 2(1 - 5^{1/2})$, $t_2 = 5 + 5^{1/2}$ 라 하세요.

그러면, 그 두개의 특정 점들은 (s_1, t_1)과 (s_2, t_2) 입니다.

이제, 점 (s_1, t_1)이 직선 P에 있다 하고 점 (s_2, t_2)는 직선 Q에 있다 하세요.
이 두개의 직선들 P 와 Q 는 L 에 평행이므로 기울기가 2 입니다. 그러면:
P의 방정식은 다음과 같습니다: $y - t_1 = 2(x - s_1)$.
Q의 방정식은 다음과 같습니다: $y - t_2 = 2(x - s_2)$.

그러면 이제, 두 개의 직선들 P 와 Q 에 있는 모든 점들 각각이 주어진 직선 L 에서 5 만큼 떨어져있는 겁니다.

$s_1 = 2(1 + 5^{1/2}) \approx 6.47$, $t_1 = 5 - 5^{1/2} \approx 2.76$, $s_2 = 2(1 - 5^{1/2}) \approx -2.47$, $t_2 = 5 + 5^{1/2} \approx 7.24$.

이제, 이것들을 모두 그래프에 넣어보기로 하죠.

그림. 8.0

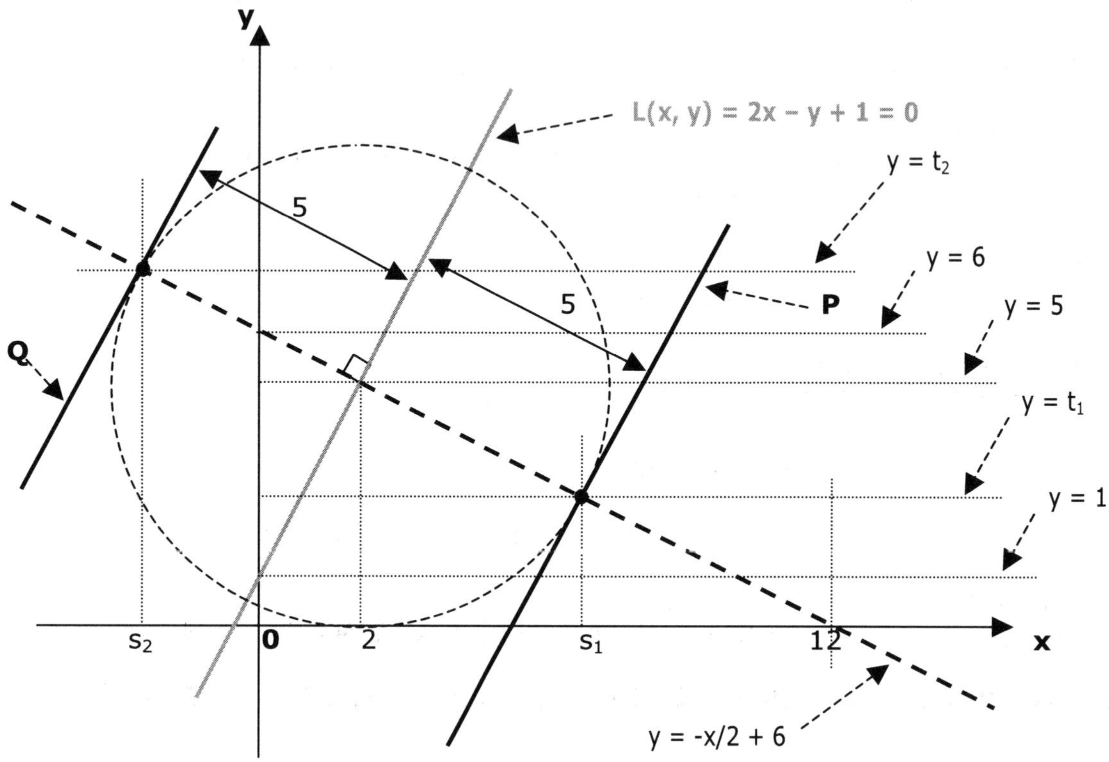

검정 색 직선들(P 와 Q)에 있는 모든 점들 각각이 회색 선(L)에서 5 만큼 떨어져있는 겁니다.

요약:

주어진 직선을 L이라 하고 두개의 점들 R_1과 R_2가 L에 있는 점 (2, 5)에서 5 만큼 떨어져있다 하세요.
또한, 두개의 직선들 P와 Q가 L에 평행이라 하고 P는 R_1을 포함하고 Q는 R_2를 포함한다고 하세요.
그러면, P 와 Q 는 이 L 로부터 제각기 5 만큼 떨어져있는 직선들이며 이 두 직선들에 있는 모든 점들 각각이 L 로부터 5 만큼 떨어져있는 겁니다.

그 두개의 점들 R_1과 R_2는 하나의 직선 N에 있는데 이 N은 직선 L에 수직이며 점 (2, 5)를 포함합니다.
그러면, 이 두 점들은 N 이 점 (2, 5)를 중심으로 하고 반지름이 5 인 원을 만나는 점들입니다.

N 의 기울기를 n 이라 하세요. 그러면, $2n = -1$ 이죠 왜냐하면 L 의 기울기가 2 이고 서로 수직인 직선들의 기울기들의 곱은 –1 이니까.
그러면, $n = -1/2$ 이고 또한, N 도 (2, 5)를 포함하므로 N 의 방정식은 $y - 5 = (-1/2)(x - 2)$.

다음으로, 그 원의 방정식은 다음과 같죠: $5^2 = (x - 2)^2 + (y - 5)^2$.

그러면 이제, 그 두개의 점들 R_1과 R_2를 구하기 위하여 다음과 같은 연립방정식을 풀어야 합니다:
$5^2 = (x - 2)^2 + (y - 5)^2$과 $y - 5 = (-1/2)(x - 2)$.

먼저, y 를 소거하고 x 를 구하기로 하죠.

$y - 5 = (-1/2)(x - 2) \Rightarrow y = -x/2 + 1 + 5 = -x/2 + 6.$

그러면:
$5^2 = (x - 2)^2 + (6 - x/2 - 5)^2 = (x - 2)^2 + (1 - x/2)^2$
$= (x - 2)^2 + \{(1/2)(2 - x)\}^2 = (x - 2)^2 + (2 - x)^2/4 = (x - 2)^2 + (x - 2)^2/4 = 5(x - 2)^2/4.$

$\Rightarrow 25 = 5(x - 2)^2/4 \Rightarrow (x - 2)^2 = 20 \Rightarrow x - 2 = 20^{1/2}$ 또는 $-(20^{1/2}) \Rightarrow x = 2 + 20^{1/2}$ 또는 $2 - 20^{1/2}$.

$20^{1/2} = 2 \times 5^{1/2} = \sqrt{20} = 2\sqrt{5}$

그러므로, $x = 2(1 + 5^{1/2})$ 또는 $2(1 - 5^{1/2})$.

그러면:
$x = 2(1 + 5^{1/2})$일 때 $t = -x/2 + 6 \Rightarrow t = -(1 + 5^{1/2}) + 6 = 5 - 5^{1/2}$.
$x = 2(1 - 5^{1/2})$일 때 $t = -x/2 + 6 \Rightarrow t = -(1 - 5^{1/2}) + 6 = 5 + 5^{1/2}$.

그럼 이제, $s_1 = 2(1 + 5^{1/2})$, $t_1 = 5 - 5^{1/2}$, $s_2 = 2(1 - 5^{1/2})$, $t_2 = 5 + 5^{1/2}$으로 놓으세요.

또한, R_1은 (s_1, t_1)에 있고 R_2는 (s_2, t_2)에 있다고 하세요.

이 두개의 직선들 P 와 Q 는 L 에 평행이므로 기울기가 2 입니다. 그러면:
P의 방정식은 $y - t_1 = 2(x - s_1)$.
Q의 방정식은 $y - t_2 = 2(x - s_2)$.

그러면 이제, 두 개의 직선들 P 와 Q 에 있는 모든 점들 각각이 주어진 직선 L 에서 5 만큼 떨어져있는 겁니다.

제안과 풀이
9 번 예제에서의 문제

점 (-2, 1)에서의 거리와 점 (3, 2)에서의 거리가 동일한 점 (s, t)를 구하세요.

물론, 그 점 (s, t)는 주어진 두 개의 점들간의 중점이라고 할 수 있죠.
그런데, 이 중점만이 답이 될 수 있는 유일한 점인가요?
아니죠, 더 많은 점들이 답이 될 수 있죠.

주어진 두개의 점들을 A 와 B 라 하세요. 그러면, 이 두 점들과 하나의 특정한 점으로 이등변삼각형을 만들면 그 특정 점은 A와 B로부터 동일한 거리만큼 떨어져있는 거죠. 즉, 그 특정 점도 (s, t)가 될 수 있는 겁니다.

그러면, 어떤 점들이 답이 될 수 있나요?
위에서 언급한 그런 이등변삼각형을 얼마나 많이 만들 수 있을까요?
무한히 많죠!

이제, A 와 B 사이의 중점을 C 라 하고 A 와 B 를 포함하는 직선이 L 이라 하세요. 또한, 그 중점 C 를 포함하고 L 에 수직인 직선이 N 이라 하세요.
그러면, 직선 N에 있는 모든 점들 하나하나가 A와 B로부터 동일한 거리만큼 떨어져있는 거죠.
그러니까, 그 N을 이루는 모든 점들 각각이 (s, t)가 될 수 있는 겁니다.
이제, 그 두 직선들 N과 L을 그 두 점들 A와 B 그리고 그 중점 M과 함께 그래프에 나타내보도록 하죠.

그림. 9.0

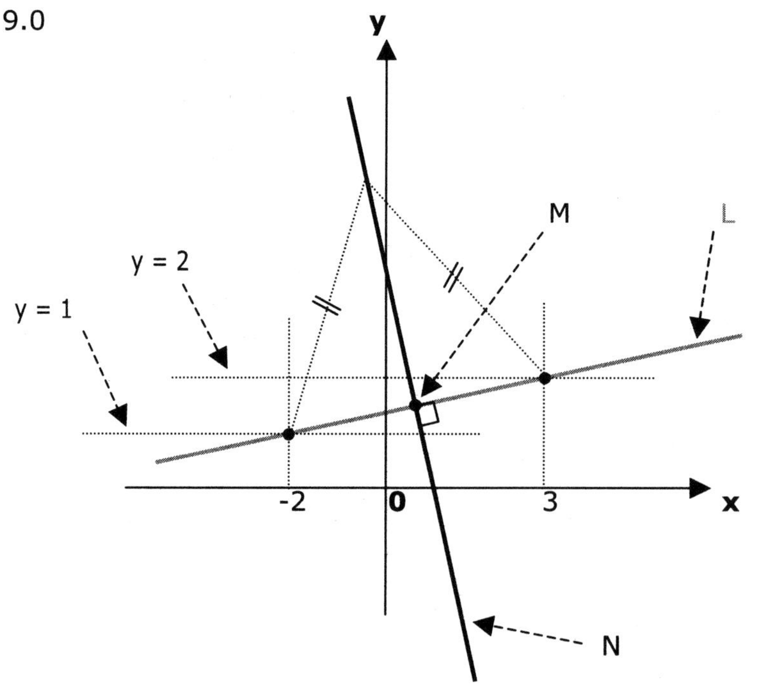

그러면 이제, 그 직선 N을 구하는 겁니다.

그 N이 그 중점 M을 포함한다는 것을 알고있죠.
그 M은 (-2, 1)과 (3, 2)의 중점이므로 {(-2 + 3)/2, (1 + 2)/2} = (1/2, 3/2)에 있죠.
그러면, N의 기울기를 구해야 하네요.
서로 수직인 두 직선들의 기울기들의 곱은 −1이죠.
그러므로, N은 L에 수직이니까 L 기울기를 구하면 N의 기울기를 바로 알 수 있죠.
L의 기울기는 = $\Delta y/\Delta x$ = (2 − 1)/{3 − (-2)} = 1/5 이네요.
그러면, N의 기울기는 −5인 거죠. 그러면 이제, N의 방정식은 y − 3/2 = -5(x − 1/2)이고 이 N에 있는 모든 점 하나하나가 (s, t)가 될 수 있어요.

요약:

(-2, 1)을 A라 하고 (3, 2)를 B라 하며 A와 B 사이의 중점을 M이라 하세요.
그러면, M은 A와 B로부터 동일한 거리만큼 떨어져있으므로 (s, t)가 될 수 있죠.
또한, A와 B를 포함하는 직선을 L이라 하고 M을 포함하며 L에 수직인 직선을 N이라 하세요.
그러면, 그 N에 있는 모든 점 하나하나가 (s, t)가 될 수 있는 겁니다.
이는 왜냐하면, A와 B 그리고 직선 N에 있는 어느 점이든 한 점은 이등변삼각형을 이룰 수 있고 여기에서 선분 AB가 그 밑변이기 때문입니다.

그 N이 그 중점 M을 포함한다는 것을 알고있죠.
그 M은 (-2, 1)과 (3, 2)의 중점이므로 {(-2 + 3)/2, (1 + 2)/2} = (1/2, 3/2)에 있죠.
그러면, 그 N의 기울기를 구해야 하네요.
서로 수직인 두 직선들의 기울기들의 곱은 −1이죠.
이제, N의 기울기를 n이라 하고 L의 기울기를 m이라 하세요.
그러면, $m = \Delta y/\Delta x = (2 - 1)/\{3 - (-2)\} = 1/5$.
따라서, $mn = (1/5)n = -1 \Rightarrow n = -5$.
그러므로, N의 방정식은 y − 3/2 = -5(x − 1/2)이고 이 N에 있는 모든 점 하나하나가 (s, t)가 될 수 있습니다

제안과 풀이
A 번 예제에서의 0 번 문제

다음과 같이 가정하세요:
x-y 평면에 여섯 개의 점들 $P_1(1, 1)$, $P_2(1, 3)$, $P_3(2, 1)$, $P_4(4, 1)$, $P_5(s, t)$, $P_6(u, v)$가 있다.
P_5 와 P_6 는 P_3 와 P_4 의 위쪽에 있고 P_2 의 왼쪽에 있다.
삼각형 $P_5P_1P_2$와 또 하나의 삼각형 $P_5P_3P_4$를 만드는데 이들의 면적의 합이 3 이다.

그러면, 점 P_5 를 명시하세요.

먼저, 아래의 그림에서와 같이 x-y 평면에 주어진 점들을 넣어보기로 하죠.

그림. A.0.0

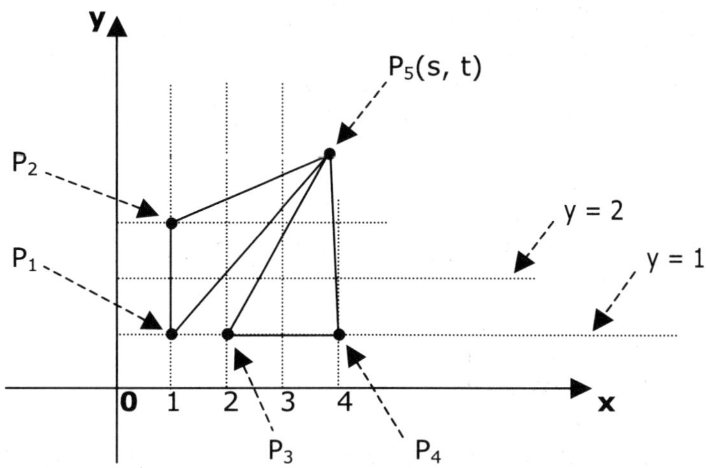

그러면, 점 P_1을 원점으로 이동시키고 나머지 점들도 마찬가지 방법으로 이동시키면 일이 훨씬 쉬워지죠. 다시 말해서, 논의의 대상이 되는 모든 점들을 평행으로 이동시키는 겁니다. 그러면, P_5를 구한 후에 이를 반대로 이동시키면 되는 거죠. 물론, 그렇게 하는 것 대신에 축들을 평행 이동시켜도 되죠. 사실, 축들을 평행 이동시키고 좌표 값들을 이에 맞게 다시 쓰는 것이 더 낫죠. 하지만, 점들의 관점에서 보면, $\Delta x = -1$ 이고 $\Delta y = -1$ 입니다. 그러면 이제, 이동시키고 나면, 다음과 같은 새로운 그래프를 갖는 겁니다:

그림. A.0.1

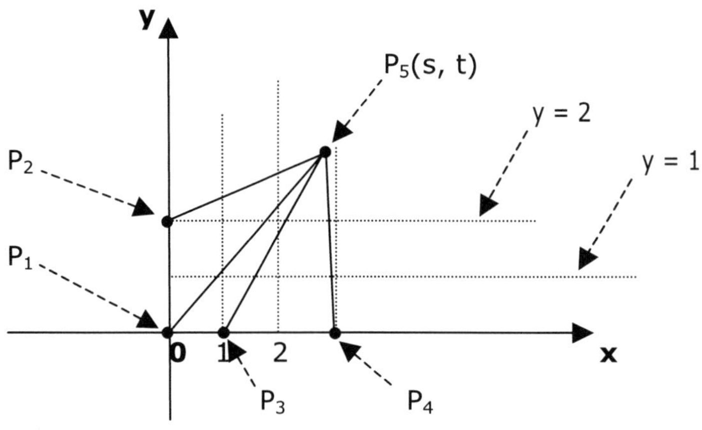

이제, P_5의 좌표들로 그 삼각형들의 면적들의 합을 나타내보기로 하죠.
A를 삼각형 $P_5P_1P_2$라 하고 B를 삼각형 $P_5P_3P_4$하 하세요. 그러면:
A의 면적 = (1/2)(2)s = s 왜냐하면 s를 그 삼각형의 높이로 할 수 있기 때문이죠.
위와 비슷한 이유로, B의 면적 = (1/2)(2)t = t.
그러면, s + t = 3 이죠. 그러면 이제, 그 면적들의 합이 3이 되는 그런 한 쌍의 삼각형들을 만들 수 있는 P_5와 같은 그런 점들이 많이 있다는 것을 알 수 있네요. 그런데, 얼마나 많죠? 사실은, 무한히 많은 거죠! 그 s와 t로 선택할 수 있는 수들은 무제한이죠.
더 정확히 말해서, 하나의 직선에 있는 모든 점들 각각이 그 (s, t)가 될 수 있는 것이고 그 직선은 x + y = 3 입니다. 다시 말해서, P_5(s, t)가 이제는 일반점이고 이 일반점은 x-y평면에 있는 거니까 그 좌표들의 이름들이 x와 y가 되도록 선택하는 겁니다. 하지만, 아직 이것으로 완전히 다된 거는 아니죠 왜냐하면 그 직선을 제대로 된 위치로 역으로 이동시켜야 하기 때문이에요. Δx는 –1 이었고 Δy도 –1 이었죠. 그럼, Δx는 이제 1 이고 Δy도 이제 1 이죠. 그럼 이제, 그 제대로 된 직선의 방정식은 (x – 1) + (y – 1) = 3 입니다.

그런데, 어찌해서 위의 방정식이 그와 같이 역으로 평행 이동된 그 직선의 방정식이 될 수 있냐고요?
그러면 이번에는, 곧이곧대로 그 방정식을 구해보기로 하죠.
점 (1, 2)는 그 방정식 x + y = 3을 만족하죠.
그러면, 이 점(1, 2)를 다음과 같이 평행 이동시키는 겁니다: Δx = 1 이고 Δy = 1.
그러면 평행 이동된 후, 그 점은 (2, 3)이 되죠.
직선 x + y = 3의 기울기는 –1 이죠. 직선을 평행 이동시키면 그 기울기는 그대로 이죠.
그러면, 그 평행 이동된 직선의 기울기도 –1 이네요.
또한, 그 평행 이동된 직선은 점 (2, 3)을 포함하는 거죠.
그러므로, 그 평행 이동된 직선의 방정식은 y – 3 = (-1)(x – 2)입니다.
그러면 이제, 위의 방정식을 다음과 같이 다시 써보기로 하죠:
y – 3 = (-1)(x – 2) ⇒ y – 3 = -x + 2 ⇒ x – 2 + y – 3 = 0 ⇒ x – 1 + y – 1 – 1 – 2 = 0
⇒ x – 1 + y – 1 = 3. 이것을 표준형으로 쓰면, y = -x + 5 이죠.

하지만 아직도 이것이 정답은 아니죠. 이는 왜냐하면 그 일반점이 그 두 점들 P_3와 P_4의 위쪽에 있어야 하고 또한, 점 P_2의 오른쪽에 있기로 되어있기 때문입니다. 그럼, 그 답은 직선이 아니라 선분인 것이죠. 그러면 이제, 그 선분의 방정식은 (x – 1) + (y – 1) = 3 이고 여기에서 1 < x < 4 이며 이 선분에 있는 모든 점들 각각이 (s, t)가 될 수 있는 겁니다.

그림. A.0.2

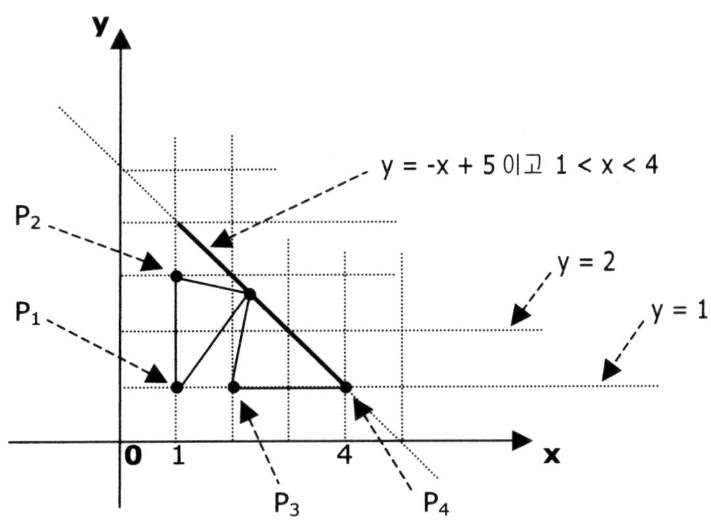

요약:

먼저, 주어진 모든 점들을 다음과 같이 평행으로 이동시키도록 하죠: $\Delta x = -1$ 이고 $\Delta y = -1$.
이동시키고 나면, 다음과 같은 그래프를 갖는 겁니다:

그림. A.0.1

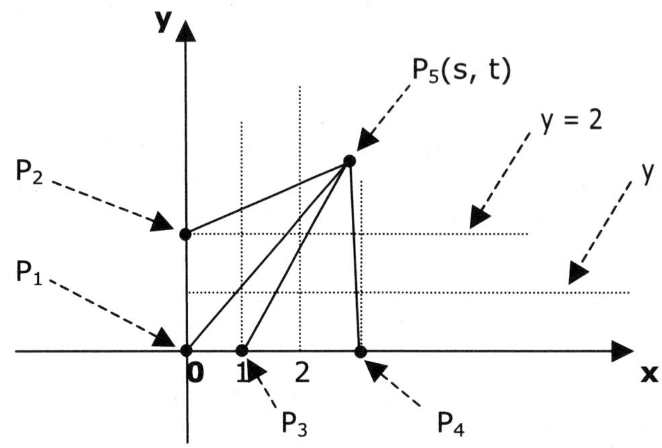

A를 삼각형 $P_5P_1P_2$라 하고 B를 삼각형 $P_5P_3P_4$라 하세요. 그러면:
A 의 면적 = (1/2)(2)s = s 왜냐하면 그 삼각형의 높이로 할 수 있기 때문이죠.
위와 비슷한 이유로, B 의 면적 = (1/2)(2)t = t.
그러면, s + t = 3 이죠.
그럼 일단, 직선에 있는 모든 점들 각각이 (s, t)가 될 수 있는 것이고 그 직선은 x + y = 3 입니다.
이제, 이 직선을 제대로 된 위치로 반대로 이동시켜야 하죠.
처음에 Δx 는 −1 이었고 Δy 도 −1 이었죠. 그럼, Δx 는 이제 1 이고 Δy 도 이제 1 이죠.
점 (1, 2)는 그 방정식 x + y = 3 을 만족하죠.
그러면, 이 점 (1, 2)를 다음과 같이 평행 이동시키는 겁니다: $\Delta x = 1$ 이고 $\Delta y = 1$.
그러면 평행 이동된 후, 그 점은 (2, 3)이 되죠.
직선 x + y = 3 의 기울기는 −1 이죠. 직선을 평행 이동시키면 그 기울기는 그대로 이죠. 그러면, 그 평행 이동된 직선의 기울기도 −1 이네요.
또한, 그 평행 이동된 직선은 (2, 3)을 포함하는 거죠.
그러므로, 그 평행 이동된 직선의 방정식은 y − 3 = (-1)(x − 2)입니다.

하지만 아직도 이것이 정답은 아니죠. 이는 왜냐하면 그 일반점이 그 두 점들 P_3와 P_4의 위쪽에 있어야 하고 또한, 점 P_2의 오른쪽에 있기로 되어있기 때문입니다. 그럼, 이는 직선이 아니라 선분인 것이죠.
그러면 이제, 그 선분의 방정식은 (x − 1) + (y − 1) = 3 이고 여기에서 $1 < x < 4$ 이며 이 선분에 있는 모든 점들 각각이 (s, t)가 될 수 있는 겁니다.

자 이제, 다음 부분(1 번)으로 넘어가보기로 하죠.

제안과 풀이
A번 예제에서의 1번 문제

다음과 같이 가정하세요:
x-y 평면에 여섯 개의 점들 $P_1(1, 1)$, $P_2(1, 3)$, $P_3(2, 1)$, $P_4(4, 1)$, $P_5(s, t)$, $P_6(u, v)$가 있다.
P_5 와 P_6 는 P_3 와 P_4 의 위쪽에 있고 P_2 의 오른쪽에 있다.
삼각형 $P_5P_1P_2$ 와 또 하나의 삼각형 $P_5P_3P_4$ 를 만드는데 이들의 면적의 합이 3 이다.

또한, 다음과 같이 가정하세요:
D_1은 P_6와 P_4 사이의 거리. D_2는P_6와 P_2 사이의 거리.
$D_1^2 - D_2^2 = 5$.

그러면 이제, 점 P_6 를 명시하세요.

그림. A.1.0

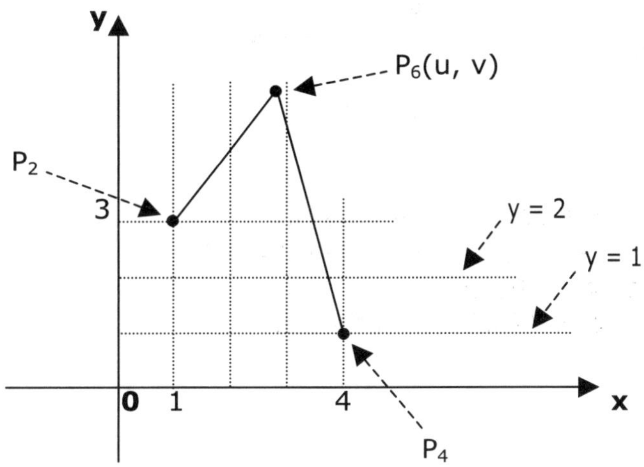

D_1은 $P_6(u, v)$와 $P_4(4, 1)$ 사이의 거리입니다.
그러면, $D_1^2 = (u - 4)^2 + (v - 1)^2$.

D_2는 $P_6(u, v)$와 $P_2(1, 3)$ 사이의 거리이죠.
그러면, $D_2^2 = (u - 1)^2 + (v - 3)^2$.

그러면 이제, $D_1^2 - D_2^2 = 5 \Rightarrow (u - 4)^2 + (v - 1)^2 - (u - 1)^2 - (v - 3)^2 = 5$
$\Rightarrow u^2 - 8u + 16 + v^2 - 2v + 1 - u^2 + 2u - 1 - v^2 + 6v - 9 = 5 \Rightarrow -6u + 4v + 7 = 5 \Rightarrow v = 3u/2 - 1/2$.

이제, 문제에서 제시된 조건들을 만족하는 P_6와 같은 그런 점들이 수없이 많이 있다는 것을 알 수 있네요. 그 좌표들 u와 v로 선택할 수 있는 수들은 무제한이죠. 다시 말해서, $P_6(u, v)$가 이제는 일반점이고 이 일반점은 x-y평면에 있는 거니까 그 좌표들의 이름들이 x와 y가 되도록 선택하는 겁니다. 그러니까, 직선 $y = 3x/2 - 1/2$ 에 있는 각각의 모든 점이 (u, v)가 될 수 있는 거죠.
하지만 아직은 이것이 완전한 답은 아니죠. 이는 왜냐하면 그 일반점은 P_4의 위쪽에 있으며 또한, P_2의 오른쪽에 있어야 하기로 되어있거든요. 그러면, 그 답은 직선이 아니고 방사선인 거죠. 방사선(또한, 반직선이라고도 함)이란 여전히 곧은 선이긴 한데 이는 한 방향으로만 진행하는 반면에 직선은 양방으로 뻗어나가는 겁니다. 예를 들면, 직선은 좌우로 진행하지만 반직선은 왼쪽으로 아니면 오른쪽으로만 늘어나죠.

그러므로, 이 반직선의 방정식은 $y = 3x/2 - 1/2$ 이며 $x > 1$ 이고 이 반직선의 모든 점들 각각이 (u, v)가 될 수 있는 겁니다.

그림. A.1.2

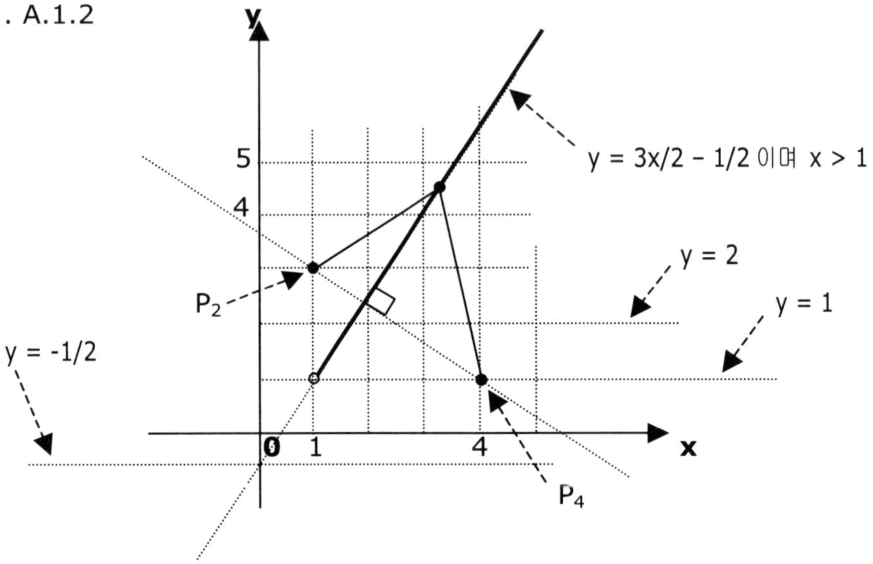

그런데, 이 반직선은 두 점들 P_2와 P_4를 포함하는 직선에 수직입니다.
그러면, 이 반직선과 이 수직선의 기울기의 곱이 −1 인지 확인해보기로 하죠.
이 수직선의 기울기는 $(3 - 1)/(1 - 4) = -2/3$ 이죠.
그러면, 기울기들의 곱이 $(-2/3)(3/2) = -1$ 이니까 서로 수직이네요.

요약:

D_1은 $P_6(u, v)$와 $P_4(4, 1)$ 사이의 거리니까 $D_1^2 = (u - 4)^2 + (v - 1)^2$.
D_2 는 $P_6(u, v)$와 $P_2(1, 3)$ 사이의 거리므로 $D_2^2 = (u - 1)^2 + (v - 3)^2$.

$D_1^2 - D_2^2 = 5 \Rightarrow (u - 4)^2 + (v - 1)^2 - (u - 1)^2 - (v - 3)^2 = 5$
$\Rightarrow u^2 - 8u + 16 + v^2 - 2v + 1 - u^2 + 2u - 1 - v^2 + 6v - 9 = 5 \Rightarrow -6u + 4v + 7 = 5 \Rightarrow v = 3u/2 - 1/2$.

그러면 일단은, 직선 $y = 3x/2 - 1/2$ 의 모든 점 하나하나가 그 (u, v)가 될 수 있는 거죠.
그런데, (u, v)는 점 P_4의 위쪽에 있으며 P_2의 오른쪽에 있어야 하기로 되어있죠.
그러므로, 그런 모든 점들은 반직선을 이루는 것이고 이 반직선의 방정식은 $y = 3x/2 - 1/2$ 이며 $x > 1$ 이고 이 반직선에서의 모든 점들 각각이 (u, v)가 될 수 있는 겁니다.

그럼 이제, 다음 예제를 처리해보기로 하죠.

제안과 풀이
B 번 예제에서의 문제

x-y평면에 있는 두 점들 P_1 과 P_2 사이의 거리가 5 라고 하세요.
또한, 다음과 같이 가정하세요:
D_1은 P_1과 또 하나의 점 P_3 사이의 거리. D_2는 P_2와 P_3 사이의 거리.
$D_1^2 - D_2^2 = 10$.

그러면, P_3를 명시하세요.

먼저, x-y 평면에 논의의 대상이 되는 점들을 놓아보기로 하죠. 또한, 그 점 P_3를 (s, t)라 하세요.

그림. B.0

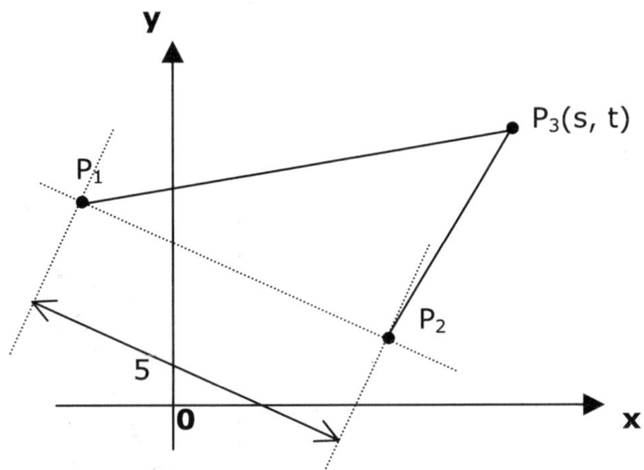

이번 예제에서도 또한, 일이 좀 더 쉽게 되도록 이 두개의 점들 P_1과 P_2를 적당한 위치에 다시 놓아보기로 하죠. 하지만 이번에는 Δx와 Δy를 고려할 필요는 없어요. 이는 왜냐하면 제한조건이 단지 해당 점들 사이의 거리에만 주어져 있기 때문이죠. 다시 말해서, 그 제한조건이 충족되는 한 그 점들이 x-y 평면에서 어디에 있든 상관이 없다는 겁니다. 그러므로, 알아내고자 하는 바는 두 개의 점들 P_1과 P_2의 위치에 대한 P_3의 상대적인 위치인 거죠. 그러면, 아래와 같이 그 점들을 놓아보기로 하죠.

그림. B.1

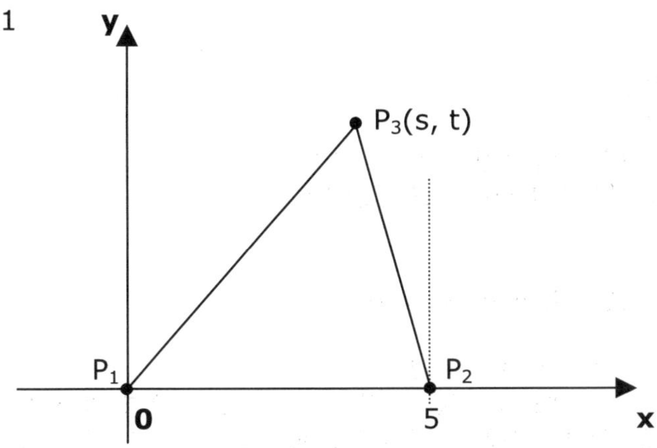

그럼 이제, P_1은 원점에 있고 P_2는 (5, 0)에 있는 겁니다.
그러면, 주어진 사실들을 모아서 그 답을 구하려면 무엇을 할 수 있을지 알아보도록 하죠.

D_1은 P_1과 $P_3(s, t)$ 사이의 거리이죠.
그러니까, $D_1^2 = (s - 0)^2 + (t - 0)^2 = s^2 + t^2$.

D_2는 P_2와 P_3 사이의 거리죠. 그러면, $D_2^2 = (s - 5)^2 + (t - 0)^2 = (s - 5)^2 + t^2$.

또한, $D_1^2 - D_2^2 = 10$ 이죠. 그러면, $s^2 + t^2 - \{(s - 5)^2 + t^2\} = 10$
$\Rightarrow s^2 + t^2 - (s^2 - 10s + 25 + t^2) = 10 \Rightarrow 10s - 25 = 10 \Rightarrow s = 7/2$.

그러면 또 다시, 이전 예제의 경우에서와 마찬가지로 문제에서 제시하는 조건을 만족하는 점들이 무한히 많네요. 이 경우에는 직선 $x = 7/2$ 에 있는 모든 점들 각각이 P_3가 될 수 있는 겁니다. 정확히 말하자면, 이 직선은P_1으로부터 3.5 만큼 그리고 P_2로부터는 1.5 만큼 떨어져있는 점을 포함하고, 또한, 이는P_1과 P_2를 포함하는 직선에 수직이죠.

그림. B.2.

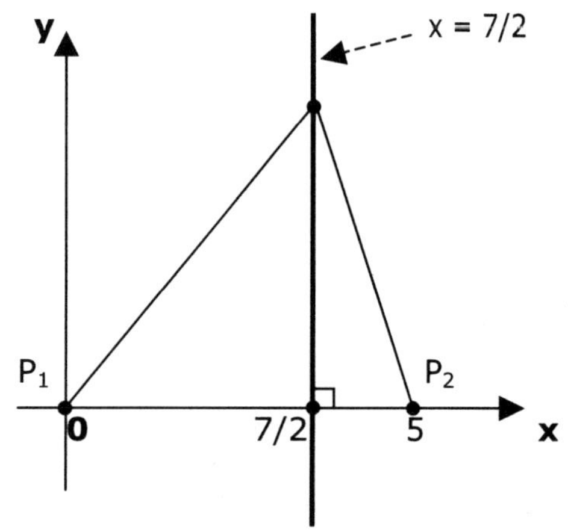

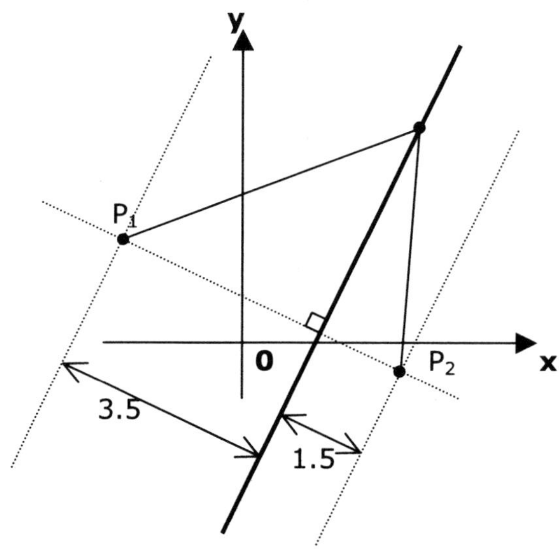

요약:

P_1은 원점에 있고 P_2는 (5, 0)에 있다고 하세요.
또한, P_3은 (s, t)에 있다고 하세요.
D_1은 P_1과 $P_3(s, t)$ 사이의 거리이므로 $D_1^2 = (s - 0)^2 + (t - 0)^2 = s^2 + t^2$.
D_2는 P_2와 P_3 사이의 거리이니까 $D_2^2 = (s - 5)^2 + (t - 0)^2 = (s - 5)^2 + t^2$.
또한, $D_1^2 - D_2^2 = 10$ 이죠.
그러면, $s^2 + t^2 - \{(s - 5)^2 + t^2\} = 10$
$\Rightarrow s^2 + t^2 - (s^2 - 10s + 25 + t^2) = 10 \Rightarrow 10s - 25 = 10 \Rightarrow s = 7/2$.

그러면, 직선 $x = 7/2$ 의 모든 점 하나하나가 P_3가 될 수 있는 겁니다.
정확히 말하자면, 이 P_3는 하나의 직선에 있는 그 어느 점도 될 수 있는데 그 직선은 점 P_1으로부터 3.5 만큼 그리고 점 P_2로부터는 1.5 만큼 떨어져있는 점을 포함하고 또한, 두 점들 P_1과 P_2를 포함하는 직선에 수직인 겁니다.

제안과 풀이
C 번 예제에서의 **0** 번 문제

x-y평면에 4 개의 점들 $P_1(-2, 3)$, $P_2(-2, -1)$, $P_3(2, 2)$, P_4 (-2, 0)가 있는데 세 개의 점들 $P_1P_2P_3$를 연결하는 삼각형 T를 만든다고 하세요. 또한, A는 P_1과 P_2를 잇는 직선이고 또 하나의 점 P_5가 A에 있다고 하세요. P_5를 포함하며 직선 A에 수직인 직선을 그으면 그 수직선은 T의 면적을 이등분합니다. 그러면 이제, P_5의 위치를 구하세요.

먼저, 주어진 점들을 그래프에 표시하고 그 삼각형 T를 만들어보도록 하죠.

그림. C.0.0

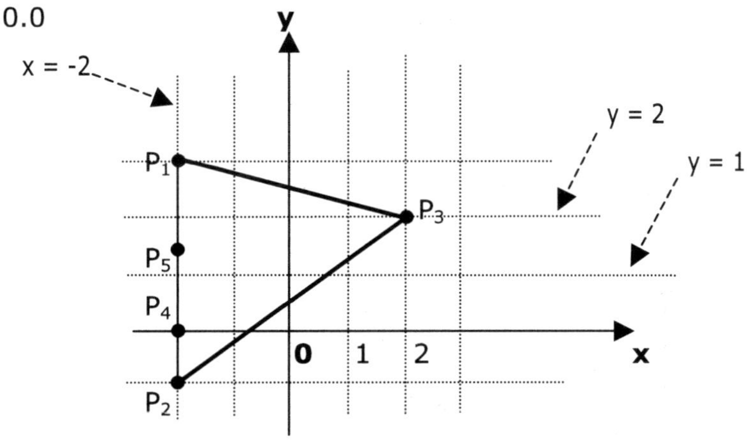

그러면 이제, P_2를 원점으로 옮기고 나머지 점들도 마찬가지 방식으로 이동시키면 문제가 훨씬 용이해진다는 것을 알 수 있죠. 그런데 이 경우, 그렇게 하지 말고 좌표축들을 이동시키는 것이 더 낫겠네요. 아무튼, 점들의 관점에서 보면, $\Delta x = 2$ 이고 $\Delta y = 1$ 입니다. 물론, 나중에 그 답을 제대로 된 위치로 이동시켜야 하는 거죠. 그러면, 이동후의 그래프의 모습은 다음과 같습니다.

그림. C.0.1

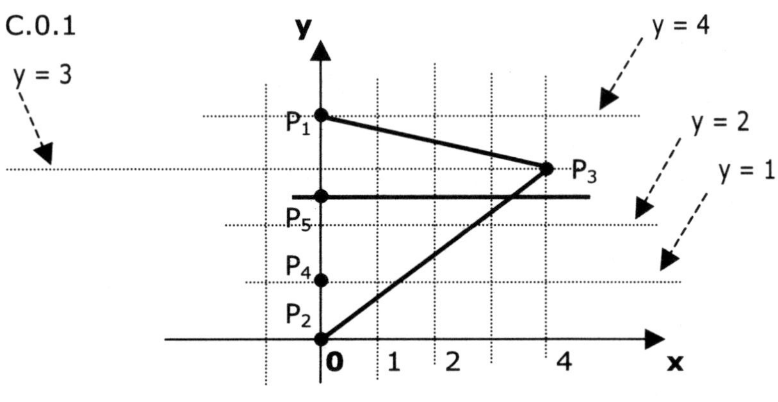

위의 그래프에서 P_5를 포함하고 y축에 수직인 직선이 P_3와 P_2를 연결하는 선분과 만나는 것을 알 수 있죠.
이제, 그 선분을 포함하는 직선을 C라 하고 이 C가 그 수직선을 만나는 점을 P_6라고 하세요.
또한, 그 작은 삼각형을 L이라 하고 P_5는 (0, t)에 있다고 하세요. 그러면, 그 수직선의 방정식은 $y = t$이죠.
그러면, P_6를 구하면 그 수직선 밑에 있는 작은 삼각형의 면적을 t로 나타낼 수 있어요.
그럼 이제, 직선 C 의 방정식부터 시작해보죠. C 의 기울기는 $\Delta y/\Delta x = (3 - 0)/(4 - 0) = 3/4$ 이네요.

따라서, C 의 방정식은 $y = (3/4)x$ 이죠 왜냐하면 이는 원점을 포함하기 때문입니다.
그럼, P_6를 구해보죠. $y = t$이고 $y = 3x/4$ 입니다. 그러면, $t = 3x/4 \Rightarrow x = 4t/3$ 이죠.
그러면 이제, x좌표는 $4t/3$ 이고 y좌표는 t네요. 그러므로, P_6는 $(4t/3, t)$에 있는 겁니다.

이제, 그 삼각형 L 의 밑변이 t 라 하세요. 그러면, 그 높이는 $4t/3$ 이죠.
그러면, L의 면적 $= (t)(4t/3)/2 = 2t^2/3$.
그럼 이제, 삼각형 L 의 면적이 삼각형 T 의 반이 되길 원하는 거죠.
삼각형 T 의 넓이 $= (4)(4)/2 = 8$.

그러면, $2t^2/3 = 8/2 = 4 \Rightarrow t^2 = 6 \Rightarrow t = 6^{1/2} \approx 2.45$ 왜냐하면 P_5가 원점 위쪽에 있기 때문입니다.
그러므로, 이 경우에 P_5는 $(0, \sqrt{6})$에 있죠. 이제, 이 P_5를 제대로 된 위치로 이동시켜야 합니다. Δx는 2 였고 Δy는 1 이었죠. 그러면, Δx는 -2 이고 Δy는 -1 이죠. 그러므로, P_5는 $(-2, \sqrt{6} - 1)$에 있는 겁니다.

요약:

P_2를 원점으로 옮기고 나머지 점들도 마찬가지 방식으로 이동시키면, 그래프의 모습은 아래와 같습니다.

그림. C.0.1

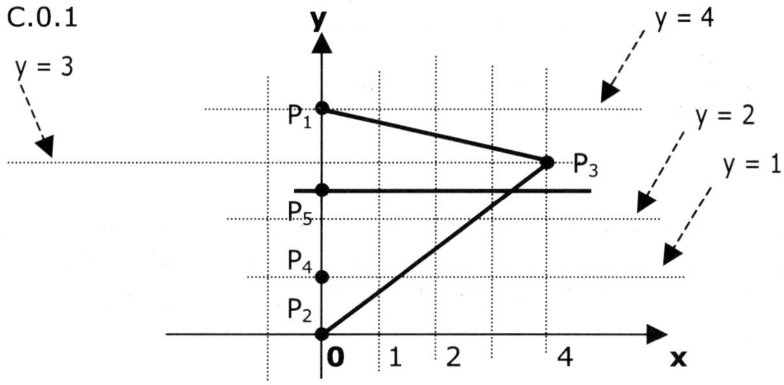

P_3와 P_2를 지나는 선을 C라 하고 P_5를 지나는 수평선이 C를 만나는 점을 P_6라 하세요.
또한, 그 수평선 밑에 있는 작은 삼각형을 L이라 하고 P_5는 $(0, t)$에 있다고 하세요.
그러면, 그 수직선의 방정식은 $y = t$ 이죠.
C 는 원점을 포함하고 그 기울기는 $\Delta y/\Delta x = (3 - 0)/(4 - 0) = 3/4$ 이므로 C 의 방정식은 $y = (3/4)x$.
그럼, P_6를 구해보죠. $y = t$이고 $y = 3x/4$ 입니다. 그러면, $t = 3x/4 \Rightarrow x = 4t/3$ 이죠.
그러면, x좌표는 $4t/3$ 이고 y좌표는 t네요. 그러므로, P_6는 $(4t/3, t)$에 있는 겁니다.

이제, 삼각형 L 의 밑변을 t 라 하세요. 그러면, 그 높이는 $4t/3$ 이죠.
그러면, L의 면적 $= (t)(4t/3)/2 = 2t^2/3$ 이고 이는 삼각형 T의 면적의 반이어야 하는 거죠.
삼각형 T 의 넓이 $= (4)(4)/2 = 8$.

그러면, $2t^2/3 = 8/2 = 4 \Rightarrow t^2 = 6 \Rightarrow t = 6^{1/2} \approx 2.45$ 왜냐하면 P_5가 원점 위쪽에 있기 때문입니다.

그러므로, 이 경우에 P_5는 $(0, \sqrt{6})$에 있죠. 이제, 이 P_5를 제대로 된 위치로 이동시켜야 합니다. Δx는 2 였고 Δy는 1 이었죠. 그러면, Δx는 -2 이고 Δy는 -1 이죠. 그러므로, P_5는 $(-2, \sqrt{6} - 1)$에 있는 겁니다.
이제, 다음 부분(1 번)으로 넘어가보도록 하죠.

제안과 풀이
C번 예제에서의 1번 문제

x-y평면에 4개의 점들 $P_1(-2, 3)$, $P_2(-2, -1)$, $P_3(2, 2)$, $P_4(-2, 0)$가 있는데 세 개의 점들 $P_1P_2P_3$를 연결하는 삼각형 T를 만든다고 하세요. 또한, A는 P_1과 P_2를 잊는 직선이고 또 하나의 점 P_5가 A에 있다고 하세요. P_4를 포함하는 직선 B를 그으면 그 직선 B는 T의 면적을 이등분합니다. 그러면 이제, 직선 B의 방정식을 구하세요.

P_2를 원점으로 옮기고 나머지 점들도 마찬가지 방식으로 이동시키면 이동후의 그래프의 모습은 다음과 같습니다. 점들의 관점에서 보면, $\Delta x = 2$ 이고 $\Delta y = 1$ 이죠.

그림. C.1.0

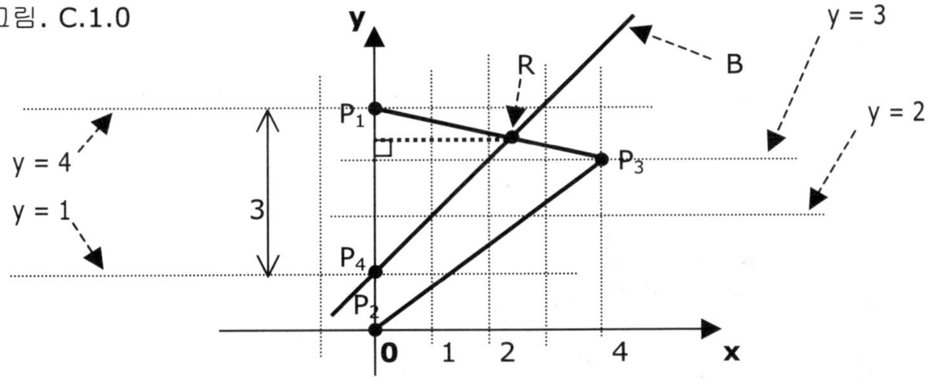

우선, 직선 B가 두 선분들 P_1P_3와 P_2P_3의 어느 것과 만나야 하는지 알아보기로 하죠.
T의 면적은 4 X 4 X 1/2 = 8 이죠.
그런데, 삼각형 $P_2P_3P_4$의 면적은 선분 P_2P_4를 밑변으로 하면 1 X 4 X 1/2 = 2 입니다.
그러면, B는 선분 P_1P_3와 만나야 하는 거죠.
이제, P_1과 P_3를 포함하는 직선을 L이라 하고 L이 B를 만나는 점을 R이라 하세요. 또한, 삼각형 P_1P_4R을 H라 하세요. 그러면, H의 면적이 T의 면적의 반 즉, 4 이어야 하죠.

그래프에서 선분 P_1P_4가 H의 밑변이고 그 길이가 3 이라는 걸 알 수 있죠.
또한, R 의 x 좌표가 H 의 높이라는 것도 알 수 있어요.

그러면 이제, B 의 방정식과 L 의 방정식을 구해서 R 의 x 좌표를 구해야 하네요.
먼저, L 의 방정식부터 시작하죠.
L은 두 점들 P_1과 P_3를 포함하므로 그 기울기는 $\Delta y/\Delta x = (3 - 4)/(4 - 0) = -1/4$ 입니다.
또한, L은 $P_1(0, 4)$를 포함하므로 L의 방정식은 $y - 4 = (-1/4)(x - 0) = -x/4 \Rightarrow y = -x/4 + 4$ 이죠.

다음으로, 직선 B 의 기울기를 m 이라 하세요.
그러면, B는 점 $P_4(0, 1)$를 포함하므로 B의 방정식은 $y - 1 = m(x - 0) \Rightarrow y = mx + 1$ 입니다.

그럼 이제, R 의 x 좌표를 구해보죠.
($y = -x/4 + 4$ 와 $y = mx + 1$) $\Rightarrow -x/4 + 4 = mx + 1 \Rightarrow -x + 16 = 4mx + 4 \Rightarrow (4m + 1)x = 12$
$\Rightarrow x = 12/(4m + 1)$. 그러므로, R 의 x 좌표는 $12/(4m + 1)$이죠.
그러면 이제, H 의 면적 $= (1/2)(3)\{12/(4m + 1)\} = 18/(4m + 1)$입니다.
그러면, $18/(4m + 1) = 4$ 이는 T 의 면적의 반이죠. $18/(4m + 1) = 4 \Rightarrow 18 = 16m + 4 \Rightarrow m = 7/8$.

그러면, B 의 방정식은 $y = 7x/8 + 1$ 그러나 아직, 이는 완전한 답이 아니죠. 이제, 이 직선 B 를 제 위치로 이동시켜야 하죠. Δx 는 2 였고 Δy 는 1 이었죠. 그럼 이제, Δx 는 −2 이고 Δy 는 −1 이죠.

($y = 7x/8 + 1$ 이고 $\Delta x = -2$ 이며 $\Delta y = -1$) $\Rightarrow y - (-1) = 7\{x - (-2)\}/8 + 1$
$\Rightarrow y + 1 = 7(x + 2)/8 + 1 = 7x/8 + 7/4 + 1$. 그러므로, B 의 방정식은 $y = 7x/8 + 7/4$ 입니다.

위의 방법을 모르면 다음과 같이 B를 이동시킬 수도 있죠. B는 $P_4(0, 1)$를 포함하므로 P_4를 먼저 이동시키죠. 그러면, P_4의 원래 위치는 $(0 - 2, 1 - 1) = (-2, 0)$이죠. 평행 이동이므로 B의 기울기는 그대로 7/8 이죠.
기울기가 a 이고 (s, t)를 포함하는 직선의 방정식은 $y - t = a(x - s)$ 이죠.
그러므로, B 의 제대로 된 방정식은 $y - 0 = (7/8)\{x - (-2)\} = (7/8)(x + 2) \Rightarrow y = (7/8)x + 7/4$ 입니다.

요약:

P_2를 원점으로 옮기고 나머지 점들도 마찬가지 방식으로 이동시키도록 하죠. 그러면, $\Delta x = 2$ 이고 $\Delta y = 1$ 이죠.

그림. C.1.0

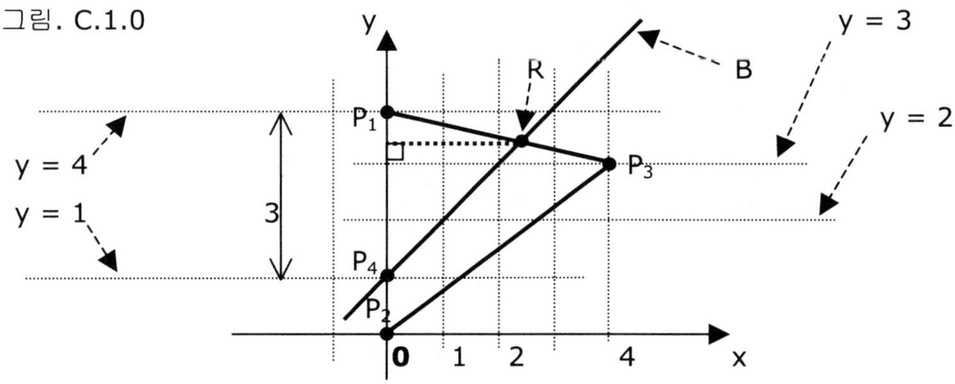

T의 면적은 4 X 4 X 1/2 = 8 이고 삼각형 $P_2P_3P_4$의 면적은 선분 P_2P_4를 밑변으로 하면 1 X 4 X 1/2 = 2 입니다.
그러면, B는 선분 P_1P_3와 만나야 하는 거죠.
L을 P_1과 P_3를 포함하는 직선이라 하고 L이 B를 만나는 점을 R이라 하며 삼각형 P_1P_4R을 H라 하세요.
그래프에서 선분 P_1P_4가 H의 밑변이고 그 길이가 3 이고 또한, R의 x좌표가 H의 높이 라는 걸 알 수 있죠.

L은 두 점들 P_1과 P_3를 포함하므로 그 기울기는 $\Delta y/\Delta x = (3 - 4)/(4 - 0) = -1/4$ 입니다.
또한, L은 $P_1(0, 4)$를 포함하므로 L의 방정식은 $y - 4 = (-1/4)(x - 0) = -x/4 \Rightarrow y = -x/4 + 4$ 이죠.
B 의 기울기를 m 이라 하세요.
그러면, B는 $P_4(0, 1)$를 포함하므로 B의 방정식은 $y - 1 = m(x - 0) \Rightarrow y = mx + 1$ 입니다.
그러면 이제, $-x/4 + 4 = mx + 1 \Rightarrow -x + 16 = 4mx + 4 \Rightarrow (4m + 1)x = 12 \Rightarrow x = 12/(4m + 1)$.
그러므로, R 의 x 좌표는 $12/(4m + 1)$이죠.
그러면, H 의 면적 $= (1/2)(3)\{12/(4m + 1)\} = 18/(4m + 1)$입니다.
그러면, $18/(4m + 1) = 4$ 이는 T 의 면적의 반이죠. $18/(4m + 1) = 4 \Rightarrow 18 = 16m + 4 \Rightarrow m = 7/8$.
그러면, B 의 방정식은 $y = 7x/8 + 1$ 그러나 아직, 이는 완전한 답이 아니죠. 이제, 이 직선 B 를 제 위치로 이동시켜야 하죠. Δx 는 2 였고 Δy 는 1 이었죠. 그럼 이제, Δx 는 −2 이고 Δy 는 −1 이죠.

B는 P_4를 포함하므로 P_4를 먼저 이동시키죠. 그러면, P_4의 원래 위치는 $(0 - 2, 1 - 1) = (-2, 0)$이죠.
평행 이동이므로 B 의 기울기는 그대로 7/8 이죠.
그러므로, B 의 제대로 된 방정식은 $y - 0 = (7/8)\{x - (-2)\} = (7/8)x + 7/4 \Rightarrow y = (7/8)x + 7/4$ 입니다.

제안과 풀이
D 번 예제에서의 문제

다음과 같은 방정식이 있다 하세요: $y = (t/4 + 1)x + t + 4$ 이며 여기에서 t 는 상수.
그러면, $x > -3$ 일 때 $y > 0$ 가 되는 t 의 값을 구하세요.

Math 에서는 사물의 행동 변화 따위 들을 그래프에로 나타낼 때가 많죠. 더 정확히 말하자면, 그런 것을 그래프에서의 커브를 사용해서 표현하는 겁니다. 여러 종류의 커브들 가운데 가장 많이 사용되는 것이 직선이죠. 직선을 가지고 뭘 할 때 주로 직선의 방정식을 사용하죠. 그 방정식에 두 가지 형이 있는데 하나는 표준형이고 나머지 하나는 일반형이죠. 그 중에서도 표준형을 자주 사용합니다. 이번 예제부터 다음 몇 개의 예제들에서는 표준형 방정식으로 표현되는 직선이 그래프에 어떻게 놓이는 가에 따라 무엇을 어떻게 할 것인지에 대하여 논의할 겁니다. 특히, 직선을 함수의 커브로 사용하는 경우에 표준형 방정식이 적합해요. 이는 그런 방정식을 함수로 즉각 전환하기가 쉽기 때문이죠. 사실, 그대로 그저 사용하면 되는 거죠.

그래프에 방정식이나 개념을 나타내면 문제가 어떻게 돌아가는지 파악하는데 도움이 되므로 그 문제에 어떻게 접근할 것인지 더 나은 판단을 할 수 있는 거죠. 문제가 말하는 바를 그저 생각만 하는 것이 아니라 눈으로 보면서 하니까 당연히 그렇죠. 그러니까, 뭔가 추상적인 것을 눈으로 볼 수 있게 하는 능력은 매우 중요한 거죠. 대부분의 경우에는 개략적인 것으로 그 그래프를 시작하는 것이 좋아요. 그렇다 하더라도, 그래프에 나타내려는 커브의 기본적인 구조를 이해하고는 있어야 되요. 예를 들어, 직선은 방향이 일단 정해지면 그 방향을 바꿀 수가 없는 거죠 하지만, 일반적으로 곡선은 계속 방향을 바꾸죠. 이러한 기본을 위반해서는 안 되는 겁니다. 일단 해당 좌표계에서 커브를 대강 나타내기 시작하면 문제가 어떻게 돌아가는지 알아챌 수 있는 경우가 흔합니다. 그러면, 무엇이 어떻게 돌아가는지 충분히 알 수 있을 때까지 계속해서 문제에서 주어지거나 암시하는 정보(사실들)를 추가해 가는 겁니다. 일단 그걸 알아내고 나서, 필요하다면 그 그래프를 더 자세하게 나타내는 거죠. 처음부터 정밀하게 할 것까지야 없다는 거죠.

또한, 그래프에 직선의 방정식을 표현하는데 직선 한 개만 달랑 넣어놓지는 마세요. 몇 개는 넣어봐야 좋습니다. 가능할 수 있다고 생각되면 극단적인 경우도 표현해 보는 것이 좋아요. 이렇게 하다 보면, 우선 그 문제를 제대로 이해할 수 있게 되고 따라서 그 답이 어디쯤에 있어야 하는지 또는 어떻게 만들어져야 하는 지를 알게 되는 경우가 많은 거죠. 그러면 이제, 이 예제에서의 문제를 풀어보기로 하죠.

주어진 방정식은 다음과 같죠: $y = (t/4 + 1)x + t + 4$ 이며 여기에서 t 는 상수.
먼저, 이 것이 무슨 방정식인지 따져봐야 합니다. 물론, 어떤 학생들에게는 척 보고 알 수도 있는 것이긴 하지만요. 그런데, 저자가 첨에 공부할 때는 척 보고 잘 모르겠더라 구요. 아무튼, 확인해서 나쁠 건 없는 거죠.
변수들은 x 와 y 이고 이것들의 지수들은 모두 1 입니다. 그런데, $y = 1/x$ 에서 x 의 지수는 1 이 아니라 -1 입니다.
그러면, $(t/4 + 1)$을 직선의 기울기로 보고 $(t + 4)$를 y 절편으로 볼 수 있죠.
따라서, 이 방정식을 직선의 방정식으로 취급할 수 있네요.

그러면 이제, 이 방정식이 무슨 직선을 말하고 있는 건지 알아보기로 하죠.

이 직선은 다음과 같은 조건을 갖고 있는 거죠: $x > -3$ 일 때 $y > 0$. 이 경우에 -3 을 x 의 임계 값이라고 하기로 하죠. 임계 값이란 특정 상태의 경계를 설명할 때 사용하는 값이라고 할 수 있어요. 즉, 이는 전환점을 나타낸다고 할 수 있죠. 그러면, 이를 경계 값이라고 하는 것이 더 낫지 않을까요? Math 에서 '경계 값'이란 말은 보통 어떤 다른 목적으로 사용 되요.

이제, 바로 위에서 언급된 그 조건이 무엇을 의미하는지 알아보기로 하죠.

먼저, 여러분이 좋아하는 x-y 좌표계에 직선들을 집어 넣으면서 어떤 일이 일어나는지 보기로 하죠.
즉, 이 문제가 말하는 바를 가시화 시키는 겁니다.

그림. D.0

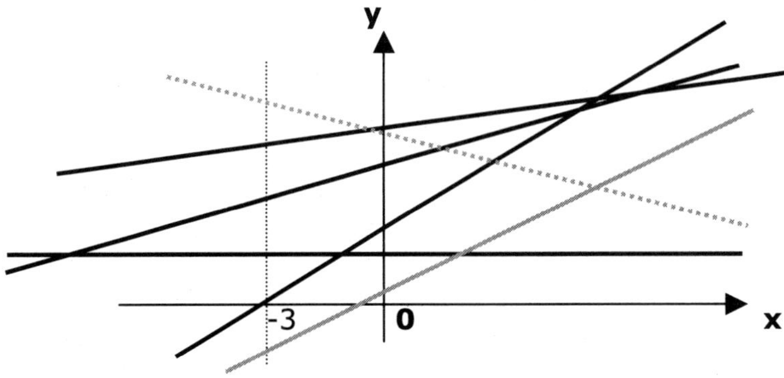

그래프를 작성할 때에는 위에서 언급한 임계 값과 같은 결정적 사항을 먼저 나타내는 것이 좋죠. 그리고 나서, 몇 개의 직선들을 집어넣다 보면 적합한 직선이 문제에서 언급(또는 암시)되어있는 구체적인 사항들(또는 조건들)에 부합하기 위해서 어떻게 놓여져야 하는가를 알아챌 수 있을 때가 많아요. 즉, 문제가 요구하는 바를 충족시키려면 그 직선이 어떻게 표현되어야만 한다는 것을 알아내는 경우가 흔하다는 거죠. 그러면 이제, x 가 –3 보다 클 때 y 가 양의 값을 갖기를 원하면 그 직선은 위의 그래프에서의 검정 직선들 가운데 하나이어야 하는 거죠. 수평선을 무시해서는 안됩니다. 하지만, 회색 점선은 그래프에서 작성된 만큼만으로는 문제에서의 조건에 부합하지만 결국 x 축 밑으로 내려가죠. 그러니까 이 회색 점선은 될 수 없는 거죠. 자 이제, 뭘 알 수 있나요?

답은 그 기울기에 달린 거죠. 우선, 그 직선의 기울기가 음이 될 수 없다는 걸 알 수 있어요. 즉, 그 기울기는 0 보다 크거나 같아야 할 필요가 있다는 겁니다. 다시 말해서, 다음 같은 두 가지 경우들을 고려해야 한다는 거죠:
"그 기울기가 양이다." 와 "그 기울기가 0 이다."

이제, 더 구체적으로 표현해보도록 하죠. 함수 표현을 사용하면 변수들간의 값들을 비교하거나 나타내기가 용이합니다. 그러면, 주어진 방정식을 함수로 나타내보도록 하죠:

이제, $y = f(x) = (t/4 + 1)x + t + 4$ 여기에서 t 는 상수라고 하세요. 물론, 이 x 와 t 는 실수들이죠.
그러면, t 의 값에 따라 f 가 결정되는 겁니다.
그러면 이제, "$x > -3$ 일 때 $f(x) > 0$."라는 걸 만족시킬 수 있는 f 의 커브는 위의 그래프에서 회색 선들을 제외한 직선들 중의 하나인 겁니다. 그 직선의 기울기는 $(t/4 + 1)$이죠.

한편, 직선에 관한 기본 원리에 따라야 하죠. 기본 원리라니 요?
일단 정해지면 직선의 기울기(또는 방향)는 변하지 않는다는 것이 그 원리죠. 그 기울기는 일정한 겁니다.
직선은 계속 같은 방식으로 올라가기만 하거나 내려가기만 하거나 또는 수평이거나 수직이기만 한 거죠.
그 기울기가 양이면 이는 올라가고 음이면 내려가죠. 그 기울기가 0 이면 이는 수평인 겁니다.
그러면, 수직인 직선은 어쩌고요? Math 에서는 수직선에는 기울기가 없다고 합니다. 그러니까, 기울기가 없다는 것이 기울기가 0 이라는 것은 아니라는 것을 유의해야 하죠.

이제, $y = ax + b$ 에서 x 의 특정 값에 대하여 y 가 양이고 그 이후 x 가 커짐에 따라 y 가 계속 증가한다고 하세요.
그러면, x 의 그 특정 값 이상에 대하여 y 는 언제든지 양인 겁니다.
무엇이 y 를 계속 증가하게 하는 거죠? 그것은 그 기울기 a 이고 이는 양이죠.
반대로, $y = ax + b$ 에서 x 의 특정 값에 대하여 y 가 음이고 그 이후 x 가 커짐에 따라 y 가 계속 감소한다고 하세요. 그러면, x 의 그 특정 값 이상에 대하여 y 는 언제든지 음인 것이고 y 가 계속 감소하게 하는 것이 바로 그 기울기 a 이고 이는 음인 것이죠.

그러면 이제, "$x > -3$ 일 때 $f(x) > 0$."라는 말은 다음과 같은 말을 의미합니다: "그 기울기가 양이고 또한 $f(-3)$이 0 보다 크거나 같다." 또는 "그 기울기가 0 이고 또한 $f(-3)$이 양이다."
즉, 이는 다음과 같은 말을 의미하는 거죠: "$t/4 + 1 > 0$ 이고 $f(-3) \geq 0$." 또는 "$t/4 + 1 = 0$ 이고 $f(-3) > 0$."

그런데, "$f(-3) \geq 0$."는 "$f(-3) > 0$."을 포함하는 것이지만 "$t/4 + 1 > 0$"와 "$t/4 + 1 = 0$."은 서로 별개의 이야기입니다. 그러니까, 먼저 $(t/4 + 1)$이 0 이 될 수 있는지 즉, 그 직선의 기울기가 0 이 될 수 있는지 알아 보아야 하죠.

그런데, y 절편 $t + 4$ 는 어쩌고요? 이것도 양이 되어야 하는 거 아닌가요?
물론 그렇죠. 위의 조건이 만족되면 $t + 4$ 는 양의 값을 갖죠.
사실, 이는 그 기울기가 0 인 경우에 포함됩니다. 왜 그런지는 곧 알게 되요.

그러면 이제, 그 t 를 명시해보도록 하죠. 먼저, 그 기울기가 0 이 될 수 있는지 알아보아야 하죠.
그러면, 그 기울기 즉, $t/4 + 1$ 이 0 일 때 $f(-3)$이 양이이어야 되죠.

한편, $t/4 + 1 = 0 \Rightarrow f(x) = (t/4 + 1)x + t + 4 = t + 4 \Rightarrow$ 모든 x 에 대하여 $f(x) = t + 4$.
이 $(t + 4)$가 바로 그 y 절편이죠. 또한, t 가 상수이니까 $(t + 4)$도 상수이죠. 그래서, 이 $f(x) = t + 4$ 와 같은 함수를 상수 함수라고 해요. 상수 함수란 모든 입력 값에 대하여 동일한 값을 갖는 겁니다.

그러면 이제, $t/4 + 1 = 0 \Rightarrow t = -4$. 또한, $t/4 + 1 = 0 \Rightarrow f(x) = t + 4 = -4 + 4 = 0$ 왜냐하면 $t = -4$ 이니까.
그러면, $f(x) = 0$ 이고 이는 이 f 가 모든 x 에 대하여 0 이라는 거죠.
따라서, $f(-3) = 0$ 이는 양이 아니죠 $\Rightarrow f(-3)$이 양이 아닙니다. 그러므로, 그 기울기는 0 이 될 수 없는 겁니다.

그러니까, 그 기울기 > 0 즉, $t/4 + 1 > 0$ 이어야 하고 또한, $f(-3) \geq 0$ 이어야 하는 겁니다.
먼저, $t/4 + 1 > 0 \Rightarrow t > -4$.
또한, $f(-3) = (t/4 + 1)(-3) + t + 4 = -3t/4 - 3 + t + 4 = t/4 + 1 \geq 0 \Rightarrow t \geq -4$.
그럼 이제, 이 두개의 조건들 "그 기울기 > 0 와 $f(-3) \geq 0$."를 동시에 고려해야 하죠.
그러면, $(t \geq -4$ 이고 $t > -4) \Rightarrow t > -4$.
그리하여, t 의 값으로 -4 보다 큰 값을 선택하면 $x > -3$ 일 때 $y > 0$ 가 되는 겁니다.
그런데, 답안을 작성할 때에는 그 기울기가 음인 경우도 설명해야 합니다.

요약:

두 변수들 x 와 y 의 지수들은 제각기 1 이고 t 가 상수이므로 이는 직선의 방정식이 될 수 있습니다.
그러면, $(t/4 + 1)$이 그 기울기입니다.
$y = f(x) = (t/4 + 1)x + t + 4$ 여기에서 t 는 상수라고 하세요.
먼저, 그 기울기가 음인 경우에는 $f(-3) > 0$ 이더라도 결국에는 그 직선이 x 축 밑으로 내려가기 때문에 이 경우를 만족하는 t 값은 없습니다. 그리하여, 다음 두 가지 경우들이 있을 수 있죠.

첫째 경우: $t/4 + 1 = 0$ 이고 $f(-3) > 0$. 먼저, $t/4 + 1 = 0 \Rightarrow t = -4$ 이고 $f(x) = t + 4$.
그러면, $t = -4 \Rightarrow t + 4 = -4 + 4 = 0$ 이는 양이 아닙니다 $\Rightarrow t/4 + 1 \neq 0$.

둘째 경우: $t/4 + 1 > 0$ 이고 $f(-3) \geq 0$. 먼저, $t/4 + 1 > 0 \Rightarrow t > -4$.
또한, $f(-3) = (t/4 + 1)(-3) + t + 4 = -3t/4 - 3 + t + 4 = t/4 + 1 \geq 0 \Rightarrow t \geq -4$.
그러면, $(t \geq -4$ 이고 $t > -4) \Rightarrow t > -4$. 그러므로, $t > -4$.

이제, 다음 예제에서 비슷한 것을 해보기로 하죠.

제안과 풀이
E번 예제에서의 문제

다음과 같은 직선의 방정식이 있다 하세요: $y = (t/4 + 1)x + t + 5$ 이며 여기에서 t는 상수.
그러면, $x > -3$ 일 때 $y > 0$ 가 되는 t의 값을 구하세요.

그림. E.0

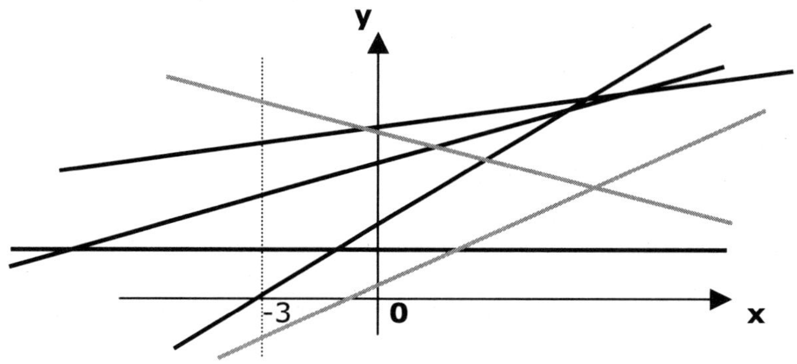

$y = f(x) = (t/4 + 1)x + t + 5$ 여기에서 t는 상수라고 하세요.
그러면, $x > -3$ 일 때 $f(x) > 0$ 이어야 합니다. 그 기울기는 $(t/4 + 1)$이죠.

직선은 그 기울기에 달려있죠. 직선은 곧 기울기이고 기울기는 곧 직선이죠.

그러면 이제, "$x > -3$ 일 때 $f(x) > 0$." 라는 말은 $x = -3$ 이후에는 f의 값이 언제나 양이어야 한다는 거죠.
즉, x의 값이 그 어떤 값이든 –3보다 큰 값에 대하여는 f의 값이 절대적으로 양이어야 하는 겁니다.

지금, f 의 커브는 직선입니다. 직선은 올라가면 항상 올라가기만 하고 내려가면 계속 내려가기만 하죠 또는, 수평이면 줄곧 수평으로만 진행하는 거죠.
그러면, x 의 값이 증가함에 따라 f(x)의 값은 계속 증가하거나 아니면 계속 감소하죠; 그것도 아니면, 이는 일정하겠네요.

그러면, -3보다 큰 모든 x에 대한 f(x)의 값이 양이어야 하면 $x = -3$ 일 때의 f의 값은 무엇이어야 하고 그 이후에 즉, -3보다 큰 x값들에 대하여 f의 값은 어때야 하나요?
즉, $x > -3$ 이고 $f(x) > 0$ 이려면 f(-3)의 값은 무엇이어야 하고 또한, $x > -3$ 일 때 f(x)의 값은 어때야 하죠?

먼저, f(-3)은 0이거나 또는 0보다 커야 하죠. 즉, $f(-3) \geq 0$ 이어야 하죠.
다음으로, $x > -3$ 일 때 f(x) 의 값이 계속 증가해야 하거나 또는, $f(x) = c > 0$ 이고 여기에서 c는 상수인 거죠.
다시 말해서, f의 기울기가 양이거나 또는 0이어야 하는 거죠. 즉, 기울기 ≥ 0 이어야 하는 겁니다.

그러면 이제, f(-3)의 값과 기울기를 함께 고려해보기로 하죠.

먼저, 기울기 = 0이면 $f(-3) > 0$ 이어야 하죠.
다음으로, 기울기 > 0이면 $f(-3) \geq 0$ 이어야 하죠.

한편, 기울기 < 0이면 $f(-3) > 0$ 이더라도 결국에는 그 직선이 x축 밑으로 내려가기 때문에 $f < 0$ 가 될 수 있죠.

그러면 이제, 그 t를 명시해보도록 하죠.

다음 두 가지 경우들이 있을 수 있죠:
기울기 = 0 이면 $f(-3) > 0$. 기울기 > 0 이면 $f(-3) \ge 0$.

먼저, 기울기가 0 이 될 수 있는지 알아보기로 하죠.
그러면, 기울기가 0 즉, $t/4 + 1 = 0$ 일 때 $f(-3)$이 양이어야 하죠.

$t/4 + 1 = 0 \Rightarrow f(x) = (t/4 + 1)x + t + 5 = t + 5$
$\Rightarrow$ 모든 x 에 대하여 $f(x) = t + 5$ 인데 이는 f 가 상수함수라는 걸 말하죠.
그러면, $t/4 + 1 = 0 \Rightarrow t = -4 \Rightarrow t + 5 = -4 + 5 = 1 > 0 \Rightarrow f(-3) > 0$.
따라서, 기울기는 0 이 될 수 있는 겁니다. 그러므로, $t = -4$ 일 수 있죠.

다음으로, 기울기가 양 즉, $t/4 + 1 > 0$ 일 때 $f(-3) \ge 0$ 이어야 하는 겁니다.

먼저, $t/4 + 1 > 0 \Rightarrow t > -4$.
또한, $f(-3) = (t/4 + 1)(-3) + t + 5 = -3t/4 - 3 + t + 5 = t/4 + 2 \ge 0 \Rightarrow t \ge -8$.
그럼 이제, 그 두개의 조건들, "기울기 > 0 와 $f(-3) \ge 0$" 를 동시에 고려해야 하죠.
그러면, ($t > -4$ 이고 $t \ge -8$) $\Rightarrow t > -4$.
그런데 또한, 기울기가 0 일 수 있으므로 t 가 -4 일 수 있죠.
그러므로, $t \ge -4$.

그러면 이제, t 에 대하여 -4 보다 크거나 같은 값을 선택하면 $x > -3$ 일 때 $y > 0$ 가 되는 겁니다.

요약:

$y = f(x) = (t/4 + 1)x + t + 5$ 여기에서 t 는 상수라고 하세요.

먼저, 기울기 < 0 이면 $f(-3) > 0$ 이더라도 결국에는 그 직선이 x 축 밑으로 내려가기 때문에 $f < 0$ 가 될 수 있죠.

그러면, 다음 두 가지 경우들이 있을 수 있죠:
기울기 = 0 이면 $f(-3) > 0$.
기울기 > 0 이면 $f(-3) \ge 0$.

먼저, 기울기 = 0 이 될 수 있는지 알아보기로 하죠.
$t/4 + 1 = 0 \Rightarrow t = -4$ 이고 또한, $f(x) = (t/4 + 1)x + t + 5 = t + 5$.
그러면, $t = -4 \Rightarrow t + 5 = -4 + 5 = 1 > 0 \Rightarrow f(-3) > 0$.
따라서, 기울기는 0 이 될 수 있는 겁니다. 그러므로, t 는 -4 일 수 있죠.

다음으로, 기울기 $t/4 + 1 > 0$ 이고 또한, $f(-3) \ge 0$ 이어야 합니다.
먼저, $t/4 + 1 > 0 \Rightarrow t > -4$.
또한, $f(-3) = (t/4 + 1)(-3) + t + 5 = -3t/4 - 3 + t + 5 = t/4 + 2 \ge 0 \Rightarrow t \ge -8$.
그러면, ($t > -4$ 이고 $t \ge -8$) $\Rightarrow t > -4$. 또한, t 가 -4 일 수도 있죠.

그러므로, $t \ge -4$.

제안과 풀이
F번 예제에서의 문제

다음과 같은 직선의 방정식이 있다 하세요: **$y = (t/4 + 1)x + t + 4$** 이며 여기에서 **t**는 상수.
그러면, **$x > -3$** 일 때 **$y \geq 0$** 가 되는 **t**의 값을 구하세요.

그림. F.0

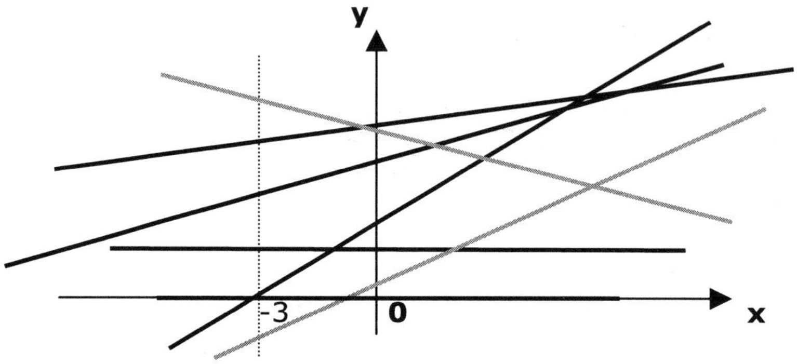

먼저, $y = f(x) = (t/4 + 1)x + t + 4$ 여기에서 t는 상수라고 하세요.
이번에는, $x > -3$ 일 때 $f(x) \geq 0$ 입니다. 또한, 그 기울기는 $t/4 + 1$ 이죠.

직선하면 뭐죠? 기울기죠! 기울기하면 뭐? 직선!

그러면 이제, "$x > -3$ 일 때 $f(x) \geq 0$." 는 $x = -3$ 이후에는 f의 값이 언제나 0이거나 양이어야 한다는 거죠.
즉, x의 값이 그 어떤 값이든 –3보다 큰 값에 대하여는 f의 값이 절대적으로 0이거나 양인 겁니다.

지금, f의 커브는 직선입니다.
그러면, x의 값이 증가함에 따라 f(x)의 값은 내내 증가하거나 아니면 계속 감소하죠; 아니면, 줄곧 일정한 거죠.

그러면, $x > -3$ 이고 $f(x) \geq 0$ 이려면 f(-3)의 값은 무엇이어야 하고 $x > -3$ 일 때 f(x)의 값은 어때야 하죠?

우선, f(-3)은 0보다 크거나 같아야 하죠. 즉, $f(-3) \geq 0$ 이어야 하죠.
또한, -3보다 큰 x값들에 대한 f(x)의 값은 f(-3)보다 크면서 계속 증가해야 하죠; 아니면, $f(x) = c = f(-3) \geq 0$ 이고 여기에서 c는 상수이어야 하죠. 즉, f의 기울기가 양이거나 아니면, 그 기울기가 0이고 $f(x) = c \geq 0$ 이며 여기에서 c는 상수인 거죠.

그러면 이제, f(-3)의 값과 기울기를 함께 고려해보기로 하죠.

먼저, 기울기 = 0이면 $f(-3) \geq 0$ 이어야 하죠.
다음으로, 기울기 > 0이면 $f(-3) \geq 0$ 이어야 하죠.

그러면 결국, 기울기 ≥ 0 이고 또한, $f(-3) \geq 0$ 이어야 하는 거네요.

한편, 기울기 < 0이면 $f(-3) > 0$ 이더라도 결국에는 그 직선이 x축 밑으로 내려가기 때문에 $f < 0$ 가 될 수 있죠.

그러면 이제, 그 t를 명시해보도록 하죠.

기울기 ≥ 0 이고 $f(-3) \geq 0$ 이어야 하죠.

그러면, $t/4 + 1 \geq 0 \Rightarrow t \geq -4$.
또한, $f(-3) = (t/4 + 1)(-3) + t + 4 = -3t/4 - 3 + t + 4 = t/4 + 1 \geq 0 \Rightarrow t \geq -4$.
그럼 이제, 이 두개의 조건들, "기울기 ≥ 0 와 $f(-3) \geq 0$" 를 동시에 고려해야 하죠.
그러면, ($t \geq -4$ 이고 $t \geq -4$) $\Rightarrow t \geq -4$.
그러므로, $t \geq -4$.

그럼 이제, t에 대하여 -4보다 크거나 같은 값을 선택하면 $x > -3$ 일 때 $y \geq 0$가 되는 겁니다.

그런데, 기울기가 0인 경우를 한번 보기로 하죠.

$t/4 + 1 = 0 \Rightarrow t = -4 \Rightarrow t + 4 = -4 + 4 = 0$.
또한, $t/4 + 1 = 0 \Rightarrow f(x) = (t/4 + 1)x + t + 4 = t + 4$.
따라서, $f(x) = 0$.

그러니까, 그 기울기가 0이면 그 직선이 x축 자체이어야 하는 거네요.
함수 f의 커브가 x축 자체이면 이 f를 0함수라고 하기도 해요.

요약:

$y = f(x) = (t/4 + 1)x + t + 4$ 여기에서 t는 상수라고 하세요.

먼저, 기울기 < 0 이면 $f(-3) > 0$ 이더라도 결국에는 그 직선이 x축 밑으로 내려가기 때문에 $f < 0$가 될 수 있죠.

그러면, 다음 두 가지 경우들이 있을 수 있죠:
기울기 $= 0$ 이면 $f(-3) \geq 0$.
기울기 > 0 이면 $f(-3) \geq 0$.

그러면 결국, 기울기 ≥ 0 이고 $f(-3) \geq 0$ 이어야 하죠.
먼저, $t/4 + 1 \geq 0 \Rightarrow t \geq -4$.
또한, $f(-3) = (t/4 + 1)(-3) + t + 4 = -3t/4 - 3 + t + 4 = t/4 + 1 \geq 0 \Rightarrow t \geq -4$.
그러면, ($t \geq -4$ 이고 $t \geq -4$) $\Rightarrow t \geq -4$.

그러므로, $t \geq -4$.

제안과 풀이
G 번 예제에서의 문제

다음과 같은 직선의 방정식이 있다 하세요: **$y = (t/4 + 1)x + t + 5$** 이며 여기에서 **t**는 상수.
그러면, **$x > -3$** 일 때 **$y \geq 0$** 가 되는 **t**의 값을 구하세요.

그림. G.0

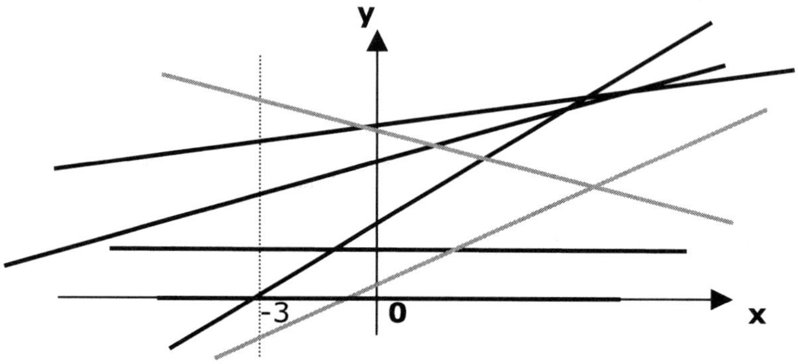

먼저, $y = f(x) = (t/4 + 1)x + t + 5$ 여기에서 t는 상수라고 하세요.
이번에도, $x > -3$ 일 때 $f(x) \geq 0$ 입니다. 그 기울기는 $t/4 + 1$ 이죠.
그러면 이제, "$x > -3$ 일 때 $f(x) \geq 0$." 는 $x = -3$ 이후에는 f가 언제나 0 보다 크거나 같다는 거죠.

그러면, $x > -3$ 이고 $f(x) \geq 0$ 이려면 $f(-3)$의 값은 무엇이어야 하고 $x > -3$ 일 때 $f(x)$의 값은 어때야 하죠?
우선, $f(-3) \geq 0$ 이어야 하죠.
또한, f의 기울기가 양이거나 아니면, 그 기울기가 0 이고 $f(x) = c \geq 0$ 이며 여기에서 c는 상수인 거죠.

그러면 이제, $f(-3)$의 값과 기울기를 함께 고려해보기로 하죠.
먼저, 기울기 = 0 이면 $f(-3) \geq 0$ 이어야 하죠. 다음으로, 기울기 > 0 이면 $f(-3) \geq 0$ 이어야 하죠.
그러면 결국, 기울기 ≥ 0 이고 $f(-3) \geq 0$ 이어야 하는 거네요.
한편, 기울기 < 0 이면 $f(-3) > 0$ 이더라도 결국에는 그 직선이 x축 밑으로 내려가기 때문에 $f < 0$ 가 될 수 있죠.

그러면 이제, 그 t를 명시해보도록 하죠. 기울기 ≥ 0 이고 $f(-3) \geq 0$ 이어야 하죠.
그러면, $t/4 + 1 \geq 0 \Rightarrow t \geq -4$.
또한, $f(-3) = (t/4 + 1)(-3) + t + 5 = -3t/4 - 3 + t + 5 = t/4 + 2 \geq 0 \Rightarrow t \geq -8$.
그럼 이제, 이 두개의 조건들, "기울기 ≥ 0 와 $f(-3) \geq 0$." 를 동시에 고려해야 하죠.
그러면, $t \geq -4$ 이고 $t \geq -8 \Rightarrow t \geq -4$. 그러므로, $t \geq -4$.

요약:

$y = f(x) = (t/4 + 1)x + t + 5$ 여기에서 t는 상수라고 하세요.
먼저, 기울기 < 0 이면 $f(-3) > 0$ 이더라도 결국에는 그 직선이 x축 밑으로 내려가기 때문에 $f < 0$ 가 될 수 있죠.
그러면, 다음 두 가지 경우들이 있을 수 있죠: 기울기 = 0 이고 $f(-3) \geq 0$. 기울기 > 0 이고 $f(-3) \geq 0$.
그러면 결국, 기울기 ≥ 0 이고 $f(-3) \geq 0$ 이죠. 먼저, $t/4 + 1 \geq 0 \Rightarrow t \geq -4$.
또한, $f(-3) = (t/4 + 1)(-3) + t + 5 = -3t/4 - 3 + t + 5 = t/4 + 2 \geq 0 \Rightarrow t \geq -8$.
그러면, $t \geq -4$ 이고 $t \geq -8 \Rightarrow t \geq -4$. 그러므로, $t \geq -4$.

제안과 풀이
H번 예제에서의 문제

다음과 같은 직선의 방정식이 있다 하세요: **$y = (t/4 + 1)x + t + 3$** 이며 여기에서 **t**는 상수.
그러면, **$x > -3$** 일 때 **$y \geq 0$** 가 되는 **t**의 값을 구하세요.

그림. H.0

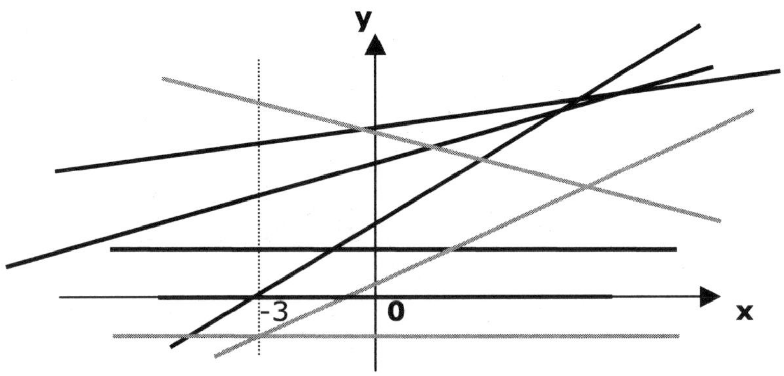

먼저, $y = f(x) = (t/4 + 1)x + t + 3$ 여기에서 t는 상수라고 하세요.
이번에도, $x > -3$ 일 때 $f(x) \geq 0$ 입니다. 그 기울기는 $t/4 + 1$ 이죠.
그러면 이제, "$x > -3$ 일 때 $f(x) \geq 0$." 는 $x = -3$ 이후에는 f가 언제나 0보다 크거나 같다는 거죠.

그러면, $x > -3$ 이고 $f(x) \geq 0$ 이려면 $f(-3)$의 값은 무엇이어야 하고 $x > -3$ 일 때 $f(x)$의 값은 어때야 하죠?
우선, $f(-3) \geq 0$ 이어야 하죠.
또한, f의 기울기가 양이거나 아니면, 그 기울기가 0이고 $f(x) = c \geq 0$ 이며 여기에서 c는 상수인 거죠.

그러면 이제, $f(-3)$의 값과 기울기를 함께 고려해보기로 하죠.
먼저, 기울기 = 0이면 $f(-3) \geq 0$ 이어야 하죠. 다음으로, 기울기 > 0이면 $f(-3) \geq 0$ 이어야 하죠.
그러면 결국, 기울기 ≥ 0 이고 $f(-3) \geq 0$ 이어야 하는 거네요.
한편, 기울기 < 0이면 $f(-3) > 0$ 이더라도 결국에는 그 직선이 x축 밑으로 내려가기 때문에 $f < 0$가 될 수 있죠.

그러면 이제, 그 t를 명시해보도록 하죠. 기울기 ≥ 0 이고 $f(-3) \geq 0$ 이어야 하죠.
그러면, $t/4 + 1 \geq 0 \Rightarrow t \geq -4$.
또한, $f(-3) = (t/4 + 1)(-3) + t + 3 = -3t/4 - 3 + t + 3 = t/4 \geq 0 \Rightarrow t \geq 0$.
그럼 이제, 이 두개의 조건들, "기울기 ≥ 0 와 $f(-3) \geq 0$." 를 동시에 고려해야 하죠.
그러면, $t \geq -4$ 이고 $t \geq 0 \Rightarrow t \geq 0$. 그러므로, $t \geq 0$.

요약:

$y = f(x) = (t/4 + 1)x + t + 3$ 여기에서 t는 상수라고 하세요.
먼저, 기울기 < 0이면 $f(-3) > 0$ 이더라도 결국에는 그 직선이 x축 밑으로 내려가기 때문에 $f < 0$가 될 수 있죠.
그러면, 다음 두 가지 경우들이 있을 수 있죠: 기울기 = 0이고 $f(-3) \geq 0$. 기울기 > 0이고 $f(-3) \geq 0$.
그러면 결국, 기울기 ≥ 0 이고 $f(-3) \geq 0$ 이죠. 먼저, $t/4 + 1 \geq 0 \Rightarrow t \geq -4$.
또한, $f(-3) = (t/4 + 1)(-3) + t + 3 = -3t/4 - 3 + t + 3 = t/4 \geq 0 \Rightarrow t \geq 0$.
그러면, $t \geq -4$ 이고 $t \geq 0 \Rightarrow t \geq 0$. 그러므로, $t \geq 0$.

제안과 풀이
I 번 예제에서의 문제

다음과 같은 직선의 방정식이 있다 하세요: **y = (t/4 + 1)x + t + 3** 이며 여기에서 **t** 는 상수.
그러면, **-2 < x < 3** 일 때 **y > 0** 가 되는 **t** 의 값을 구하세요.

그림. I.0

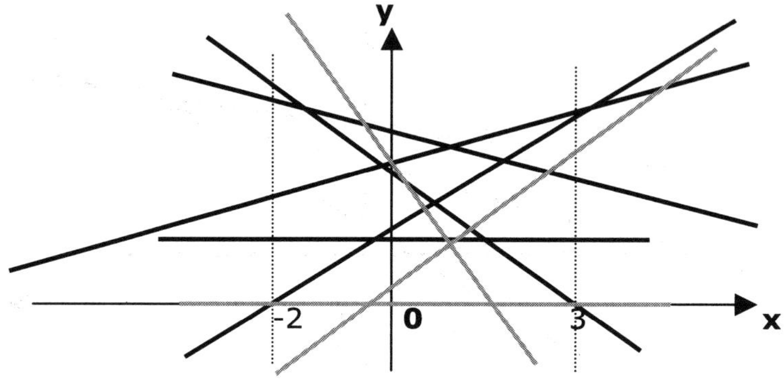

먼저, y = f(x) = (t/4 + 1)x + t + 3 이며 여기에서 t 는 상수라고 하세요.
그러면, -2 < x < 3 일 때 y > 0 이므로 f(x) > 0 이고 여기에서 -2 < x < 3 이죠. 또한, t/4 + 1 이 기울기이죠.

그런데, 이 예제와 이전에 나온 예제들 사이의 다른 점은 뭘까요?
앞서 나왔던 예제들에서는 답이 되는 직선의 기울기가 음이 될 수 없었죠. 하지만 이 예제에서는 그 기울기가 음이 될 수도 있는 거죠. 그러니까 그 직선의 기울기는 심하게 따질 필요는 없는 것 같네요. 그래도, 기울기는 따져 주어야 합니다. 그럼, 뭐 더 따질 거 없나요? 임계 값에서의 f 값들이죠. 두개의 임계 값들이 있어요. 하나는 -2 이고 나머지 하나는 3 입니다.

먼저, 기울기를 따져보기로 하죠.

그러면, "-2 < x < 3 일 때 f(x) > 0." 는 x 의 값이 -2 와 3 사이에 있을 때는 f 의 값이 언제나 양이어야 한다는 거죠.

그러면, -2 < x < 3 이고 f > 0 이려면 f(-2)의 값은 무엇이어야 하고 -2 < x < 3 일 때 f 의 값들은 어때야 하죠?

먼저, f(-2) ≥ 0 이어야 하죠.
또한, -2 보다 큰 x 값에 대한 f 의 값은 f(-2)보다 크면서 계속 증가하거나 아니면, f(x) = c = f(-2) > 0 이고 여기에서 c 는 상수이어야 하는 거죠.
그러니까, 이 경우에는 기울기 ≥ 0 이네요. 하지만, 이게 다는 아니죠.

다음과 같은 또 하나의 상황이 있을 수 있어요.
먼저, f(-2) > 0 이어야 하죠.
또한, -2 보다 큰 x 값에 대한 f 의 값은 f(-2)보다 작으면서 계속 감소하지만 x 가 3 이 되기 바로 전까지는 f 의 값이 양이어야 하죠. 즉, 우선 f(-2) > 0 이고 그리고 나서, x > -2 일 때 f(x) < f(-2)이지만 f(3) ≥ 0 이어야 하죠.
이 경우에는 기울기 < 0 이네요.

그러면 이제, 기울기는 음, 양 또는 0 이 될 수도 있는 거죠.

이런 식으로 일일이 기울기를 따져서 이 문제를 해결하려면 위에서 볼 수 있다시피 꽤나 번거로운 계산을 귀찮게도 해야 할 겁니다.

그런데, 그래프를 조사해보면 직선이 어떻게 놓여야 하는지 알 수 있어요.

"$-2 < x < 3$ 일 때 $f(x) > 0$." 는 x 의 값이 -2 와 3 사이에 있을 때는 f 의 값이 언제나 양이어야 한다는 거죠.

그러면, x 의 값이 -2 와 3 사이에 있을 때는 직선이 x 축 위쪽에 있어야 하는 거죠.
그러면, $f(-2) \geq 0$ 이고 $f(3) \geq 0$ 이어야 하죠.
하지만, 임계 값들 –2 와 3 둘 다에서 f 가 0 이 될 수도 있으니 f 의 커브가 x 축 자체가 될 수도 있다는 것을 유의해야 합니다. 즉, f 가 0 함수가 될 수도 있는 겁니다. 그렇다면, $f(x) > 0$ 이어야 하므로 그 경우의 t 값은 제외시켜야 하죠.

그러면 이제, t 의 값을 명시해보도록 하죠.

$y = f(x) = (t/4 + 1)x + t + 3$ 입니다.

우선, f 가 0 함수가 될 수 있는지 알아보기로 하죠.
$t/4 + 1 = 0 \Rightarrow f(x) = (t/4 + 1)x + t + 3 = t + 3$.
$t/4 + 1 = 0 \Rightarrow t = -4 \Rightarrow t + 3 = -4 + 3 = -1 \Rightarrow$ f 가 0 함수는 될 수 없네요.

그러면, $f(-2) = (t/4 + 1)(-2) + t + 3 = -t/2 - 2 + t + 3 = t/2 + 1 \geq 0 \Rightarrow t \geq -2$.
다음으로, $f(3) = (t/4 + 1)(3) + t + 3 = 3t/4 + 3 + t + 3 = 7t/4 + 6 \geq 0 \Rightarrow t \geq -24/7 = -3 - 3/7$.

그럼 이제, 이 두개의 조건들 $f(-2) \geq 0$ 와 $f(3) \geq 0$ 를 동시에 고려해야 하죠.
그러면, $t \geq -2$ 이고 $t \geq -24/7 \Rightarrow t \geq -2$.
그러므로, $t \geq -2$.

그리하여, t 에 대하여 -2 보다 크거나 같은 값을 선택하면 $-2 < x < 3$ 일 때 $f(x) > 0$ 가 되는 겁니다.

요약:

$y = f(x) = (t/4 + 1)x + t + 3$ 이며 여기에서 t 는 상수라고 하세요.
그러면 이제, "$-2 < x < 3$ 일 때 $f(x) > 0$." 는 x 의 값이 -2 와 3 사이에 있을 때는 f 의 값이 언제나 양이어야 한다는 거죠.
그러면, x 의 값이 -2 와 3 사이에 있을 때는 직선이 x 축 위쪽에 있어야 하는 거죠.
그러면, $f(-2) \geq 0$ 이고 $f(3) \geq 0$ 이어야 하죠.

하지만, f 가 0 함수가 되서는 안되죠. 그러면, 확인해보기로 하죠.
$t/4 + 1 = 0 \Rightarrow t = -4 \Rightarrow t + 3 = -4 + 3 = -1 \Rightarrow$ f 가 0 함수는 될 수 없다.

먼저, $f(-2) = (t/4 + 1)(-2) + t + 3 = -t/2 - 2 + t + 3 = t/2 + 1 \geq 0 \Rightarrow t \geq -2$.
또한, $f(3) = (t/4 + 1)(3) + t + 3 = 3t/4 + 3 + t + 3 = 7t/4 + 6 \geq 0 \Rightarrow t \geq -24/7 = -3 - 3/7$.

그러면, ($t \geq -2$ 이고 $t \geq -24/7 \Rightarrow t \geq -2$. 그러므로, $t \geq -2$.

제안과 풀이
J번 예제에서의 문제

다음과 같은 직선의 방정식이 있다 하세요: **$y = (t/4 + 1)x + t + 4$** 이며 여기에서 **t** 는 상수.
그러면, **$-2 < x < 3$** 일 때 **$y > 0$** 가 되는 **t** 의 값을 구하세요.

그림. J.0

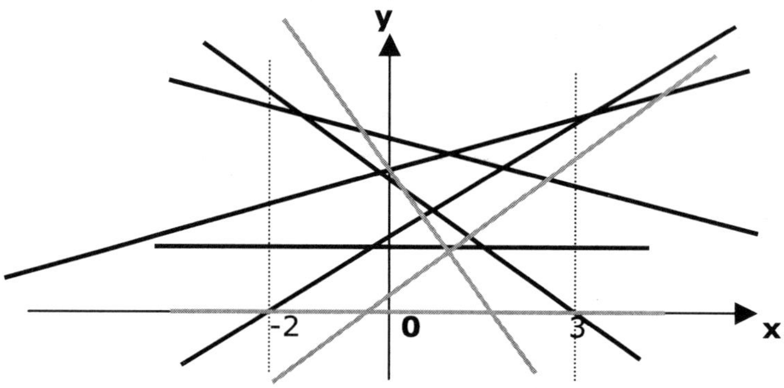

먼저, $y = f(x) = (t/4 + 1)x + t + 4$ 이며 여기에서 t 는 상수라고 하세요.
그러면, $f(x) > 0$ 이고 여기에서 $-2 < x < 3$ 이죠. 또한, $t/4 + 1$ 이 기울기이죠.
이 예제는 앞서 나왔던 예제와 별로 다를 것이 없죠. 그러면 이제, $-2 < x < 3$ 일 때 $f(x) > 0$ 라는 말은 x 의 값이 두개의 임계 값들 -2 와 3 사이에 있을 때는 f 의 값이 언제나 양이어야 한다는 거죠.
이전 예제에서와 마찬가지로, 그래프를 조사해보면 직선이 어떻게 놓여야 하는지 알 수 있죠.
그러면, x 의 값이 -2 와 3 사이에 있을 때는 직선이 x 축 위쪽에 있어야 하는 거죠.
그러면, $f(-2) \geq 0$ 이고 $f(3) \geq 0$ 이어야 하죠.
하지만, f 가 0 함수가 되서는 안되니까 먼저, f 가 0 함수가 될 수 있는지 알아보기로 하죠.
$t/4 + 1 = 0 \Rightarrow f(x) = (t/4 + 1)x + t + 4 = t + 4.$
$t/4 + 1 = 0 \Rightarrow t = -4 \Rightarrow t + 4 = -4 + 4 = 0 \Rightarrow t = -4$ 이면 f 는 0 함수가 될 수 있네요.
먼저, $f(-2) = (t/4 + 1)(-2) + t + 4 = -t/2 - 2 + t + 4 = t/2 + 2 \geq 0 \Rightarrow t \geq -4.$
또한, $f(3) = (t/4 + 1)(3) + t + 4 = 3t/4 + 3 + t + 4 = 7t/4 + 7 \geq 0 \Rightarrow t \geq -4.$
그럼 이제, 이 두개의 조건들, "$f(-2) \geq 0$ 와 $f(3) \geq 0$." 를 동시에 고려해야 하죠.
그러면, $t \geq -4$ 이고 $t \geq -4 \Rightarrow t \geq -4$. 그러나, $t \neq -4$ 왜냐하면 $t = -4$ 이면 $f(x) = 0$ 이기 때문이죠.
그러므로, $t > -4$.
그리하여, t 에 대하여 -4 보다 큰 값을 선택하면 $-2 < x < 3$ 일 때 $f(x) > 0$ 가 되는 겁니다.

요약:

$y = f(x) = (t/4 + 1)x + t + 4$ 이며 여기에서 t 는 상수라고 하세요. 그러면, $-2 < x < 3$ 일 때 $f(x) > 0$.
그러면, x 의 값이 -2 와 3 사이에 있을 때는 직선이 x 축 위쪽에 있어야 하는 거죠.
그러면, $f(-2) \geq 0$ 이고 $f(3) \geq 0$ 이어야 하죠. 하지만, f 가 0 함수가 되서는 안되니까 될 수 있는지 알아보아야 하죠.
$t/4 + 1 = 0 \Rightarrow t = -4 \Rightarrow t + 4 = -4 + 4 = 0 \Rightarrow t = -4$ 이면 f 는 0 함수가 될 수 있네요.
먼저, $f(-2) = (t/4 + 1)(-2) + t + 4 = -t/2 - 2 + t + 4 = t/2 + 2 \geq 0 \Rightarrow t \geq -4.$
또한, $f(3) = (t/4 + 1)(3) + t + 4 = 3t/4 + 3 + t + 4 = 7t/4 + 7 \geq 0 \Rightarrow t \geq -4.$
그러면, $t \geq -4$ 이고 $t \geq -4 \Rightarrow t \geq -4$. 그러나, $t \neq -4$ 왜냐하면 $t = -4$ 이면 $f(x) = 0$.
그러므로, $t > -4$.

제안과 풀이
K번 예제에서의 문제

다음과 같은 직선의 방정식이 있다 하세요**: y = (t/4 + 1)x + t + 4** 이며 여기에서 **t**는 상수**.**
그러면**, -5 < x < -3** 일 때 **y**가 양이 되는 **t**의 값을 구하세요**.**

그림. K.0

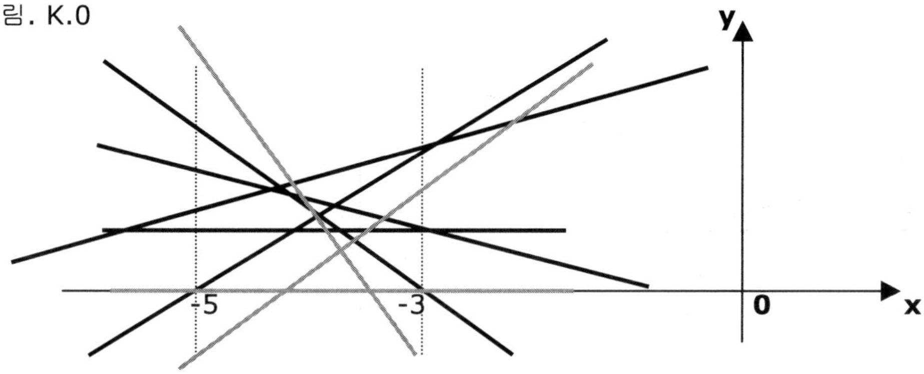

먼저, y = f(x) = (t/4 + 1)x + t + 4 이며 여기에서 t는 상수라고 하세요.
그러면, f(x) > 0 이고 여기에서 -5 < x < -3 이죠. 또한, t/4 + 1 이 기울기이죠.
이 예제도 앞서 나왔던 예제와 별로 다를 것이 없죠. 그러면 이제, "-5 < x < -3 일 때 f(x) > 0." 는 x의 값이 두 개의 임계 값들 -5와 -3 사이에 있을 때는 f의 값이 언제나 양이어야 한다는 거죠.
이전 예제에서와 마찬가지로, 그래프를 조사해보면 직선이 어떻게 놓여야 하는지 알 수 있죠.
그러면, x의 값이 -5와 -3 사이에 있을 때는 직선이 x축 위쪽에 있어야 하는 거죠.
그러면, f(-5) ≥ 0 이고 f(-3) ≥ 0 이어야 하죠. 하지만, f가 0 함수가 되면 안되니까 될 수 있는지 알아보아야 하죠.
t/4 + 1 = 0 ⇒ f(x) = (t/4 + 1)x + t + 4 = t + 4.
t/4 + 1 = 0 ⇒ t = -4 ⇒ t + 4 = -4 + 4 = 0 ⇒ t = -4 이면 f는 0 함수가 될 수 있네요.
그러면, f(-5) = (t/4 + 1)(-5) + t + 4 = -5t/4 − 5 + t + 4 = -t/4 − 1 ≥ 0 ⇒ t ≤ -4.
다음으로, f(-3) = (t/4 + 1)(-3) + t + 4 = -3t/4 − 3 + t + 4 = t/4 + 1 ≥ 0 ⇒ t ≥ -4.
그럼 이제, 이 두개의 조건들, "f(-5) ≥ 0 and f(-3) ≥ 0." 를 동시에 고려해야 하죠.
그러면, (t ≤ -4 이고 t ≥ -4) ⇒ t = -4. 그러나, t ≠ -4 왜냐하면 t = -4 이면 f(x) = 0 이기 때문이죠.
따라서, 이 경우에서는 t에 대한 값은 없는 거죠.
그리니까, t에 대해서 어떤 값을 선택하든 -5 < x < -3 일 때 f(x) > 0 가 되지않는 겁니다.

요약:

y = f(x) = (t/4 + 1)x + t + 4 이며 여기에서 t는 상수라고 하세요. 그러면, -5 < x < -3 일 때 f(x) > 0.
그러면, x의 값이 -5와 -3 사이에 있을 때는 직선이 x축 위쪽에 있어야 하는 거죠.
그러면, f(-5) ≥ 0 이고 f(-3) ≥ 0 이어야 하죠. 먼저, f가 0 함수가 되면 안되니 될 수 있는지 알아보기로 하죠.
t/4 + 1 = 0 ⇒ t = -4 ⇒ t + 4 = -4 + 4 = 0 ⇒ t = -4 이면 f는 0 함수가 될 수 있네요.
그러면, f(-5) = (t/4 + 1)(-5) + t + 4 = -5t/4 − 5 + t + 4 = -t/4 − 1 ≥ 0 ⇒ t ≤ -4.
다음으로, f(-3) = (t/4 + 1)(-3) + t + 4 = -3t/4 − 3 + t + 4 = t/4 + 1 ≥ 0 ⇒ t ≥ -4.
그러면, t ≤ -4 이고 t ≥ -4 ⇒ t = -4. 그러나, t ≠ -4 왜냐하면 t = -4 이면 f(x) = 0.

그러므로, 이 문제에서 t에 대한 값은 없습니다.

제안과 풀이
L 번 예제에서의 문제

다음과 같은 직선의 방정식이 있다 하세요**: $y = (t/4 + 1)x + t + 4$** 이며 여기에서 **t** 는 상수**.**
그러면**, $-5 \le x \le -3$** 일 때 **y** 가 양이 되는 **t** 의 값을 구하세요**.**

그림. L.0

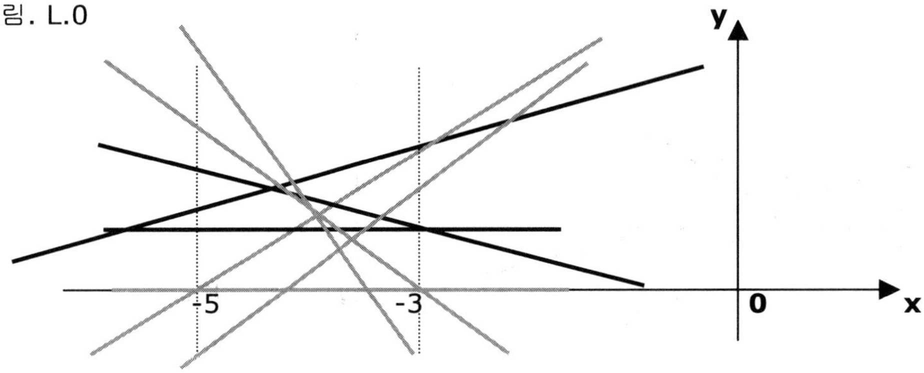

먼저, $y = f(x) = (t/4 + 1)x + t + 4$ 이며 여기에서 t 는 상수라고 하세요.
그러면, $f(x) > 0$ 이고 여기에서 $-5 \le x \le -3$ 이죠. 또한, $t/4 + 1$ 이 기울기이죠.
이 예제도 앞서 나왔던 예제와 별로 다를 것이 없죠. 그러면 이제, "$-5 \le x \le -3$ 일 때 $f(x) > 0$." 는 x 가 –5 에서 –3 까지 취할 때는 f 의 값이 언제나 양이어야 한다는 거죠.
이전 예제에서와 마찬가지로, 그래프를 조사해보면 직선이 어떻게 놓여야 하는지 알 수 있죠.
그러면, x 가 –5 에서 –3 까지의 값을 취할 때는 직선이 x 축 위쪽에 있어야 하는 거죠.
그러니 이번에는, $f(-5) > 0$ 이고 $f(-3) > 0$ 이어야 하죠.

이 경우에는, f 가 위의 조건을 만족하면 0 함수가 될 수 없죠.

먼저, $f(-5) = (t/4 + 1)(-5) + t + 4 = -5t/4 - 5 + t + 4 = -t/4 - 1 > 0 \Rightarrow t < -4$.
또한, $f(-3) = (t/4 + 1)(-3) + t + 4 = -3t/4 - 3 + t + 4 = t/4 + 1 > 0 \Rightarrow t > -4$.
그럼 이제, 이 두개의 조건들, "$f(-5) \ge 0$ 와 $f(-3) \ge 0$." 를 동시에 고려해야 하죠.
그러면, $t < -4$ 이고 $t > -4 \Rightarrow$ 이 경우에서의 t 값은 없네요.
그러므로, 이 문제를 만족시키는 t 값은 없는 겁니다.
그러면 이제, t 의 값으로 어떤 값을 선택하든 $-5 \le x \le -3$ 일 때 $f(x) > 0$ 가 되지않는 겁니다.

요약:

$y = f(x) = (t/4 + 1)x + t + 4$ 이며 여기에서 t 는 상수라고 하세요. 그러면, $-5 \le x \le -3$ 일 때 $f(x) > 0$.
그러면, x 가 –5 에서 –3 까지 값을 취할 때는 직선이 x 축 위쪽에 있어야 하는 거죠.
그러면, $f(-5) > 0$ 이고 $f(-3) > 0$ 이어야 하죠.
먼저, $f(-5) = (t/4 + 1)(-5) + t + 4 = -5t/4 - 5 + t + 4 = -t/4 - 1 > 0 \Rightarrow t < -4$.
또한, $f(-3) = (t/4 + 1)(-3) + t + 4 = -3t/4 - 3 + t + 4 = t/4 + 1 > 0 \Rightarrow t > -4$.
그러면, $t < -4$ 이고 $t > -4 \Rightarrow$ 이 경우에서의 t 값은 없습니다.

그러므로, 이 문제를 만족시키는 t 값은 없습니다.

제안과 풀이
M 번 예제에서의 문제

다음과 같은 직선의 방정식이 있다 하세요: **$y = (t/4 + 1)x + t + 4$** 이며 여기에서 **t**는 상수.
그러면, **$-5 < x < -3$** 일 때 **$y \ge 0$**가 되는 **t**의 값을 구하세요.

그림. M.0

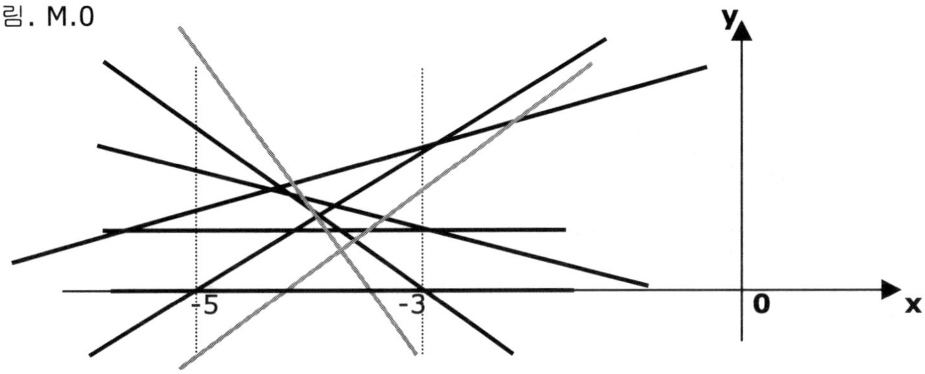

먼저, $y = f(x) = (t/4 + 1)x + t + 4$ 이며 여기에서 t는 상수라고 하세요.
그러면, $f(x) \ge 0$ 이고 여기에서 $-5 < x < -3$ 이죠. 또한, $t/4 + 1$ 이 기울기이죠.
이 예제도 앞서 나왔던 예제와 비슷하죠. 그러면 이제, $-5 < x < -3$ 일 때 $f(x) \ge 0$ 라는 말은 x의 값이 두개의 임계 값들 –5와 –3 사이에 있을 때는 f의 값이 0보다 크거나 같다는 거죠.
이전 예제에서와 마찬가지로, 그래프를 조사해보면 직선이 어떻게 놓여야 하는지 알 수 있죠.
먼저, x의 값이 -5와 -3 사이에 있을 때는 직선이 x축 위쪽에 있어야 하는 거죠.
또한, $f(-5) = 0$ 이거나 $f(-3) = 0$ 이어도 되는 거죠. 그러니 이번에는, $f(-5) \ge 0$ 이고 $f(-3) \ge 0$ 이어야 하죠.
따라서 이번에는, f가 0함수가 되도 되죠 그러니까 그 검사를 해보아야 하죠.
$t/4 + 1 = 0 \Rightarrow f(x) = (t/4 + 1)x + t + 4 = t + 4.$
$t/4 + 1 = 0 \Rightarrow t = -4 \Rightarrow t + 4 = -4 + 4 = 0 \Rightarrow t = -4$ 이면 f는 0함수가 될 수 있죠.
먼저, $f(-5) = (t/4 + 1)(-5) + t + 4 = -5t/4 - 5 + t + 4 = -t/4 - 1 \ge 0 \Rightarrow t \le -4.$
또한, $f(-3) = (t/4 + 1)(-3) + t + 4 = -3t/4 - 3 + t + 4 = t/4 + 1 \ge 0 \Rightarrow t \ge -4.$
그럼 이제, 이 두개의 조건들, "$f(-5) \ge 0$ 와 $f(-3) \ge 0$." 를 동시에 고려해야 하죠.
그러면, ($t \le -4$ 이고 $t \ge -4$) $\Rightarrow t = -4.$
그러므로, t 값은 -4만 되는 겁니다. 그러면, $f(x) = 0$ 즉, f는 0함수만 될 수 있죠.
그러면 이제, t의 값으로 -4를 선택하면 $-5 < x < -3$ 일 때 $f(x) \ge 0$가 되는 겁니다.

요약:

$y = f(x) = (t/4 + 1)x + t + 4$ 이며 여기에서 t는 상수라고 하세요. 그러면, $-5 < x < -3$ 일 때 $f(x) \ge 0$.
먼저, x의 값이 -5와 -3 사이에 있을 때는 직선이 x축 위쪽에 있어야 하는 거죠.
또한, $f(-5) = 0$ 이거나 $f(-3) = 0$ 이어도 되는 거죠. 그러니, $f(-5) \ge 0$ 이고 $f(-3) \ge 0$ 이어야 하죠.
따라서, f가 0함수가 되도 되니까 그 검사를 해보아야 하죠.
$t/4 + 1 = 0 \Rightarrow t = -4 \Rightarrow t + 4 = -4 + 4 = 0 \Rightarrow t = -4$ 이면 f는 0함수가 될 수 있죠.
먼저, $f(-5) = (t/4 + 1)(-5) + t + 4 = -5t/4 - 5 + t + 4 = -t/4 - 1 \ge 0 \Rightarrow t \le -4.$
또한, $f(-3) = (t/4 + 1)(-3) + t + 4 = -3t/4 - 3 + t + 4 = t/4 + 1 \ge 0 \Rightarrow t \ge -4.$
그러면, ($t \le -4$ 이고 $t \ge -4$) $\Rightarrow t = -4.$

제안과 풀이

N 번 예제에서의 문제

다음과 같은 직선의 방정식이 있다 하세요: **$y = (t/4 + 1)x + t + 4$** 이며 여기에서 **t** 는 상수.
그러면, **$-5 \le x \le -3$** 일 때 **$y \ge 0$** 가 되는 **t** 의 값을 구하세요.

그림. N.0

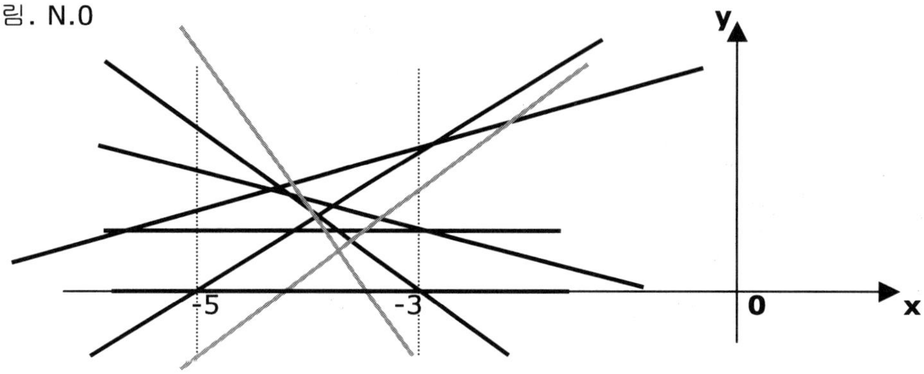

먼저, $y = f(x) = (t/4 + 1)x + t + 4$ 이며 여기에서 t 는 상수라고 하세요.
그러면, $f(x) \ge 0$ 이고 여기에서 $-5 \le x \le -3$ 이죠. 또한, $t/4 + 1$ 이 기울기이죠.
이 예제도 앞서 나왔던 예제와 비슷하죠. 그러면 이제, "$-5 \le x \le -3$ 일 때 $f(x) \ge 0$." 는 x 가 −5 에서 −3 까지 값을 취할 때는 f 의 값이 0 보다 크거나 같다는 거죠.
이전 예제에서와 마찬가지로, 그래프를 조사해보면 직선이 어떻게 놓여야 하는지 알 수 있죠.
먼저, x 의 값이 -5 와 -3 사이에 있을 때는 직선이 x 축 위쪽에 있어야 하는 거죠.
또한, $f(-5) = 0$ 이거나 $f(-3) = 0$ 이어도 되는 거죠. 그러니, $f(-5) \ge 0$ 이고 $f(-3) \ge 0$ 이어야 하죠.
따라서, f 가 0 함수가 되도 되니까 그 검사를 해보아야 하죠.
$t/4 + 1 = 0 \Rightarrow f(x) = (t/4 + 1)x + t + 4 = t + 4.$
$t/4 + 1 = 0 \Rightarrow t = -4 \Rightarrow t + 4 = -4 + 4 = 0 \Rightarrow t = -4$ 이면 f 는 0 함수가 될 수 있죠.
먼저, $f(-5) = (t/4 + 1)(-5) + t + 4 = -5t/4 - 5 + t + 4 = -t/4 - 1 \ge 0 \Rightarrow t \le -4.$
또한, $f(-3) = (t/4 + 1)(-3) + t + 4 = -3t/4 - 3 + t + 4 = t/4 + 1 \ge 0 \Rightarrow t \ge -4.$
그럼 이제, 이 두개의 조건들 $f(-5) \ge 0$ 와 $f(-3) \ge 0$ 를 동시에 고려해야 하죠.
그러면, $t \le -4$ 이고 $t \ge -4 \Rightarrow t = -4.$
그러므로, t 값은 -4 만 되는 겁니다. 그러면, $f(x) = 0$ 즉, f 는 0 함수만 될 수 있죠.
그러면 이제, t 의 값으로 -4 를 선택하면 $-5 \le x \le -3$ 일 때 $f(x) \ge 0$ 가 되는 겁니다.

요약:

$y = f(x) = (t/4 + 1)x + t + 4$ 이며 여기에서 t 는 상수라고 하세요. 그러면, $-5 \le x \le -3$ 일 때 $f(x) \ge 0$.
먼저, x 의 값이 −5 와 −3 사이에 있을 때는 직선이 x 축 위쪽에 있어야 하는 거죠.
또한, $f(-5) = 0$ 이거나 $f(-3) = 0$ 이어도 되는 거죠. 그러니, $f(-5) \ge 0$ 이고 $f(-3) \ge 0$ 이어야 하죠.
또한, f 가 0 함수가 되어도 되니까 그 검사를 해보아야 하죠.
$t/4 + 1 = 0 \Rightarrow t = -4 \Rightarrow t + 4 = -4 + 4 = 0 \Rightarrow t = -4$ 이면 f 는 0 함수가 될 수 있죠.
먼저, $f(-5) = (t/4 + 1)(-5) + t + 4 = -5t/4 - 5 + t + 4 = -t/4 - 1 \ge 0 \Rightarrow t \le -4.$
또한, $f(-3) = (t/4 + 1)(-3) + t + 4 = -3t/4 - 3 + t + 4 = t/4 + 1 \ge 0 \Rightarrow t \ge -4.$
그러면, $t \le -4$ 이고 $t \ge -4 \Rightarrow t = -4.$

제안과 풀이
O 번 예제에서의 문제

다음과 같은 직선의 방정식이 있다 하세요: $y = (t/4 + 2)x + t + 3$ 이며 여기에서 t 는 상수.
그러면, $-5 \le x \le -3$ 일 때 y 가 음, 0, 그리고 양도 되는 t 의 값을 구하세요.

그림. O.0

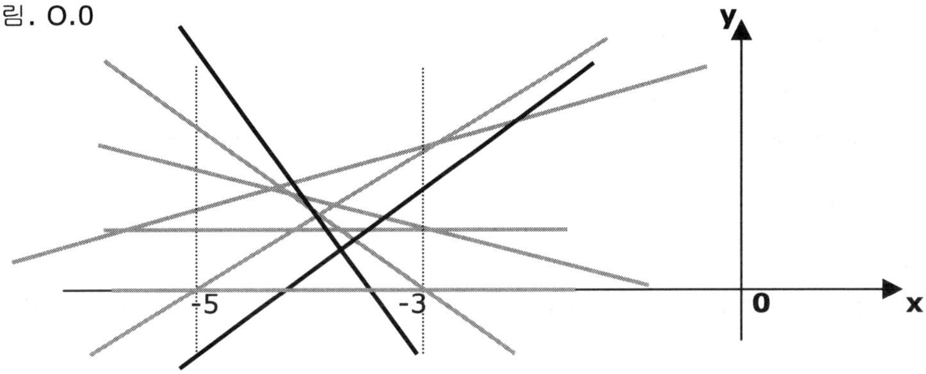

이 예제는 앞서 나왔던 예제와 비슷한 것 같지만 꽤 다르죠.

먼저, $y = f(x) = (t/4 + 2)x + t + 3$ 이며 여기에서 t 는 상수라고 하세요.
그러면, $f(x) \ge 0$ 또는 $f(x) < 0$ 이고 여기에서 $-5 \le x \le -3$ 이죠. 또한, $t/4 + 2$ 가 기울기이죠.

그러면 이제, "$-5 \le x \le -3$ 일 때 $f(x) \ge 0$ 또는 $f(x) < 0$." 는 x 가 -5 에서 -3 까지 값을 취할 때는 f 의 값들이 양과 0 그리고 음이 될 수다는 거죠.
즉, 그 값들 중의 하나가 0 이고 그 나머지들은 양의 값들과 음의 값들인 겁니다.

이전 예제에서와 마찬가지로, 그래프를 조사해보면 직선이 어떻게 놓여야 하는지 알 수 있죠.
그러면, x 축에서 -5 와 -3 사이에서 직선이 x 축을 통과해야 하는 거죠.
그러면, 다음과 같은 두 가지 경우들이 있을 수 있네요:

$f(-5) < 0$ 이고 $f(-3) > 0$.
$f(-5) > 0$ 이고 $f(-3) < 0$.

그러면 이제, 첫번째 경우부터 시작하도록 하죠: $f(-5) < 0$ 이고 $f(-3) > 0$.
$f(x) = (t/4 + 2)x + t + 3$ 입니다.
그러면, $f(-5) = (t/4 + 2)(-5) + t + 3 = -5t/4 - 10 + t + 3 = -t/4 - 7 < 0 \Rightarrow t > -28$.
또한, $f(-3) = (t/4 + 2)(-3) + t + 3 = -3t/4 - 6 + t + 3 = t/4 - 3 > 0 \Rightarrow t > 12$.
그럼 이제, 이 두개의 조건들, "$f(-5) < 0$ 와 $f(-3) > 0$."를 동시에 고려해야 하죠.
그러면, ($t > -28$ 이고 $t > 12$) $\Rightarrow t > 12$.

다음으로, 그 두 번째 경우로 넘어가기로 하죠: $f(-5) > 0$ 이고 $f(-3) < 0$.
그러면, $f(-5) = -t/4 - 7 > 0 \Rightarrow t < -28$.
또한, $f(-3) = t/4 - 3 < 0 \Rightarrow t < 12$.
그럼 이제, 이 두개의 조건들 $f(-5) > 0$ 와 $f(-3) < 0$ 를 동시에 고려해야 하죠.
그러면, ($t < -28$ 이고 $t < 12$) $\Rightarrow t < -28$.

그러면 이제, 그 두개의 경우들을 합쳐야 하죠.
그러면, $t > 12$ 또는 $t < -28$.

그러므로, t 의 값으로 12 보다 크거나 −28 보다 작은 값을 선택하면:
$-5 \le x \le -3$ 일 때 $f(x) \ge 0$ 또는 $f(x) < 0$ 가 되는 겁니다.

그런데 또 하나의 방법을 사용해볼 수도 있어요.
다음과 같은 두 가지 경우들이 있죠:
$f(-5) < 0$ 이고 $f(-3) > 0$. $f(-5) > 0$ 이고 $f(-3) < 0$.

그러면, 그 둘 다의 경우들에서, f(-5)와 f(-3)은 서로 반대의 부호를 갖고 있는 거죠.
그러므로, 그 둘의 경우는 다음과 같은 하나의 식으로 표현할 수도 있는 겁니다: $f(-5)f(-3) < 0$.

$f(-5) = -t/4 - 7$ 이고 $f(-3) = t/4 - 3$ 이죠.

그러면, $f(-5)f(-3) = (-t/4 - 7)(t/4 - 3) = -(t/4 + 7)(t/4 - 3) < 0$ 이죠.
따라서, $-(t/4 + 7)(t/4 - 3) < 0 \Rightarrow (t/4 + 7)(t/4 - 3) > 0 \Rightarrow (t + 28)(t - 12) > 0 \Rightarrow t > 12$ 또는 $t < -28$.

요약:

먼저, $y = f(x) = (t/4 + 2)x + t + 3$ 이며 여기에서 t 는 상수라고 하세요.
그러면, $f(x) \ge 0$ 또는 $f(x) < 0$ 이고 여기에서 $-5 \le x \le -3$ 이죠.
그러면, x 축에서 -5 와 -3 사이에서 직선이 x 축을 통과해야 하는 거죠.
그러면, 다음과 같은 두 가지 경우들이 있을 수 있네요:
$f(-5) < 0$ 이고 $f(-3) > 0$.
$f(-5) > 0$ 이고 $f(-3) < 0$.

그러면, 위의 둘 다의 경우들에서, f(-5)와 f(-3)은 서로 반대의 부호를 갖고 있는 거죠.
그러므로, 위의 둘의 경우는 다음과 같은 하나의 식으로 표현할 수도 있는 겁니다: $f(-5)f(-3) < 0$.

그러면, $f(-5) = (t/4 + 2)(-5) + t + 3 = -5t/4 - 10 + t + 3 = -t/4 - 7$.
다음으로, $f(-3) = (t/4 + 2)(-3) + t + 3 = -3t/4 - 6 + t + 3 = t/4 - 3$.

그러면, $f(-5)f(-3) = (-t/4 - 7)(t/4 - 3) = -(t/4 + 7)(t/4 - 3) < 0$ 이죠.
따라서, $-(t/4 + 7)(t/4 - 3) < 0 \Rightarrow (t/4 + 7)(t/4 - 3) > 0 \Rightarrow (t + 28)(t - 12) > 0 \Rightarrow t > 12$ 또는 $t < -28$.

제안과 풀이
P 번 예제에서의 문제

다음과 같은 직선의 방정식이 있다 하세요: $y = (t/4 + 2)x + t + 3$ 이며 여기에서 t는 상수.
그러면, $1 \le t \le 5$ 일 때 $y > 0$가 되는 x의 범위를 구하세요.

이번에는 상수의 범위가 주어지고 변수의 특정 범위를 구해야 하는군요.
상수는 변수와는 본질적으로 다르지만 변수의 성질도 가지고 있죠. 변수는 보통, 값을 취하는 범위를 갖고있죠. 이는 상수에 대하여도 마찬가집니다. 그래서, 상수를 변수처럼 사용하기도 하죠 하지만 상수는 변수가 아니죠. 그러면 이제, 상수 t의 범위사용해서 x의 범위를 결정해야 하죠.

그러면, y가 어느 x값에서 양이 되기 시작하는지 어떻게 알죠?
그 직선의 x절편을 알아야 해요. 그런데, 그 x절편만 알아서는 충분하지 않죠. 왜냐하면 그 x절편만 알아 가지고는 x의 값이 그 x절편보다 커지면 그 y가 양이 되는지 아니면 음이 되는지 알 수 없으니까요.
그럼, 무엇을 더 알아야 하나요?
그 직선의 기울기도 알아야 하는 거죠. 그런데, 그 기울기도 상수 t의 범위에 따라 결정되죠.

그러면 이제, 기울기와 x절편의 범위를 구하는데 기울기부터 시작해보기로 하죠.
그 기울기는 $t/4 + 2$이죠. $a = t/4 + 2$라 하세요. $1 \le t \le 5$입니다. 그러면, t가 증가함에 따라 기울기 a도 증가하네요. 즉, t가 증가함에 따라 그 직선은 더욱 가파르게 되는 거죠. 그러면, a의 범위는 다음과 같아요:
$1 \le t \le 5 \Rightarrow 1/4 \le t/4 \le 5/4 \Rightarrow 9/4 \le t/4 + 2 \le 13/4 \Rightarrow 9/4 \le a \le 13/4$.

다음으로, x절편의 범위를 구해보기로 하죠. 그런데, 두 가지 방법들을 사용할 수 있겠어요.
첫째 방법은 직접, x절편을 구해서 그 범위를 구하는 겁니다.
둘째 방법은 먼저, y절편의 범위를 구하고 그리고 나서 x절편의 범위를 구하는 거죠.

둘째 방법부터 시작해보기로 하죠.
그 y절편은 $t + 3$이죠. $b = t + 3$이라고 하세요. $1 \le t \le 5$입니다. 그러면, t가 증가함에 따라 y절편 b도 증가하네요. 즉, t가 증가함에 따라 그 직선은 y축을 따라 올라가는 것처럼 보이게 되는 거죠. 그러면, b의 범위는 다음과 같아요: $1 \le t \le 5 \Rightarrow 4 \le t + 3 \le 8 \Rightarrow 4 \le b \le 8$.
이제, 그 x절편은 어떻게 구하나요?
그것은 $y = 0$일 때 x의 값이니까 직선의 방정식에서 $y = 0$으로 놓고 x에 대하여 풀고 그 답을 x절편으로 하면 되죠. 이제, 그 x절편을 s라고 하세요.
우선, $t = 1$일 때 $y = (9/4)x + 4$이죠. 그러면, $0 = (9/4)x + 4 \Rightarrow x = -16/9 \Rightarrow s = -16/9$.
다음으로, $t = 5$일 때 $y = (13/4)x + 8$이죠. 그러면, $0 = (13/4)x + 8 \Rightarrow x = -32/13 \Rightarrow -32/13$.
그러므로, $-32/13 \le s \le -16/9$이죠.

다음으로, 첫째 방법은 다음과 같아요. $y = (t/4 + 2)x + t + 3$이죠. 그러면:
$0 = (t/4 + 2)x + t + 3 \Rightarrow x = -(t + 3)/(t/4 + 2) = -4(t + 3)/(t + 8)$.
$(t + 3)/(t + 8) = (t + 3 + 5 - 5)/(t + 8) = (t + 8 - 5)/(t + 8) = 1 - 5/(t + 8)$.
그러므로, $x = -4\{1 - 5/(t + 8)\} = -4 + 20/(t + 8)$. 그러면, x절편 $s = 20/(t + 8) - 4$이네요.
$1 \le t \le 5$ 이죠. 그러면, t가 커짐에 따라 s는 작아지는군요. 분모가 계속 커지기 때문이죠.
이제, 실제로 s의 범위를 구해보기로 하죠.
$t = 1$일 때 $s = 20/(t + 8) - 4 = 20/9 - 4 = -16/9 = -1 - 7/9$.
$t = 5$일 때 $s = 20/(t + 8) - 4 = 20/13 - 4 = (20 - 52)/13 = -32/13 = -2 - 6/13$.
따라서, $-32/13 \le s \le -16/9$이죠.

그러면 이제, t = 1 일 때 a = 9/4 이고 s = -16/9, 또한, t = 5 일 때 a = 13/4 이고 s = -32/13.
그러면, 1 ≤ t ≤ 5 일 때 기울기는 항상 양이니까 x 의 값이 그 x 절편(s)보다 크면 y 는 언제나 양이 되는 거죠.
그러면, 1 ≤ t ≤ 5 일 때 가장 큰 x 절편(s)이 −16/9 이니까 x > -16/9 이면 y > 0 이네요.
그러면 이제, 그 그래프는 대략 다음과 같죠:

그림. P.0

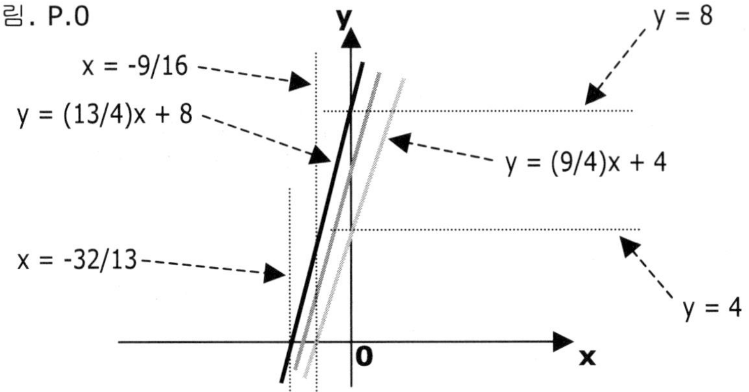

그러먼 이제, t가 1 일 때, 그 직신은 옅은 회색이고 그 직선의 방정식은 y = 9x/4 + 4 여기에시 x 절편은 −9/16 입니다. 또한, t가 5 이면 그 직선은 검정 색이고 그 직선의 방정식은 y = 13x/4 + 8 여기에서 x 절편은 −32/13 이죠. 그러면, t가 1 에서 5 로 변할 때 그 직선은 옅은 회색직선에서 검정직선으로 변하죠.
그러면 이제, 1 ≤ t ≤ 5 일 때 f가 항상 0 보다 크려면 x 의 값이 −16/9 보다 커야 하는 겁니다.
반면에, 1 ≤ t ≤ 5 일 때 f가 항상 0 보다 적으려면 x 의 값이 −32/13 보다 작아야 하는 거죠.
그러므로, 1 ≤ t ≤ 5 이고 x > -16/9 일 때 y = (t/4 + 2)x + t + 3 > 0 가 되는 겁니다.

그런데, 1 < t ≤ 5 라면 어떨까요? 그 때는 x 의 값이 −16/9 이어도 되죠. 왜냐하면 옅은 회색직선은 t = 1 일 때 만들어지는데 이 경우에는 t가 1 이 될 수 없기 때문이죠. 그러므로, 이 경우에서는 x ≥ -16/9 인 것이죠.
그러면 이제, 이전 예제들에 대하여 했던 방법으로 이 경우에 대한 답에 접근해보도록 하죠.
먼저, 주어진 직선의 방정식에서 t를 변수로 취급하고 x를 상수로 취급하고 그 방정식을 변형하는 겁니다.
그리고 나서, x 의 범위를 구하는 거죠. 물론, 상수는 변수와 다른 겁니다. 상수와 변수 둘 다 범위를 가질 수 있다는 면을 이용하는 거죠. 이제, y = f(x) = (t/4 + 2)x + t + 3 여기에서 t는 상수라고 하세요.
그러면, 이 f를 함수 g로 변형합니다. 이것들은 좌표계 이름만 다를 뿐이지 서로 같은 것들이죠.
(t/4 + 2)x + t + 3 = xt/4 + 2x + t + 3 = (x/4 + 1)t + 2x + 3.
이제, y = g(t) = (x/4 + 1)t + 2x + 3 여기에서 x는 상수라고 하세요.
그러면, 1 < t ≤ 5 일 때 g(t) > 0 가 되도록 x 의 범위를 구하는 겁니다.

그림. P.1

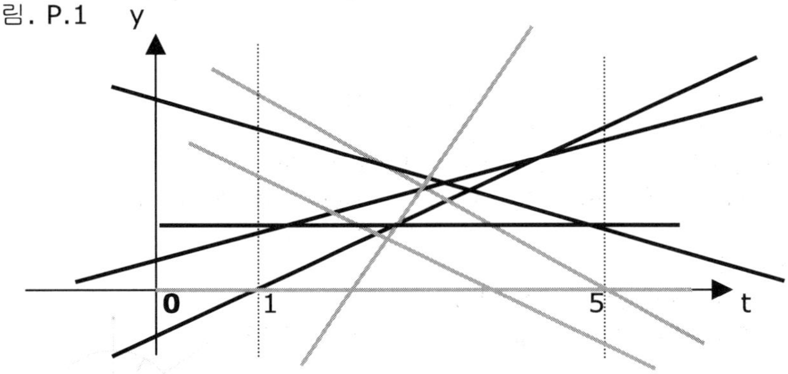

위의 그래프를 조사해보면 직선이 어떻게 놓여야 하는지 알 수 있죠.
그러면, t 의 값이 1 과 5 사이에 있을 때는 직선이 t 축 위쪽에 있어야 하는 거죠.
또한, $g(1) = 0$ 이어도 되지만 $g(5) \neq 0$ 이어야 하는 거죠. 그러니까 $g(1) \geq 0$ 이고 $g(5) > 0$ 이어야 하죠.
그러면, $g(1) = (x/4 + 1)(1) + 2x + 3 = x/4 + 1 + 2x + 3 = 9x/4 + 4 \geq 0 \Rightarrow x \geq -16/9$.
다음으로, $g(5) = (x/4 + 1)(5) + 2x + 3 = 5x/4 + 5 + 2x + 3 = 13x/4 + 8 > 0 \Rightarrow x > -32/13$.
그럼 이제, 이 두개의 조건들, "$g(1) \geq 0$ 와 $g(5) > 0$." 를 동시에 고려해야 하죠.
그러면, ($x \geq -16/9$ 이고 $x > -32/13$) $\Rightarrow x \geq -16/9$.
그러면 이제, x 의 값이 $-16/9$ 보다 크거나 같으면 $1 < t \leq 5$ 일 때 $f(x) > 0$ 가 되는 겁니다.

요약:

a 는 기울기이고 s 는 x 절편이라 하세요. 그러면, $a = t/4 + 2$. 이제, s 를 구합니다.
$0 = (t/4 + 2)x + t + 3 \Rightarrow x = -(t + 3)/(t/4 + 2) = -4(t + 3)/(t + 8)$.
$(t + 3)/(t + 8) = (t + 3 + 5 - 5)/(t + 8) = (t + 8 - 5)/(t + 8) = 1 - 5/(t + 8)$.
그러므로, $x = -4\{1 - 5/(t + 8)\} = -4 + 20/(t + 8)$. 그러면, $s = 20/(t + 8) - 4$.

그러면 이제, a 와 s 의 범위들을 구합니다.
우선, t 가 증가함에 따라 a 도 증가.
$1 \leq t \leq 5 \Rightarrow 1/4 \leq t/4 \leq 5/4 \Rightarrow 9/4 \leq t/4 + 2 \leq 13/4 \Rightarrow 9/4 \leq a \leq 13/4$.
다음으로, t 가 증가함에 따라 s 는 감소.
$t = 1$ 일 때 $s = 20/(t + 8) - 4 = 20/9 - 4 = -16/9 = -1 - 7/9$.
$t = 5$ 일 때 $s = 20/(t + 8) - 4 = 20/13 - 4 = (20 - 52)/13 = -32/13 = -2 - 6/13$.
그러므로, $-32/13 \leq s \leq -16/9$.

그러면:
$t = 1$ 일 때 $a = 9/4$ 이고 x 절편은 $-16/9$. $t = 5$ 일 때 $a = 13/4$ 이고 x 절편은 $-32/13$.
기울기가 양일 때는 x 의 값이 그 x 절편보다 크면 y 가 양이 되죠.
따라서, $1 \leq t \leq 5$ 이고 $x > -16/9$ 일 때 $y = (t/4 + 2)x + t + 3 > 0$ 가 됩니다.
그러므로, $x > -16/9$.

다른 방법:

먼저, 주어진 직선의 방정식에서 t 를 변수로 취급하고 x 를 상수로 취급하고 그 방정식을 변형하는 겁니다.
$y = f(x) = (t/4 + 2)x + t + 3$ 여기에서 t 는 상수라고 하세요.
그러면, 이 f 를 다음과 같이 함수 g 로 변형합니다:
$(t/4 + 2)x + t + 3 = xt/4 + 2x + t + 3 = (x/4 + 1)t + 2x + 3$.

이제, $y = g(t) = (x/4 + 1)t + 2x + 3$ 여기에서 x 는 상수라고 하세요.
그러면, $1 \leq t \leq 5$ 일 때 $g(t) > 0$ 가 되도록 x 의 범위를 구하는 겁니다.
그러면, t 의 값이 1 에서 5 까지 변할 때는 직선이 t 축 위쪽에 있어야 하죠.
그러니까 $g(1) > 0$ 이고 $g(5) > 0$ 이어야 하죠.
그러면, $g(1) = (x/4 + 1)(1) + 2x + 3 = x/4 + 1 + 2x + 3 = 9x/4 + 4 > 0 \Rightarrow x > -16/9$.
또한, $g(5) = (x/4 + 1)(5) + 2x + 3 = 5x/4 + 5 + 2x + 3 = 13x/4 + 8 > 0 \Rightarrow x > -32/13$.
그러면, ($x > -16/9$ 이고 $x > -32/13$) $\Rightarrow x > -16/9$.
따라서, x 의 값이 $-16/9$ 보다 크면 $1 \leq t \leq 5$ 일 때 $f(x) > 0$ 가 됩니다.
그러므로, $x > -16/9$.

제안과 풀이
Q 번 예제에서의 문제

다음과 같은 직선의 방정식이 있다 하세요: $y = 3(t - 1)x + 3 - 2t$ 이며 여기에서 t 는 상수.
그러면, $0 \le t \le 2$ 일 때 $y > 0$ 가 되는 x 의 값을 구하세요.

이번에도 상수의 범위가 주어지고 변수의 특정 범위를 구해야 하죠.
상수는 변수와는 본질적으로 다르지만 변수의 성질도 가지고 있죠. 변수는 보통, 값을 취하는 범위를 갖고있죠. 이는 상수에 대하여도 마찬가집니다. 그래서, 상수를 변수처럼 사용하기도 하죠. 하지만 상수는 변수가 아니죠.
그러면 이제, 상수 t 의 범위는 결국 기울기와 x 절편과 y 절편의 범위들을 결정하게 되네요.
그러면, 기울기를 a, x 절편을 s, y 절편을 b 라 하고 이것들의 범위들을 구해보기로 하죠.

그러면 이제, 기울기 a 부터 시작해보기로 하죠.
그 기울기 $a = 3(t - 1)$이고 $0 \le t \le 2$ 입니다. 그러면, t 가 증가함에 따라 기울기 a 도 증가하네요. 즉, t 가 증가함에 따라 그 직선은 더욱 가파르게 되는 거죠. 그러면, a 의 범위는 다음과 같네요:
$0 \le t \le 2 \Rightarrow -1 \le t - 1 \le 1 \Rightarrow -3 \le 3(t - 1) \le 3 \Rightarrow -3 \le a \le 3$, 즉, $|a| \le 3$.

다음으로, x 절편 s 의 범위를 구합니다.
그런데, 먼저, y 절편 b 의 범위를 구하고 그리고 나서 s 의 범위를 구해보기로 하죠.

그 y 절편 $b = 3 - 2t$ 이고 $0 \le t \le 2$ 입니다. 그러면, t 가 증가함에 따라 y 절편 b 는 감소하네요. 즉, t 가 증가함에 따라 그 직선은 y 축을 따라 내려가는 것처럼 보이게 되는 거죠. 그럴 것 같죠? 어쨌든, b 의 범위는 다음과 같아요: $0 \le t \le 2 \Rightarrow 0 \ge -2t \ge -4 \Rightarrow 3 \ge 3 - 2t \ge -1 \Rightarrow -1 \le b \le 3$.

이제, 그 x 절편의 범위를 구하기로 하죠. 그 x 절편은 $y = 0$ 일 때 x 의 값이니까 직선의 방정식에서 $y = 0$ 으로 놓고 x 에 대하여 풀고 그 답을 x 절편으로 하면 되죠.
우선, $t = 0$ 일 때 $y = -3x + 3$ 이죠. 그러면, $0 = -3x + 3 \Rightarrow x = 1 \Rightarrow s = 1$.
다음으로, $t = 2$ 일 때 $y = 3x - 1$ 이죠. 그러면, $0 = 3x - 1 \Rightarrow x = 1/3 \Rightarrow s = 1/3$.
그러므로, $1/3 \le s \le 1$ 입니다. 그런데, 과연 그럴까요?

한편, x 절편 s 를 직접 구하면 다음과 같아요. $y = 3(t - 1)x + 3 - 2t$ 이죠.
그러면, $0 = 3(t - 1)x + 3 - 2t \Rightarrow 3x = (2t - 3)/(t - 1) = (2t - 2 - 1)/(t - 1) = (2t - 2)/(t - 1) - 1/(t - 1) = 2 - 1/(t - 1) = 2 + 1/(1 - t)$.
그러므로, $x = (1/3)\{2 + 1/(1 - t)\}$. 그러면, x 절편 $s = (1/3)\{2 + 1/(1 - t)\}$이네요.

그러면, t 가 커짐에 따라 s 가 어떻게 변하는지 알아보기로 하죠. $0 \le t \le 2$ 이죠.
그런데 우선, $1/(1 - t)$이 어떻게 변하는지 볼 겁니다. 그러면, $c = 1/(1 - t)$이라 하세요.
그러면, $t = 0 \Rightarrow c = 1/(1 - t) = 1/1 = 1$ 이죠. 그러면, $s = (1/3)(2 + 1) = 1$ 이네요.
그러면, t 가 0 에서 1 을 향하여 커감에 따라 c 도 커지긴 하는데 얼마까지 커지나요?
분모가 0 에 가까워지니까 이는 무한히 커지게 되네요.
그런데, $t = 1$ 이면 $1 - t = 0$ 인데 이는 분모이니까 허용될 수 없는 거죠.
아무튼, t 의 값이 1 과 거의 같으면 결국 s 도 무한히 커지게 되죠.
그러면, $0 \le t < 1$ 일 때에는 $s \ge 1$ 입니다.
사실, $t = 1$ 이면 그 기울기 $a = 3(t - 1) = 0$ 이므로 그 직선은 수평선이고 이 경우에 x 절편은 없는 겁니다.

그러면 이제, $t = 2 \Rightarrow c = 1/(1 - t) = 1/(-1) = -1$ 입니다. 그러면, $s = (1/3)(2 - 1) = 1/3$ 이네요.

그러면, t 가 2 에서 1 을 향하여 감소함에 따라 c 도 감소하긴 하는데 얼마까지 작아지나요?
분모가 0 에 가까워지고 또한 음이므로 음의 무한대로 작아지네요.
그러면, $1 < t \leq 2$ 일 때에는 $s \leq 1/3$ 입니다.
그 t 가 1 이면 그 직선은 수평선이라는 걸 알고 있죠.

그럼 이제, 실제 s 의 범위는 다음과 같죠: $s \leq 1/3$ 또는 $s \geq 1$.
그러니까, s 는 1/3 과 1 사이의 값은 가질 수가 없는 거죠. 따라서, 위에서 구한 $1/3 \leq s \leq 1$ 은 잘못된 겁니다. 그러면, 그저 y 절편의 범위만 갖고 x 절편의 범위를 구하는 것은 좋은 생각이 아니죠.

그럼 이제, $0 \leq t < 1$ 와 $1 < t \leq 2$ 의 두 개의 경우들 각각에 대하여 그 기울기 $a = 3(t - 1)$의 범위들을 알아보도록 하죠. 그러면:
$0 \leq t < 1 \Rightarrow -1 \leq t - 1 < 0 \Rightarrow -3 \leq 3(t - 1) < 0$.
$1 < t \leq 2 \Rightarrow 0 < t - 1 \leq 1 \Rightarrow 0 < 3(t - 1) \leq 3$.

그러면:
$0 \leq t < 1 \Rightarrow -3 \leq a < 0$ 이고 $s \geq 1$. $1 < t \leq 2 \Rightarrow 0 < a \leq 3$ 이고 $s \leq 1/3$.

기울기가 음이고 x 가 x 절편보다 작으면 y 가 양이고 또한, 기울기가 양이고 x 가 x 절편보다 크면 y 가 양이죠.
그러면, $s < 1$ 이고 $s > 1/3$ 이어야 하죠. 그러므로, $1/3 < s < 1$ 입니다.
물론, $t = 1$ 일 때에는 $a = 0$ 이어서 그 직선은 수평선이죠.
또한, $t = 1 \Rightarrow y = 3(t - 1)x + 3 - 2t = 1$ 이니까 y 는 모든 x 에 대하여 양이네요.
그러므로, $0 \leq t \leq 2$ 일 때 $1/3 < x < 1$ 이면 $y > 0$ 입니다. 그러면 이제, 그 그래프는 대략 다음과 같죠:

그림. Q.0

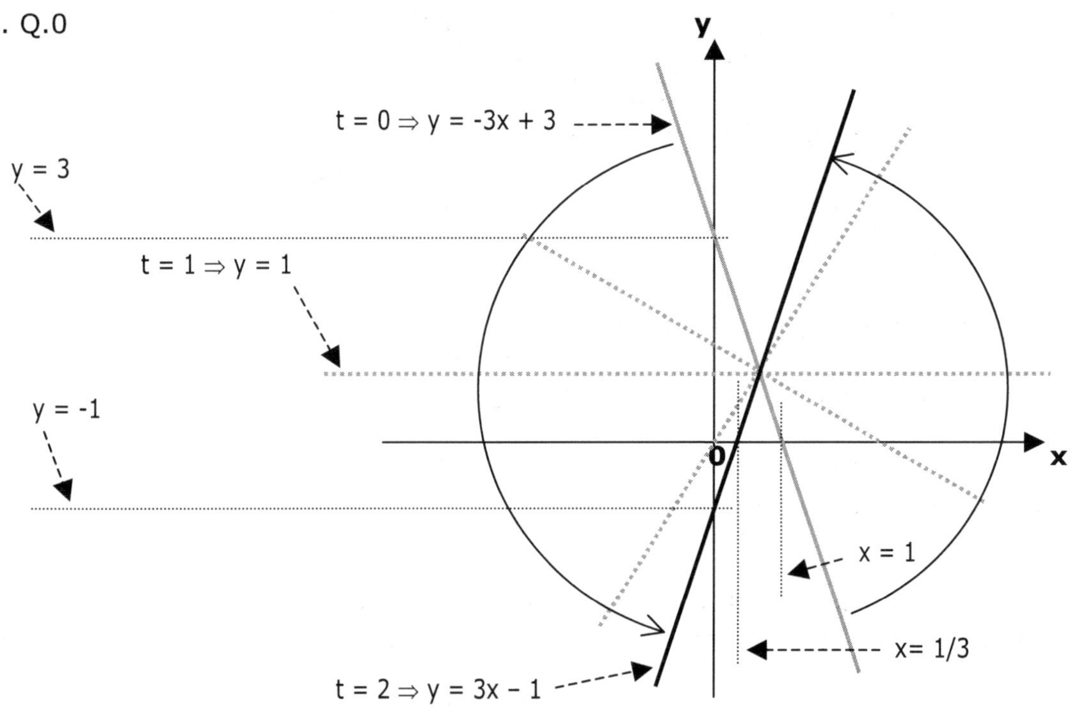

그러면, 회색과 검정 색 직선들이 서로 만나는 점을 구해보기로 하죠. x 좌표부터 시작하도록 하죠.
$-3x + 3 = 3x - 1 \Rightarrow 6x = 4 \Rightarrow x = 2/3 \Rightarrow y = 3x - 1 = 3(2/3) - 1 = 1$.

그러므로, 그 두 직선들은 점 (2/3, 1)에서 서로 만나네요.
사실, 방정식 $y = 3(t - 1)x + 3 - 2t$ 여기에서 $0 \le t \le 2$로 나타낼 수 있는 그 모든 직선들이 점 (2/3, 1)에서 모두 함께 서로 만납니다. 어째서 요? 다음 절에서 알게 될 겁니다.

그러니까, 직선이 y축을 따라서 내려가는 것이 아니라 하나의 점을 중심으로 회전하는 거네요.

그런데, $0 < t < 2$라면 어떨까요? 그 때는 $1/3 < x < 1$가 되죠. 왜냐하면 회색과 검정 직선들은 t가 0과 2일 때 만들어지는데 이 경우에는 t가 0과 2가 될 수 없기 때문이죠.

그러면 이제, 앞서 나왔던 예제들에 대하여 했던 방법으로 이 경우를 처리해보도록 하죠.
먼저, 주어진 직선의 방정식에서 t를 변수로 취급하고 x를 상수로 취급하고 그 방정식을 변형하는 겁니다.
그리고 나서, 상수로서의 x의 범위를 구하는 거죠. 물론, 상수와 변수는 서로 다른 겁니다. 상수와 변수 둘 다 범위를 가질 수 있다는 면을 이용하는 거죠. 이제, $y = f(x) = 3(t - 1)x + 3 - 2t$ 여기에서 t는 상수라고 하세요.
그러면, 이 f를 함수 g로 변형합니다. 이것들은 좌표계만 다를 뿐이지 서로 같은 것들이죠.

$3(t - 1)x + 3 - 2t = 3xt - 3x + 3 - 2t = (3x - 2)t + 3 - 3x.$

이제, $y = g(t) = (3x - 2)t + 3 - 3x$ 여기에서 x는 상수라고 하세요.
그러면, $0 < t < 2$일 때 $g(t) > 0$가 되도록 x의 범위를 구하는 겁니다.

그림. Q.1

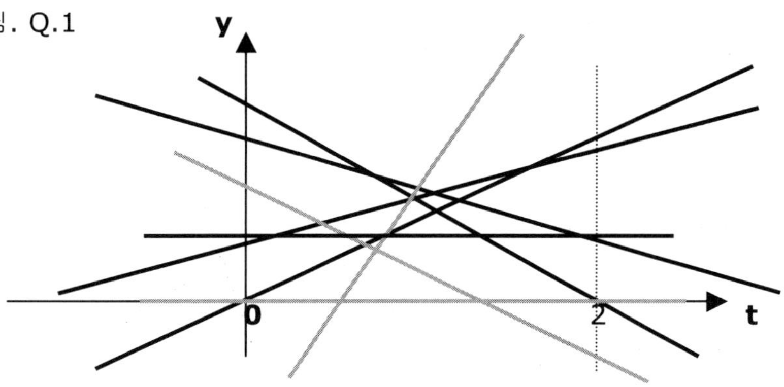

위의 그래프를 조사해보면 직선이 어떻게 놓여야 하는지 알 수 있죠.
그러면, t의 값이 0과 2사이에 있을 때는 직선이 t축 위쪽에 있어야 하는 거죠.
또한, $g(0) = 0$이고 $g(2) = 0$이어도 되는 거죠. 그러니까 $g(0) \ge 0$이고 $g(2) \ge 0$이어야 하죠.

먼저, g가 0함수가 되면 안되니까 그 검사를 해보아야 하죠. 그 기울기는 $3x - 2$입니다.
$3x - 2 = 0 \Rightarrow g(t) = (3x - 2)t + 3 - 3x = 3 - 3x.$
$3x - 2 = 0 \Rightarrow x = 2/3 \Rightarrow 3 - 3x = 3 - 2 = 1 \ne 0 \Rightarrow$ g는 0함수가 될 수 없죠.

그러면, $g(0) = (3x - 2)(0) + 3 - 3x = 3 - 3x \ge 0 \Rightarrow x \le 1.$
다음으로, $g(2) = (3x - 2)(2) + 3 - 3x = 6x - 4 + 3 - 3x = 3x - 1 \ge 0 \Rightarrow x \ge 1/3.$

그럼 이제, 이 두개의 조건들, "$g(0) \ge 0$와 $g(2) \ge 0$." 를 동시에 고려해야 하죠.
그러면, ($x \le 1$이고 $x \ge 1/3$) $\Rightarrow 1/3 \le x \le 1.$
그러면 이제, x의 값이 1/3보다 크거나 같고 1보다 작거나 같으면 $0 < t < 2$일 때 $f(x) > 0$가 되는 겁니다.

요약:

a 는 기울기이고 s 는 x 절편이라 하세요. 그러면, a = 3(t − 1). 이제, s 를 구합니다.
0 = 3(t − 1)x + 3 − 2t ⇒ 3x = (2t − 3)/(t − 1) = (2t − 2 − 1)/(t − 1) = (2t − 2)/(t − 1) − 1/(t − 1)
= 2 − 1/(t − 1) = 2 + 1/(1 − t) ⇒ x = (1/3){2 + 1/(1 − t)} ⇒ s = (1/3){2 + 1/(1 − t)}.

그러면 이제, a 와 s 의 범위들을 구합니다. 우선, t 가 증가함에 따라 a 도 증가.
0 ≤ t ≤ 2 ⇒ -1 ≤ t − 1 ≤ 1 ⇒ -3 ≤ 3(t − 1) ≤ 3 ⇒ -3 ≤ a ≤ 3.

다음으로, 먼저 c = 1/(1 − t)이라 하세요.
그러면, t = 0 ⇒ c = 1/(1 − t) = 1/1 = 1 ⇒ s = (1/3)(2 + 1) = 1.
그러면, t 가 0 에서 1 을 향하여 커감에 따라 분모가 0 에 가까워지니까 c 는 무한히 커지게 되네요.
따라서, 이 경우에 t 의 값이 1 과 거의 같으면 결국 s 도 무한히 커지게 되죠. 그러면, 0 ≤ t < 1 일 때에는 s ≥ 1.
사실, t = 1 이면 그 기울기 a = 3(t − 1) = 0 이므로 그 직선은 수평이고 이 경우에 x 절편은 없는 겁니다.

그러면 이제, t = 2 ⇒ c = 1/(1 − t) = 1/(-1) = -1. 그러면, s = (1/3)(2 − 1) = 1/3.
그러면, t 가 2 에서 1 을 향하여 감소함에 따라 분모가 0 에 가까워지고 또한 음이므로 c 는 음의 무한대로 작아지네요. 따라서, s 도 음의 무한대로 작아지죠. 그러면, 1 < t ≤ 2 일 때에는 s ≤ 1/3 입니다.
또한, t 가 1 이면 그 직선은 수평이라는 걸 알고 있죠.
그럼 이제, s 의 범위는 다음과 같아요: s ≤ 1/3 또는 s ≥ 1.

그러면:
0 ≤ t < 1 ⇒ -1 ≤ t − 1 < 0 ⇒ -3 ≤ 3(t − 1) < 0 ⇒ -3 ≤ a < 0 이고 s ≥ 1.
1 < t ≤ 2 ⇒ 0 < t − 1 ≤ 1 ⇒ 0 < 3(t − 1) ≤ 3 ⇒ 0 < a ≤ 3 이고 s ≤ 1/3.

기울기가 음이고 x 가 x 절편보다 작으면 y 가 양이고 또한, 기울기가 양이고 x 가 x 절편보다 크면 y 가 양이죠.
그러면, s < 1 이고 s > 1/3 이어야 하죠. 그러므로, 1/3 < s < 1 입니다.
물론, t = 1 일 때에는 a = 0 이어서 그 직선은 수평이죠.
또한, t = 1 ⇒ y = 3(t − 1)x + 3 − 2t = 1 이니까 y 는 모든 x 에 대하여 양이네요.
따라서, 0 ≤ t ≤ 2 일 때 1/3 < x < 1 이면 y > 0 입니다. 그러므로, 1/3 < x < 1.

다른 방법:

먼저, 주어진 직선의 방정식에서 t 를 변수로 취급하고 x 를 상수로 취급하고 그 방정식을 변형하는 겁니다.
y = f(x) = 3(t − 1)x + 3 − 2t 여기에서 t 는 상수라고 하세요.
그러면, 이 f 를 다음과 같이 함수 g 로 변형합니다:
3(t − 1)x + 3 − 2t = 3xt − 3x + 3 − 2t = (3x − 2)t + 3 − 3x.

이제, y = g(t) = (3x − 2)t + 3 − 3x 여기에서 x 는 상수라고 하세요.
그러면, 0 ≤ t ≤ 2 일 때 g(t) > 0 가 되도록 x 의 범위를 구하는 겁니다.
그러면, t 의 값이 0 에서 2 까지 변할 때는 직선이 t 축 위쪽에 있어야 하죠.
그러니까 g(0) > 0 이고 g(2) > 0 이어야 하죠.
그러면, g(0) = (3x − 2)(0) + 3 − 3x = 3 − 3x > 0 ⇒ x < 1.
또한, g(2) = (3x − 2)(2) + 3 − 3x = 6x − 4 + 3 − 3x = 3x − 1 > 0 ⇒ x > 1/3.
그러면, (x < 1 이고 x > 1/3) ⇒ 1/3 < x < 1.
그러면 이제, x 의 값이 1/3 보다 크고 1 보다 작으면 0 ≤ t ≤ 2 일 때 f(x) > 0 가 됩니다.
그러므로, 1/3 < x < 1.

제안과 풀이
R 번 예제에서의 문제

$-1/2 \le x \le 2$ 일 때 **$y = f(x) = 2x + 1$** 이고 **$2 < x \le 9/2$** 일 때 **$y = f(x) = 9 - 2x$** 이며 **$g(x) = (f \bullet f)(x)$** 라 하세요. 그러면, **$g(x)$**의 그래프를 작성하세요.

유의사항**:** 이 책과 **Runmath** 시리즈(seires)에서의 모든 책에서는 하나의 함수 표현에서 그 수식부분을 그 함수의 메커니즘(mechanism)이라고 합니다. 예를 들어, **$f(x) = 2x^3 + 3/x$** 이면 **f**의 메커니즘은 **$2x^3 + 3/x$**이죠. 또한, '**f**'는 함수의 **호출명**이라 하고 **f(x)**를 함수의 **이름**이라고 해요.

직선을 함수의 커브로 하여 그 함수를 사용하는 경우가 많습니다. 함수 f 에서는 정의역이 두 부분으로 되어있어요. 하나는 $-1/2 \le x \le 2$ 이고 나머지 하나는 $2 < x \le 9/2$ 입니다. 또한, 이 f 는 2 개의 메커니즘들을 갖고있죠.
그러면 이제, g 는 간단한 합성 함수인데 이는 f 로만 합성되어있죠. 그러면, 함수 g 에서도 정의역은 f 의 것과 같죠. 먼저, 정의역의 각 부분에 대하여 g 의 메커니즘을 구하기로 하죠.

$y = f(x) = 2x + 1$ 이고 여기에서 $-1/2 \le x \le 2$
$\Rightarrow g(x) = (f \bullet f)(x) = f\{f(x)\} = 2(2x + 1) + 1 = 4x + 3$ 이고 여기에서 $-1/2 \le x \le 2$.

$y = f(x) = 9 - 2x$ 이고 여기에서 **$2 < x \le 9/2$**
$\Rightarrow y = g(x) = (f \bullet f)(x) = f\{f(x)\} = 9 - 2(9 - 2x) = 9 - 18 + 4x = 4x - 9$ 이고 여기에서 **$2 < x \le 9/2$.**

그림. R.0.

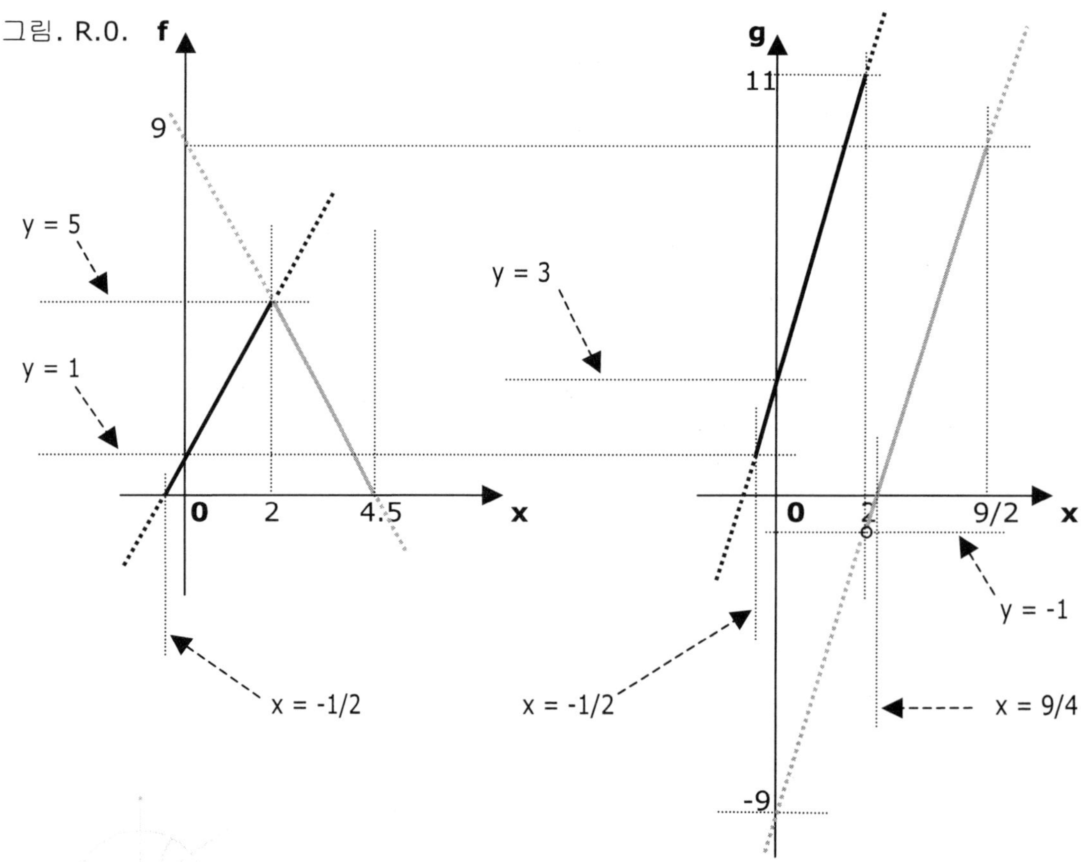

제안과 풀이
S 번 예제에서의 0 번 문제

주어진 방정식은 $y = |3 - 2x| - 1$ 이죠.
이런 방정식은 구간별로 표현할 수도 있어요. 각 구간은 $|3 - 2x|$으로 알 수가 있죠. $|a|$를 처리할 때는 $a \geq 0$ 와 $a < 0$ 인 두개의 경우들을 처리합니다. 그러니까, a 자체가 갖을 수 있는 값들을 부호(+ 또는 −)별로 구분하는 겁니다. 보통, 0 은 양인 경우와 함께 처리하죠.
그러면, $3 - 2x$ 에 대하여 두개의 경우들이 있을 수 있죠. 첫째 경우에서는 $3 - 2x$ 가 0 보다 크거니 같죠. 두 번째에서는 $3 - 2x$ 가 0 보다 작은 겁니다. 그러면 이제, 그 절대기호를 제거하면 다음과 같다는 것을 알 수 있어요:

$3 - 2x \geq 0 \Rightarrow 2x \leq 3 \Rightarrow x \leq 3/2 \Rightarrow y = 3 - 2x - 1 = 2 - 2x$ 이며 $x \leq 3/2$.
$3 - 2x < 0 \Rightarrow 2x > 3 \Rightarrow x > 3/2 \Rightarrow y = -(3 - 2x) - 1 = -3 + 2x - 1 = 2x - 4$ 이며 $x > 3/2$.

그런데, 어째서 $3 - 2x < 0$ 일 때 $|3 - 2x| = -(3 - 2x)$가 되죠?
이는 $|3 - 2x|$가 $3 - 2x$ 의 크기이므로 $3 - 2x$ 가 음이면 그 때는 이것을 양으로 만들어야 하기 때문이죠 크기라는 것은 0 이거나 양이니까.
그러므로, $y = 2 - 2x$ 이고 $x \leq 3/2$ 또는 $y = 2x - 4$ 이고 $x > 3/2$ 입니다.

이제, $y = |3 - 2x| - 1$ 을 그래프로 나타내보도록 하죠.

그림. S.0.0 $y = 2 - 2x$ 이고 $x \leq 3/2$ 또는 $y = 2x - 4$ 이고 $x > 3/2$

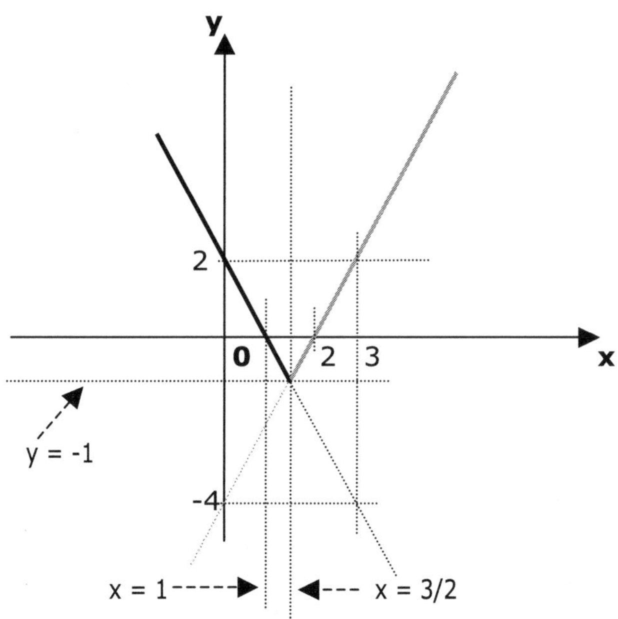

회색과 검정 직선들은 직선 $x = 3/2$ 에 관하여 서로 대칭이라는 것을 알 수 있죠.
검정직선은 점 (3/2, -1)을 포함하지만 회색직선은 이 점을 포함하지 않습니다.
이것들은 사실, 반직선 들이죠.

제안과 풀이
S 번 예제에서의 **1** 번 문제

주어진 방정식은 $y = |2x - 3| + 3x + 1$ 이죠.
이런 방정식은 구간별로 표현할 수도 있는데 각 구간은 $|3 - 2x|$으로 알 수 있죠.
그러면, $2x - 3$ 가 갖을 수 있는 값들을 음 또는 양의 부호별로 구분하죠. 그러면, $2x - 3$ 에 대하여 두개의 경우들이 있을 수 있죠. 첫째 경우에서는 $2x - 3$ 가 0 보다 크거나 같죠. 두 번째에서는 $2x - 3$ 가 0 보다 작은 겁니다. 그러면 이제, 그 절대기호를 제거해보도록 하죠. 그러면:

$2x - 3 \geq 0 \Rightarrow 2x \geq 3 \Rightarrow x \geq 3/2$
$\Rightarrow y = 2x - 3 + 3x + 1 = 5x - 2$ 이고 $x \geq 3/2$.

$2x - 3 < 0 \Rightarrow 2x < 3 \Rightarrow x < 3/2$
$\Rightarrow y = -(2x - 3) + 3x + 1 = 3 - 2x + 3x + 1 = x + 4$ 이고 $x < 3/2$.

그러므로, $y = 5x + 2$ 이고 $x \geq 3/2$ 또는, $y = x + 4$ 이고 $x < 3/2$ 이죠.

그럼 이제, $y = |2x - 3| + 3x + 1$ 을 그래프에 나타내보기로 하죠.

그림. S.1.0 $y = 5x + 2$ 이고 $x \geq 3/2$ 또는 $y = x + 4$ 이며 $x < 3/2$

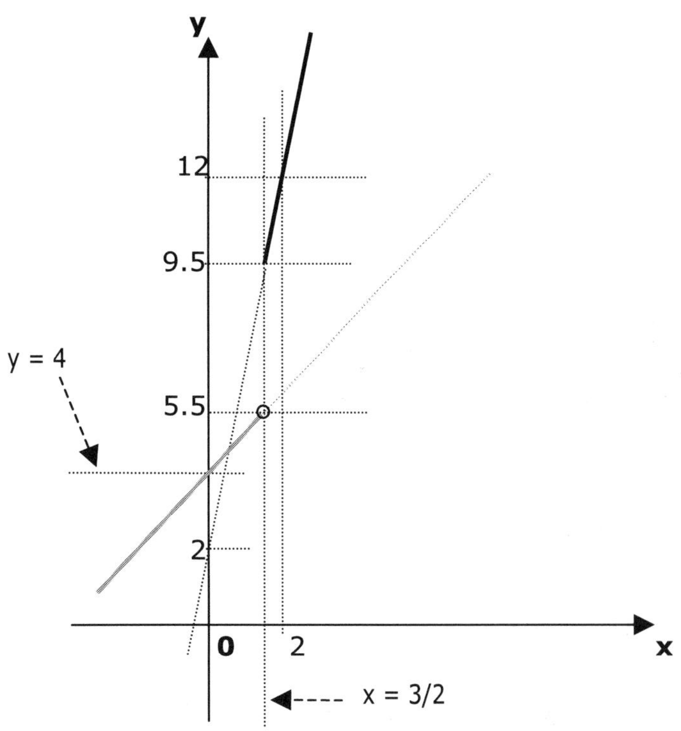

검정직선은 점 **(3/2, 9.5)**를 포함하지만 회색직선은 점 **(3/2, 5.5)**를 포함하지 않아요.

제안과 풀이
S번 예제에서의 2번 문제

주어진 방정식은 $y = x - |3x - 2|/2$ 이죠.
이런 방정식을 구간별로 표현할 수도 있는데 각 구간은 $|3x - 2|$로 알 수가 있죠.
그러면, $3x - 2$가 갖을 수 있는 값들을 + 또는 –의 부호별로 구분해보기로 하죠.
첫째 경우에서는 $3x - 2$가 0보다 크거나 같죠. 두 번째에서는 $3x - 2$가 0보다 작은 거죠.
그러면 이제, 그 절대기호를 제거하기로 하죠. 그러면:

$3x - 2 \geq 0 \Rightarrow 3x \geq 2 \Rightarrow x \geq 3/2$
$\Rightarrow y = x - (3x - 2)/2 = \{2x - (3x - 2)\}/2 = (-x + 2)/2 = 1 - x/2$ 이고 $x \geq 3/2$.

$3x - 2 < 0 \Rightarrow x < 3/2$
$\Rightarrow y = x - \{-(3x - 2)\}/2 = x + (3x - 2)/2 = (5x - 2)/2 = 5x/2 - 1$ 이고 $x < 3/2$.

그러므로, $y = 1 - x/2$ 이고 $x \geq 3/2$ 또는 $y = 5x/2 - 1$ 이고 $x < 3/2$.

그러면 이제, $y = x - |3x - 2|/2$를 그래프에 나타내보기로 하죠.

그림. S.2.0 $y = 1 - x/2$ 이고 $x \geq 3/2$ 또는 $y = 5x/2 - 1$ 이고 $x < 3/2$

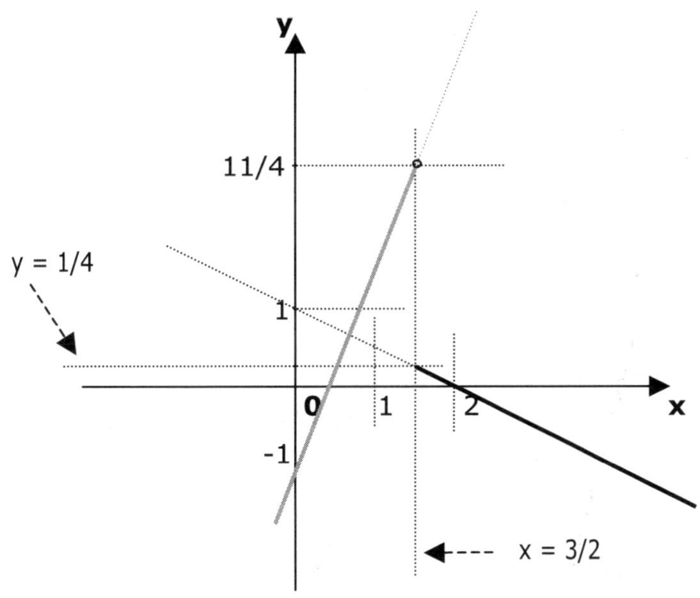

점 **(3/2, 11/4)**는 검정직선에 속하는 겁니다.

제안과 풀이
S번 예제에서의 3번 문제

주어진 방정식은 $2x + |y/2| = 2$ 이고 여기에서 x 뿐만이 아니라 y 도 변수입니다.
그러면 이제, y 가 그 어떤 값을 갖든 $|y/2|$는 y 가 갖는 값의 크기의 반을 나타내는 거죠.
그럼, $|y|/2$ 는 $|y/2|$하고 같은 건가요?
그렇죠. 그러니까, $2x + |y/2| = 2$ 는 $2x + |y|/2 = 2$ 와 마찬가지죠.
그럼, $|-x|$는 $|x|$하고도 같은가요?
그렇죠. 그럼 이제, $x \geq 0$ 이면 $|-x| = x$ 이고; 한편, $x < 0$ 이면 $|-x| = |x| = -x$ 입니다.

먼저, $y/2 \geq 0$ 인 경우부터 시작해보죠.
그러면, $y/2 \geq 0 \Rightarrow y \geq 0 \Rightarrow 2x + y/2 = 2 \Rightarrow y = 4 - 4x$ 이고 $y \geq 0$.
그러면, x 의 범위를 구해보죠.
$y \geq 0 \Rightarrow 4 - 4x \geq 0 \Rightarrow 4x \leq 4 \Rightarrow x \leq 1$. 따라서, $y = 4 - 4x$ 이고 $x \leq 1$.
이 x 를 y 에 관한 식으로 표현해보기로 하죠.
그러면, $y = 4 - 4x \Rightarrow 4x = 4 - y \Rightarrow x = 1 - y/4$ 이고 $y \geq 0$.

이제, $y/2 < 0$ 인 경우로 넘어가보기로 하죠.
그럼, $y/2 < 0 \Rightarrow y < 0 \Rightarrow 2x + \{-(y/2)\} = 2 \Rightarrow 2x - y/2 = 2 \Rightarrow y = 4x - 4$ 이고 $y < 0$.
그럼, x 의 범위를 구해보죠.
$y < 0 \Rightarrow 4x - 4 < 0 \Rightarrow 4x < 4 \Rightarrow x < 1$. 따라서, $y = 4x - 4$ 이고 $x < 1$.
이제, 이 x 를 y 에 관한 식으로 표현해보기로 하죠.
그러면, $y = 4x - 4 \Rightarrow 4x = y + 4 \Rightarrow x = y/4 + 1$ 이고 $y < 0$.

그러면, y 는 모든 실수가 다 되지만 x 는 1 보다 클 수가 없다는 것을 알 수 있죠.
이제, $y = 4 - 4x$ 이고 $x \leq 1$ 또는 $y = 4x - 4$ 이고 $x < 1$ 이죠.
또한 마찬가지로, $x = 1 - y/4$ 이고 $y \geq 0$ 또는 $x = y/4 + 1$ 이고 $y < 0$ 입니다.
그러면 이제, $2x + |y/2| = 2$ 를 그래프로 표현해보기로 하죠.

그림. S.3.0

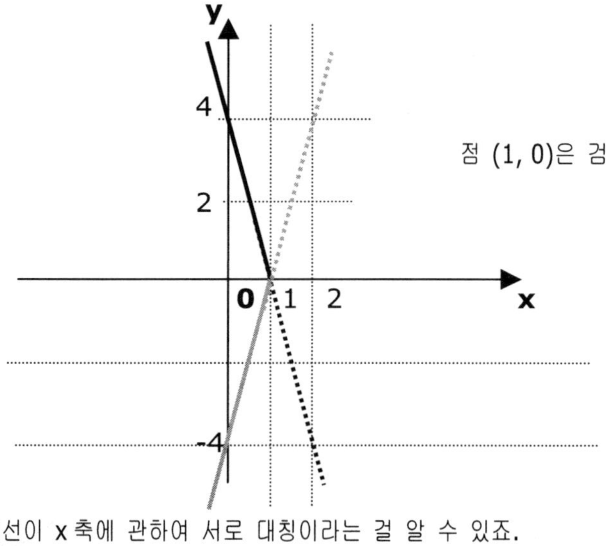

점 (1, 0)은 검정직선에 속하는 겁니다.

회색직선과 검정직선이 x 축에 관하여 서로 대칭이라는 걸 알 수 있죠.

제안과 풀이
S번 예제에서의 4번 문제

주어진 방정식은 $|2x| + y/2 = 2$ 이고 이는 구간별로 표현될 수 있는데 각 구간은 $|2x|$로 알 수가 있죠.
$|2x|$가 무슨 뜻이죠?
$|2x|$는 x가 갖는 값의 두 배의 크기입니다. 또한, x가 양이면 2x도 양이고 x가 음이면 2x도 음이죠.
그러니까, $|2x| = 2|x|$입니다.
그러면, 두개의 구간이 있는데, 첫째 구간에서는 x가 0보다 크거나 같죠. 두 번째에서는 x가 0보다 작은 거죠.

먼저, 주어진 방정식은 다음과 같이 변형할 수 있어요.
$|2x| + y/2 = 2 \Rightarrow 2|x| + y/2 = 2 \Rightarrow |x| + y/4 = 1$.

그런데, y = 어쩌고저쩌고 하는 표현 형식이 더 편하면 그렇게 하면 됩니다. 이는 그 표준형과 비슷한 거죠.
그러면, $|x| + y/4 = 1 \Rightarrow y/4 = 1 - |x| \Rightarrow y = 4 - 4|x|$.

그러면 이제, 그 절대기호를 제거하도록 하죠.
$x \geq 0 \Rightarrow y = 4 - 4x$. $x < 0 \Rightarrow y = 4 - 4(-x) = 4 + 4x$.

그러면 이제, $y = 4 - 4x$ 이고 $x \geq 0$, 또는 $y = 4 + 4x$ 이고 $x < 0$.

그러면, $|2x| + y/2 = 2$ 의 그래프는 다음과 같습니다:

그림. S.4.0 $y = 4 - 4x$ 이고 $x \geq 0$, 또는 $y = 4 + 4x$ 이고 $x < 0$

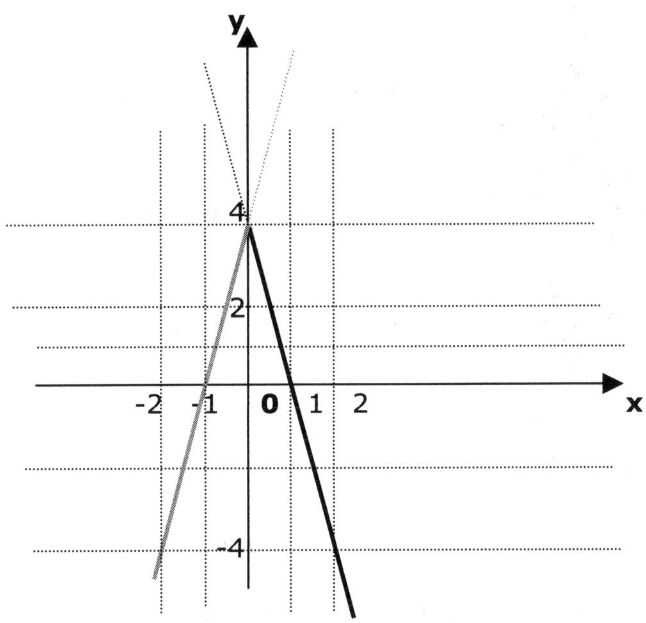

회색직선과 검정직선이 y축에 관하여 서로 대칭이라는 걸 알 수 있죠.

제안과 풀이
S번 예제에서의 5번 문제

주어진 방정식은 $|2x| + |y/2| = 2$ 이죠. 이 경우엔 x와 y 둘 다 절대기호를 갖고있네요.
먼저, 주어진 방정식을 다음과 같이 변형해보도록 하죠.
$|2x| + |y/2| = 2 \Rightarrow 2|x| + |y|/2 = 2 \Rightarrow |y| = 4 - 4|x|$.

그러면, 이 방정식을 구간별로 표현해야 하는데 각 구간은 $|x|$와 $|y|$로 알 수 있죠.
두 변수들 x와 y 각각에 대하여 2개의 구간들이 있으니까 4개의 경우들을 고려해야 합니다.

$x \geq 0$ 이고 $y \geq 0 \Rightarrow y = 4 - 4x$ 여기에서 $x \geq 0$ 이고 $y \geq 0$.
$x \geq 0$ 이고 $y < 0 \Rightarrow -y = 4 - 4x \Rightarrow y = 4x - 4$ 여기에서 $x \geq 0$ 이고 $y < 0$.
$x < 0$ 이고 $y \geq 0 \Rightarrow y = 4 - 4(-x) = 4 + 4x \Rightarrow y = 4x + 4$ 여기에서 $x < 0$ 이고 $y \geq 0$.
$x < 0$ 이고 $y < 0 \Rightarrow -y = 4 - 4(-x) = 4 + 4x \Rightarrow y = -4x - 4$ 여기에서 $x < 0$ 이고 $y < 0$.

그럼 이제, $|2x| + |y/2| = 2$를 그래프로 나타내보기로 하죠.

그림. S.5.0

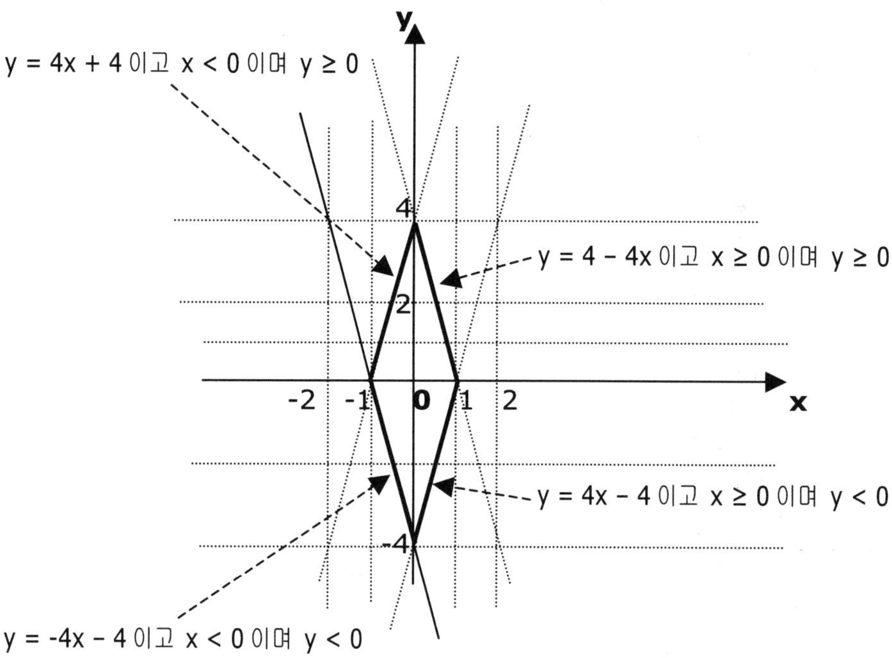

각 선분은 나머지 선분들 각각과 x축, y축, 또는 원점에 관하여 서로 대칭이라는 걸 알 수 있죠.

선분 $y = 4 - 4x$ 이고 $x \geq 0$ 이며 $y \geq 0$ 은 양 끝점들 (0, 4)와 (1, 0)을 포함합니다.
선분 $y = 4x - 4$ 이고 $x \geq 0$ 이며 $y < 0$ 은 점 (0, -4)는 포함하지만 (1, 0)은 포함하지않죠.
선분 $y = -4x - 4$ 이고 $x < 0$ 이며 $y < 0$ 은 양 끝점들을 포함하지 않아요.
선분 $y = 4x + 4$ 이고 $x < 0$ 이며 $y \geq 0$ 은 (-1, 0)은 포함하지만 (0, 4)는 포함하지않습니다.

제안과 풀이
S번 예제에서의 6번 문제

주어진 방정식은 y = $(x^2)^{1/2}$ = $\sqrt{x^2}$ 이죠. 이 경우엔 x와 y 둘 다 절대기호를 갖고있지 않네요. 하지만, 제곱근호는 절대기호로 취급하는 것이 좋습니다. 예를 들어, 근호 안에 변수의 지수가 짝수이면 근호를 제거할 때 절대기호를 제거하듯 하는 것이죠.

일반적으로 $\sqrt{x^2} \geq 0$ 입니다. 그러면 다음과 같은 예를 하나 들어보기로 하죠.

$\sqrt{4} = \sqrt{2^2} = \sqrt{(-2)^2} = 2$ 라는 것을 알고있죠.

그러면, $\sqrt{x^2}$ 에서 x = -2 이면 $\sqrt{x^2} \neq x$ 이죠 왜냐하면 x = -2 이고 $\sqrt{x^2} > 0$ 이기 때문입니다.

또한, $\sqrt{x^2} = \sqrt{4}$ 라는 것도 알고있죠. 그러므로, 이 경우에는 $\sqrt{x^2}$ = -x 인 것이죠.

마찬가지로, 그저 $\sqrt{(x-2)^2}$ = x – 2 라고 하면 안되죠.

이는 왜냐하면 $\sqrt{(x-2)^2} \geq 0$ 인데 x – 2 < 0 일 수도 있기 때문입니다.

그러므로, $\sqrt{(x-2)^2}$ = x – 2 여기에서 x ≥ 2 라고 해야 하는 거죠.

그러면 이제, y = $(x^2)^{1/2}$ = $\sqrt{x^2}$ 입니다. 그러면, $\sqrt{x^2} \geq 0$ 이므로 y ≥ 0 이라는 걸 알고있죠.
그러므로, 그저 y = x 라고 할 수는 없죠 왜냐하면 x 가 음이 될 수도 있으니까.

그러면, 다음과 같이 근호를 제거할 수 있네요:
y = x 여기에서 x ≥ 0. y = -x 여기에서 x < 0.
이제, y = $(x^2)^{1/2}$ = $\sqrt{x^2}$ 의 그래프를 작성해보도록 하죠.

그림. S.6.0 y = x 이고 x ≥ 0, 또는 y = -x 이고 x < 0

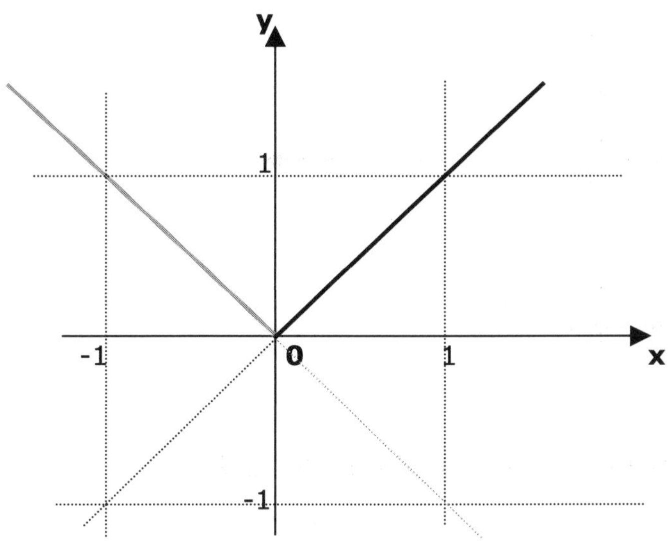

검정직선은 원점을 포함하지만 회색직선은 원점을 포함하지 않습니다.
y = $(x^2)^{1/2}$ = |x|라는 것을 알 수 있죠.

제안과 풀이
S번 예제에서의 7번 문제

주어진 방정식은 $y = (x^2)^{1/2}/x = \sqrt{x^2}\,/\,x$ 이죠. 이 경우에도 x와 y 둘 다 절대기호를 갖고있지 않네요. 하지만, 제곱근호는 절대기호처럼 처리하는 것이 좋죠. 일반적으로 $\sqrt{x^2} \geq 0$ 입니다.

그런데, 분모는 0 절대 0 이 될 수 없다는 것을 알고있죠. 그러니까, $x \neq 0$ 이어야 하고 그저, $x \geq 0$ 라고 할 수는 없는 거죠. 따라서, x 의 범위를 아래와 같은 방법으로 정해야 합니다.

$x > 0 \Rightarrow y = x/x = 1$ 이고 $x > 0$.
$x < 0 \Rightarrow y = -x/x = -1$ 이고 $x < 0$.

그러면, $y = \sqrt{x^2}\,/\,x$ 를 다음과 같이 표현할 수 있어요:
$y = 1$ 이고 $x > 0$, 또는 $y = -1$ 이고 $x < 0$.

이제, $y = (x^2)^{1/2}/x = \sqrt{x^2}\,/\,x$ 의 그래프를 작성해보도록 하죠.

그림. S.7.0 $y = -1$ 이고 $x > 0$, 또는 $y = 1$ 이고 $x < 0$

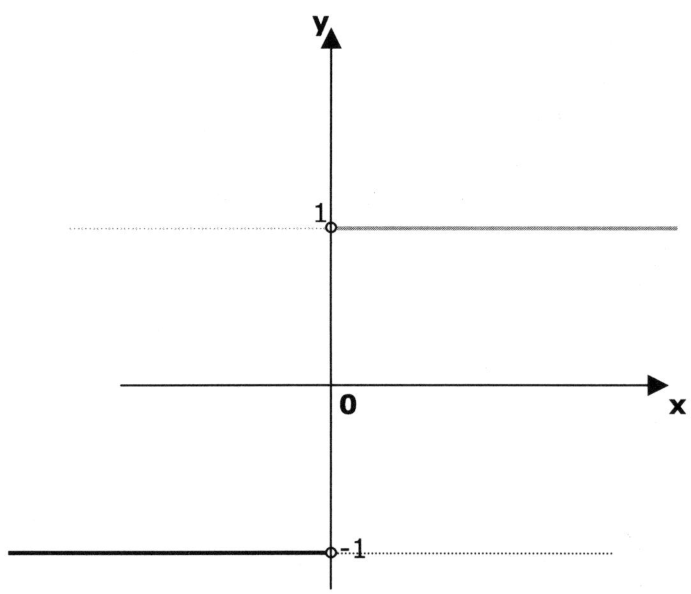

회색직선은 점 (0, 1)을 포함하지 않고 검정직선은 점 (0, -1)을 포함하지 않죠.

제안과 풀이
S번 예제에서의 8번 문제

주어진 방정식은 $y = (x^2)^{1/2}/|x| = \sqrt{x^2}\ /\ |\ x\ |$이죠. 이 경우에는 x가 절대기호를 갖고있네요. 제곱근호는 절대기호처럼 처리하는 것이 좋죠. 일반적으로 $\sqrt{x^2} \geq 0$ 입니다.

그런데, 분모는 0 절대 0 이 될 수 없다는 것을 알고있죠. 그러니까, $x \neq 0$ 이어야 하고 그저, $x \geq 0$ 라고 할 수는 없는 거죠. 따라서, x 의 범위를 아래와 같은 방법으로 설정해야 합니다.

$x > 0 \Rightarrow y = x/x = 1$ 이고 $x > 0$.
$x < 0 \Rightarrow y = -x/(-x) = 1$ 이고 $x < 0$.

그러면, $y = \sqrt{x^2}\ /\ |\ x\ |$를 다음과 같이 표현할 수 있어요:
$y = 1$ 이고 x 는 0 을 제외한 모든 실수.

이제, $y = (x^2)^{1/2}/|x| = \sqrt{x^2}\ /\ |\ x\ |$를 그래프에 나타내보도록 하죠.

그림. S.8.0 0 을 제외한 모든 실수에 대하여 $y = 1$.

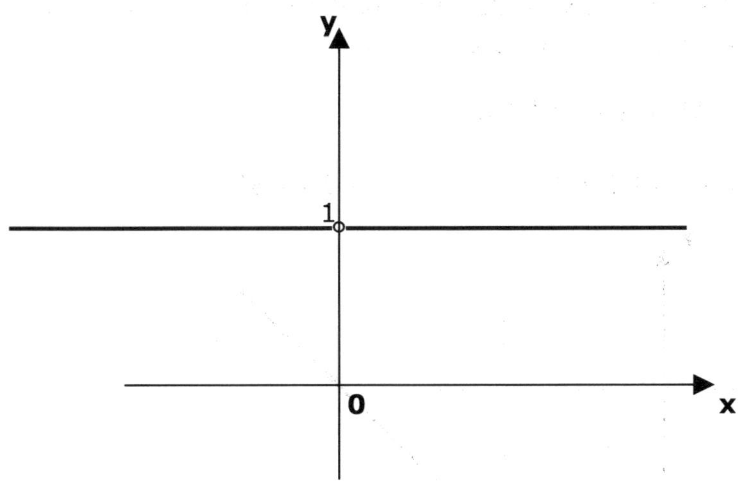

위의 직선은 점 (0, 1)을 포함하지않는 거죠.

제안과 풀이
S번 예제에서의 9번 문제

주어진 방정식은 $y = (x^2 - 5x + 25/4)^{1/2} = \sqrt{x^2 - 5x + 25/4}$ 이죠. 제곱근호는 절대기호처럼 처리하는 것이 좋죠. 예를 들어, 근호 안에 변수의 지수가 짝수이면 근호를 제거할 때 절대기호를 제거하듯 하는 것이죠. 일반적으로 $\sqrt{x^2} \geq 0$ 이죠.

그런데, 근호 안에 있는 것이 완전 제곱식 같네요. 그럼, 그런지 확인해보도록 하죠.
$x^2 - 5x + 25/4 = x^2 - 2x(5/2) + (5/2)^2 = (x - 5/2)^2$. 그러네요.

그러면, $y = \{(x - 5/2)^2\}^{1/2} = |x - 5/2|$이고 이는 직선의 방정식이군요.
그런데, 왜 절대기호인가요?
이는 $(x^2 - 5x + 25/4)^{1/2} \geq 0$ 이니까 y가 0보다 크거나 같아야 하기 때문이죠.
$x - 5/2$는 어떤 x 값들에 대하여는 음이 될 수도 있다는 것을 알고있죠.

자 이제, 절대기호도 제거하기로 하죠.

$x - 5/2 \geq 0 \Rightarrow x \geq 5/2 \Rightarrow y = x - 5/2$ 이고 $x \geq 5/2$.
$x - 5/2 < 0 \Rightarrow x < 5/2 \Rightarrow y = -(x - 5/2) = 5/2 - x$ 이고 $x < 5/2$.

그리하여, $y = (x^2 - 5x + 25/4)^{1/2} = \sqrt{x^2 - 5x + 25/4}$ 를 다음과 같은 형태로 나타낼 수 있죠:
$y = x - 5/2$ 이고 $x \geq 5/2$, 또는 $y = 5/2 - x$ 이고 $x < 5/2$.

이제, $y = (x^2 - 5x + 25/4)^{1/2}$를 그래프에 놓아보기로 하죠.

그림. S.9.0 $y = x - 5/2$ 이고 $x \geq 5/2$, 또는 $y = 5/2 - x$ 이고 $x < 5/2$

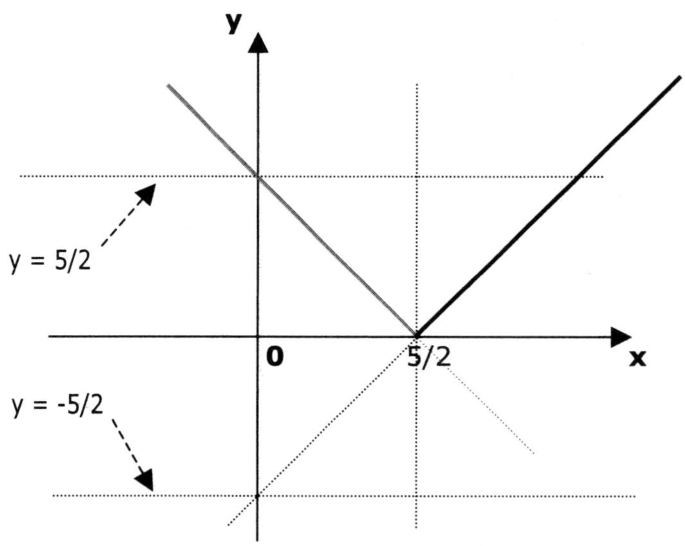

이 두 직선들은 직선 $x = 5/2$ 에 관하여 대칭입니다.
회색직선은 점 (5/2, 0)을 포함하지 않지만 검정직선은 이를 포함하는 거죠.

제안과 풀이
S번 예제에서의 **A**번 문제

주어진 방정식은 $y = 2x\{|x^2 - 1|/(x^2 - 1)\}$이죠.

절대기호 때문에 다음과 같이 두개의 경우를 고려해야 하죠.

첫째, $x^2 - 1 \ge 0 \Rightarrow (x + 1)(x - 1) \ge 0 \Rightarrow x \ge 1$ 이거나 $x \le -1$
$\Rightarrow y = 2x(x^2 - 1)/(x^2 - 1) = 2x$이고 $x \ge 1$ 또는 $x \le -1$.

다음으로, $x^2 - 1 < 0 \Rightarrow (x + 1)(x - 1) < 0 \Rightarrow -1 < x < 1$
$\Rightarrow y = 2x\{-(x^2 - 1)\}/(x^2 - 1) = -2x(x^2 - 1)\}/(x^2 - 1) = -2x$이고 $-1 < x < 1$.

그러므로, $y = 2x\{|x^2 - 1|/(x^2 - 1)\}$를 다음과 같이 나타낼 수 있죠:

$y = 2x$ 이고 $x \ge 1$ 이거나 $x \le -1$, 또는 $y = -2x$ 이고 $-1 < x < 1$.

이제, $y = 2x\{|x^2 - 1|/(x^2 - 1)\}$를 그래프로 나타내보기로 하죠.

그림. S.A.0 $y = 2x$ 이고 $x \ge 1$ 이거나 $x \le -1$, 또는 $y = -2x$ 이고 $-1 < x < 1$

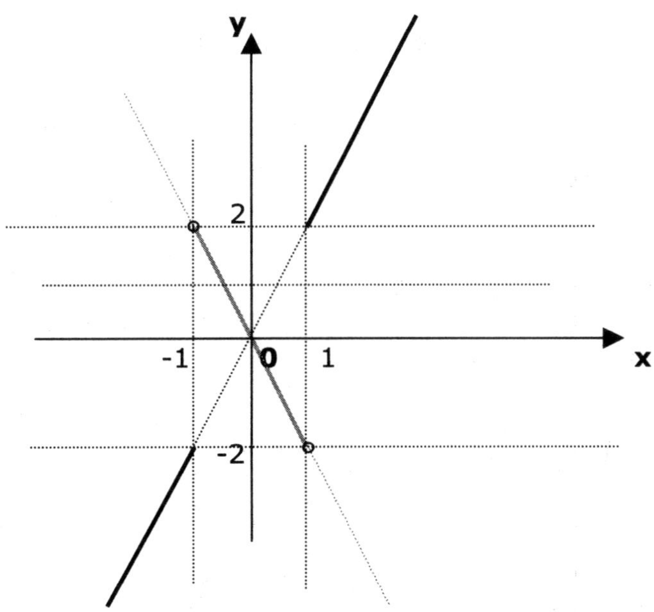

검정직선은 두 점들 (-1, -2)와 (1, 2)를 포함하지만 회색직선은 양 끝점들 (-1, 2)와 (1, -2)를 포함하지 않습니다.

제안과 풀이
S번 예제에서의 B번 문제

주어진 방정식은 $y = ((2x + 1)^2)^{1/2} + ((2x - 1)^2)^{1/2} = \sqrt{(2x+1)^2} + \sqrt{(2x-1)^2}$ 이죠.
근호를 제거할 때는 그 결과가 0 보다 크거나 같은지를 확인해야 합니다. 이는 실수의 제곱근은 0 보다 크거나 같기 때문이죠. 그러니까, 근호와 지수를 그냥 제거해서는 안 되는 거죠. 하나의 근호에 두 개의 경우들이 있고 근호가 둘이므로, 다음과 같이 네 경우들을 고려해야 합니다:

$(2x + 1 \geq 0$ 이고 $2x - 1 \geq 0) \Rightarrow (x \geq -1/2$ 이고 $x \geq 1/2) \Rightarrow x \geq 1/2$
$\Rightarrow y = 2x + 1 + 2x - 1 = 4x$ 이며 $x \geq 1/2$.

$(2x + 1 \geq 0$ 이고 $2x - 1 < 0) \Rightarrow (x \geq -1/2$ 이고 $x < 1/2) \Rightarrow -1/2 \leq x < 1/2$
$\Rightarrow y = 2x + 1 + \{-(2x - 1)\} = 2x + 1 - 2x + 1 = 2$ 이며 $-1/2 \leq x < 1/2$.

$(2x + 1 < 0$ 이고 $2x - 1 \geq 0) \Rightarrow (x < -1/2$ 이고 $x \geq 1/2) \Rightarrow$ 없습니다.

$(2x + 1 < 0$ 이고 $2x - 1 < 0) \Rightarrow (x < -1/2$ 이고 $x < 1/2) \Rightarrow x < -1/2$
$\Rightarrow y = -(2x + 1) + \{-(2x - 1)\} = -2x - 1 - 2x + 1 = -4x$ 이며 $x < -1/2$.

그리하여, $y = \sqrt{(2x+1)^2} + \sqrt{(2x-1)^2}$ 를 다음과 같은 형태로 표현할 수 있어요:
$y = 4x$ 이고 $x \geq 1/2$, $y = 2$ 이고 $-1/2 \leq x < 1/2$, 또는 $y = -4x$ 이고 $x < -1/2$.

그러면 이제, $y = ((2x + 1)^2)^{1/2} + ((2x - 1)^2)^{1/2} = \sqrt{(2x+1)^2} + \sqrt{(2x-1)^2}$ 를 그래프로 나타내보기로 하죠.

Fig. S.B.0 **$y = 4x$ 이고 $x \geq 1/2$, $y = 2$ 이고 $-1/2 \leq x < 1/2$, 또는 $y = -4x$ 이고 $x < -1/2$**

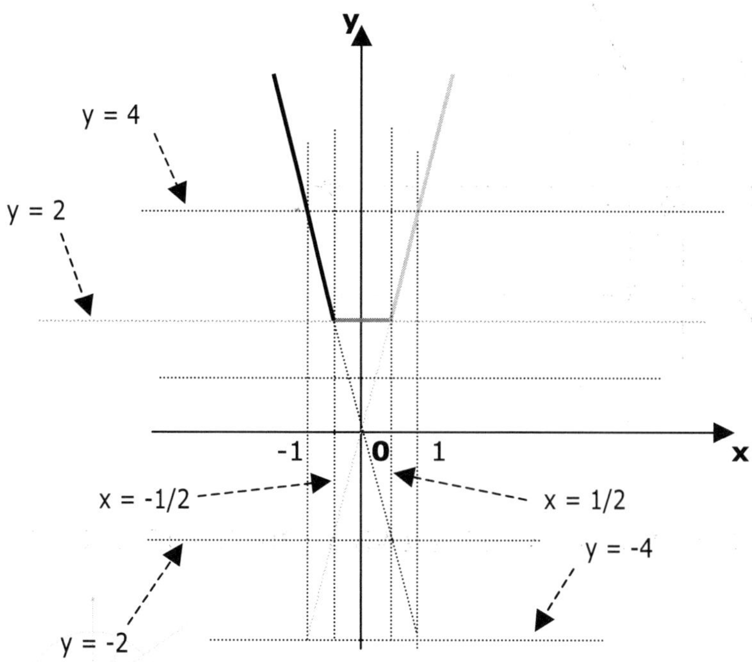

제안과 풀이
S번 예제에서의 **C**번 문제

주어진 방정식은 $||2x| - |3y|| = 1$ 이죠.
절대기호 내부에 있는 것이 음인 경우와 0보다 크거나 같은 경우를 고려해야 하죠.
그러니까, $|x|$와 $|y|$뿐만 아니라 $|2x| - |3y| \geq 0$ 와 $|2x| - |3y| < 0$ 도 고려해야 합니다.

그러면, 다음과 같이 8개의 경우들을 고려해야 하죠.

($x \geq 0$ 이고 $y \geq 0$ 이며 $2x \geq 3y$) $\Rightarrow$ ($x \geq 0$ 이고 $y \geq 0$ 이며 $y \leq 2x/3$) $\Rightarrow 2x - 3y = 1$.
($x \geq 0$ 이고 $y \geq 0$ 이며 $2x < 3y$) $\Rightarrow$ ($x \geq 0$ 이고 $y \geq 0$ 이며 $y > 2x/3$) $\Rightarrow -(2x - 3y) = 3y - 2x = 1$.

($x \geq 0$ 이고 $y < 0$ 이며 $2x \geq -3y$) $\Rightarrow$ ($x \geq 0$ 이고 $y < 0$ 이며 $y \geq -2x/3$) $\Rightarrow 2x - (-3y) = 2x + 3y = 1$.
그런데, 어째서 "$2x \geq -3y$"에서 $-3y$ 인가요?
이는 $y < 0$ 이니까 $|3y|$가 절대기호가 제거되면 $-3y$ 가 되기 때문이죠. 그러면, 다음으로 진행하도록 하죠.

($x \geq 0$ 이고 $y < 0$ 이며 $2x < -3y$) $\Rightarrow$ ($x \geq 0$ 이고 $y < 0$ 이며 $y < -2x/3$) $\Rightarrow -\{2x - (-3y)\} = -2x - 3y = 1$.

($x < 0$ 이고 $y \geq 0$ 이며 $-2x \geq 3y$) $\Rightarrow$ ($x < 0$ 이고 $y \geq 0$ 이며 $y \leq -2x/3$) $\Rightarrow -2x - 3y = 1$.
($x < 0$ 이고 $y \geq 0$ 이며 $-2x < 3y$) $\Rightarrow$ ($x < 0$ 이고 $y \geq 0$ 이며 $y > -2x/3$) $\Rightarrow -(-2x - 3y) = 2x + 3y = 1$.

($x < 0$ 이고 $y < 0$ 이며 $-2x \geq -3y$) $\Rightarrow$ ($x < 0$ 이고 $y < 0$ 이며 $y \geq 2x/3$) $\Rightarrow -2x - (-3y) = 3y - 2x = 1$.
($x < 0$ 이고 $y < 0$ 이며 $-2x < -3y$) $\Rightarrow$ ($x < 0$ 이고 $y < 0$ 이며 $y < 2x/3$) $\Rightarrow -\{-2x - (-3y)\} = 2x - 3y = 1$.

그러면 이제:
($x \geq 0$ 이고 $y \geq 0$ 이며 $y \leq 2x/3$) $\Rightarrow 2x - 3y = 1 \Rightarrow y = 2x/3 - 1/3$.
($x \geq 0$ 이고 $y \geq 0$ 이며 $y > 2x/3$) $\Rightarrow 3y - 2x = 1 \Rightarrow y = 2x/3 + 1/3$.
($x \geq 0$ 이고 $y < 0$ 이며 $y \geq -2x/3$) $\Rightarrow 2x + 3y = 1 \Rightarrow y = -2x/3 + 1/3$.
($x \geq 0$ 이고 $y < 0$ 이며 $y < -2x/3$) $\Rightarrow -2x - 3y = 1 \Rightarrow y = -2x/3 - 1/3$.
($x < 0$ 이고 $y \geq 0$ 이며 $y \leq -2x/3$) $\Rightarrow -2x - 3y = 1 \Rightarrow y = -2x/3 - 1/3$.
($x < 0$ 이고 $y \geq 0$ 이며 $y > -2x/3$) $\Rightarrow 2x + 3y = 1 \Rightarrow y = -2x/3 + 1/3$.
($x < 0$ 이고 $y < 0$ 이며 $y \geq 2x/3$) $\Rightarrow 3y - 2x = 1 \Rightarrow y = 2x/3 + 1/3$.
($x < 0$ 이고 $y < 0$ 이며 $y < 2x/3$) $\Rightarrow 2x - 3y = 1 \Rightarrow y = 2x/3 - 1/3$.

그러므로:
$x \geq 0$ 이고 $y \geq 0$ 이며 $y \leq 2x/3$ 일 때 또는 $x < 0$ 이고 $y < 0$ 이며 $y < 2x/3$ 일 때 $y = 2x/3 - 1/3$.
$x \geq 0$ 이고 $y \geq 0$ 이며 $y > 2x/3$ 일 때 또는 $x < 0$ 이고 $y < 0$ 이며 $y \geq 2x/3$ 일 때 $y = 2x/3 + 1/3$.
$x \geq 0$ 이고 $y < 0$ 이며 $y \geq -2x/3$ 일 때 또는 $x < 0$ 이고 $y \geq 0$ 이며 $y > -2x/3$ 일 때 $y = -2x/3 + 1/3$.
$x \geq 0$ 이고 $y < 0$ 이며 $y < -2x/3$ 일 때 또는 $x < 0$ 이고 $y \geq 0$ 이며 $y \leq -2x/3$ 일 때 $y = -2x/3 - 1/3$.

이제, $||2x| - |3y|| = 1$ 을 그래프로 나타내어보죠.

그림. S.C.0

$x \ge 0$ 이고 $y \ge 0$ 이며 $y > 2x/3$ 일 때 또는 $x < 0$ 이고 $y < 0$ 이며 $y \ge 2x/3$ 일 때 $y = 2x/3 + 1/3$.

$x \ge 0$ 이고 $y < 0$ 이며 $y \ge -2x/3$ 일 때 또는 $x < 0$ 이고 $y \ge 0$ 이며 $y > -2x/3$ 일 때 $y = -2x/3 + 1/3$.

$x \ge 0$ 이고 $y \ge 0$ 이며 $y \le 2x/3$ 일 때 또는 $x < 0$ 이고 $y < 0$ 이며 $y < 2x/3$ 일 때 $y = 2x/3 - 1/3$.

$x \ge 0$ 이고 $y < 0$ 이며 $y < -2x/3$ 일 때 또는 $x < 0$ 이고 $y \ge 0$ 이며 $y \le -2x/3$ 일 때 $y = -2x/3 - 1/3$.

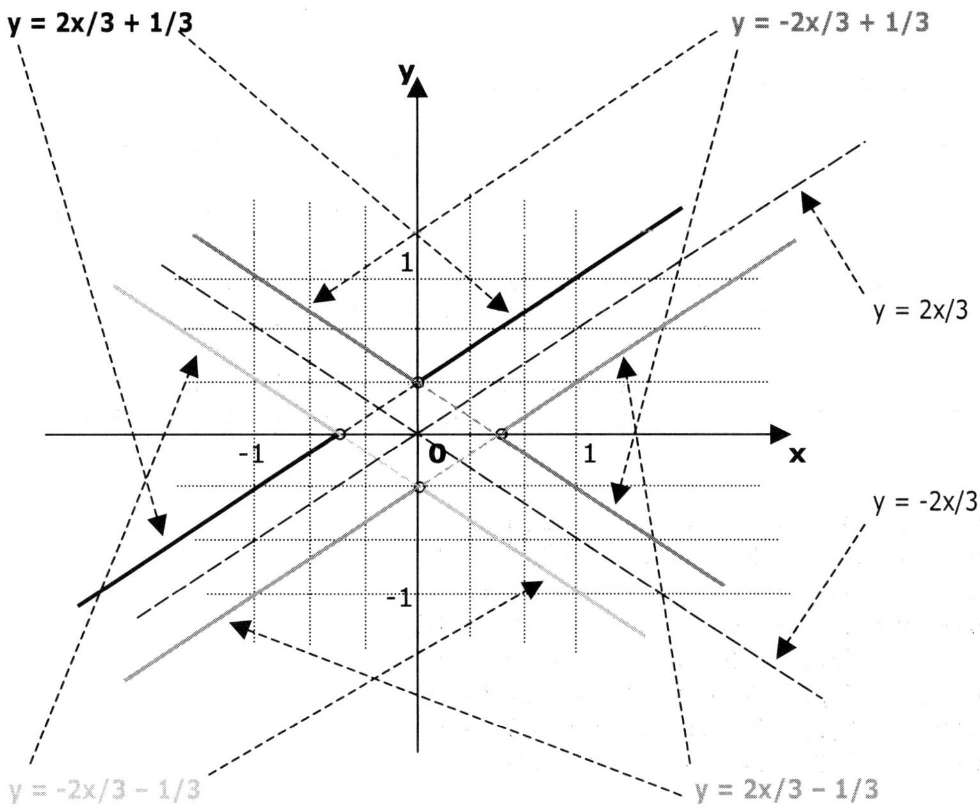

제안과 풀이
S번 예제에서의 D번 문제

주어진 방정식은 $y = |2x - 1| + |3x + 2|$이죠. 네 가지 경우들을 고려해야 하죠.

$(2x - 1 \geq 0$ 이고 $3x + 2 \geq 0) \Rightarrow (x \geq 1/2$ 이고 $x \geq -2/3) \Rightarrow x \geq 1/2$
$\Rightarrow y = 2x - 1 + 3x + 2 = 5x + 1$ 이고 $x \geq 1/2 \Rightarrow y = 5x + 1$ 이고 $x \geq 1/2$.

$(2x - 1 \geq 0$ 이고 $3x + 2 < 0) \Rightarrow (x \geq 1/2$ 이고 $x < -2/3) \Rightarrow$ 없는 거죠.
$(2x - 1 < 0$ 이고 $3x + 2 \geq 0) \Rightarrow (x < 1/2$ 이고 $x \geq -2/3) \Rightarrow -2/3 \leq x < 1/2$
$\Rightarrow y = -(2x - 1) + 3x + 2 = -2x + 1 + 3x + 2 = x + 3$ 이고 $-2/3 \leq x < 1/2$.
$\Rightarrow y = x + 3$ 이고 $-2/3 \leq x < 1/2$.

$(2x - 1 < 0$ 이고 $3x + 2 < 0) \Rightarrow (x < 1/2$ 이고 $x < -2/3) \Rightarrow x < -2/3$
$\Rightarrow y = -(2x - 1) + \{-(3x + 2)\} = -2x + 1 - 3x - 2 = -5x - 1$ 이고 $x < -2/3$.
$\Rightarrow y = -5x - 1$ 이고 $x < -2/3$.

이제, $y = |2x - 1| + |3x + 2|$를 그래프에 나타내보기로 하죠.

그림. S.D.0

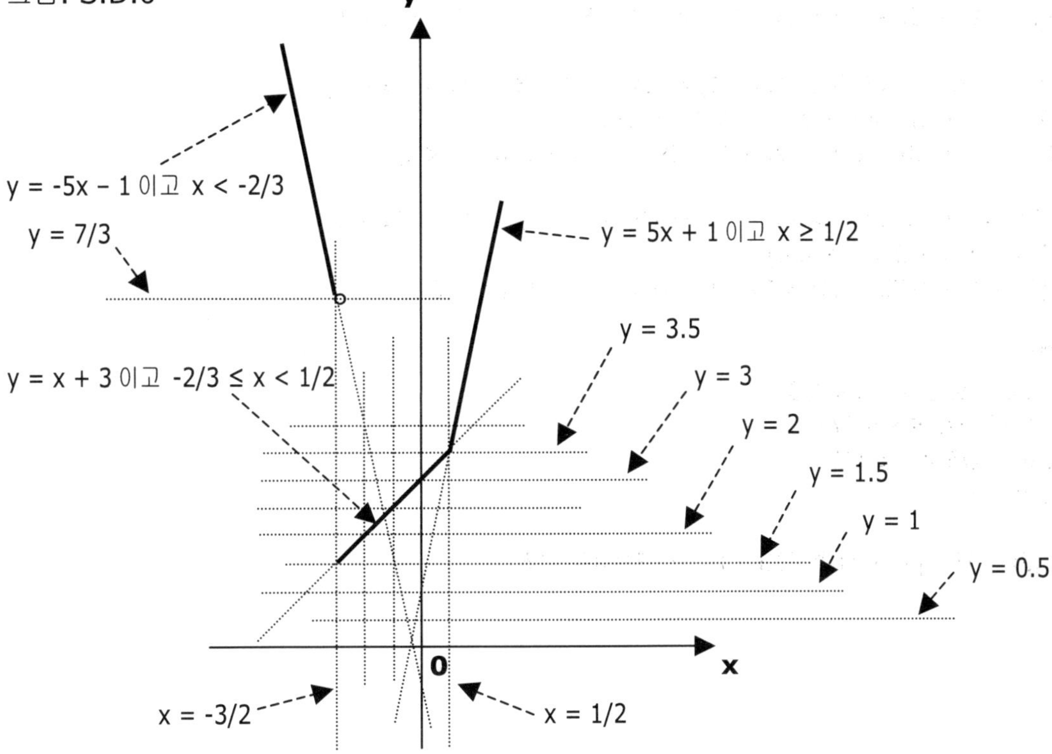

$y = -5x - 1$ 이고 $x < -2/3$ 은 점 $(-3/2, 7/3)$를 포함하죠.
$y = x + 3$ 이고 $-2/3 \leq x < 1/2$ 은 점 $(-3/2, 3/2)$를 포함하지만 점 $(1/2, 7/2)$는 포함하지 않죠.
$y = 5x + 1$ 이고 $x \geq 1/2$ 은 점 $(1/2, 7/2)$를 포함합니다.

제안과 풀이
S 번 예제에서의 **E** 번 문제

주어진 방정식은 $y = ||2x - 1| - |3x - 1||$이죠. 여덟 개의 경우들을 고려해야 하죠.

($2x - 1 \geq 0$ 이고 $3x - 1 \geq 0$ 이며 $2x - 1 \geq 3x - 1$) ⇒ ($x \geq 1/2$ 이고 $x \geq 1/3$ 이며 $x \leq 0$) ⇒ 없음.

($2x - 1 \geq 0$ 이고 $3x - 1 \geq 0$ 이며 $2x - 1 < 3x - 1$) ⇒ ($x \geq 1/2$ 이고 $x \geq 1/3$ 이며 $x > 0$) ⇒ $x \geq 1/2$
⇒ $y = -\{2x - 1 - (3x - 1)\} = -2x + 1 + 3x - 1 = x$ 이며 $x \geq 1/2$.

($2x - 1 \geq 0$ 이고 $3x - 1 < 0$ 이며 $2x - 1 \geq -(3x - 1) = -3x + 1$) ⇒ ($x \geq 1/2$ 이고 $x < 1/3$ 이며 $x \geq 2/5$)
⇒ 없음.
$2x - 1 \geq 0$ 이고 $3x - 1 < 0$ 이며 $2x - 1 < -(3x - 1)$ ⇒ 없음.

($2x - 1 < 0$ 이고 $3x - 1 \geq 0$ 이며 $-(2x - 1) = -2x + 1 \geq 3x - 1$) ⇒ ($x < 1/2$ 이고 $x \geq 1/3$ 이며 $x \leq 2/5$)
⇒ $1/3 \leq x \leq 2/5$
⇒ $y = -(2x - 1) - (3x - 1) = -2x + 1 - 3x + 1 = -5x + 2$ 이고 $1/3 \leq x \leq 2/5$.

($2x - 1 < 0$ 이고 $3x - 1 \geq 0$ 이며 $-(2x - 1) = -2x + 1 < 3x - 1$) ⇒ ($x < 1/2$ 이고 $x \geq 1/3$ 이며 $x > 2/5$)
⇒ $2/5 < x < 1/2$
⇒ $y = -\{-(2x - 1) - (3x - 1)\} = 2x - 1 + 3x - 1 = 5x - 2$ 이며 $2/5 < x < 1/2$.

($2x - 1 < 0$ 이고 $3x - 1 < 0$ 이며 $-(2x - 1) = -2x + 1 \geq -(3x - 1) = -3x + 1$)
⇒ ($x < 1/2$ 이고 $x < 1/3$ 이며 $x \geq 0$) ⇒ $0 \leq x < 1/3$
⇒ $y = -(2x - 1) - \{-(3x - 1)\} = -2x + 1 + 3x - 1 = x$ 이며 $0 \leq x < 1/3$.

($2x - 1 < 0$ 이고 $3x - 1 < 0$ 이며 $-(2x - 1) = -2x + 1 < -(3x - 1) = -3x + 1$)
⇒ ($x < 1/2$ 이고 $x < 1/3$ 이며 $x < 0$)
⇒ $y = -\{-(2x - 1) - \{-(3x - 1)\} = 2x - 1 - 3x + 1 = -x$ 이고 $x < 0$.

그러면 이제:
$y = x$ 이고 $x \geq 0$ 또는 $0 \leq x < 1/3$.
$y = -5x + 2$ 이고 $1/3 \leq x \leq 2/5$.
$y = 5x - 2$ 이고 $2/5 < x < 1/2$.
$y = -x$ 이고 $x < 0$.

이제, $y = ||2x - 1| - |3x - 1||$를 그래프에 나타내보기로 하죠.

그림. S.E.0.

$y = x$ 이고 $0 \le x < 1/3$ 또는 $x \ge 1/2$

$y = -x$ 이고 $x < 0$

$y = 5x - 2$ 이고 $2/5 < x < 1/2$

$y = -5x + 2$ 이고 $1/3 \le x \le 2/5$

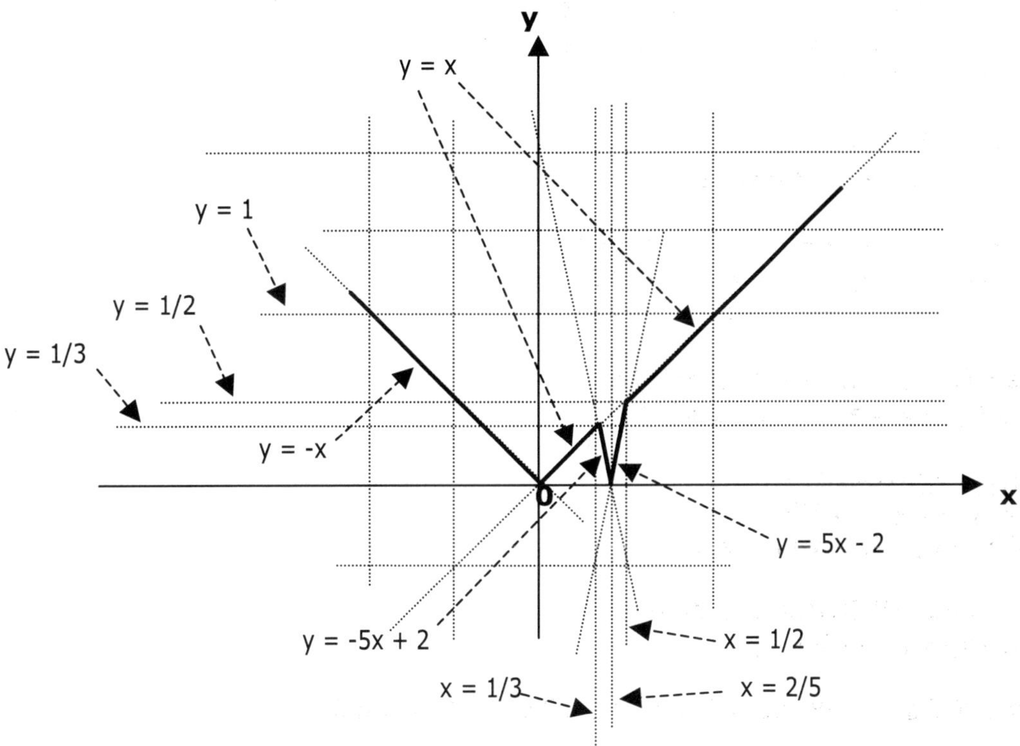

제안과 풀이
T번 예제에서의 문제

x-y평면에서의 세 개의 점들 $P_1(x_1, y_1)$, $P_2(x_2, y_2)$, $P_2(x_3, y_3)$가 하나의 직선에 있다고 하고 이들 좌표들간의 관계식을 구하세요.

그 직선의 방정식이 $y = ax + b$ 이며 여기에서 a는 기울기이고 b는 y 절편이라 하세요.
그러면, 먼저 두 점들 P_1과 P_2를 사용해서 그 기울기 a를 구해보죠.
그러면, $a = \Delta y/\Delta x = (y_2 - y_1)/(x_2 - x_1)$.

그럼 이제, P_1을 사용해서b를 명시합니다.
$y = ax + b \Rightarrow y_1 = ax_1 + b \Rightarrow b = y_1 - ax_1$.

그러면, $y = ax + b \Rightarrow y = ax + y_1 - ax_1 = a(x - x_1) + y_1$.
그러면, $y = a(x - x_1) + y_1 \Rightarrow y - y_1 = a(x - x_1)$이죠.
점 P_3이 그 직선에 있으니까 $y_3 - y_1 = a(x_3 - x_1)$이네요.

그 기울기 $a = (y_2 - y_1)/(x_2 - x_1)$라는 걸 알고있죠.
그러면, 이 a를 $y_3 - y_1 = a(x_3 - x_1)$에 대입합니다.

$y_3 - y_1 = \{(y_2 - y_1)/(x_2 - x_1)\}(x_3 - x_1)$
$\Rightarrow y_3 - y_1 = (y_2 - y_1)(x_3 - x_1)/(x_2 - x_1)$
$\Rightarrow (y_3 - y_1)(x_2 - x_1) = (y_2 - y_1)(x_3 - x_1)$.

그러면 이제, 다음과 같은 관계식을 갖는 것이죠:
$(y_3 - y_1)(x_2 - x_1) = (x_3 - x_1)(y_2 - y_1)$.

위의 관계식은 세 점들이 하나의 직선에 있을 수 있는 조건이라고 할 수 있네요.

간단한 예를 들면, 세 점들 $(x_1, y_1) = (1, 1)$, $(x_2, y_2) = (2, 2)$, $(x_3, y_3) = (3, 3)$은 하나의 직선에 있는 거죠.
물론, 이 직선은 $y = x$ 입니다.
그러면, $(y_3 - y_1)(x_2 - x_1) = (x_3 - x_1)(y_2 - y_1) \Rightarrow (3 - 1)(2 - 1) = (3 - 1)(2 - 1) \Rightarrow 2 = 2$.

예를 하나 더 들어, 직선 $y = 2x + 1$ 에 있는 다음과 같은 세 점들을 사용해보기로 하죠:
$(x_1, y_1) = (1, 3)$, $(x_2, y_2) = (3, 7)$, $(x_3, y_3) = (5, 11)$.

그러면, $(y_3 - y_1)(x_2 - x_1) = (x_3 - x_1)(y_2 - y_1) \Rightarrow (11 - 3)(3 - 1) = (5 - 1)(7 - 3) \Rightarrow 16 = 16$.

제안과 풀이
U 번 예제에서의 문제

점 P_1(2, 1)에서 하나의 특정 직선까지의 거리가 또 하나의 점 P_2(-2, -3)에서 그 특정 직선까지의 거리와 같다고 하고 그 특정 직선의 방정식을 구하세요.

Math 에서 거리라는 것은 수직 거리이죠.
그 특정 직선을 직선 A 라고 하세요.
그러면, 그 두개의 점들 P_1과 P_2는 직선 A에 수직인 직선에 있는 겁니다.
또한, 그 두 거리들이 같으니까 서로 수직인 두 직선들이 만나는 점은 그 두 점들 P_1과 P_2의 중점이죠.
그러니까, 그 중점은 직선 A 에 있는 거네요.
그러면, 직선 A 의 기울기와 그 중점을 구하면 직선 A 의 방정식을 구할 수 있는 거죠.

서로 수직인 두 직선들의 기울기들의 곱이 −1 이니까 A 의 기울기는 A 에 수직인 직선의 기울기로부터 구할 수 있죠.
그 중점의 좌표들은 P_1과 P_2의 좌표들의 제각기 평균값들이죠.

그러면 이제, A 에 수직인 직선의 기울기부터 시작하기로 하죠.
이 수직선은 P_1(2, 1)과 P_2(-2, -3)를 지나니까:
그 기울기는 $\Delta y/\Delta x = (1 - (-3))/(2 - (-2)) = 4/4 = 1$ 이네요.
그러면, A 의 기울기는 −1 이죠.

다음으로, 그 중점의 x 좌표 $= (2 + (-2))/2 = 0$ 이고 또한, y 좌표 $= (1 + (-3))/2 = -1$ 이죠.
따라서, 그 중점 = (0, -1)이네요.

기울기가 a 이고 점 (m, n)을 지나는 직선의 방정식은 $y - n = a(x - m)$이죠.

그러니까, 직선 A 의 방정식은 $y - (-1) = (-1)(x - 0) \Rightarrow y + 1 = -x \Rightarrow y = -x - 1$ 이네요.

요약:

그 특정 직선을 A라 하고 A에 수직이고 P_1과 P_2를 지나는 직선을 B라 하세요.

그러면, B는 P_1(2, 1)과 P_2(-2, -3)를 지나니까:
그 기울기는 $\Delta y/\Delta x = (1 - (-3))/(2 - (-2)) = 4/4 = 1$.
그러면, A 의 기울기는 −1.

또한, A는 P_1과 P_2 사이의 중점을 포함.
그 중점의 x 좌표 $= (2 + (-2))/2 = 0$ 이고 또한, y 좌표 $= (1 + (-3))/2 = -1$.
따라서, 그 중점 = (0, -1).

그러므로, 직선 A 의 방정식은 $y - (-1) = (-1)(x - 0) \Rightarrow y + 1 = -x \Rightarrow y = -x - 1$.

제안과 풀이
V번 예제에서의 문제

다음과 같이 가정하세요:

T: (x, y) ⟶ (3x − y, x − y). y = f(x) = ax 여기에서 a 는 상수. T•f = f.

그러면, a 의 값을 구하세요.

방정식 $y = ax$ 의 커브를 갖는 함수 f 를 변환 T 를 사용해서 변환시켜도 변하지 않도록 a 의 값을 정해야 하죠.
$g = T \cdot f$ 이고 $t = g(s)$라 하세요. 그러면, $g = f$ 이고 또한 다음과 같아요:
T: (x, y) ⟶ (s, t) = (3x − y, x − y).
따라서, $s = 3x - y$ 이고 $t = x - y$ 이죠. 그러면, 이것들을 x 와 y 에 관한 연립방정식으로 하는 거죠.
그러면, 이를 풀면, x 와 y 각각을 s 와 t 에 관하여 나타낼 수 있죠.

$s - t = 3x - y - (x - y) = 3x - y - x + y = 2x \Rightarrow s - t = 2x \Rightarrow x = (s - t)/2$. 이제, x 는 s 와 t 의 식으로 나타내어진 것이죠. 그러면, 이 x 를 $s = 3x - y$ 에 대입하여 y 를 s 와 t 의 식으로 나타낼 수 있어요.
$s = 3(s - t)/2 - y = (3s - 3t)/2 - y \Rightarrow y = (3s - 3t)/2 - s = (s - 3t)/2$.
그럼 이제, $x = (s - t)/2$ 이고 $y = (s - 3t)/2$ 이네요.

그러면, 위의 식들을 원래 함수 f 에 적용해서 그 새로운 함수 g 를 구해야 하죠.
그 원래 함수는 $y = f(x) = ax$ 이죠. 그러면, $(s - 3t)/2 = f\{(s - t)/2\} = a\{(s - t)/2\}$.
그러면, $(s - 3t)/2 = a(s - t)/2 \Rightarrow s - 3t = a(s - t) \Rightarrow at - 3t = as - s = (a - 1)s \Rightarrow (a - 3)t = (a - 1)s$
$\Rightarrow t = (a - 1)s/(a - 3)$. 따라서, $t = g(s) = (a - 1)s/(a - 3)$.
그런데, 함수 g 는 x-y 좌표계에 있으므로 s 와 t 를 제각기 x 와 y 로 바꾸어야 하죠.
그러면, $y = g(x) = (a - 1)x/(a - 3) = \{(a - 1)/(a - 3)\}x$ 이죠.

그러면 이제, $f(x) = ax$ 이고 $f(x) = g(x)$이니까 $(a - 1)/(a - 3) = a$ 이고 여기에서 $a \neq 3$.
따라서, $a - 1 = a(a - 3) = a^2 - 3a \Rightarrow a^2 - 4a + 1 = 0$.
그러면, $a^2 - 4a + 1 = a^2 - 4a + 4 - 4 + 1 = (a - 2)^2 - 3 = 0$.
그러므로, $a - 2 = \pm 3^{1/2} \Rightarrow a = 2 \pm \sqrt{3}$.

요약:

$g = T \cdot f$ 이고 $t = g(s)$라 하세요. 그러면, $g = f$ 이고 또한 $s = 3x - y$ 이고 $t = x - y$.
$s - t = 3x - y - (x - y) = 3x - y - x + y = 2x \Rightarrow s - t = 2x \Rightarrow x = (s - t)/2$.
$s = 3x - y = 3(s - t)/2 - y = (3s - 3t)/2 - y \Rightarrow y = (3s - 3t)/2 - s = (s - 3t)/2$.
그러므로, $x = (s - t)/2$ 이고 $y = (s - 3t)/2$.
$y = f(x) = ax$ 이니까 $(s - 3t)/2 = f\{(s - t)/2\} = a\{(s - t)/2\}$.
그러면, $(s - 3t)/2 = a(s - t)/2 \Rightarrow s - 3t = a(s - t) \Rightarrow at - 3t = as - s = (a - 1)s \Rightarrow (a - 3)t = (a - 1)s$
$\Rightarrow t = (a - 1)s/(a - 3)$. 따라서, $t = g(s) = (a - 1)s/(a - 3)$.
그러면, $g(x) = (a - 1)x/(a - 3) = \{(a - 1)/(a - 3)\}x$.
그러면 이제, $f(x) = ax$ 이고 $f(x) = g(x)$이니까 $(a - 1)/(a - 3) = a$ 이고 여기에서 $a \neq 3$.
따라서, $a - 1 = a(a - 3) = a^2 - 3a \Rightarrow a^2 - 4a + 1 = 0$.
그러면, $a^2 - 4a + 1 = a^2 - 4a + 4 - 4 + 1 = (a - 2)^2 - 3 = 0$.
그러므로, $a - 2 = \pm 3^{1/2} \Rightarrow a = 2 \pm \sqrt{3}$.

제안과 풀이
W 번 예제에서의 문제

다음과 같이 가정하세요:
T: (x, y) ⟶ (3x – 2y, x + y). y = f(x) = ax 여기에서 a 는 상수. T•f = f.

그러면, a 의 값을 구하세요.

이번에도 y = ax 의 커브를 갖는 함수 f 를 변환 T 를 사용해서 변환시켜도 변하지 않도록 a 의 값을 정해야 하는 거죠. g = T•f 이고 t = g(s)라 하세요. 그러면, g = f 이고 또한 다음과 같은 거죠:
T: (x, y) ⟶ (s, t) = (3x – 2y, x + y).
따라서, s = 3x – 2y 이고 t = x + y 이죠.

그러면, 이것들을 x 와 y 에 관한 연립방정식으로 해서 풀면 x 와 y 각각을 s 와 t 에 관하여 나타낼 수 있죠.
먼저, s + 2t = 3x – 2y + 2(x + y) = 3x – 2y + 2x + 2y = 5x ⇒ s + 2t = 5x ⇒ x = (s + 2t)/5.
다음으로, x = (s + 2t)/5 ⇒ s = 3x – 2y = 3(s + 2t)/5 – 2y = (3s + 6t)/5 – 2y ⇒ 2y = (3s + 6t)/5 – s = (-2s + 6t)/5 ⇒ y = (3t – s)/5.

그럼 이제, x = (s + 2t)/5 이고 y = (3t – s)/5 이네요.
그러면, 위의 식들을 원래 함수 f 에 적용해서 그 새로운 함수 g 를 구해야 하죠.
그 원래 함수는 y = f(x) = ax 이죠. 그러면, (3t – s)/5 = f{(s + 2t)/5} = a{(s + 2t)/5}.
그러면, (3t – s)/5 = a(s + 2t)/5 ⇒ 3t – s = a(s + 2t) ⇒ 2at – 3t = -as – s = -(a + 1)s
⇒ (2a – 3)t = -(a + 1)s ⇒ t = (a + 1)s/(3 – 2a).
따라서, t = g(s) = (a + 1)s/(3 – 2a).
그런데, 함수 g 는 x-y 좌표계에 있으므로 s 와 t 를 제각기 x 와 y 로 바꾸어야 하죠.
그러면, y = g(x) = (a + 1)x/(3 – 2a) = {(a + 1)/(3 – 2a)}x 이죠.

그러면 이제, f(x) = ax 이고 f(x) = g(x)이니까 (a + 1)/(3 – 2a) = a 이고 여기에서 a ≠ 3/2.
따라서, a + 1 = a(3 – 2a) = $-2a^2$ + 3a ⇒ $2a^2$ – 2a + 1 = 0.
그런데, $2a^2$ – 2a + 1 = 2(a^2 – a + 1/4 – 1/4) + 1 = 2$(a – 1/2)^2$ + 1/2 인데 이는 a가 실수이니까 0 이 될 수 없죠. 그러니까, 답이 없는 거죠. 즉, a의 값이 없는 겁니다.

요약:

g = T•f 이고 t = g(s)라 하세요. 그러면, g = f 이고 또한 s = 3x – 2y 이고 t = x + y.
s + 2t = 3x – 2y + 2(x + y) = 3x – 2y + 2x + 2y = 5x ⇒ s + 2t = 5x ⇒ x = (s + 2t)/5.
s = 3(s + 2t)/5 – 2y = (3s + 6t)/5 – 2y ⇒ 2y = (3s + 6t)/5 – s = (-2s + 6t)/5 ⇒ y = (3t – s)/5.
그러므로, x = (s + 2t)/5 이고 y = (3t – s)/5.
y = f(x) = ax 이니까 (3t – s)/5 = f{(s + 2t)/5} = a{(s + 2t)/5}.
그러면, (3t – s)/5 = a(s + 2t)/5 ⇒ 3t – s = a(s + 2t) ⇒ 2at – 3t = -as – s = -(a + 1)s
⇒ (2a – 3)t = -(a + 1)s ⇒ t = (a + 1)s/(3 – 2a). 따라서, t = g(s) = (a + 1)s/(3 – 2a).
그러면, g(x) = (a + 1)x/(3 – 2a) = {(a + 1)/(3 – 2a)}x.
그러면 이제, f(x) = ax 이고 f(x) = g(x)이니까 (a + 1)/(3 – 2a) = a 이고 여기에서 a ≠ 3/2.
따라서, a + 1 = a(3 – 2a) = $-2a^2$ + 3a ⇒ $2a^2$ – 2a + 1 = 0.
그런데, $2a^2$ – 2a + 1 = 2(a^2 – a + 1/4 – 1/4) + 1 = 2$(a – 1/2)^2$ + 1/2 인데 이는 a가 실수이니까 0 이 될 수 없음. 그러므로, a의 값은 없음.

제안과 풀이
X번 예제에서의 문제

다음과 같이 가정하세요:
T: (x, y) ⟶ (2x + y + 3, 3x + 2y – 2).
y = f(x) = ax + b 여기에서 **a**와 **b**는 상수.
T•f = f.

그러면, **a**와 **b**의 값을 구하세요.

방정식 y = ax + b의 커브를 갖는 함수 f를 변환 T를 사용해서 변환시켜도 변하지 않도록 a와 b의 값들을 정해야 하는 거죠. g = T•f이고 t = g(s)라 하세요. 그러면, g = f이고 또한 다음과 같은 거죠:
T: (x, y) ⟶ (s, t) = (2x + y + 3, 3x + 2y – 2).
따라서, s = 2x + y + 3이고 t = 3x + 2y – 2이죠.

그러면, 이것들을 x와 y에 관한 연립방정식으로 해서 x와 y 각각을 s와 t에 관하여 나타내는 거죠.

먼저, 2s – t = 4x + 2y + 6 – (3x + 2y – 2) = x + 8 ⇒ 2s – t = x + 8 ⇒ x = 2s – t – 8.
다음으로, x = 2s – t – 8 ⇒ s = 2(2s – t – 8) + y + 3 = 4s – 2t – 16 + y + 3 ⇒ y = -3s + 2t + 13.
그럼 이제, x = 2s – t – 8이고 y = -3s + 2t + 13이네요.

y = f(x) = ax + b이죠. 그러면, -3s + 2t + 13 = f(2s – t – 8) = a(2s – t – 8) + b.
그러면, -3s + 2t + 13 = 2as – at – 8a + b ⇒ (2 + a)t = (2a + 3)s – 8a + b – 13
⇒ t = (2a + 3)s/(a + 2) + (b – 8a – 13)/(a + 2).
그러므로, t = g(s) = (2a + 3)s/(a + 2) + (b – 8a – 13)/(a + 2).

그런데, 함수 g는 x-y 좌표계에 있으므로 s와 t를 제각기 x와 y로 바꾸어야 하죠.
그러면, y = g(x) = (2a + 3)x/(a + 2) + (b – 8a – 13)/(a + 2)이죠.
그러면 이제, f(x) = ax이고 f(x) = g(x)이니까:
(2a + 3)/(a + 2) = a 이고 (b – 8a – 13)/(a + 2) = b 여기에서 a ≠ -2.
먼저, a의 값을 구하기로 하죠. 그러면, 2a + 3 = a(a + 2) = a^2 + 2a ⇒ a^2 = 3. 따라서, a = ±$3^{1/2}$.
그러면 이제, b의 값을 구해야 하죠.
(b – 8a – 13)/(a + 2) = b ⇒ b – 8a – 13 = b(a + 2) = ab + 2b ⇒ b(a + 1) = -8a – 13
⇒ b = -(8a + 13)/(a + 1). 그러면:

$a = \sqrt{3} \Rightarrow b = -(8\sqrt{3} + 13)/(\sqrt{3} + 1) = -(8\sqrt{3} + 13)(\sqrt{3} - 1)/(\sqrt{3} + 1)(\sqrt{3} - 1)$
$= -(8\sqrt{3} + 13)(\sqrt{3} - 1)/2 = -(24 + 5\sqrt{3} - 13)/2 = -(11 + 5\sqrt{3})/2.$

$a = -\sqrt{3} \Rightarrow b = -(-8\sqrt{3} + 13)/(1 - \sqrt{3}) = (8\sqrt{3} - 13)/(1 - \sqrt{3}) = (13 - 8\sqrt{3})/(\sqrt{3} - 1)$
$= (13 - 8\sqrt{3})(\sqrt{3} + 1)/2 = (5\sqrt{3} - 24 + 13)/2 = (5\sqrt{3} - 11)/2.$

그러므로, $a = \sqrt{3}$이고 $b = -(11 + 5\sqrt{3})/2$, 또는 $a = -\sqrt{3}$이고 $b = (5\sqrt{3} - 11)/2$.

그러면 이제, 직선 $y = \sqrt{3}x - (11 + 5\sqrt{3})/2$을 T에 의하여 변환한다 하더라도 동일한 직선을 갖는 것이고 또한, 이는 직선 $y = -\sqrt{3}x + (5\sqrt{3} - 11)/2$에 대하여도 마찬가지죠.

0.1. 여러 개의 직선들

보통, 하나의 방정식으로 하나의 직선을 나타내죠. 하지만, 앞서 언급했던 바와 같이 여러 개의 직선들을 하나의 방정식으로 나타낼 수도 있어요. 하나의 방정식으로 두개 이상의 직선들을 나타낼 때는 일반형 방정식을 사용합니다: $ax + by + c = 0$ 여기에서 a, b, c 는 상수.

그런데, 하나의 방정식으로 여러 개의 직선들을 어떻게 나타내죠?
우선, 두개의 직선들을 하나의 방정식으로 나타내보도록 하죠.
하나의 직선을 다음과 같은 일반형 방정식으로 나타낼 수 있죠: $ax + by + c = 0$ 여기에서 a, b, c 는 상수.
또 하나의 직선을 다음과 같이 일반형으로 나타내보죠: $px + qy + r = 0$ 여기에서 p, q, r 은 상수.
그러면 이제, 이 두개의 직선들을 다음과 같이 하나의 방정식으로 나타낼 수 있는 겁니다:

$$(ax + by + c)(px + qy + r) = 0.$$

어떻게 요?
다음과 같은 방정식이 있다고 하세요: $AB = 0$.
그러면, $A = 0$ 또는 $B = 0$ 이죠.
마찬가지로, $(ax + by + c)(px + qy + r) = 0 \Rightarrow ax + by + c = 0$ 또는 $px + qy + r = 0$ 이네요.
따라서, $(ax + by + c)(px + qy + r) = 0$ 은 두 개의 직선들을 나타내는 것이고 하나는 $ax + by + c = 0$ 이고 나머지 하나는 $px + qy + r = 0$ 이죠.

마찬가지의 개념이 셋 이상의 직선들의 방정식에도 적용되는 겁니다.

이제, $L_1L_2L_3 \ldots L_n = 0$ 이라 하세요.
또한, $L_k = a_kx + b_ky + c_k$ 여기에서 a_k, b_k, c_k는 상수이고 $k = 1, 2, \ldots n$이라 하세요.
즉, $L_1 = a_1x + b_1y + c_1, L_2 = a_2x + b_2y + c_2, \ldots L_n = a_nx + b_ny + c_n$

그러면, $L_1L_2L_3 \ldots L_n = 0 \Rightarrow (a_1x + b_1y + c_1)(a_2x + b_2y + c_2) \ldots (a_nx + b_ny + c_n) = 0$.

그러면, $L_1 = 0, L_2 = 0, \ldots$ 또는 $L_n = 0$ 입니다.
즉, $L_k = 0$ 여기에서 $k = 1, 2, \ldots n$이죠.
다시 말해서, $a_1x + b_1y + c_1 = 0, a_2x + b_2y + c_2 = 0 \ldots$ 또는 $a_nx + b_ny + c_n = 0$ 인 것이죠.

그러므로, $L_1L_2L_3 \ldots L_n = 0$ 은 n개의 직선들을 나타내는 하나의 방정식인 겁니다.

직선은 점들의 집합이라는 것을 알고있죠.
그러면, 두개의 직선을 다음과 같은 점들의 집합 A 로 나타낼 수 있어요:

$A = \{(x, y) | (ax + by + c)(px + qy + r) = 0$ 여기에서 a, b, c, p, q, r 은 상수$\}$
물론, a 와 b 는 동시에 0 이 아니고 p 와 q 도 동시에 0 이 아닙니다.

그러면, 셋 이상의 직선을 다음과 같은 점들의 집합으로 나타낼 수 있는 거죠:

$B = \{(x, y) | (a_1x + b_1y + c_1)(a_2x + b_2y + c_2) \ldots (a_nx + b_ny + c_n) = 0$ 여기에서 a_k, b_k, c_k는 상수$\}$
물론, $k = 1, 2, \ldots n$이고 a_k와 b_k는 동시에 0 이 아니죠.

주로, 직선을 방정식이나 또는 점의 집합으로 나타내죠.
직선을 만들려면 적어도 두개의 점들이 있어야 하죠. 그러면, 그 방정식을 구할 수 있죠.
이제, 그 두 개의 점들이 (s, t)와 (u, v)라 하세요.
그러면, 그 직선의 방정식은 $y - t = \{(t - v)/(s - u)\}(x - s)$ 또는 $y - v = \{(t - v)/(s - u)\}(x - u)$이고 여기에서 $\{(t - v)/(s - u)\}$가 그 기울기이죠. 물론, 위의 두 개의 방정식들은 서로 같은 겁니다. 그렇지만, 예를 하나 만들어서 확인해보세요.
물론 또한, 그 기울기와 그 직선에서의 한 점을 알면 그 직선의 방정식을 구할 수 있죠.
그 기울기가 a 이고 그 한 점이 (m, n)이라 하세요. 그러면, 그 방정식은 $y - n = a(x - m)$이죠.

그런데, 직선을 만드는 또 하나의 방법이 있습니다. 직선으로 직선을 만드는 겁니다.
서로 다른 두개의 직선들로 또 하나의 직선을 만들 수 있어요.
그 두개의 직선들을 원래의 직선들이라 해보죠.
그러면, 두개의 원래의 직선들로 하나의 새로운 직선을 만드는 겁니다.

그런데, 그 원래의 직선들이 서로 평행이면 그 새로운 그 직선도 평행입니다.
그렇지 않으면 즉, 그 원래의 직선들이 한 점에서 만나면, 그 새로운 직선도 그 점을 지나갑니다.
즉, 그 세 직선들은 모두 서로 평행이거나 아니면 한 점에서 만나는 거죠.

그러면, 어떻게 그 새로운 직선을 만드나요?
그 원래의 직선들의 평균을 구하면 그 평균이 그 새로운 직선인 겁니다.

그런데, 직선들의 평균이라니 요?
그 평균이란 그 원래의 직선들의 방정식들의 합을 둘로 나눈 몫이죠.

예를 들어, 다음과 같은 직선들의 방정식들이 있다고 하세요: $y = a_1x + b_1$과 $y = a_2x + b_2$. 그러면:
$y + y = a_1x + b_1 + (a_2x + b_2) \Rightarrow 2y = (a_1 + a_2)x + (b_1 + b_2) \Rightarrow y = (a_1 + a_2)x/2 + (b_1 + b_2)/2$.

그러면 이제, 그 새로운 직선의 방정식은 $y = (a_1 + a_2)x/2 + (b_1 + b_2)/2$ 인 거죠.

사실, 그 새로운 직선이 그 평균이 아니라도 됩니다. 이제, 일반형으로 해보죠.
$ax + by + c = 0$ 이고 $ux + vy + w = 0$ 라고 하세요.
그러면, $(a + u)x + (b + v)y + c + w = 0$ 이고 이는 새로운 직선의 방정식입니다.
또한, $2(ax + by + c) = 0$ 도 되죠. 그러면, $(2a + u)x + (2b + v)y + 2c + w = 0$ 이 되고 이것도 새로운 직선의 방정식이죠.

이제, 두개의 직선들이 $y = ax + b$ 이고 $y = ax + c$ 라고 하세요.
그럼, 이 두 직선들은 서로 평행이죠. 그러면:
$\{y = ax + b\} + \{y = ax + c\} \Rightarrow 2y = 2ax + b + c \Rightarrow y = ax + (b + c)/2$ 이는 원래의 두 직선들과 평행이죠.

이번에는 원래의 직선들이 한 점에서 서로 만나는 간단한 예를 들어보죠.
두 개의 직선들 $y = x + 2$ 와 $y = -3x + 2$ 이 있다고 하세요.
먼저, 이 원래의 직선들이 서로 만나는 점을 구해보죠.
그러면, $x + 2 = -3x + 2 \Rightarrow 4x = 0 \Rightarrow x = 0 \Rightarrow y = x + 2 = 2 \Rightarrow (0, 2)$.
그러므로, 그 원래의 직선들은 점 (0, 2)에서 만나죠. 이제, 그 새로운 직선을 구해보죠.
$\{y = x + 2\} + \{y = -3x + 2\} \Rightarrow 2y = -2x + 4 \Rightarrow y = -x + 2$.
그럼 이제, 이 새로운 직선이 그 점 (0, 2)를 포함하는지 확인해보죠.
$y = -x + 2 \Rightarrow 2 = 0 + 2$. 포함하네요.

그림. 0.1.0

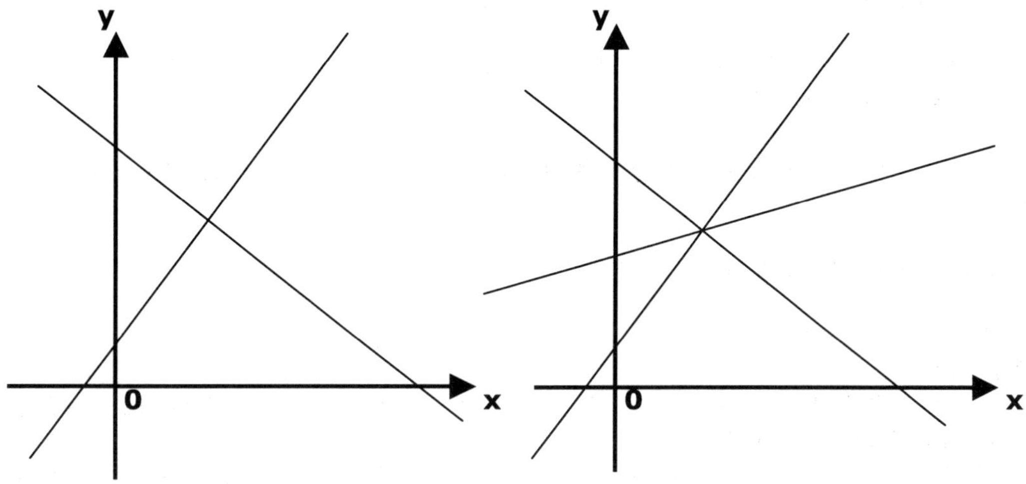

또한, 그 세 직선들 중에 그 어느 두개를 가지고 새로운 직선을 만들어도 그 새로운 직선은 그 점 (0, 2)를 지납니다. 예를 들어, y = -3x + 2 와 y = -x + 2 를 가지고 새로운 걸 만들어보죠. 그러면:
{y = -3x + 2} + {y = -x + 2} ⇒ 2y = -4x + 4 ⇒ y = -2x + 2.
그럼 이제, 이 새로운 직선이 그 점 (0, 2)를 지나가는지 확인해보죠.
y = -2x + 2 ⇒ 2 = 0 + 2. 지나가네요. 그러니까, 한 점에서 서로 만나는 두개의 직선들이 있으면 그 점을 지나는 무한히 많은 새로운 직선들을 만들어낼 수 있는 거죠.

그림. 0.1.1

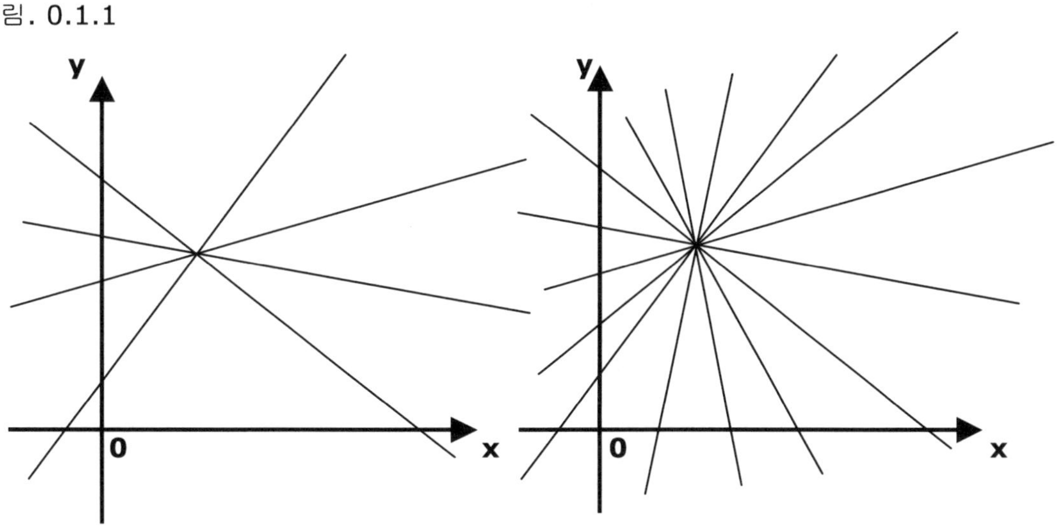

이는 서로 평행한 직선들에 대하여도 마찬가집니다.
서로 평행인 두 직선들로 또 하나의 평행 직선을 만들 수 있고 또한, 그 세 직선들 중에 그 어느 두개를 가지고 새로운 직선을 만들어도 그 새로운 직선도 평행한 거죠.
그럼, 그 세 직선들을 모두 합치면 어떨까요?
그 새로운 직선도 평행이죠.
앞에 나온 예를 들면, y = ax + b, y = ax + c, y = ax + (b + c)/2 이죠.

그러면, $\{y = ax + b\} + \{y = ax + c\} + \{y = ax + (b + c)/2\} \Rightarrow 3y = 3ax + 3(b + c)/2$
$\Rightarrow y = ax + (b + c)/2$ 이는 이전에 구한 그 세 번째 직선이네요.

그럼 이번에는 다음과 같이 서로 평행인 세 직선을 가지고 하나의 새로운 직선을 만들어보죠:
$y = ax + b$, $y = ax + c$, $y = ax + d$.
그러면:
$\{y = ax + b\} + \{y = ax + c\} + \{y = ax + d\} \Rightarrow 3y = 3a + (b + c + d) \Rightarrow y = a + (b + c + d)/3$ 이것도 평행한 직선이네요.

이는 한 점에서 서로 만나는 직선들에 대하여도 마찬가집니다.

그럼, 위의 이야기들의 요점은 무엇이죠?
그 요점이 점과 기울기이죠.
우선, 그 요점은 하나의 특정 점을 지나는 무수히 많은 직선들이 있다는 것이고 이들 각각은 두 개의 직선들로 만들어 지는데 이것들도 그 특정 점을 지나는 거죠.
다음으로, 그 요점은 서로 평행한 무수히 많은 직선들이 있다는 것이고 이들 각각은 두 개의 직선들로 만들어지는데 이것들도 평행인 거죠.

그런데, 사실은 그 원래의 직선들의 방정식들을 합하면 그 합이 그 새로운 직선을 나타내는 겁니다.
더 정확히 이야기하면, 그 원래의 직선들의 방정식들 각각에 0 이 아닌 수를 곱하고 나서 그 곱들을 모두 합하면 그 합이 하나의 새로운 직선을 나타내는데 이는 그 원래의 직선들이 서로 만나는 하나의 특정 점을 지나는 것이죠.
또한, 그 원래의 직선들이 서로 평행이면 그 새로운 직선도 그 원래의 직선들에 평행한 겁니다.

그럼 이제, 예제를 좀 다루어보도록 하죠.

예제

여러 개의 직선들

0. 두개의 직선들 $y = x$ 와 $y = -x$ 가 있다 하고 이것들을 하나의 방정식으로 나타내세요.

1. 두개의 직선들 $y = -x + 1$ 과 $y = x$ 가 있다 하고 이것들을 하나의 방정식으로 나타내세요.

2. 방정식 $2x^2 - 2y^2 - 3xy - x + 2y = 0$ 은 두개의 직선들을 나타냅니다. 그러면, 그 두개의 직선들을 구하세요.

3. 아래의 방정식은 두개의 직선을 나타낸다 하고 t 의 값을 구하세요:
$2x^2 + 4xy - ty^2 + 4x + 2y - 2 = 0$ 여기에서 t는 상수.

4. 다음과 같은 두개의 직선들이 있다 하세요: $2x - y - 1 = 0$ 과 $3x + 2y - 2 = 0$.
그러면, 이 두 직선들 각각으로부터 동일한 거리만큼 떨어져있는 점 $P(s, t)$를 구하세요.

5. 다음과 같은 직선의 방정식들이 있다고 하세요:
$ax + by + c = 0$ 이고 $ux + vy + w = 0$ 이며 여기에서 a, b, c, u, v, w 는 상수이고 $abc \neq 0$ 이며 $uvw \neq 0$.

5.0. 또한, 두 방정식들은 사실, 동일한 직선을 나타낸다 하고 상수들간의 관계식을 구하세요.

5.1. 또한, 두 직선들은 서로 평행이다 하고 상수들간의 관계식을 구하세요.

5.2. 또한, 두 직선들은 한 점에서 서로 만난다 하고 상수들간의 관계식을 구하세요.

5.3. 또한, 두 직선들은 서로 수직이다 하고 상수들간의 관계식을 구하세요.

5.4. 그러면, 두 직선들이 서로 만나는 점을 지나는 직선의 방정식을 구하세요.

6. 직선들의 모임을 나타내는 방정식이 있는데 그 방정식은 다음과 같다고 하세요:
$3(k - 1)x + (2k + 3)y + 4k + 5 = 0$ 여기에서 k 는 상수.

또한, 그 모임의 모든 직선들은 하나의 특정한 점 $P(s, t)$에서 만난다 하고 s 와 t 의 값을 구하세요.

7. 직선들의 모임을 나타내는 방정식이 있는데 그 방정식은 다음과 같다고 하세요:
$y = ax - b$ 여기에서 a 와 b 는 상수이고 $3a + b = 2$ 를 만족.

또한, 그 모든 직선들은 하나의 특정한 점 $P(s, t)$에서 만난다 하고 s 와 t 의 값을 구하세요.

8. 직선들의 모임을 나타내는 방정식이 있는데 그 방정식은 다음과 같다고 하세요:
$2x/a + y/2b = 1$ 여기에서 a 와 b 는 상수 이고 $1/2a + 1/b = 1/5$ 를 만족.

또한, 그 모임의 모든 직선들은 하나의 특정한 점 $P(u, v)$에서 만난다 하고 u 와 v 의 값을 구하세요.

9. 직선들의 모임을 나타내는 방정식이 있는데 그 방정식은 다음과 같다고 하세요:
$5ax - (3b - 2)y + 2b = 8$ 여기에서 a 와 b 는 상수이고 $2a - b = 2$ 를 만족하며 $ab \neq 0$.

9.0. 또한, 그 모임의 모든 직선들이 하나의 특정한 점 $P(s, t)$에서 만난다 하고 그 점 P 를 구하세요.

9.1. 그러면, 직선 $2x + 5y = 10$ 에 평행이고 그 모임에 속하는 직선을 구하세요.

A. 두개의 직선들이 점 $P(s, t)$에서 만나고 $s > 0$ 이며 $t > 0$ 이라 하고 그 직선들의 방정식은 다음과 같다고 하세요: $2x + y - 2 = 0$ 과 $2kx - y + k - 1 = 0$ 여기에서 k 는 상수.

그러면, k 의 범위를 구하세요.

B. 다음과 같이 가정하세요:
$(2k - 1)x - (3k + 1)y + 5 + 2k = 0$ 여기에서 k 는 상수.
$y = 2x$ 이고 $|x| \leq 1$.

그러면, 위의 두 직선들이 서로 만날 수 있도록 k 값을 정하세요.

제안과 풀이
0 번 예제에서의 문제

두개의 직선들 **$y = x$** 와 **$y = -x$** 가 있다 하고 이것들을 하나의 방정식으로 나타내세요.

$y = -x$ 또는 $y = x \Rightarrow x + y = 0$ 또는 $x - y = 0 \Rightarrow (x + y)(x - y) = 0$.

그러면, 그 두개의 직선들을 하나의 방정식 $(x + y)(x - y) = 0$ 으로 나타낼 수 있네요.

그러면 이제, 역으로 해보죠.

$(x + y)(x - y) = 0 \Rightarrow x^2 - xy + xy - y^2 = 0 \Rightarrow x^2 - y^2 = 0 \Rightarrow y^2 = x^2 \Rightarrow y = \pm x$.

$\Rightarrow y = x$ 또는 $y = -x$.

그림. 0.0

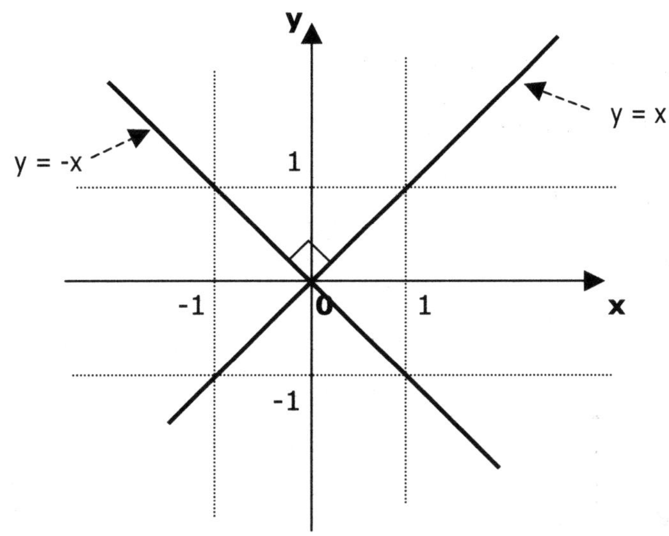

유의 사항:
다항식의 인수분해에 대하여 잘 모르면 방정식을 푸는데 많은 제한을 받습니다. 잘 모르면, 다항식의 인수분해 관련 서적이나 **Runmath** 시리즈의 **인수분해 하세요** 를 참고 하세요.

제안과 풀이

1번 예제에서의 문제

두개의 직선들 **y = -x + 1**과 **y = x**가 있다 하고 이것들을 하나의 방정식으로 나타내세요.

$y = -x + 1$ 또는 $y = x \Rightarrow x + y - 1 = 0$ 또는 $x - y = 0 \Rightarrow (x + y - 1)(x - y) = 0$.

그러면, 그 두개의 직선들을 하나의 방정식 $(x + y - 1)(x - y) = 0$으로 나타낼 수 있네요.

그러면 이제, 역으로 해보죠.

$(x + y - 1)(x - y) = 0 \Rightarrow x^2 - xy + xy - y^2 - x + y = 0$

$\Rightarrow x^2 - y^2 - x + y = 0 \Rightarrow y^2 - y = x^2 - x$

$\Rightarrow y^2 - y + 1/4 - 1/4 = x^2 - x + 1/4 - 1/4 \Rightarrow (y - 1/2)^2 - 1/4 = (x - 1/2)^2 - 1/4$
$\Rightarrow (y - 1/2)^2 = (x - 1/2)^2$

$\Rightarrow y - 1/2 = \pm(x - 1/2) \Rightarrow y = \pm(x - 1/2) + 1/2$

$\Rightarrow y = (x - 1/2) + 1/2$ 또는 $y = -(x - 1/2) + 1/2$

$\Rightarrow y = x$ 또는 $y = -x + 1/2 + 1/2 = -x + 1$

$\Rightarrow y = x$ 또는 $y = -x + 1$.

그림. 1.0

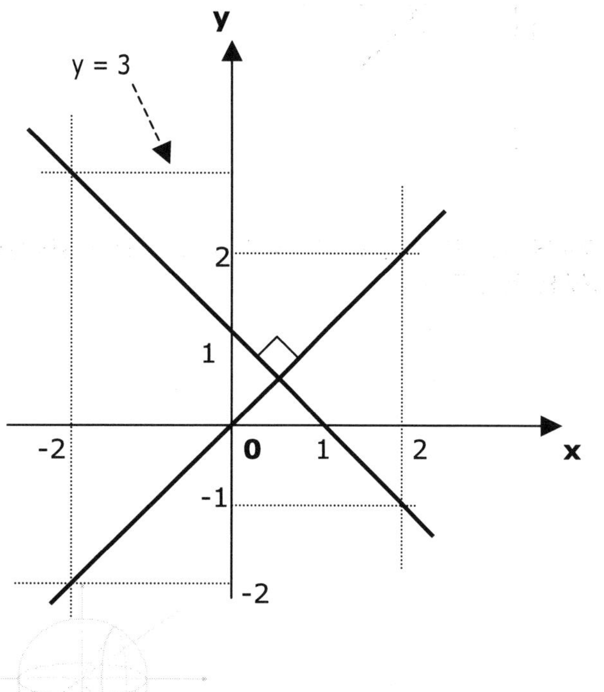

제안과 풀이

2번 예제에서의 문제

방정식 $2x^2 - 2y^2 - 3xy - x + 2y = 0$은 두개의 직선들을 나타냅니다. 그러면, 그 두개의 직선들을 구하세요.

이 문제는 인수분해 문제죠. 사실, 대부분의 방정식 문제들은 다항식의 인수분해 문제입니다. 이 경우에는 인수분해가 보통의 인수분해보다 훨씬 쉬운 겁니다. 인수의 형태가 이미 알려져 있기 때문이죠. 그 인수들 각각은 직선의 방정식이죠. 그러니까, 예를 들면 그 인수들이 $ax + by + c$와 $dx + ey + f$가 되는 것으로 간주할 수 있는 거죠.

그러면 이제, 주어진 방정식을 인수분해 하여 보기로 하죠.

$$2x^2 - 2y^2 - 3xy - x + 2y = 2x^2 - 3xy - x - 2y^2 + 2y = x(2x - 3y - 1) - 2y(y - 1)$$
$$= x(2x - 4y + y - 1) - 2y(y - 1) = x(2x - 4y) + x(y - 1) - 2y(y - 1)$$
$$= 2x(x - 2y) + (x - 2y)(y - 1) = (x - 2y)(2x + y - 1).$$

그러므로, 그 직선들은 $2x + y - 1 = 0$과 $x - 2y = 0$입니다.

그런데, 방정식에 있는 다항식을 인수분해 하지 못하면 어쩌죠?
물론, 인수분해 그 자체가 일반적으로는 쉽지 않아요 왜냐하면 직선들의 방정식들 중 많은 것들이 분수 또는 심지어는 무리수로 된 계수 또는 상수들을 갖을 수 있기 때문이죠.
그런데, 그러한 방정식에서의 다항식에 있는 지수들을 살펴보면 그 다항식이 2차식이라는 걸 알 수 있죠.
그러면, 완전 제곱식 방법을 사용하는 것이 좋겠어요.
그러면 이제, 완전 제곱식 방법을 사용해서 풀어보도록 하죠.

$$2x^2 - 2y^2 - 3xy - x + 2y = 2x^2 - (1 + 3y)x - 2y^2 + 2y = 2\{x^2 - (1 + 3y)x/2\} - 2y^2 + 2y$$
$$= 2\{x^2 - (1 + 3y)x/2 + (1 + 3y)^2/16 - (1 + 3y)^2/16\} - 2y^2 + 2y$$
$$= 2\{x - (1 + 3y)/4\}^2 - (1 + 3y)^2/8 - 2y^2 + 2y$$

$$-(1 + 3y)^2/8 - 2y^2 + 2y =-(1 + 6y + 9y^2)/8 - 2y^2 + 2y = -1/8 - 3y/4 - 9y^2/8 - 2y^2 + 2y$$
$$= -25y^2/8 + 5y/4 - 1/8 = (-1/8)(25y^2 - 10y + 1) = (-1/8)(5y - 1)^2$$

그러면, $2x^2 - 2y^2 - 3xy - x + 2y = 2\{x - (1 + 3y)/4\}^2 - 1/8(5y - 1)^2$
$= (1/8)[16\{x - (1 + 3y)/4\}^2 - (5y - 1)^2].$

$A^2 - B^2 = (A + B)(A - B)$라는 것을 알고있죠. 그러면:

$$16\{x - (1 + 3y)/4\}^2 - (5y - 1)^2$$
$$= \{4(x - (1 + 3y)/4) + (5y - 1)\}\{4(x - (1 + 3y)/4) - (5y - 1)\}$$
$$= (4x - 1 - 3y + 5y - 1)(4x - 1 - 3y - 5y + 1)$$
$$= (4x + 2y - 2)(4x - 8y) = 2(2x + y - 1)4(x - 2y) = 8(2x + y - 1)(x - 2y)$$

그러면, $(1/8)[16\{x - (1 + 3y)/4\}^2 - (5y - 1)^2] = (1/8)8(2x + y - 1)(x - 2y).$
그러므로, $2x^2 - 2y^2 - 3xy - x + 2y = (2x + y - 1)(x - 2y).$

요약:

$$2x^2 - 2y^2 - 3xy - x + 2y = 2x^2 - 3xy - x - 2y^2 + 2y = x(2x - 3y - 1) - 2y(y - 1)$$
$$= x(2x - 4y + y - 1) - 2y(y - 1) = x(2x - 4y) + x(y - 1) - 2y(y - 1)$$
$$= 2x(x - 2y) + (x - 2y)(y - 1) = (x - 2y)(2x + y - 1) \Rightarrow 2x + y - 1 = 0 \text{과 } x - 2y = 0.$$

제안과 풀이
3번 예제에서의 문제

아래의 방정식은 두개의 직선을 나타낸다 하고 t의 값을 구하세요:
$2x^2 + 4xy - ty^2 + 4x + 2y - 2 = 0$ 여기에서 t는 상수.

두개의 직선들을 하나의 방정식 $(ax + by + c)(ux + vy + w) = 0$ 으로 나타낼 수 있죠.
다시 말해서, 그 다항식을 인수분해하고 그 개별적인 직선들을 분리하는 거죠.
그럼 이제, 그 두개의 인수들 각각이 x와 y의 일차식인 간단한 다항식이 되는 겁니다.
그런데, 어떻게 인수분해 하죠?
완전 제곱식을 사용해볼 수 있어요.
이제, 인수분해를 시작해보기로 하죠.
먼저, $P = 2x^2 + 4xy - ty^2 + 4x + 2y - 2$ 라 하세요. 그러면:

$P = 2x^2 + (4y + 4)x - ty^2 + 2y - 2$
$= 2\{x^2 + 2(y + 2)x\} - ty^2 + 2y - 2$
$= 2\{x^2 + 2(y + 2)x + (y + 2)^2 - (y + 2)^2\} - ty^2 + 2y - 2$
$= 2\{x + (y + 2)\}^2 - 2(y + 2)^2 - ty^2 + 2y - 2$
$= 2\{x + (y + 2)\}^2 - \{2(y + 2)^2 + ty^2 - 2y + 2\}.$

그럼 이제, 그 다항식 P 전체를 $(A + B)(A - B)$로 쉽게 인수분해 될 수 있는 $A^2 - B^2$와 같은 형태로 표현할 수 있도록 $2(y + 2)^2 + ty^2 - 2y + 2$를 완전 제곱식으로 변형해야 하죠.

$2(y + 2)^2 + ty^2 - 2y + 2 = 2(y^2 + 4y + 4) + ty^2 - 2y + 2 = (t + 2)y^2 + 6y + 10.$

그런데, 다항식 $(t + 2)y^2 + 6y + 10$이 완전 제곱식으로 변형될 수 있으려면 그 판별식이 0이어야만 하는 거죠.
그 판별식이란 이차방정식의 근의 공식에서의 그 근호 안에 있는 부분입니다.

예를 들어, $ax^2 + bx + c = 0$이라 하세요.

그러면, 그 근의 공식은 $x = (-b \pm (b^2 - 4ac)^{1/2})/2a = \dfrac{-b \pm \sqrt{b^2 - 4ac}}{2a}$ 이죠.
그러면 이제, 그 판별식은 $b^2 - 4ac$이죠.
그 판별식이 0이면 그 답은 $-b/2a$뿐이고 따라서, 그 다항식 $ax^2 + bx + c$은 완전 제곱식으로 될 수 있는 거죠.
실제로, 그 판별식 $b^2 - 4ac = 0$이면 $ax^2 + bx + c = (x + b/2a)^2$ 입니다.

그럼 이제, 다항식 $(t + 2)y^2 + 6y + 10$의 판별식이 0이어야만 하죠.

그 판별식 $= 6^2 - 4(t + 2)10 = 36 - 40(t + 2) = 36 - 40t - 80 = -40t - 44.$
그러면, $-40t - 44 = 0 \Rightarrow t = -44/40 = -11/10.$

그럼 이제, t가 $-11/10$이면 그 두 직선들의 방정식에서의 다항식 **$2x^2 + 4xy - ty^2 + 4x + 2y - 2$**가 두 직선들에 대한 개별적인 두개의 방정식들에서의 다항식들로 인수분해 될 수 있는 것이죠.

그러므로, $2x^2 + 4xy + 11y^2/10 + 4x + 2y - 2 = 0$은 두 직선들의 방정식인 겁니다.

이제, 그 두개의 개별적인 방정식들이 어떻게 생겼는지 보도록 하죠. 그러면:

$t = -11/10 \Rightarrow (t + 2)y^2 + 6y + 10 = 9y^2/10 + 6y + 10 = (1/10)(9y^2 + 60 + 100) = (1/10)(3y + 10)^2$.

그러면 이제, $2x^2 + 4xy + 11y^2/10 + 4x + 2y - 2 = 2\{x + (y + 2)\}^2 - (1/10)(3y + 10)^2$.

$A^2 - B^2 = (A + B)(A - B)$라는 걸 알고있죠. 그러면:

$2\{x + (y + 2)\}^2 - (1/10)(3y + 10)^2$
$= \{2^{1/2}(x + y + 2) + (1/10)^{1/2}(3y + 10)\}\{2^{1/2}(x + y + 2) - (1/10)^{1/2}(3y + 10)\}$

$2^{1/2}(x + y + 2) + (1/10^{1/2})(3y + 10)$
$= 2^{1/2}x + 2^{1/2}y + 2^{3/2} + (3/10^{1/2})y + 10^{1/2}$
$= 2^{1/2}x + \{2^{1/2} + (3/10^{1/2})\}y + 2^{3/2} + 10^{1/2}$
$= 2^{1/2}x + \{2^{1/2} + 3(10^{1/2})/10\}y + 2^{3/2} + 10^{1/2}$.
$= \sqrt{2}x + (\sqrt{2} + 3\sqrt{10}/10)y + \sqrt{2^3} + \sqrt{10}$
$= \sqrt{2}x + (\sqrt{2} + 3\sqrt{10}/10)y + 2\sqrt{2} + \sqrt{10}$

$2^{1/2}(x + y + 2) - (1/10)^{1/2}(3y + 10)$
$= 2^{1/2}x + 2^{1/2}y + 2^{3/2} - (3/10^{1/2})y - 10^{1/2}$
$= 2^{1/2}x + \{2^{1/2} - (3/10^{1/2})\}y + 2^{3/2} - 10^{1/2}$
$= 2^{1/2}x + \{2^{1/2} - 3(10^{1/2})/10\}y + 2^{3/2} - 10^{1/2}$.
$= \sqrt{2}x + (\sqrt{2} - 3\sqrt{10}/10)y + \sqrt{2^3} - \sqrt{10}$
$= \sqrt{2}x + (\sqrt{2} - 3\sqrt{10}/10)y + 2\sqrt{2} - \sqrt{10}$

그러므로:
그 두 직선들 중 하나는 $\sqrt{2}x + (\sqrt{2} + 3\sqrt{10}/10)y + 2\sqrt{2} + \sqrt{10} = 0$.
그 나머지 하나는 $\sqrt{2}x + (\sqrt{2} - 3\sqrt{10}/10)y + 2\sqrt{2} - \sqrt{10} = 0$.

요약:

$P = 2x^2 + 4xy - ty^2 + 4x + 2y - 2$ 라 하세요. 그러면:
$P = 2x^2 + (4y + 4)x - ty^2 + 2y - 2 = 2\{x^2 + 2(y + 2)x\} - ty^2 + 2y - 2$
$= 2\{x^2 + 2(y + 2)x + (y + 2)^2 - (y + 2)^2\} - ty^2 + 2y - 2$
$= 2\{x + (y + 2)\}^2 - 2(y + 2)^2 - ty^2 + 2y - 2$
$= 2\{x + (y + 2)\}^2 - \{2(y + 2)^2 + ty^2 - 2y + 2\}$.

그럼 이제, 그 다항식 P 전체를 $(A + B)(A - B)$로 쉽게 인수분해 될 수 있는 $A^2 - B^2$와 같은 형태로 표현할 수 있도록 $2(y + 2)^2 + ty^2 - 2y + 2$를 완전 제곱식으로 변형해야 하죠.

$2(y + 2)^2 + ty^2 - 2y + 2 = 2(y^2 + 4y + 4) + ty^2 - 2y + 2 = (t + 2)y^2 + 6y + 10$.

그러면, 다항식 $(t + 2)y^2 + 6y + 10$의 판별식이 0이어야만 하죠.
그 판별식 $= 6^2 - 4(t + 2)10 = 36 - 40(t + 2) = 36 - 40t - 80 = -40t - 44$.

그러므로, $-40t - 44 = 0 \Rightarrow t = -44/40 = -11/10$.

제안과 풀이
4 번 예제에서의 문제

다음과 같은 두개의 직선들이 있다 하세요: **$2x - y - 1 = 0$** 와 **$3x + 2y - 2 = 0$.**
그러면, 이 두 직선들 각각으로부터 동일한 거리만큼 떨어져있는 점 **P(s, t)**를 구하세요.

Math 에서는 두개의 사물들간의 거리를 잴 때는 그 둘을 잇는 선분의 길이를 재는데 이 선분은 그 두 사물들에 수직이죠. 직선 $2x - y - 1 = 0$을 A라 하고 직선 $3x + 2y - 2 = 0$을 B라 하세요.
그러면, 점 P와 직선 A 사이의 거리를 구하려면 먼저, 점 P를 지나고 그 A에 수직인 직선의 방정식을 구해야 하는 거죠. 또한, 이는 점 P와 직선 B 사이의 거리에 대하여도 마찬가지죠.
이제, C는 A에 수직인 직선이고 D는 B에 수직인 직선이라 하세요. 그러면, 그 점 P는 그 수직선 둘 다에 포함되는 거죠. 그러면, C와 D는 어디서 서로 만나나요? 점 P에서 이죠.

그림. 4.0

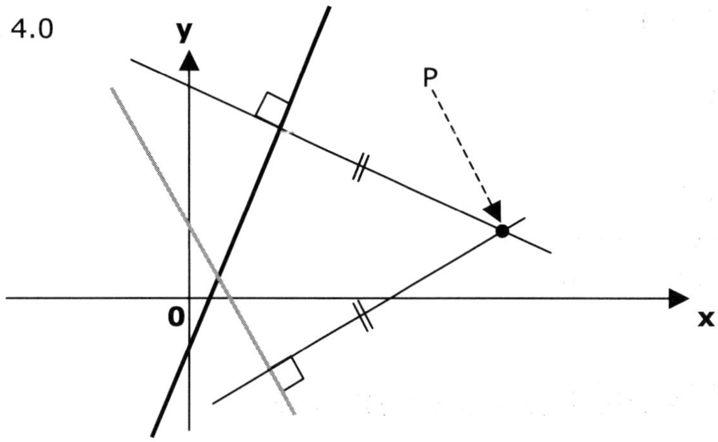

그런데, 또 한 점이 그 두 직선들로부터 동일한 거리만큼 떨어져있네요.

그림. 4.1

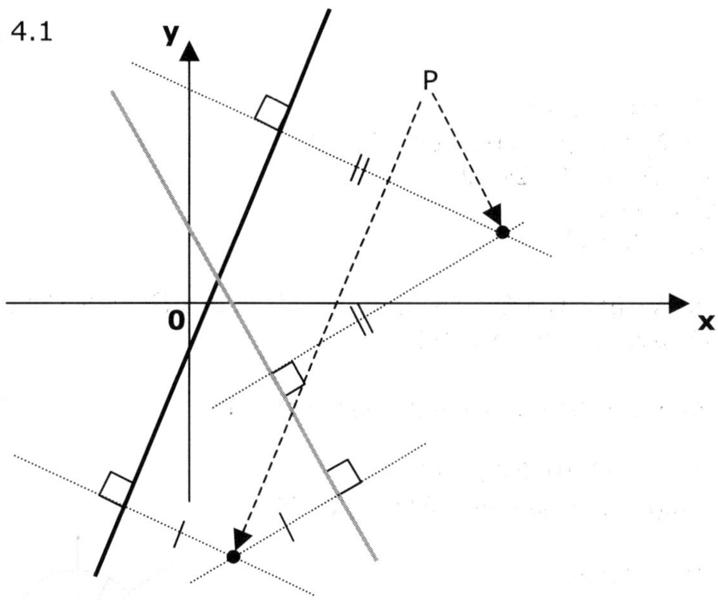

사실은, 무한정 많은 점들 각각이 그 두 직선들로부터 동일한 거리만큼 떨어져있는 겁니다. 그러한 모든 점들은 아래에 나타낸 바와 같이 점선으로 된 두개의 직선들을 이루고 있어요.

그림. 4.2

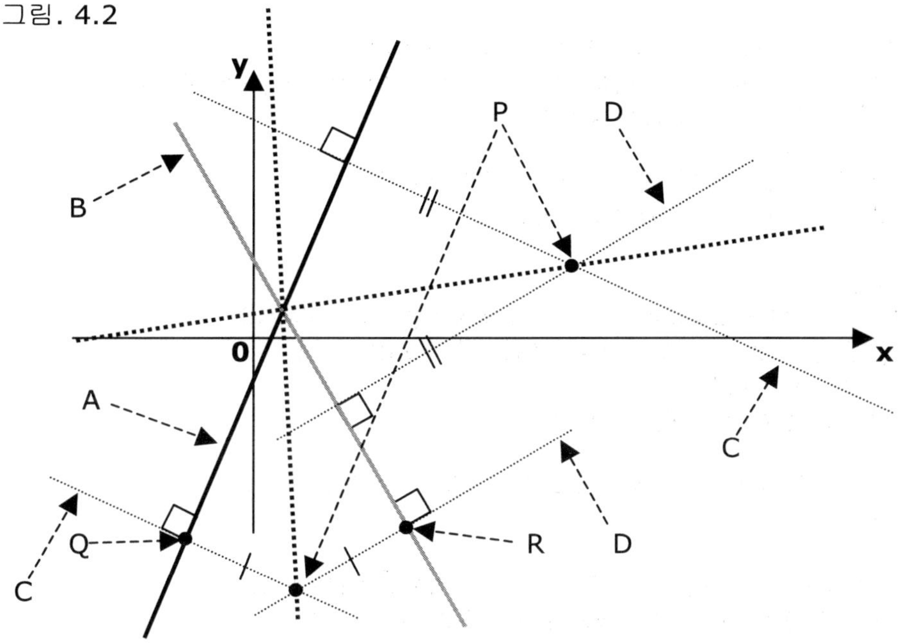

그러면, 점 P(s, t)는 그저 하나의 특정한 점이 아니고 그 두개의 점선으로 된 직선들 각각에 있는 그 모든 점들을 대표하는 일반점인 거죠. 그러면 이제, 이 일반점 P 를 가지고 그 두개의 점선으로 된 직선들 각각을 구해보기로 하죠. 즉, P를 두 번 사용하는 겁니다.

먼저, 그 두 직선들 C와 D부터 시작해야 해요. 그리고 나서, 두 직선들 A와 C가 서로 만나는 점과 두 직선들 B와 D가 서로 만나는 점을 구해야 하죠. 그러면 이제, 그 두개의 점선으로 된 직선들을 구할 수 있는 거죠.

직선 C가 직선 A에 수직이고 직선 D는 직선 B에 수직이죠.
두 직선들이 서로 수직이면 그 기울기들의 곱이 −1이죠.
그러면 먼저, 주어진 일반형 방정식들을 표준형으로 바꾸어보도록 하죠. 다음과 같습니다:
직선 A: $2x - y - 1 = 0 \Rightarrow y = 2x - 1$.
직선 B: $3x + 2y - 2 = 0 \Rightarrow y = -3x/2 + 1$.

이제, 직선 C의 방정식이 $y = a_1x + b_1$이라고 하세요.
그럼, A의 기울기가 2이므로 $a_1 = -1/2$이죠 왜냐하면 기울기들의 곱이 −1이니까.
그러면, $y = -x/2 + b_1$이네요.
또한, C는 점 P(s, t)를 지나니까 $t = -s/2 + b_1 \Rightarrow b_1 = t + s/2$.
그러므로, C의 방정식은 $y = -x/2 + t + s/2$입니다.

이번에는, 직선 D의 방정식은 $y = a_2x + b_2$라고 하세요.
그럼, B의 기울기가 −3/2이므로 $a_2 = 2/3$이죠 왜냐하면 기울기들의 곱이 −1이니까.
그러면, $y = 2x/3 + b_2$네요.
또한, D도 점 P(s, t)를 지나니까 $t = 2s/3 + b_2 \Rightarrow b_2 = t - 2s/3$.
그러므로, D의 방정식은 $y = 2x/3 + t - 2s/3$입니다.

다음으로, 두 직선 A와 C가 서로 만나는 점을 Q라 하고 이 Q를 구해보기로 하죠.
직선 A는 $y = 2x - 1$이고 직선 C는 $y = -x/2 + t + s/2$이죠.
그러면, Q의 x 좌표부터 시작합니다.
$2x - 1 = -x/2 + t + s/2$.
그러면, $2x + x/2 = t + s/2 + 1 \Rightarrow 5x/2 = t + s/2 + 1 \Rightarrow x = (2t + s + 2)/5$. 따라서:
$y = 2x - 1 = 2(2t + s + 2)/5 - 1 = (4t + 2s + 4)/5 - 1 = (4t + 2s + 4 - 5)/5 = (4t + 2s - 1)/5$.
그러므로, Q는 $((2t + s + 2)/5, (4t + 2s - 1)/5)$에 있는 거죠.

다음으로, 두 직선 B와 D가 만나는 점을 R이라 하고 이 R을 구해보기로 하죠.
직선 B는 $y = -3x/2 + 1$이고 직선 D는 $y = 2x/3 + t - 2s/3$이죠.
그러면, R의 x 좌표부터 시작합니다.
$-3x/2 + 1 = 2x/3 + t - 2s/3$.
그러면, $(2/3 + 3/2)x = 1 - t + 2s/3 \Rightarrow 13x/6 = 1 - t + 2s/3 \Rightarrow x = (6 - 6t + 4s)/13$.
따라서, $y = -3x/2 + 1 = (-3/2)(6 - 6t + 4s)/13 + 1 = (-3)(3 - 3t + 2s)/13 + 1$
$= (9t - 6s - 9)/13 + 1 = (9t - 6s + 4)/13$.
그러므로, R은 $((6 - 6t + 4s)/13, (9t - 6s + 4)/13)$에 있죠.

그러면 이제, 여러분이 좋아하는 그 거리 공식(보통, 피타고라스 정리라고 하죠)을 사용해서 그 두개의 점선으로 된 직선들을 구해보기로 하죠.

P와 Q사이의 거리와 P와 R 사이의 거리는 동일하고 거리 공식은 $d^2 = (\Delta x)^2 + (\Delta y)^2$이죠.
먼저, P와 Q사이의 거리를 d_1 그리고 P와 R사이의 거리를 d_2라 하고 d_1부터 시작하죠.

$\Delta x = s - (2t + s + 2)/5 = (4s - 2t - 2)/5$.
$\Delta y = t - (4t + 2s - 1)/5 = (t - 2s + 1)/5$.
그러면, $d_1^2 = (4s - 2t - 2)^2/5^2 + (t - 2s + 1)^2/5^2 = \{(4s - 2t - 2)^2 + (t - 2s + 1)^2\}/5^2$.

다음으로, d_2로 넘어가보죠.

$\Delta x = s - (6 - 6t + 4s)/13 = (6t + 9s - 6)/13$.
$\Delta y = t - (9t - 6s + 4)/13 = (4t + 6s - 4)/13$.
그럼, $d_2^2 = (6t + 9s - 6)^2/13^2 + (4t + 6s - 4)^2/13^2 = \{(6t + 9s - 6)^2 + (4t + 6s - 4)^2\}/13^2$.

그럼 이제, $\{(4s - 2t - 2)^2 + (t - 2s + 1)^2\}/5^2 = \{(6t + 9s - 6)^2 + (4t + 6s - 4)^2\}/13^2$.

사실, 위의 방정식은 두개의 직선들의 방정식이고 그 직선들은 그 두개의 점선으로 된 직선들이죠.
그러면, 그 두 직선들을 어떻게 생겼는지 알 수 있도록 분리할 수 있어야 하네요. 인수분해로 해볼 수 있죠. 즉, 위의 방정식을 다음과 같은 형태로 인수분해 하는 겁니다:
$(ax + by + c)(dx + ey + f) = 0$이고 여기에서 a, b, c, d, e, f는 상수.
그런데, 계산이 꽤나 복잡해 보이네요. 그렇더라도 좀 더 진행해보도록 하죠.

$\{(4s - 2t - 2)^2 + (t - 2s + 1)^2\}/5^2 = \{(6t + 9s - 6)^2 + (4t + 6s - 4)^2\}/13^2$
$\Rightarrow 13^2\{(4s - 2t - 2)^2 + (t - 2s + 1)^2\} = 5^2\{(6t + 9s - 6)^2 + (4t + 6s - 4)^2\}$
$\Rightarrow 13^2(4s - 2t - 2)^2 + 13^2(t - 2s + 1)^2 = 5^2(6t + 9s - 6)^2 + 5^2(4t + 6s - 4)^2$
$\Rightarrow 13^2(4s - 2t - 2)^2 - 5^2(6t + 9s - 6)^2 + 13^2(t - 2s + 1)^2 - 5^2(4t + 6s - 4)^2 = 0$.

$13^2(4s - 2t - 2)^2 - 5^2(6t + 9s - 6)^2$
$= \{13(4s - 2t - 2) + 5(6t + 9s - 6)\}\{13(4s - 2t - 2) - 5(6t + 9s - 6)\}$

$= (52s - 26t - 26 + 30t + 45s - 30)(52s - 26t - 26 - 30t - 45s + 30)$
$= (97s + 4t - 56)(7s - 56t + 4)$
$= 679s^2 - 5432st + 388s + 28st - 224t^2 + 16t - 392s + 3136t - 224$
$= 679s^2 - 224t^2 - 5404st - 4s + 3152t - 224$

$13^2(t - 2s + 1)^2 - 5^2(4t + 6s - 4)^2$
$= \{13(t - 2s + 1) + 5(4t + 6s - 4)\}\{13(t - 2s + 1) - 5(4t + 6s - 4)\}$
$= (13t - 26s + 13 + 20t + 30s - 20)(13t - 26s + 13 - 20t - 30s + 20)$
$= (33t + 4s - 7)(-7t - 56s + 33)$
$= -231t^2 - 1848st + 1089t - 28st - 224s^2 + 132s + 49t + 392s - 231$
$= -224s^2 - 231t^2 - 1876st + 524s + 1138t - 231$

$13^2(4s - 2t - 2)^2 - 5^2(6t + 9s - 6)^2 + 13^2(t - 2s + 1)^2 - 5^2(4t + 6s - 4)^2$
$= 679s^2 - 224t^2 - 5404st - 4s + 3152t - 224 - 224s^2 - 231t^2 - 1876st + 524s + 1138t - 231$
$= 455s^2 - 455t^2 - 7280st + 520s + 4290t - 455$
$= 0$

이제, 마지막 등식의 양변을 65로 나누면 다음과 같습니다:
$7s^2 - 7t^2 - 112st + 8s + 66t - 7 = 0.$

이 계산 정말 복잡하네요!
자 이제, 이 예제를 다른 방법으로 해결해보는 것이 좋겠어요.

한 점에서 직선까지의 거리를 구해야 할 때가 많죠. 그래서, 이에 대한 공식이 있어요. 그럼 이제, 그 공식을 사용하기로 하죠. 그런데, 먼저 그 공식이 어떻게 만들어지는지 알아보도록 하겠습니다.

직선의 방정식의 일반형은 다음과 같죠: $ax + by + c = 0$ 여기에서 a, b, c는 상수.
이제, 직선 $ax + by + c = 0$을 L이라 하고 한 점 P(s, t)에서 직선 L까지의 거리를 구한다 하세요.
그러면, 평소와 마찬가지로 그 수직선을 구하고 그 두 직선들이 서로 만나는 점을 구하고 그리고 나서 그 유명한 거리 공식을 사용하는 겁니다.

그러면, T를 그 직선 L에 수직인 직선이라 하고 그 T를 먼저 구하기로 하죠.
T가 점 P(s, t)를 지난다는 것을 알고있죠. T의 기울기가 m이라 하세요.
그러면, T의 방정식은 다음과 같습니다: $y - t = m(x - s)$. 이제, 그 m을 명시하기로 하죠.
우선, L의 방정식을 표준형으로 변형하고 그 기울기를 구합니다.
$ax + by + c = 0 \Rightarrow y = -ax/b - c/b \Rightarrow$ L의 기울기는 $-a/b$입니다.
그러면, b/a가 T의 기울기네요 왜냐하면 그 기울기들의 곱이 -1이니까.

그러면, $y - t = m(x - s) \Rightarrow y - t = (b/a)(x - s)$이죠.
따라서, $b(x - s) - a(y - t) = 0$입니다. 그러면:

직선 L: $ax + by + c = 0$. 직선 T: $b(x - s) - a(y - t) = 0$.

그러면, 이 두 직선들이 어디서 만나는지 알아내야 하죠.
즉, 위의 방정식들을 x와 y에 관한 연립방정식으로 해서 그 답을 구하는 겁니다.
L의 방정식에 a를 곱하고 T의 방정식에는 b를 곱해보죠. 그러면:
$a(ax + by + c) = a^2x + aby + ac = 0.$
$b\{b(x - s) - a(y - t)\} = b^2(x - s) - ab(y - t) = 0.$

그럼 이제, 이 둘을 합하면 y가 소거될 겁니다.
$(a^2x + aby + ac) + \{b^2(x - s) - ab(y - t)\} = a^2x + aby + ac + b^2x - b^2s - aby + abt$
$= a^2x + ac + b^2x - b^2s + abt = (a^2 + b^2)x - b^2s + abt + ac = 0.$
그러면, $(a^2 + b^2)x = b^2s - abt - ac.$
그러므로, $x = (b^2s - abt - ac)/(a^2 + b^2).$

이번에는, L의 방정식에 b를 곱하고 T의 방정식에는 a를 곱해보죠. 그러면:
$b(ax + by + c) = abx + b^2y + bc = 0.$
$a\{b(x - s) - a(y - t)\} = ab(x - s) - a^2(y - t) = 0.$

그럼 이제, 첫번째 방정식식에서 두 번째를 빼면 x가 소거될 겁니다.
$(abx + b^2y + bc) - \{ab(x - s) - a^2(y - t)\} = abx + b^2y + bc - abx + abs + a^2y - a^2t$
$= b^2y + bc + abs + a^2y - a^2t = (a^2 + b^2)y - a^2t + abs + bc = 0.$
그러면, $(a^2 + b^2)y = a^2t - abs - bc.$
그러므로, $y = (a^2t - abs - bc)/(a^2 + b^2).$

따라서, 두 직선들 L과 T는 $((b^2s - abt - ac)/(a^2 + b^2), (a^2t - abs - bc)/(a^2 + b^2))$에서 만나죠.

이제, 그 두 직선들 L과 T가 점 Q(u, v)에서 만나고 d가 점 P(s, t)와 점 Q(u, v) 사이의 거리라고 하세요.

그러면, $u = (b^2s - abt - ac)/(a^2 + b^2)$이고 $v = (a^2t - abs - bc)/(a^2 + b^2)$네요.

따라서, $(a^2 + b^2)u = b^2s - abt - ac$이고 $(a^2 + b^2)v = a^2t - abs - bc$이죠.

$d^2 = (u - s)^2 + (v - t)^2$ 라는 걸 알고있죠.
그러면, 다음과 같은 식들로부터 u - s와 v - t를 구하는 겁니다:
$(a^2 + b^2)u = b^2s - abt - ac$와 $(a^2 + b^2)v = a^2t - abs - bc.$

$(a^2 + b^2)u = b^2s - abt - ac = a^2s - a^2s + b^2s - abt - ac = (a^2 + b^2)s - a^2s - abt - ac$
$\Rightarrow (a^2 + b^2)u = (a^2 + b^2)s - a^2s - abt - ac$
$\Rightarrow (a^2 + b^2)u - (a^2 + b^2)s = -(a^2s + abt + ac)$
$\Rightarrow (a^2 + b^2)(u - s) = -(a^2s + abt + ac)$
$\Rightarrow u - s = -(a^2s + abt + ac)/(a^2 + b^2) = -a(as + bt + c)/(a^2 + b^2).$

$(a^2 + b^2)v = a^2t - abs - bc = a^2t + b^2t - b^2t - abs - bc = (a^2 + b^2)t - (b^2t + abs + bc)$
$\Rightarrow (a^2 + b^2)(v - t) = -(b^2t + abs + bc).$
$\Rightarrow v - t = -(b^2t + abs + bc)/(a^2 + b^2) = -b(bt + as + c)/(a^2 + b^2).$

자 이제, 다음과 같다는 것을 알 수 있는 거죠:

$d^2 = (u - s)^2 + (v - t)^2$
$= a^2(as + bt + c)^2/(a^2 + b^2)^2 + b^2(as + bt + c)^2/(a^2 + b^2)^2$
$= (a^2 + b^2)(as + bt + c)^2/(a^2 + b^2)^2$
$= (as + bt + c)^2/(a^2 + b^2).$

그러면, $d = \pm(as + bt + c)/(a^2 + b^2)^{1/2}.$
그런데, d는 거리이니까 $d > 0$ 이죠.
그러므로, $d = |as + bt + c|/(a^2 + b^2)^{1/2}.$

그럼 이제, 점 P(s, t)에서 직선 $ax + by + c = 0$ 까지 거리를 나타내는 공식을 구한 겁니다.
그러면, 이 공식을 가지고 이 예제에서의 문제로 돌아가보죠. 현재, 두 직선 $2x - y - 1 = 0$ 와 $3x + 2y - 2 = 0$ 모두로부터 동일한 거리만큼 떨어져있는 점 P(s, t)에 대해서 하고있는 거죠.

직선 A: $2x - y - 1 = 0$. 직선 B: $3x + 2y - 2 = 0$.

그러면, 그 공식에 의하여:
직선 A와 점 P 사이의 거리는 $|2s - t - 1|/(2^2 + (-1)^2)^{1/2}$.
직선 B와 점 P 사이의 거리는 $|3s + 2t - 2|/(3^2 + 2^2)^{1/2}$.

그러면, $|2s - t - 1|/(2^2 + (-1)^2)^{1/2} = |3s + 2t - 2|/(3^2 + 2^2)^{1/2}$이네요.
따라서, $(2s - t - 1)/5^{1/2} = \pm(3s + 2t - 2)/13^{1/2}$.

그런데, 어째서 ±이죠?
이는 절대기호 안에 있는 것이 양 또는 음이 될 수도 있기 때문입니다. 즉, 이는 두 개의 직선들을 의미하는 거죠.
무슨 직선들?
이는 앞에서 언급된 점선으로 된 두개의 직선들인 것이죠.
그 점선으로 된 두개의 직선들을 이루는 모든 점들 각각이 P(s, t)가 될 수 있다는 것은 앞에서 이미 알아냈죠.
즉, P가 그 점선으로 된 직선들 각각에서의 일반점이죠.
이제, 그 직선들은 x-y 좌표계에 있으므로 s 와 t 를 제각기 x 와 y 로 교체하는 겁니다.
그러면 이제, 그 점선으로 된 두개의 직선들 중의 하나는 다음과 같아요:

$$(2x - y - 1)/5^{1/2} = (3x + 2y - 2)/13^{1/2}$$
$$\Rightarrow (2/5^{1/2} - 3/13^{1/2})x - (1/5^{1/2} + 2/13^{1/2})y + 2/13^{1/2} - 1/5^{1/2} = 0$$

$$\Rightarrow (2X13^{1/2} - 3X5^{1/2})x/65^{1/2} - (13^{1/2} + 2X5^{1/2})y/65^{1/2} + (2X5^{1/2} - 13^{1/2})/65^{1/2} = 0$$

$$\Rightarrow (2X13^{1/2} - 3X5^{1/2})x - (13^{1/2} + 2X5^{1/2})y + 2X5^{1/2} - 13^{1/2}$$
$$= (2\sqrt{13} - 3\sqrt{5})x - (\sqrt{13} + 2\sqrt{5})y + 2\sqrt{5} - \sqrt{13} = 0.$$

다음으로 그 나머지 직선은 다음과 같죠:

$$(2x - y - 1)/5^{1/2} = -(3x + 2y - 2)/13^{1/2}$$
$$\Rightarrow (2/5^{1/2} + 3/13^{1/2})x + (2/13^{1/2} - 1/5^{1/2})y - 1/5^{1/2} - 2/13^{1/2} = 0.$$

$$\Rightarrow (2X13^{1/2} + 3X5^{1/2})x/65^{1/2} + (2X5^{1/2} - 13^{1/2})y/65^{1/2} - (13^{1/2} + 2X5^{1/2})/65^{1/2} = 0$$

$$\Rightarrow (2X13^{1/2} + 3X5^{1/2})x + (2X5^{1/2} - 13^{1/2})y - 13^{1/2} - 2X5^{1/2}$$
$$= (2\sqrt{13} + 3\sqrt{5})x + (2\sqrt{5} - \sqrt{13})y - \sqrt{13} - 2\sqrt{5} = 0.$$

그러면 이제, 위의 두 직선들을 이루는 모든 점들 하나하나가 두 직선들 A 와 B 모두로부터 동일한 거리만큼 떨어져있는 겁니다.

요약:

다음과 같이 가정하세요:
직선 A: $2x - y - 1 = 0$. 직선 B: $3x + 2y - 2 = 0$.

그러면, 그 거리공식에 의하여:
직선 A와 점 P 사이의 거리는 $|2s - t - 1|/(2^2 + (-1)^2)^{1/2}$.
직선 B와 점 P 사이의 거리는 $|3s + 2t - 2|/(3^2 + 2^2)^{1/2}$.

그러면, $|2s - t - 1|/(2^2 + (-1)^2)^{1/2} = |3s + 2t - 2|/(3^2 + 2^2)^{1/2}$이네요.
따라서, $(2s - t - 1)/5^{1/2} = \pm(3s + 2t - 2)/13^{1/2}$.

그러면 이제, 위의 수식은 두 개의 직선들을 나타내는 방정식이고 이 직선들을 이루는 모든 점들 하나하나가 두 직선들 A 와 B 로부터 동일한 거리만큼 떨어져있는 겁니다. 그러므로, P(s, t)는 그 직선들 각각에서의 일반점이죠.

이제, 그 직선들은 x-y 좌표계에 있으므로 s 와 t 를 제각기 x 와 y 로 교체하도록 하죠.

그러면 이제, 그 두개의 직선들 중의 하나는 다음과 같아요:

$$(2x - y - 1)/5^{1/2} = (3x + 2y - 2)/13^{1/2}$$
$$\Rightarrow (2/5^{1/2} - 3/13^{1/2})x - (1/5^{1/2} + 2/13^{1/2})y + 2/13^{1/2} - 1/5^{1/2} = 0$$

$$\Rightarrow (2X13^{1/2} - 3X5^{1/2})x/65^{1/2} - (13^{1/2} + 2X5^{1/2})y/65^{1/2} + (2X5^{1/2} - 13^{1/2})/65^{1/2} = 0$$

$$\Rightarrow (2X13^{1/2} - 3X5^{1/2})x - (13^{1/2} + 2X5^{1/2})y + 2X5^{1/2} - 13^{1/2}$$
$$= (2\sqrt{13} - 3\sqrt{5})x - (\sqrt{13} + 2\sqrt{5})y + 2\sqrt{5} - \sqrt{13} = 0.$$

다음으로 그 나머지 직선은 다음과 같죠:

$$(2x - y - 1)/5^{1/2} = -(3x + 2y - 2)/13^{1/2}$$
$$\Rightarrow (2/5^{1/2} + 3/13^{1/2})x + (2/13^{1/2} - 1/5^{1/2})y - 1/5^{1/2} - 2/13^{1/2} = 0.$$

$$\Rightarrow (2X13^{1/2} + 3X5^{1/2})x/65^{1/2} + (2X5^{1/2} - 13^{1/2})y/65^{1/2} - (13^{1/2} + 2X5^{1/2})/65^{1/2} = 0$$

$$\Rightarrow (2X13^{1/2} + 3X5^{1/2})x + (2X5^{1/2} - 13^{1/2})y - 13^{1/2} - 2X5^{1/2}$$
$$= (2\sqrt{13} + 3\sqrt{5})x + (2\sqrt{5} - \sqrt{13})y - \sqrt{13} - 2\sqrt{5} = 0.$$

그런데, 내친김에 공식을 하나 마저 더 만들어보기로 하죠.
삼각형의 면적을 구해야 할 때가 많아요. 그 때는 보통, 그 면적이 밑변과 높이의 곱의 반이라는 간단한 공식을 사용하죠. 하지만, 그 높이를 구하는데 곤란을 겪을 때가 더러 있죠. 그런데, 높이를 모르고도 그 면적을 구할 수 있는 공식들이 몇 개 있어요. 그것들 중에, 지금 막 만들어보려는 것이 x-y 평면과 같은 직교(또는, 데카르트) 좌표평면에 명시되어있는 삼각형에 대하여 뭘 할 때 꽤나 소용이 될 수도 있죠.

이제, 세 개의 점들 $P_1(x_1, y_1)$, $P_2(x_2, y_2)$, $P_3(x_3, y_3)$가 주어져있고 이 점들이 삼각형을 이룬다고 하세요.
그러면, 아래의 그림에서와 같이 그 점들을 x-y 평면에 놓고 삼각형을 만들어보기로 하죠.

그림. 4.3

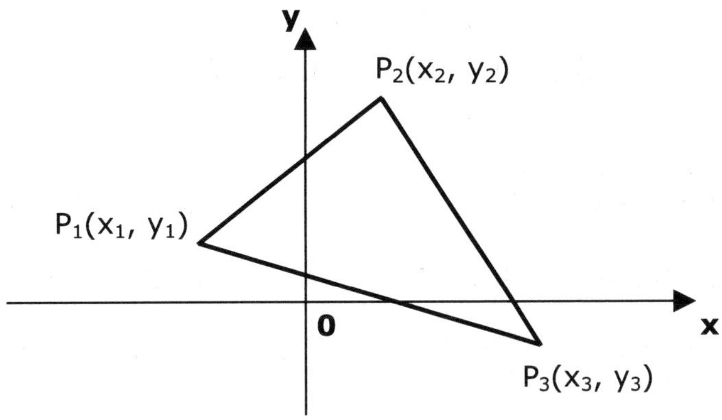

하지만, 여전히 면적 = (밑변 X 높이)/2 라는 그 기본 공식을 사용해야 합니다.
이제, 선분 P_1P_2가 밑변이라고 하세요. 그러면, 점 P_3에서 그 밑변까지의 수직거리가 그 높이죠.
이제, 두 점들 P_1과 P_2를 지나는 직선의 방정식부터 시작하기로 하죠.
먼저, 그 직선의 기울기는 $(y_2 - y_1)/(x_2 - x_1)$이죠.
또한, 그 직선은 P_1을 지나므로, 그 방정식은 다음과 같은 겁니다: $y - y_1 = \{(y_2 - y_1)/(x_2 - x_1)\}(x - x_1)$.
그럼 이제, 이 방정식을 일반형으로 변형해보기로 하죠.

그러면, $y - y_1 = (y_2 - y_1)(x - x_1)/(x_2 - x_1)$이죠. 따라서, $(y - y_1)(x_2 - x_1) = (y_2 - y_1)(x - x_1)$.
한편:
$(y - y_1)(x_2 - x_1) = y(x_2 - x_1) - y_1(x_2 - x_1)$. $(y_2 - y_1)(x - x_1) = x(y_2 - y_1) - x_1(y_2 - y_1)$.

그러면, $(y - y_1)(x_2 - x_1) = (y_2 - y_1)(x - x_1) \Rightarrow (y - y_1)(x_2 - x_1) - (y_2 - y_1)(x - x_1) = 0$
$\Rightarrow y(x_2 - x_1) - y_1(x_2 - x_1) - \{x(y_2 - y_1) - x_1(y_2 - y_1)\} = 0$
$\Rightarrow y(x_2 - x_1) - y_1(x_2 - x_1) - x(y_2 - y_1) + x_1(y_2 - y_1)\} = 0$
$\Rightarrow y(x_2 - x_1) - x(y_2 - y_1) - y_1x_2 + y_1x_1 + x_1y_2 - x_1y_1 = y(x_2 - x_1) - x(y_2 - y_1) - y_1x_2 + x_1y_2 = 0$
$\Rightarrow (y_2 - y_1)x - (x_2 - x_1)y - x_1y_2 + y_1x_2 = 0$ 이는 직선의 일반형 방정식이죠.

그럼 이제, 방금 앞에서 만들어낸 거리공식을 사용하여 이 직선과 점 P_3 사이의 거리를 구할 수 있죠.
그러면, 그 거리가 그 삼각형의 높이가 될 겁니다. 그러므로:
그 높이 = $|(y_2 - y_1)x_3 - (x_2 - x_1)y_3 - x_1y_2 + y_1x_2|/\{(y_2 - y_1)^2 + (x_2 - x_1)^2\}^{1/2}$.
또한, P_1과 P_2 사이의 거리 즉, 그 밑변의 길이는 $\{(y_2 - y_1)^2 + (x_2 - x_1)^2\}^{1/2}$ 입니다.

그러면, 밑변 X 높이
$= \{(y_2 - y_1)^2 + (x_2 - x_1)^2\}^{1/2}|(y_2 - y_1)x_3 - (x_2 - x_1)y_3 - x_1y_2 + y_1x_2|/\{(y_2 - y_1)^2 + (x_2 - x_1)^2\}^{1/2}$
$= |(y_2 - y_1)x_3 - (x_2 - x_1)y_3 - x_1y_2 + y_1x_2|$.

그러므로, 그 삼각형의 면적은 $|(y_2 - y_1)x_3 - (x_2 - x_1)y_3 - x_1y_2 + y_1x_2|/2$ 인 것이죠.
절대기호 내부를 좀더 보기 좋게 해보죠.
$(y_2 - y_1)x_3 - (x_2 - x_1)y_3 - x_1y_2 + y_1x_2 = (x_1 - x_2)y_3 + y_2x_3 - y_1x_3 - x_1y_2 + y_1x_2$
$= (x_1 - x_2)y_3 + (x_2 - x_3)y_1 + (x_3 - x_1)y_2$.

그럼 이제, 삼각형 $P_1P_2P_3$의 면적 = $|(x_1 - x_2)y_3 + (x_2 - x_3)y_1 + (x_3 - x_1)y_2|/2$.

제안과 풀이
5 번 예제에서의 0 번 문제

다음과 같은 직선의 방정식들이 있다고 하세요:
$ax + by + c = 0$ 이고 $ux + vy + w = 0$ 이며 여기에서 a, b, c, u, v, w 는 상수이고 $abc \neq 0$ 이며 $uvw \neq 0$.

또한, 두 방정식들은 사실, 동일한 직선을 나타낸다 하고 상수들간의 관계식을 구하세요.

어떤 방정식들은 달라보이기는 해도 실제로는 동일한 직선을 나타낼 수도 있어요.
그럼, 그것들이 사실 동일한 방정식들이라는 걸 어떻게 알아내죠?
시험을 좀 해 보아야 하는 거죠.

주어진 방정식들의 상수들을 흔히 계수라고도 합니다.
그런데, 엄밀히 말하면 c 와 w 는 계수라고 하지는 않죠.
이 계수들간의 관계를 조사해보면 그 문제의 방정식들이 과연 동일한 것들인가를 확인할 수 있는 겁니다.

먼저, 동일한 방정식들이 어떻게 해서 달라보일 수 있는지 알아보기로 하죠.
예를 들면, $2x + 3y - 5 = 0$, $4x + 6y - 10 = 0$, $-2x - 3y + 5 = 0$, $x + 3y/2 - 5/2 = 0$ 은 모두 동일한 직선을 나타내는 방정식들입니다. 왜냐하면 이것들이 서로 곱이 되기 때문이죠.
첫번째 방정식에 2 를 곱하면 두 번째 것이 되도 또한, 첫번째 것에 -1 을 곱하면 세 번째 것이 됩니다.
첫번째 것에 2 를 곱하면 그 상수들 각각에 2 가 곱해지고 그 곱들은 제각기 두 번째 것에서의 상수들이 되는 거죠.
또한, 두 번째에 $-1/2$ 을 곱하면 세 번째가 되고 그 세 번째에 있는 상수들 각각에 $-1/2$ 를 다시 곱하면 네 번째 방정식을 갖게 되는 거죠.
그러니까, 서로 대응하는 상수들간의 비들이 동일하면 그 방정식들은 동일한 직선을 나타내는 겁니다.
위에서 예로든 것들 중 처음 두개의 방정식들을 비교해보도록 하죠.
그것들은 $2x + 3y - 5 = 0$ 과 $4x + 6y - 10 = 0$ 입니다.
그러면, 그 대응하는 상수들간의 비들은 다음과 같죠: $2/4 = 3/6 = -5/-10$.
그 비들이 모두 $1/2$ 이고 서로 다 같은 것들이네요.

이제, 그 일반형 방정식들을 표준형 방정식들로 표현해보기로 하죠. 그러면:
$ax + by + c = 0 \Rightarrow y = (-a/b)x - c/b$
$ux + vy + w = 0 \Rightarrow y = (-u/v)x - w/v$

그러면 이제, 이 두 방정식들이 동일하다면 그 기울기들이 동일해야 하고 그 y 절편들도 같아야 하죠.
그러므로, $-a/b = -u/v$ 이고 $-c/b = -w/v$ 이어야 하는 거죠.

그러면:
$-a/b = -u/v \Rightarrow a/b = u/v \Rightarrow av = bu \Rightarrow a/u = b/v$.
$-c/b = -w/v \Rightarrow c/b = w/v \Rightarrow cv = bw \Rightarrow b/v = c/w$.
따라서, $a/u = b/v$ 이고 $b/v = c/w$ 이면 $a/u = b/v = c/w$.

그러므로, $ax + by + c = 0$ 과 $ux + vy + w = 0$ 이 동일한 직선을 나타낸다면 $a/u = b/v = c/w$ 입니다.

즉, 직선의 일반형 방정식들을 비교할 때 그것들이 동일한 직선을 나타내는지를 알아내려면 대응하는 상수들 간의 비들이 모두 동일한지 확인해야 하는 거죠.

제안과 풀이
5번 예제에서의 1번 문제

다음과 같은 직선의 방정식들이 있다고 하세요:
$ax + by + c = 0$ 이고 $ux + vy + w = 0$ 이며 여기에서 a, b, c, u, v, w 는 상수이고 $abc \neq 0$ 이며 $uvw \neq 0$.

또한, 두 직선들은 서로 평행이다 하고 상수들간의 관계식을 구하세요.

서로 평행인 두 직선들은 동일한 기울기를 갖고 있죠.
그러니까, 두개의 방정식들이 서로 평행인 두 직선들을 나타낸다면 그 기울기들이 같아야 하죠.
그러면, 무엇이 달라야 하나요?
그 y 절편들이나 x 절편들이 달라야 합니다.
그러니까, 두 방정식들에서 기울기들은 같지만 y 절편들(또는, x 절편들)이 다르면 그 두 방정식들은 서로 평행인 두개의 다른 직선들을 나타내는 것이죠.

이제, 그 일반형 방정식들을 표준형 방정식들로 표현해보기로 하죠. 그러면:
$ax + by + c = 0 \Rightarrow y = (-a/b)x - c/b$
$ux + vy + w = 0 \Rightarrow y = (-u/v)x - w/v$

그러면, $-a/b = -u/v$ 이지만 $-c/b \neq -w/v$ 이면 그 두 방정식들은 서로 평행인 두개의 다른 직선들을 나타내는 겁니다. 그러면 이제:

$-a/b = -u/v \Rightarrow a/b = u/v \Rightarrow av = bu \Rightarrow a/u = b/v$.
$-c/b \neq -w/v \Rightarrow c/b \neq w/v \Rightarrow cv \neq bw \Rightarrow b/v \neq c/w$.

따라서, $a/u = b/v$ 이지만 $b/v \neq c/w$ 이면 $a/u = b/v \neq c/w$.

그러므로, 두개의 방정식들 $ax + by + c = 0$ 과 $ux + vy + w = 0$ 이 서로 평행인 두개의 다른 직선들을 나타낸다면 $a/u = b/v \neq c/w$ 입니다.

즉, 두개의 일반형 직선의 방정식들을 비교할 때 그것들이 서로 평행인 두개의 다른 직선들을 나타내는지를 알아내려면 x 의 계수들간의 비가 y 의 계수들 간의 비와 동일한지 확인해야 하고 그리고 나서, 동일한 비면, 그 비가 그 상수 항들 간의 비와 다른지 확인해야 하는 거죠.

그런데, 그 모든 비들이 다 같으면 그 방정식들은 동일한 직선을 나타내는 거죠.

제안과 풀이
5번 예제에서의 2번 문제

다음과 같은 직선의 방정식들이 있다고 하세요:
$ax + by + c = 0$ 이고 $ux + vy + w = 0$ 이며 여기에서 a, b, c, u, v, w 는 상수이고 $abc \neq 0$ 이며 $uvw \neq 0$.

또한, 두 직선들은 한 점에서 만난다 하고 상수들간의 관계식을 구하세요.

두 직선들이 서로 평행 하다면 그 기울기들이 같은 거고 그 직선들은 서로 교차하지 않는 거죠.
두 직선들의 기울기들이 다르면 그 직선들은 서로 교차하는 거죠.
그러니까, 두 직선들이 한 점에서 서로 만나려면 그 기울기들이 서로 달라야 하는 겁니다.

이제, 그 일반형 방정식들을 표준형 방정식들로 표현해보기로 하죠. 그러면:
$ax + by + c = 0 \Rightarrow y = (-a/b)x - c/b$
$ux + vy + w = 0 \Rightarrow y = (-u/v)x - w/v$

그러면, $-a/b \neq -u/v$ 이면 그 두 방정식들은 한 점에서 서로 만나는 두개의 다른 직선들을 나타내는 겁니다.

그러면 이제, $-a/b \neq -u/v \Rightarrow a/b \neq u/v \Rightarrow av \neq bu \Rightarrow a/u \neq b/v$.

그러므로, 두개의 방정식들 $ax + by + c = 0$ 과 $ux + vy + w = 0$ 이 한 점에서 서로 만나는 두개의 다른 직선들을 나타낸다면 $a/u \neq b/v$ 입니다.

즉, 두개의 일반형 직선의 방정식들을 비교할 때 그것들이 한 점에서 서로 만나는 두 직선들을 나타내는지를 알아내려면 x 의 계수들 간의 비가 y 의 계수들 간의 비와 다른지 확인해야 하는 거죠.

그런데, 두 직선들의 기울기들이 다르면 그 y 절편들도 달라야 하나요?
아니죠. 그 y 절편들은 서로 같을 수도 있어요.
어째서 요?
그 직선들이 y 축에 있는 한 점에서 만날 수도 있거든요.
또한, 기울기가 다른 직선들이 x 축에 있는 한 점에서 만날 수도 있으니까 x 절편들도 서로 동일할 수도 있는 겁니다.

제안과 풀이
5번 예제에서의 3번 문제

다음과 같은 직선의 방정식들이 있다고 하세요:
ax + by + c = 0 이고 **ux + vy + w = 0** 이며 여기에서 **a, b, c, u, v, w** 는 상수이고 **abc ≠ 0** 이며 **uvw ≠ 0.**

또한, 두 직선들은 서로 수직이다 하고 상수들간의 관계식을 구하세요.

두개의 직선들이 서로 수직이 되는 조건은 무엇인가요?
그 기울기들의 곱이 –1 이죠.

이제, 그 일반형 방정식들을 표준형 방정식들로 표현해보기로 하죠. 그러면:
ax + by + c = 0 ⇒ y = (-a/b)x – c/b
ux + vy + w = 0 ⇒ y = (-u/v)x – w/v

그러면, (-a/b)(-u/v) = -1 이면 그 두개의 방정식들은 서로 수직인 두개의 직선들을 나타내는 겁니다.

그러면 이제, (-a/b)(-u/v) = -1 ⇒ au/bv = -1 ⇒ au = -bv ⇒ au + bv = 0.

그러므로, 두개의 방정식들 ax + by + c = 0 과 ux + vy + w = 0 이 서로 수직인 두개의 직선들을 나타낸다면 au + bv = 0 입니다.

즉, 일반형 방정식들을 비교할 때 그것들이 서로 수직인 다른 직선들을 나타내는지를 알아내려면 그 기울기들의 곱을 확인하는 대신에 x 와 y 의 계수들 각각의 곱들의 합이 0 인지를 확인해도 되는 거죠.

두 직선들이 서로 수직이면 어째서 그 기울기들의 곱이 –1 이 되는지는 이전 절에서의 4 번 예제에서 설명합니다.

제안과 풀이
5번 예제에서의 4번 문제

다음과 같은 직선의 방정식들이 있다고 하세요:
$ax + by + c = 0$ 이고 $ux + vy + w = 0$ 이며 여기에서 a, b, c, u, v, w 는 상수이고 $abc \neq 0$ 이며 $uvw \neq 0$.
그러면, 두 직선들이 만나는 점을 지나는 직선의 방정식을 구하세요.

두개의 직선들이 만나는 점을 지나는 직선은 얼마나 많을까요?
무한히 많죠.
그렇다고 해서 아무 직선이나 다 되는 건 아니죠.

위의 두개의 직선들이 하나의 점 (s, t)에서 만난다고 하세요.
그러면, 그 두개의 직선들이 점 (s, t)을 지나가니까 $as + bt + c = 0$ 이고 $us + vt + w = 0$ 이죠.
그러면, $m(as + bt + c) + n(us + vt + w) = 0$ 여기에서 m 과 n 은 상수라는 것을 알 수 있네요.

그러면 이제, $m(ax + by + c) + n(ux + vy + w) = 0$ 도 하나의 직선의 방정식이라는 것도 알 수 있죠.

그 이유는 다음과 같습니다:

우선, $m(ax + by + c) + n(ux + vy + w) = (am + nu)x + (bm + nv)y + cm + nw = 0$ 이죠.
다음으로, $(am + nu)x + (bm + nv)y + cm + nw = 0$ 에서 $am + nu$, $bm + nv$, $cm + nw$ 는 상수들입니다.
그러므로, $(am + nu)x + (bm + nv)y + cm + nw = 0$ 은 하나의 직선의 방정식이죠.

그러면 이제, $m(as + bt + c) + n(us + vt + w) = 0$ 이면 그 직선 $m(ax + by + c) + n(ux + vy + w) = 0$ 은 그 점 (s, t)를 지나가는 것이죠.

그러므로, $m(ax + by + c) + n(ux + vy + w) = 0$ 은 그 두개의 직선들 $ax + by + c = 0$ 과 $ux + vy + w = 0$ 이 서로 만나는 그 점 (s, t)를 지나가는 직선의 방정식인 겁니다.

특히 이 경우에, 그 두개의 방정식들 $ax + by + c = 0$ 과 $ux + vy + w = 0$ 을 기본 방정식들이라고 할 수 있겠죠.
즉, 이것들은 기저(基底) 같은 것들이죠.
그러면, 그 두개의 기본 방정식들은 두개의 기본 직선들을 나타내는 겁니다.
그 두개의 기본 직선들은 하나의 특정한 점에서 만나는데 그 방정식 $m(ax + by + c) + n(ux + vy + w) = 0$ 으로 표현될 수 있는 모든 직선들도 그 특정한 점을 지나는 것이죠.
사실, 방정식 $m(ax + by + c) + n(ux + vy + w) = 0$ 으로 표현될 수 있는 모든 직선들 중의 아무거나 두 직선들을 무작위로 선택해도 그것들도 기본 직선들이 될 수 있죠.

그러면 이제, 이 기본 직선들에 0 이 아닌 수를 제각기 곱해서 더하면 그 합이 또 하나의 직선의 방정식이 되는데 이는 그 점 (s, t)를 지나가는 겁니다.

또한, 이는 서로 평행인 직선들에 대하여도 마찬가집니다.
이제, 두 직선들 $ax + by + c = 0$ 과 $ux + vy + w = 0$ 이 서로 평행이라고 하세요.
그러면, $m(ax + by + c) + n(ux + vy + w) = 0$ 은 두 직선 $ax + by + c = 0$ 과 $ux + vy + w = 0$ 모두에 평행인 직선의 방정식인 겁니다.

제안과 풀이

6번 예제에서의 문제

직선들의 모임을 나타내는 방정식이 있는데 그 방정식은 다음과 같다고 하세요:
$3(k-1)x + (2k+3)y + 4k + 5 = 0$ 여기에서 k는 상수.

또한, 그 모임의 모든 직선들은 하나의 특정한 점 P(s, t)에서 만난다 하고 s와 t의 값을 구하세요.

다음과 같이 가정하세요:
방정식 $m(ax + by + c) + n(ux + vy + w) = 0$은 A.
방정식 $ax + by + c = 0$은 B. 방정식 $ux + vy + w = 0$은 C.

그러면, A는 그 두개의 직선들 B와 C가 서로 만나는 그 점을 지나가는 그 어느 직선이든 그 직선의 방정식이죠. 즉, A는 직선의 방정식인데 이는 그 두개의 직선들 B와 C가 서로 만나는 그 점을 지나가는 모든 직선들을 대표하는 겁니다. 그러니까, B와 C는 그 기본 방정식들인 거죠.

그러면 이제, 방정식 $3(k-1)x + (2k+3)y + 4k + 5 = 0$은 두개의 기본 직선들의 방정식들로 이루어진 것이죠. 그러면, 그 두 기본 직선들을 추출해내어 그 기본 직선들이 만나는 점을 구하면 그 점이 그 특정한 점 P(s, t)인 것이죠. 그런데, 그것들을 어떻게 추출해내나요?
주어진 방정식 $3(k-1)x + (2k+3)y + 4k + 5 = 0$을 다음과 같은 형태로 변형시키는 겁니다:
$m(ax + by + c) + n(ux + vy + w) = 0$. 어떻게 요?

$3(k-1)x + (2k+3)y + 4k + 5 = 0 \Rightarrow 3kx - 3x + 2ky + 3y + 4k + 5 = 0$
$\Rightarrow 3kx + 2ky + 4k - 3x + 3y + 5 = 0 \Rightarrow k(3x + 2y + 4) - (3x - 3y - 5) = 0$인데
이는 $m(ax + by + c) + n(ux + vy + w) = 0$의 형태로 표현되어있는 겁니다.

그러므로, $m = k$이고 $n = -1$인 것이죠; 또한, $a = 3$, $b = 2$, $c = 4$, $u = 3$, $v = -3$, $w = -5$.

그러면, 방정식 $3(k-1)x + (2k+3)y + 4k + 5 = 0$은 두개의 직선들 $3x + 2y + 4 = 0$과 $3x - 3y - 5 = 0$이 서로 만나는 그 점을 지나가는 모든 직선들 각각을 나타내는 겁니다.
그러면 이제, 위의 두 개의 직선의 방정식들을 x와 y에 관한 연립방정식으로 하고 그 답을 s와 t의 값들로 하면 되죠.
($3x + 2y + 4 = 0$이고 $3x - 3y - 5 = 0$) $\Rightarrow 3x + 2y + 4 - (3x - 3y - 5) = 0 \Rightarrow 5y + 9 = 0 \Rightarrow y = -9/5$.
그러면, $y = -9/5 \Rightarrow 3x + 2y + 4 = 3x - 18/5 + 4 = 0 \Rightarrow 3x = -2/5 \Rightarrow x = -2/15$.
그러면, 그 두개의 기본 직선들 뿐만 아니라 그 직선들의 모임에 있는 나머지 모든 직선들도 (-2/15, -9/5)를 지나는 거죠. 그러므로, $s = -2/15$이고 $t = -9/5$입니다.

요약:

$3(k-1)x + (2k+3)y + 4k + 5 = 0 \Rightarrow 3kx - 3x + 2ky + 3y + 4k + 5 = 0$
$\Rightarrow 3kx + 2ky + 4k - 3x + 3y + 5 = 0 \Rightarrow k(3x + 2y + 4) - (3x - 3y - 5) = 0$.

($3x + 2y + 4 = 0$이고 $3x - 3y - 5 = 0$) $\Rightarrow 3x + 2y + 4 - (3x - 3y - 5) = 0 \Rightarrow 5y + 9 = 0 \Rightarrow y = -9/5$.

그러면, $y = -9/5 \Rightarrow 3x + 2y + 4 = 3x - 18/5 + 4 = 0 \Rightarrow 3x = -2/5 \Rightarrow x = -2/15$.
그러므로, $s = -2/15$이고 $t = -9/5$.

제안과 풀이
7 번 예제에서의 문제

직선들의 모임을 나타내는 방정식이 있는데 그 방정식은 다음과 같다고 하세요:
$y = ax - b$ 여기에서 a 와 b 는 상수이고 $3a + b = 2$ 를 만족.
또한, 그 모든 직선들은 하나의 특정한 점 $P(s, t)$에서 만난다 하고 s 와 t 의 값을 구하세요.

먼저, 이 문제가 하는 말을 들어보기로 하죠.
"관계식 $3a + b = 2$ 에서의 a 와 b 에 대하여 여러 쌍의 수들을 선택하면 그 선택한 만큼 많은 직선들이 만들어지는데, 그 만들어진 모든 직선들 각각은 방정식 $y = ax - b$ 가 나타내는 거죠. 그러면, 그 만들어진 모든 직선들 각각은 하나의 특정한 점 $P(s, t)$를 지나갑니다. 그러면 이제, s 와 t 의 값을 구하세요."

그럼 이제, 무엇을 어떻게 해야 할지 따져보기로 하죠.
하나의 특정한 점에서 서로 만나는 모든 직선들 각각은 두개의 기본 직선들에 의해서 만들어지죠.
그러면, 이 특정 점에서 그 두개의 기본 직선들이 서로 만나는 겁니다.
그럼 이제, 그 두개의 기본 직선들을 구해야 하고 그 기본 직선들이 서로 만나는 점을 구해야 하는 거죠.
그 두개의 기본 직선들을 나타내는 방정식들은 $y = ax - b$ 가 가지고 있죠.

그러면 이제, 방정식 $y = ax - b$ 를 $m(px + qy + r) + n(ux + vy + w) = 0$ 의 형태로 변형시키는 겁니다.
그러면, $y = ax - b \Rightarrow ax - y - b = 0$.
이제, 그 관계식을 사용해서 a 아니면 b 를 소거해야 하겠네요. a 를 소거하기로 하죠.

그러면, $3a + b = 2 \Rightarrow a = (2 - b)/3 \Rightarrow y = ax - b = (2 - b)x/3 - b = 2x/3 - bx/3 - b$
$\Rightarrow y - 2x/3 + bx/3 + b = 0 \Rightarrow y - 2x/3 + b(x/3 + 1) = 0$. 그러면, 위의 형태에서 $m = 1$ 이고 $n = b$ 라고 할 수 있겠네요; 또한, $p = -2/3$, $q = 1$, $r = 0$, $u = 1/3$, $v = 0$, $w = 1$.

그러니까, 두개의 방정식들 $y - 2x/3 = 0$ 과 $x/3 + 1 = 0$ 이 그 두개의 기본 직선들을 나타내는 거네요.
그럼, 이 기본 직선들이 서로 만나는 점을 구해야 하죠.

$y - 2x/3 = 0 \Rightarrow y = 2x/3$.
$x/3 + 1 = 0 \Rightarrow x = -3 \Rightarrow y = 2x/3 = -2$.

그러면, 그 두개의 기본 직선들은 점 $(-3, -2)$에서 서로 만나네요.
그러므로, $s = -3$ 이고 $t = -2$ 입니다.

그런데, b 를 소거하면 어떨까요? 마찬가지 결과가 되어야 하겠죠. 어디 그런지 한번 봅시다.

$3a + b = 2 \Rightarrow b = 2 - 3a \Rightarrow y = ax - b = ax - (2 - 3a) = ax - 2 + 3a = a(x + 3) - 2$
$\Rightarrow a(x + 3) - 2 - y = 0 \Rightarrow a(x + 3) - (2 + y) = 0$.

그러니까, 두개의 방정식들 $x + 3 = 0$ 과 $2 + y = 0$ 이 그 두개의 기본 직선들을 나타내는 거네요.
그럼, 이 기본 직선들이 서로 만나는 점을 구해야 하죠.

$x + 3 = 0 \Rightarrow x = -3$. $2 + y = 0 \Rightarrow y = -2$.

그러면, 그 두개의 기본 직선들은 점 $(-3, -2)$에서 서로 만나네요. 그러므로, $s = -3$ 이고 $t = -2$ 입니다.

그림. 7.0

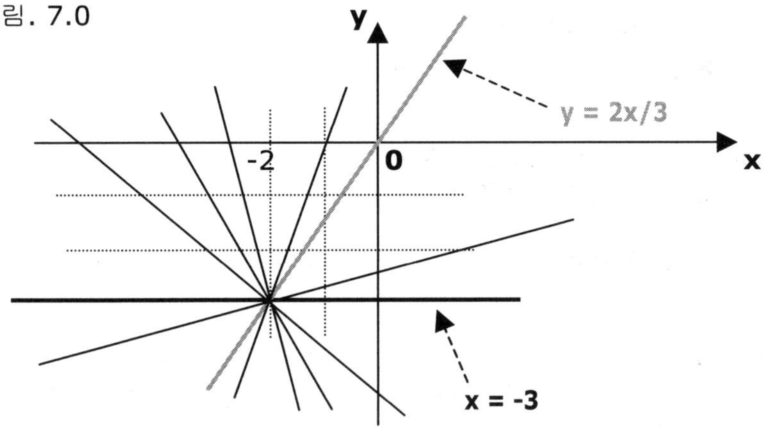

그림. 7.1

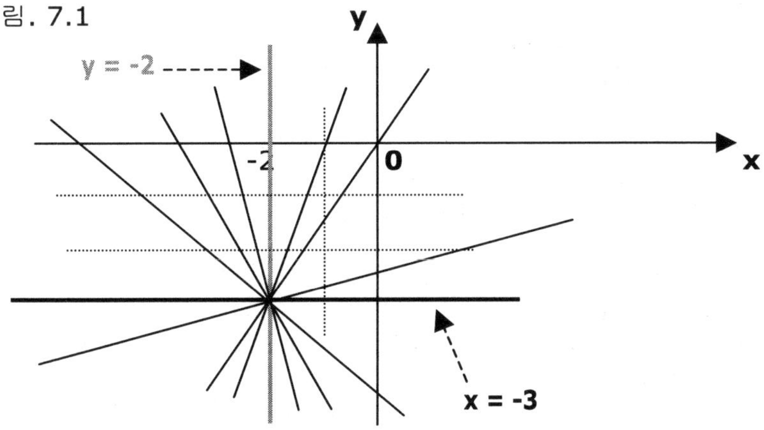

요약:

먼저, 방정식 y = ax – b 을 m(ax + by + c) + n(ux + vy + w) = 0 의 형태로 변형시킵니다.
y = ax – b ⇒ ax – y – b = 0. 주어진 관계식을 사용해서 a 를 소거하기로 하죠.

그러면, 3a + b = 2 ⇒ a = (2 – b)/3 ⇒ y = ax – b = (2 – b)x/3 – b = 2x/3 – bx/3 – b
⇒ y – 2x/3 + bx/3 + b = 0 ⇒ y – 2x/3 + b(x/3 + 1) = 0.

그러므로, 두개의 직선들 y – 2x/3 = 0 과 x/3 + 1 = 0 이 서로 만나는 점이 그 모든 직선들이 모두 함께 서로 만나는 점입니다.
그러면, 그 두 직선들이 서로 만나는 점을 구해야 합니다.

y – 2x/3 = 0 ⇒ y = 2x/3.
x/3 + 1 = 0 ⇒ x = -3 ⇒ y = 2x/3 = -2.

그러면, 그 직선들은 점 (-3, -2)에서 서로 만납니다.
그러므로, s = -3 이고 t = -2.

제안과 풀이
8 번 예제에서의 문제

직선들의 모임을 나타내는 방정식이 있는데 그 방정식은 다음과 같다고 하세요:
$2x/a + y/2b = 1$ 여기에서 a 와 b 는 상수이고 $1/2a + 1/b = 1/5$ 를 만족.

또한, 그 모임의 모든 직선들은 하나의 특정한 점 P(u, v)에서 만난다 하고 u 와 v 의 값을 구하세요.

이 문제가 하는 말은 다음과 같죠:
"관계식 $1/2a + 1/b = 1/5$ 를 만족하는 a 와 b 에 대하여 여러 쌍의 수들을 선택하면 그 선택한 만큼의 직선들이 만들어지는데 그 만들어진 직선들 각각은 방정식 $2x/a + y/2b = 1$ 이 나타냅니다. 그러면, 그 만들어진 모든 직선들 각각은 하나의 특정한 점 P(u, v)를 지납니다. 그러면 이제, u 와 v 의 값을 구하세요."

하나의 특정한 점에서 만나는 모든 직선들 각각은 두개의 기본 직선들에 의해서 만들어지죠.
그러면, 이 특정 점에서 그 두개의 기본 직선들이 만나는 겁니다.
그럼 이제, 그 두개의 기본 직선들을 구해야 하고 그 직선들이 만나는 점을 구해야 하는 거죠.
그 두개의 기본 직선들을 나타내는 방정식들은 $2x/a + y/2b = 1$ 가 가지고 있죠.

그러면 이제, $2x/a + y/2b = 1$ 을 $m(ax + by + c) + n(ux + vy + w) = 0$ 의 형태로 변형시키는 겁니다.
먼저, $t = 1/a$ 그리고 $s = 1/b$ 로 놓기로 하죠. 그러면, t 와 s 도 상수들이고 또한:
$2x/a + y/2b = 1 \Rightarrow 2tx + sy/2 = 1 \Rightarrow 2tx + sy/2 - 1 = 0 \Rightarrow 4tx + sy - 2 = 0.$

이제, 그 관계식을 사용해서 s 아니면 t 를 소거해야 하겠네요. s 를 소거하기로 하죠. 그러면:
$1/2a + 1/b = 1/5 \Rightarrow t/2 + s = 1/5 \Rightarrow s = 1/5 - t/2$
$\Rightarrow 4tx + sy - 2 = 4tx + (1/5 - t/2)y - 2 = 4tx + y/5 - ty/2 - 2 = 0 \Rightarrow t(4x - y/2) + y/5 - 2 = 0.$
그러면, 위의 형태에서 $m = t$ 이고 $n = 1$ 이라 할 수 있네요.
그러니까, 두개의 방정식들 $4x - y/2 = 0$ 과 $y/5 - 2 = 0$ 이 그 두개의 기본 직선들을 나타내는 거네요.

그럼, 이 직선들이 만나는 점을 구해야 하죠.
$y/5 - 2 = 0 \Rightarrow y = 10.\ 4x - y/2 = 0 \Rightarrow x = y/8 = 10/8 = 5/4.$
그러면, 그 두개의 기본 직선들은 점 (5/4, 10)에서 서로 만나네요.
그러므로, $u = 5/4$ 이고 $v = 10$ 입니다.

요약:

$2x/a + y/2b = 1$ 을 $m(ax + by + c) + n(ux + vy + w) = 0$ 의 형태로 변형시킵니다.
$t = 1/a$ 그리고 $s = 1/b$ 라 하세요. 그러면:
$2x/a + y/2b = 1 \Rightarrow 2tx + sy/2 = 1 \Rightarrow 2tx + sy/2 - 1 = 0 \Rightarrow 4tx + sy - 2 = 0.$
t 를 소거하기로 하죠.
$1/2a + 1/b = 1/5 \Rightarrow t/2 + s = 1/5 \Rightarrow t = 2/5 - 2s$
$\Rightarrow 4tx + sy - 2 = 4(2/5 - 2s)x + sy - 2 = 8x/5 - 8sx + sy - 2 = 8x/5 - 2 + s(y - 8x) = 0.$
그러면, $8x/5 - 2 = 0$ 과 $y - 8x = 0$ 이 서로 만나는 점을 구해야 하죠.
$8x/5 - 2 = 0 \Rightarrow x = 10/8 = 5/4 \Rightarrow x = 5/4.$
$y - 8x = 0 \Rightarrow y = 8x = 8(5/4) = 10.$
그러므로, $u = 5/4$ 이고 $v = 10.$

제안과 풀이
9번 예제에서의 0번 문제

직선들의 모임을 나타내는 방정식이 있는데 그 방정식은 다음과 같다고 하세요:
$5ax - (3b - 2)y + 2b = 8$ 여기에서 **a**와 **b**는 상수이고 **$2a - b = 2$**를 만족하며 **$ab \neq 0$.**
또한, 그 모임의 모든 직선들이 하나의 특정한 점 **P(s, t)**에서 만난다 하고 그 점 **P**를 구하세요.

방정식 $m(ax + by + c) + n(dx + ey + f) = 0$을 $ax + by + c = 0$과 $dx + ey + f = 0$의 선형조합이라고 합니다. 그러면, 두 개의 직선들이 하나의 특정한 점에서 서로 만나면 그 두개의 직선들의 선형조합은 그 특정한 점을 지나는 거죠. 다시 말해서, 그 직선들에 어떤 수를 곱해서 합하더라도 그 합은 그 특정한 점을 지나죠. 즉, 두 상수들 m과 n에 대하여 그 어떤 수들을 선택하더라도 그 합이 나타내는 직선은 그 특정한 점을 지나는 겁니다.

먼저, A를 그 주어진 방정식 $5ax - (3b - 2)y + 2b = 8$이라 하세요.
그러면, A가 그 특정 점 P(s, t)를 지나는 모든 직선들 각각을 나타내는 방정식이죠.
그러면 이제, A로부터 그 두개의 기본 직선들을 추출해서 그 두 직선들이 만나는 점을 구해야 하죠.
그러면, 바로 그 점이 그 A가 나타내는 모든 직선들 각각이 지나는 그 특정 점 P(s, t)입니다.

그럼 이제, A를 $m(ax + by + c) + n(ux + vy + w) = 0$의 형태로 해야 하죠.

그러면, $5ax - (3b - 2)y + 2b = 8 \Rightarrow 5ax - 3by + 2y + 2b - 8 = 0$.
이제, 주어진 관계식을 사용해서 a 아니면 b를 소거해야 하겠네요. a를 소거하기로 하죠.

그러면, $2a - b = 2 \Rightarrow a = (2 + b)/2 \Rightarrow 5ax - 3by + 2y + 2b - 8 = 5(2 + b)x/2 - 3by + 2y + 2b - 8$
$= 5x + 5bx/2 - 3by + 2y + 2b - 8 = b(5x/2 - 3y + 2) + 5x + 2y - 8 = 0$.

그러면, 위의 형태에서 $m = b$이고 $n = 1$이라 할 수 있네요.
그러니까, 두개의 방정식들 $5x/2 - 3y + 2 = 0$과 $5x + 2y - 8 = 0$이 그 두개의 기본 직선들을 나타내는 거네요.
그럼, 이 두 직선들이 서로 만나는 점을 구해야 하죠.

$5x/2 - 3y + 2 = 0 \Rightarrow 5x - 6y + 4 = 0 \Rightarrow 5x = 6y - 4$.
$5x + 2y - 8 = 0 \Rightarrow 6y - 4 + 2y - 8 = 8y - 12 = 0 \Rightarrow y = 3/2$.
그러면, $5x = 6y - 4 = 6(3/2) - 4 = 5 \Rightarrow x = 1$.

그러면, 그 두개의 기본 직선들은 점 (1, 3/2)에서 서로 만나네요.
따라서, 이 점 (1, 3/2)가 그 A가 나타내는 모든 직선들 각각이 지나는 그 특정 점 P(s, t)이죠.
그러므로, $s = 1$이고 $t = 3/2$입니다.

요약:

A가 방정식 $5ax - (3b - 2)y + 2b = 8$이라 하세요.
그러면, A를 $m(ax + by + c) + n(ux + vy + w) = 0$의 형태로 나타냅니다.
$5ax - (3b - 2)y + 2b = 8 \Rightarrow 5ax - 3by + 2y + 2b - 8 = 0$.
주어진 관계식으로 b를 소거하기로 하죠. 그러면:
$2a - b = 2 \Rightarrow b = 2a - 2 \Rightarrow 5ax - 3by + 2y + 2b - 8 = 5ax - 3(2a - 2)y + 2y + 2(2a - 2) - 8$
$= 5ax - 6ay + 6y + 2y + 4a - 4 - 8 = a(5x - 6y + 4) + 8y - 12 = 0$.

그러면, $5x - 6y + 4 = 0$과 $8y - 12 = 0$이 서로 만나는 점을 구해야 하죠.
$8y - 12 = 0 \Rightarrow y = 3/2$
$5x - 6y + 4 = 0 \Rightarrow 5x = 6y - 4 = 6(3/2) - 4 = 5 \Rightarrow x = 1$.
그러면, 그 두 직선들은 점 (1, 3/2)에서 서로 만나네요.
그러므로, s = 1 이고 t = 3/2 입니다.

그림. 9.0.0

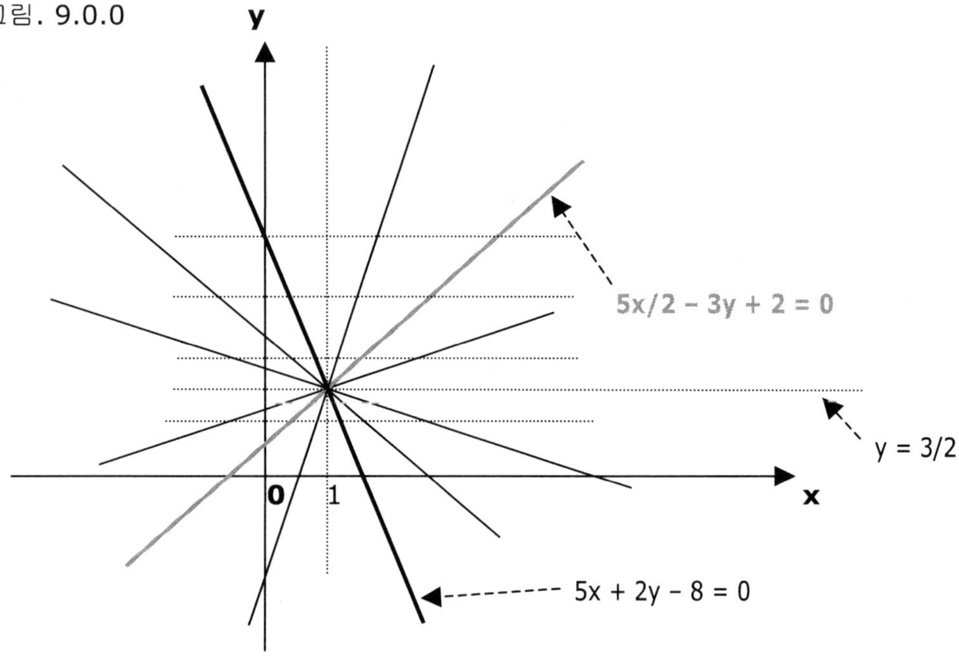

그림. 9.0.1

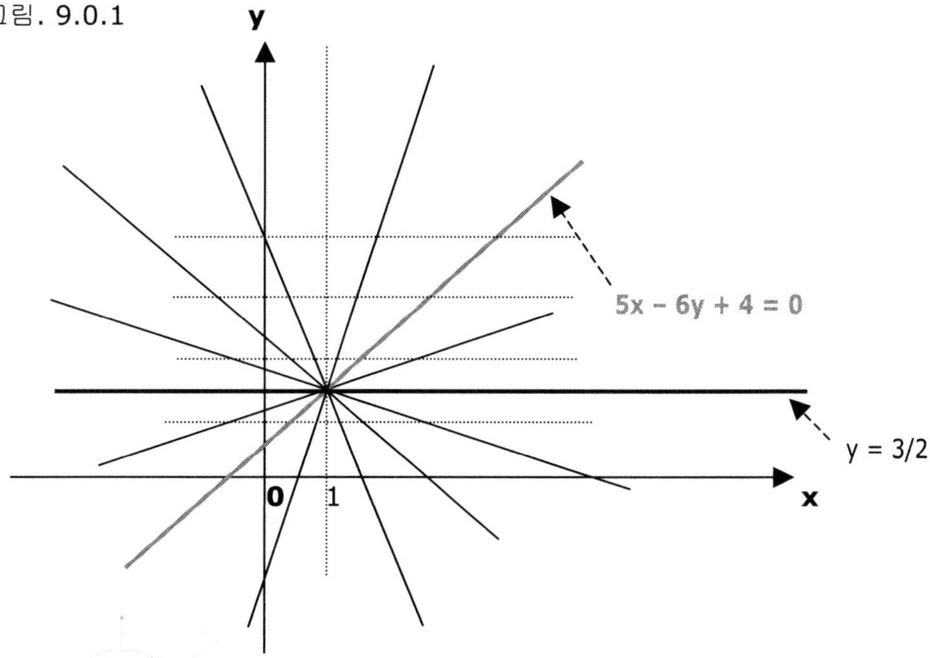

제안과 풀이
9 번 예제에서의 1 번 문제

직선들의 모임을 나타내는 방정식이 있는데 그 방정식은 다음과 같다고 하세요:
$5ax - (3b - 2)y + 2b = 8$ 여기에서 a 와 b 는 상수이고 $2a - b = 2$ 를 만족하며 $ab \neq 0$.
그러면, 직선 $2x + 5y = 10$ 에 평행이고 그 모임에 속하는 직선을 구하세요.

"직선들의 모임을 나타내는 방정식이 있는데 그 방정식은 다음과 같다고 하세요:
$5ax - (3b - 2)y + 2b = 8$ 여기에서 a 와 b 는 상수이고 $2a - b = 2$ 를 만족하며 $ab \neq 0$." 라는 말은 "관계식 $2a - b = 2$ 를 만족하며 $ab \neq 0$ 만족하는 a 와 b 에 대하여 많은 수들을 선택하면 그 선택한 만큼의 직선들이 만들어지는데 그 만들어진 직선들 각각은 방정식 $5ax - (3b - 2)y + 2b = 8$ 이 나타냅니다." 라는 말이죠.

그러니, 위에서 언급한 직선들 중에 직선 $2x + 5y = 10$ 에 평행한 직선을 구해야 하는 거죠. 즉, 그러한 직선을 결정하는 a 와 b 의 값을 찾아야 하는 거죠. 그리고 나서, 그 값들을 주어진 직선의 방정식 $5ax - (3b - 2)y + 2b = 8$ 에 대입하면 그러한 직선의 방정식을 갖게 되는 겁니다. 그러면 이제, 그 값들을 찾아서 대입해야 하죠.

두개의 직선들이 서로 평행하면 그 기울기들이 동일하죠. 따라서, 찾고있는 직선의 기울기는 직선 $2x + 5y = 10$ 의 기울기와 같은 거죠. 그러면, 주어진 방정식들을 표준형으로 표현해서 그 기울기들을 구해보죠.
$2x + 5y = 10 \Rightarrow y = -2x/5 + 2 \Rightarrow$ 기울기 $= -2/5$.
$5ax - (3b - 2)y + 2b = 8 \Rightarrow y = 5ax/(3b - 2) + (8 - 2b)/(3b - 2) \Rightarrow$ 기울기 $= 5a/(3b - 2)$.

사실, 일반형 방정식에서 y 의 계수에 대한 x 의 계수의 비의 음의 값이 그 직선의 기울기죠.
그러면, 찾고있는 a 와 b 의 값들은 다음과 같은 관계식들을 만족시켜야 하는 거죠:
$5a/(3b - 2) = -2/5$ 와 $2a - b = 2$.

그럼 이제, 위의 식들을 a 와 b 에 관한 연립방정식으로 하고 풀어서 a 와 b 의 값을 구해야 하죠.
$2a - b = 2 \Rightarrow a = (2 + b)/2 \Rightarrow 5a = 5(2 + b)/2$. 한편, $5a/(3b - 2) = -2/5 \Rightarrow 5a = 2(2 - 3b)/5$.
그러면, $5(2 + b)/2 = 2(2 - 3b)/5 \Rightarrow 25(2 + b) = 4(2 - 3b) \Rightarrow 50 + 25b = 8 - 12b \Rightarrow 37b = -42$
$\Rightarrow b = -42/37$. 그러면, $a = (2 + b)/2 = \{2 - (42/37)\}/2 = 1 - 42/37/2 = 1 - 21/37 = 16/37$.

그러면 이제, $5ax - (3b - 2)y + 2b = 8 \Rightarrow 80x/37 - (-126/37 - 2)y - 84/37 = 8$
$\Rightarrow 80x/37 + 200y/37 - 380/37 = 0 \Rightarrow 80x + 200y - 380 = 0 \Rightarrow 4x + 10y - 19 = 0$.

그러므로, 직선 $2x + 5y = 10$ 에 평행인 직선은 $4x + 10y - 19 = 0$ 입니다.

요약:
$2x + 5y = 10 \Rightarrow$ 기울기 $= -2/5$. $5ax - (3b - 2)y + 2b = 8 \Rightarrow$ 기울기 $= 5a/(3b - 2)$.
그럼 이제, 다음의 식들을 a 와 b 에 관한 연립방정식으로 하고 풀어서 a 와 b 의 값을 구해야 하죠:
$5a/(3b - 2) = -2/5$ 와 $2a - b = 2$.

$2a - b = 2 \Rightarrow a = (2 + b)/2 \Rightarrow 5a = 5(2 + b)/2$. 한편, $5a/(3b - 2) = -2/5 \Rightarrow 5a = 2(2 - 3b)/5$.
그러면, $5(2 + b)/2 = 2(2 - 3b)/5 \Rightarrow 25(2 + b) = 4(2 - 3b) \Rightarrow 50 + 25b = 8 - 12b \Rightarrow 37b = -42$
$\Rightarrow b = -42/37$. 그러면, $a = (2 + b)/2 = \{2 - (42/37)\}/2 = 1 - 42/37/2 = 1 - 21/37 = 16/37$.

그러면 이제, $5ax - (3b - 2)y + 2b = 8 \Rightarrow 80x/37 - (-126/37 - 2)y - 84/37 = 8$
$\Rightarrow 80x/37 + 200y/37 - 380/37 = 0 \Rightarrow 80x + 200y - 380 = 0 \Rightarrow 4x + 10y - 19 = 0$.

제안과 풀이
A 번 예제에서의 문제

두개의 직선들이 점 **P(s, t)**에서 만나고 **s > 0** 이며 **t > 0** 이라 하고 그 직선들의 방정식은 다음과 같다고 하세요**: 2x + y - 2 = 0** 와 **2kx - y + k - 1 = 0** 여기에서 **k**는 상수**.** 그러면**, k**의 범위를 구하세요**.**

먼저, 그 답이 어디쯤에 있을 건지 보다 잘 알 수 있도록 $2x + y - 2 = 0$의 그래프를 작성해보죠.
$x = 0 \Rightarrow y = 2$. $y = 0 \Rightarrow x = 1$.

그림. A.0.

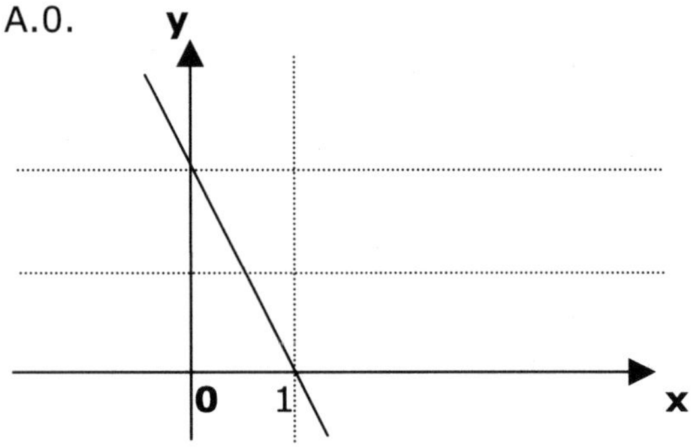

P(s, t)에서 $s > 0$ 이며 $t > 0$ 이므로 제 1 사분면에 있는 선분만 해당되죠.

그럼 이제, 두 번째 방정식은 한 집단의 직선들을 나타내니까 그 집단이 어디에 있는지 알 수 있도록 그 두개의 기본 직선들을 구해보기로 하죠.
$2kx - y + k - 1 = 0 \Rightarrow k(2x + 1) - y - 1 = 0$.

그러면, 그 두개의 기본 직선들은 $2x + 1 = 0$과 $-y - 1 = 0$이죠.
$-y - 1 = 0 \Rightarrow y = -1$. $2x + 1 = 0 \Rightarrow x = -1/2$.
그러므로, 그 집단에 있는 모든 직선들은 (-1/2, -1)에서 서로 만나네요. 이제 이것들도 그 그래프에 넣어보기로 하죠.

그림. A.1

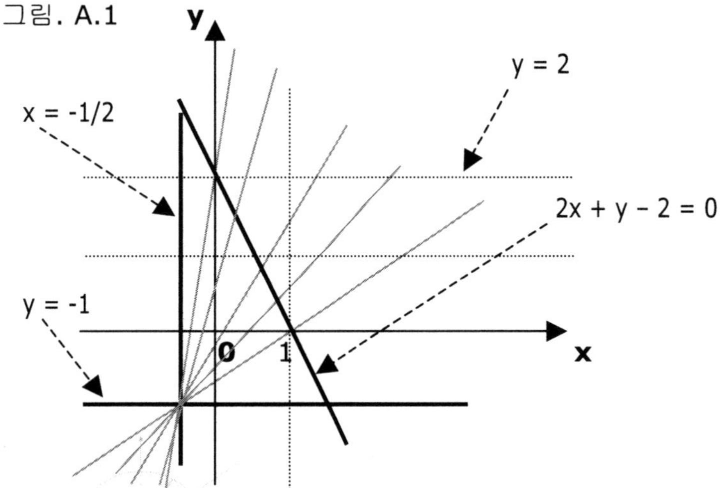

자 이제, 회색 직선들이 직선 $2x + y - 2 = 0$를 어떻게 만나야 한다는 것을 알 수 있죠.
가장 위에 있는 회색선의 y 절편은 2 이고 가장 아래에 있는 회색선의 x 절편은 1 이죠.
또한, y 절편이 2 일 때 그 기울기 2k 가 최대이고 x 절편이 1 이 될 때까지 줄곧 감소하여 최소가 되죠.

$2kx - y + k - 1 = 0 \Rightarrow$ 기울기 $= 2k$.

가장 위에 있는 회색선의 기울기 $= \{2 - (-1)\}/\{0 - (-1/2)\} = 3/(1/2) = 6$.
가장 아래에 있는 회색선의 기울기 $= \{0 - (-1)\}/\{1 - (-1/2)\} = 1/(3/2) = 2/3$.

그러므로, $2/3 < 2k < 6 \Rightarrow 1/3 < k < 3$. 점 P(s, t)에서 $s > 0$ 이고 $t > 0$ 이므로 경계 값들은 제외되죠.

요약:

$2kx - y + k - 1 = 0$은 하나의 특정한 점을 지나는 모든 직선들 각각을 나타냅니다.
$2kx - y + k - 1 = 0 \Rightarrow k(2x + 1) - (y + 1) = 0$.
그러면, 직선들 $2x + 1 = 0$과 $y + 1 = 0$이 서로 만나는 점에서 그 모든 직선들이 모두 함께 서로 만납니다.

$y + 1 = 0 \Rightarrow y = -1$. $2x + 1 = 0 \Rightarrow x = -1/2$.
그러므로, 그 특정한 점은 (-1/2, -1)입니다

그림. A.1

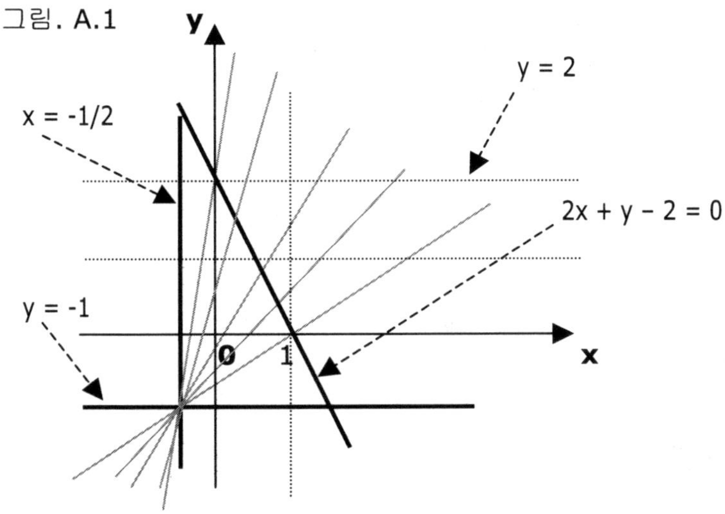

가장 위에 있는 회색선의 y 절편은 2 이고 가장 아래에 있는 회색선의 x 절편은 1.
또한, y 절편이 2 일 때 그 기울기 2k 가 최대이고 x 절편이 1 이 될 때까지 줄곧 감소하여 최소.
$2kx - y + k - 1 = 0 \Rightarrow$ 기울기 $= 2k$.

가장 위에 있는 회색선의 기울기 $= \{2 - (-1)\}/\{0 - (-1/2)\} = 3/(1/2) = 6$.
가장 아래에 있는 회색선의 기울기 $= \{0 - (-1)\}/\{1 - (-1/2)\} = 1/(3/2) = 2/3$.

그러므로, $2/3 < 2k < 6 \Rightarrow 1/3 < k < 3$. 점 P(s, t)에서 $s > 0$ 이고 $t > 0$ 이므로 경계 값들은 제외.

제안과 풀이
B 번 예제에서의 문제

다음과 같이 가정하세요:
$(2k - 1)x - (3k + 1)y + 5 + 2k = 0$ 여기에서 k 는 상수.
$y = 2x$ 이고 $|x| \le 1$.
그러면, 위의 두 직선들이 만나도록 k 값을 구하세요.

다음 방정식은 직선들의 모임을 나타내죠: $(2k - 1)x - (3k + 1)y + 5 + 2k = 0$ 여기에서 k 는 상수.
다음 방정식은 하나의 선분을 나타내죠: $y = 2x$ 이고 $|x| \le 1$.
그러니까, 그 모임에 있는 모든 직선들이 그 선분과 만나도록 k 의 범위를 구해야 하는 겁니다.
먼저, 그 기본 직선들부터 구하기로 하죠.
$(2k - 1)x - (3k + 1)y + 5 + 2k = 0 \Rightarrow k(2x - 3y + 2) - (x + y - 5) = 0$
그러면, 그 기본 직선들은 $2x - 3y + 2 = 0$ 과 $x + y - 5 = 0$.

이제, 그 기본 직선들이 서로 만나는 점을 구해야 하죠.
두 번째 방정식에 -2 를 곱하고 그 곱을 첫번째 방정식에 더하여 x 를 소거하고 y 의 값을 구합니다.

$-2(x + y - 5) = 0 \Rightarrow -2x - 2y + 10 = 0$.
$2x - 3y + 2 + (-2x - 2y + 10) = 0 \Rightarrow -5y + 12 = 0 \Rightarrow y = 12/5$.
그러면, $y = 12/5 \Rightarrow x + y - 5 = x + 12/5 - 5 = 0 \Rightarrow x = 13/5$.
그리하여, 그 두 기본 직선들은 그 나머지 모든 직선들과 함께 (13/5, 12/5)에서 만납니다.
그 기본 직선들과 선분을 그래프에 나타내면 그 답이 어디쯤에 있는지 보다 잘 알 수 있죠.

그림. B.0

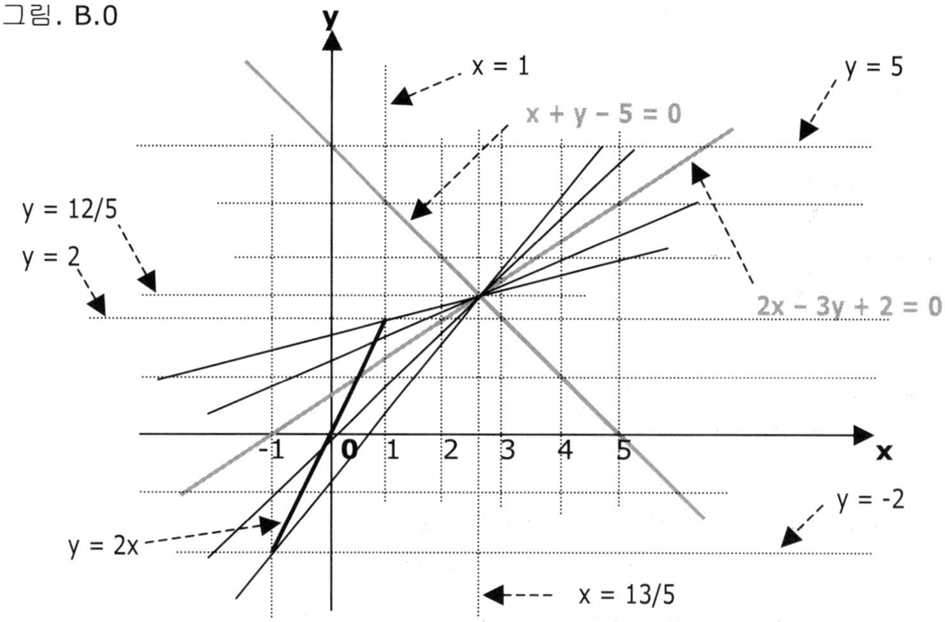

위의 그래프에서 그 선분과 만나는 검정 직선들을 볼 수 있죠.
가장 위에 있는 검정 직선은 두개의 점들 (1, 2)와 (13/5, 12/5)를 지납니다.
가장 아래에 있는 검정 직선은 두개의 점들 (-1, -2)와 (13/5, 12/5)를 지납니다.

방정식 $(2k - 1)x - (3k + 1)y + 5 + 2k = 0$ 으로부터 위의 검정 직선들 각각의 기울기가 $(2k - 1)/(3k + 1)$으로 나타내어질 수 있는 것을 알 수 있죠.

가장 위에 있는 직선의 기울기 $= \Delta y/\Delta x = (12/5 - 2)/(13/5 - 1) = (2/5)/(8/5) = 1/4$.
그러면, $(2k - 1)/(3k + 1) = 1/4 \Rightarrow 2k - 1 = 4(3k + 1) = 12k + 4 \Rightarrow 10k = -5 \Rightarrow k = -1/2$.

가장 아래에 있는 직선의 기울기 $= \Delta y/\Delta x = (12/5 - (-2))/(13/5 - (-1)) = (22/5)/(18/5) = 11/9$.
그러면, $(2k - 1)/(3k + 1) = 11/9 \Rightarrow 9(2k - 1) = 11(3k + 1) \Rightarrow 18k - 9 = 33k + 11$
$\Rightarrow 15k = -20 \Rightarrow k = -4/3$.

그러므로, $-4/3 \le k \le -1/2$.

요약:

다음 방정식은 직선들의 모임을 나타내죠: $(2k - 1)x - (3k + 1)y + 5 + 2k = 0$ 여기에서 k는 상수.
$(2k - 1)x - (3k + 1)y + 5 + 2k = 0 \Rightarrow k(2x - 3y + 2) - (x + y - 5) = 0$
그러면, 그 기본 직선들은 $2x - 3y + 2 = 0$과 $x + y - 5 = 0$.
$-2(x + y - 5) = 0 \Rightarrow -2x - 2y + 10 = 0$.
$2x - 3y + 2 + (-2x - 2y + 10) = 0 \Rightarrow -5y + 12 = 0 \Rightarrow y = 12/5$.
$y = 12/5 \Rightarrow x + y - 5 = x + 12/5 - 5 = 0 \Rightarrow x = 13/5$.
그리하여, 그 두 기본 직선들은 (13/5, 12/5)에서 서로 만납니다.

그림. B.0

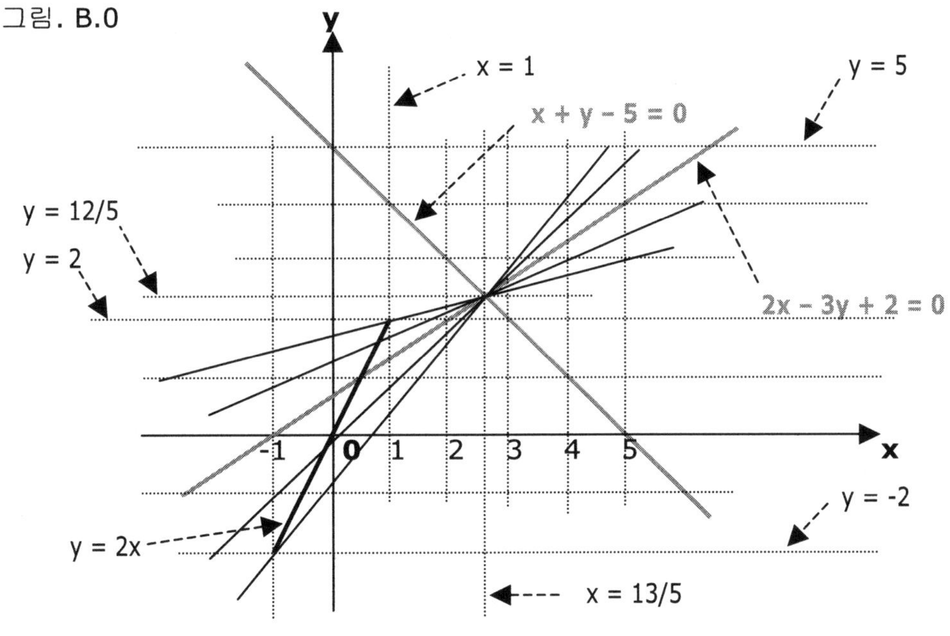

가장 위에 있는 직선의 기울기 $= \Delta y/\Delta x = (12/5 - 2)/(13/5 - 1) = (2/5)/(8/5) = 1/4$.
그러면, $(2k - 1)/(3k + 1) = 1/4 \Rightarrow 2k - 1 = 4(3k + 1) = 12k + 4 \Rightarrow 10k = -5 \Rightarrow k = -1/2$.
가장 아래에 있는 직선의 기울기 $= \Delta y/\Delta x = (12/5 - (-2))/(13/5 - (-1)) = (22/5)/(18/5) = 11/9$.
그러면, $(2k - 1)/(3k + 1) = 11/9 \Rightarrow 9(2k - 1) = 11(3k + 1) \Rightarrow 18k - 9 = 33k + 11$
$\Rightarrow 15k = -20 \Rightarrow k = -4/3$. 그러므로, $-4/3 \le k \le -1/2$.

1. 포물선(抛物線)이란 무엇인가요?

포물선이란 하나의 선이죠 하지만 곡선입니다. 이는 하나의 커브인데 이 커브는 원뿔의 옆면과 평행한 평면이 그 원뿔과 교차해서 생기는 거죠. 하나의 원뿔을 그 원뿔의 옆면과 평행한 평면으로 자른다고 하세요. 그러면, 그 단면이 포물선인 겁니다. 하나의 직선은 점들의 모임이죠. 직선에서는 모든 점들이 동일한 방향을 갖고 있어요. 포물선도 점들의 모임인데, 하지만 그 모든 점들은 제각기 모두 다른 방향들을 갖고있는 거죠. 하지만, 그 방향들은 아무렇게나 향하고 있는 것이 아니라 그 커브가 대칭이 되도록 향하고있는 겁니다. 무엇에 관해서 대칭이죠? 다음 절들을 공부하면서 그걸 알아내게 될 겁니다.

포물선이란 직선의 경우에서와 마찬가지로 하나의 개념이고 이에는 두께나 폭이 없어요. 길이는 요? 일반 math 에서는 좀처럼 포물선의 길이에 관해서는 이야기하지 않죠. 좀 더 정확히 말하면, 포물선에서의 궤적의 크기라고 해요. 포물선에서의 궤적이란 포물선분(抛物線分)입니다. 미적분학에서의 도구를 가지고 포물선분의 크기를 계산할 수 있어요. Math 에서 길이란 주로 거리를 의미하는데 이는 선분의 크기죠. 궤적의 길이를 측정할 때는 궤적이란 것도 사실은 아주 작은 선분들을 합쳐놓은 겁니다.

포물선을 많이 사용합니다. 이는 궤적과 관련이 많죠. 포물선분이 단순 발사 체의 경로로 자주 사용되거든요. 예를 들어, 공을 던지면 그 공은 포물선인 경로를 따라 공중을 이동해요. 야구공을 치면 보통 그런 식으로 날아가는 거죠. 포물선을 찾을 수 있는 경우가 이것이 다는 아닙니다. 예를 들어, 위성방송 TV 를 볼 때 접시 안테나를 사용하죠. 실제로, 그 접시의 단면이 포물선이죠. 이는 회중전등의 반사경에 대해서도 마찬가집니다.

직선에서의 경우와 마찬가지로 math 에서 포물선에 대해서 뭘 해야 할 때 포물선을 방정식으로 표현합니다. 물론, 이를 그래프로 나타내기도 하죠. 원한다면 두개 이상의 포물선들을 하나의 방정식으로 나타낼 수도 있어요. 그럼 이제, 그 방정식이 무엇이고, 어떻게 만들어지며, 어떻게 돌아가는지 알아보도록 하죠.

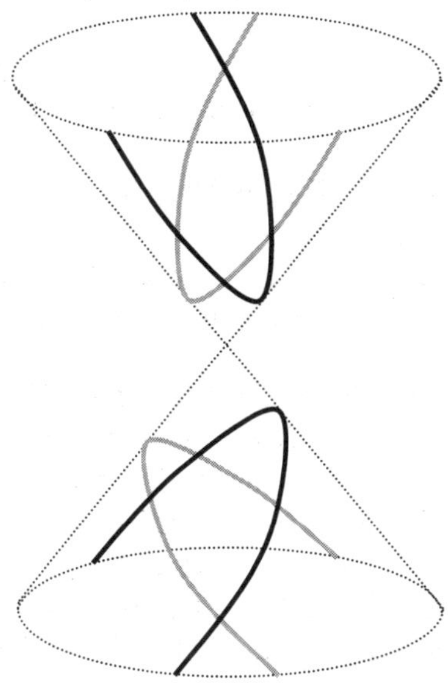

포물선이란 점들의 집합이죠. 그럼, 무슨 점들의 집합인가요? 포물선에 있는 한 점을 예로 들어보죠. 그러면, 이 점은 하나의 특정한 점과 하나의 특정한 직선 둘 다로부터 동일한 거리만큼 떨어져있는 겁니다. 즉, 그 점과 그 특정 점 사이의 거리가 그 점과 그 특정 직선 사이의 거리와 같은 거죠. 이는 그 포물선에 있는 다른 모든 점들 각각에 대하여도 마찬가집니다. 그러면 이제, 그런 점들의 집합이 포물선인 거죠. 이제, 하나의 포물선이 x-y 좌표계에 있다고 하세요. 그러면, 그 점들 각각은 한 쌍의 좌표들을 갖는 것이고 그 포물선을 다음과 같이 점들의 집합으로 나타낼 수 있습니다. 먼저, 표준형을 보기로 하죠.

$A = \{(x, y) | y^2 = 4px$ 여기에서 p는 상수.$\}$
$B = \{(x, y) | x^2 = 4py$ 여기에서 p는 상수.$\}$
$C = \{(x, y) | (y - v)^2 = 4p(x - u)$ 여기에서 p, u, v는 상수.$\}$
$D = \{(x, y) | (x - u)^2 = 4p(y - v)$ 여기에서 p, u, v는 상수.$\}$

$A = \{(x, y) | y^2 = 4px$ 여기에서 p는 상수.$\}$라는 말은 집합 A가 x-y좌표계에서의 점들로 이루어져 있는데 각 점의 좌표들 사이의 관계식은 일반적으로 $y^2 = 4px$이고 여기에서 p는 상수이다 라는 것을 말하고 있는 겁니다. 그러니까, 이러한 점들의 집합을 포물선이라 하고 이 포물선의 이름이 A인 거죠. 이 포물선 A의 방정식을 써라 하면 $A(x, y) = y^2 - 4px = 0$ 또는 그저, $y^2 = 4px$라고 쓰면 되는 거죠. 그러니까, 이 경우에 A(x, y)는 함수가 아니고 방정식인 겁니다.

다음으로, 일반형입니다.

$P = \{(x, y) | ax^2 + by^2 + cxy + dx + ey + f = 0$ 여기에서 a, b, c, d, e, f는 상수이고 또한, $ab > 0$; 아니면, c가 0 이고 또한, a 아니면 b가 0 이다.$\}$

위의 집합 P 는 일반적 포물선의 포괄적 표현이죠. 즉, 그 어떠한 포물선이든 다 해당되는 겁니다. 그런데, 다음과 같은 형태가 널리 사용되고 있어요:

$Q = \{(x, y) | ax^2 + bx + c + dy = 0$ 여기에서 a, b, c, d는 상수이고 또한, $ad \neq 0\}$
$R = \{(x, y) | ay^2 + by + c + dx = 0$ 여기에서 a, b, c, d는 상수이고 또한, $ad \neq 0\}$

아래의 집합 S 는 Q 와 R 을 포괄적으로 나타내는 겁니다:
$S = \{(x, y) | ax^2 + by^2 + cx + dy + e = 0$ 여기에서 a, b, c, d, e는 상수이고 또한, a아니면 b가 0 이다.$\}$

일반 math 에서는 주로, 포물선의 일반형 방정식들은 다음과 같이 쓰죠:

$Q(x, y) = ax^2 + bx + c + dy = 0$ 또는 그저, $ax^2 + bx + c + dy = 0$.
$R(x, y) = ay^2 + by + c + dx = 0$ 또는 그저, $ay^2 + by + c + dx = 0$.

직선에서의 경우와 마찬가지로, math 에서 포물선에 대하여 뭘 할 때 그 포물선을 방정식으로 표현합니다. 즉, 포물선의 방정식을 작성해서 사용하는 거죠. 그런데, 직선에서와 마찬가지로, 방정식의 두 가지 형태가 있는데 하나는 일반형이고 나머지 하나는 표준형입니다. 이 두 가지 형태의 목적도 직선에서의 경우와 마찬가지죠. 즉, 일반형은 모든 포물선을 나타낼 수 있고 표준형은 주로 포물선의 기하학적 요소를 나타내기 위한 것이다라고 할 수도 있죠. 표준형을 사용하면 꼭지점과 초점, 준선과 대칭축 그리고 좌표축과 만나는 점(근 또는 절편)을 알아보기가 용이하기 때문이죠. 또한, 이 표준형은 즉각적으로 함수로 사용할 수 있고요. 한편, 일반형은 일반적으로 모든 포물선을 표현하기에 적합하죠. 그럼, 그런 방정식을 어떻게 만드는지 알아보기로 하죠. 그럼 이제, 표준형부터 시작합니다.

1.0. 포물선의 표준방정식

이제, x-y 평면에 하나의 직선 D 가 있다 하고 이 D 의 방정식은 x = d 라고 하세요. 그러면, 점들의 집합을 하나 만드는데, 각 점에서 원점까지의 거리는 그 점에서 직선 D 까지의 거리와 같은 겁니다. 그러면, 그 점들이 어떤 점들인지 구체적으로 나타내보기로 하죠. 그런데, 그 집합은 무한히 많은 점들로 되어있어요. 그러면, 그 많은 점들을 일일이 설명할 수는 없으니까 그 모든 점들을 대표할 수 있는 점을 하나 만들어서 그걸 가지고 하도록 하죠. 그 점을 일반점이라고 해요. 또한, 이런 점을 변점 또는 점변수라고 하기도 하죠. 이제, 그 일반점을 P(s, t)라고 하세요. 그럼, 이 일반점 P 는 원점과 직선 D 로부터 동일한 거리만큼 떨어져있는 겁니다. 자 이제, 이 일반점 P(s, t)가 x-y 평면에서 그 집합의 모든 점들을 망라하면서 움직이고 있는 겁니다. 그러면, P 가 움직이는 동안 한 순간을 포착하면 다음과 같아요.

그림. 1.0.0

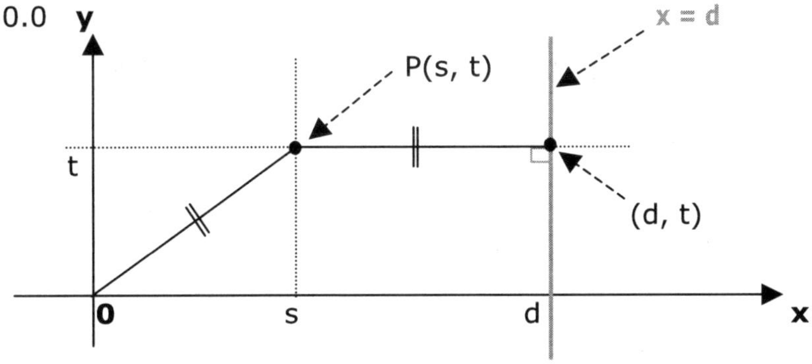

이제, P의 자취를 추적해보기로 하죠. P는 원점과 직선 D로부터 동일한 거리를 유지합니다. 거리가 Δ라 하면 그 거리공식으로부터 $\Delta^2 = (\Delta x)^2 + (\Delta y)^2$가 되죠. 이제, P(s, t)와 원점사이의 거리를 U라 하고 P(s, t)와 직선 D 사이의 거리를 V라고 하세요. 그러면, $U^2 = (s - 0)^2 + (t - 0)^2$이고 $V^2 = (d - s)^2 + (t - t)^2$이 되죠. 그러므로, $U^2 = s^2 + t^2$이고 $V^2 = (d - s)^2$이네요. 그러면 이제, P는 원점과 직선 D로부터 제각기 동일한 거리만큼을 유지해야 하니까 $U^2 = V^2$이어야 하죠. 그러니까, $U^2 = V^2 \Rightarrow s^2 + t^2 = (d - s)^2 = d^2 - 2ds + s^2 \Rightarrow \mathbf{t^2 = d^2 - 2ds}$. 그러면 이제, 이 P가 x-y평면에서 만드는 자취를 그래프에 나타내보도록 하죠.

그림. 1.0.1

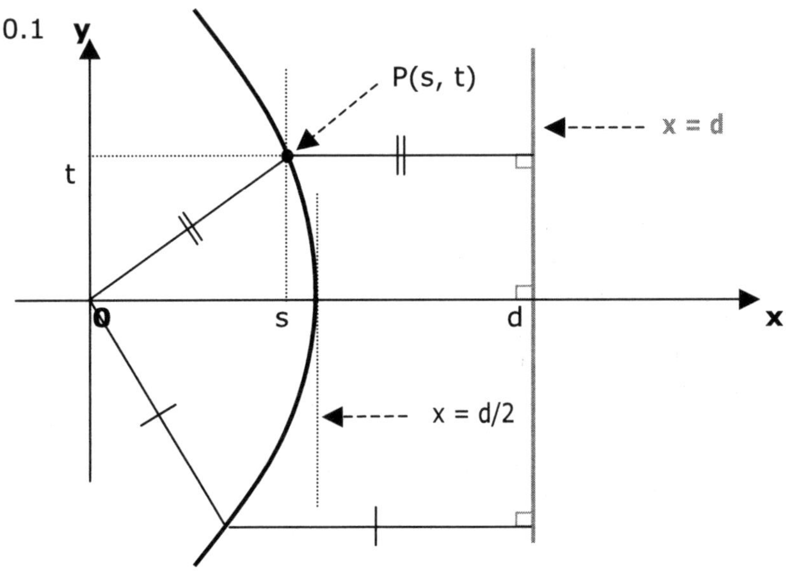

그러면, 점 P(s, t)가 만드는 자취는 하나의 포물선입니다. 다시 말해서, P는 포물선을 따라 움직이는 겁니다.
위의 그래프에서 그 자취 즉, 그 포물선은 x축에 관하여 대칭이라는 걸 알 수 있죠.
그러면 이제, 이 P(s, t)는 x-y 평면에서의 일반점입니다.
그러면, 이 s와 t는 제각기 x-y 좌표계에서 x축과 y축의 값들을 갖는 겁니다.
x-y 좌표계에서의 점들을 명시하는데 x축과 y축에서의 값들을 그 좌표 값들로 사용하죠.
그러니까, s와 t는 변수들이죠. 또한, x-y 좌표계에서의 변수들로서는 x와 y를 사용하죠.
또한, 위에서 나타낸 바와 같이, 이 s와 t의 관계식은 $t^2 = d^2 - 2ds$이죠.
그러면, 그 관계식 $t^2 = d^2 - 2ds$에서 s와 t를 제각기 x와 y로 바꾸면 그 관계식은 바로 x와 y사이의 관계식이죠.
그러면, 포물선을 이루는 모든 점들 각각의 좌표들 간의 관계식은 $y^2 = d^2 - 2dx$가 되네요.
그러면 이제, 이 관계식을 포물선의 방정식이라고 하는 겁니다.

그런데, 그 직선 D는 뭔가요?
그건 준선이라고 해요. 일반적으로 이는 포물선 뿐만이 아니라 타원과 쌍곡선을 작성하는 데에도 사용되죠. 그러한 커브에서는, 각 점에서 준선까지의 거리와 그 점에서 특정 점까지의 거리 사이의 비가 일정한데, 그 비가 포물선의 경우에는 1이죠. 또한, 그 특정 점은 초점이라고 하는데, 지금 하고 있는 이 경우에 이는 원점입니다. 포물선의 경우, 준선은 포물선의 오리엔테이션을 나타내주는데 이는 그 포물선이 열린 방향을 말해주는 겁니다. 초점이란 예를 들면, 회중전등에서 전구가 놓이는 위치인 것이죠. 이것들은 잠시 후에 설명될 겁니다.

자 이제, 그 준선 D를 바꾸어보도록 하죠. 이번에는, 아래 나타낸 바와 같이 수직선 $x = d$ 대신에 수평선 $y = c$를 그래프에 놓고 그 방정식에 어떤 일이 일어나는 지 알아보기로 하죠.

그림. 1.0.2

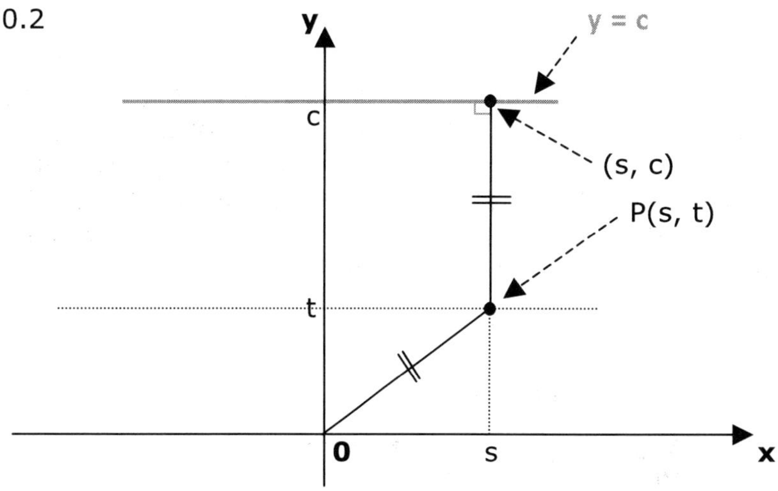

이번에도, P(s, t)와 원점사이의 거리를 U라 하고 P(s, t)와 직선 D 사이의 거리를 V라고 하세요.
그러면, $U^2 = (s - 0)^2 + (t - 0)^2$이고 $V^2 = (s - s)^2 + (c - t)^2$이 되죠.
그러므로, $U^2 = s^2 + t^2$이고 $V^2 = (c - t)^2$이네요.
그러면 이제, P는 원점과 직선 D로부터 동일한 거리만큼을 유지해야 하는 거니까 $U^2 = V^2$이어야 하는 거죠.
그러니까, $U^2 = V^2 \Rightarrow s^2 + t^2 = (c - t)^2 = c^2 - 2ct + t^2 \Rightarrow \mathbf{s^2 = c^2 - 2ct}$.
그러면 이제, 이 s와 t의 관계식은 $s^2 = c^2 - 2ct$이죠.
또한, 이 P(s, t)는 x-y 평면에서의 일반점입니다. 그러면, 이 s와 t는 제각기 x-y 좌표계에서 x축과 y축의 값들을 갖는 겁니다. 그러니까, 이것들은 변수들이죠. 또한, x-y 좌표계에서의 변수들로서 x와 y를 사용하죠.
x-y 좌표계에서의 점들을 명시하기 위해서는 x축과 y축에서의 값들을 그 좌표 값들로 사용하죠. 그러면, 그 관계식 $s^2 = c^2 - 2ct$에서 s와 t를 제각기 x와 y로 바꾸면 그 관계식은 바로 x와 y사이의 관계식이죠.

그러면, 포물선을 이루는 모든 점들 각각의 좌표들 간의 관계식은 $x^2 = c^2 - 2cy$가 되네요.
그러면 이제, 이 P가 x-y 평면에서 만드는 자취를 그래프에 나타내보도록 하죠.

그림. 1.0.3

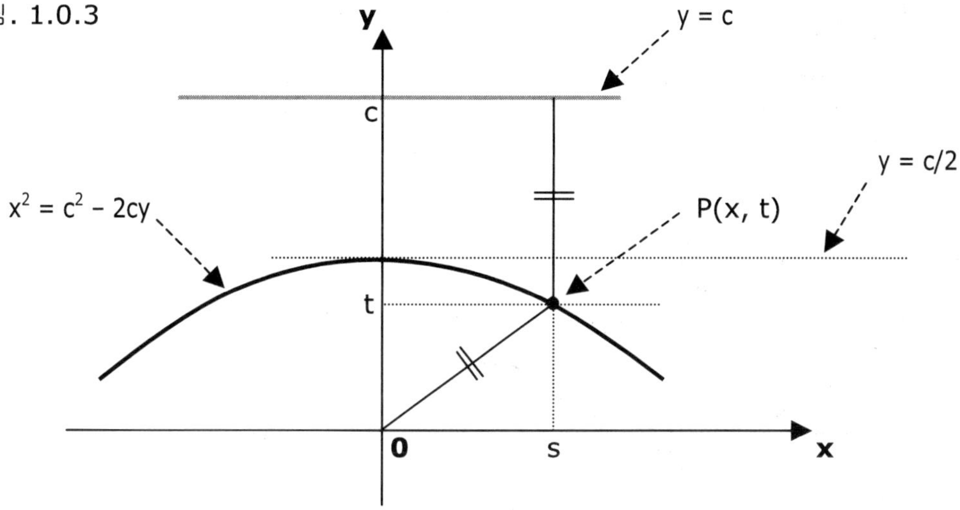

위의 그래프에서 그 자취 즉, 그 포물선은 y 축에 관하여 대칭이라는 걸 알 수 있죠.
그런데, 이 초점을 다른 위치로 옮길 수 있고 준선의 위치도 변경할 수 있어요. 수직선 또는 수평선을 그 준선으로 잡는 이유는 계산이 간단지기 때문입니다. 포물선이 초점쪽으로 굽어있다는 걸 알 수 있죠. 이제, 광선들이 포물선으로 들어온다고 하세요. 그러면, 그 광선들은 그 초점에서 모두 집중되는 겁니다. 역으로, 초점에 광원(예를 들어, 전구)을 놓으면 모든 광선들은 준선에 수직이 되는 겁니다. 준선을 평행 이동하면 아래 그림의 점선 수직선과 같죠.

그림. 1.0.4

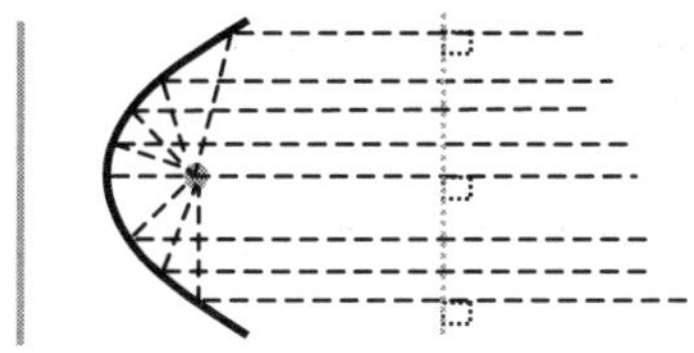

자 이제, 그 초점이 (0, f)에 있다 하고 준선 D가 직선 y = -f 라고 해보죠.

그림. 1.0.5

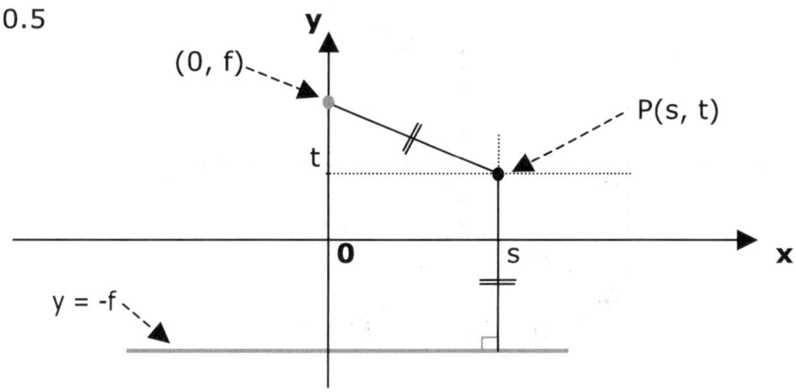

P(s, t)와 초점사이의 거리를 U 라 하고 P(s, t)와 준선 D 사이의 거리를 V 라고 하세요.
그러면, $U^2 = (s - 0)^2 + (t - f)^2$이고 $V^2 = (s - s)^2 + \{t - (-f)\}^2$이 되죠.
그러므로, $U^2 = s^2 + (t - f)^2$이고 $V^2 = (t + f)^2$이네요.

그러면 이제, P는 초점과 준선 D로부터 동일한 거리만큼을 유지해야 하는 거니까 $U^2 = V^2$이어야 하는 거죠.
그러니까, $U^2 = V^2 \Rightarrow s^2 + (t - f)^2 = (t + f)^2 \Rightarrow s^2 + t^2 - 2ft + f^2 = t^2 + 2ft + f^2 \Rightarrow \mathbf{s^2 = 4ft}$.
그러면 이제, 이 s와 t의 관계식은 $s^2 = 4ft$이죠.
또한, 이 P(s, t)는 x-y 평면에서의 일반점이니까 이 s와 t는 제각기 x-y 좌표계에서 x 축과 y 축의 값들을 갖는 변수들이고 x-y 좌표계에서는 변수들로서 x와 y를 사용하죠.
또한, x-y 좌표계에서의 점들을 명시하는데 x 축과 y 축에서의 값들을 그 좌표 값들로 사용하죠.
그러면, 그 관계식 $s^2 = 4ft$에서 s와 t를 제각기 x와 y로 바꾸면 그 관계식은 바로 x와 y사이의 관계식이죠.
그러면, 포물선을 이루는 점들 각각에서의 좌표들의 관계식은 $\mathbf{x^2 = 4fy}$가 되네요.

그림. 1.0.6

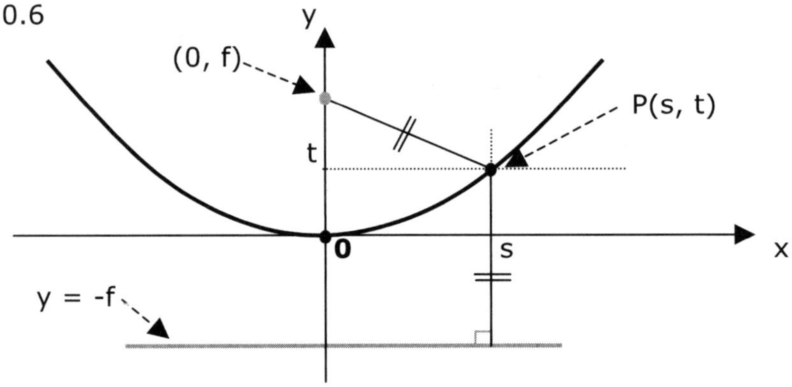

이제, f = 1/4 라고 해보세요. 그러면, $y = x^2$을 갖게 되죠.
그런데, 이런 커브에 또 하나의 특별한 점이 있는데 이를 그 포물선의 꼭지점이라고 합니다. 위의 경우에는, 그 꼭지점이 원점에 있네요. 꼭지점은 초점과 준선 사이의 중점이죠. 또한, 이 포물선은 y 축에 관하여 대칭이죠. 그러므로, 대칭축은 준선에 수직이고 그 꼭지점을 지납니다. 따라서, 위의 경우에는 대칭축이 y 축 즉, 직선 x = 0 이죠.
또한, 이 포물선의 입구는 포물선의 위쪽입니다.

이번에는 준선을 직선 y = f 라 하고 초점은 (0, -f)에 있다고 해보죠.

그림. 1.0.7

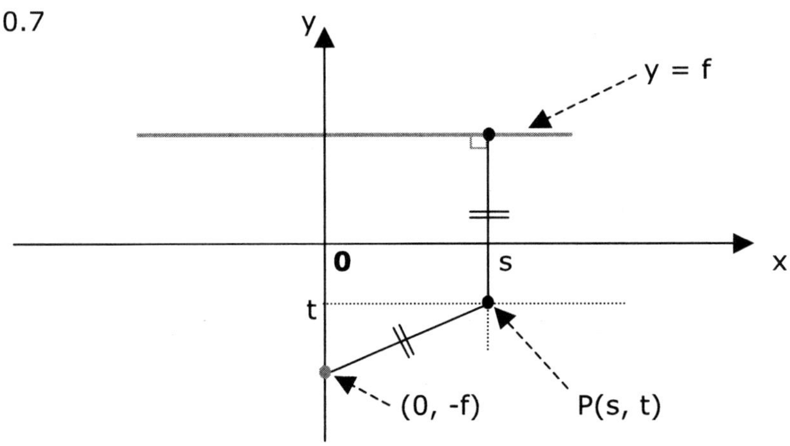

P(s, t)와 초점사이의 거리를 U라 하고 P(s, t)와 준선 사이의 거리를 V라고 하세요.
그러면, $U^2 = (s - 0)^2 + \{t - (-f)\}^2$이고 $V^2 = (s - s)^2 + (t - f)^2$이 되죠.
그러므로, $U^2 = s^2 + (t + f)^2$이고 $V^2 = (t - f)^2$이네요.

그러면 이제, P는 초점과 준선으로부터 동일한 거리만큼을 유지해야 하니까 $U^2 = V^2$이어야 하죠.
그러니까, $U^2 = V^2 \Rightarrow s^2 + (t + f)^2 = (t - f)^2 \Rightarrow s^2 + t^2 + 2ft + f^2 = t^2 - 2ft + f^2 \Rightarrow \mathbf{s^2 = -4ft}$.
그러면 이제, 이 s와 t의 관계식은 $s^2 = -4ft$이죠.
또한, 이 P(s, t)는 x-y 평면에서의 일반점이니까 이 s 와 t 는 제각기 x-y 좌표계에서 x 축과 y 축의 값들을 갖는 변수들이고 x-y 좌표계에서는 변수들로서 x 와 y 를 사용하죠.
또한, x-y 좌표계에서의 점들을 명시하는데 x축과 y축에서의 값들을 그 좌표 값들로 사용하죠. 그러면, 그 관계식 $s^2 = -4ft$에서 s와 t를 x와 y로 바꾸면 포물선을 이루는 점들의 좌표 값들의 관계식은 $x^2 = -4fy$가 되네요.
물론, 이를 $y = -x^2/4f$로 놓아도 되죠.

그림. 1.0.8

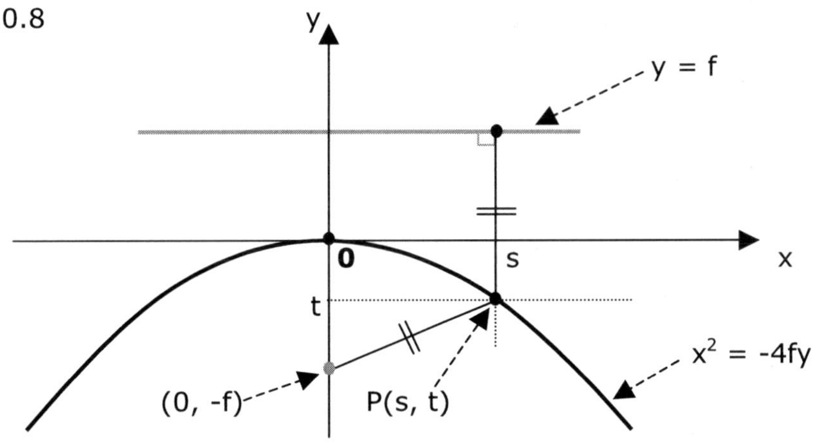

또한, f = 1/4 라 놓아보죠. 그러면, $y = -x^2$가 되는 거죠.
위의 포물선에도 그 꼭지점이 원점에 있고 그 대칭축도 y 축 즉, 직선 x = 0 이죠.
하지만, 그 포물선의 입구는 아래에 있네요.

이번에는 초점이 (f, 0)에 있고 준선은 직선 x = -f 라 해보죠.

그림. 1.0.9

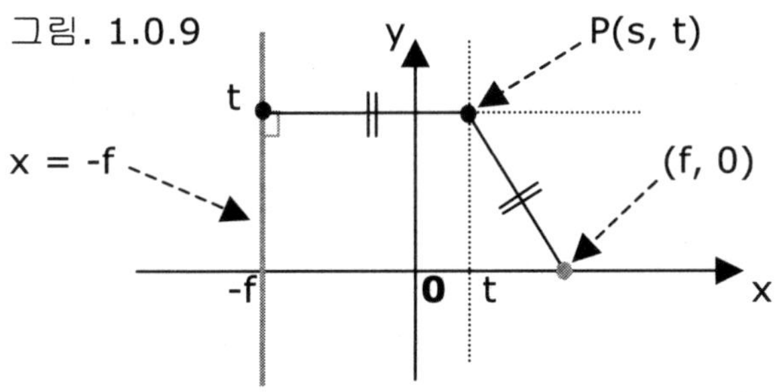

P(s, t)와 초점사이의 거리를 U 라 하고 P(s, t)와 준선 D 사이의 거리를 V 라고 하세요.
그러면, $U^2 = (s - f)^2 + (t - 0)^2$이고 $V^2 = \{s - (-f)\}^2 + (t - t)^2$이 되죠.
그러므로, $U^2 = (s - f)^2 + t^2$이고 $V^2 = (s + f)^2$이네요.
그러면 이제, P는 초점과 준선으로부터 동일한 거리만큼을 유지해야 하니까 $U^2 = V^2$이어야 하죠.
그러니까, $U^2 = V^2 \Rightarrow (s - f)^2 + t^2 = (s + f)^2 \Rightarrow s^2 - 2fs + f^2 + t^2 = s^2 + 2fs + f^2 \Rightarrow \mathbf{t^2 = 4fs}$.
그러면 이제, 이 s와 t의 관계식은 $t^2 = 4fs$이죠.
또한, 이 P(s, t)는 x-y 평면에서의 일반점이니까 이 s 와 t 는 제각기 x-y 좌표계에서 x 축과 y 축의 값들을 갖는 변수들이고 x-y 좌표계에서는 변수들로서 x 와 y 를 사용하죠.

또한, x-y 좌표계에서의 점들을 명시하는데 x축과 y축에서의 값들을 그 좌표 값들로 사용하죠. 그러면, 그 관계식 $t^2 = 4fs$에서 s와 t를 x와 y로 바꾸면 포물선을 이루는 점들 각각에서의 좌표들 간의 관계식은 $\mathbf{y^2 = 4fx}$가 되네요. 물론, 이를 $x = y^2/4f$로 놓아도 되죠. 그러면, $f = 1/4$ 이라면 이 포물선의 방정식은 $x = y^2$이 되는 거죠.

그림. 1.0.A

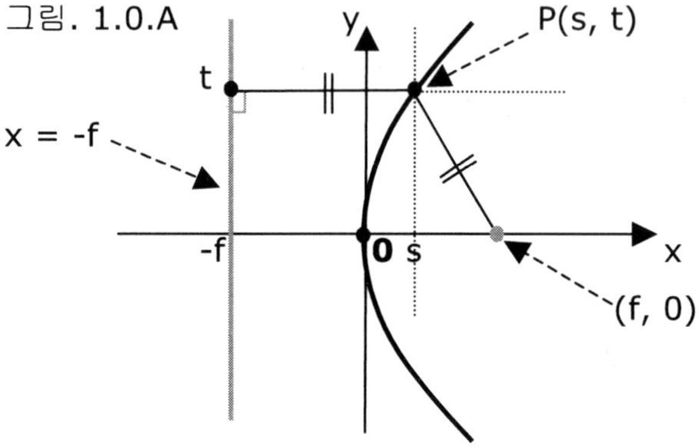

위의 포물선에도 그 꼭지점이 원점에 있지만 그 대칭축은 y 축이 아니고 x 축 즉, 직선 $y = 0$ 이네요.
이번에는, 그 포물선의 입구가 오른쪽에 있죠.

이번에는 초점이 (0, v)에 있다고 해보죠.

그러면, 준선을 어디에 놓나요? 그 준선의 방정식은 $y = -v$ 인가요?
꼭 그렇지는 않아요. 즉, 준선은 놓고 싶은 곳이면 어디에 놓아도 되는 겁니다. 하지만, 간단하게 할 수 있으니까 일반 math 에서는 준선을 좌표축들 중의 하나와 평행하게 놓는 것뿐이죠.

물론, 초점과 준선 사이의 중점도 원점이어야 할 필요는 없는 거죠.
중점이라니 요?
그건 포물선의 꼭지점이고 이는 포물선에서 매우 중요한 점이죠. 사실, 포물선에 관련한 문제 풀이에서 반드시 알아내야 하는 점인 경우가 많아요.

이 책에서는 직각 포물선을 공부합니다. 직각 포물선에서는 준선이 좌표축들 중 하나에 평행한 겁니다. 그러니, 직각 포물선의 대칭축은 그 좌표축에 수직인 거죠. 대칭축이나 준선이 좌표축에 비스듬하게 기울어진 포물선은 **제 2 권: 그래프 조작**을 참고하세요.

그러면 이제, 포물선에서 유의해야 할 구성 요소가 꼭지점과 초점입니다.
준선은 요?
물론, 준선도 중요하죠. 그런데, 이는 꼭지점과 초점을 알면 알 수 있는 거죠. 꼭지점이 초점과 준선 사이의 중점이니까. 그러니까, 또한 꼭지점과 초점이 준선에 수직인 직선에 있는 거죠. 그 직선이 뭐죠? 대칭축이죠.

꼭지점 V가 (u, v)에 있으며 대칭축이 $x = u$ 라 하고 초점 F가 (@, f)에 있다 하고 @는 아직 모른다고 하세요.
그러면, @는 뭐가 될 수 있나요?
그 @는 u 이어야 하는 거죠 꼭지점과 초점은 대칭축에 있고 대칭축이 직선 $x = u$ 이니까.
이제, 초점 F가 (u, f)에 있다 하고 이 포물선의 방정식을 구한다 하세요.
먼저, 그 일반점 P(s, t)와 함께 그 꼭지점과 초점을 그래프에 넣어보기로 하죠.
그러면, 그 방정식을 도출해내기 위해서 무엇을 할 수 있는지 보다 잘 알 수 있죠.

그림. 1.0.B

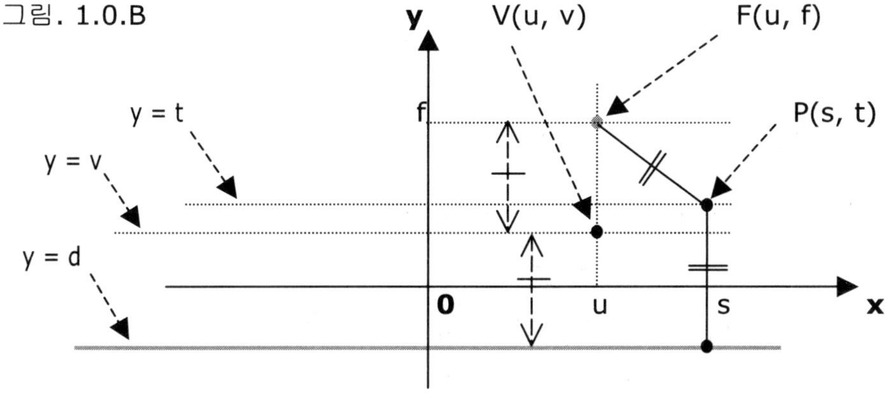

이제, 준선이 직선 $y = d$ 라 하세요.
그러면, V가 준선과 초점 사이의 중점이니까 v는 f와 d의 평균이네요.
그러면, $v = (f + d)/2 \Rightarrow d = 2v - f$. 따라서, 준선은 직선 $y = 2v - f$ 네요.

P(s, t)와 초점사이의 거리를 Q라 하고 P(s, t)와 준선 D 사이의 거리를 R이라고 하세요.

그러면, $Q^2 = (s - u)^2 + (t - f)^2$이고 $R^2 = (s - s)^2 + \{t - (2v - f)\}^2 = \{t - (2v - f)\}^2$이 되죠.
그러면 이제, P는 초점과 준선으로부터 동일한 거리만큼을 유지해야 하니까 $Q^2 = R^2$이어야 하죠.
그러니까, $Q^2 = R^2 \Rightarrow (s - u)^2 + (t - f)^2 = \{t - (2v - f)\}^2$
$\Rightarrow (s - u)^2 = \{t - (2v - f)\}^2 - (t - f)^2 = \{t - (2v - f) + (t - f)\}\{t - (2v - f) - (t - f)\}$
$= (t - 2v + f + t - f)(t - 2v + f - t + f) = 2(t - v)2(f - v)$.

$\Rightarrow (s - u)^2 = 4(t - v)(f - v)$.

그러면 이제, 이 s와 t의 관계식은 $(s - u)^2 = 4(f - v)(t - v)$이죠.
또한, 이 P(s, t)는 x-y평면에서의 일반점이니까 이 s와 t는 제각기 x-y좌표계에서 x축과 y축의 값들을 갖는 변수들이고 x-y좌표계에서는 변수들로서 x와 y를 사용하죠. 또한, x-y 좌표계에서의 점들을 명시하는데 x축과 y축에서의 값들을 그 좌표 값들로 사용하죠. 그러면, 그 관계식 $(s - u)^2 = 4(f - v)(t - v)$에서 s와 t를 제각기 x와 y로 바꾸면 포물선을 이루는 모든 점들 각각에서의 좌표들 간의 관계식은 $(x - u)^2 = 4(f - v)(y - v)$가 되네요.

그림. 1.0.C

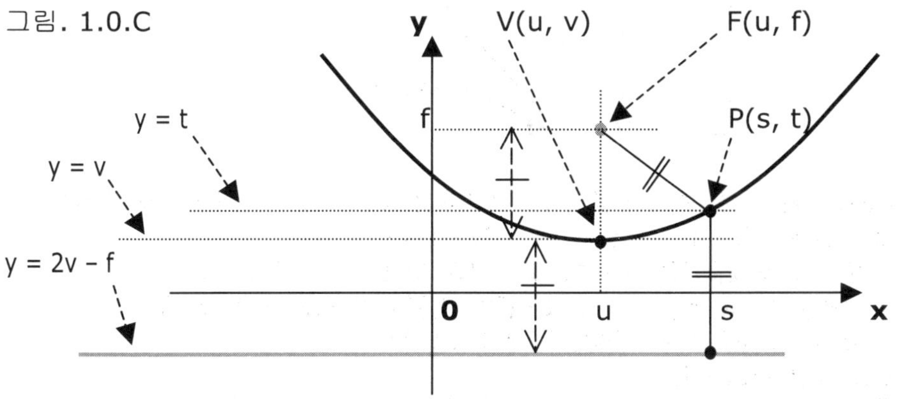

또한, 위의 커브는 아래에 나타낸 바와 같이 회색점선커브를 x 축 방향으로 u 만큼 y 축 방향으로 v 만큼 평행이동한 결과로 할 수도 있어요.

그림. 1.0.D

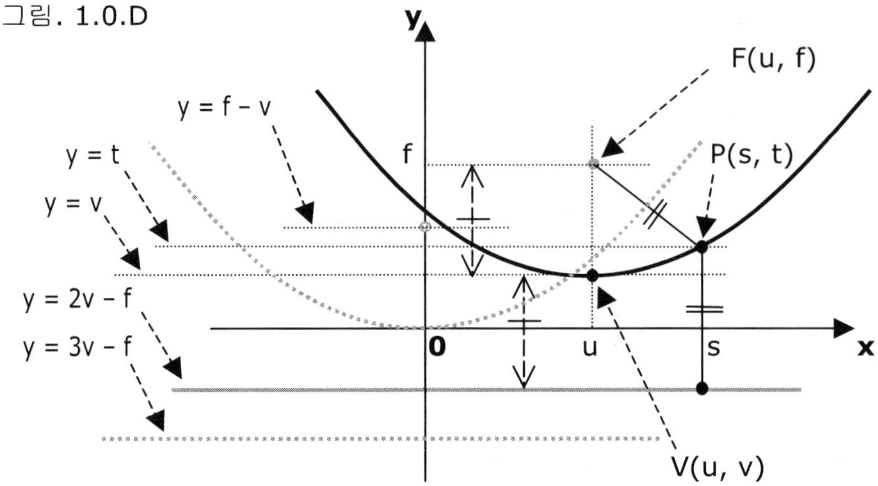

회색점선포물선의 방정식은 $x^2 = 4(f - v)y$이고 꼭지점은 원점에 그리고 초점은 $(0, f - v)$에 있는 거죠.
검정포물선의 대칭축은 직선 $x = u$ 입니다.

이번에는 꼭지점 V가 (u, v)에 있고 초점 F는 $(u, -f)$에 있다고 하세요.
그러면, 대칭축은 여전히 그 직선 $x = u$ 라는 걸 즉각 알 수 있죠.

그림. 1.0.E

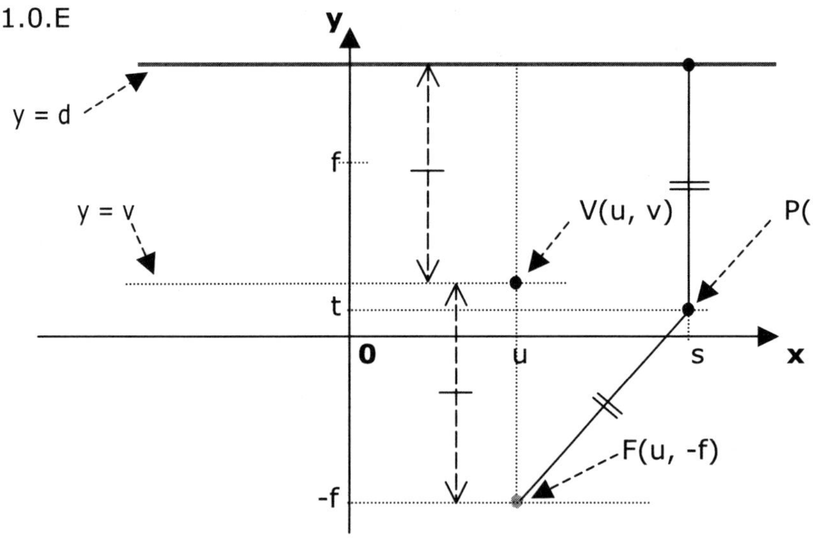

다음으로, 준선이 직선 $y = d$ 라 하세요. 그러면, V가 준선과 초점 사이의 중점이니까 v는 -f와 d의 평균이죠.
그러면, $v = (-f + d)/2 \Rightarrow d = 2v + f$. 따라서, 준선은 직선 $y = 2v + f$ 네요.
이제, $P(s, t)$와 초점사이의 거리를 Q라 하고 $P(s, t)$와 준선 D 사이의 거리를 R이라고 하세요.
그러면, $Q^2 = (s - u)^2 + \{t - (-f)\}^2$이고 $R^2 = (s - s)^2 + \{t - (2v + f)\}^2 = \{t - (2v + f)\}^2$이 되죠.
그러면 이제, P는 초점과 준선으로부터 동일한 거리만큼을 유지해야 하니까:
$Q^2 = R^2 \Rightarrow (s - u)^2 + (t + f)^2 = \{t - (2v + f)\}^2$
$\Rightarrow (s - u)^2 = \{t - (2v + f)\}^2 - (t + f)^2 = \{t - (2v + f) + (t + f)\}\{t - (2v + f) - (t + f)\}$
$= (t - 2v - f + t + f)(t - 2v - f - t - f) = 2(t - v)(-2)(f + v) \Rightarrow (s - u)^2 = -4(t - v)(f + v)$.
그러면 이제, 이 s와 t의 관계식은 $(s - u)^2 = -4(f + v)(t - v)$이죠.

또한, 이 P(s, t)는 x-y평면에서의 일반점이니까 이 s와 t는 제각기 x-y좌표계에서 x축과 y축의 값들을 갖는 변수들이고 x-y좌표계에서는 변수들로서 x와 y를 사용하죠. 또한, x-y좌표계에서의 점들을 명시하는데 x축과 y축에서의 값들을 그 좌표 값들로 사용하죠. 그러면, 그 관계식 $(s - u)^2 = -4(f + v)(t - v)$에서 s와 t를 제각기 x와 y로 바꾸면 포물선을 이루는 모든 점들 각각의 좌표들 간의 관계식은 $(x - u)^2 = -4(f + v)(y - v)$가 되네요.

그림. 1.0.F

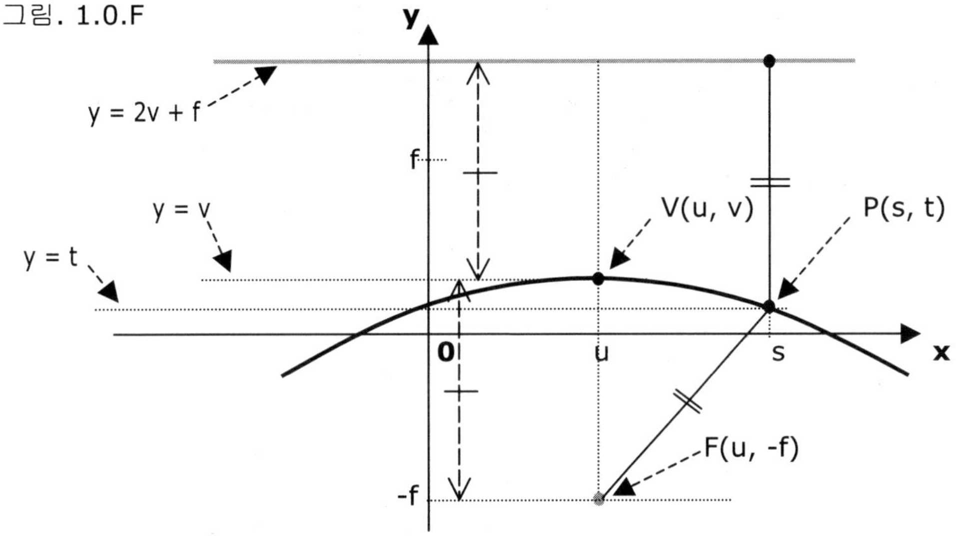

또한, 위의 커브는 아래에 나타낸 바와 같이 회색점선커브를 x 축 방향으로 u 만큼 y 축 방향으로 v 만큼 평행이동한 결과로 할 수도 있어요.

그림. 1.0.G

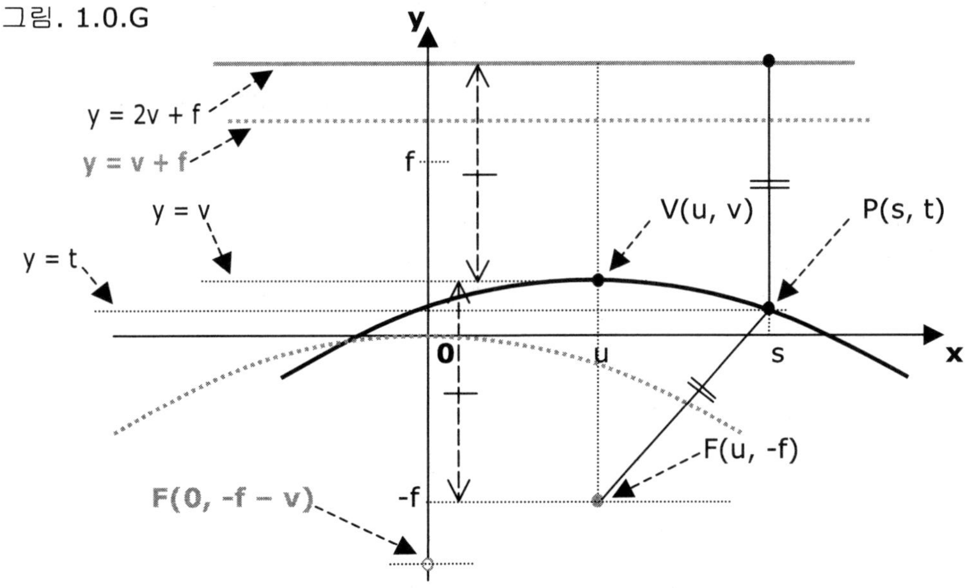

회색점선포물선의 방정식은 $x^2 = 4(f + v)y$이며 초점은 (0, -f − v)에 있고 대칭축은 직선 x = 0 이죠. 검정포물선의 방정식은 $(x - u)^2 = -4(f + v)(y - v)$이며 초점은 (u, -f)에 있고 대칭축은 직선 x = u입니다.

그러면 이제, 다음 두 포물선들의 그래프들을 서로 비교해보기로 하죠:

$(x - u)^2 = -4(f + v)(y - v)$와 $(x - u)^2 = 4(f - v)(y - v)$.

그림. 1.0.C

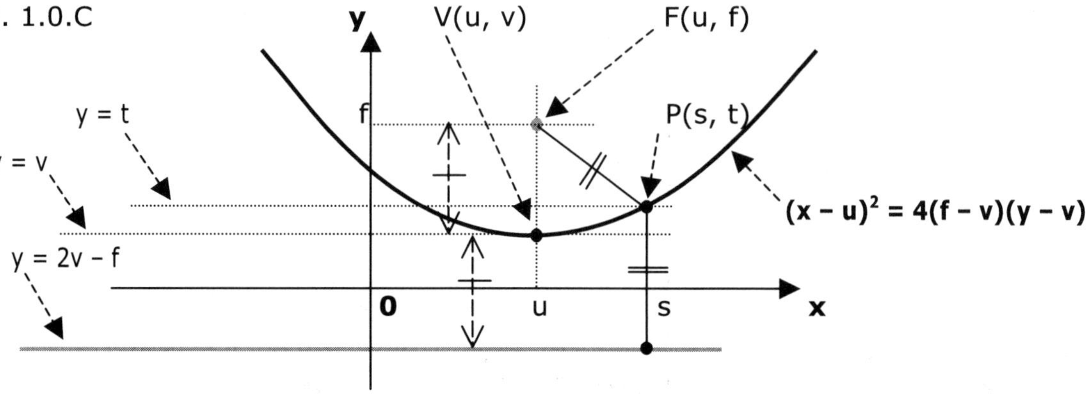

그림. 1.0.F

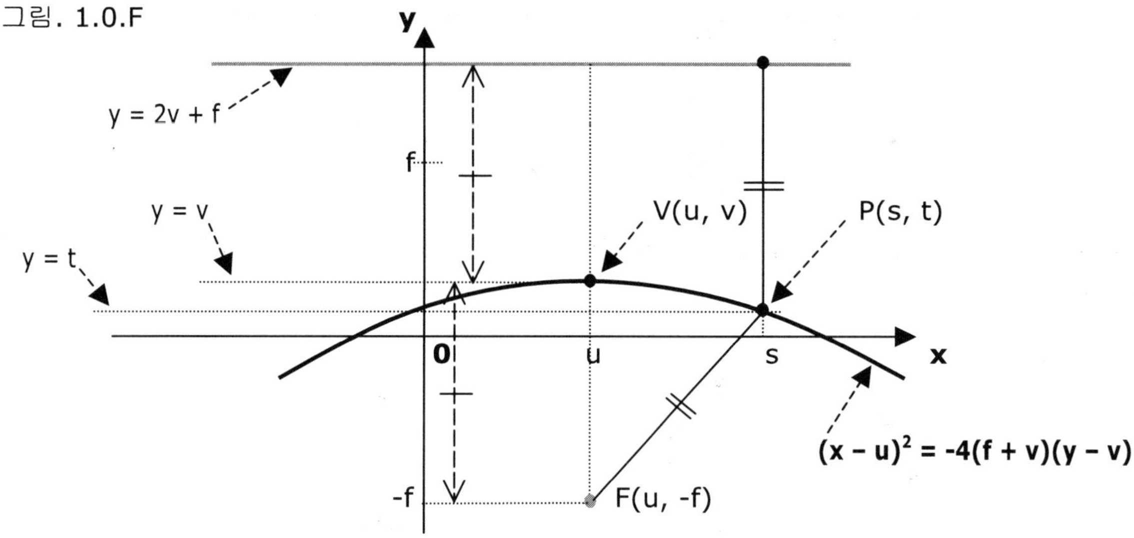

이번에는, 꼭지점 V가 (u, v)에 그리고 초점 F는 (f, v)에 있다 하세요.
그러면, 대칭축이 직선 y = v라는 걸 즉각 알 수 있죠.
먼저, 그 일반점 P(s, t)와 함께 그 꼭지점과 초점을 그래프에 넣어보기로 하죠. 꼭지점은 초점과 준선 사이의 중점이라는 걸 알고있죠. 이제, 준선을 직선 x = d라 하세요. 그러면:

그림. 1.0.H

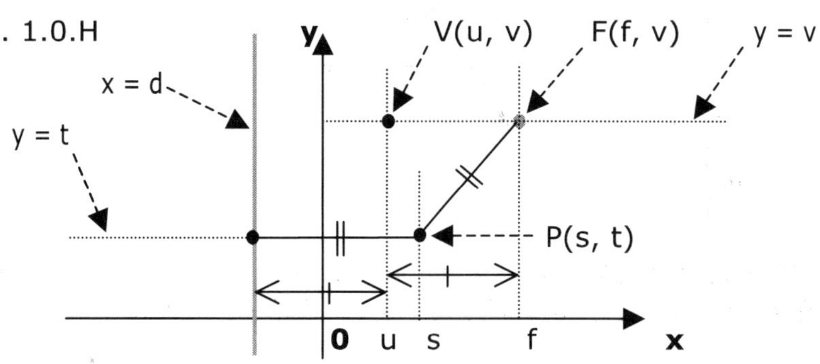

그러면, 위의 그래프에서 볼 수 있는 바와 같이 u는 f와 d의 평균이죠.

그러면, $u = (f + d)/2 \Rightarrow d = 2u - f$. 그러므로, 준선은 직선 $x = 2u - f$네요.
P(s, t)와 초점사이의 거리를 Q라 하고 P(s, t)와 준선 D 사이의 거리를 R이라고 하세요.
그러면, $Q^2 = (s - f)^2 + (t - v)^2$이고 $R^2 = \{s - (2u - f)\}^2 + (t - t)^2 = \{s - (2u - f)\}^2$이 되죠.

그러면 이제, P는 초점과 준선으로부터 동일한 거리만큼을 유지하니까:
$Q^2 = R^2 \Rightarrow (s - f)^2 + (t - v)^2 = \{s - (2u - f)\}^2$
$\Rightarrow (t - v)^2 = \{s - (2u - f)\}^2 - (s - f)^2 = \{s - (2u - f) + (s - f)\}\{s - (2u - f) - (s - f)\}$
$= (s - 2u + f + s - f)(s - 2u + f - s + f) = 2(s - u)2(f - u) \Rightarrow (t - v)^2 = 4(f - u)(s - u)$.

그러면 이제, 이 s와 t의 관계식은 $(t - v)^2 = 4(f - u)(s - u)$이죠.
또한, 이 P(s, t)는 x-y평면에서의 일반점이니까 이 s와 t는 제각기 x-y좌표계에서 x축과 y축의 값들을 갖는 변수들이고 x-y좌표계에서는 변수들로서 x와 y를 사용하죠. 또한, x-y좌표계에서의 점들을 명시하는데 x축과 y축에서의 값들을 그 좌표 값들로 사용하죠. 그러면, 위의 관계식에서 s와 t를 x와 y로 바꾸면 포물선을 이루는 모든 점들 각각에서의 좌표들 간의 관계식은 $(y - v)^2 = 4(f - u)(x - u)$가 되네요.

그림. 1.0.I

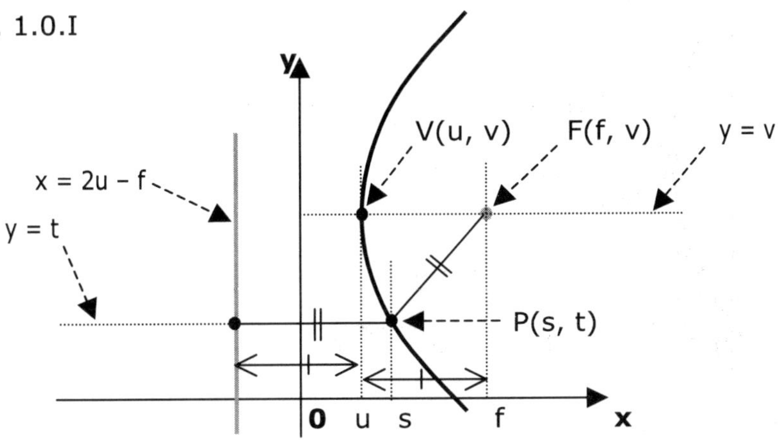

다음으로, 꼭지점 V는 (u, v)에 그리고 초점 F는 (-f, v)에 있다 하세요.
그러면, 대칭축은 여전히 $y = v$ 이죠. 이제, 준선을 직선 $x = d$ 라 하세요. 그러면, 꼭지점은 초점과 준선 사이의 중점이니까 꼭지점과 초점을 준선과 함께 그래프에 넣으면 다음과 같죠:

그림. 1.0.J

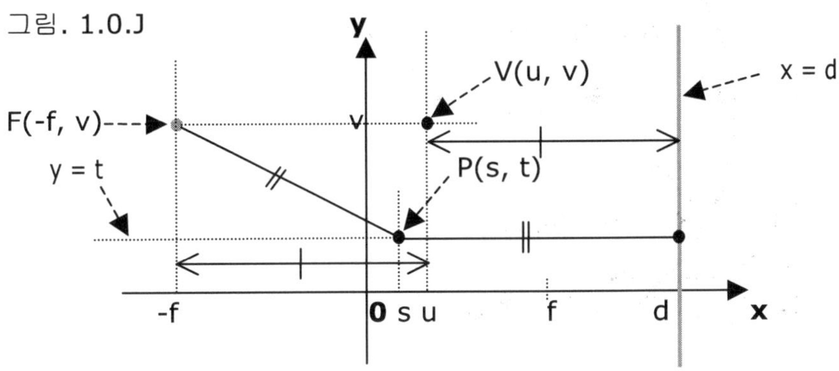

그러면, u가 -f와 d의 평균이니까, $u = (-f + d)/2 \Rightarrow d = 2u + f$.
따라서, 준선은 직선 $x = 2u + f$.

P(s, t)와 초점사이의 거리를 Q 라 하고 P(s, t)와 준선 D 사이의 거리를 R 이라고 하세요.
그러면, $Q^2 = (s + f)^2 + (t - v)^2$이고 $R^2 = \{s - (2u + f)\}^2 + (t - t)^2 = \{s - (2u + f)\}^2$이 되죠.
그러면 이제, $Q^2 = R^2 \Rightarrow (s + f)^2 + (t - v)^2 = \{s - (2u + f)\}^2$
$\Rightarrow (t - v)^2 = \{s - (2u + f)\}^2 - (s + f)^2 = \{s - (2u + f) + (s + f)\}\{s - (2u + f) - (s + f)\}$
$= (s - 2u - f + s + f)(s - 2u - f - s - f) = 2(s - u)(-2)(f + u)$
$\Rightarrow (t - v)^2 = -4(s - u)(f + u)$.

그러면 이제, 이 s와 t의 관계식은 $(t - v)^2 = -4(f + u)(s - u)$이죠.
또한, 이 P(s, t)는 x-y평면에서의 일반점이니까 이 s와 t는 제각기 x-y좌표계에서 x축과 y축의 값들을 갖는 변수들이고 x-y좌표계에서는 변수들로서 x와 y를 사용하죠. 또한, x-y좌표계에서의 점들을 명시하는데 x축과 y축에서의 값들을 그 좌표 값들로 사용하죠. 그러면, 위의 관계식에서 s와 t를 제각기 x와 y로 바꾸면 포물선을 이루는 모든 점들 각각에서의 좌표들 간의 관계식은 $(y - v)^2 = -4(f + u)(x - u)$가 되죠.

그림. 1.0.K

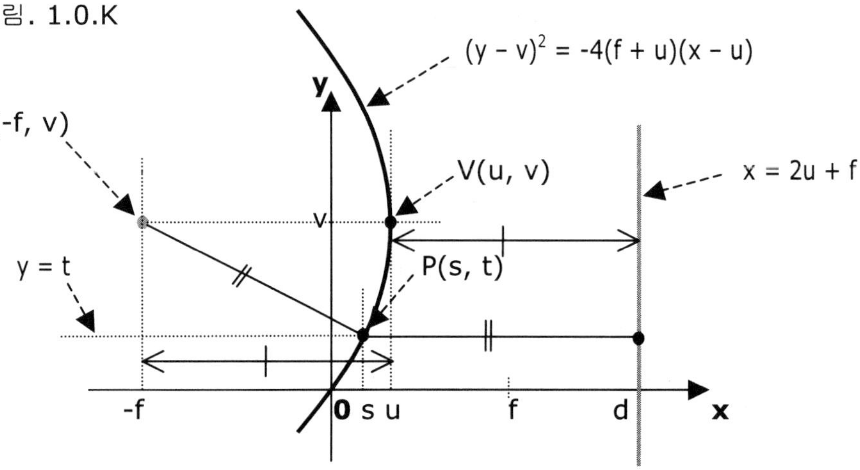

위의 포물선은 직선 y = v 에 관하여 대칭이죠.
그러면, 위의 그래프와 아래 다시 나타낸 그 이전 그래프를 서로 비교해보도록 하죠.

그림. 1.0.I

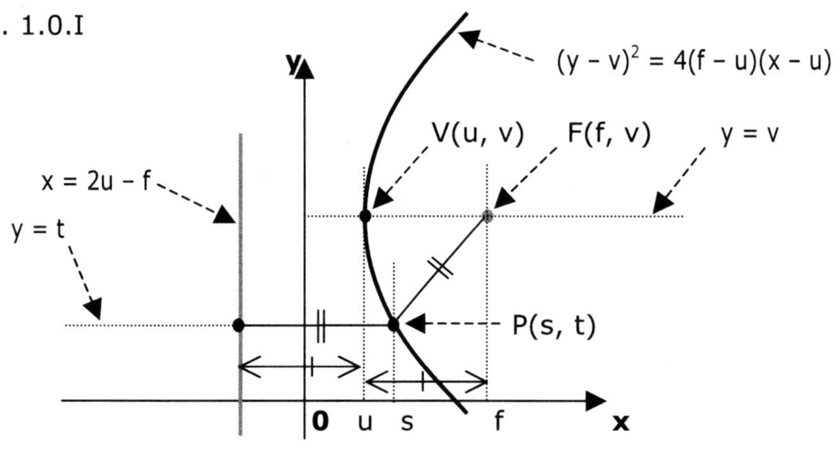

자 이제, 포물선의 표준형 방정식에 대하여 이야기해보기로 하죠.
위에서 작성한 것들이 표준형 방정식들의 예라고 할 수 있어요.
V가 꼭지점이고 F는 초점이며 D는 준선이라 하세요. 그러면:

$y^2 = d^2 - 2dx$ (V: (d/2, 0), F: 원점, D: x = d)
$x^2 = c^2 - 2cy$ (V: (0, c/2), F: 원점, D: y = c)
$x^2 = 4fy$ (V: 원점, F: (0, f), D: y = -f)
$x^2 = -4fy$ (V: 원점, F: (0, -f), D: y = f)
$y^2 = 4fx$ (V: 원점, F: (f, 0), D: x = -f)
$(x - u)^2 = 4(f - v)(y - v)$ (V: (u, v), F: (u, f)), D: y = 2v - f)
$x^2 = 4(f - v)y$ (V: 원점, F: (0, f - v), D: y = 3v - f)
$(x - u)^2 = -4(f + v)(y - v)$ (V: (u, v), F: (u, -f), D: y = 2v + f)
$x^2 = 4(f + v)y$ (V: 원점, F: (0, -f - v), D: y = v + f)
$(y - v)^2 = 4(f - u)(x - u)$ (V: (u, v), F: (f, v), D: x = 2u - f)
$(y - v)^2 = -4(f + u)(x - u)$ (V: (u, v), F: (-f, v), D: x = 2u + f)

위의 방정식들 중에서 굵은 글씨로 되어있는 것들을 표준형에서의 기본형들이라고 할 수 있어요.
이 기본형들 중에서도 특히, $y^2 = 4fx$와 $x^2 = 4fy$는 단순 기본형들인데 실제로는 이것들을 포물선의 표준형 방정식이라고 합니다. 그런데, 일반적으로 표준형을 나타낼 때는 통상적으로 문자 f대신에 p를 사용해요.
그러므로, 포물선의 표준형 방정식은 다음과 같습니다:

$x^2 = 4py$ 여기에서 꼭지점은 원점, 초점은 (0, p)이고 준선은 직선 y = -p.
$y^2 = 4px$ 여기에서 꼭지점은 원점, 초점은 (p, 0)이고 준선은 직선 x = -p.

그런데, 다음과 같은 방정식들이 있죠:
$(x - u)^2 = 4(f - v)(y - v)$ (꼭지점: (u, v), 초점: (u, f), 준선: y = 2v - f)
$(y - v)^2 = 4(f - u)(x - u)$ (꼭지점: (u, v), 초점: (f, v), 준선: x = 2u - f)

바로 위의 첫번째 방정식에서 $f - v = p$ 라고 놓으세요.
그러면, $f - v = p \Rightarrow f = p + v$.
따라서, 준선의 경우에, $y = 2v - f = 2v - (p + v) = 2v - p - v = v - p$.
그러므로, $(x - u)^2 = 4p(y - v)$이며 여기에서 꼭지점은 (u, v), 초점은 (u, v + p)이고 준선은 y = v - p.

두 번째 방정식에서 $f - u = p$ 라고 놓으세요.
그러면, $f - u = p \Rightarrow f = p + u$.
따라서, 준선의 경우에, $x = 2u - f = 2u - (p + u) = 2u - p - u = u - p$.
그러므로, $(y - v)^2 = 4p(x - u)$이고 여기에서 꼭지점은 (u, v), 초점은 (u + p, v)이며 준선은 x = u - p.

따라서, 다음의 방정식들은 좀더 일반화 시킨 표준형이라고 해도 될 겁니다:
$(x - u)^2 = 4p(y - v)$ 여기에서 꼭지점은 (u, v), 초점은 (u, v + p)이고 준선은 y = v - p.
$(y - v)^2 = 4p(x - u)$ 여기에서 꼭지점은 (u, v), 초점은 (u + p, v)이고 준선은 x = u - p.

그러면 이제, 각 표준형에 따른 커브의 오목한 상태를 보도록 하죠.

먼저, $p > 0$ 라고 하세요. 그러면, 다음과 같은 커브들은 위로 오목(우리식으로는, 아래로 볼록)입니다:
$x^2 = 4py$ 여기에서 꼭지점은 원점, 초점은 (0, p)이고 준선은 직선 y = -p.
$(x - u)^2 = 4p(y - v)$ 여기에서 꼭지점은 (u, v) 초점은 (u, v + p)이고 준선은 y = v - p.

그림. 1.0.L

다음으로, 다음과 같은 커브들은 아래로 오목(위로 볼록)입니다:
$x^2 = -4py$ 여기에서 꼭지점은 원점, 초점은 $(0, -p)$이고 준선은 직선 $y = p$.
$(x - u)^2 = -4p(y - v)$ 여기에서 꼭지점은 (u, v), 초점은 $(u, v - p)$이고 준선은 $y = v + p$.

그림. 1.0.M

다음으로, 다음과 같은 커브들은 오른쪽으로 오목(왼쪽으로 볼록)입니다:
$y^2 = 4px$ 여기에서 꼭지점은 원점, 초점은 $(p, 0)$이고 준선은 직선 $x = -p$.
$(y - v)^2 = 4p(x - u)$ 여기에서 꼭지점은 (u, v), 초점은 $(u + p, v)$이고 준선은 $x = u - p$.

그림. 1.0.N

다음으로, 다음과 같은 커브들은 왼쪽으로 오목(오른쪽으로 볼록)입니다:
$y^2 = -4px$ 여기에서 꼭지점은 원점, 초점은 $(-p, 0)$이고 준선은 직선 $x = p$.
$(y - v)^2 = -4p(x - u)$ 여기에서 꼭지점은 (u, v), 초점은 $(u - p, v)$이고 준선은 $x = u + p$.

그림. 1.0.O

그림. 1.0.P p = 1/4 일 때 표준형

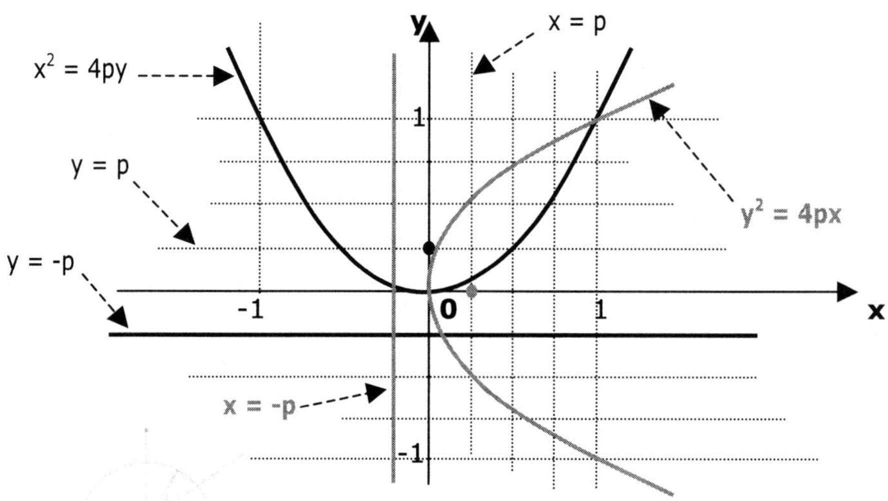

그림. 1.0.Q $x^2 = 4py$와 $(x - u)^2 = 4p(y - v)$

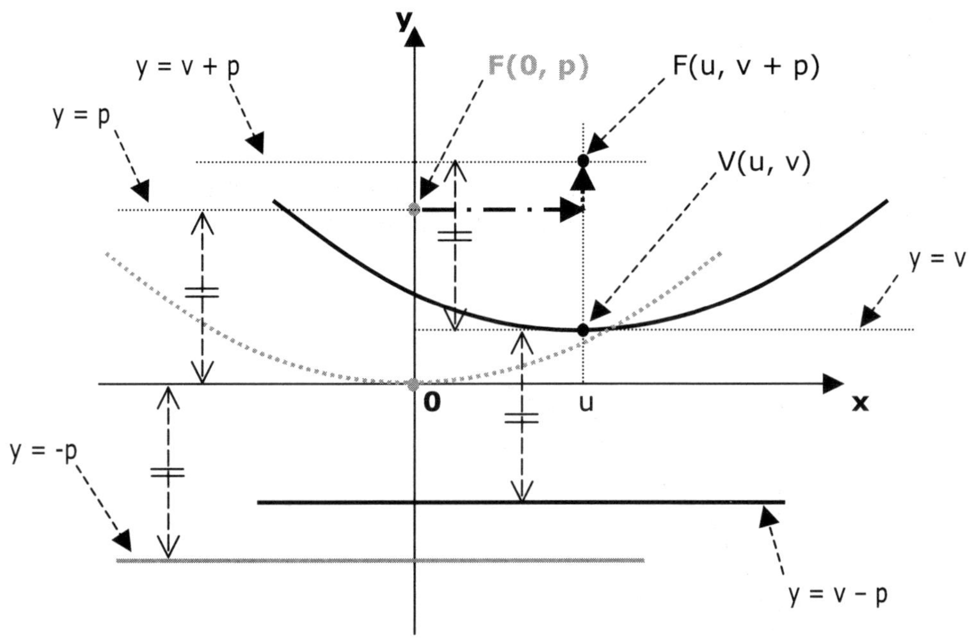

그림. 1.0.R $y^2 = 4px$와 $(y - v)^2 = 4p(x - u)$

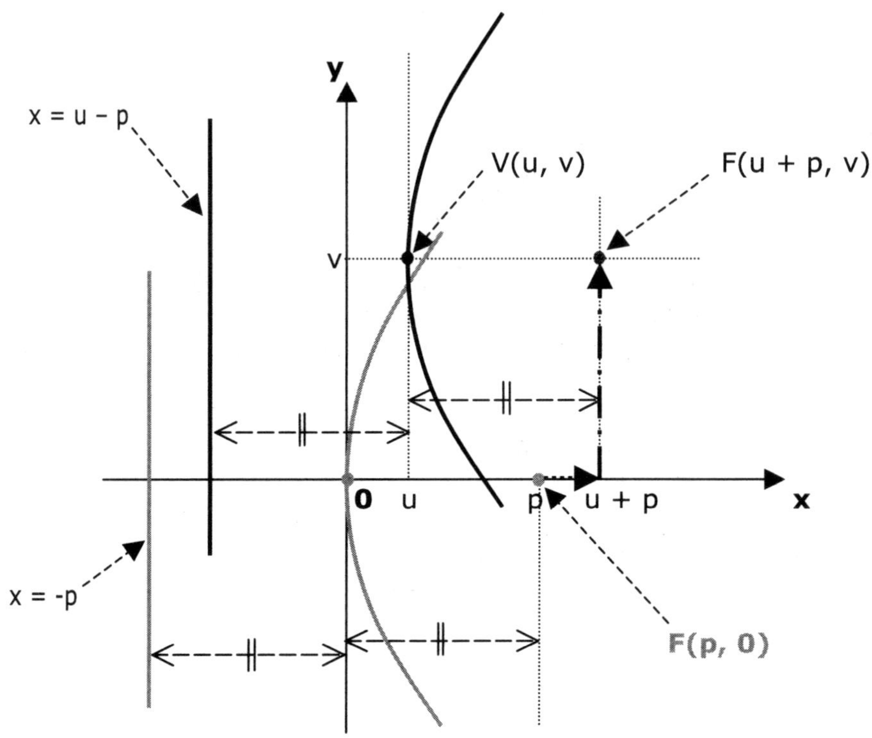

포물선은 점들의 집합이죠. 이제, 하나의 포물선이 x-y 좌표계에 있다고 하세요. 그러면, 그 점들 각각은 한 쌍의 좌표들을 갖는 것이고 그 포물선을 표준형 방정식을 사용해서 다음과 같이 점들의 집합으로 나타낼 수 있죠:

A = {(x, y)| $y^2 = 4px$ 여기에서 p는 상수.}
B = {(x, y)| $x^2 = 4py$ 여기에서 p는 상수.}
C = {(x, y)| $(y - v)^2 = 4p(x - u)$ 여기에서 p, u, v는 상수.}
D = {(x, y)| $(x - u)^2 = 4p(y - v)$ 여기에서 p, u, v는 상수.}

포물선 A는 x-y좌표계에서의 점들의 집합인데 각 점의 좌표들 간의 관계는 $y^2 = 4px$ 여기에서 p는 상수이다 이죠. 포물선 A의 방정식은 $A(x, y) = y^2 - 4px = 0$ 또는 그저, $y^2 = 4px$라고 쓸 수 있어요.
이는 위의 나머지 다른 포물선들에 대하여도 마찬가지죠.

포물선의 몇 가지 특성은 다음과 같습니다:

그림. 1.0.S

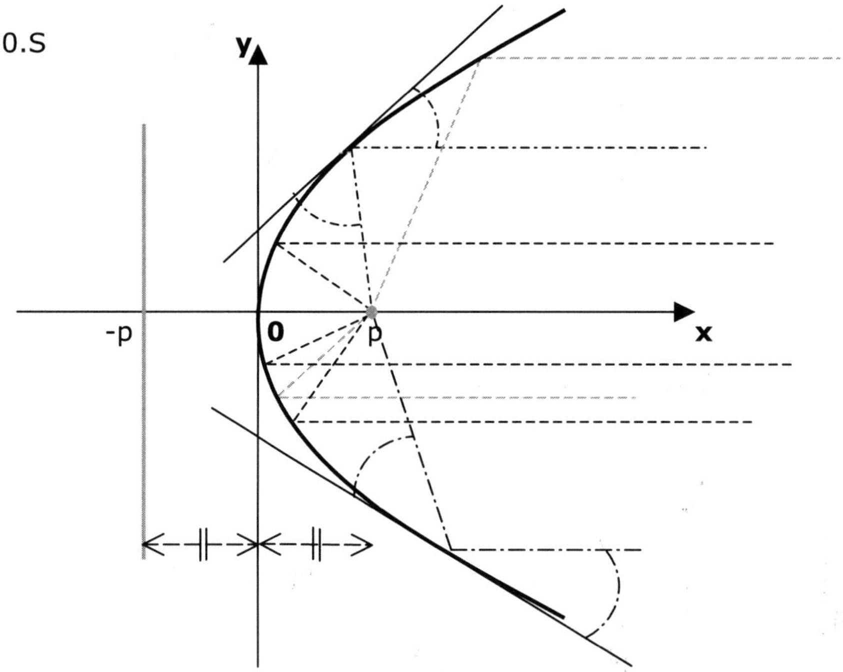

포물선으로 들어오는 준선에 수직인 방사선들은 초점에 모이고 초점에서 나가는 방사선들도 준선에 수직입니다. 이는 방사선이 포물선에 반사될 때의 입사각과 반사각이 같기 때문이죠. 방사선은 빛이나 소리와 같은 파동이 될 수 있죠. 이제, 포물선의 입구에서 포물선까지의 방사선의 길이를 A 라 하고 포물선에서 초점까지의 방사선의 길이를 B 라 하세요. 그러면, A + B 는 일정합니다.

그림. 1.0.T

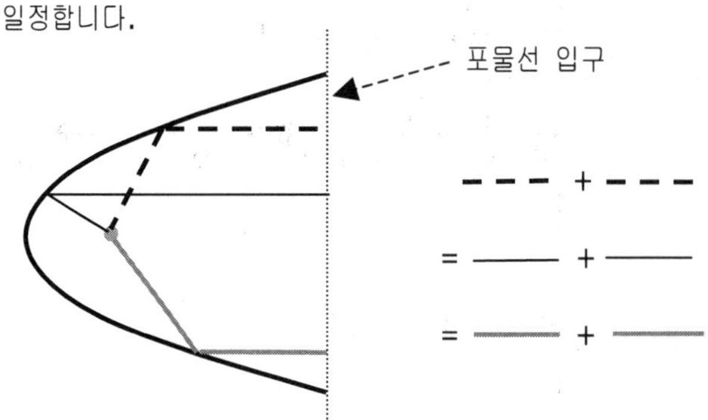

아래와 같은 케플러의 포물선 작성법은 위의 성질을 이용하는 겁니다.
먼저, 아래에 나타낸 바와 같이 초점의 위치를 정하고 포물선의 준선 역할을 하는 수평막대 위에 수직막대를 놓습니다. 또한, 수직막대의 길이와 동일한 길이의 줄이 필요합니다. 다음으로, 핀(pin)으로 줄의 한쪽 끝을 초점에 고정시키고 다른 한쪽은 수직막대 끝에다가 고정시키세요. 줄은 회색입니다. 연필을 줄 위에 놓고 막대에 줄이 팽팽하게 밀착시키세요. 그리고 나서, 수직막대를 수평막대를 따라 미끄러지게 이동시키는 겁니다. 그러면, 포물선이 만들어지는 거죠.

그림. 1.0.U 케플러(Kepler)의 포물선 작성법

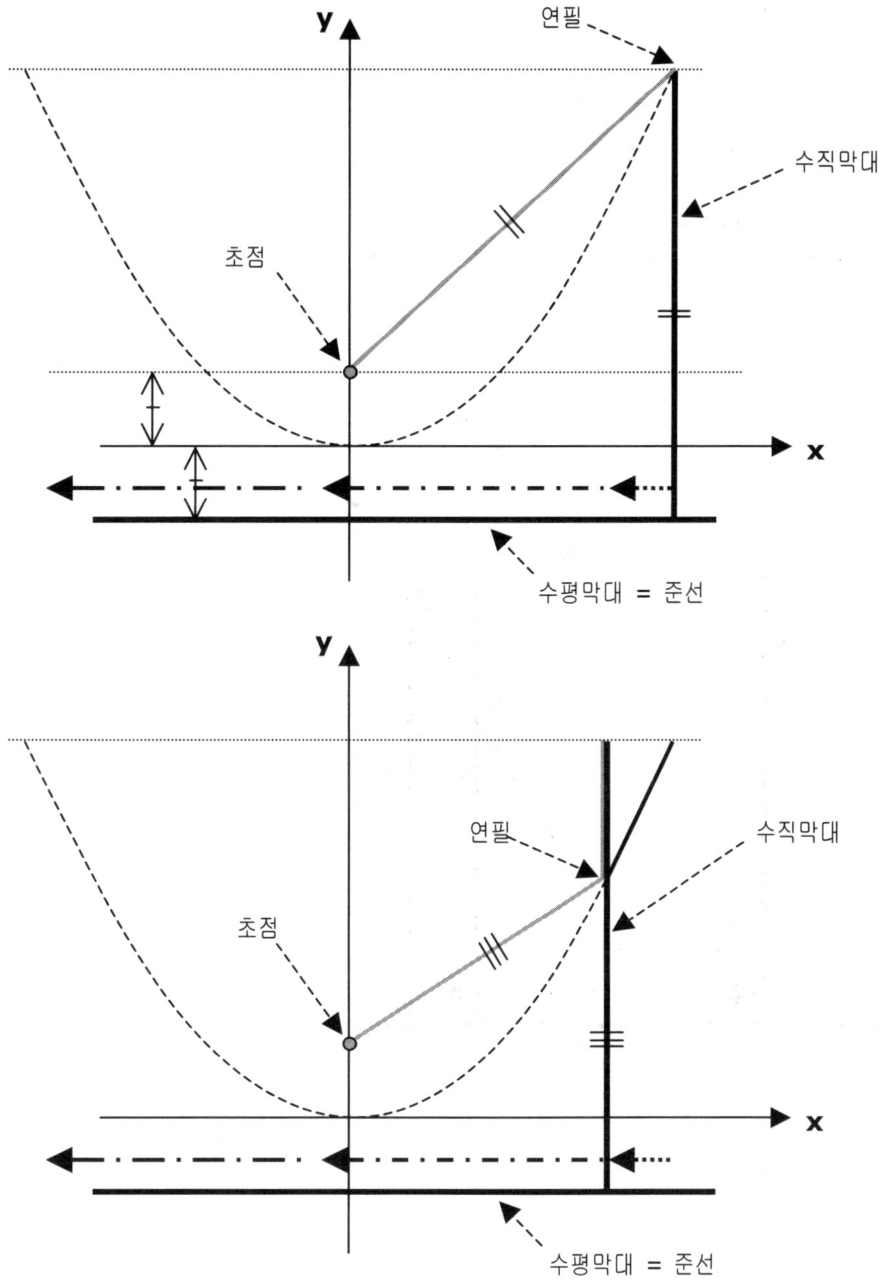

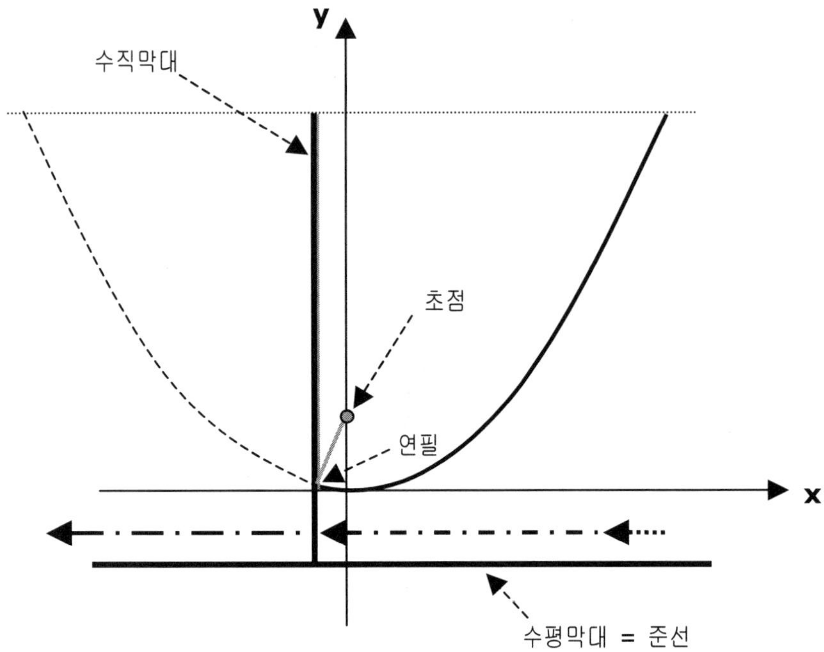

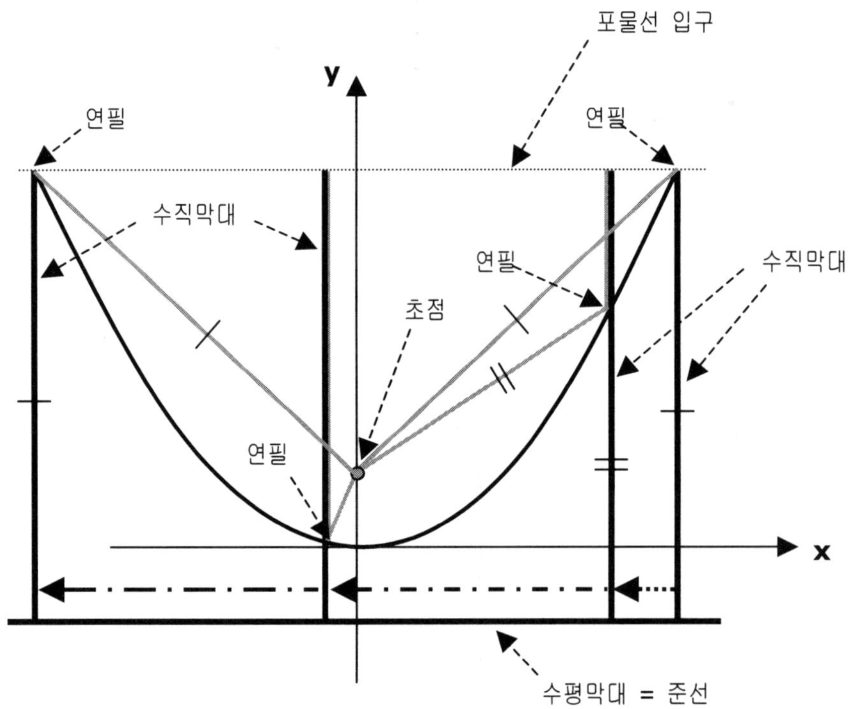

그럼 이제, 예제를 좀 해보기로 하죠.

예제

포물선의 표준방정식

V를 꼭지점, F는 초점 그리고 D는 준선이라 하고 다음과 같은 포물선의 방정식을 구하세요:

0. V(3, 0), F(0, 0)

1. V(0, -1), F(3, -1)

2. V(1, 0), F(3, 0)

3. V(1, -1), F(3, -1)

4. V(0, 0), F(0, 3)

5. V(-1, 0), F(-1, 3)

6. V(0, 1), F(0, 3)

7. V(-1, 1), F(-1, 3)

8. V(3, 1), D: x = -2

제안과 풀이
0 번 예제에서의 문제

꼭지점은 원점이고 초점이 (3, 0).

포물선에서의 각 점은 초점과 준선으로부터 동일한 거리만큼 떨어져있다는 것을 알고있죠.
또한, 꼭지점은 준선과 초점 사이의 중점이고 꼭지점과 초점은 대칭축에 있는 거죠.
꼭지점은 (0, 0)이고 초점이 (3, 0)이니까 y 좌표들이 0 으로 0 동일하니까 대칭축은 직선 $y = 0$ 즉, x 축이네요.
따라서, 준선은 대칭축에 수직이므로 x 축에 수직 또는 y 축에 평행이어야 하죠.
그러면 이제, 준선을 직선 $x = d$ 라고 하세요. 그러면, 그 포물선의 그래프는 아래의 것과 비슷하겠죠.

그림. 0.0

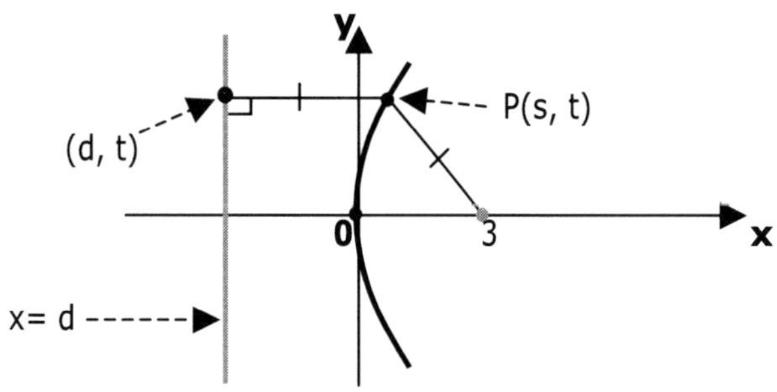

그러면, 꼭지점은 준선과 초점 사이의 중점이니까:
$0 = (d + 3)/2 \Rightarrow d = -3$.
그러므로, 준선은 직선 $x = -3$ 이죠.

이제, 포물선의 일반점을 P(s, t)라 하세요. 그러면:
(점 P와 준선 사이의 거리)$^2 = \{s - (-3)\}^2 + (t - t)^2 = (s + 3)^2$.
(점 P와 초점 사이의 거리)$^2 = (s - 3)^2 + (t - 0)^2 = (s - 3)^2 + t^2$.

두 거리들이 같으므로 $(s + 3)^2 = (s - 3)^2 + t^2$
$\Rightarrow t^2 = (s + 3)^2 - (s - 3)^2 = (s + 3 + s - 3)(s + 3 - s + 3) = 2s(6) = 12s$
$\Rightarrow t^2 = 12s$.

그런데, P(s, t)는 x-y좌표계에 있으므로 그 포물선의 방정식은 $y^2 = 12x$입니다.

꼭지점이 원점이고 초점이 (3, 0)이므로 표준형 방정식 $y^2 = 4px$에서 $p = 3$ 이죠. 이를 외우고 있다면 즉각 답을 구할 수 있겠네요. 외우기도 해야 하겠지만 항상 외우고 있는 것도 쉬운 일은 아니죠.

요약:
준선을 직선 $x = d$ 라고 하세요.
그러면, $0 = (d + 3)/2 \Rightarrow d = -3$.
포물선의 일반점을 P(s, t)라 하세요. 그러면:
$\{s - (-3)\}^2 + (t - t)^2 = (s + 3)^2 = (s - 3)^2 + (t - 0)^2 = (s - 3)^2 + t^2 \Rightarrow (s + 3)^2 = (s - 3)^2 + t^2$
$\Rightarrow t^2 = (s + 3)^2 - (s - 3)^2 = (s + 3 + s - 3)(s + 3 - s + 3) = 2s(6) = 12s \Rightarrow t^2 = 12s$.
그런데, P(s, t)는 x-y좌표계에 있으므로 그 포물선의 방정식은 $y^2 = 12x$.

제안과 풀이
1 번 예제에서의 문제

꼭지점은 **(0, -1)**이고 초점이 **(3, -1).**

포물선이란 점들의 집합인데 각 점은 초점과 준선으로부터 동일한 거리만큼 떨어져있다는 것을 알고있죠.
또한, 꼭지점은 준선과 초점 사이의 중점이고 꼭지점과 초점은 대칭축에 있는 거죠.
꼭지점은 (0, -1)이고 초점이 (3, -1)이니까 y 좌표들이 −1 로 동일하니까 대칭축은 직선 y = -1 이네요.
따라서, 준선은 대칭축에 수직이므로 x 축에 수직 또는 y 축에 평행이어야 하죠.
그러면 이제, 준선을 직선 x = d 라고 하세요. 그러면, 그 포물선의 그래프는 아래의 것과 비슷하겠죠.

그림. 1.0

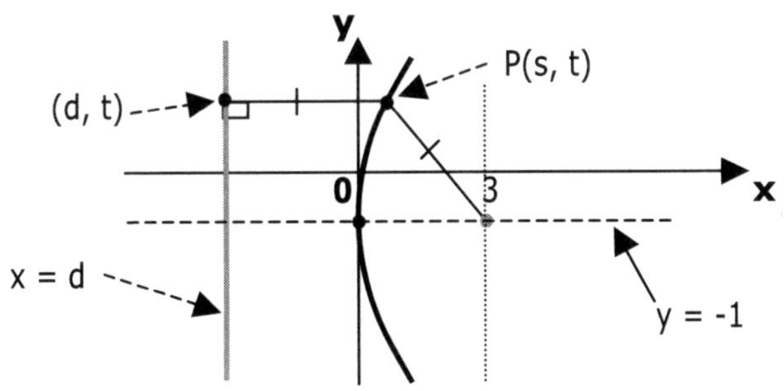

그러면, 꼭지점은 준선과 초점 사이의 중점이니까:
$0 = (d + 3)/2 \Rightarrow d = -3$.
그러므로, 준선은 x = -3 이죠.

이제, 포물선의 일반점을 P(s, t)라 하세요. 그러면:
(점 P와 준선 사이의 거리)$^2 = \{s - (-3)\}^2 + (t - t)^2 = (s + 3)^2$.
(점 P와 초점 사이의 거리)$^2 = (s - 3)^2 + \{t - (-1)\}^2 = (s - 3)^2 + (t + 1)^2$.

두 거리들이 같으므로 $(s + 3)^2 = (s - 3)^2 + (t + 1)^2$
$\Rightarrow (t + 1)^2 = (s + 3)^2 - (s - 3)^2 = (s + 3 + s - 3)(s + 3 - s + 3) = 2s(6) = 12s$.

그런데, P(s, t)는 x-y좌표계에 있으므로 그 포물선의 방정식은 $(y + 1)^2 = 12x$이죠.

요약:

준선을 직선 x = d 라고 하세요.
$0 = (d + 3)/2 \Rightarrow d = -3$.
포물선의 일반점을 P(s, t)라 하세요. 그러면:
$\{s - (-3)\}^2 + (t - t)^2 = (s + 3)^2 = (s - 3)^2 + \{t - (-1)\}^2 = (s - 3)^2 + (t + 1)^2$
$\Rightarrow (t + 1)^2 = (s + 3)^2 - (s - 3)^2 = (s + 3 + s - 3)(s + 3 - s + 3) = 2s(6) = 12s$
$\Rightarrow (t + 1)^2 = 12s$.

그런데, P(s, t)는 x-y좌표계에 있으므로 그 포물선의 방정식은 $(y + 1)^2 = 12x$.

제안과 풀이
2 번 예제에서의 문제

꼭지점은 **(1, 0)**이고 초점이 **(3, 0).**

포물선이란 점들의 집합인데 각 점은 초점과 준선으로부터 동일한 거리만큼 떨어져있죠.
또한, 꼭지점은 준선과 초점 사이의 중점이고 꼭지점과 초점은 대칭축에 있는 거죠.
꼭지점은 (1, 0)이고 초점이 (3, 0)이니까 y 좌표들이 일치하고 0 이니까 대칭축은 직선 $y = 0$ 즉, x 축이네요.
따라서, 준선은 대칭축에 수직이므로 x 축에 수직 또는 y 축에 평행이어야 하죠.
그러면 이제, 준선을 직선 $x = d$ 라고 하세요. 그러면, 그 포물선의 그래프는 아래의 것과 비슷하겠죠.

그림. 2.0

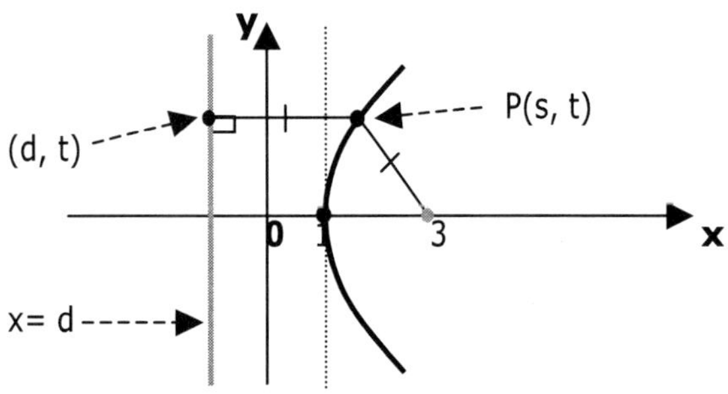

그러면, 꼭지점은 준선과 초점 사이의 중점이니까:
$1 = (d + 3)/2 \Rightarrow d = -1.$
그러므로, 준선은 직선 $x = -1$ 이죠.

이제, 포물선의 일반점을 P(s, t)라 하세요. 그러면:
(점 P와 준선 사이의 거리)$^2 = \{s - (-1)\}^2 + (t - t)^2 = (s + 1)^2$.
(점 P와 초점 사이의 거리)$^2 = (s - 3)^2 + (t - 0)^2 = (s - 3)^2 + t^2$.

두 거리들이 같으므로 $(s + 1)^2 = (s - 3)^2 + t^2$
$\Rightarrow t^2 = (s + 1)^2 - (s - 3)^2 = (s + 1 + s - 3)(s + 1 - s + 3) = 2(s - 1)(4) = 8(s - 1).$

그런데, P(s, t)는 x-y좌표계에 있으므로 그 포물선의 방정식은 $y^2 = 8(x - 1)$.

요약:

준선을 직선 $x = d$ 라고 하세요.
$1 = (d + 3)/2 \Rightarrow d = -1.$
포물선의 일반점을 P(s, t)라 하세요. 그러면:
$\{s - (-1)\}^2 + (t - t)^2 = (s + 1)^2 = (s - 3)^2 + (t - 0)^2 = (s - 3)^2 + t^2$
$\Rightarrow t^2 = (s + 1)^2 - (s - 3)^2 = (s + 1 + s - 3)(s + 1 - s + 3) = 2(s - 1)(4) = 8(s - 1)$
$\Rightarrow t^2 = 8(s - 1).$

그런데, P(s, t)는 x-y좌표계에 있으므로 그 포물선의 방정식은 $y^2 = 8(x - 1)$.

제안과 풀이

3번 예제에서의 문제

꼭지점은 **(1, -1)**이고 초점이 **(3, -1).**

포물선이란 점들의 집합인데 각 점은 초점과 준선으로부터 동일한 거리만큼 떨어져있죠.
또한, 꼭지점은 준선과 초점 사이의 중점이고 꼭지점과 초점은 대칭축에 있는 거죠.
꼭지점은 (1, -1)이고 초점이 (3, -1)이니까 y 좌표들이 일치하고 -1 이니까 대칭축은 직선 y = -1 이네요.
따라서, 준선은 대칭축에 수직이므로 x 축에 수직 또는 y 축에 평행이어야 하죠.
그러면 이제, 준선을 직선 x = d 라고 하세요. 그러면, 그 포물선의 그래프는 아래의 것과 비슷하겠죠.

그림. 3.0

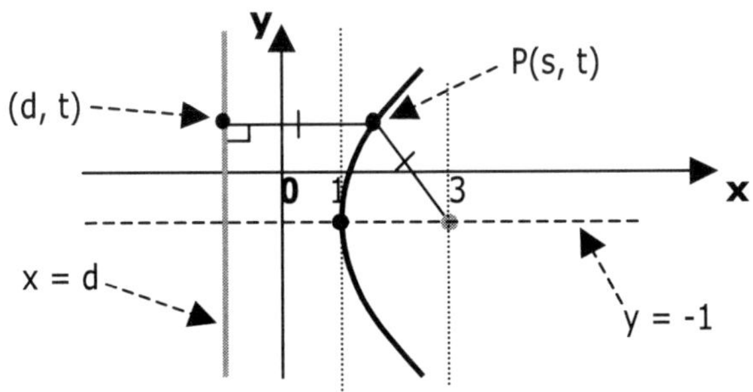

그러면, 꼭지점은 준선과 초점 사이의 중점이니까:
$1 = (d + 3)/2 \Rightarrow d = -1$.
그러므로, 준선은 x = -1 이죠.

이제, 포물선의 일반점을 P(s, t)라 하세요. 그러면:
(점 P와 준선 사이의 거리)$^2 = \{s - (-1)\}^2 + (t - t)^2 = (s + 1)^2$.
(점 P와 초점 사이의 거리)$^2 = (s - 3)^2 + \{t - (-1)\}^2 = (s - 3)^2 + (t + 1)^2$.

두 거리들이 같으므로 $(s + 1)^2 = (s - 3)^2 + (t + 1)^2$
$\Rightarrow (t + 1)^2 = (s + 1)^2 - (s - 3)^2 = (s + 1 + s - 3)(s + 1 - s + 3) = 2(s - 1)(4) = 8(s - 1)$
$\Rightarrow (t + 1)^2 = 8(s - 1)$.

그런데, P(s, t)는 x-y좌표계에 있으므로 그 포물선의 방정식은 $(y + 1)^2 = 8(x - 1)$.

요약:

준선을 직선 x = d 라고 하세요.
그러면, $1 = (d + 3)/2 \Rightarrow d = -1$.
포물선의 일반점을 P(s, t)라 하세요. 그러면:
$\{s - (-1)\}^2 + (t - t)^2 = (s + 1)^2 = (s - 3)^2 + \{t - (-1)\}^2 = (s - 3)^2 + (t + 1)^2$
$\Rightarrow (t + 1)^2 = (s + 1)^2 - (s - 3)^2 = (s + 1 + s - 3)(s + 1 - s + 3) = 2(s - 1)(4) = 8(s - 1)$
$\Rightarrow (t + 1)^2 = 8(s - 1)$.
그런데, P(s, t)는 x-y좌표계에 있으므로 그 포물선의 방정식은 $(y + 1)^2 = 8(x - 1)$.

제안과 풀이
4 번 예제에서의 문제

꼭지점은 원점이고 초점이 **(0, 3).**

포물선에서의 각 점은 초점과 준선으로부터 동일한 거리만큼 떨어져있죠.
또한, 꼭지점은 준선과 초점 사이의 중점이고 꼭지점과 초점은 대칭축에 있는 거죠.
꼭지점은 (0, 0)이고 초점이 (0, 3)이니까 x 좌표들이 0 으로 동일하니까 대칭축은 x = 0 즉, y 축이네요.
따라서, 준선은 대칭축에 수직이므로 y 축에 수직 또는 x 축에 평행이어야 하죠.
그러면 이제, 준선을 직선 y = d 라고 하세요. 그러면, 그 포물선의 그래프는 아래의 것과 비슷하겠죠.

그림. 4.0

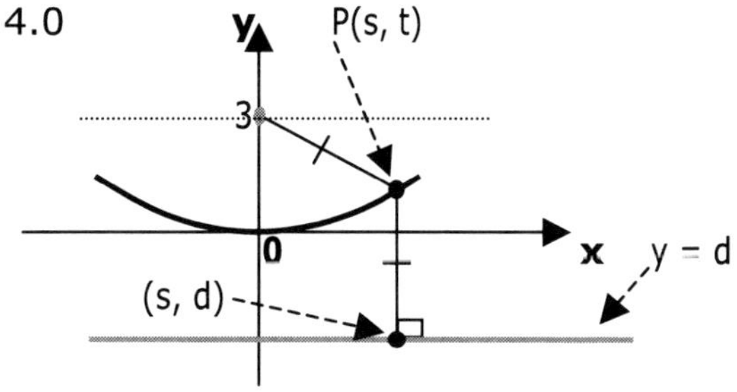

그러면, 꼭지점은 준선과 초점 사이의 중점이니까:
$0 = (d + 3)/2 \Rightarrow d = -3.$
그러므로, 준선은 직선 $y = -3$ 이죠.

이제, 포물선의 일반점을 P(s, t)라 하세요. 그러면:
(점 P와 준선 사이의 거리)$^2 = (s - s)^2 + \{t - (-3)\}^2 = (t + 3)^2$.
(점 P와 초점 사이의 거리)$^2 = (s - 0)^2 + (t - 3)^2 = s^2 + (t - 3)^2$.

두 거리들이 같으므로 $(t + 3)^2 = (t - 3)^2 + s^2$
$\Rightarrow s^2 = (t + 3)^2 - (t - 3)^2 = (t + 3 + t - 3)(t + 3 - t + 3) = 2t(6) = 12t$
$\Rightarrow s^2 = 12t.$

그런데, P(s, t)는 x-y좌표계에 있으므로 그 포물선의 방정식은 $x^2 = 12y$.

꼭지점이 원점이고 초점이 (3, 0)이므로 표준형 방정식 $x^2 = 4py$에서 p = 3 이죠. 이를 외우고 있다면 즉각 답을 구할 수 있겠네요. 외우기도 해야 하겠지만 항상 외우고 있는 것도 쉬운 일은 아니죠.

요약:
준선을 직선 y = d 라고 하세요.
그러면, $0 = (d + 3)/2 \Rightarrow d = -3$.
포물선의 일반점을 P(s, t)라 하세요. 그러면:
$(s - s)^2 + \{t - (-3)\}^2 = (t + 3)^2 = (s - 0)^2 + (t - 3)^2 = s^2 + (t - 3)^2 \Rightarrow (t + 3)^2 = (t - 3)^2 + s^2$
$\Rightarrow s^2 = (t + 3)^2 - (t - 3)^2 = (t + 3 + t - 3)(t + 3 - t + 3) = 2t(6) = 12t$
$\Rightarrow s^2 = 12t.$

그런데, P(s, t)는 x-y좌표계에 있으므로 그 포물선의 방정식은 $x^2 = 12y$.

제안과 풀이
5 번 예제에서의 문제

꼭지점은 (-1, 0)이고 초점이 (-1, 3).

포물선에서의 각 점은 초점과 준선으로부터 동일한 거리만큼 떨어져있죠.
또한, 꼭지점은 준선과 초점 사이의 중점이고 꼭지점과 초점은 대칭축에 있는 거죠.
꼭지점은 (-1, 0)이고 초점이 (-1, 3)이니까 x 좌표들이 -1로 동일하니까 대칭축은 $x = -1$ 이네요.
따라서, 준선은 대칭축에 수직이므로 y 축에 수직 또는 x 축에 평행이어야 하죠.
그러면 이제, 준선을 $y = d$ 라고 하세요. 그러면, 그 포물선의 그래프는 아래의 것과 비슷하겠죠.

그림. 5.0

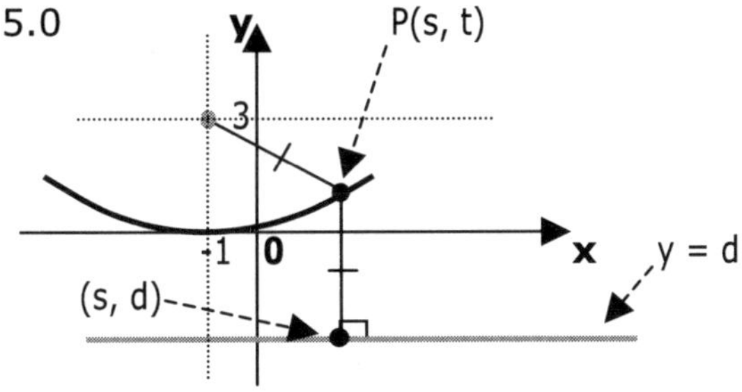

그러면, 꼭지점은 준선과 초점 사이의 중점이니까:
$0 = (d + 3)/2 \Rightarrow d = -3.$
그러므로, 준선은 $y = -3$ 이죠.

이제, 포물선의 일반점을 P(s, t)라 하세요. 그러면:
$(\text{점 P와 준선 사이의 거리})^2 = (s - s)^2 + \{t - (-3)\}^2 = (t + 3)^2.$
$(\text{점 P와 초점 사이의 거리})^2 = \{s - (-1)\}^2 + (t - 3)^2 = (s + 1)^2 + (t - 3)^2.$

두 거리들이 같으므로 $(t + 3)^2 = (t - 3)^2 + (s + 1)^2$
$\Rightarrow (s + 1)^2 = (t + 3)^2 - (t - 3)^2 = (t + 3 + t - 3)(t + 3 - t + 3) = 2t(6) = 12t$
$\Rightarrow (s + 1)^2 = 12t.$

그런데, P(s, t)는 x-y좌표계에 있으므로 그 포물선의 방정식은 $(x + 1)^2 = 12y.$

요약:

준선을 $y = d$ 라고 하세요.
그러면, $0 = (d + 3)/2 \Rightarrow d = -3.$
포물선의 일반점을 P(s, t)라 하세요. 그러면:
$(s - s)^2 + \{t - (-3)\}^2 = (t + 3)^2 = \{s - (-1)\}^2 + (t - 3)^2 = (s + 1)^2 + (t - 3)^2$
$\Rightarrow (t + 3)^2 = (t - 3)^2 + (s + 1)^2$
$\Rightarrow (s + 1)^2 = (t + 3)^2 - (t - 3)^2 = (t + 3 + t - 3)(t + 3 - t + 3) = 2t(6) = 12t$
$\Rightarrow (s + 1)^2 = 12t.$

그런데, P(s, t)는 x-y좌표계에 있으므로 그 포물선의 방정식은 $(x + 1)^2 = 12y.$

제안과 풀이
6번 예제에서의 문제

꼭지점은 **(0, 1)**이고 초점이 **(0, 3).**

포물선에서의 각 점은 초점과 준선으로부터 동일한 거리만큼 떨어져있죠.
또한, 꼭지점은 준선과 초점 사이의 중점이고 꼭지점과 초점은 대칭축에 있는 거죠.
꼭지점은 (0, 1)이고 초점이 (0, 3)이니까 x 좌표들이 0 으로 동일하니까 대칭축은 $x = 0$ 즉, y 축이네요.
따라서, 준선은 대칭축에 수직이므로 y 축에 수직 또는 x 축에 평행이어야 하죠.
그러면 이제, 준선을 $y = d$ 라고 하세요. 그러면, 그 포물선의 그래프는 아래의 것과 비슷하겠죠.

그림. 6.0

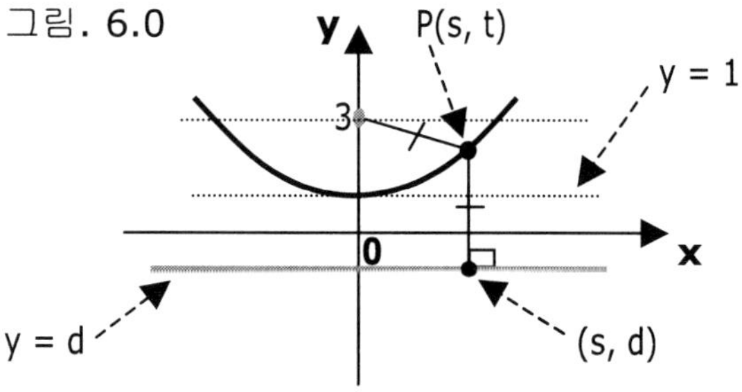

그러면, 꼭지점은 준선과 초점 사이의 중점이니까:
$1 = (d + 3)/2 \Rightarrow d = -1$.
그러므로, 준선은 $y = -1$ 이죠.

이제, 포물선의 일반점을 P(s, t)라 하세요. 그러면:
(점 P와 준선 사이의 거리)$^2 = (s - s)^2 + \{t - (-1)\}^2 = (t + 1)^2$.
(점 P와 초점 사이의 거리)$^2 = (s - 0)^2 + (t - 3)^2 = s^2 + (t - 3)^2$.

두 거리들이 같으므로 $(t + 1)^2 = (t - 3)^2 + s^2$
$\Rightarrow s^2 = (t + 1)^2 - (t - 3)^2 = (t + 1 + t - 3)(t + 1 - t + 3) = 2(t - 1)(4) = 8(t - 1)$
$\Rightarrow s^2 = 8(t - 1)$.

그런데, P(s, t)는 x-y좌표계에 있으므로 그 포물선의 방정식은 $x^2 = 8(y - 1)$.

요약:

준선을 $y = d$ 라고 하세요.
그러면, $1 = (d + 3)/2 \Rightarrow d = -1$.
포물선의 일반점을 P(s, t)라 하세요. 그러면:
$(s - s)^2 + \{t - (-1)\}^2 = (t + 1)^2 = (s - 0)^2 + (t - 3)^2 = s^2 + (t - 3)^2 \Rightarrow (t + 1)^2 = (t - 3)^2 + s^2$
$\Rightarrow s^2 = (t + 1)^2 - (t - 3)^2 = (t + 1 + t - 3)(t + 1 - t + 3) = 2(t - 1)(4) = 8(t - 1)$
$\Rightarrow s^2 = 8(t - 1)$.

그런데, P(s, t)는 x-y좌표계에 있으므로 그 포물선의 방정식은 $x^2 = 8(y - 1)$.

제안과 풀이
7 번 예제에서의 문제

꼭지점은 (-1, 1)이고 초점이 (-1, 3).

포물선에서의 각 점은 초점과 준선으로부터 동일한 거리만큼 떨어져있죠.
또한, 꼭지점은 준선과 초점 사이의 중점이고 꼭지점과 초점은 대칭축에 있는 거죠.
꼭지점은 (-1, 1)이고 초점이 (-1, 3)이니까 x 좌표들이 일치하고 -1 이니까 대칭축은 $x = -1$ 이네요.
따라서, 준선은 대칭축에 수직이므로 y 축에 수직 또는 x 축에 평행이어야 하죠.
그러면 이제, 준선을 $y = d$ 라고 하세요. 그러면, 그 포물선의 그래프는 아래의 것과 비슷하겠죠.

그림. 7.0

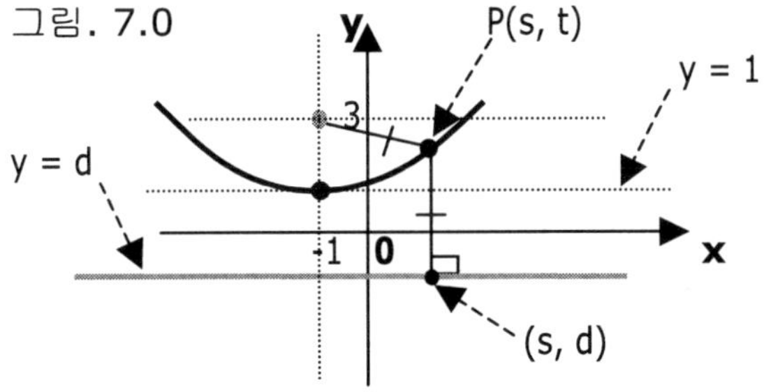

그러면, 꼭지점은 준선과 초점 사이의 중점이니까:
$1 = (d + 3)/2 \Rightarrow d = -1$.
그러므로, 준선은 $y = -1$ 이죠.

이제, 포물선의 일반점을 P(s, t)라 하세요. 그러면:
(점 P와 준선 사이의 거리)$^2 = (s - s)^2 + \{t - (-1)\}^2 = (t + 1)^2$.
(점 P와 초점 사이의 거리)$^2 = \{s - (-1)\}^2 + (t - 3)^2 = (s + 1)^2 + (t - 3)^2$.

두 거리들이 같으므로 $(t + 1)^2 = (t - 3)^2 + (s + 1)^2$
$\Rightarrow (s + 1)^2 = (t + 1)^2 - (t - 3)^2 = (t + 1 + t - 3)(t + 1 - t + 3) = 2(t - 1)(4) = 8(t - 1)$
$\Rightarrow (s + 1)^2 = 8(t - 1)$.

그런데, P(s, t)는 x-y좌표계에 있으므로 그 포물선의 방정식은 $(x + 1)^2 = 8(y - 1)$.

요약:

준선을 $y = d$ 라고 하세요.
그러면, $1 = (d + 3)/2 \Rightarrow d = -1$.
포물선의 일반점을 P(s, t)라 하세요. 그러면:
$(s - s)^2 + \{t - (-1)\}^2 = (t + 1)^2 = \{s - (-1)\}^2 + (t - 3)^2 = (s + 1)^2 + (t - 3)^2$
$\Rightarrow (t + 1)^2 = (t - 3)^2 + (s + 1)^2$
$\Rightarrow (s + 1)^2 = (t + 1)^2 - (t - 3)^2 = (t + 1 + t - 3)(t + 1 - t + 3) = 2(t - 1)(4) = 8(t - 1)$
$\Rightarrow (s + 1)^2 = 8(t - 1)$.

그런데, P(s, t)는 x-y좌표계에 있으므로 그 포물선의 방정식은 $(x + 1)^2 = 8(y - 1)$.

제안과 풀이
8번 예제에서의 문제

꼭지점은 **(3, 1)**이고 준선이 **x = -2.**

포물선이란 점들의 집합인데 각 점은 초점과 준선으로부터 동일한 거리만큼 떨어져있죠.
또한, 꼭지점은 준선과 초점 사이의 중점이고 꼭지점과 초점은 대칭축에 있는 거죠.
준선이 $x = -2$이니까 대칭축은 x축에 평행이죠.
또한, 대칭축은 꼭지점 (3, 1)을 지나므로 직선 $y = 1$이네요.
꼭지점과 초점은 대칭축에 있으니까 초점의 y좌표는 1이죠.
그러면 이제, 초점을 (f, 1)이라고 하세요. 그러면, 그 포물선의 그래프는 아래의 것과 비슷하겠죠.

그림. 8.0

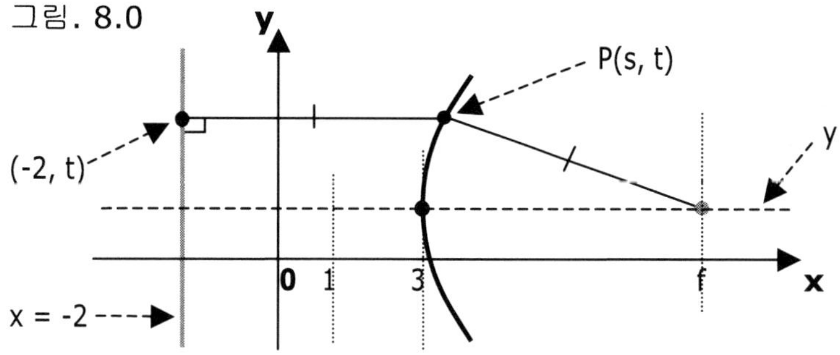

그러면, 꼭지점은 준선과 초점 사이의 중점이니까:
$3 = (-2 + f)/2 \Rightarrow f = 8.$
그러므로, 초점은 (8, 1)이죠.

이제, 포물선의 일반점을 P(s, t)라 하세요. 그러면:
(점 P와 준선 사이의 거리)$^2 = \{s - (-2)\}^2 + (t - t)^2 = (s + 2)^2$.
(점 P와 초점 사이의 거리)$^2 = (s - 8)^2 + (t - 1)^2 = (s - 8)^2 + (t - 1)^2$.

두 거리들이 같으므로 $(s + 2)^2 = (s - 8)^2 + (t - 1)^2$
$\Rightarrow (t - 1)^2 = (s + 2)^2 - (s - 8)^2 = (s + 2 + s - 8)(s + 2 - s + 8) = 2(s - 3)(10) = 20(s - 3)$
$\Rightarrow (t - 1)^2 = 20(s - 3)$.

그런데, P(s, t)는 x-y좌표계에 있으므로 그 포물선의 방정식은 $(y - 1)^2 = 20(x - 3)$.

요약:
초점을 (f, 1)이라고 하세요.
그러면, $3 = (-2 + f)/2 \Rightarrow f = 8$.
포물선의 일반점을 P(s, t)라 하세요. 그러면:
$\{s - (-2)\}^2 + (t - t)^2 = (s + 2)^2 = (s - 8)^2 + (t - 1)^2 = (s - 8)^2 + (t - 1)^2$
$\Rightarrow (t - 1)^2 = (s + 2)^2 - (s - 8)^2 = (s + 2 + s - 8)(s + 2 - s + 8) = 2(s - 3)(10) = 20(s - 3)$
$\Rightarrow (t - 1)^2 = 20(s - 3)$.
그런데, P(s, t)는 x-y좌표계에 있으므로 그 포물선의 방정식은 $(y - 1)^2 = 20(x - 3)$.

1.1. 포물선의 일반방정식

포물선의 일반방정식이란 일반형으로 된 포물선의 방정식인데 이는 모든 포물선을 나타낼 수 있는 방정식입니다. 그러니까, 그런 방정식은 그 어떤 포물선도 나타낼 수 있는 거죠. 그런데, $y^2 = 4px$와 같은 표준형 방정식으로는 직각 포물선을 표현합니다. 직각 포물선에서는 대칭축이 그 좌표계의 어느 한 좌표축과 나란하죠. 예를 들면, $(y - 1)^2 = 4x$에서는 대칭축이 $y = 1$ 이고 이는 x축과 나란한 거죠. 표준방정식은 꼭지점과 초점 그리고 준선과 그 근들을 즉각 알 수 있는 장점이 있어요. 물론, 포물선의 일반방정식으로도 직각 포물선을 나타낼 수 있죠. 또한, 일반형을 표준형으로 바꿀 수도 있는데 이를 완전 제곱식 변형 법이라고 하죠. 그러면 이제, 이 일반방정식은 직각 포물선 뿐만 아니라 대칭축이 좌표축에 비스듬한 포물선도 나타낼 수 있는 거니까 이러한 방정식으로 그 어떤 포물선도 나타낼 수 있는 겁니다.

포물선은 점들의 집합이죠. 그런데, 무슨 점들인가요? 포물선이 x-y 좌표계에 있다고 하세요. 그러면, 그 점들 각각은 한 쌍의 좌표들을 갖고있는데 포물선을 일반적으로 나타낼 때 다음과 같이 나타낼 수 있어요:
$P = \{(x, y)|\ ax^2 + by^2 + cxy + dx + ey + f = 0$ 여기에서 a, b, c, d, e, f는 상수이고 $ab > 0$; 아니면, c가 0 이고 또한, a와 b 중의 하나가 0 이다.$\}$

유의사항: "$ab > 0$" 는 a 와 b 의 부호들이 같다는 것도 의미하죠.

위의 집합 P 는 모든 포물선의 포괄적 표현이죠. 즉, 그 어떠한 포물선이든 다 해당되는 거죠. 그러니까, 포물선의 일반방정식은 다음과 같다고 할 수 있죠:
$P(x, y) = ax^2 + by^2 + cxy + dx + ey + f = 0$ 또는 단순히, **$ax^2 + by^2 + cxy + dx + ey + f = 0$.**

그런데, 이런 방정식은 그저 모든 포물선을 포괄적으로 나타내는 것뿐이지 실제로 자주 사용하지는 않습니다.
일반 math 에서는 주로 직각 포물선을 다루죠. 그러면, 다음과 같은 점들의 집합을 보기로 하죠:
$A = \{(x, y)|\ ax^2 + by^2 + cx + dy + e = 0$ 여기에서 a, b, c, d, e는 상수이고 a 아니면 b가 0 이다.$\}$

A 는 모든 직각 포물선을 포괄적으로 나타내는 겁니다. 그런데, 이 A 는 다음과 같은 두 종류의 포물선들을 동시에 나타내는 것이라고 할 수 있어요:
$B = \{(x, y)|\ ax^2 + bx + c + dy = 0$ 여기에서 a, b, c, d는 상수이고 또한, $ad \neq 0\}$.
$C = \{(x, y)|\ ay^2 + by + c + dx = 0$ 여기에서 a, b, c, d는 상수이고 또한, $ad \neq 0\}$.

유의사항: $ad \neq 0$ 이란 a 도 d 도 0 이 아니다 라는 것도 의미합니다.

B 와 C 는 둘 다 직각 포물선들인데 B 의 대칭축은 y 축에 평행이고 C 의 대칭축은 x 축에 평행이죠.
그러므로, 일반 math 에서는 주로, 직각 포물선을 사용하는데 직각 포물선의 일반방정식들은 다음과 같이 쓸 수 있어요:
$B(x, y) = ax^2 + bx + c + dy = 0$ 또는 그저, **$ax^2 + bx + c + dy = 0$.**
$C(x, y) = ay^2 + by + c + dx = 0$ 또는 그저, **$ay^2 + by + c + dx = 0$.**

그런데 흔히, 다음과 같은 준(準) 일반형을 사용하죠: **$y = ax^2 + bx + c$**. 이런 형태는 포물선을 함수의 커브로 취급할 때 편리해요. 예를 들면, $y = f(x) = 2x^2 + 3x + 5$ 와 $y = g(x) = (x - 1)^{1/2} = \sqrt{x-1}$ 입니다. 함수 f의 커브는 방정식 $y = 2x^2 + 3x + 5$ 의 커브와 같고 함수 g의 커브는 반쪽 포물선의 방정식 $y = (x - 1)^{1/2} = \sqrt{x-1}$ 의 커브와 같아요. 이 커브는 $y^2 = x - 1$ 의 커브의 반쪽인데, 이는 아래와 같은 포물선에서 x축 위에 있는 부분과 같죠.

그림. 1.1.0

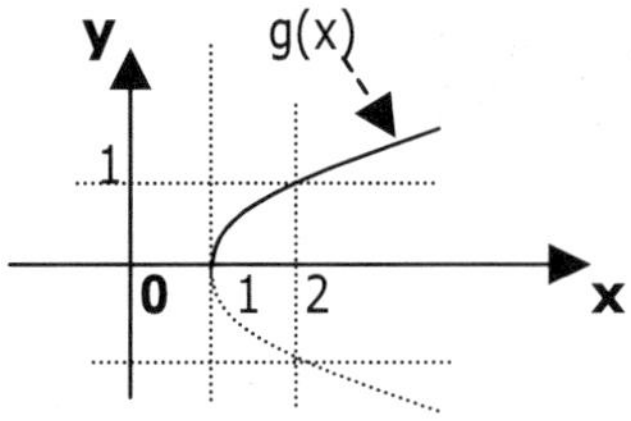

또한, 방정식 B(x, y) 또는 C(x, y)에서 d = 0 으로 놓으면 그 등식은 하나의 2 차 방정식이 되는 거죠.
즉, $ax^2 + bx + c = 0$ 또는 $ay^2 + by + c = 0$. 이와 같은 방정식을 미지수 방정식이라고 합니다.
즉, 수식 $ax^2 + bx + c$나 수식 $ay^2 + by + c$의 값을 0 으로 만드는 x나 y의 값을 찾는 것이고 이 x또는 y를 그 미지수라고 하죠.

일반 math에서는 흔히 이런 준 일반형을 사용하는데 이를 보통은 그저 포물선의 일반방정식이라고 해요. 그러면, 이러한 일반형에 대해서 알아보기로 하죠. 이는 직각 포물선을 표현하는 데 사용합니다. 직각 포물선이란 예를 들면, x-y좌표계에서 그 대칭축이 그 두개의 축들 중 어느 하나 즉, x축 아니면 y축에 평행한 포물선이죠. 그러면 이제, 대칭축이 y축에 평행한 경우부터 시작하도록 하죠. 즉, 방정식 $\mathbf{y = ax^2 + bx + c}$에 대해서 알아보려는 겁니다.

우선, 방정식 $\mathbf{y = ax^2 + bx + c}$의 좌변은 하나의 항으로, 우변은 세 개의 항들로 되어있죠. 변수가 있는 항을 변수 항이라 하고 상수만 있는 항을 상수 항이라고 해요. 개별적으로는, $\mathbf{ax^2}$을 $\mathbf{x^2}$항 그리고 **bx**를 **x**항이라고 하죠. 그런데 흔히, 변수와 함께 있는 상수를 계수라고도 해요. 그러니까, x^2의 계수는 a이고 x의 계수는 b이죠.

직선의 방정식의 경우에서와 마찬가지로 포물선의 방정식도 기울기와 절편들을 나타내줍니다. 물론, 직선의 경우에는 표준형 방정식이 그렇죠. 그런데, 포물선의 일반방정식인 $\mathbf{y = ax^2 + bx + c}$를 보면 당장 알 수 있는 것이 기울기와 y절편입니다. 이 방정식에서 x^2의 계수인 a가 기울기이고 c가 y절편이죠. 직선의 경우에서처럼, y절편 c는 포물선이 y축과 만나는 점에서의 y좌표의 값이죠. 그런데, 그 기울기 a는 직선의 경우와는 좀 달라요. 우선, 이 기울기 a는 포물선의 오목한 상태를 알려주는 거죠. 즉, 어느쪽으로 포물선이 굽어있는가를 말하죠. 예를 들면, 대칭축이 y축에 평행한 경우에 이 a가 양이면 포물선이 아래로 볼록한 것이고 음이면 위로 볼록한 것이죠. 또한, 이 a의 크기 즉, 절대값이 크면 그 포물선이 많이 굽어 있는 것이에요.

그런데, math 에서의 기울기란 커브에 있는 특정 점의 방향을 말하는 것이고 이는 그 특정 점을 포함하는 직선의 기울기를 나타내는 거죠. 즉, 예를 들면, 이는 그 직선이 x 축에 대해서 어떻게 기울어져 있는가를 말하죠. 하나의 직선에서는 그 모든 점들의 방향들이 동일하기 때문에 그 중 어느 한 점의 방향을 그저 그 직선의 기울기라고 할 수 있는 겁니다. 하지만, 포물선과 같이 휘어져있는 커브에서의 점들은 모두 동일한 방향을 갖는 것이 아니죠. 사실, 포물선에서는 모든 점들이 다른 방향들을 갖고 있어요. 그러면, 기울기가 도대체 뭐냐고요? 기울기의 정확한 정의는 미분학을 공부할 때 다룹니다. 그러니, 엄밀한 정의는 그때 논의하기로 하죠. 하지만, 기울기 자체가 갖고있는 특성은 직선을 공부할 때 설명되었죠. 우선, 기울기란 한 점에서의 기울기를 말하며 이는 그 점이 갖고 있는 방향입니다. 다음으로, 커브에 하나의 직선이 접한다고 하세요. 즉, 그 커브를 가로지르지 않고 그 직선이 그 커브와 만나는 거죠. 그러면 보통은, 그 직선은 그 커브와 한 점에서 접합니다. 그러면 이제, 그 커브에서 그 점에서의 기울기는 그 접선의 기울기와 같은 겁니다. 일단은 여기까지만 해두기로 하죠.

포물선의 준 일반형 방정식에서 x^2항의 계수 즉, a를 그 포물선의 기울기라 하는데 이는 그 포물선의 오목한 상태를 나타낸다고 하면 되겠어요. 또한, 그 크기가 크면, 그 포물선은 많이 휘어져 있는 거죠. 그 y절편은 그 포물선이 y축과 만나는 점의 y좌표 값인데 이는 그 상수 항 즉, c의 값인 거죠. 그러면, x절편은 어떤가요? 그 x절편이란 포물선이 x축과 만나는 점에서의 x좌표 값이니까 이는 y = 0 으로 놓고$ax^2 + bx + c = 0$을 풀어서 구해야 하죠. 그러면, 그 x항의 계수인 b는 별로 쓸모가 없는 건가요? 그렇지는 않아요. 이는 나중에 판별식이라는 걸 설명할 때 의미 있다는 걸 알게 될 겁니다.

그 x절편을 구하려면 그 일반방정식에서 y를 0 으로 놓고 그 미지수 방정식을 풀어야 하죠. 그 미지수는 물론, x이고 그 방정식을 2 차 방정식이라고 하죠. 이때 그 근의 공식(영어로는 흔히, 2 차 공식이라고 해요)을 이용해서 그 근들을 구하면 되죠. 그 공식을 잊었다면 대신에 완전 제곱식 법을 사용할 수 있어요. 즉, 그 일반방정식을 표준형으로 바꾸는 거죠: $ax^2 + bx + c = 0 \rightarrow a(x + p)^2 + q = 0$. 그리고 나서, 그 표준형을 x에 대하여 풀면 그 근들을 구할 수 있죠. 이는 사실, 그 근의 공식을 유도하는 거나 다름이 없죠. 그런데 우선, 그 포물선이 x축과 만나야 x절편이 있는 거죠. 그 포물선이 x축과 만나면 이는 한 점 아니면 두개의 점들에서 만나죠. 포물선이 그 x절편 즉, 2 차 방정식의 근을 갖는지 그리고 갖으면 몇 개를 갖는지는 판별식을 이용해서 알 수 있어요. 그 판별식은 $b^2 - 4ac$인데 이것이 음이면 근이 없는 거죠. 그 판별식이 0 이면 근이 하나인데 이를 중 근이라 하고 x축이 그 포물선에 접하는 겁니다. 그 판별식이 양이면 근이 둘 즉, x절편이 두개인 겁니다.

자 이제, 이 일반형에 대하여 좀 더 구체적으로 알아보기로 하죠. 그러면, 가장 간단한 포물선 $y = x^2$부터 시작합니다. 이는 가장 기본적인 포물선인데 **Runmath** 시리즈에서는 이를 원형이라고 합니다. 표준형 방정식 $x^2 = 4py$를 알고있죠. 이 원형 $y = x^2$에서는 $p = 1/4$ 이죠. 이 방정식에서, 대칭축이 y축이고 초점이 (0, 1/4)이며 준선은 직선 $y = -1/4$ 이라는 걸 쉽게 알 수 있어요. 또한, 그 기울기는 1 이고 이는 양이므로 아래로 볼록하네요. 이 원형에서는 c가 0 이고 따라서, y절편은 0 입니다. 또한, $y = 0$ 으로 놓으면 $x^2 = 0$ 이고 따라서, x절편도 0 입니다. 그러면 이제, x절편과 y절편 둘 다 0 이므로 꼭지점이 원점(0, 0)이라는 걸 바로 알 수 있죠.

그림.1.1.1 $y = x^2 \Leftrightarrow x^2 = 4(1/4)y$

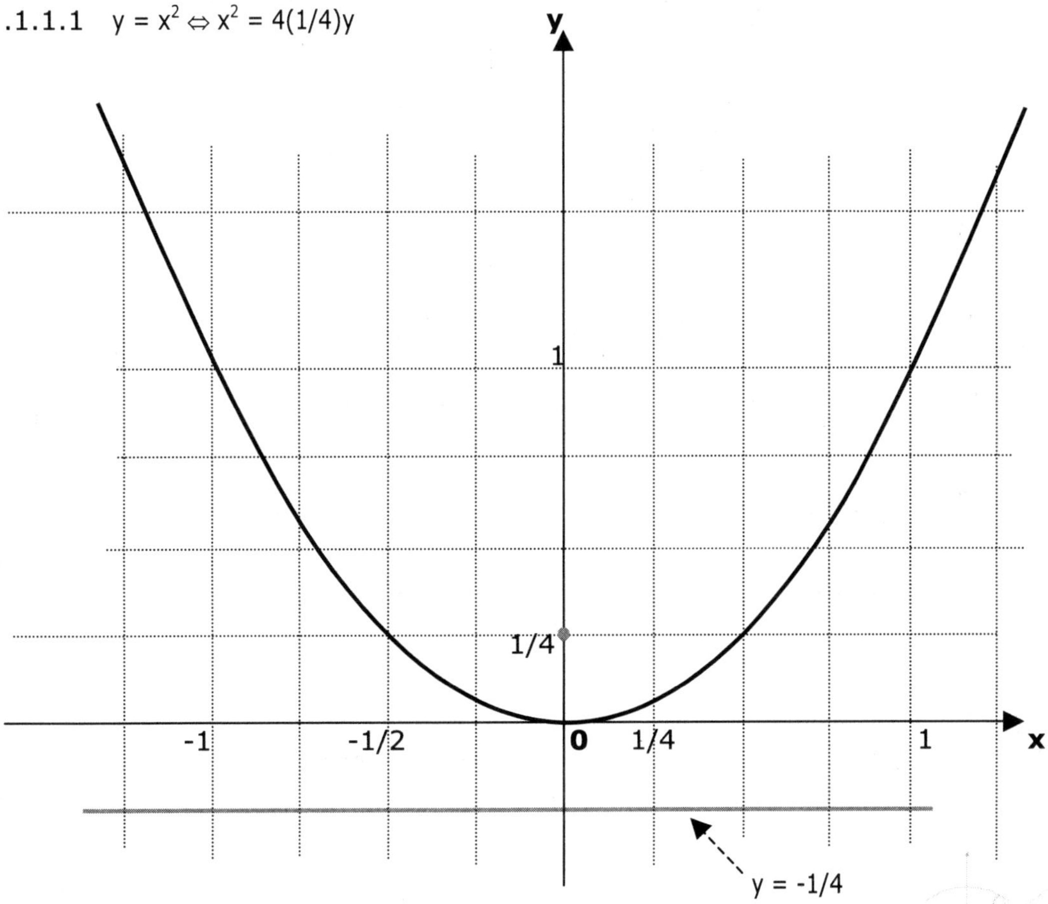

자 이제, $y = ax^2$의 형태로 되어있는 포물선들을 보기로 하죠.
예를 들어, 다음과 같은 포물선들을 고려해보죠. 그러면:

$y = (1/2)x^2 \Rightarrow x^2 = 2y = 4(1/2)y \Rightarrow$ 초점 = (0, 1/2), 준선: y = -1/2, 꼭지점 = (0, 0).
$y = 2x^2 \Rightarrow x^2 = (1/2)y = 4(1/8)y \Rightarrow$ 초점 = (0, 1/8), 준선: y = -1/8, 꼭지점 = (0, 0).
$y = 4x^2 \Rightarrow x^2 = y/4 = 4(1/16)y \Rightarrow$ 초점 = (0, 1/16), 준선: y = -1/16, 꼭지점 = (0, 0).

그러면 이제, 기울기 a 가 커짐에 따라 초점과 준선이 꼭지점에 더 가까워지는 걸 알 수 있죠. 또한, 포물선은 더욱 굽어지는 거죠.

그림. 1.1.2

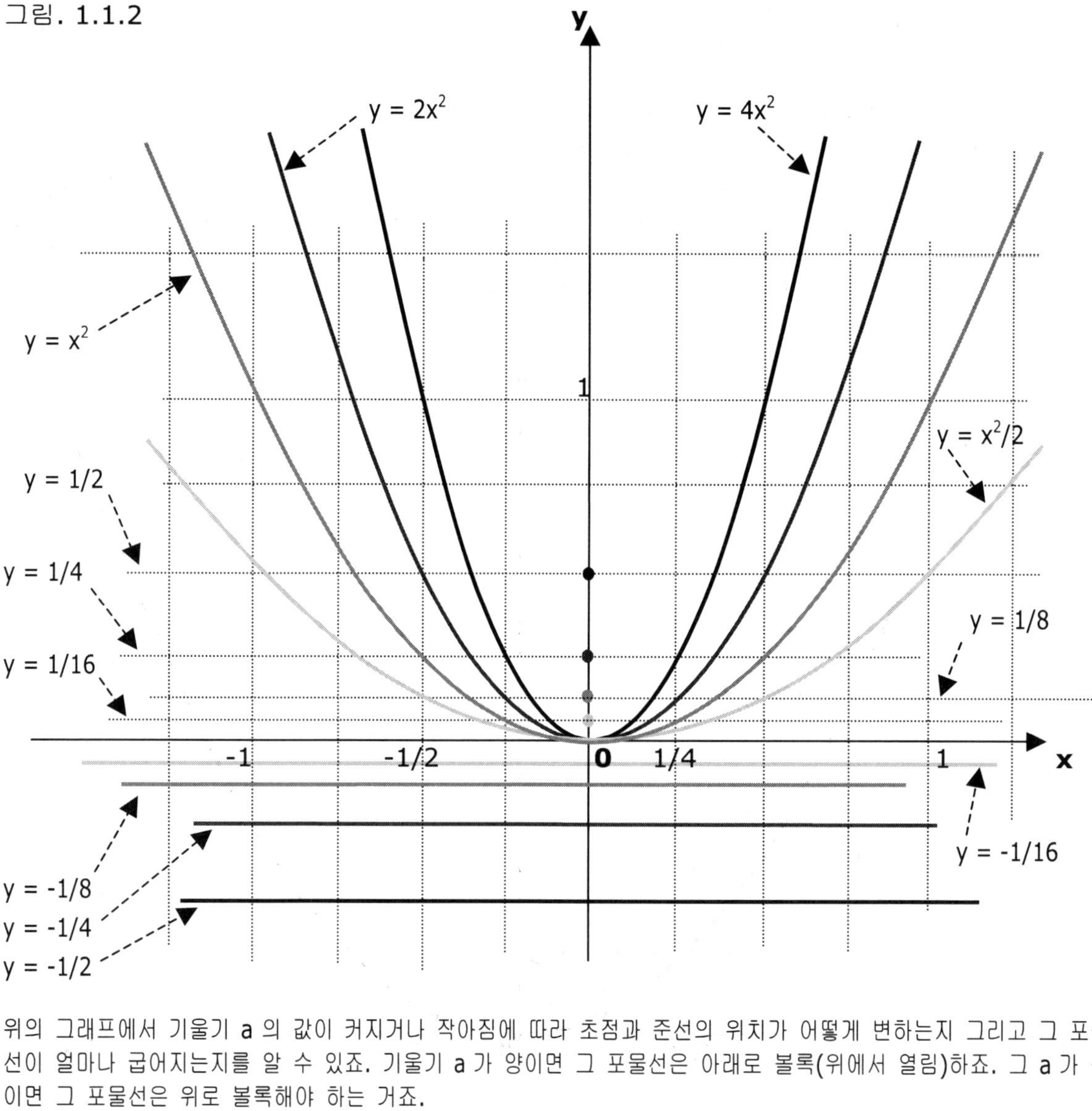

위의 그래프에서 기울기 a 의 값이 커지거나 작아짐에 따라 초점과 준선의 위치가 어떻게 변하는지 그리고 그 포물선이 얼마나 굽어지는지를 알 수 있죠. 기울기 a 가 양이면 그 포물선은 아래로 볼록(위에서 열림)하죠. 그 a 가 음이면 그 포물선은 위로 볼록해야 하는 거죠.

이제, 포물선 $y = ax^2$을 x축을 따라 m만큼 y축을 따라서는 n만큼 이동시킨다고 하세요. 즉, 그 포물선을 평행 이동시키는 겁니다. 그러면, 그 꼭지점과 초점 등도 모두 동일하게 이동되는 거죠. 그러면, 새로운 포물선을 갖게 되는데 이 새로운 포물선은 원래의 포물선과 동일한 것이고 단지 위치만 바뀐 거죠. 그 새로운 포물선의 방정식은 다음과 같아요: $y - n = a(x - m)^2$.

그림. 1.1.3

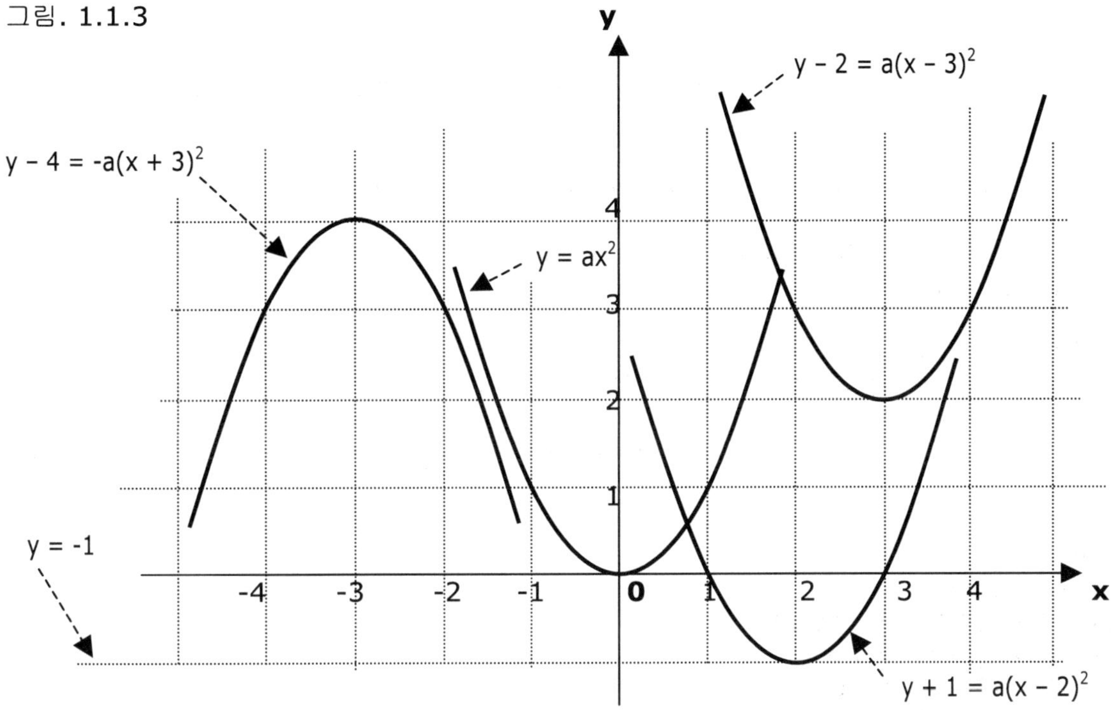

마찬가지의 개념이 포물선들 $x = ay^2$과 $x - n = a(y - m)^2$에도 적용되는 거죠. 그 포물선들은 옆으로(왼쪽으로 또는 오른쪽으로) 볼록하죠. 그 준선들은 y축에 평행합니다.

$y = ax^2 \Rightarrow x^2 = (1/a)y = 4(1/4a)y \Rightarrow x^2 = 4(1/4a)y$. 그러므로:

$y = ax^2 \Rightarrow x^2 = 4(1/4a)y$.

그러면, 포물선 $y = ax^2$에서는 다음과 같죠:

꼭지점 = (0, 0).
초점 = (0, 1/4a).
준선: $y = -1/4a$.

그런데, $y - n = a(x - m)^2 \Rightarrow (x - m)^2 = (1/a)(y - n) = 4(1/4a)(y - n)$이죠.
그러므로, $y - n = a(x - m)^2 \Rightarrow (x - m)^2 = 4(1/4a)(y - n)$이네요.

그러면, 그 포물선 $y = ax^2$의 모든 것이 x축을 따라 m만큼 y축을 따라서는 n만큼 이동되는 거니까 다음과 같은 거죠:

꼭지점 = (0 + m, 0 + n) = (m, n).
초점 = (0 + m, n + 1/4a) = (m, 1/4a + n).
준선: $y = -1/4a + n = n - 1/4a$.

그런데, 방정식들이 다르면 그 포물선의 모양들이나 오목한 상태들도 다를까요?
꼭 그렇지는 않아요! 포물선의 방정식들에서 기울기들이 같으면 그 모양들이나 오목한 상태들도 모두 같은 겁니다. 예를 들면, $y = x^2$, $y = x^2 - x + 1$, $y = x^2 + 2x + 5$, $y = x^2 + x/3 + \sqrt{5}$ 등등과 같은 것들은 모두 동일한 오목한 상태를 갖고 그 모양 자체들도 같은 겁니다. 다시 말해서, 동일한 포물선들이 다른 위치들에 놓여있는 것 뿐인 거죠.
그런데, 그냥 x^2 항의 계수를 기울기라고 하면 안되죠. 먼저, 주어진 방정식을 그 일반형이나 표준형으로 표현해놓고 나서 x^2 항의 계수를 기울기로 해야 하는 거죠.

그러면 이제, 방정식 $y - n = a(x - m)^2$를 전개해서 자세한 내용을 보도록 하죠.

$y - n = a(x - m)^2 = a(x^2 - 2mx + m^2) = ax^2 - 2amx + am^2$.
$\Rightarrow y = ax^2 - 2amx + am^2 + n$.

이제, $-2am = b$이고 $am^2 + n = c$라 놓으세요.
그러면, 이 a와 m과 n은 상수들이죠. 그러므로, b와 c도 상수들입니다.
그러면, $y = ax^2 + bx + c$가 되는 거죠.
자 이제, 하고자 하는 바는 $y = ax^2 + bx + c$의 형태를 갖는 방정식을 $y - n = a(x - m)^2$ 형태를 갖도록 변형시키는 겁니다.

$ax^2 + bx + c$
$= a\{x^2 + (b/a)x\} + c$
$= a\{x^2 + (b/a)x + (b/2a)^2 - (b/2a)^2\} + c$
$= a\{x + (b/2a)\}^2 - a(b/2a)^2 + c$
$= a\{x + (b/2a)\}^2 - b^2/4a + c$
$= a\{x + (b/2a)\}^2 - (b^2 - 4ac)/4a$

그러면, $y + (b^2 - 4ac)/4a = a\{x + (b/2a)\}^2$이 되네요.

이제, $n = -(b^2 - 4ac)/4a$이고 $m = -b/2a$이라고 놓기로 하죠.
그러면, $y - n = a(x - m)^2$이 되는 거죠.

그러면, 꼭지점 = (m, n)입니다. 대칭축은 y축에 평행이고 꼭지점을 지나가죠.
그러면, 대칭축은 수직선 $x = m$이죠. 또한, $m = -b/2a$입니다.
그러면, 대칭축은 직선 $x = -b/2a$이네요.

이제, 예제들을 다루어보기로 하죠.

예제

포물선의 일반방정식

다음 포물선들 각각의 꼭지점, 초점, 준선을 구하세요:

0. $y = x^2 - x + 1$.

1. $y = 2x^2 - 4x + 2$ 와 $2x^2 - 3x + 2y - 1 = 0$.

2. $x = 3y^2 - 6y + 3$ 와 $2y^2 - 3x - 2y - 1 = 0$.

제안과 풀이
0번 예제에서의 문제

다음 포물선의 꼭지점, 초점, 준선을 구하세요: $y = x^2 - x + 1$.

주어진 방정식은 일반형이죠. 이를 표준형으로 나타내면 꼭지점과 초점 그리고 준선을 쉽게 볼 수 있을 겁니다. 그럼 이제, 바꾸어보기로 하죠.

$y = x^2 - x + 1 = x^2 - x + (1/2)^2 - (1/2)^2 + 1 = (x - 1/2)^2 - 1/4 + 1 = (x - 1/2)^2 - 3/4$.

그러면, 다음과 같은 거죠:
$y = x^2 - x + 1 \Rightarrow y = (x - 1/2)^2 - 3/4 \Rightarrow y + 3/4 = (x - 1/2)^2 \Rightarrow y - (-3/4) = (x - 1/2)^2$
$\Rightarrow (x - 1/2)^2 = 4(1/4)\{y - (-3/4)\}$.

그 표준방정식이 다음과 같다는 것을 알고있죠:
$x^2 = 4py$와 $(x - m)^2 = 4p(y - n)$.

포물선 $x^2 = 4py$에는 꼭시섬 = (0, 0)이고 초점 = (0, p)이며 준선은 y = -p입니다.
이 포물선을 x축 방향으로 m만큼 y축 방향으로는 n만큼 이동시키면 포물선 $(x - m)^2 = 4p(y - n)$을 갖게 되죠.
따라서, 꼭지점 = (0 + m, 0 + n) = (m, n)이고 초점 = (0 + m, p + n) = (m, p + n)이며
준선은 y = -p + n = n - p 입니다.

그럼 이제, $y = x^2 - x + 1 \Rightarrow (x - 1/2)^2 = 4(1/4)\{y - (-3/4)\}$이죠.

그러면, p = 1/4 이고 m = 1/2 이며 n = -3/4 이죠. 따라서:
그 꼭지점 = (m, n) = (1/2, -3/4)이고 초점 = (m, p + n) = (1/2, 1/4 - 3/4) = (1/2, -1/2)이며
준선은 y = n - p = -3/4 - 1/4 = -1 이네요.

그러므로, 꼭지점 = (1/2, -3/4), 초점 = (1/2, -1/2), 준선: y = -1.

요약:

$(x - m)^2 = 4p(y - n) \Rightarrow$ 꼭지점 = (m, n)이고 초점 = (m, p + n)이며 준선은 y = n - p.

$y = x^2 - x + 1 = x^2 - x + (1/2)^2 - (1/2)^2 + 1 = (x - 1/2)^2 - 1/4 + 1 = (x - 1/2)^2 - 3/4$.
$\Rightarrow y = (x - 1/2)^2 - 3/4 \Rightarrow y + 3/4 = (x - 1/2)^2 \Rightarrow y - (-3/4) = (x - 1/2)^2$
$\Rightarrow (x - 1/2)^2 = 4(1/4)\{y - (-3/4)\}$.

그러므로, $y = x^2 - x + 1 \Rightarrow (x - 1/2)^2 = 4(1/4)\{y - (-3/4)\}$.
따라서, p = 1/4, m = 1/2, n = -3/4. 그러면:

꼭지점 = (m, n) = (1/2, -3/4)이고 초점 = (0 + m, p + n) = (1/2, 1/4 - 3/4) = (1/2, -1/2)이며
준선은 y = n - p = -3/4 - 1/4 = -1.

그러므로, 꼭지점 = (1/2, -3/4), 초점 = (1/2, -1/2), 준선: y = -1.

그림. 0.0 $y = x^2 - x + 1 \Leftrightarrow (x - 1/2)^2 = 4(1/4)\{y - (-3/4)\}$

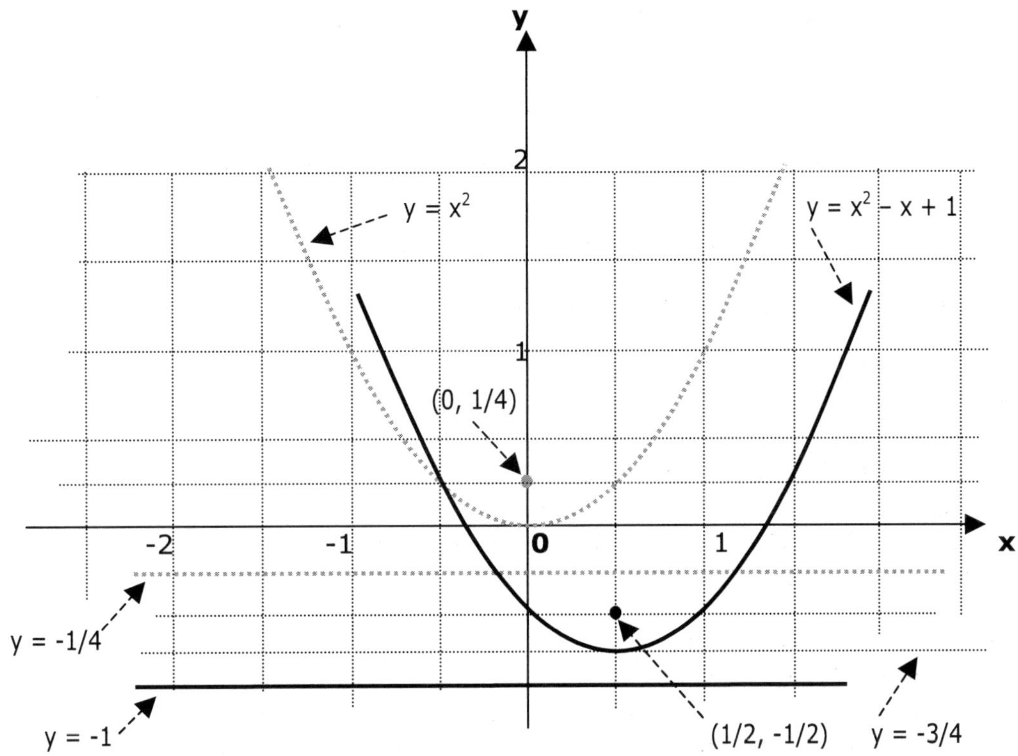

제안과 풀이
1 번 예제에서의 문제

다음 포물선들의 꼭지점들과 초점들 그리고 준선들을 구하세요: $y = 2x^2 - 4x + 2$ 와 $2x^2 - 3x + 2y - 1 = 0$.

주어진 방정식들은 일반형이죠. 이를 표준형으로 바꾸면 꼭지점과 초점 그리고 준선을 쉽게 볼 수 있을 겁니다.
그럼 이제, 바꾸어보기로 하죠.
$y = 2x^2 - 4x + 2 = 2(x^2 - 2x + 1) = 2(x - 1)^2 \Rightarrow (x - 1)^2 = y/2 = 4(1/8)y$.
그러면, 다음과 같은 거죠:
$y = 2x^2 - 4x + 2 \Rightarrow y = 2(x - 1)^2 \Rightarrow (x - 1)^2 = 4(1/8)y$.

그 표준방정식이 다음과 같다는 것을 알고있죠:
$x^2 = 4py$와 $(x - m)^2 = 4p(y - n)$.

포물선 $x^2 = 4py$에는 꼭지점 = (0, 0)이고 초점 = (0, p)이며 준선은 y = -p입니다.
이 포물선을 x축 방향으로 m만큼 y축 방향으로는 n만큼 이동시키면 포물선 $(x - m)^2 = 4p(y - n)$을 갖게 되죠.
따라서, 꼭지점 = (0 + m, 0 + n) = (m, n)이고 초점 = (0 + m, p + n) = (m, p + n)이며
준선은 y = -p + n = n - p 입니다.

그럼 이제, $y = 2x^2 - 4x + 2 \Rightarrow (x - 1)^2 = 4(1/8)y$이죠.
그러면, p = 1/8 이고 m = 1 이며 n = 0 이죠. 따라서:
그 꼭지점 = (m, n) = (1, 0)이고 초점 = (0 + m, p + n) = (0 + 1, 1/8 + 0) = (1, 1/8)이며
준선은 y = n - p = 0 - 1/8 = -1/8 이네요. 그러므로, 꼭지점 = (1, 0), 초점 = (1, 1/8), 준선: y = -1/8.

그림. 1.0

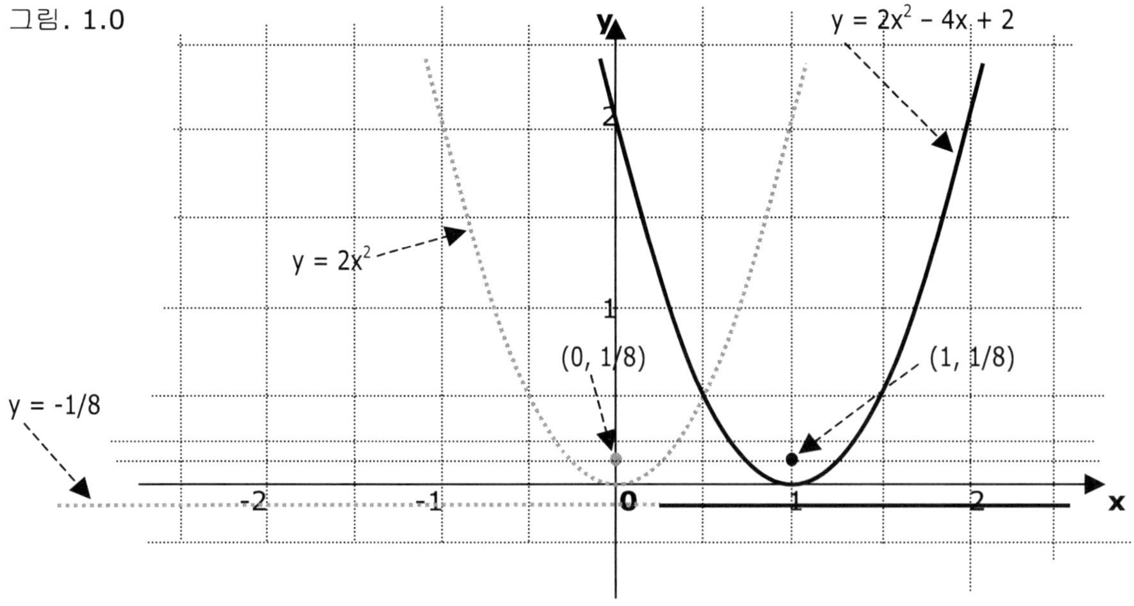

자 이제, 다음 포물선으로 넘어가보죠: $2x^2 - 3x + 2y - 1 = 0$.
그러면, $2x^2 - 3x + 2y - 1 = 0 \Rightarrow 2y = -2x^2 + 3x + 1 \Rightarrow y = -x^2 + 3x/2 + 1/2$.
$-x^2 + 3x/2 + 1/2 = -(x^2 - 3x/2) + 1/2 = -\{x^2 - 3x/2 + (3/4)^2 - (3/4)^2\} + 1/2$
$= -(x - 3/4)^2 + (3/4)^2 + 1/2 = -(x - 3/4)^2 + 17/16$.

따라서, $2x^2 - 3x + 2y - 1 = 0 \Rightarrow y = -(x - 3/4)^2 + 17/16 \Rightarrow y - 17/16 = -(x - 3/4)^2$
$\Rightarrow (x - 3/4)^2 = -(y - 17/16) = 4(-1/4)(y - 17/16)$.

그러면, 다음과 같은 거죠: $2x^2 - 3x + 2y - 1 = 0 \Rightarrow (x - 3/4)^2 = 4(-1/4)(y - 17/16)$.

그 표준방정식이 다음과 같다는 것을 알고있죠: $x^2 = 4py$와 $(x - m)^2 = 4p(y - n)$.
포물선 $x^2 = 4py$에는 꼭지점 = (0, 0)이고 초점 = (0, p)이며 준선은 $y = -p$입니다.
이 포물선을 x축 방향으로 m만큼 y축 방향으로는 n만큼 이동시키면 포물선 $(x - m)^2 = 4p(y - n)$을 갖게 되죠.
따라서, 꼭지점 = (0 + m, 0 + n) = (m, n)이고 초점 = (0 + m, p + n) = (m, p + n)이며
준선은 $y = -p + n = n - p$ 입니다.

그럼 이제, $2x^2 - 3x + 2y - 1 = 0 \Rightarrow (x - 3/4)^2 = 4(-1/4)(y - 17/16)$이죠.
그러면, $p = -1/4$ 이고 $m = 3/4$ 이며 $n = 17/16$ 이죠. 따라서:
그 꼭지점 = (m, n) = (3/4, 17/16)이고 초점 = (0 + m, p + n) = (0 + 3/4, -1/4 + 17/16) = (3/4, 13/16)
이며 준선은 $y = n - p = 17/16 + 1/4 = 21/16$ 이네요.
그러므로, 꼭지점 = (3/4, 17/16), 초점 = (3/4, 13/16), 준선: $y = 21/16$.

그림. 1.1 $2x^2 - 3x + 2y - 1 = 0 \Leftrightarrow (x - 3/4)^2 = 4(-1/4)(y - 17/16)$

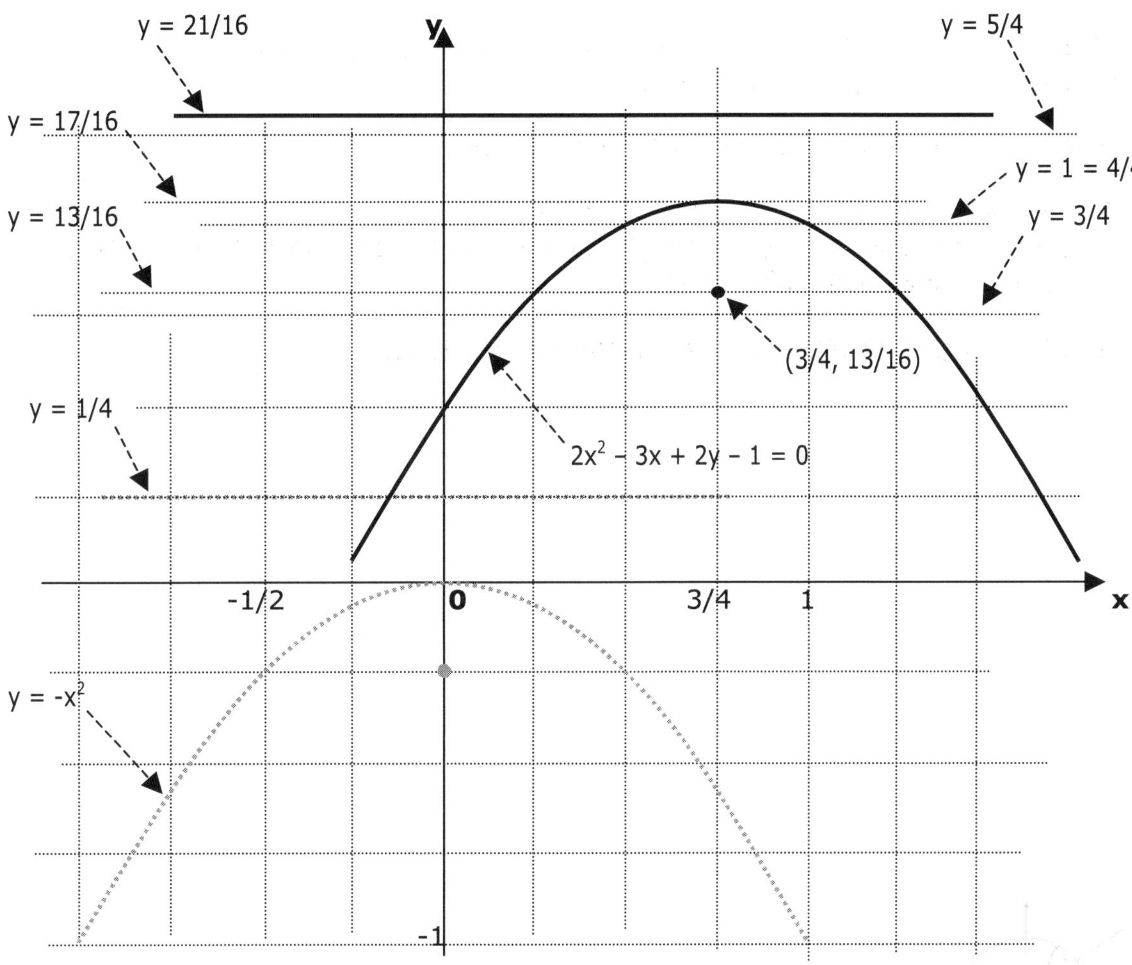

요약:

$(x - m)^2 = 4p(y - n) \Rightarrow$ 꼭지점 = (m, n)이고 초점 = (m, p + n)이며 준선은 y = n - p.

$y = 2x^2 - 4x + 2 = 2(x^2 - 2x + 1) = 2(x - 1)^2 \Rightarrow (x - 1)^2 = y/2 = 4(1/8)y$.

그러면, $y = 2x^2 - 4x + 2 \Rightarrow (x - 1)^2 = 4(1/8)y$.

그러면, p = 1/8 이고 m = 1 이며 n = 0 이죠. 따라서:
그 꼭지점 = (m, n) = (1, 0)이고 초점 = (0 + m, p + n) = (0 + 1, 1/8 + 0) = (1, 1/8)이며
준선은 y = n - p = 0 - 1/8 = -1/8 이네요.

그러므로, 꼭지점 = (1, 0), 초점 = (1, 1/8), 준선: y = -1/8.

다음으로, $2x^2 - 3x + 2y - 1 = 0$ 입니다.
그러면, $2x^2 - 3x + 2y - 1 = 0 \Rightarrow 2y = -2x^2 + 3x + 1 \Rightarrow y = -x^2 + 3x/2 + 1/2$.

$-x^2 + 3x/2 + 1/2 = -(x^2 - 3x/2) + 1/2 = -\{x^2 - 3x/2 + (3/4)^2 - (3/4)^2\} + 1/2$
$= -(x - 3/4)^2 + (3/4)^2 + 1/2 = -(x - 3/4)^2 + 17/16$.

따라서, $2x^2 - 3x + 2y - 1 = 0 \Rightarrow y = -(x - 3/4)^2 + 17/16 \Rightarrow y - 17/16 = -(x - 3/4)^2$
$\Rightarrow (x - 3/4)^2 = -(y - 17/16) = 4(-1/4)(y - 17/16)$.

그러면, $2x^2 - 3x + 2y - 1 = 0 \Rightarrow (x - 3/4)^2 = 4(-1/4)(y - 17/16)$.

그러면, p = -1/4 이고 m = 3/4 이며 n = 17/16 이죠. 따라서:
그 꼭지점 = (m, n) = (3/4, 17/16)이고 초점 = (0 + m, p + n) = (0 + 3/4, -1/4 + 17/16) = (3/4, 13/16)
이며 준선은 y = n - p = 17/16 + 1/4 = 21/16 이네요.

그러므로, 꼭지점 = (3/4, 17/16), 초점 = (3/4, 13/16), 준선: y = 21/16.

제안과 풀이

2 번 예제에서의 문제

다음 포물선들의 꼭지점들과 초점들 그리고 준선들을 구하세요: $x = 3y^2 - 6y + 3$ 과 $2y^2 + x - y - 1 = 0$.

주어진 방정식들을 표준형으로 바꾸면 꼭지점과 초점 그리고 준선을 쉽게 볼 수 있죠.
그럼 이제, 바꾸어보기로 하죠.

$x = 3y^2 - 6y + 3 = 3(y^2 - 2y + 1) = 3(y - 1)^2 \Rightarrow (y - 1)^2 = x/3$.

그러면, 다음과 같죠: $x = 3y^2 - 6y + 3 \Rightarrow (y - 1)^2 = 4(1/12)x$.

그 표준방정식이 다음과 같다는 것을 알고있죠: $y^2 = 4px$와 $(y - n)^2 = 4p(x - m)$.

포물선 $y^2 = 4px$에는 꼭지점 $= (0, 0)$이고 초점 $= (p, 0)$이며 준선은 $x = -p$입니다.
이 포물선을 x축 방향으로 m만큼 y축 방향으로는 n만큼 이동시키면 포물선 $(y - n)^2 = 4p(x - m)$을 갖게 되죠.
따라서, 꼭지점 $= (0 + m, 0 + n) = (m, n)$이고 초점 $= (p + m, n) = (p + m, n)$이며
준선은 $x = -p + m = m - p$ 입니다.

그럼 이제, $x = 3y^2 - 6y + 3 \Rightarrow (y - 1)^2 = 4(1/12)x$이죠.

그러면, $p = 1/12$ 이고 $m = 0$ 이며 $n = 1$ 이죠. 따라서:
그 꼭지점 $= (m, n) = (0, 1)$이고 초점 $= (p + m, n) = (1/12 + 0, 1) = (1/12, 1)$이며
준선은 $x = m - p = 0 - 1/12 = -1/12$ 이네요.

그러므로, 꼭지점 $= (0, 1)$, 초점 $= (1/12, 1)$, 준선: $x = -1/12$.

준선이 y 축에 평행이고 기울기가 양이므로 그 포물선은 왼쪽으로 볼록합니다.

요약:

$(y - n)^2 = 4p(x - m) \Rightarrow$ 꼭지점 $= (m, n)$이고 초점 $= (p + m, n)$이며 준선은 $x = m - p$.

$x = 3y^2 - 6y + 3 = 3(y^2 - 2y + 1) = 3(y - 1)^2 \Rightarrow (y - 1)^2 = x/3$.

그러면, $x = 3y^2 - 6y + 3 \Rightarrow (y - 1)^2 = 4(1/12)x$.

그러면, $p = 1/12$ 이고 $m = 0$ 이며 $n = 1$ 이죠. 따라서:
그 꼭지점 $= (m, n) = (0, 1)$이고 초점 $= (p + m, n) = (1/12 + 0, 1) = (1/12, 1)$이며
준선은 $x = m - p = 0 - 1/12 = -1/12$ 이네요.

그러므로, 꼭지점 $= (0, 1)$, 초점 $= (1/12, 1)$, 준선: $x = -1/12$.

그림. 2.0

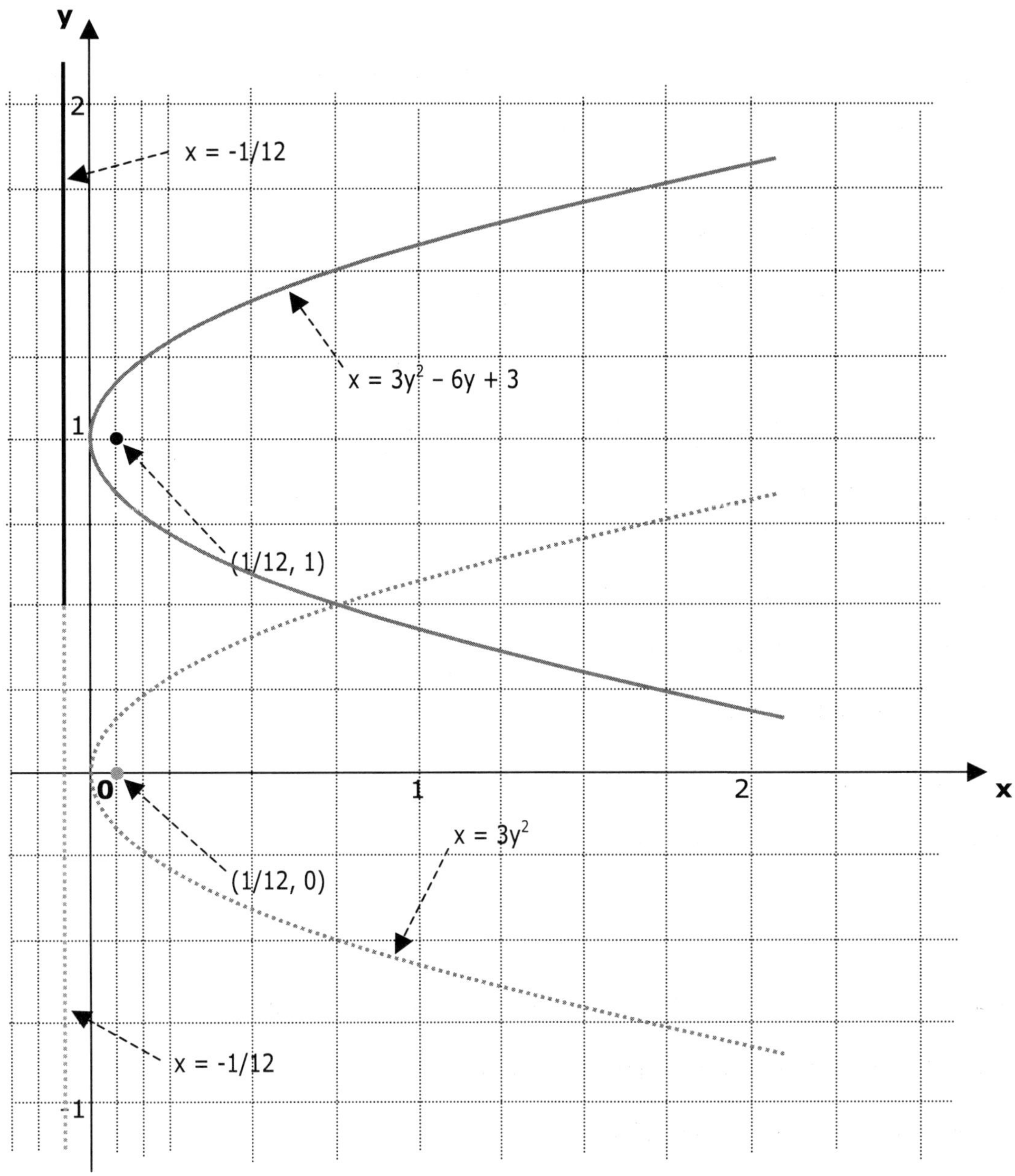

다음으로, $2y^2 + x - y - 1 = 0$ 입니다.

그러면, $2y^2 + x - y - 1 = 0 \Rightarrow x = -2y^2 + y + 1$.

$-2y^2 + y + 1 = -2(y^2 - y/2) + 1 = -2\{y^2 - y/2 + (1/4)^2 - (1/4)^2\} + 1 = -2(y - 1/4)^2 + 2(1/16) + 1$
$= -2(y - 1/4)^2 + 9/8$.

그러면, $2y^2 + x - y - 1 = 0 \Rightarrow x = -2(y - 1/4)^2 + 9/8 \Rightarrow x - 9/8 = -2(y - 1/4)^2$
$\Rightarrow (y - 1/4)^2 = (-1/2)(x - 9/8) = 4(-1/8)(x - 9/8)$.

그러면, 다음과 같은 거죠: $2y^2 + x - y - 1 = 0 \Rightarrow (y - 1/4)^2 = 4(-1/8)(x - 9/8)$.

그 표준방정식이 다음과 같다는 것을 알고있죠: $y^2 = 4px$와 $(y - n)^2 = 4p(x - m)$.
포물선 $y^2 = 4px$에는 꼭지점 = (0, 0)이고 초점 = (p, 0)이며 준선은 $x = -p$입니다.
이 포물선을 x축 방향으로 m만큼 y축 방향으로는 n만큼 이동시키면 포물선 $(y - n)^2 = 4p(x - m)$을 갖게 되죠.
따라서, 꼭지점 = (0 + m, 0 + n) = (m, n)이고 초점 = (p + m, n) = (p + m, n)이며
준선은 $x = -p + m = m - p$ 입니다.

그럼 이제, $2y^2 + x - y - 1 = 0 \Rightarrow (y - 1/4)^2 = 4(-1/8)(x - 9/8)$이죠.
그러면, $p = -1/8$ 이고 $m = 9/8$ 이며 $n = 1/4$ 이죠. 따라서:
그 꼭지점 = (m, n) = (9/8, 1/4)이고 초점 = (p + m, n) = (-1/8 + 9/8, 1/4) = (1, 1/4)이며
준선은 $x = m - p = 9/8 + 1/8 = 5/4$ 이네요.
그러므로, 꼭지점 = (9/8, 1/4), 초점 = (1, 1/4), 준선: $x = 5/4$.

그런데, 준선이 y 축에 평행하고 기울기가 −2 즉, 음이므로 이 포물선은 오른쪽으로 볼록하죠.

요약:

$(y - n)^2 = 4p(x - m) \Rightarrow$ 꼭지점 = (m, n)이고 초점 = (p + m, n)이며 준선은 $x = m - p$.

그러면, $2y^2 + x - y - 1 = 0 \Rightarrow x = -2y^2 + y + 1$.

$-2y^2 + y + 1 = -2(y^2 - y/2) + 1 = -2\{y^2 - y/2 + (1/4)^2 - (1/4)^2\} + 1 = -2(y - 1/4)^2 + 2(1/16) + 1$
$= -2(y - 1/4)^2 + 9/8$.

그러면, $2y^2 + x - y - 1 = 0 \Rightarrow x = -2(y - 1/4)^2 + 9/8 \Rightarrow x - 9/8 = -2(y - 1/4)^2$
$\Rightarrow (y - 1/4)^2 = (-1/2)(x - 9/8) = 4(-1/8)(x - 9/8)$.

그러면, $2y^2 + x - y - 1 = 0 \Rightarrow (y - 1/4)^2 = 4(-1/8)(x - 9/8)$.

그러면, $p = -1/8$ 이고 $m = 9/8$ 이며 $n = 1/4$. 따라서:
그 꼭지점 = (m, n) = (9/8, 1/4)이고 초점 = (p + m, n) = (-1/8 + 9/8, 1/4) = (1, 1/4)이며
준선은 $x = m - p = 9/8 + 1/8 = 5/4$.

그러므로, 꼭지점 = (9/8, 1/4), 초점 = (1, 1/4), 준선: $x = 5/4$.

그림. 2.1

$x = -2y^2 \Leftrightarrow y^2 = 4(-1/8)x$

$2y^2 + x - y - 1 = 0 \Leftrightarrow x = -2y^2 + y + 1 \Leftrightarrow (y - 1/4)^2 = 4(-1/8)(x - 9/8)$

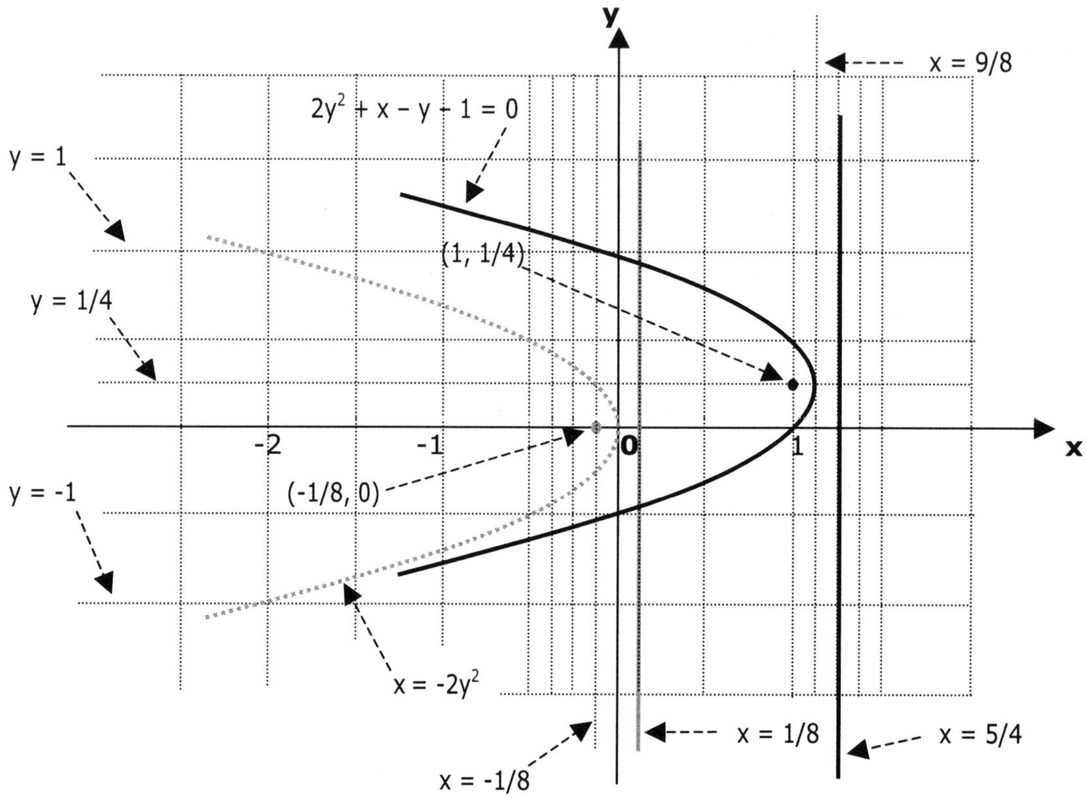

1.2. 포물선의 방정식을 구하려면 무엇이 필요한가요?

문제를 풀다 보면 직선이나 포물선과 같은 커브의 방정식을 구해야 하는 경우가 많죠. 다시 말해서, 그런 커브의 방정식을 작성해야 하는 겁니다. 물론, 그냥 그걸 작성할 수야 없는 거죠. 그러면, 뭘 가지고 그걸 작성하나요? 문제에 대한 답은 그 문제 안에 있습니다. 그러니까, 그 문제의 내용으로부터 필요한 사항(사실)을 추출해 내어야 하죠. 따라서, 우선 그 문제 자체를 제대로 이해해야 하는 거죠. 어떤 때는 그런 사항이 문제의 내용에 명시적으로 드러나 있기도 하지만 보통은 그런 사항의 일부 또는 전부가 감추어져 있어요. 어떤 다른 경우에는 그런 사항이 암시적으로 내재되어있죠. 그런 사항이 명시적으로 드러나 있지 않으면 추론해서 찾아내어야 하는 겁니다. 그러면, 구체적으로 어떤 사항을 속아내어야 하는지 알아 보기로 하죠. 먼저, 직선의 방정식을 작성하는 데 필요 사항부터 시작해보기로 하죠.

직선의 방정식을 작성하는 경우, 다음과 같은 두 가지의 사항 중 어느 한가지가 필요해요:

그 직선에 있는 두 개의 점들
그 직선에 있는 하나의 점과 그 기울기

위의 사항 중 어느 한가지를 가지고 다음과 같은 표준형을 사용해 볼 수 있겠죠: **$y = ax + b$**.
그러면, a 와 b 를 구해야 하네요. 즉, a 와 b 가 미지수들인 거죠. 그러면, 미지수가 두개이니까 두개의 정보가 필요한 겁니다. 그런데, 위의 두 가지 사항 각각은 두개의 정보로 구성되어 있죠. 한가지는 두 개의 점들 즉, 두 쌍의 좌표들로 되어있고 다른 한가지는 하나의 점과 기울기입니다. 그래서, 두개의 방정식들을 세울 수 있는 거죠. 물론, 그것들은 a 와 b 에 관한 방정식들입니다.

그러면 이제, 포물선의 방정식을 작성하려면 무엇이 필요한가요?
다음과 같은 세가지 사항 중에 어느 한가지가 있어야 하죠:

그 포물선에 있는 세 개의 점들
그 포물선에 있는 두개의 점들과 그 기울기나 y 절편 등등과 같은 또 하나의 정보
그 포물선의 꼭지점과 기울기

위의 처음 두 가지 중 어느 한가지를 갖는 경우에는 다음과 같은 두개의 형태들 중 어느 한 개를 사용해볼 수 있겠어요:

$y = ax^2 + bx + c$
$x = ay^2 + by + c$.

사실, 위의 두개의 형태들 중 어느 하나를 선택하려면 또 하나의 정보가 필요한데 이는 그 대칭축이나 준선의 놓임새를 말해주는 거죠. 예를 들면, 준선이 x 축에 평행한지를 알아야 하는 거죠.

그러면 이제, a 와 b 그리고 c 를 구해야 하네요. 즉, a, b, c 가 미지수들인 거죠. 그러면, 미지수가 세 개이니까 세 개의 정보가 필요하죠. 그런데, 위의 두 가지 사항들 각각은 세 개의 정보로 구성되어 있어요. 한가지는 세 개의 점들 즉, 세 쌍의 좌표들로 되어있고 다른 한가지는 두 개의 점들과 기울기입니다. 그래서, 세 개의 방정식들을 세울 수 있는 거죠. 물론, 그것들은 a 와 b 그리고 c 에 관한 방정식들입니다.

다음으로, 그 포물선의 기울기와 꼭지점을 알고있는 경우에는 다음과 같은 두 가지 형태들 중의 어느 하나를 사용해볼 수 있겠어요: $y - n = a(x - m)^2$과 $x - m = a(y - n)^2$.

그런데, 미지수가 세 개 아닌가요? 물론, 그렇죠.
그러면, 또 하나의 정보가 있어야 하는 거 아닌가요? 아니죠, 없어도 되요.
어째서 요? 그 꼭지점이 (m, n)이거든요 그러니까, 사실, 단 하나의 미지수 만을 구해야 하는데 이는 그 기울기 a 이죠.

또한, 위의 두개의 형태들 중 어느 하나를 선택하려면 또 하나의 정보가 필요한데 이는 그 대칭축이나 준선의 놓임새를 말해주는 겁니다. 예를 들면, 대칭축이 x 축에 수식인지를 알아야 하는 거죠.

그러면 이제, 직선이나 포물선의 방정식을 구하는 문제를 풀 때 문제의 내용으로부터 위에서 설명한 그런 정보를 어떻게 해서든 추출해낼 수 있어야 하는 겁니다.

이제, 예제를 좀 다루어보도록 하죠.

예제

포물선의 방정식을 구하려면 무엇이 필요한가요?

0. 포물선이 세 개의 점들 $P_1(1, 6)$, $P_2(-1, 4)$, $P_3(-2, 9)$를 포함하고 그 준선은 x축에 평행하다 하세요. 그러면, 그 포물선의 방정식을 구하세요.

1. 포물선이 세 개의 점들 $P_1(1, 6)$, $P_2(-1, 4)$, $P_3(-2, 9)$를 포함하고 그 대칭축은 x축에 평행하다 하세요. 그러면, 그 포물선의 방정식을 구하세요.

2. 포물선 $y = 2x^2 + bx + c$가 있다 하고 그 포물선은 또 하나의 포물선 $y = -2x^2 + 3x/2 + 9/4$가 하나의 직선 $y = x/2 - 3/4$를 만나는 두개의 점들을 포함한다 하세요. 그러면, b와 c의 값들을 구하세요.

3. 하나의 포물선 $y = ax^2 - x/2 + c$가 있다 하고 그 포물선은 또 하나의 포물선 $y = -2x^2 + 3x/2 + 9/4$가 직선 $y = x/2 - 3/4$를 만나는 두개의 점들을 포함한다 하세요. 그러면, a와 c의 값들을 구하세요.

4. 포물선 $x = ay^2 - y/2 + c$가 있다 하고 그 포물선은 또 하나의 포물선 $y = 2x^2 + x - 3$이 직선 $y = 3x + 1$을 만나는 두개의 점들을 포함한다 하세요. 그러면, a와 c의 값들을 구하세요.

5. 하나의 포물선 $y/2 = ax^2 - 2x + c/2$가 $(1/2, -1/2)$에 있는 점을 포함하고 그 꼭지점은 직선 $y = x - 1$이 지난다 하세요. 그러면, a와 c의 값들을 구하세요.

6. 하나의 포물선 $y = 6x^2 + bx + c$가 $(2, 2)$에 있는 점을 포함하고 그 꼭지점은 직선 $y = x - 1$이 지난다 하세요. 그러면, b와 c의 값들을 구하세요.

7. 하나의 포물선 $y = ax^2 + bx + c$가 두 개의 점들 $(1, 1)$과 $(2, 3)$을 포함하고 그 꼭지점은 직선 $y = 2x - 1$에 있다 하세요. 그러면, a와 b 그리고 c의 값들을 구하세요.

8. 하나의 포물선 $y = ax^2 + bx + c$가 두 개의 점들 $(1, 1)$과 $(2, 4)$을 포함하고 그 꼭지점은 직선 $y = 2x - 1$에 있다 하세요. 그러면, a와 b 그리고 c의 값들을 구하세요.

9. 포물선 A의 방정식이 $y = -4x^2 + 8mx - n(n + 2)$라 하고 포물선 B의 방정식은 $y = 8x^2 - 2x + 5/4$라 하세요. 또한, A의 꼭지점이 B에 있다고 하세요. 그러면, m과 n의 값들을 구하세요.

A. 포물선 A의 방정식이 $y = x^2 + bx + c$라 하고 포물선 B의 방정식은 $y = -2x^2 - x + 5$라 하세요. 또한, A의 꼭지점이 B에 있고 B의 꼭지점도 A에 있다고 하세요. 그러면, b와 c의 값들을 구하세요.

B. 포물선 A의 방정식이 $y = 2x^2 + bx + c$라 하고 포물선 B의 방정식은 $y = -2x^2 + 8x - 5$라 하세요. 또한, A의 꼭지점이 B에 있고 B의 꼭지점도 A에 있다고 하세요. 그러면, b와 c의 값들을 구하세요.

C. 포물선 A의 방정식이 $y = x^2 + bx + c$라 하고 포물선 B의 방정식은 $y = -2x^2 + 8x - 5$라 하세요. 또한, A의 꼭지점이 B에 있고 B의 꼭지점도 A에 있다고 하세요. 그러면, b와 c의 값들을 구하세요.

D. 다음 예제들에서는 포물선의 방정식을 그래프에 나타내는 연습을 좀 하겠어요.

유의사항:
아래의 수식들에 있는 한 쌍의 수직 막대들은 절대 기호라고 하죠. 절대 기호가 있는 수식을 처리할 때에는 그 절대 기호를 제거하고 나서 진행해야 해요. 그런데, 그 절대 기호들을 제거하기 전에 두 가지 경우들을 고려해야 하는 겁니다. 하나는 그 절대 기호 안에 있는 것이 0 보다 크거나 같은 경우이고 다른 하나는 절대 기호 내부에 있는 것이 0 보다 작은 경우이죠. 그러면, 그것이 0 보다 작은 경우에는 그 절대 기호들을 제거하고 나서 그것에 (-) 부호를 붙여주어야 하죠. 예를 들면:

실수인 s에 대하여 $t = |s|$라 하세요. 그러면:
$s \ge 0$이면 $t = s$; 즉, $|s| = s$. $s < 0$이면 $t = -s$; 즉, $|s| = -s$ 왜냐하면 $|s|$는 양이니까.

그러면 이제, 다음 방정식들의 그래프들을 작성하세요.

0. $y = 2(x - 1)^2 - 2|x - 2| + 1$

1. $y = (|2x| - 1)(x + 1)$

2. $y = |x^2 + 2x - 3|$

3. $y = |x|^2 + 2|x| - 3$

4. $|y| = x^2 + 2x - 3$

5. $|y| = |x|^2 + 2|x| - 3$

6. $|y| = x^2 + |2x - 3|$

7. $y = (x^2 - 2x + 1)^{1/2} =$

8. $y = \{(x^2 - 4x + 4)^2\}^{1/2} =$

9. $y = \{(x^2 - 2x + 2)^2\}^{1/2} =$

A. $y = \{(x^2 - 2x)^2\}^{1/2} = \sqrt{(x^2 - 2x)^2}$

B. $y = \{(-x^2 + 2x)^2\}^{1/2} = \sqrt{(-x^2+2x)^2}$

C. $y = \{(-x^2 + 2x - 1)^2\}^{1/2} = \sqrt{(-x^2+2x-1)^2}$

D. $y = \{(-x^2 + 2x - 2)^2\}^{1/2} = \sqrt{(-x^2+2x-2)^2}$

E. $y = f(x) = 2x^2 + bx + c$ 이고 $0 \le x \le 1$ 이며 $0 \le y \le 2$ 라고 하고 b와 c의 값들을 구하세요.

F. $y = f(x) = ax^2 + bx + 1$ 이고 $0 \le x \le 1$ 이며 $0 \le y \le 2$ 라고 하고 a와 b의 값들을 구하세요.

G. $y = f(x) = ax^2 + bx + c$ 이고 $0 \le x \le 1$ 이며 $0 \le y \le 2$ 라고 하고 a와 b 그리고 c의 값들을 구하세요.

H. $y = f(x) = (2x^2 + 3)/7$ 이고 $m \le x \le n$이며 $m \le y \le n$이고 여기에서 m 과 n은 상수라 하고 m과 n의 값들을 구하세요.

I. $y = f(x) = (2x^2 - 3)/5$ 이고 $m \le x \le n$이며 $m \le y \le n$이고 여기에서 m과 n은 상수라 하고 m과 n의 값들을 구하세요.

J. $y = f(x) = (2x^2 + c)/5$ 이고 $m \le x \le n$이며 $m \le y \le n$이고 여기에서 c와 m 그리고 n은 상수라 하고 c와 m 그리고 n의 값들을 구하세요.

K. 다음과 같은 포물선의 방정식이 있다고 하세요: $2x^2 + (2c + 1)x - (c + 1)y + c + 1 = 0$ 여기에서 c는 상수. 그러면, 그 포물선은 그 어느 c값에 대하여도 두개의 특정한 점들을 지난다 하고 그 두 점들을 구하세요.

L. 다음과 같은 포물선의 방정식이 있다고 하세요: $(c + 2)x^2 + (2c + 1)x - (c + 1)y + 3c + 1 = 0$ 여기에서 c는 상수. 그러면, 그 포물선이 그 어느 c값에 대하여도 두개의 특정한 점들을 지난다 하고 그 두 점들을 구하세요.

M. 다음과 같은 포물선의 방정식이 있다고 하세요: $cx^2 + 2cx - (c + 1)y + 3c + 3 = 0$ 여기에서 c는 상수. 그러면, 그 포물선이 그 어느 c값에 대하여도 두개의 특정한 점들을 지난다 하고 그 두 점들을 구하세요.

N. 다음과 같은 포물선의 방정식이 있다고 하세요: $cx^2 + (2c + 1)x - cy + 3c + 1 = 0$ 여기에서 c는 상수. 그러면, 그 포물선이 그 어느 c값에 대하여도 하나의 특정한 점을 지난다 하고 그 특정한 점을 구하세요.

O. 다음과 같은 포물선의 방정식이 있다고 하세요: $(c + 2)x^2 + (2c + 1)x - (c + 1)y + 3c + 1 = 0$ 여기에서 c는 상수. 그러면, 그 포물선이 그 어느 c값에 대하여도 두개의 특정한 점들을 지난다 하고 그 두 점들을 구하세요.

다음 예제들에서는 함수의 변환에 대하여 연습을 좀 해 보죠. 물론, 그 함수들의 커브들은 포물선들입니다. 변환에 대한 자세한 내용은 **제 2 권: 그래프 조작**을 참고하세요.

P. 다음과 같이 가정하세요: $y = f(x) = 2(x + 1)^2$ 이고 $f(x) = T_3 \bullet T_2 \bullet T_1 \bullet f$.
T_1: $(x, y) \longrightarrow (3 - x, -2 - y)$.
T_2: $(x, y) \longrightarrow (-x, -y)$.
T_3: $(x, y) \longrightarrow (x + u, y + v)$ 여기에서 u와 v는 상수.

그러면, u 와 v 의 값들을 구하세요.

Q. 다음과 같이 가정하세요: $y = f(x) = 2x^2 - 2x + 3$ 이고 $g = T_2 \bullet T_1 \bullet f$.
T_1: $(x, y) \longrightarrow (x - 3, y - 2)$.
T_2: $(x, y) \longrightarrow (y, x)$.

그러면, g 를 구하세요.

R. 다음과 같이 가정하세요: $y = f(x) = -2x^2 + 8x - 5$ 이고 $y = g(x) = 2x^2 - 6x + 5/2$ 이며 $g = T \bullet f$
T: $(x, y) \longrightarrow (u - x, v - y)$

그러면, u 와 v 의 값들을 구하세요.

S. 다음과 같이 가정하세요: $x = f(y)= 2(y - 1)^2 + 2$ 이고 $y = g(x) = (x - 3/2)^2 + 2$ 이며 $g = T_2 \bullet T_1 \bullet f$.
T_1: $(x, y) \longrightarrow (cy, 2x)$ 여기에서 c는 상수
T_2: $(x, y) \longrightarrow (x + u, y + v)$ 여기에서 u와 v는 상수

그러면, c 와 u 그리고 v 의 값들을 구하세요.

T. 다음과 같이 가정하세요:
$f(x) = 2x^2 - 2cx + c$ 여기에서 c는 상수이고 $c \geq -1$. $y = f(x)$의 커브는 포물선인데 그 꼭지점은 (s, t)이다.
T: $c \longrightarrow (s, t)$.
$t = g(s)$.

그러면, g 를 명시하세요.

U. 다음과 같이 가정하세요:
포물선 $x = y^2 - 2(c + 1)y + 3c^2$의 꼭지점은$(s, t)$인데 여기에서 c는 상수이고 $c \geq -1$.
T: $c \longrightarrow (s, t)$.
$t = g(s)$.

그러면, g 를 구하세요.

제안과 풀이

0 번 예제에서의 문제

포물선이 세 개의 점들 $P_1(1, 6)$, $P_2(-1, 4)$, $P_3(-2, 9)$를 포함하고 그 준선은 x축에 평행하다 하세요. 그러면, 그 포물선의 방정식을 구하세요.

준선이 x 축에 평행하니까 다음과 같은 방정식의 형태를 사용하는 것이 좋겠어요:
$y = ax^2 + bx + c$ 여기에서 a와 b 그리고 c는 상수.

그러면 이제, 주어진 세 개의 점들을 이용해서 a 와 b 그리고 c 에 관한 방정식들을 세워야 하죠. 이는 연립방정식이 되겠어요.

$y = ax^2 + bx + c$이죠. 그러면:

$P_1(1, 6) \Rightarrow 6 = a + b + c$
$P_2(-1, 4) \Rightarrow 4 = a - b + c$
$P_3(-2, 9) \Rightarrow 9 = 4a - 2b + c$

자 이제, 계산 좀 해야 하죠. 먼저, b 를 소거합니다.
$(a + b + c) + (a - b + c) = 6 + 4 \Rightarrow 2a + 2c = 10 \Rightarrow a + c = 5.$
$2(a + b + c) + (4a - 2b + c) = 12 + 9 \Rightarrow 6a + 3c = 21 \Rightarrow 2a + c = 7.$

다음으로, c 를 소거하고 a 를 구하죠.
$(2a + c) - (a + c) = 7 - 5 \Rightarrow a = 2.$

이제, c 를 구하면 되죠.
$(a + c = 5$ 이고 $a = 2) \Rightarrow 2 + c = 5 \Rightarrow c = 3.$

그러면, b 를 구해야 하죠. $a = 2$ 이고 $c = 3$ 이니까:
$6 = a + b + c \Rightarrow 6 = 2 + b + 3 \Rightarrow b = 1.$

그러니까, 그 방정식은 $y = 2x^2 + x + 3$ 이네요.

이제, 그 포물선을 그래프에 나타내보기로 하죠.
그러면, 그 방정식을 완전 제곱으로 변형해서 꼭지점을 드러내보는 것이 좋겠어요.

$y = 2x^2 + x + 3 = 2(x^2 + x/2) + 3$
$= 2\{x^2 + x/2 + (1/4)^2 - (1/4)^2\} + 3 = 2(x + 1/4)^2 - 1/8 + 3 = 2(x + 1/4)^2 + 23/8.$
$\Rightarrow y - 23/8 = 2(x + 1/4)^2.$

그러면 이제, 꼭지점은 $(-1/4 , 23/8)$에 있네요.

그림. 0.0

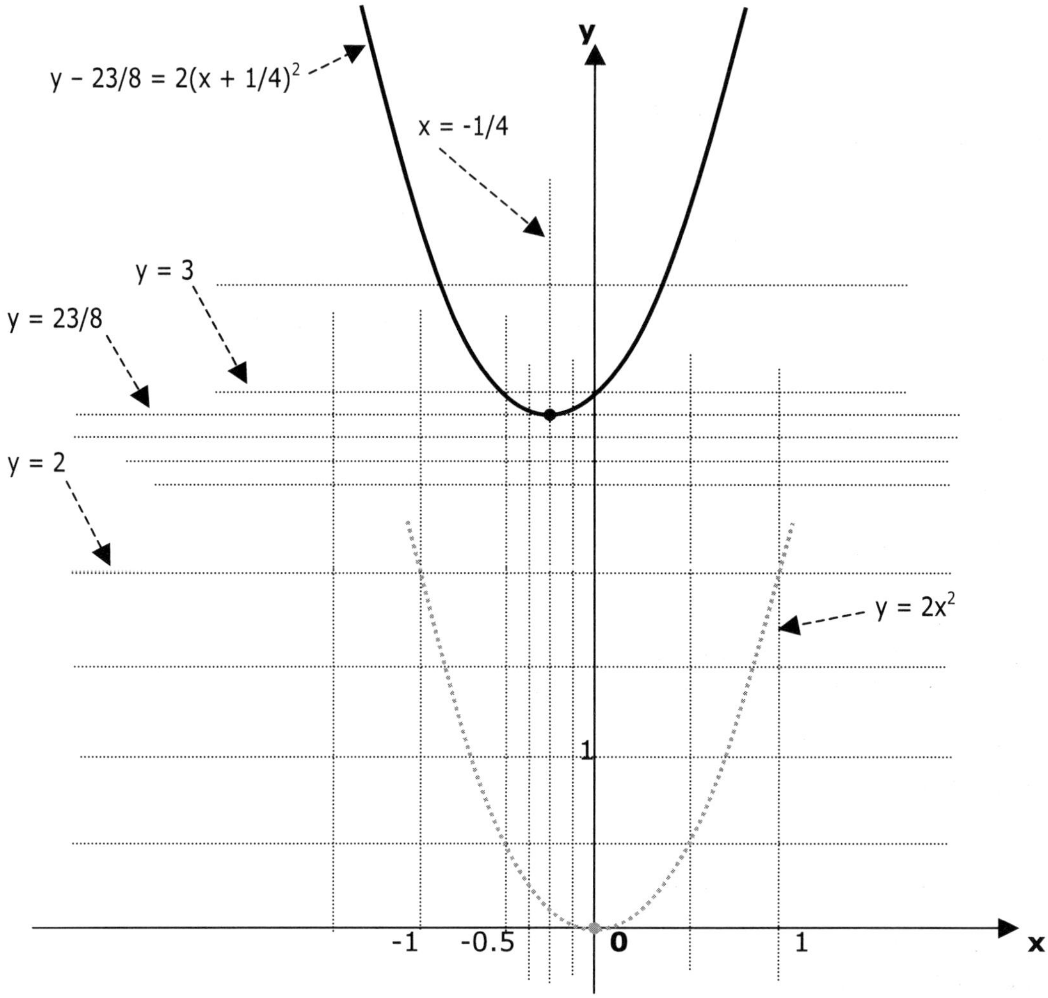

요약:

준선이 x 축에 평행이니까 다음과 같은 형태를 사용할 수 있죠:

$y = ax^2 + bx + c$ 여기에서 a와 b 그리고 c는 상수. 그러면:

$P_1(1, 6) \Rightarrow 6 = a + b + c$

$P_2(-1, 4) \Rightarrow 4 = a - b + c$

$P_3(-2, 9) \Rightarrow 9 = 4a - 2b + c$

먼저, b를 소거합니다.

$(a + b + c) + (a - b + c) = 6 + 4 \Rightarrow 2a + 2c = 10 \Rightarrow a + c = 5.$

$2(a + b + c) + (4a - 2b + c) = 12 + 9 \Rightarrow 6a + 3c = 21 \Rightarrow 2a + c = 7.$

다음으로, c를 소거하고 a를 구하죠.

$(2a + c) - (a + c) = 7 - 5 \Rightarrow a = 2.$

이제, c를 구하면 되죠.

$a = 2 \Rightarrow a + c = 2 + c = 5 \Rightarrow c = 3.$

그러면, b를 구해야 하죠.

$(6 = a + b + c, a = 2, c = 3) \Rightarrow 6 = 2 + b + 3 \Rightarrow b = 1.$

그러니까, 그 방정식은 $y = 2x^2 + x + 3.$

제안과 풀이

1번 예제에서의 문제

포물선이 세 개의 점들 $P_1(1, 6)$, $P_2(-1, 4)$, $P_3(-2, 9)$를 포함하고 그 대칭축은 x축에 평행하다 하세요. 그러면, 그 포물선의 방정식을 구하세요.

대칭축이 x축에 평행하니까 다음과 같은 방정식의 형태를 사용하는 것이 좋겠죠: $x = ay^2 + by + c$ 여기에서 a와 b 그리고 c는 상수. 그러면 이제, 주어진 세 개의 점들을 이용해서 a와 b 그리고 c에 관한 세 개로 된 연립방정식을 세워서 그 연립방정식을 풀어야 하죠. 그러면:

$P_1(1, 6) \Rightarrow 1 = 36a + 6b + c$
$P_2(-1, 4) \Rightarrow -1 = 16a + 4b + c$
$P_3(-2, 9) \Rightarrow -2 = 81a + 9b + c$

$(36a + 6b + c) - (16a + 4b + c) = 1 - (-1) \Rightarrow 20a + 2b = 2 \Rightarrow 10a + b = 1.$
$(81a + 9b + c) - (16a + 4b + c) = -2 - (-1) \Rightarrow 65a + 5b = -1 \Rightarrow 13a + b = -1/5.$
$(13a + b) - (10a + b) = -1/5 - 1 \Rightarrow 3a = -6/5 \Rightarrow a = -2/5.$

$a = -2/5 \Rightarrow 10a + b = -20/5 + b = 1 \Rightarrow b = 1 + 4 = 5.$

$a = -2/5$ 이고 $b = 5$ 이므로:
$-1 = 16a + 4b + c = -32/5 + 20 + c \Rightarrow c = (32 - 105)/5 = -73/5.$

그러므로, 그 방정식은 $x = -2y^2/5 + 5y - 73/5$ 이네요.

요약:

대칭축이 x축에 평행하니까: $x = ay^2 + by + c$ 여기에서 a와 b 그리고 c는 상수. 그러면:
$P_1(1, 6) \Rightarrow 1 = 36a + 6b + c$, $P_2(-1, 4) \Rightarrow -1 = 16a + 4b + c$, $P_3(-2, 9) \Rightarrow -2 = 81a + 9b + c$.

$(36a + 6b + c) - (16a + 4b + c) = 1 - (-1) \Rightarrow 20a + 2b = 2 \Rightarrow 10a + b = 1.$
$(81a + 9b + c) - (16a + 4b + c) = -2 - (-1) \Rightarrow 65a + 5b = -1 \Rightarrow 13a + b = -1/5.$
$(13a + b) - (10a + b) = -1/5 - 1 \Rightarrow 3a = -6/5 \Rightarrow a = -2/5.$

$a = -2/5 \Rightarrow 10a + b = -20/5 + b = 1 \Rightarrow b = 1 + 4 = 5.$
$a = -2/5$ 이고 $b = 5$ 이므로 $-1 = 16a + 4b + c = -32/5 + 20 + c \Rightarrow c = (32 - 105)/5 = -73/5.$
그러므로, 그 방정식은 $x = -2y^2/5 + 5y - 73/5$ 이네요.

이제, 그 포물선을 그래프에 넣어보기로 하죠. 먼저, 완전 제곱식 법으로 꼭지점 V를 나타내보이기로 하죠.
$x = -2y^2/5 + 5y - 73/5$
$\Rightarrow 5x = -2y^2 + 25y - 73 = -2(y^2 - 25y/2) - 73 = -2\{y^2 - 25y/2 + (25/4)^2 - (25/4)^2\} - 73$
$= -2(y - 25/4)^2 + 2(25/4)^2 - 73 = -2(y - 25/4)^2 + 25^2/8 - 73 = -2(y - 25/4)^2 + (625 - 584)/8$
$= -2(y - 25/4)^2 + 41/8$
$\Rightarrow 5x = -2(y - 25/4)^2 + 41/8 \Rightarrow x = (-2/5)(y - 25/4)^2 + 41/40 \Rightarrow x - 41/40 = (-2/5)(y - 25/4)^2.$
그러면 이제, 그 꼭지점 V는 (41/40, 25/4)에 있네요.

그림. 1.0

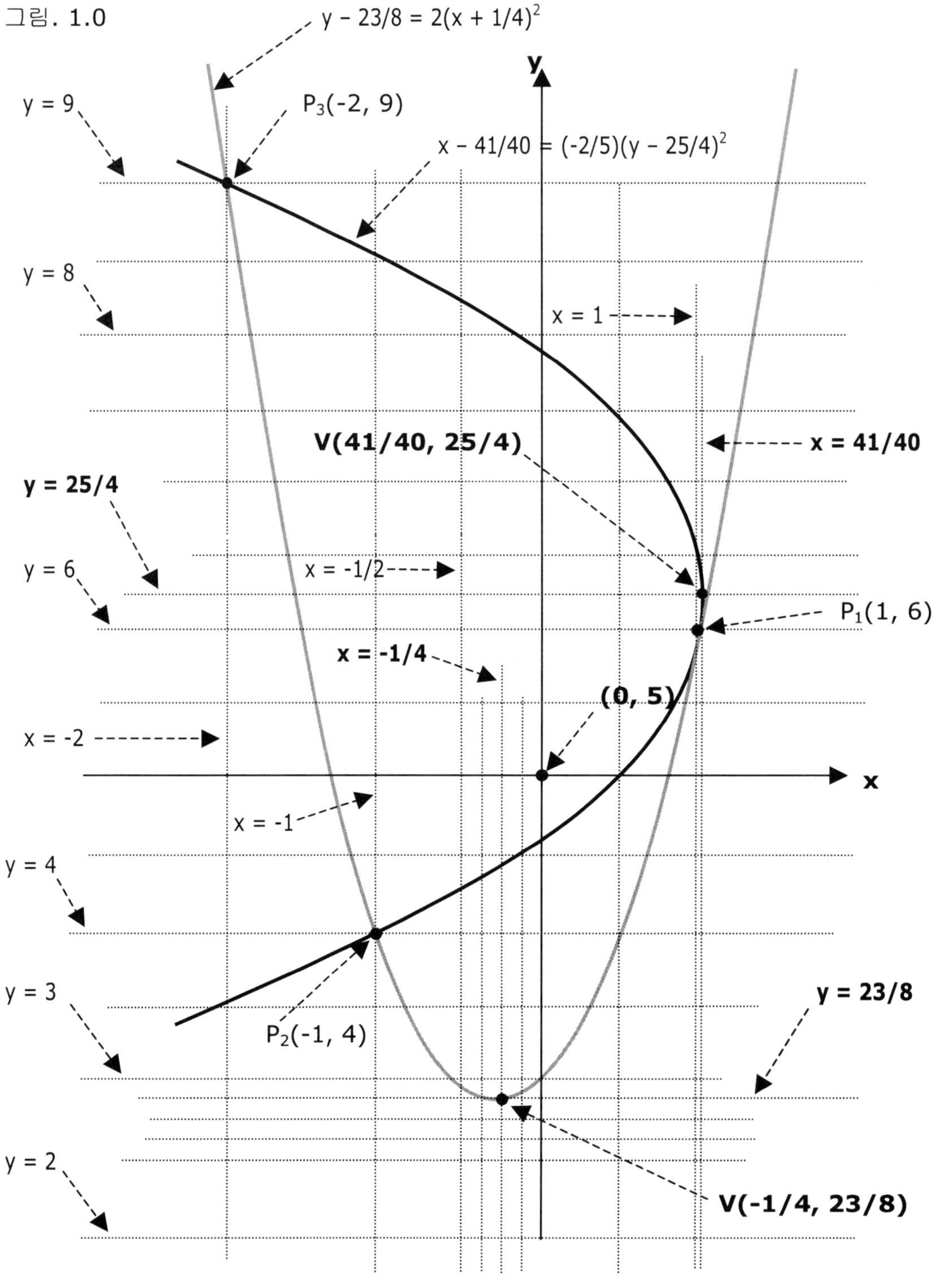

유의사항:

원점은 위의 그래프에는 나타나있지는 않지만 직선 $y = 2$ 아래 어디쯤에 있는 것은 물론이죠.

검정 포물선이 이 문제의 답인데, 회색 포물선은 0 번 예제에서 구한 것이고 이것도 주어진 그 세 점들을 지납니다.

제안과 풀이
2 번 예제에서의 문제

포물선 $y = 2x^2 + bx + c$가 있다 하고 그 포물선은 또 하나의 포물선 $y = -2x^2 + 3x/2 + 9/4$가 하나의 직선 $y = x/2 - 3/4$를 만나는 두개의 점들을 포함한다 하세요. 그러면, b와 c의 값들을 구하세요.

포물선 $y = 2x^2 + bx + c$를 A라 하세요.
그러면, A 에 대해서는 대칭축이 y 축에 평행이고 그 포물선은 두개의 점들을 지나는데 이 두 점들에서 문제에서 설명한 그 다른 하나의 포물선과 그 직선이 서로 만나는 거죠. 그러면, 그 두 점들을 찾아야 하네요.
그 두 커브들 $y = -2x^2 + 3x/2 + 9/4$와 $y = x/2 - 3/4$이 서로 만나는 그 두 점들에서는 y의 값들이 같아야 하죠. 그러면, x에 관한 방정식 $-2x^2 + 3x/2 + 9/4 = x/2 - 3/4$를 갖게 되네요. 그러면 이제:
$-2x^2 + 3x/2 + 9/4 = x/2 - 3/4 \Rightarrow 2x^2 - x - 3 = 0 \Rightarrow 2x^2 - x - 3 = (2x - 3)(x + 1) = 0 \Rightarrow x = 3/2$ 또는 -1.
그러면, $x = 3/2 \Rightarrow y = x/2 - 3/4 = 3/4 - 3/4 = 0$. $x = -1 \Rightarrow y = x/2 - 3/4 = -1/2 - 3/4 = -5/4$.
그러니까, 그 다른 포물선과 직선은 (3/2, 0)과 (-1, -5/4)에서 서로 만나네요.

그러면 이제, 그 포물선 $y = 2x^2 + bx + c$은 위의 두 점들을 지나는 겁니다. 그러면:
$0 = 2(3/2)^2 + (3/2)b + c = 9/2 + 3b/2 + c \Rightarrow 3b + 2c + 9 = 0 \Rightarrow 3b + 2c = -9$.
$-5/4 = 2(-1)^2 + (-1)b + c = 2 - b + c \Rightarrow b - c = 2 + 5/4 = 13/4 \Rightarrow b - c = 13/4$.

$(3b + 2c) + 2(b - c) = -9 + 26/4 = -10/4 \Rightarrow 5b = -10/4 \Rightarrow b = -1/2$.

그러면, $b = -1/2 \Rightarrow b - c = -1/2 - c = 13/4 \Rightarrow c = -1/2 - 13/4 = -15/4$.
따라서, 그 포물선 A는 $y = 2x^2 - x/2 - 15/4$이고 그러니까, $b = -1/2$이며 $c = -15/4$입니다.

요약:
그 다른 포물선 $y = -2x^2 + 3x/2 + 9/4$와 그 직선 $y = x/2 - 3/4$이 서로 만나는 그 두 점들을 찾아야 합니다.
$-2x^2 + 3x/2 + 9/4 = x/2 - 3/4 \Rightarrow 2x^2 - x - 3 = 0 \Rightarrow 2x^2 - x - 3 = (2x - 3)(x + 1) = 0 \Rightarrow x = 3/2$ 또는 -1.
그러면: $x = 3/2 \Rightarrow y = x/2 - 3/4 = 3/4 - 3/4 = 0$. $x = -1 \Rightarrow y = x/2 - 3/4 = -1/2 - 3/4 = -5/4$.
그러니까, 그 다른 포물선과 직선은 (3/2, 0)과 (-1, -5/4)에서 서로 만나네요.
그러면 이제, 그 포물선 $y = 2x^2 + bx + c$은 위의 두 점들을 지나는 겁니다. 그러면:
$0 = 2(3/2)^2 + (3/2)b + c = 9/2 + 3b/2 + c \Rightarrow 3b + 2c + 9 = 0 \Rightarrow 3b + 2c = -9$.
$-5/4 = 2(-1)^2 + (-1)b + c = 2 - b + c \Rightarrow b - c = 2 + 5/4 = 13/4 \Rightarrow b - c = 13/4$.
그러므로, 먼저 $(3b + 2c) + 2(b - c) = -9 + 26/4 = -10/4 \Rightarrow 5b = -10/4 \Rightarrow b = -1/2$.
다음으로, $b = -1/2 \Rightarrow b - c = -1/2 - c = 13/4 \Rightarrow c = -1/2 - 13/4 = -15/4$.

이제, 그 포물선들과 직선을 그래프로 나타내보기로 하죠.
$y = 2x^2 - x/2 - 15/4$
$\Rightarrow 4y = 8x^2 - 2x - 15 = 8(x^2 - x/4) - 15 = 8\{x^2 - x/4 + (1/8)^2 - (1/8)^2\} - 15$
$= 8(x - 1/8)^2 - 8(1/8)^2 - 15 = 8(x - 1/8)^2 - 1/8 - 15 = 8(x - 1/8)^2 - 121/8$.
$\Rightarrow y = 2(x - 1/8)^2 - 121/32 \Rightarrow y + 121/32 = 2(x - 1/8)^2$.
$\Rightarrow$ 꼭지점 = (1/8, -121/32) = (0.125, -3.78125).

$y = -2x^2 + 3x/2 + 9/4 = -2\{x^2 - 3x/4 + (3/8)^2 - (3/8)^2\} + 9/4 = -2(x - 3/8)^2 + 2(3/8)^2 + 9/4$
$= -2(x - 3/8)^2 + 9/32 + 9/4 = -2(x - 3/8)^2 + 81/32 \Rightarrow y - 81/32 = -2(x - 3/8)^2$.
$\Rightarrow$ 꼭지점 = (3/8, 81/32) = (0.375, 2.53125).

기울기들의 크기들이 같으니까 그 포물선들의 모양들이 같죠. 하지만 그 오목한 상태들은 서로 정반대죠 왜냐하면 부호가 반대니까. 그런데, 포물선 $y = 2x^2 - x/2 - 15/4$ 만이 그 두 점들을 지나는 유일한 포물선인가요?
물론, 아니죠! 그 두 점들을 지나는 포물선들은 무수히 많이 있을 수 있는 거죠.
하지만, 그 많은 포물선들 중에 기울기가 2 이며 대칭축이 y 축에 평행이고 그 두 점들을 포함하는 포물선은 단 하나밖에는 없는 겁니다, 그리고 그 단 하나의 포물선이 바로 그 포물선 A 인 거죠.

그림. 2.0

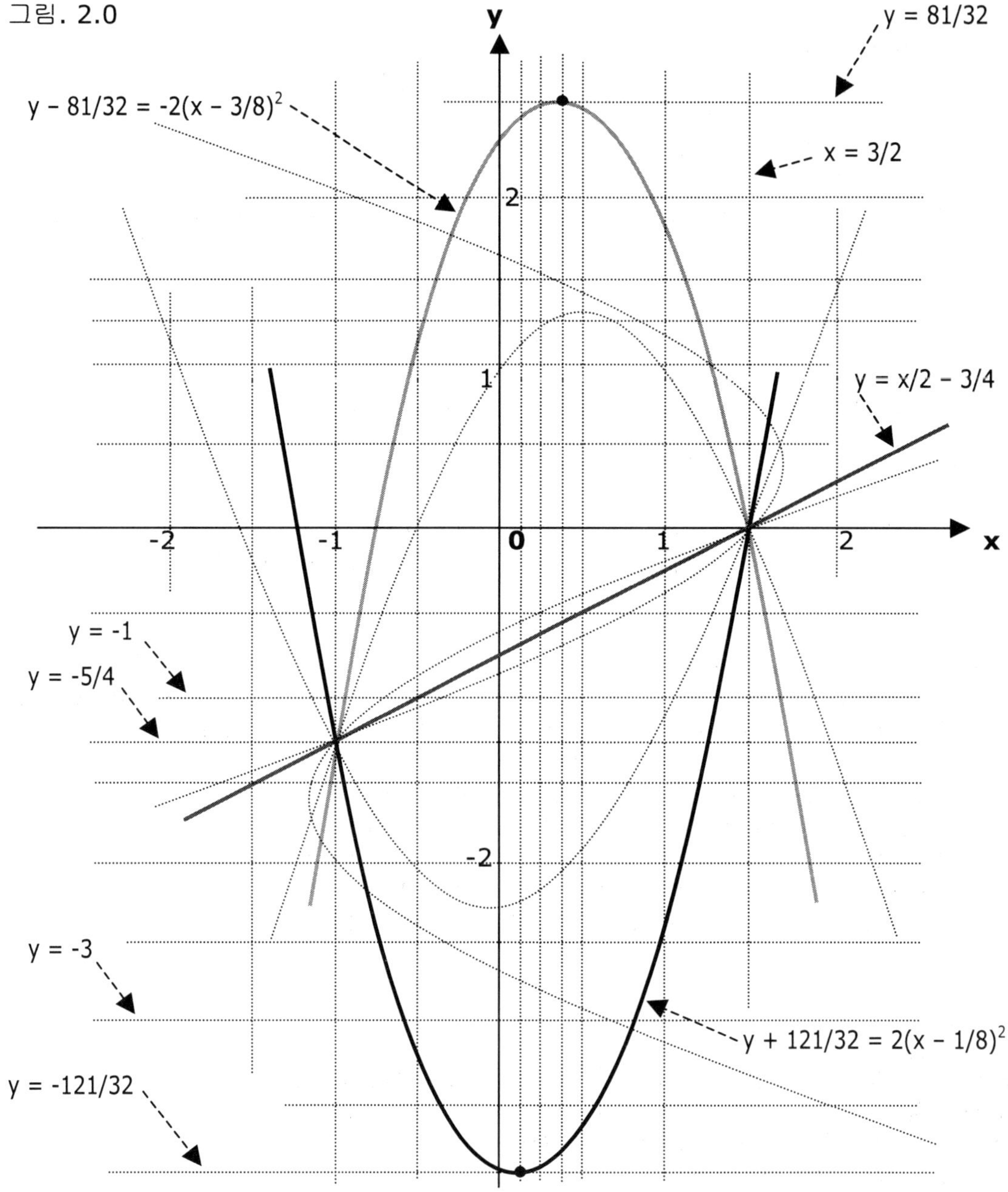

그럼 이제, 다음 예제에서 비슷한 것을 해보기로 하죠.

제안과 풀이
3번 예제에서의 문제

하나의 포물선 $y = ax^2 - x/2 + c$가 있다 하고 그 포물선은 또 하나의 포물선 $y = -2x^2 + 3x/2 + 9/4$가 직선 $y = x/2 - 3/4$를 만나는 두개의 점들을 포함한다 하세요. 그러면, a와 c의 값들을 구하세요.

포물선 $y = ax^2 - x/2 + c$를 B라 하세요. 그러면, B에 대해서는 대칭축이 y축에 평행이고 B는 두개의 점들을 지나는데 여기에서 문제에서 설명한 그 다른 하나의 포물선과 그 직선이 서로 만나는 거죠.
그런데, 그 두 점들은 (3/2, 0)와 (-1. -5/4)라는 걸 이전 예제에서 이미 알아냈죠.

또한, x 항의 계수가 −1/2 이고 이는 이전 예제에서의 그 포물선 A 의 방정식에서의 x 항의 계수와 같은 거죠.
그러면, 포물선 B 는 이전 예제에서의 포물선 A 와 같은 건가요?
다시 말해서, a 의 값이 2 와 같고 c 의 값은 −15/4 와 같은 건가요?
아마도 같을 겁니다. A 와 B 가 서로 같은 건지 알아보기로 하죠.

주어진 포물선 B의 방정식은 $y = ax^2 - x/2 + c$이고 이 포물선은 두 점들 (3/2, 0)와 (-1, -5/4)를 지나는 걸 알고있죠. 그러면:

$0 = a(3/2)^2 - (3/2)/2 + c = 9a/4 - 3/4 + c \Rightarrow 9a + 4c - 3 = 0 \Rightarrow 9a + 4c = 3.$
$-5/4 = a(-1)^2 - (-1)/2 + c = a + 1/2 + c \Rightarrow a + c = -1/2 - 5/4 = -7/4 \Rightarrow a + c = -7/4.$
$(9a + 4c) - 4(a + c) = 3 - (-7) = 10 \Rightarrow 5a = 10 \Rightarrow a = 2.$
그러면, $a = 2 \Rightarrow a + c = 2 + c = -7/4 \Rightarrow c = -7/4 - 2 = -15/4.$

그러면 이제, A 와 B 는 서로 동일한 거네요.

주어진 방정식이 B대신에 $y = ax^2 + bx - 15/4$ 라면 어떨까요?
그것도 동일한 방정식이 될 겁니다.
이는 왜냐하면 하나의 특정 포물선에 대하여는 단 하나의 특정한 방정식만이 있기 때문이죠.
몇 개의 동일한 방정식들이 달라보일 수 있어요.
예를 들면, $y = 2x^2 + 3x + 5/2$ 는 $4x^2 + 6x - 2y + 5 = 0$ 과 같은 겁니다.

그럼 이제, 다음 예제에서 또 하나의 유사한 것을 해보기로 하죠.

제안과 풀이
4번 예제에서의 문제

포물선 $x = ay^2 - y/2 + c$가 있다 하고 그 포물선은 또 하나의 포물선 $y = 2x^2 + x - 3$이 직선 $y = 3x + 1$을 만나는 두개의 점들을 포함한다 하세요. 그러면, a와 c의 값들을 구하세요.

포물선 $x = ay^2 - y/2 + c$를 C라 하세요. 그러면, C에 대해서는 대칭축이 x축에 평행이고 C는 두개의 점들을 지나는데 이 두 점들에서 문제에서 설명한 그 다른 포물선과 그 직선이 서로 만나는 거죠. 그러면 이제, 그 두 점들을 찾아야 하죠. 두 커브들 $y = 2x^2 + x - 3$과 $y = 3x + 1$에 있는 그 두 점들에서는 y의 값들이 같아야 하죠. 그러면, x에 관한 방정식 $2x^2 + x - 3 = 3x + 1$을 갖게 되네요. 그러면 이제:

$2x^2 + x - 3 = 3x + 1 \Rightarrow 2x^2 - 2x - 4 = 0 \Rightarrow 2(x + 1)(x - 2) = 0 \Rightarrow x = -1$ 또는 2.

그러면, $x = -1 \Rightarrow y = 3x + 1 = -3 + 1 = -2$. $x = 2 \Rightarrow y = 3x + 1 = 6 + 1 = 7$.

그러니까, 그 다른 포물선과 직선은 (-1, -2)와 (2, 7)에서 서로 만나네요.
그러면 이제, 그 포물선 $x = ay^2 - y/2 + c$는 위의 두 점들을 지나는 겁니다. 그러면:
$-1 = a(-2)^2 - (-2)/2 + c = 4a + 1 + c \Rightarrow 4a + c + 1= -1 \Rightarrow 4a + c = -2$.
$2 = 7^2a - 7/2 + c = 49a + c - 7/2 \Rightarrow 49a + c = 2 + 7/2 = 11/2 \Rightarrow 49a + c = 11/2$.

$(49a + c) - (4a + c) = 11/2 - (-2) = 15/2 \Rightarrow 45a = 15/2 \Rightarrow a = 1/6$.

그러면, $a = 1/6 \Rightarrow 4a + c = 4/6 + c = -2 \Rightarrow c = -16/6 = -8/3$.
따라서, 그 포물선 C는 $x = y^2/6 - y/2 - 8/3$이니까 $a = 1/6$이며 $c = -8/3$입니다.

요약:
포물선 $y = 2x^2 + x - 3$과 직선 $y = 3x + 1$이 서로 만나는 그 두 점들을 찾아야 하죠.
$2x^2 + x - 3 = 3x + 1 \Rightarrow 2x^2 - 2x - 4 = 0 \Rightarrow 2(x + 1)(x - 2) = 0 \Rightarrow x = -1$ 또는 2.
그러면, $x = -1 \Rightarrow y = 3x + 1 = -3 + 1 = -2$. $x = 2 \Rightarrow y = 3x + 1 = 6 + 1 = 7$.
그러니까, 그 다른 포물선과 직선은 (-1, -2)와 (2, 7)에서 서로 만나네요.

그러면 이제, 그 포물선 $x = ay^2 - y/2 + c$는 위의 두 점들을 지나는 겁니다. 그러면:
$-1 = a(-2)^2 - (-2)/2 + c = 4a + 1 + c \Rightarrow 4a + c + 1= -1 \Rightarrow 4a + c = -2$.
$2 = 7^2a - 7/2 + c = 49a + c - 7/2 \Rightarrow 49a + c = 2 + 7/2 = 11/2 \Rightarrow 49a + c = 11/2$.
$(49a + c) - (4a + c) = 11/2 - (-2) = 15/2 \Rightarrow 45a = 15/2 \Rightarrow a = 1/6$.
그러면, $a = 1/6 \Rightarrow 4a + c = 4/6 + c = -2 \Rightarrow c = -16/6 = -8/3$.

이제, 그 포물선들과 직선을 그래프로 나타내보기로 하죠.

$x = y^2/6 - y/2 - 8/3 \Rightarrow 6x = y^2 - 3y - 16 = y^2 - 3y + (3/2)^2 - (3/2)^2 - 16$
$= (y - 3/2)^2 - 9/4 - 16 = (y - 3/2)^2 - 73/4 \Rightarrow 6x = (y - 3/2)^2 - 73/4 \Rightarrow x = (1/6)(y - 3/2)^2 - 73/24$
$\Rightarrow x + 73/24 = (y - 3/2)^2/6$.
$\Rightarrow$ 꼭지점 $= (-73/24, 3/2) \approx (-3.04, 1.5)$.

$y = 2x^2 + x - 3 = 2\{x^2 + x/2 + (1/4)^2 - (1/4)^2\} - 3 = 2(x + 1/4)^2 - 2(1/4)^2 - 3 = 2(x + 1/4)^2 - 1/8 - 3$
$= 2(x + 1/4)^2 - 25/8 \Rightarrow y + 25/8 = 2(x + 1/4)^2 \Rightarrow$ 꼭지점 $= (-1/4, -25/8) = (-0.25, -3.125)$.

그런데, 포물선 $x = y^2/6 - y/2 - 8/3$ 만이 그 두 점들을 지나는 유일한 포물선인가요?
물론, 아니죠! 그 두 점들을 지나는 포물선들은 무수히 많이 있을 수 있는 거죠.
하지만, 그 많은 포물선들 중에 기울기가 1/6 이며 대칭축이 x 축에 평행이고 그 두 점들을 포함하는 포물선은 단 하나밖에는 없는 겁니다, 그리고 그 단 하나의 포물선이 바로 그 포물선 C 인 거죠.

그림. 4.0.

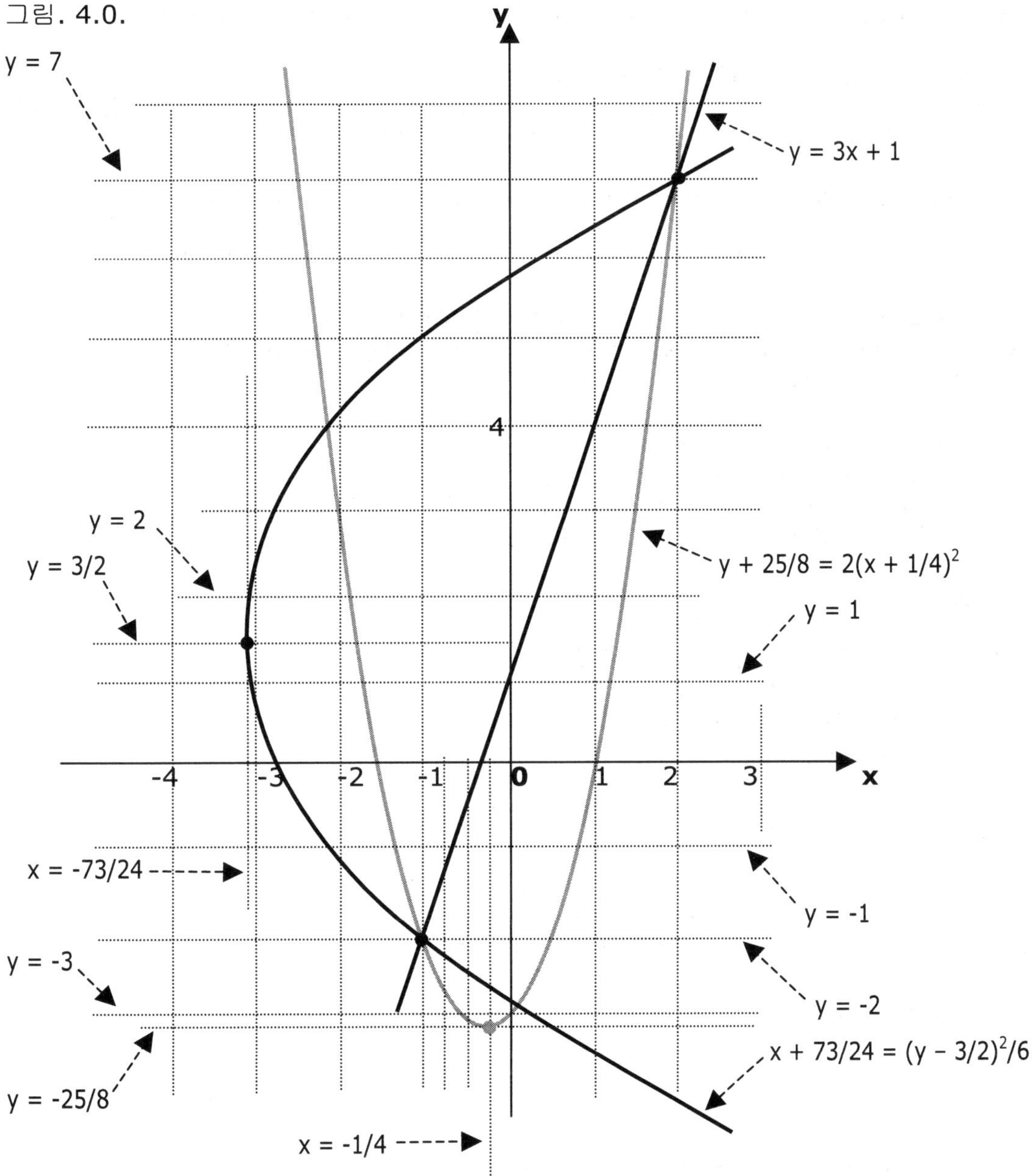

그럼 이제, 다음 예제를 해보기로 하죠.

제안과 풀이
5번 예제에서의 문제

하나의 포물선 $\mathbf{y/2 = ax^2 - 2x + c/2}$가 **(1/2, -1/2)**에 있는 점을 포함하고 그 꼭지점은 직선 $\mathbf{y = x - 1}$이 지난다 하세요. 그러면, **a**와 **c**의 값들을 구하세요.

포물선 $y/2 = ax^2 - 2x + c/2$가 (1/2, -1/2)에 있는 점을 지나니까:
$(-1/2)/2 = -1/4 = a(1/2)^2 - 2(1/2) + c/2 = a/4 - 1 + c/2 \Rightarrow -1 = a - 4 + 2c \Rightarrow a + 2c = 3.$

미지수가 a 와 c 두개니까 방정식을 하나 더 구해야 하죠.
그런데, 그 꼭지점이 그 직선에 있죠. 그러니까, 그 꼭지점의 좌표들을 a와 c로 나타내면 이것들에 대한 방정식을 하나 더 갖게 되는 거죠. 포물선 $y = ax^2 + bx + c$의 꼭지점은 다음과 같아요:

$y = ax^2 + bx + c = a(x + b/2a)^2 - (b^2 - 4ac)/4a \Rightarrow$ 그 꼭지점은 $\{-b/2a, -(b^2 - 4ac)/4a\}$.

흔히 이를 꼭지점의 공식이라고 하는데, 그 공식이라는 게 다음과 같은 완전 제곱 법의 결과일 뿐이죠.

$ax^2 + bx + c = a(x^2 + bx/a) + c = a\{x^2 + bx/a + (b/2a)^2 - (b/2a)^2\} + c$
$= a(x + b/2a)^2 - ab^2/4a^2 + c = a(x + b/2a)^2 - b^2/4a + c = a(x + b/2a)^2 - (b^2/4a - c)$
$= a(x + b/2a)^2 - (b^2 - 4ac)/4a.$

그러면 이제, 주어진 방정식에 완전 제곱 법을 바로 적용해보기로 하죠:
$y/2 = ax^2 - 2x + c/2 \Rightarrow y = 2ax^2 - 4x + c = 2a(x^2 - 2x/a + 1/a^2 - 1/a^2) + c$
$= 2a(x^2 - 2x/a + 1/a^2) - 2a/a^2 + c = 2a(x^2 - 2x/a + 1/a^2) - 2/a + c = 2a(x - 1/a)^2 - 2/a + c$
$\Rightarrow y + 2/a - c = 2a(x - 1/a)^2 \Rightarrow y - (-2/a + c) = 2a(x - 1/a)^2 \Rightarrow$ 꼭지점 $= (1/a, c - 2/a)$.

그 꼭지점 (1/a, -2/a + c)가 직선 $y = x - 1$에 있어야 하는 거죠.
그럼 이제, 그 꼭지점을 그 직선에 놓아보도록 하죠.
어떻게 그걸 직선에 놓나요? 그 방정식에 대입하는 거죠.
그러면, 그 꼭지점 (1/a, -2/a + c)이 직선 $y = x - 1$에 있으므로:
$y = x - 1 \Rightarrow -2/a + c = 1/a - 1 \Rightarrow 3/a - c = 1 \Rightarrow 3 - ac = a \Rightarrow a + ac = 3$
$\Rightarrow a(1 + c) = 3 \Rightarrow a = 3/(1 + c).$

자 이제, 두 개의 미지수 a 와 c에 대한 두개의 방정식들이 되었죠:
$a + 2c = 3$과 $a = 3/(1 + c)$.

그러면, $a + 2c = 3/(1 + c) + 2c = 3 \Rightarrow 3 + 2c(1 + c) = 3(1 + c) = 3 + 3c$
$\Rightarrow 3 + 3c = 3 + 2c + 2c^2 \Rightarrow 2c^2 - c = 0 \Rightarrow c(2c - 1) = 0 \Rightarrow c = 0$ 또는 1/2.

따라서, $c = 0 \Rightarrow a + 2c = a + 0 = 3 \Rightarrow a = 3$. $c = 1/2 \Rightarrow a + 2c = a + 1 = 3 \Rightarrow a = 2$.

요약:

포물선 $y/2 = ax^2 - 2x + c/2$가 (1/2, -1/2)에 있는 점을 지나니까:
$(-1/2)/2 = -1/4 = a(1/2)^2 - 2(1/2) + c/2 = a/4 - 1 + c/2 \Rightarrow -1 = a - 4 + 2c \Rightarrow a + 2c = 3.$
$y/2 = ax^2 - 2x + c/2 \Rightarrow y = 2ax^2 - 4x + c = 2a(x^2 - 2x/a + 1/a^2 - 1/a^2) + c$
$= 2a(x^2 - 2x/a + 1/a^2) - 2a/a^2 + c = 2a(x^2 - 2x/a + 1/a^2) - 2/a + c = 2a(x - 1/a)^2 - 2/a + c$

$\Rightarrow y + 2/a - c = 2a(x - 1/a)^2 \Rightarrow y - (-2/a + c) = 2a(x - 1/a)^2 \Rightarrow$ 꼭지점 = (1/a, c - 2/a).

그 꼭지점 (1/a, -2/a + c)가 직선 y = x - 1 에 있어야 하는 거죠. 그러므로, -2/a + c = 1/a - 1
$\Rightarrow 3/a - c = 1 \Rightarrow 3 - ac = a \Rightarrow a + ac = 3 \Rightarrow a(1 + c) = 3 \Rightarrow a = 3/(1 + c)$.
자 이제, a + 2c = 3 과 a = 3/(1 + c).
그러면, $a + 2c = 3/(1 + c) + 2c = 3 \Rightarrow 3 + 2c(1 + c) = 3(1 + c) = 3 + 3c$
$\Rightarrow 3 + 3c = 3 + 2c + 2c^2 \Rightarrow 2c^2 - c = 0 \Rightarrow c(2c - 1) = 0 \Rightarrow c = 0$ 또는 1/2.
따라서, $c = 0 \Rightarrow a + 2c = a + 0 = 3 \Rightarrow a = 3$. $c = 1/2 \Rightarrow a + 2c = a + 1 = 3 \Rightarrow a = 2$.
그러므로, a = 3 이고 c = 0 또는 a = 2 이고 c = 1/2.

그러면 이제, 두개의 포물선들이 답이 될 수 있는 거네요. $y/2 = ax^2 - 2x + c/2 \Rightarrow y = 2ax^2 - 4x + c$이죠.

먼저, a = 3 이고 c = 0 이면 $y = 2ax^2 - 4x + c \Rightarrow y = 6x^2 - 4x = 6\{x^2 - (2/3)x + 1/9 - 1/9\}$
$= 6(x - 1/3)^2 - 2/3. \Rightarrow y + 2/3 = 6(x - 1/3)^2 \Rightarrow y - (-2/3) = 6(x - 1/3)^2$.

다음으로, a = 2 이고 c = 1/2 이면 $y = 2ax^2 - 4x + c \Rightarrow y = 4x^2 - 4x + 1/2 = 4(x^2 - x) + 1/2$
$= 4(x^2 - x + 1/4 - 1/4) + 1/2 = 4(x - 1/2)^2 - 1 + 1/2 = 4(x - 1/2)^2 - 1/2$.
$\Rightarrow y + 1/2 = 4(x - 1/2)^2 \Rightarrow y - (-1/2) = 4(x - 1/2)^2$.

그림. 5.0

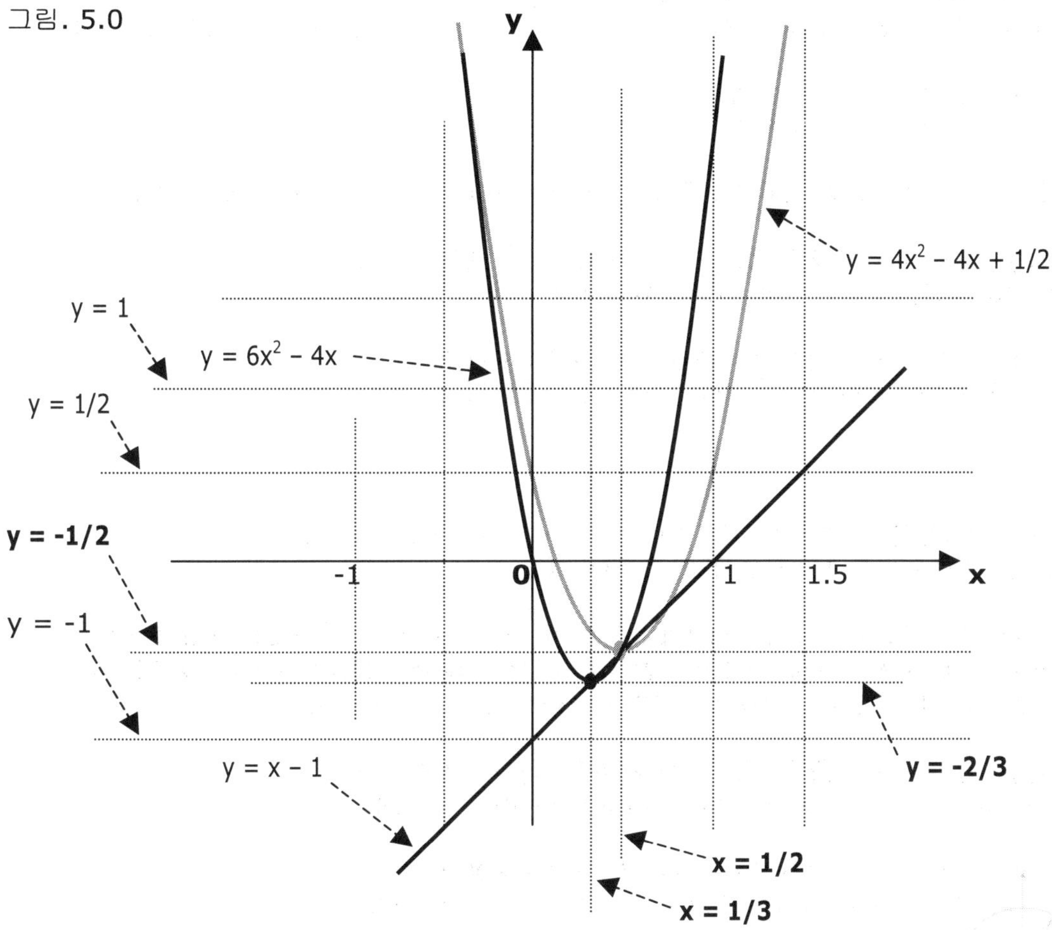

제안과 풀이
6 번 예제에서의 문제

하나의 포물선 $y = 6x^2 + bx + c$가 (2, 2)에 있는 점을 포함하고 그 꼭지점은 직선 $y = x - 1$이 지난다 하세요. 그러면, b와 c의 값들을 구하세요.

이번 예제에서는 다음과 같은 형태를 사용하는 것이 좋겠죠:
$y - n = a(x - m)^2$ 여기에서 a는 기울기이고 (m, n)은 꼭지점.

기울기가 6 이고 점 (2, 2)가 포물선에 있는 거죠.
그러면, $y - n = 6(x - m)^2$ 이고 이는 이제 m과 n에 관한 방정식이 되는 겁니다.

또한, 꼭지점에 관한 내용을 알고있죠.
꼭지점 (m, n)은 직선 $y = x - 1$ 에 있는 거니까 $n = m - 1$ 이 되는 거네요.

그러면 이제, 그 포물선의 방정식을 구할 채비가 다 된 겁니다.
$2 - n = 6(2 - m)^2$ 이고 $n = m - 1$ 이죠.

그러면, $2 - (m - 1) = 6(2 - m)^2 \Rightarrow \mathbf{2 - m + 1} = 6(2 - m)^2 = 6(4 - 4m + m^2) = 24 - 24m + 6m^2$
$\Rightarrow \mathbf{3 - m} = 24 - 24m + 6m^2 \Rightarrow 6m^2 - 23m + 21 = 0 \Rightarrow (2m - 3)(3m - 7) = 0 \Rightarrow m = 3/2$ 또는 7/3.

$m = 3/2 \Rightarrow n = m - 1 = 3/2 - 1 = 1/2$.
$m = 7/3 \Rightarrow n = m - 1 = 7/3 - 1 = 4/3$.

따라서, 다음과 같은 두개의 포물선들을 갖는 겁니다: $y - 1/2 = 6(x - 3/2)^2$와 $y - 4/3 = 6(x - 7/3)^2$.

그러면 이제, $y - 1/2 = 6(x - 3/2)^2 = 6(x^2 - 3x + 9/4) = 6x^2 - 18x + 27/2$
$\Rightarrow y = 6x^2 - 18x + 27/2 + 1/2 = 6x^2 - 18x + 14 \Rightarrow b = -18$ 이고 $c = 14$.

$y - 4/3 = 6(x - 7/3)^2 = 6(x^2 - 14x/3 + 49/9) = 6x^2 - 28x + 98/3$
$\Rightarrow y = 6x^2 - 28x + 98/3 + 4/3 = 6x^2 - 28x + 34 \Rightarrow b = -28$ 이고 $c = 34$.

요약:

기울기가 6 이고 점 (2, 2)가 포물선에 있으므로 $y - n = 6(x - m)^2$ 이라고 놓으세요.
꼭지점 (m, n)은 직선 $y = x - 1$ 에 있는 거니까 $n = m - 1$.

그러면, $2 - (m - 1) = 6(2 - m)^2 \Rightarrow \mathbf{2 - m + 1} = 6(2 - m)^2 = 6(4 - 4m + m^2) = 24 - 24m + 6m^2$
$\Rightarrow \mathbf{3 - m} = 24 - 24m + 6m^2 \Rightarrow 6m^2 - 23m + 21 = 0 \Rightarrow (2m - 3)(3m - 7) = 0 \Rightarrow m = 3/2$ 또는 7/3.
$m = 3/2 \Rightarrow n = m - 1 = 3/2 - 1 = 1/2$. $m = 7/3 \Rightarrow n = m - 1 = 7/3 - 1 = 4/3$.

그러면 이제, $y - 1/2 = 6(x - 3/2)^2 = 6(x^2 - 3x + 9/4) = 6x^2 - 18x + 27/2$
$\Rightarrow y = 6x^2 - 18x + 27/2 + 1/2 = 6x^2 - 18x + 14 \Rightarrow b = -18$ 이고 $c = 14$.

$y - 4/3 = 6(x - 7/3)^2 = 6(x^2 - 14x/3 + 49/9) = 6x^2 - 28x + 98/3$
$\Rightarrow y = 6x^2 - 28x + 98/3 + 4/3 = 6x^2 - 28x + 34 \Rightarrow b = -28$ 이고 $c = 34$.

그림. 6.0

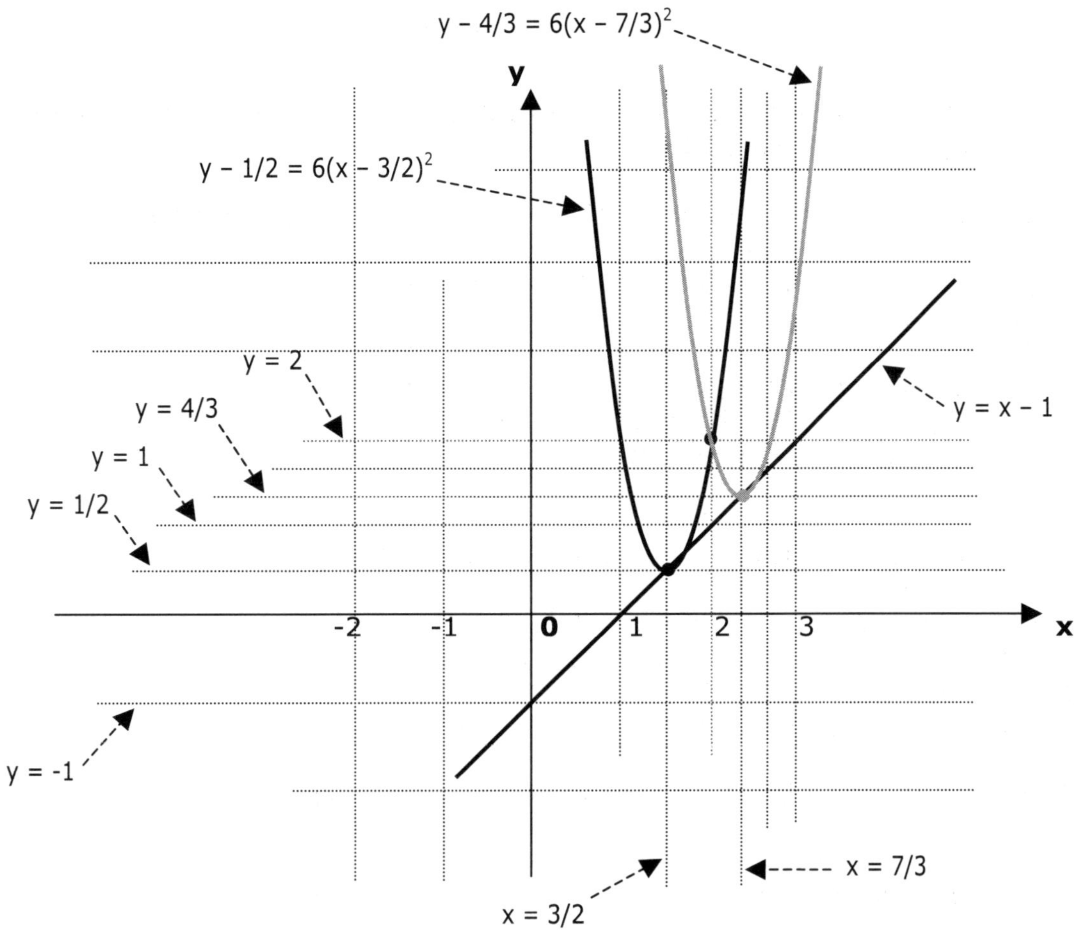

이제, 다음 예제에서 비슷한 걸 해보기로 하죠.

제안과 풀이
7번 예제에서의 문제

하나의 포물선 $y = ax^2 + bx + c$가 두 개의 점들 (1, 1)과 (2, 3)을 포함하고 그 꼭지점은 직선 $y = 2x - 1$에 있다 하세요. 그러면, a와 b 그리고 c의 값들을 구하세요.

포물선 $y = ax^2 + bx + c$의 꼭지점을 구하는 공식을 기억하면 그 꼭지점을 $\{-b/2a, -(b^2 - 4ac)/4a\}$라고 즉시 표현할 수 있죠. 기억하지 못하면 완전 제곱식으로 그걸 유도하면 되죠.
그러면 이제, 그 꼭지점이 직선 $y = 2x - 1$에 있으니까 다음과 같은 방정식을 갖는 거죠:
$-(b^2 - 4ac)/4a = 2(-b/2a) - 1$.
또한, 두 개의 점들 (1, 1)과 (2, 3)이 그 포물선에 있으니까:
$(1, 1) \Rightarrow 1 = a(1^2) + b(1) + c = a + b + c$ 이고 또한, $(2, 3) \Rightarrow 3 = 4a + 2b + c$.
자 이제, 계산하는 것만 남았네요. 다음과 같은 연립방정식을 풀어야 하죠:
$-(b^2 - 4ac)/4a = 2(-b/2a) - 1,\ a + b + c = 1,\ 4a + 2b + c = 3$.

먼저, 첫째 방정식을 간단히 해보죠:
$-(b^2 - 4ac)/4a = 2(-b/2a) - 1 = -b/a - 1 = -(b + a)/a \Rightarrow b^2 - 4ac = 4(b + a)$.
한편, $a + b + c = 1 \Rightarrow a + b = 1 - c$.
그러면, $b^2 - 4ac = 4(b + a) = 4(1 - c) \Rightarrow b^2 - 4ac = 4(1 - c)$.
그런데, $4a + 2b + c = 3 \Rightarrow 4ac + 2bc + c^2 = 3c \Rightarrow 4ac = 3c - 2bc - c^2$.
그러면, $b^2 - 4ac = 4(1 - c) \Rightarrow b^2 - (3c - 2bc - c^2) = b^2 - 3c + 2bc + c^2 = 4(1 - c) = 4 - 4c$.
따라서, $b^2 - 3c + 2bc + c^2 = 4 - 4c \Rightarrow b^2 + 2bc + c^2 + c - 4 = 0$.

그런데, $4a + 2b + c = 3$이고 $a + b + c = 1$이니까 $4a + 2b + c - 4(a + b + c) = 3 - 4 \Rightarrow -2b - 3c = -1$.

한편, $b^2 + 2bc + c^2 + c - 4 = 0 \Rightarrow (b + c)^2 + c - 4 = 0$.
또한, $-2b - 3c = -1 \Rightarrow -2b - 2c - c = -2(b + c) - c = -1 \Rightarrow 2(b + c) = 1 - c \Rightarrow b + c = (1 - c)/2$.

그러면, $(b + c)^2 + c - 4 = 0$이고 $b + c = (1 - c)/2$이니까 $(1 - c)^2/4 + c - 4 = 0$이 되죠.
따라서, $1 - 2c + c^2 + 4c - 16 = 0 \Rightarrow c^2 + 2c - 15 = 0 \Rightarrow (c + 5)(c - 3) = 0 \Rightarrow c = -5$ 또는 3.

이제, 다음 연립방정식에 c의 값을 대입하기로 하죠: $a + b + c = 1$이고 $4a + 2b + c = 3$.
그런데, 이 연립방정식을 다음과 같이 놓기로 하죠: $a + b = 1 - c$이고 $4a + 2b = 3 - c$.
그럼 이제, $c = -5$이면 $a + b = 1 + 5 = 6 \Rightarrow a + b = 6$이죠 또한, $4a + 2b = 3 + 5 = 8 \Rightarrow 2a + b = 4$.
그러면, $2a + b - (a + b) = 4 - 6 = -2 \Rightarrow a = -2 \Rightarrow a + b = -2 + b = 6 \Rightarrow b = 8$.
다음으로, $c = 3$일 때 $a + b = 1 - 3 = -2 \Rightarrow a + b = -2$ 이고 또한, $4a + 2b = 3 - 3 = 0 \Rightarrow 2a + b = 0$.
그러면, $2a + b - (a + b) = 0 - (-2) = 2 \Rightarrow a = 2 \Rightarrow a + b = 2 + b = -2 \Rightarrow b = -4$.
그러니까, 다음과 같은 두개의 포물선들을 갖게 되는 거네요:
$y = -2x^2 + 8x - 5 = -2(x^2 - 4x) - 5 = -2(x^2 - 4x + 4 - 4) - 5 = -2(x - 2)^2 + 3 \Rightarrow y - 3 = -2(x - 2)^2$.
$y = 2x^2 - 4x + 3 = 2(x^2 - 2x) + 3 = 2(x - 1)^2 + 1 \Rightarrow y - 1 = 2(x - 1)^2$.

요약:
포물선 $y = ax^2 + bx + c$의 꼭지점은 $\{-b/2a, -(b^2 - 4ac)/4a\}$이고 이는 직선 $y = 2x - 1$에 있으니까:
$-(b^2 - 4ac)/4a = 2(-b/2a) - 1$.
또한, 두 개의 점들 (1, 1)과 (2, 3)이 그 포물선에 있으니까:
$(1, 1) \Rightarrow 1 = a(1^2) + b(1) + c = a + b + c$ 이고 $(2, 3) \Rightarrow 3 = 4a + 2b + c$.

그러면 이제: $-(b^2 - 4ac)/4a = 2(-b/2a) - 1$, $a + b + c = 1$, $4a + 2b + c = 3$.
우선:
$-(b^2 - 4ac)/4a = 2(-b/2a) - 1 = -b/a - 1 = -(b + a)/a \Rightarrow b^2 - 4ac = 4(b + a)$.
$a + b + c = 1 \Rightarrow a + b = 1 - c$.
그러면, $b^2 - 4ac = 4(b + a) = 4(1 - c) \Rightarrow b^2 - 4ac = 4(1 - c)$.
그런데, $4a + 2b + c = 3 \Rightarrow 4ac + 2bc + c^2 = 3c \Rightarrow 4ac = 3c - 2bc - c^2$.
그러면, $b^2 - 4ac = 4(1 - c) \Rightarrow b^2 - (3c - 2bc - c^2) = b^2 - 3c + 2bc + c^2 = 4(1 - c) = 4 - 4c$.
따라서, $b^2 - 3c + 2bc + c^2 = 4 - 4c \Rightarrow b^2 + 2bc + c^2 + c - 4 = 0$.
그런데, $4a + 2b + c - 4(a + b + c) = 3 - 4 \Rightarrow -2b - 3c = -1$.
한편, $b^2 + 2bc + c^2 + c - 4 = 0 \Rightarrow (b + c)^2 + c - 4 = 0$.
또한, $-2b - 3c = -1 \Rightarrow -2b - 2c - c = -2(b + c) - c = -1 \Rightarrow 2(b + c) = 1 - c \Rightarrow b + c = (1 - c)/2$.
그러면, $(b + c)^2 + c - 4 = 0$ 이고 $b + c = (1 - c)/2$ 이니까 $(1 - c)^2/4 + c - 4 = 0$.
따라서, $1 - 2c + c^2 + 4c - 16 = 0 \Rightarrow c^2 + 2c - 15 = 0 \Rightarrow (c + 5)(c - 3) = 0 \Rightarrow c = -5$ 또는 3.
그럼 이제, $a + b + c = 1$ 이고 $4a + 2b + c = 3$ 이므로: $a + b = 1 - c$ 이고 $4a + 2b = 3 - c$.

먼저, $c = -5$ 일 때 $a + b = 1 + 5 = 6 \Rightarrow a + b = 6$ 이죠 또한, $4a + 2b = 3 + 5 = 8 \Rightarrow 2a + b = 4$.
그러면, $2a + b - (a + b) = 4 - 6 = -2 \Rightarrow a = -2 \Rightarrow a + b = -2 + b = 6 \Rightarrow b = 8$.
그러므로, $a = -2$, $b = 8$, $c = -5$.
다음으로, $c = 3$ 일 때 $a + b = 1 - 3 = -2 \Rightarrow a + b = -2$ 이고 또한, $4a + 2b = 3 - 3 = 0 \Rightarrow 2a + b = 0$.
그러면, $2a + b - (a + b) = 0 - (-2) = 2 \Rightarrow a = 2 \Rightarrow a + b = 2 + b = -2 \Rightarrow b = -4$.
그러므로, $a = 2$, $b = -4$, $c = 3$.

그림. 7.0

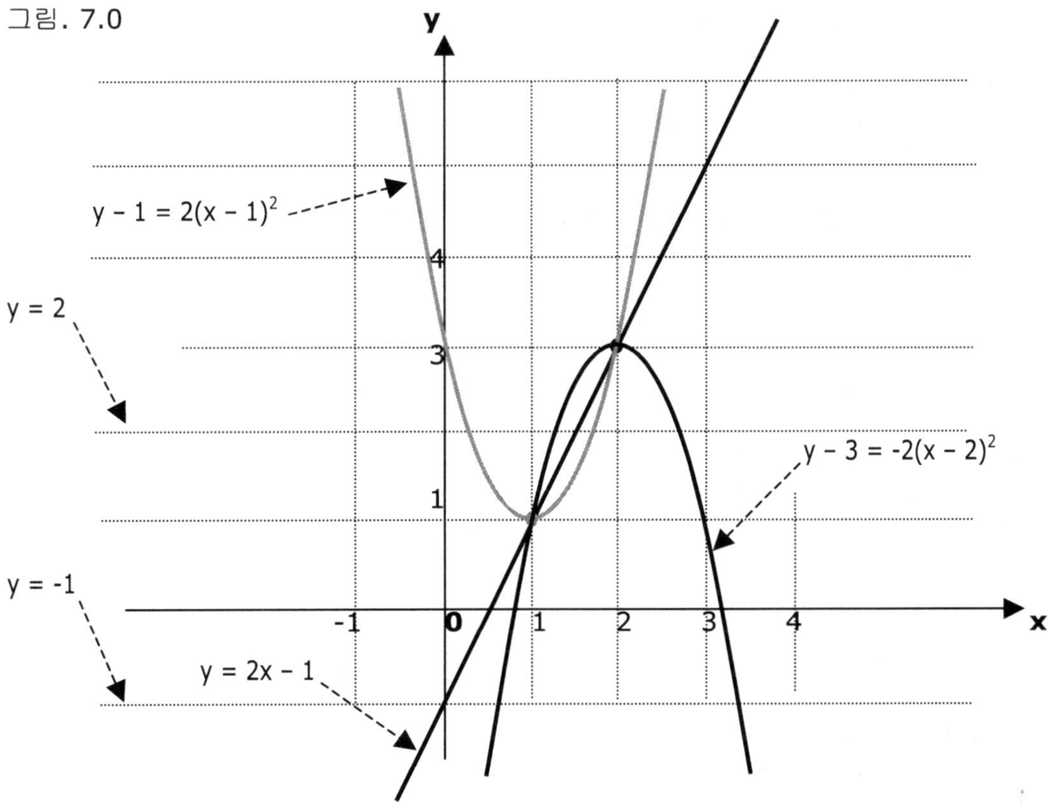

제안과 풀이
8 번 예제에서의 문제

하나의 포물선 $y = ax^2 + bx + c$가 두 개의 점들 (1, 1)과 (2, 4)을 포함하고 그 꼭지점은 직선 $y = 2x - 1$ 에 있다 하세요. 그러면, a와 b 그리고 c의 값들을 구하세요.

이번 예제에서 이전 예제와 다른 것은 그 두 점들 중 하나의 위치 뿐이네요. 이전 예제에서는 (2, 4)대신에 (2, 3)이었죠. 포물선 $y = ax^2 + bx + c$의 꼭지점은 $\{-b/2a, -(b^2 - 4ac)/4a\}$이죠. 잘 모르면 유도하는 것이 좋겠죠.
그러면 이제, 그 꼭지점이 직선 $y = 2x - 1$ 에 있으니까 다음과 같은 방정식을 갖는 거죠:
$-(b^2 - 4ac)/4a = 2(-b/2a) - 1$.
또한, 두 개의 점들 (1, 1)과 (2, 4)가 그 포물선에 있으니까:
$(1, 1) \Rightarrow 1 = a(1^2) + b(1) + c = a + b + c$ 이고 또한, $(2, 4) \Rightarrow 4 = 4a + 2b + c$.
자 이제, 계산하는 것만 남았네요. 다음과 같은 연립방정식을 풀어야 하죠:
$-(b^2 - 4ac)/4a = 2(-b/2a) - 1,\ a + b + c = 1,\ 4a + 2b + c = 4$.

먼저, 첫째 방정식을 간단히 해보죠:
$-(b^2 - 4ac)/4a = 2(-b/2a) - 1 = -b/a - 1 = -(b + a)/a \Rightarrow b^2 - 4ac = 4(b + a)$.
한편, $a + b + c = 1 \Rightarrow a + b = 1 - c$.
그러면, $b^2 - 4ac = 4(b + a) = 4(1 - c) \Rightarrow b^2 - 4ac = 4(1 - c)$.
그런데, $4a + 2b + c = 4 \Rightarrow 4ac + 2bc + c^2 = 4c \Rightarrow 4ac = 4c - 2bc - c^2$.
그러면, $b^2 - 4ac = 4(1 - c) \Rightarrow b^2 - (4c - 2bc - c^2) = b^2 - 4c + 2bc + c^2 = 4(1 - c) = 4 - 4c$.
따라서, $b^2 - 4c + 2bc + c^2 = 4 - 4c \Rightarrow b^2 + 2bc + c^2 = 4 \Rightarrow (b + c)^2 = 4 \Rightarrow b + c = 2$ 또는 -2.

그럼 이제, $a + b + c = 1$ 이니까:
$b + c = 2 \Rightarrow a + b + c = a + 2 = 1 \Rightarrow a = -1$.
$b + c = -2 \Rightarrow a + b + c = a - 2 = 1 \Rightarrow a = 3$.

그러면, 먼저 $a = -1$ 이면 $4a + 2b + c = 4 \Rightarrow 2b + c = 4 - 4a = 4 - 4(-1) = 8 \Rightarrow 2b + c = 8$.
또한, $b + c = 2$ 이죠. 그러면, $2b + c - (b + c) = 8 - 2 = 6 \Rightarrow b = 6 \Rightarrow b + c = 6 + c = 2 \Rightarrow c = -4$.

다음으로, $a = 3$ 이면 $4a + 2b + c = 4 \Rightarrow 2b + c = 4 - 4a = 4 - 4(3) = -8 \Rightarrow 2b + c = -8$.
또한, $b + c = -2$ 이죠.
그러면, $2b + c - (b + c) = -8 - (-2) = -8 + 2 = -6 \Rightarrow b = -6 \Rightarrow b + c = -6 + c = -2 \Rightarrow c = 4$.

그러니까, 다음과 같은 두개의 포물선을 갖게 되는 거네요:
$y = -x^2 + 6x - 4 = -(x^2 - 6x) - 4 = -(x - 3)^2 + 5 \Rightarrow y - 5 = -(x - 3)^2$.
$y = 3x^2 - 6x + 4 = 3(x^2 - 2x) + 4 = 3(x - 1)^2 + 1 \Rightarrow y - 1 = 3(x - 1)^2$.

요약:

포물선 $y = ax^2 + bx + c$의 꼭지점은 $\{-b/2a, -(b^2 - 4ac)/4a\}$이고 이는 직선 $y = 2x - 1$ 에 있습니다.
그러므로, $-(b^2 - 4ac)/4a = 2(-b/2a) - 1$.
또한, 두 개의 점들 (1, 1)과 (2, 4)가 그 포물선에 있으므로:
$(1, 1) \Rightarrow 1 = a(1^2) + b(1) + c = a + b + c$ 이고 또한, $(2, 4) \Rightarrow 4 = 4a + 2b + c$.
그러면 이제, 다음과 같은 연립방정식을 풀어야 합니다:
$-(b^2 - 4ac)/4a = 2(-b/2a) - 1,\ a + b + c = 1,\ 4a + 2b + c = 4$.

$-(b^2 - 4ac)/4a = 2(-b/2a) - 1 = -b/a - 1 = -(b + a)/a \Rightarrow b^2 - 4ac = 4(b + a)$.
$a + b + c = 1 \Rightarrow a + b = 1 - c$.

그러면, $b^2 - 4ac = 4(b + a) = 4(1 - c) \Rightarrow b^2 - 4ac = 4(1 - c)$.
그런데, $4a + 2b + c = 4 \Rightarrow 4ac + 2bc + c^2 = 4c \Rightarrow 4ac = 4c - 2bc - c^2$.
그러면, $b^2 - 4ac = 4(1 - c) \Rightarrow b^2 - (4c - 2bc - c^2) = b^2 - 4c + 2bc + c^2 = 4(1 - c) = 4 - 4c$.
따라서, $b^2 - 4c + 2bc + c^2 = 4 - 4c \Rightarrow b^2 + 2bc + c^2 = 4 \Rightarrow (b + c)^2 = 4 \Rightarrow b + c = 2$ 또는 -2.

그럼 이제, $a + b + c = 1$ 이므로:
$b + c = 2 \Rightarrow a + b + c = a + 2 = 1 \Rightarrow a = -1$.
$b + c = -2 \Rightarrow a + b + c = a - 2 = 1 \Rightarrow a = 3$.

먼저, $a = -1$ 이고 $b + c = 2$ 일 때:
$4a + 2b + c = 4 \Rightarrow 2b + c = 4 - 4a = 4 - 4(-1) = 8 \Rightarrow 2b + c = 8$.
그러면, $2b + c - (b + c) = 8 - 2 = 6 \Rightarrow b = 6 \Rightarrow b + c = 6 + c = 2 \Rightarrow c = -4$.
그러므로, $a = -1$, $b = 6$, $c = -4$.

다음으로, $a = 3$ 이고 $b + c = -2$ 일 때:
$4a + 2b + c = 4 \Rightarrow 2b + c = 4 - 4a = 4 - 4(3) = -8 \Rightarrow 2b + c = -8$.
그러면, $2b + c - (b + c) = -8 - (-2) = -8 + 2 = -6 \Rightarrow b = -6 \Rightarrow b + c = -6 + c = -2 \Rightarrow c = 4$.
그러므로, $a = 3$, $b = -6$, $c = 4$.

그림. 8.0

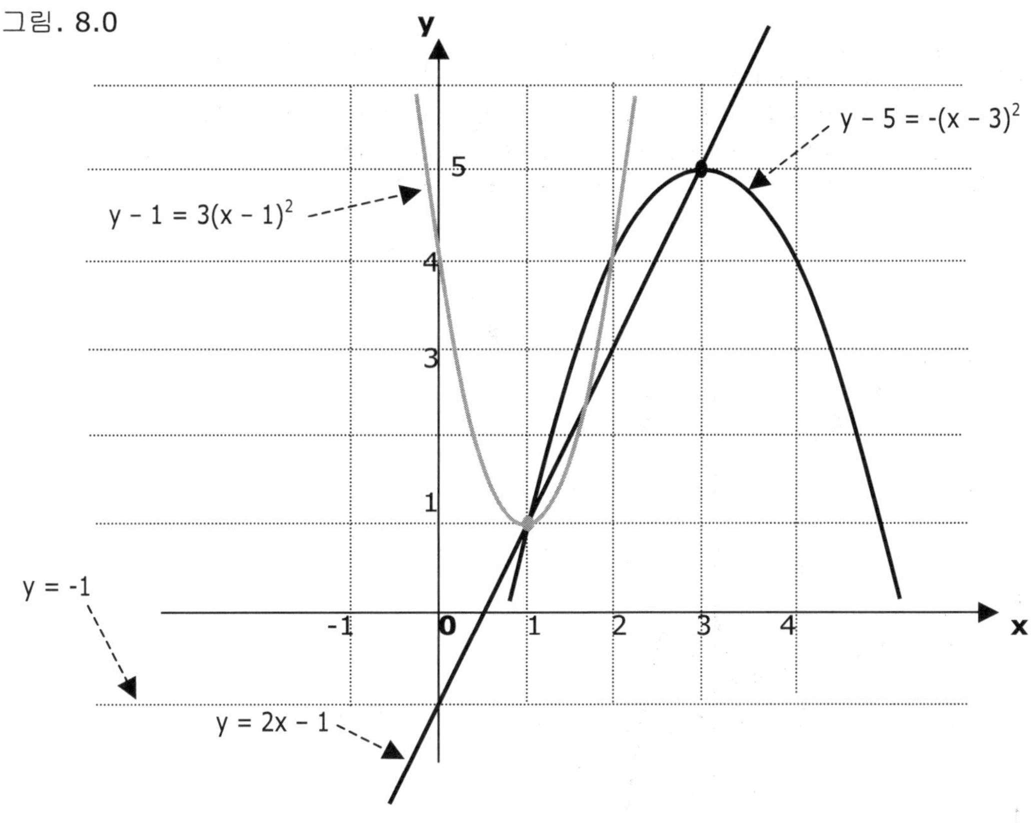

제안과 풀이
9번 예제에서의 문제

포물선 A의 방정식이 $y = -4x^2 + 8mx - n(n + 2)$라 하고 포물선 B의 방정식은 $y = 8x^2 - 2x + 5/4$라 하세요. 또한, A의 꼭지점이 B에 있다고 하세요. 그러면, m과 n의 값들을 구하세요.

먼저, 포물선 A의 꼭지점을 구해서 포물선 B에 대입해보도록 하죠.
그러면, $y = -4x^2 + 8mx - n(n + 2) = -4(x^2 - 2mx) - n(n + 2) = -4(x - m)^2 + 4m^2 - n(n + 2)$.
따라서, 그 꼭지점 = $\{m, 4m^2 - n(n + 2)\}$이네요. 그럼 이제, 이를 포물선 B에 대입해야 하죠.
그러면, $4m^2 - n(n + 2) = 8m^2 - 2m + 5/4 \Rightarrow 4m^2 - 2m + n(n + 2) + 5/4 = 0$.

그런데, 위의 식을 조사해보면 n^2항이 있다는 걸 알 수 있죠. 그러면, 이는 제각기 m과 n에 관한 완전 제곱식들의 합이 될 수도 있는 거죠. 그러면, 완전 제곱식을 시도해보기로 하죠.
$4m^2 - 2m + n(n + 2) + 5/4 = 4m^2 - 2m + 1/4 + n^2 + 2n + 1 = (2m - 1/2)^2 + (n + 1)^2 = 0$.

그러네요. 그럼 이제, 다음과 같은 사실을 고려해야 하는 겁니다: $A^2 + B^2 = 0$이면 A와 B 둘 다 0이다.
따라서, $2m - 1/2 = 0$이고 또한, $n + 1 = 0$이어야 하죠. 그러므로, $m = 1/4$이고 $n = -1$.

요약:

$y = -4x^2 + 8mx - n(n + 2) = -4(x^2 - 2mx) - n(n + 2) = -4(x - m)^2 + 4m^2 - n(n + 2)$.
따라서, 그 꼭지점 = $\{m, 4m^2 - n(n + 2)\}$. 이 꼭지점은 포물선 B에 있죠.
그러니까, $4m^2 - n(n + 2) = 8m^2 - 2m + 5/4 \Rightarrow 4m^2 - 2m + n(n + 2) + 5/4 = 0$. 그런데:
$4m^2 - 2m + n(n + 2) + 5/4 = 4m^2 - 2m + 1/4 + n^2 + 2n + 1 = (2m - 1/2)^2 + (n + 1)^2 = 0$.
따라서, $2m - 1/2 = 0$이고 또한, $n + 1 = 0$. 그러므로, $m = 1/4$이고 $n = -1$.

Fig. 9.0

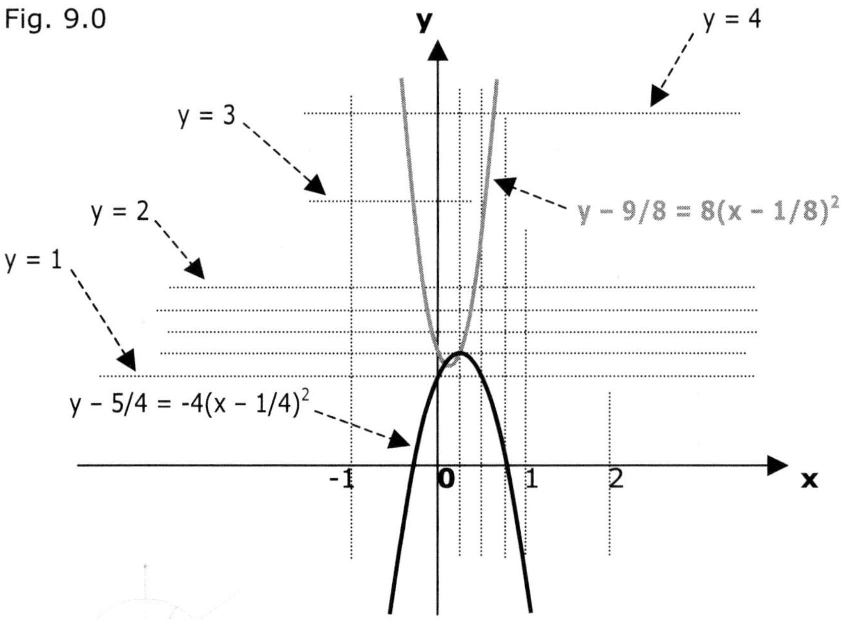

제안과 풀이
A번 예제에서의 문제

포물선 A의 방정식이 $y = x^2 + bx + c$ 라 하고 포물선 B의 방정식은 $y = -2x^2 - x + 5$ 라 하세요. 또한, A의 꼭지점이 B에 있고 B의 꼭지점도 A에 있다고 하세요. 그러면, b와 c의 값들을 구하세요.

A 와 B 의 기울기들의 부호가 서로 반대이므로 이들의 오목한 상태들도 서로 반대이죠. A 가 아래의 그림에서 회색 포물선들 중의 하나라 하고 B 는 검정 포물선이라 하세요. 그러면, 그 두개의 포물선들이 서로의 꼭지점을 지나는 데에는 세 가지 방법들이 있을 수 있죠. 그러니까, 답이 두개이던지 아니면 하나가 되어야 하겠네요. A 와 B 가 하나의 점을 꼭지점으로서 공유하면 답은 하나이며 A 는 회색 점선 포물선이고; 그렇지 않으면, 답은 둘이고 A 는 두 개의 회색 실선 포물선들 중 하나가 되어야 하는 거죠.

그림. A.0

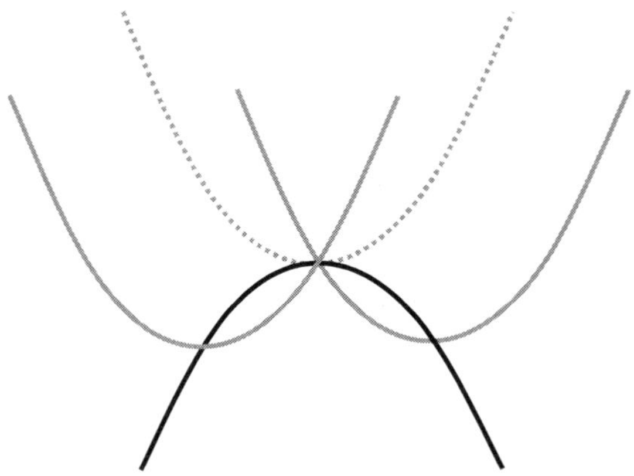

자 이제, 이 A 의 방정식에서의 b 와 c 값들을 구한다는 건 바로 이 포물선 A 를 구하는 거죠. 그러면, 그 포물선 A 를 구해보도록 하죠. 하지만 먼저, 포물선 A 의 꼭지점을 구하고 그리고 나서, 그 꼭지점을 포물선 B 에 대입할 겁니다.

$y = x^2 + bx + c = x^2 + bx + (b/2)^2 - (b/2)^2 + c = (x + b/2)^2 - (b/2)^2 + c = (x + b/2)^2 - (b^2 - 4c)/4.$

그러니까, A의 꼭지점은 $\{-b/2, -(b^2 - 4c)/4\}$이네요.

이제, 이 꼭지점을 B에 대입해보죠. B는 $y = -2x^2 - x + 5$ 이죠. 그러면:

$\{-b/2, -(b^2 - 4c)/4\} \Rightarrow -(b^2 - 4c)/4 = -2(-b/2)^2 - (-b/2) + 5 = -b^2/2 + b/2 + 5$

$\Rightarrow b^2 - 4c = 2b^2 - 2b - 20 \Rightarrow b^2 - 2b + 4c - 20 = 0.$

또한, 포물선 B 의 꼭지점도 포물선 A 에 있어야 하니까 그 꼭지점도 구해서 A 에 대입해야 하죠. 그러면:

$y = -2x^2 - x + 5 = -2(x^2 + x/2) + 5 = -2(x + 1/4)^2 + (-2)(-1/16) + 5$

$= -2(x + 1/4)^2 + 1/8 + 5 = -2(x + 1/4)^2 + 41/8.$

그러니까, B의 꼭지점은 (-1/4, 41/8)이네요. 이제, 이를 A에 대입하는데 A는 $y = x^2 + bx + c$이죠.

$41/8 = (-1/4)^2 + (-1/4)b + c = 1/16 - b/4 + c$

$\Rightarrow 41 = 1/2 - 2b + 8c \Rightarrow 8c - 2b = 81/2 \Rightarrow 4c - b = 81/4.$

자 이제, 다음과 같은 연립방정식을 풀어야 하죠: $b^2 - 2b + 4c - 20 = 0$ 과 $4c - b = 81/4$.

$4c - b = 81/4 \Rightarrow 4c = b + 81/4$. 그러면:

$b^2 - 2b + \mathbf{4c} - 20 = 0 \Rightarrow b^2 - 2b + \mathbf{b + 81/4} - 20 = b^2 - b + 1/4 = 0 \Rightarrow (b - 1/2)^2 = 0 \Rightarrow b = 1/2.$

그러면, $b = 1/2 \Rightarrow 4c - b = 4c - 1/2 = 81/4 \Rightarrow 4c = 81/4 + 1/2 = 83/4 \Rightarrow c = 83/16$.
그러니까, 포물선 A의 방정식은 $y = x^2 + x/2 + 83/16$ 입니다.

그런데, 답이 하나뿐이니까 이 A는 B를 아래와 같이 만나야 하는 거죠.

그림. A.1

다시 말해서, 그 꼭지점들이 서로 같은 겁니다.
이제, 그 꼭지점들이 과연 서로 같은지 확인해보죠. 포물선 A의 꼭지점은 $\{-b/2, -(b^2 - 4c)/4\}$이니까:

$b = 1/2 \Rightarrow -b/2 = -1/4$.
$b = 1/2$ 이고 $c = 83/16$ 이니까 $-(b^2 - 4c)/4 = -\{1/4 - 4(83/16)\}/4 = -(1/4 - 83/4)/4 = 82/4/4 = 41/8$ 이죠..

그러니까, A의 꼭지점은 $(-1/4, 41/8)$이고 A와 B의 꼭지점은 동일한 거죠.

요약:

먼저, 포물선 A의 꼭지점을 구하고 그리고 나서, 그 꼭지점을 포물선 B에 대입합니다.
$y = x^2 + bx + c = x^2 + bx + (b/2)^2 - (b/2)^2 + c = (x + b/2)^2 - (b/2)^2 + c = (x + b/2)^2 - (b^2 - 4c)/4$.

그러므로, 그 꼭지점은 $\{-b/2, -(b^2 - 4c)/4\}$.
이제, 이 꼭지점을 B에 대입합니다. B는 $y = -2x^2 - x + 5$. 그러면:
$\{-b/2, -(b^2 - 4c)/4\} \Rightarrow -(b^2 - 4c)/4 = -2(-b/2)^2 - (-b/2) + 5 = -b^2/2 + b/2 + 5$
$\Rightarrow b^2 - 4c = 2b^2 - 2b - 20 \Rightarrow b^2 - 2b + 4c - 20 = 0$.

또한, 포물선 B의 꼭지점도 포물선 A에 있어야 하니까 그것도 구해서 A에 대입해야 하죠.
그러면, $y = -2x^2 - x + 5 = -2(x^2 + x/2) + 5 = -2(x + 1/4)^2 + (-2)(-1/16) + 5$
$= -2(x + 1/4)^2 + 1/8 + 5 = -2(x + 1/4)^2 + 41/8$.
그러므로, B의 꼭지점은 $(-1/4, 41/8)$. 이제, 이를 A에 대입하는데 A는 $y = x^2 + bx + c$이죠.
$41/8 = (-1/4)^2 + (-1/4)b + c = 1/16 - b/4 + c$
$\Rightarrow 41 = 1/2 - 2b + 8c \Rightarrow 8c - 2b = 81/2 \Rightarrow 4c - b = 81/4$.

자 이제, 다음과 같은 연립방정식을 풀어야 하죠: $b^2 - 2b + 4c - 20 = 0$ 과 $4c - b = 81/4$.

$4c - b = 81/4 \Rightarrow 4c = b + 81/4$. 그러면:
$b^2 - 2b + \mathbf{4c} - 20 = 0 \Rightarrow b^2 - 2b + \mathbf{b + 81/4} - 20 = b^2 - b + 1/4 = 0 \Rightarrow (b - 1/2)^2 = 0 \Rightarrow b = 1/2$.

그러면, $b = 1/2 \Rightarrow 4c - b = 4c - 1/2 = 81/4 \Rightarrow 4c = 81/4 + 1/2 = 83/4 \Rightarrow c = 83/16$.

제안과 풀이
B번 예제에서의 문제

포물선 A의 방정식이 $y = 2x^2 + bx + c$ 라 하고 포물선 B의 방정식은 $y = -2x^2 + 8x - 5$ 라 하세요. 또한, A의 꼭지점이 B에 있고 B의 꼭지점도 A에 있다고 하세요. 그러면, b와 c의 값들을 구하세요.

이 예제는 앞서 나온 예제와 거의 마찬가지네요.
먼저, 포물선 A의 꼭지점을 구하고 그리고 나서, 그 꼭지점을 포물선 B에 대입할 겁니다.
$y = 2x^2 + bx + c = 2\{x^2 + bx/2 + (b/4)^2 - (b/4)^2\} + c = 2(x + b/4)^2 - 2(b/4)^2 + c$
$= 2(x + b/4)^2 - (b^2 - 8c)/8$.
그러면, 그 꼭지점은 $\{-b/4, -(b^2 - 8c)/8\}$이네요. 이제, 이 꼭지점을 B에 대입해보죠.
B는 $y = -2x^2 + 8x - 5$ 이죠. 그러면:
$-(b^2 - 8c)/8 = -2(-b/4)^2 + 8(-b/4) - 5 = -b^2/8 - 2b - 5$
$\Rightarrow b^2 - 8c = b^2 + 16b + 40 \Rightarrow 16b + 8c + 40 = 0 \Rightarrow 2b + c + 5 = 0$.

또한, 포물선 B의 꼭지점도 포물선 A에 있어야 하니까 그것도 구해서 A에 대입해야 하죠. 그러면:
$y = -2x^2 + 8x - 5 = -2(x^2 - 4x) - 5 = -2(x - 2)^2 + 3$.
그러면, B의 꼭지점은 (2, 3)이네요. 이제, 이를 A에 대입하는데 A는 $y = 2x^2 + bx + c$이죠.
$3 = 2(2^2) + 2b + c = 8 + 2b + c \Rightarrow 2b + c = -5$.

그런데, 미지수는 둘인데 방정식은 하나뿐이네요: $2b + c = -5$ 는 $2b + c + 5 = 0$ 와 같은 거죠.
그럼 이제, 어쩌죠?
어쩌긴, 답이 무한히 많은 거죠. 다시 말해서, 관계식 $2b + c + 5 = 0$을 만족하는 b와 c의 모든 값들이 답이 될 수 있는 겁니다. 즉, A가 될 수 있는 포물선이 무수히 많은데 이 A에서 b와 c의 값들은 $2b + c = -5$를 만족하는 거죠.

Fig. B.0

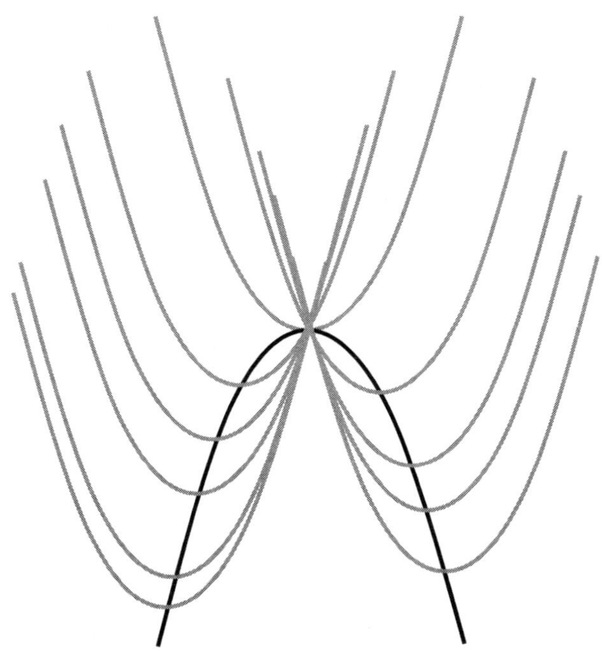

제안과 풀이
C 번 예제에서의 문제

포물선 A의 방정식이 $y = x^2 + bx + c$ 라 하고 포물선 B의 방정식은 $y = -2x^2 + 8x - 5$ 라 하세요. 또한, A의 꼭지점이 B에 있고 B의 꼭지점도 A에 있다고 하세요. 그러면, b와 c의 값들을 구하세요.

이 예제도 앞서 나온 것들과 거의 마찬가지죠.
먼저, 포물선 A 의 꼭지점을 구하고 그리고 나서, 그 꼭지점을 포물선 B 에 대입할 겁니다.

$y = x^2 + bx + c = (x + b/2)^2 - (b^2 - 4c)/4$. 그러면, 그 꼭지점은 $\{-b/2, -(b^2 - 4c)/4\}$이죠.
이 꼭지점을 B에 대입해보죠. B는 $y = -2x^2 + 8x - 5$ 이죠. 그러면:
$-(b^2 - 4c)/4 = -2(-b/2)^2 + 8(-b/2) - 5 = -b^2/2 - 4b - 5 \Rightarrow b^2 - 4c = 2b^2 + 16b + 20$
$\Rightarrow b^2 + 16b + 4c + 20 = 0$.

또한, 포물선 B 의 꼭지점도 포물선 A 에 있어야 하니까 그 꼭지점도 구해서 A 에 대입해야 하죠.

그러면, $y = -2x^2 + 8x - 5 = -2(x - 2)^2 + 3$. 그러면, B의 꼭지점은 (2, 3)이죠.
이제, 이를 A에 대입하는데 A는 $y = x^2 + bx + c$이죠. 그러면:
$3 = (2^2) + 2b + c = 4 + 2b + c \Rightarrow 2b + c = -1$.

자 이제, 다음과 같은 연립방정식을 풀어야 하죠: $b^2 + 16b + 4c + 20 = 0$ 과 $2b + c = -1$.

$2b + c = -1 \Rightarrow c = -1 - 2b \Rightarrow 4c = -4 - 8b$. 그러면:
$b^2 + 16b + \mathbf{4c} + 20 = 0 \Rightarrow b^2 + 16b \mathbf{- 4 - 8b} + 20 = b^2 + 8b + 16 = (b + 4)^2 = 0 \Rightarrow b = -4$.
그러면, $b = -4 \Rightarrow 2b + c = -8 + c = -1 \Rightarrow c = 7$. 그러니까, A의 방정식은 $y = x^2 - 4x + 7$ 이죠.

그런데, $y = x^2 - 4x + 7 = (x - 2)^2 + 3 \Rightarrow$ 꼭지점 = (2, 3) 이는 B의 꼭지점과 같네요.

자 이제, 다음과 같이 정리를 해볼 수도 있겠네요.

다음과 같이 가정하세요:
하나의 특정 포물선 B 가 있고 또 하나의 포물선 A 를 찾으려 한다.
포물선 A 의 꼭지점이 포물선 B 에 있고 또한 B 의 꼭지점도 A 에 있다.
이것들의 대칭축들은 서로 평행하다.
이것들의 기울기들의 크기들이 동일하다.

그러면, 무한히 많은 포물선들이 A 가 될 수 있는데 그러나, 그 포물선들 각각은 포물선 B 와 꼭지점을 공유하지 않아도 된다.

이제, 다음과 같이 가정하세요:
포물선 B 가 있고 또 하나의 포물선 A 를 찾으려 한다.
포물선 A 의 꼭지점이 포물선 B 에 있고 또한 B 의 꼭지점도 A 에 있다.
하지만 이것들의 기울기들의 크기들은 서로 다르다.

그러면, 무한히 많은 포물선들이 A 가 될 수 있는데 이는 B 와 꼭지점을 공유한다; 즉, A 와 B 의 꼭지점들은 동일하다.

제안과 풀이

D 번 예제에서의 0 번 문제

방정식은 다음과 같습니다**: $y = 2(x - 1)^2 - 2|x - 2| + 1$.**

절대 기호의 정의에 의해서 $|x - 2| \geq 0$ 라는 걸 알고있죠.
또한, $x - 2 \geq 0$ 와 $x - 2 < 0$ 의 두 가지 경우들을 고려해야 하는 걸 알고있죠.

$x - 2 \geq 0$ 이면 $|x - 2| = x - 2$.
$x - 2 < 0$ 이면 $|x - 2| = -(x - 2) = -x + 2$.

$x - 2 \geq 0 \Rightarrow x \geq 2$
$\Rightarrow y = 2(x - 1)^2 - 2(x - 2) + 1 = 2x^2 - 4x + 2 - 2x + 4 + 1 = 2x^2 - 6x + 7$
$= 2(x^2 - 3x) + 7 = 2(x - 3/2)^2 - 2(3/2)^2 + 7 = 2(x - 3/2)^2 + 5/2 \Rightarrow y - 5/2 = 2(x - 3/2)^2$이고 $x \geq 2$.

$x - 2 < 0 \Rightarrow x < 2$
$\Rightarrow y = 2(x - 1)^2 - \{-2(x - 2)\} + 1 = 2x^2 - 4x + 2 + 2x - 4 + 1 = 2x^2 - 2x - 1$
$= 2(x^2 - x) - 1 = 2(x - 1/2)^2 - 2(1/2)^2 - 1 = 2(x - 1/2)^2 - 3/2 \Rightarrow y + 3/2 = 2(x - 1/2)^2$이고 $x < 2$.

그림. D.0.0

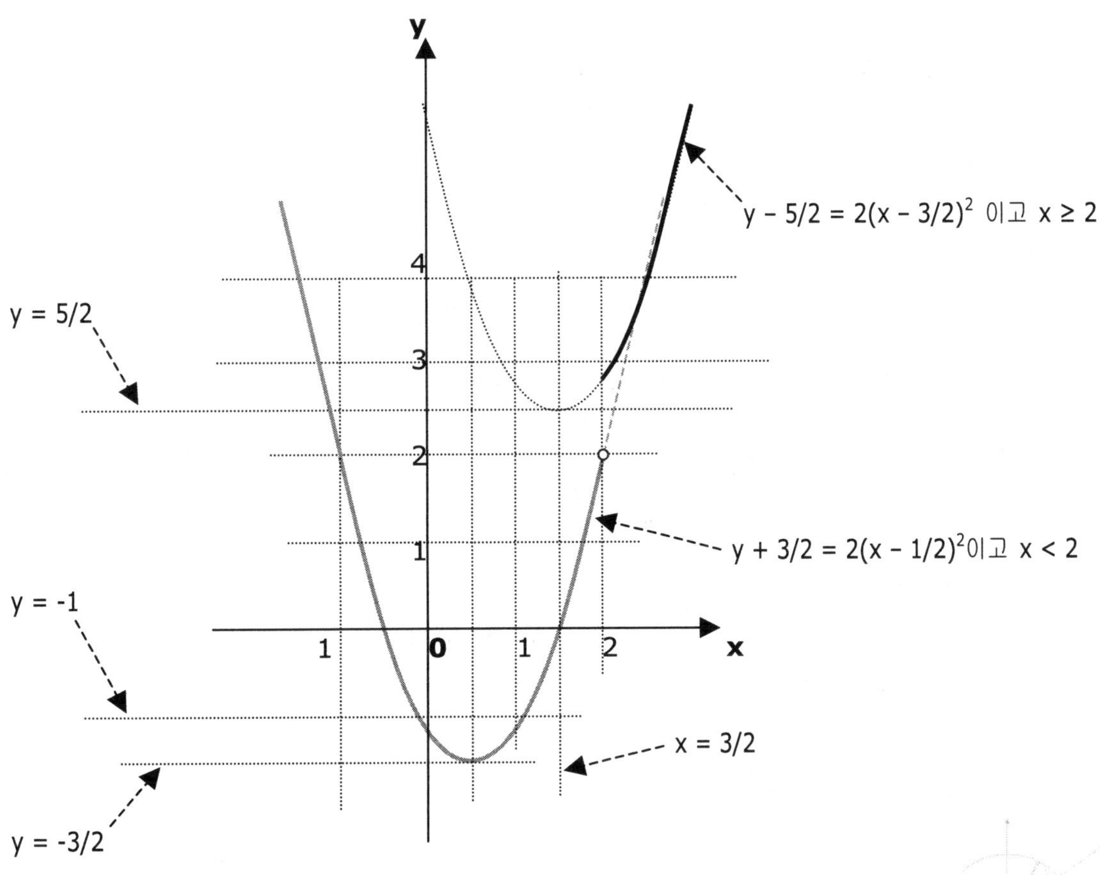

제안과 풀이
D번 예제에서의 1번 문제

방정식은 다음과 같습니다**: $y = (|2x| - 1)(x + 1)$.**

절대 기호의 정의에 의해서 $|2x| \geq 0$ 라는 걸 알고있죠.
또한, $x \geq 0$ 와 $x < 0$ 의 두 가지 경우들을 고려해야 하는 걸 알고있죠.
그런데, 왜 2는 빠진 거죠? 이는 2가 양이니까 음이나 양의 부호에 영향을 미치지않기 때문이죠.

$x \geq 0$ 이면 $|2x| = 2x$. $x < 0$ 이면 $|2x| = -2x$.

$x \geq 0 \Rightarrow y = (|2x| - 1)(x + 1) = (2x - 1)(x + 1)$이고 $x \geq 0$ 인데 여기에서 x절편들을 알 수 있고 이것들이 그 근들인데 1/2과 −1이죠.

$x < 0 \Rightarrow y = (|2x| - 1)(x + 1) = (-2x - 1)(x + 1) = -(2x + 1)(x + 1)$이고 $x < 0$ 여기에서 x절편들이 -1/2과 −1이라는 걸 알 수 있죠.

그림. D.1.0

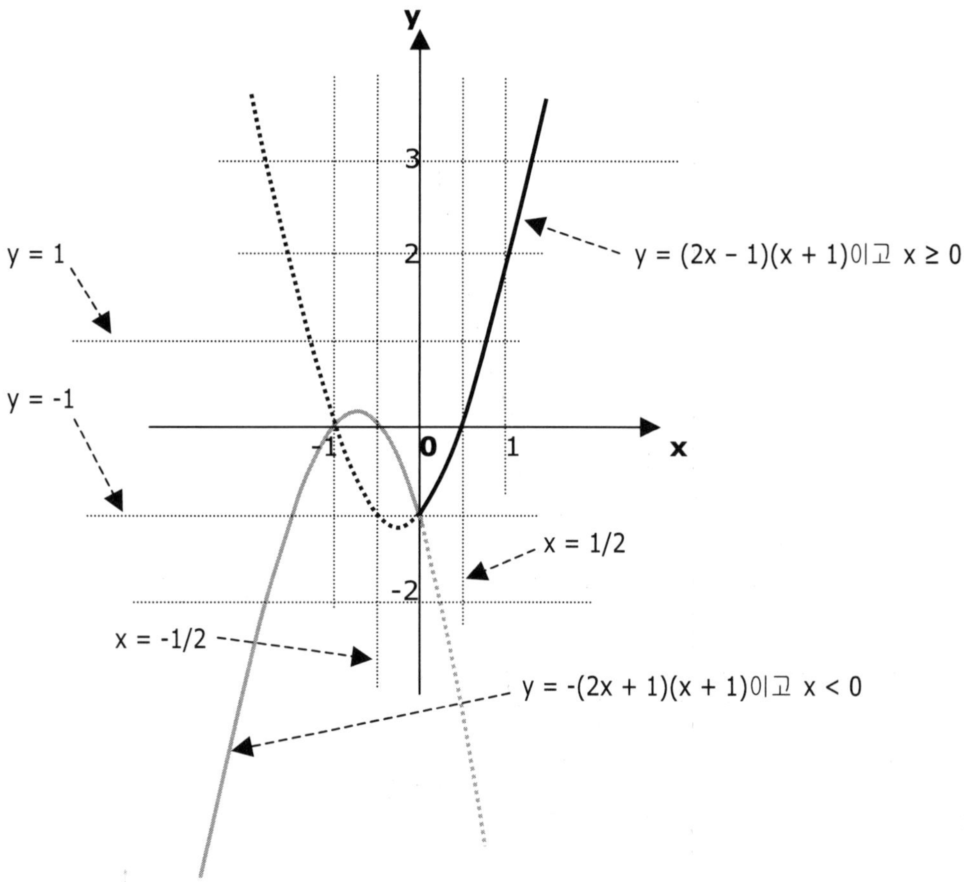

제안과 풀이
D번 예제에서의 2번 문제

방정식은 다음과 같습니다: $y = |x^2 + 3x + 2|$.

절대 기호의 정의에 의해서 $|x^2 + 3x + 2| \geq 0$ 라는 걸 알고있죠.

$x^2 + 3x + 2 = (x + 1)(x + 2) \geq 0 \Rightarrow x \geq -1$ 또는 $x \leq -2$.
그러면, $y = |x^2 + 3x + 2| = x^2 + 3x + 2 = (x + 1)(x + 2)$이며 $x \geq -1$ 또는 $x \leq -2$ 이고 여기에서 x절편들이 −1과 −2라는 걸 알 수 있죠.

$x^2 + 3x + 2 = (x + 1)(x + 2) < 0 \Rightarrow -2 < x < -1$.
그러면, $y = |x^2 + 3x + 2| = -(x^2 + 3x + 2) = -(x + 1)(x + 2)$이며 $-2 < x < -1$ 이고 여기에서 x절편들이 −1과 −2라는 걸 알 수 있죠.

그림. D.2.0

$y = (x + 1)(x + 2)$이고 $x \geq -1$ or $x \leq -2$

$y = -(x + 1)(x + 2)$이고 $-2 < x < -1$

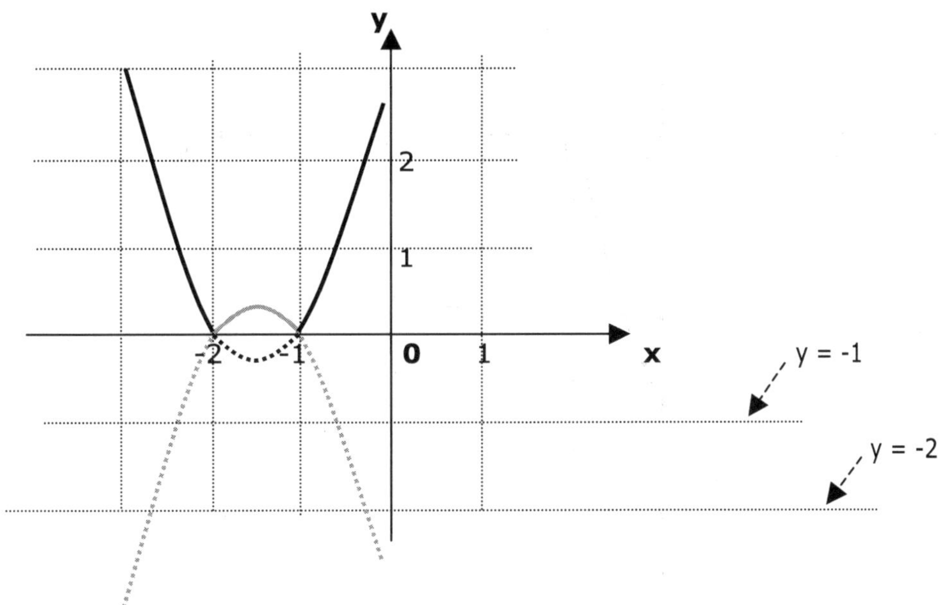

제안과 풀이
D 번 예제에서의 3 번 문제

방정식은 다음과 같습니다**: $y = |x|^2 + 2|x| - 3$.**

$x \geq 0 \Rightarrow y = |x|^2 + 2|x| - 3 = x^2 + 2x - 3 = (x - 1)(x + 3)$이고 $x \geq 0$ 여기에서 x절편들이 1 과 −3 인 것을 알 수 있죠.

$x < 0 \Rightarrow y = |x|^2 + 2|x| - 3 = x^2 - 2x - 3 = (x + 1)(x - 3)$이고 $x< 0$ 여기에서 x절편들이 −1 과 3 인 것을 알 수 있죠.

그림. D.3.0

$y = (x - 1)(x + 3)$이고 $x \geq 0$
$y = (x + 1)(x - 3)$이고 $x< 0$

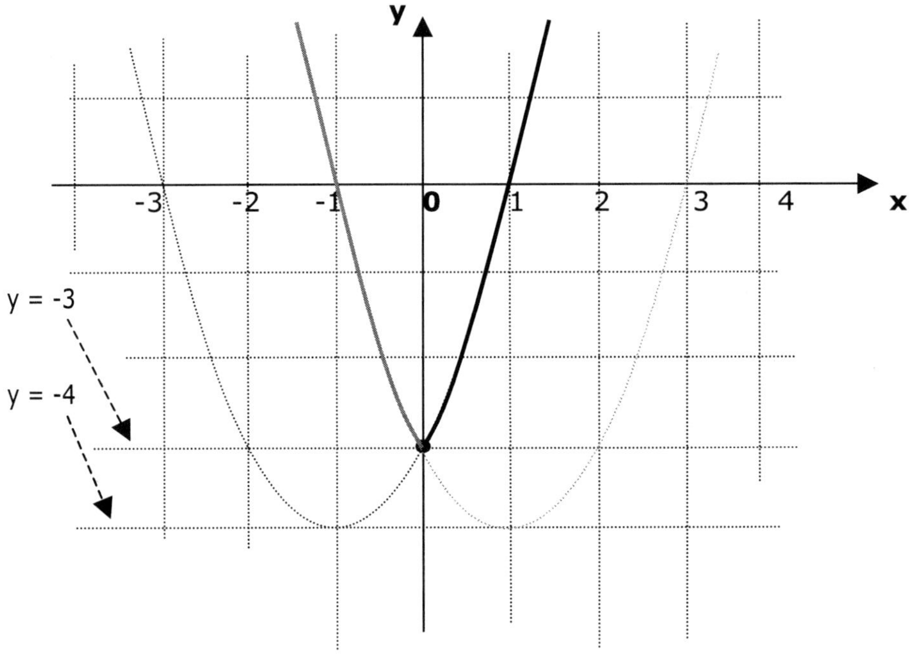

제안과 풀이

D 번 예제에서의 4 번 문제

방정식은 다음과 같습니다: $|y| = x^2 + 2x - 3$.

절대 기호의 정의에 의해서 $|y| \geq 0$ 라는 걸 알고있죠.
또한, $y \geq 0$ 와 $y < 0$ 의 두 가지 경우들을 고려해야 하는 걸 알고있죠.

$y \geq 0$ 이면 $|y| = y$ 이고 $y < 0$ 이면 $|y| = -y$.

$y \geq 0$
$\Rightarrow y = x^2 + 2x - 3 = (x + 3)(x - 1)$이고 $y \geq 0$ 이며 여기에서 x절편들이 1 과 –3 이라는 걸 알 수 있죠.

$y < 0$
$\Rightarrow -y = x^2 + 2x - 3 \Rightarrow y = -(x^2 + 2x - 3) = -(x + 3)(x - 1)$이고 $y < 0$ 여기에서 x절편들이 1 과 –3 이라는 걸 알 수 있죠.

그림. D.4.0 **$y = (x + 3)(x - 1)$이고 $y \geq 0$** **$y = -(x + 3)(x - 1)$이고 $y < 0$**

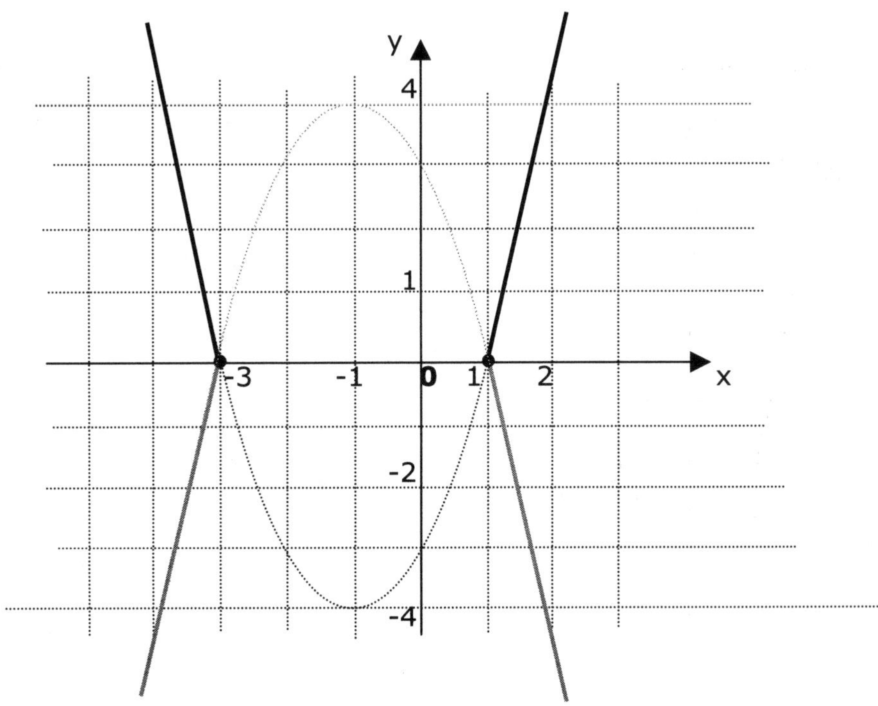

제안과 풀이
D 번 예제에서의 5 번 문제

방정식은 다음과 같습니다: **$|y| = |x|^2 + 2|x| - 3$.**

x 와 y 둘 다 절대 기호 안에 있으니까 4 개의 경우들을 고려해야 하죠.

$x \geq 0$ 이며 $y \geq 0 \Rightarrow y = x^2 + 2x - 3 = (x - 1)(x + 3)$이고 $x \geq 0$ 이며 $y \geq 0$.

$x \geq 0$ 이고 $y < 0 \Rightarrow -y = x^2 + 2x - 3 \Rightarrow y = -(x^2 + 2x - 3) = -(x - 1)(x + 3)$이고 $x \geq 0$ 이며 $y < 0$.

$x < 0$ 이고 $y \geq 0 \Rightarrow y = x^2 - 2x - 3 = (x + 1)(x - 3)$이고 $x < 0$ 이며 $y \geq 0$.

$x < 0$ 이고 $y < 0 \Rightarrow -y = x^2 - 2x - 3 \Rightarrow y = -(x^2 - 2x - 3) = -(x + 1)(x - 3)$이고 $x < 0$ 이며 $y < 0$.

그러면 이제, 방정식 $|y| = |x|^2 + 2|x| - 3$ 은 다음과 같이 확장해서 나타낼 수 있죠:
$y = (x - 1)(x + 3)$이고 $x \geq 0$ 이며 $y \geq 0$. $y = -(x - 1)(x + 3)$이고 $x \geq 0$ 이며 $y < 0$
$y = (x + 1)(x - 3)$이고 $x < 0$ 이며 $y \geq 0$. $y = -(x + 1)(x - 3)$이고 $x < 0$ 이며 $y < 0$.

그림. D.5.0

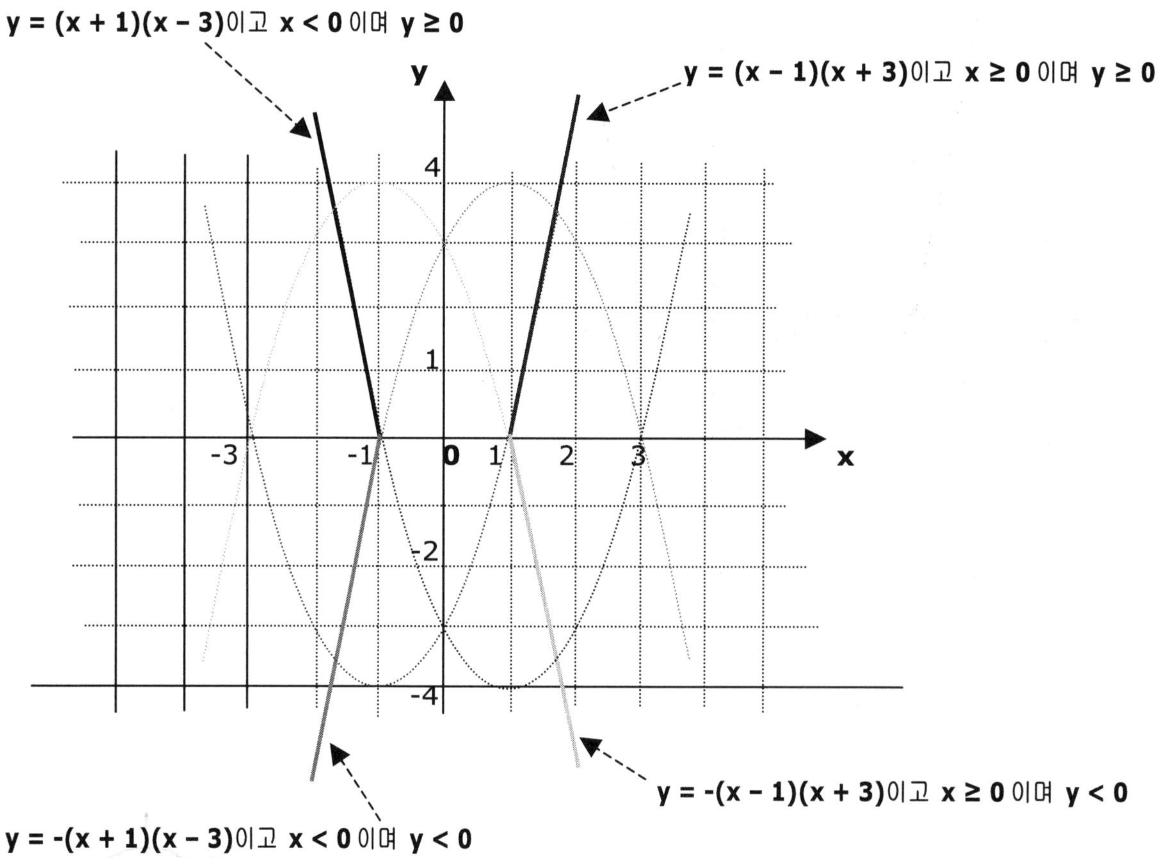

제안과 풀이
D번 예제에서의 6번 문제

방정식은 다음과 같습니다: **$|y| = x^2 + |2x - 3|$.**

'2x - 3'과 'y'가 절대기호 안에 있으니까 각각 두개의 경우씩 즉, 모두 4개의 경우들을 고려해야 하죠.

$2x - 3 \geq 0$ 이고 $y \geq 0 \Rightarrow x \geq 3/2$ 이고 $y \geq 0$
$\Rightarrow y = x^2 + 2x - 3 = (x - 1)(x + 3)$이고 $x \geq 3/2$ 이며 $y \geq 0$.

$2x - 3 \geq 0$ 이고 $y < 0 \Rightarrow x \geq 3/2$ 이고 $y < 0$
$\Rightarrow -y = x^2 + 2x - 3 \Rightarrow y = -(x^2 + 2x - 3) = -(x - 1)(x + 3)$이고 $x \geq 3/2$ 이며 $y < 0$.

$2x - 3 < 0$ 이고 $y \geq 0 \Rightarrow x < 3/2$ 이고 $y \geq 0$
$\Rightarrow y = x^2 - (2x - 3) = x^2 - 2x + 3 = (x - 1)^2 - 1 + 3 = (x - 1) + 2$ 이고 $x < 3/2$ 이며 $y \geq 0$.

$2x - 3 < 0$ 이고 $y < 0 \Rightarrow x < 3/2$ 이고 $y < 0 \Rightarrow -y = x^2 - (2x - 3) = x^2 - 2x + 3$
$\Rightarrow y = -(x^2 - 2x + 3) = -\{(x - 1) + 2\} = -(x - 1)^2 - 2$ 이고 $x < 0$ 이며 $y < 0$.

그럼 이제, $|y| = x^2 + |2x - 3|$는 다음과 같이 확장해서 나타낼 수도 있죠:

$y = (x - 1)(x + 3)$이고 $x \geq 3/2$ 이며 $y \geq 0$
$y = -(x - 1)(x + 3)$이고 $x \geq 3/2$ 이며 $y < 0$
$y = (x - 1)^2 + 2$ 이고 $x < 3/2$ 이며 $y \geq 0$
$y = -(x - 1)^2 - 2$ 이고 $x < 3/2$ 이며 $y < 0$.

이제, 이 방정식을 그래프에 나타내보도록 하죠.

그림. D.6.0

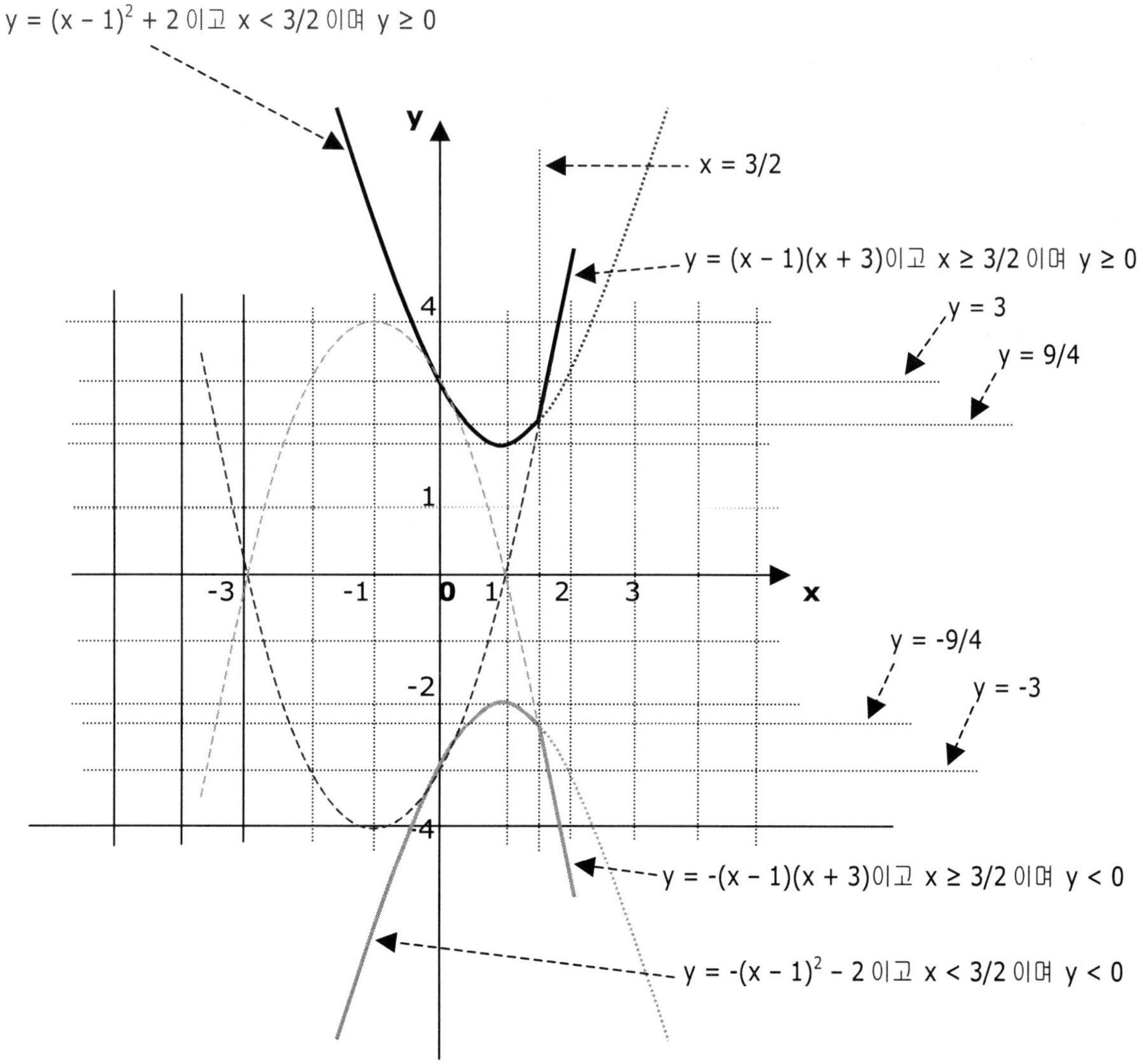
$y = (x - 1)^2 + 2$ 이고 $x < 3/2$ 이며 $y \ge 0$
y
x = 3/2
$y = (x - 1)(x + 3)$이고 $x \ge 3/2$ 이며 $y \ge 0$
y = 3
y = 9/4
4
1
-3
-1
0
1
2
3
x
y = -9/4
y = -3
-2
-4
$y = -(x - 1)(x + 3)$이고 $x \ge 3/2$ 이며 $y < 0$
$y = -(x - 1)^2 - 2$ 이고 $x < 3/2$ 이며 $y < 0$

제안과 풀이
D 번 예제에서의 7 번 문제

방정식은 다음과 같습니다: $y = (x^2 - 2x + 1)^{1/2} = \sqrt{x^2 - 2x + 1}$.

합당한 실수의 제곱근이 0 보다 크거나 같다는 것이 자명한 것이긴 해도 가끔 실수를 하는 경향이 있더군요. 이는 무엇이든 너무 가까이 있으면 좀처럼 알아 볼 수가 없어서 보통은 그냥 지나치기 때문이죠. 그러니까, 이 경우에 y 가 0 보다 크거나 같은지 확인하는 게 좋습니다. 그런데, 합당한 실수라니, 뭘 말하자는 거죠? 이중근호 안에 있는 것이 실수라면 이는 0 보다 크거나 같아야 하는 거죠.
이제, $g(x) = x^2 - 2x + 1$ 라 하세요. 그러면, g의 커브는 다음과 같죠:

그림. D.7.0

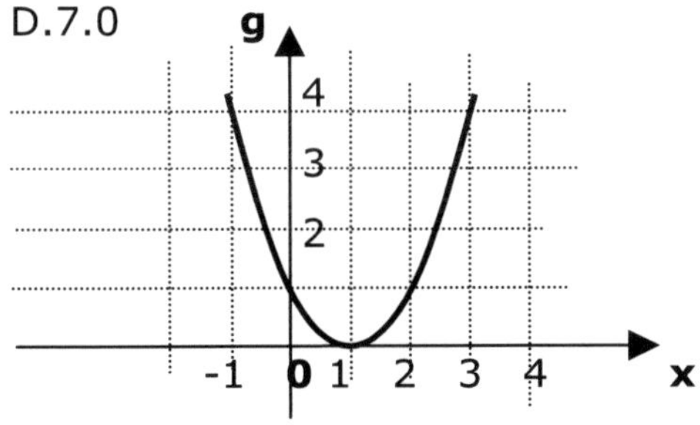

그러면, 모든 x 값에 대하여 g(x)는 0 보다 크거나 같은 값들을 만들어내죠. 그러니까, 모든 x에 대하여 그 근호 안에 있는 것은 합당한 겁니다. 그런데, $g(x) = x^2 - 2x + 1 = (x - 1)^2$ 인데 이는 완전 제곱이네요.
그러면, $y = (x^2 - 2x + 1)^{1/2} = \{(x - 1)^2\}^{1/2}$인 거죠. 그런데, $\{(x - 1)^2\}^{1/2}$은 $x - 1$ 이 아니고 $|x - 1|$이어야 한다는 것을 유의해야 합니다. 이는 왜냐하면 실수의 제곱근 즉, $\sqrt{x^2 - 2x + 1}$ 은 0 보다 크거나 같아야 하기 때문이죠. 그럼 이제, 다음과 같은 두 가지의 경우들을 고려해야 하죠:
$x - 1 \geq 0 \Rightarrow x \geq 1 \Rightarrow y = x - 1$ 이고 $x \geq 1$.
$x - 1 < 0 \Rightarrow x < 1 \Rightarrow y = -(x - 1) = -x + 1$ 이고 $x < 1$.

이제, 방정식 $y = (x^2 - 2x + 1)^{1/2}$를 그래프에 나타내보도록 하죠.

그림. D.7.1 **$y = -x + 1$ 이고 $x < 1$** **$y = x - 1$ 이고 $x \geq 1$**

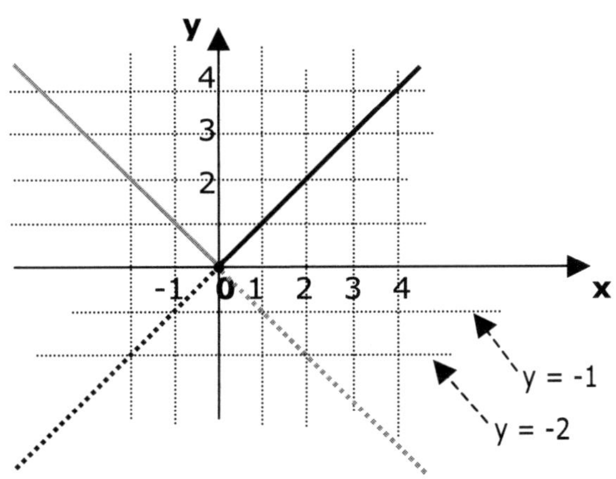

제안과 풀이
D번 예제에서의 **8**번 문제

방정식은 다음과 같습니다: $y = \{(x^2 - 4x + 4)^2\}^{1/2} = \sqrt{(x^2 - 4x + 4)^2}$.

이중근호 안에 있는 것이 0 보다 크거나 같아야 한다는 것을 알고있죠.
그러면, 이중근호 안에 뭐가 있는지 알아보기로 하죠.

이제, $g(x) = x^2 - 4x + 4$ 라고 하세요. 그러면, g의 커브는 다음과 같죠:

그림. D.8.0

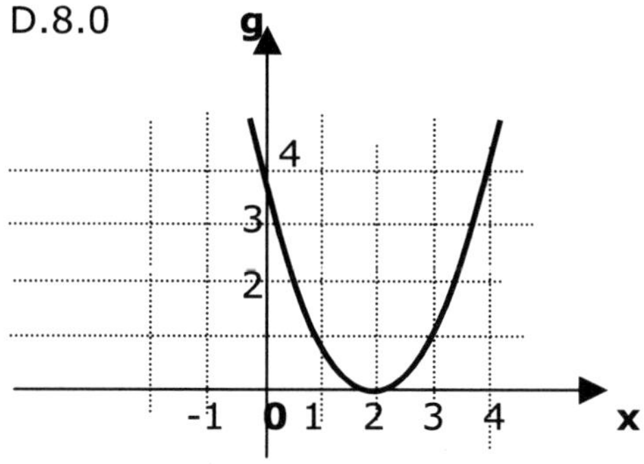

그러면, 그 어떤 실수 x에 대하여도 g(x)는 0 보다 크거나 같은 값들을 만들어내죠. 그러니까, 모든 x에 대하여 그 근호 안에 있는 것은 합당한 겁니다. 그런데, $g(x) = x^2 - 4x + 4 = (x - 2)^2$ 인데 이는 완전 제곱이네요
그러면, 다음과 같은 거죠:

$y = \{(x^2 - 4x + 4)^2\}^{1/2} = \{(x - 2)^4\}^{1/2} = (x - 2)^2 = g(x)$ 왜냐하면 the $g \ge 0$.

그러니까, x 의 범위에 대하여 여러 경우들을 고려할 필요가 없는 거죠.

그러면 이제, 주어진 방정식의 그래프는 아래와 같아요.

그림. D.8.1 $y = \{(x - 4x + 4)^2\}^{1/2} = \sqrt{(x^2 - 4x + 4)^2}$

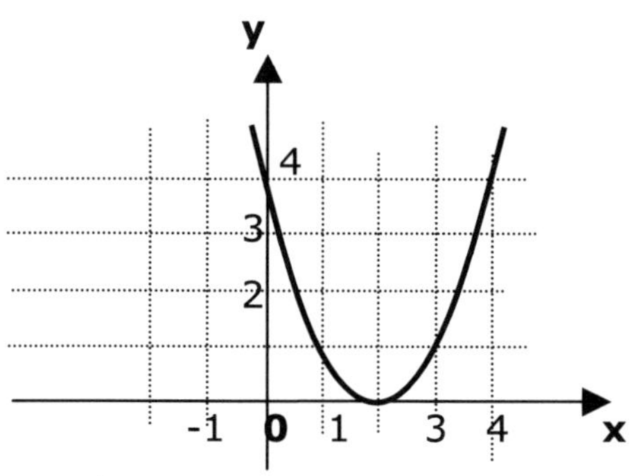

제안과 풀이
D 번 예제에서의 9 번 문제

방정식은 다음과 같습니다: $y = \{(x^2 - 2x + 2)^2\}^{1/2} = \sqrt{(x^2-2x+2)^2}$.

합당한 실수의 제곱근이 0 보다 크거나 같다는 것을 알고있죠.
또한, 이중근호 안에 있는 것도 0 보다 크거나 같아야 한다는 것도 알고있죠.

그러면, 먼저 이중근호 안에 뭐가 있는지 알아보기로 하죠.
그건 $(x^2 - 2x + 2)^2$이고 이는 0 보다 크거나 같죠 그러니까, x가 어떤 실수 값을 갖든 괜찮을 것 같네요.

그럼 이제, 이중근호를 제거하고 난 후 어떻게 해야 될지 알아보기로 하죠.

이제, $g(x) = x^2 - 2x + 2$ 라 하세요.
그러면, $g(x) = (x - 1)^2 + 1$ 이고 g의 커브는 아래와 같죠:

그림. D.9.0

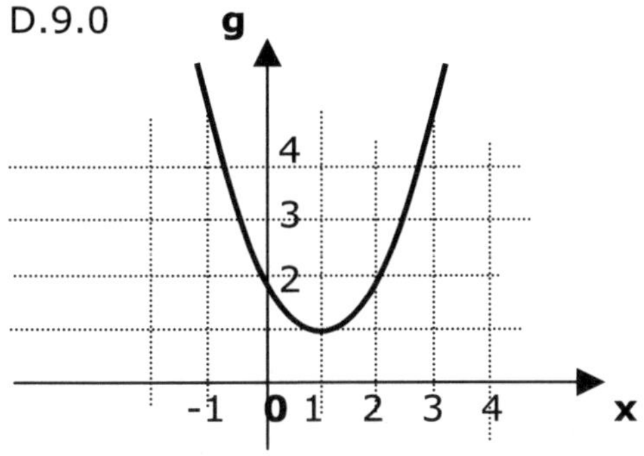

그러면, 모든 실수 x 에 대하여 g(x)는 0 보다 크거나 같은 값들을 만들어내죠. 그러니까, 모든 x 에 대하여 그 근호 안에 있는 것은 합당한 겁니다. 그러면, 다음과 같은 거죠:

$y = \{(x^2 - 2x + 2)^2\}^{1/2} = x^2 - 2x + 2 = g(x)$ 왜냐하면 $g > 0$.

그러니까, x 의 범위에 관하여 여러 경우들을 고려할 필요가 없는 거죠.

그러면 이제, 주어진 방정식의 그래프는 아래와 같죠.

그림. D.9.1 $y = \{(x^2 - 2x + 2)^2\}^{1/2} = \sqrt{(x^2-2x+2)^2}$

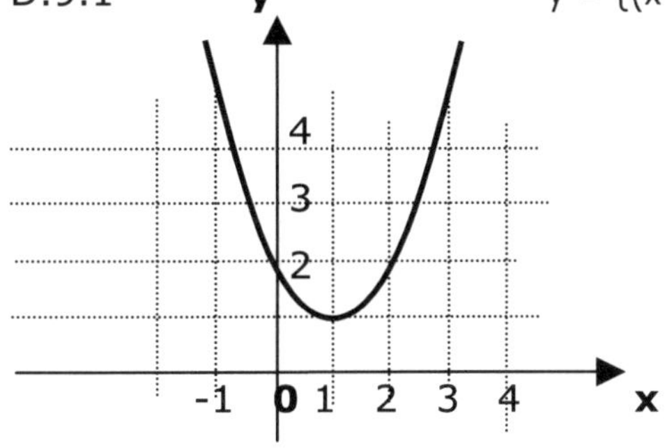

제안과 풀이
D 번 예제에서의 **A** 번 문제

방정식은 다음과 같습니다: $\mathbf{y = \{(x^2 - 2x)^2\}^{1/2}} = \sqrt{(x^2-2x)^2}$

합당한 실수의 제곱근(이중근)이 0 보다 크거나 같다는 것이 분명하긴 해도 가끔 실수를 하는 경향이 있죠. 이는 너무 당연하면 좀처럼 알아 볼 수가 없어서 보통은 그냥 지나치기 때문이죠. 그러니까, 그 제곱근 기호를 제거 했을 때 그 결과도 0 보다 크거나 같은지 확인하는 게 좋죠. 그런데, 제곱근호 안의 합당한 실수라는 건 제곱근호 안에 있는 것이 실수이면 0 보다 크거나 같아야 한다는 거죠. 그러면, 제곱근호 안에 뭐가 있는지 알아보기로 하죠.

그건 $(x^2 - 2x)^2$ 이고 이는 실수인 어느 x에 대하여도 0 보다 크거나 같으니까 그 제곱근호와는 별 문제가 없네요. 그러면 이제, 그 제곱근호를 제거하고 난 후에 어떻게 해야될지 알아보기로 하죠.
이제, $g(x) = x^2 - 2x$라 하세요.
그러면, $g(x) = x^2 - 2x + 1 - 1 = (x - 1)^2 - 1 = (x - 2)x$이고 g의 커브는 아래와 같죠.

그림. D.A.0

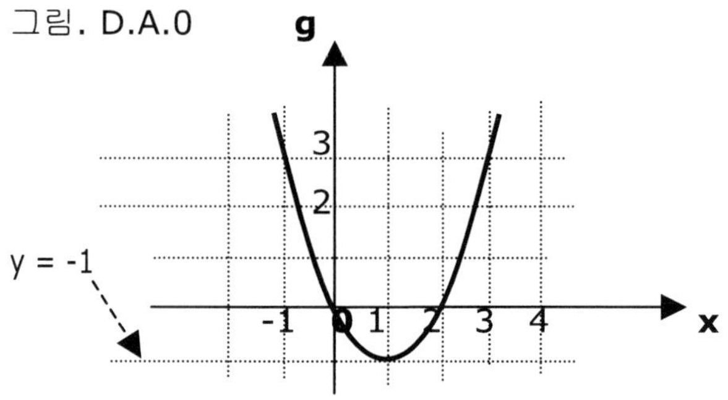

그러면, g 는 모든 실수인 x 에 대해서 항상 0 보다 크거나 같은 값을 만드는 것이 아니네요.
즉, 그저 $y = g(x) = x^2 - 2x$라고 할 수는 없는 거죠. 그러면, 이 경우에는 제곱근호 안에 있는 것이 밖으로 나올 때는 중요한 의미를 갖는 거죠. 즉, 이 경우에는 그 지수 2 와 함께 제곱근호를 그저 제거할 수는 없는 겁니다.
그런데, 합당한 실수의 제곱근은 0 보다 크거나 같으니까 $y \geq 0$ 이죠. 그러면, x 의 범위에 관하여 두 가지 경우들을 고려해야 하죠. 위의 그래프에서 다음과 같다는 것을 알 수 있죠:

$y = x^2 - 2x$이고 $x \leq 0$ 또는 $x \geq 2$; 그러나, $y = -(x^2 - 2x) = -x^2 + 2x$이고 $0 < x < 2$.

그럼 이제, 주어진 방정식의 그래프를 보기로 하죠.

그림. D.A.1

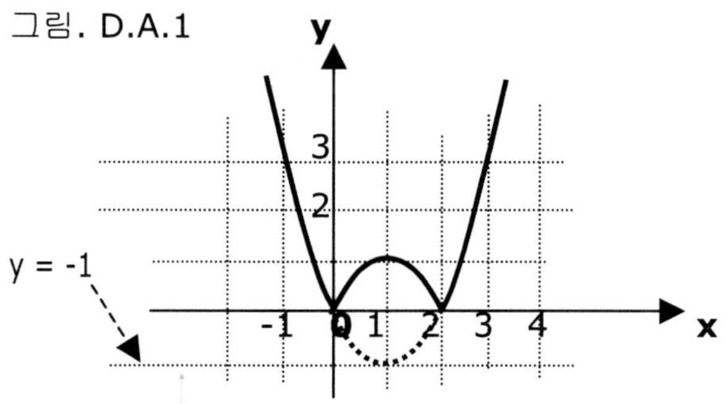

제안과 풀이
D 번 예제에서의 **B** 번 문제

방정식은 다음과 같습니다**:** $y = \{(-x^2 + 2x)^2\}^{1/2} = \sqrt{(-x^2+2x)^2}$.

이중근호 안에 있는 것부터 시작하기로 하죠.
그러면, 그건 $(-x^2 + 2x)^2$인데 이는 실수인 모든 x에 대하여 0 보다 크거나 같으니까 이중근호와는 문제가 없는 것 같네요.

그런데, 이중근호 안에 있는 것이 밖으로 나올 때 이것이 중요한 의미를 갖는 거죠.
그러면 이제, 이중근호를 제거한 후에 어떻게 되는지 알아보아야 하죠.

이제, $g(x) = -x^2 + 2x$라 하세요.
그러면, $g(x) = -x^2 + 2x = -(x^2 - 2x + 1 - 1) = -(x - 1)^2 + 1 = (2 - x)x$이고 g의 커브는 다음과 같죠:

그림. D.B.0

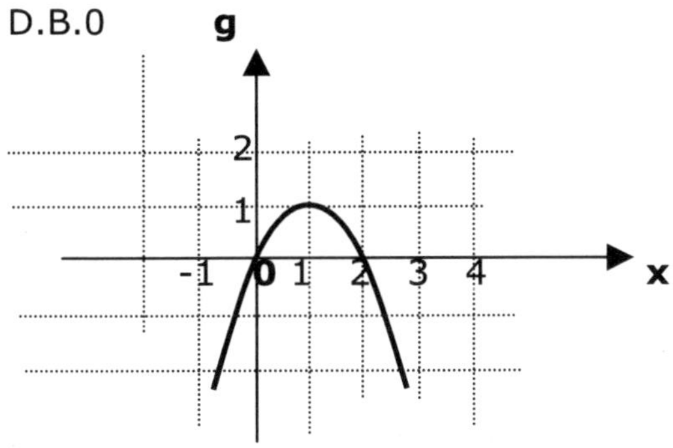

그러면, x 가 취하는 모든 실수에 대해서 g(x)가 항상 0 보다 크거나 같은 값을 만드는 건 아니네요.
즉, 그저 $y = g(x) = -x^2 + 2x$라고 할 수는 없는 거죠.

합당한 실수의 이중근은 0 보다 크거나 같은 거니까 $y \geq 0$ 라는 걸 알고있죠.

그러면, 위의 그래프로부터 다음과 같아야 한다는 것을 알 수 있죠:
$y = -x^2 + 2x$이고 $0 \leq x \leq 2$ 이죠; 하지만, $y = -(-x^2 + 2x) = x^2 - 2x$이고 $x < 0$ 또는 $x > 2$ 입니다.

그러니까, 주어진 방정식을 그래프로 다음과 같이 나타낼 수 있는 거죠:

그림. D.B.1

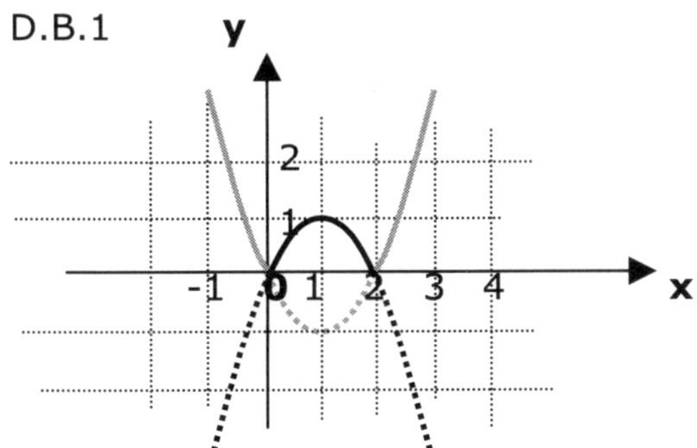

제안과 풀이
D번 예제에서의 **C**번 문제

방정식은 다음과 같습니다: $\mathbf{y = \{(-x^2 + 2x - 1)^2\}^{1/2}} = \sqrt{(-x^2+2x-1)^2}$.

이중근호 안에 있는 것부터 시작하기로 하죠.
그러면, 그건 $(-x^2 + 2x - 1)^2$인데 이는 실수인 모든 x에 대하여 0 보다 크거나 같으니까 이는 이중근호와는 합당하죠.

그런데, 이중근호 안에 있는 것이 밖으로 나올 때 이것이 중요한 의미를 갖을 수 있는 거죠.
그러면 이제, 이중근호를 제거한 후에 어떻게 해야 되는지 알아보아야 하죠.

이제, $g(x) = -x^2 + 2x - 1$ 이라 하세요.
그러면, $g(x) = -x^2 + 2x - 1 = -(x^2 - 2x + 1) = -(x - 1)^2$ 이고 g의 커브는 다음과 같죠:

그림. D.C.0

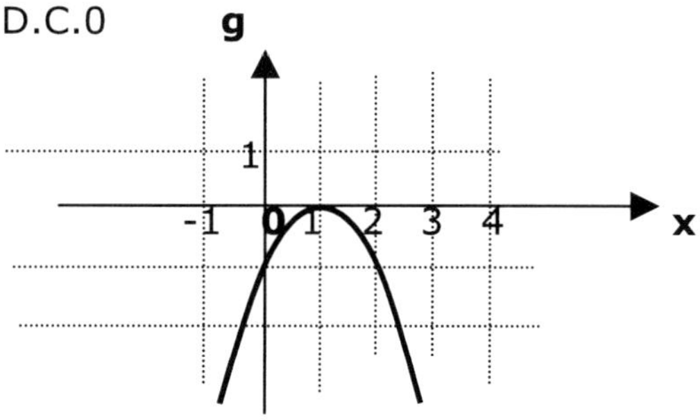

그러면, x가 취하는 그 어떤 실수에 대해서도 g(x)는 그 어떤 양수도 만들지않죠; 하지만, x = 1 일 때는 g가 0 을 만들긴 하네요. 다시 말해서, $y = -x^2 + 2x - 1 \le 0$ 이고 x는 실수인 겁니다.
합당한 실수의 이중근은 0 보다 크거나 같은 거니까 $y \ge 0$ 이죠.
그러니까, y = g(x)에서의 y가 갖는 모든 값의 부호를 바꾸어야 하는 거죠.
그러면, 등호의 좌변에 있는 수식의 기호를 바꾸기로 하죠. 등호의 어느 한쪽의 기호를 바꾸어도 되기는 하지만요.
그러면, $-y = -x^2 + 2x - 1 \Rightarrow y = -(-x^2 + 2x - 1) = x^2 - 2x + 1 = (x - 1)^2 \ge 0$ 이고 x는 실수입니다.

그럼 이제, 주어진 방정식의 그래프는 다음과 같아요:

그림. D.C.1

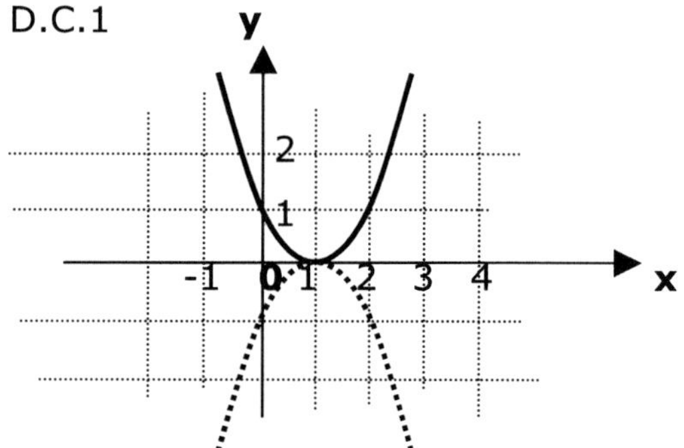

제안과 풀이

D번 예제에서의 D번 문제

방정식은 다음과 같습니다: **$y = \{(-x^2 + 2x - 2)^2\}^{1/2}$ =** $\sqrt{(-x^2+2x-2)^2}$.

이중근호 안에 있는 것부터 시작하기로 하죠. 그러면, 그건 $(-x^2 + 2x - 2)^2$인데 이는 실수인 모든 x에 대하여 0 보다 크거나 같으니까 이는 이중근호와 별 문제 없네요. 그런데, 이중근호 안에 있는 것이 밖으로 나올 때 이것이 중요한 의미를 갖는 거죠.
그러면 이제, 이중근호를 제거한 후에 어떻게 해야 되는지 알아보아야 하죠. 이제, $g(x) = -x^2 + 2x - 2$라 하세요. 그러면, $g(x) = -x^2 + 2x - 2 = -(x^2 - 2x + 2) = -(x^2 - 2x + 1 + 1) = -(x - 1)^2 - 1$이고 g의 커브는 다음과 같죠:

그림. D.D.0

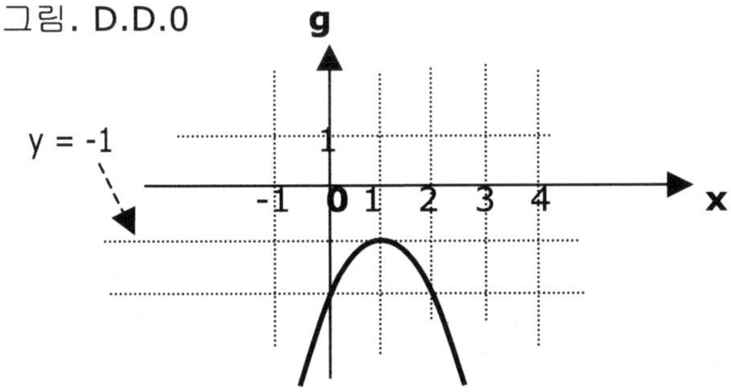

그러면, x가 취하는 모든 실수에 대해서 g(x)는 음수만 만들죠. 즉, $y = -x^2 + 2x - 2 < 0$이고 x는 실수이죠.
합당한 실수의 이중근은 0 보다 크거나 같은 거니까 $y \geq 0$ 이죠. 그러니까, $y = g(x)$에서의 y 가 갖는 모든 값의 부호를 바꾸어야 하는 거죠. 그러면, 등호의 우변에 있는 수식의 기호를 바꾸기로 하죠. 물론, 등호의 어느 한쪽의 기호를 바꾸어도 되죠.
그러면, $y = -(-x^2 + 2x - 2) = x^2 - 2x + 2 = (x - 1)^2 + 1 > 0$이고 x는 실수입니다.
그럼 이제, 주어진 방정식의 그래프는 다음과 같습니다:

그림. D.D.1

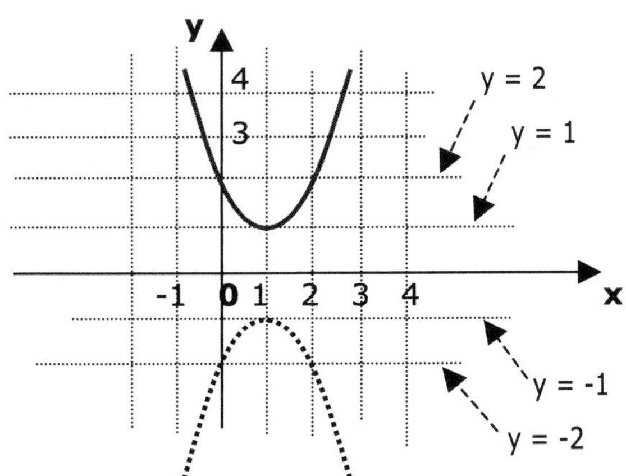

제안과 풀이
E번 예제에서의 문제

$y = f(x) = 2x^2 + bx + c$ 이고 $0 \le x \le 1$ 이며 $0 \le y \le 2$ 라고 하고 b와 c의 값들을 구하세요.

함수 f가 $0 \le x \le 1$ 인 x에 대해서 0부터 2까지의 값들을 만들어내게 해야 하죠.
그러면, 이 함수 f가 $0 \le x \le 1$ 와 $0 \le y \le 2$ 로 명시된 활동영역에서 어떻게 활동해야 하는지 알아보아야겠어요. 함수의 활동은 그 커브로 알 수 있고 f 의 커브가 포물선이라는 걸 알고있죠. 그러면, 이 커브를 x-y 좌표계에서의 그래프에 놓고 그 활동영역에서의 커브를 보도록 하죠. 아래의 x-y 평면에서 옅은 회색 영역이 그 활동영역입니다.

그림. E.0

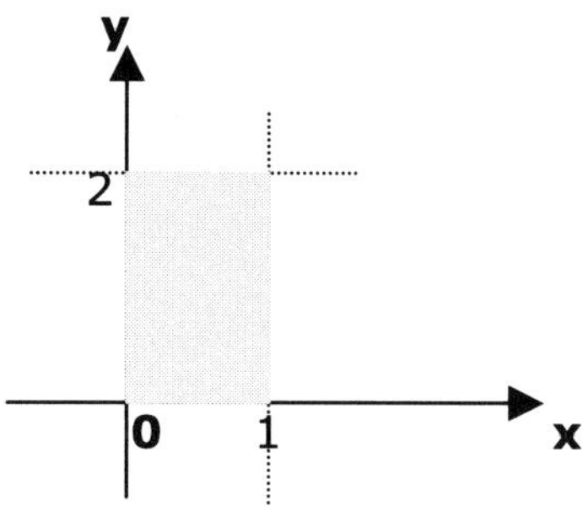

하지만, 그 커브를 그래프에 그저 놓는 건 아니죠. 커브를 놓는데 그 커브의 일부(포물선 선분)가 그 영역에 있도록 하는 겁니다. 좀 더 자세히 하자면, 그 일부가 x 축을 따라 0 에서 1 까지, y 축을 따라서는 0 에서 2 까지 걸쳐서 그 영역을 채울 수 있도록 해야 하는 겁니다. 이제, 그래프에 커브들을 몇 개 놓아보도록 하죠.

그림. E.1

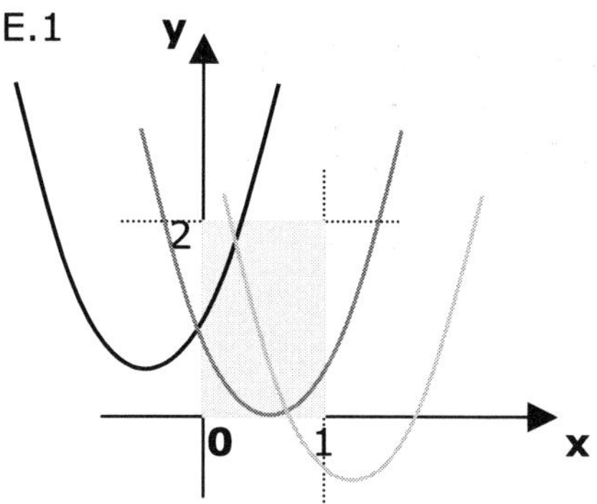

위와 같이 놓인 그 어느것도 되지않는 거네요. 왜냐하면 이것들 그 어느 것도 위에서 명시한 양 만큼에 걸쳐서 그 활동영역을 채우지않기 때문이죠. 이제, 다시 시도해보기로 하죠.

그림. E.2

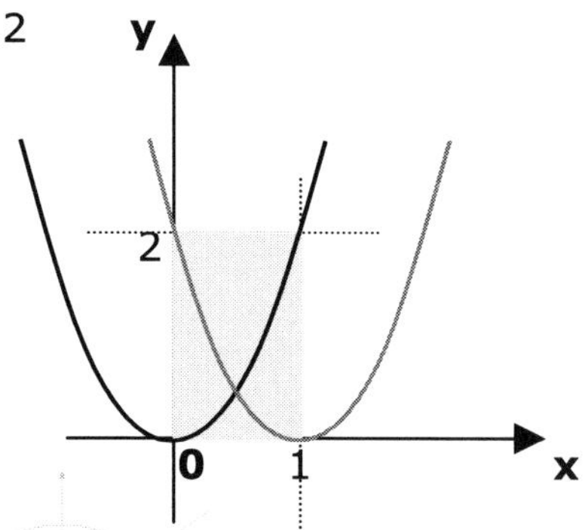

이번에는 제대로 됐네요. 이 커브들 둘 다 제자리에 놓여서 제각기 제대로 된 양 만큼에 걸쳐서 그 활동영역을 채우고있네요. 그러면 이제, 이 두개의 포물선들이 맞는 것이고 그 포물선들의 방정식들을 작성하기 시작할 수 있겠어요.
잠깐만 요! 아래에 나타낸 포물선들은 어떤가요?

그림. E.3

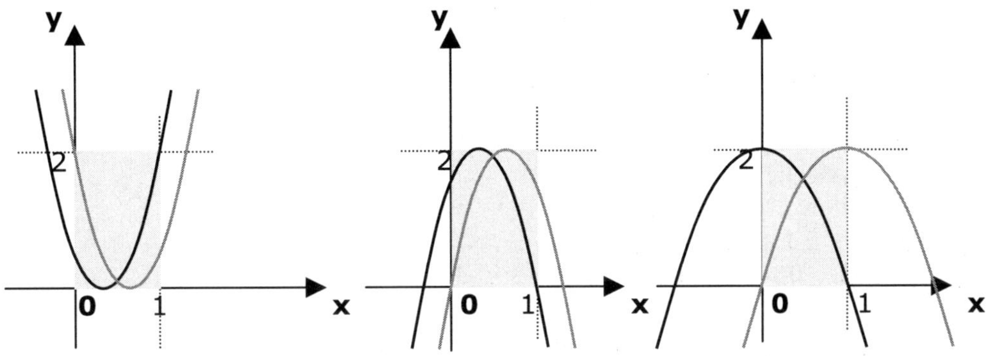

물론, 위와 같은 그런 포물선들도 될 수 있죠; 하지만, 그것들 각각의 기울기에 문제가 있어요. 다시 말해서, 그것들은 맞지 않는다는 거죠. 주어진 방정식에서 포물선의 기울기가 2 라는 걸 알 수 있는데 위에 나타낸 포물선들의 기울기는 2 가 아니죠. 하지만, 그 기울기가 하나의 상수이어서 따져야 할 대상이라면 위와 같이 놓일 수 있는 포물선들도 고려할 수 있어야 해요. 그렇다면 또한, 아래에 나타낸 포물선들도 고려해야 하는 겁니다.

그림. E.4

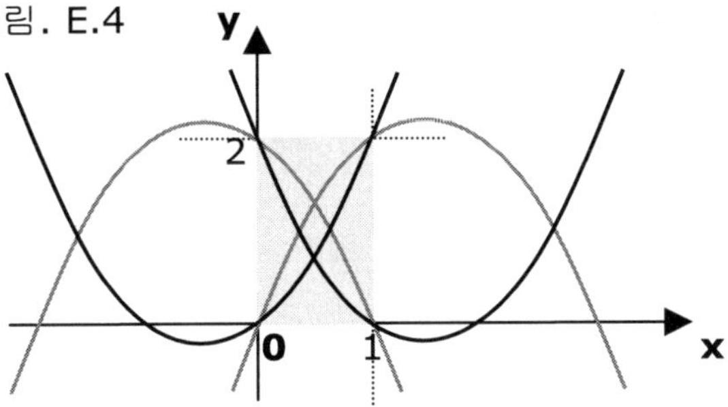

이제, 제대로 된 포물선들로 돌아가보기로 하죠. 그것들은 다음과 같습니다.

그림. E.2

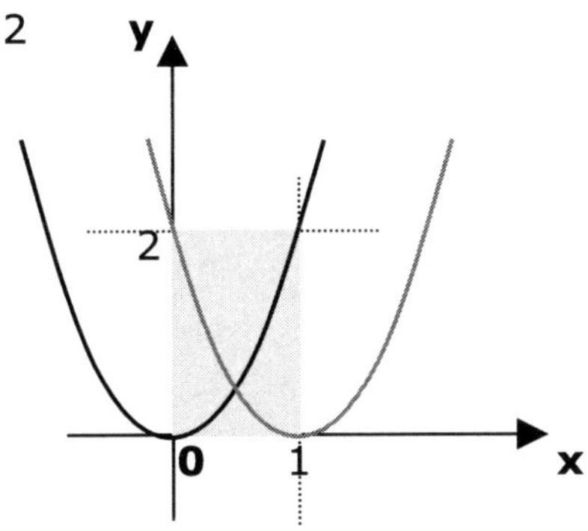

방정식은 $y = 2x^2 + bx + c$이고 $0 \le x \le 1$ 이며 $0 \le y \le 2$ 이죠.
그럼 이제, b 와 c 의 값들은 찾아야 하죠. 즉, 그 두개의 포물선들의 방정식들을 구해야 하는 겁니다. 위의 그래프에서, 구하고자 하는 그 방정식들 각각을 작성하는 데 사용할 수 있는 두개의 점들을 볼 수 있어요.

왼쪽에 있는 포물선에 대하여는, 사용될 그 점들이 원점과 (1, 2)이죠.
$y = 2x^2 + bx + c$입니다. 그러면:

$(0, 0) \Rightarrow 0 = 0 + 0 + c = c \Rightarrow c = 0.$
$((1, 2)$와 $c = 0) \Rightarrow 2 = 2 + b + 0 = 2 + b \Rightarrow b = 0.$

그러니까, 그 포물선의 방정식은 $y = 2x^2$입니다.

오른쪽에 있는 포물선에 대하여는, 사용될 그 점들이 (0, 2)와 (1, 0)이죠.
$y = 2x^2 + bx + c$입니다. 그러면:

$(0, 2) \Rightarrow 2 = 0 + 0 + c = c \Rightarrow c = 2.$
$((1, 0)$과 $c = 2) \Rightarrow 0 = 2 + b + 2 = 4 + b \Rightarrow b = -4.$

그러니까, 그 포물선의 방정식은 $y = 2x^2 - 4x + 2 = 2(x - 1)^2$이죠.

그런데, 그래프를 주의 깊게 보기만 해도 그 방정식들은 알아낼 수 있는 학생도 있겠죠.

요약:

방정식은 $y = 2x^2 + bx + c$이고 $0 \le x \le 1$ 이며 $0 \le y \le 2$ 이죠.
그러면, 그 포물선은 다음과 같은 두개 중 어느 하나이어야 합니다.

그림. E.2

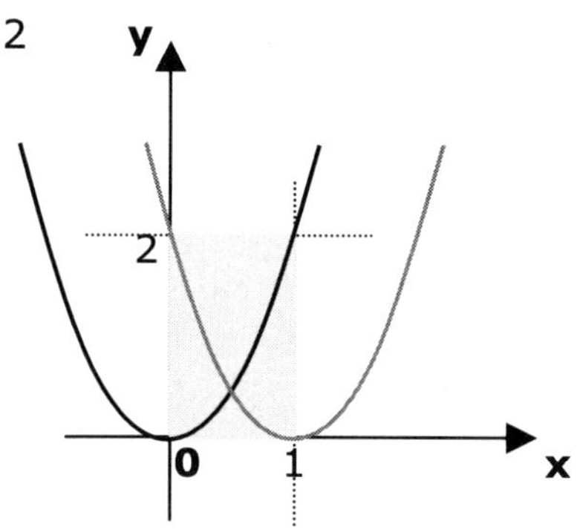

왼쪽에 있는 포물선은 원점과 (1, 2)를 지나야 하죠.
$y = 2x^2 + bx + c$입니다. 그러면:
$(0, 0) \Rightarrow 0 = 0 + 0 + c = c \Rightarrow c = 0.$
$((1, 2)$와 $c = 0) \Rightarrow 2 = 2 + b + 0 = 2 + b \Rightarrow b = 0.$

오른쪽에 있는 포물선은 (0, 2)와 (1, 0)을 지나야 합니다.
$y = 2x^2 + bx + c$입니다. 그러면:
$(0, 2) \Rightarrow 2 = 0 + 0 + c = c \Rightarrow c = 2.$
$((1, 0)$과 $c = 2) \Rightarrow 0 = 2 + b + 2 = 4 + b \Rightarrow b = -4.$

그러므로, $b = 0$ 이고 $c = 0$ 또는, $b = -4$ 이고 $c = 2$.

제안과 풀이
F번 예제에서의 문제

$y = f(x) = ax^2 + bx + 1$ 이고 $0 \le x \le 1$ 이며 $0 \le y \le 2$ 라고 하고 a와 b의 값들을 구하세요.

이 예제에서도 함수 f가 $0 \le x \le 1$ 인 x에 대해서 0부터 2까지의 값들을 만들어내게 해야 하죠.
그러면, 이 함수 f가 $0 \le x \le 1$ 와 $0 \le y \le 2$ 로 명시된 활동영역에서 어떻게 활동해야 하는지 알아보아야 하죠. 함수의 활동은 그 커브로 알 수 있고 이 f의 커브가 포물선이라는 걸 알고있죠. 그러면, 이 커브를 x-y 좌표계에서의 그래프에 놓고 그 활동영역에서의 커브를 보도록 하죠. 아래의 x-y 평면에서 옅은 회색 영역이 그 활동영역이죠.

그림. F.0

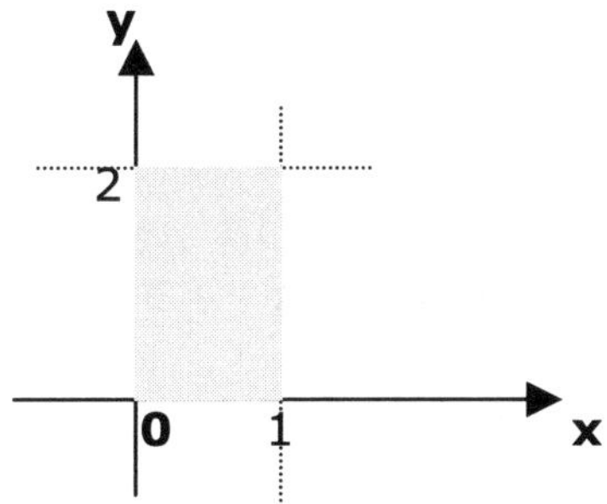

그럼 이제, 그 그래프에 커브를 놓을 건데 그 커브의 일부(포물선 선분)가 그 영역에 있도록 하는 겁니다. 좀 더 자세히 하자면, 그 일부가 x축을 따라 0에서 1까지 y축을 따라서는 0에서 2까지 걸쳐서 그 영역을 채울 수 있도록 해야 하는 겁니다. 이제, 그래프에 커브들을 몇 개 놓아보도록 하죠.

그림. F.1

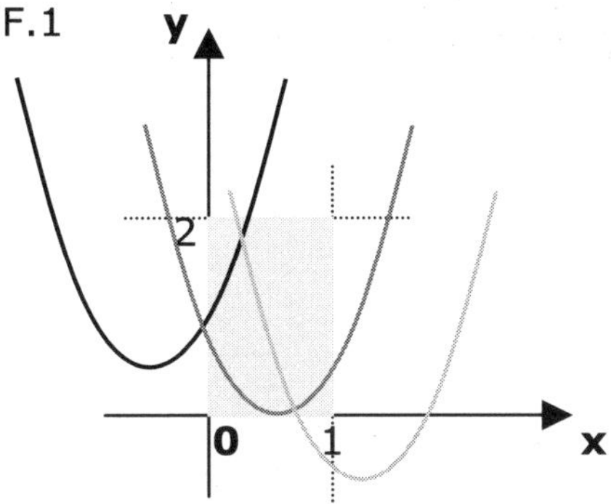

위와 같이 놓인 그 어느것도 맞지않는 거네요. 왜냐하면 이것들 그 어느 것도 위에서 명시한 양 만큼에 걸쳐서 그 활동영역을 채우지않기 때문이죠. 이제, 다시 시도해보기로 하죠.

그림. F.2

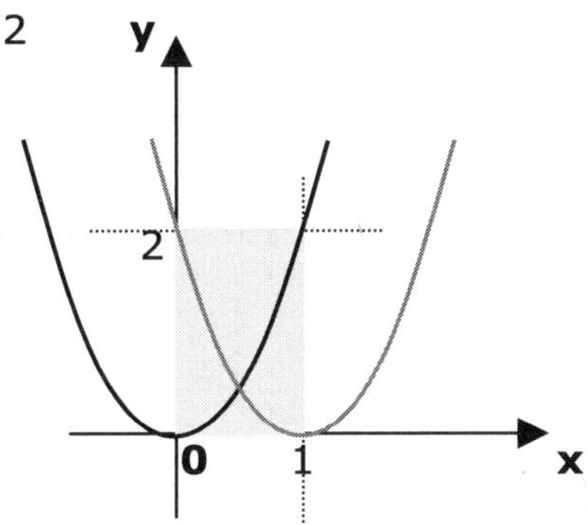

안됐지만 또 잘못됐네요.
이 커브들 둘 다 제각기 제대로 된 양 만큼에 걸쳐서 그 활동공간을 채우고는 있죠; 하지만, y절편들이 맞지 않는 군요. 주어진 방정식 $y = ax^2 + bx + 1$ 에서는 y절편이 1 이죠. 그럼 이제, 다시 시도해보기로 하죠.

그림. F.3

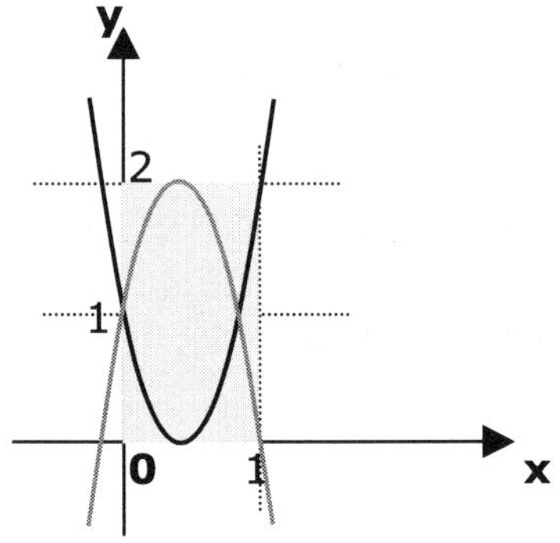

이번에는 제대로 됐어요. 이 커브들 둘 다 제자리에 놓여서 제각기 제대로 된 양 만큼에 걸쳐서 그 활동공간을 채우고 있고 y 절편들도 맞는군요. 그러면 이제, 이 두개의 포물선들이 맞는 것이고 그 방정식들을 작성하기 시작할 수 있겠죠.

방정식은 $y = ax^2 + bx + 1$ 이고 $0 \le x \le 1$ 이며 $0 \le y \le 2$ 이죠.
그럼 이제, b 와 c 의 값들은 찾아야 하죠. 즉, 그 두개의 포물선들의 방정식들을 구해야 하는 겁니다. 위의 그래프를 보면, 구하고자 하는 그 방정식들 각각을 작성하는 데 두개의 점들을 바로 사용할 수 있을 것 같죠. 하지만, 그 두 점들 중 하나는 그 방정식들을 작성하는 데 도움이 되지않는 겁니다. 도움이 되는 점이 (1, 0)에만 있는 거고 (0, 1)에 있는 점은 아니죠. 왜냐하면 예를 들어, 그 기울기가 음인 포물선에 대하여는:
그 방정식은 $y = ax^2 + bx + 1$ 이니까 $(0, 1) \Rightarrow 1 = 0 + 0 + 1$ 이고 이는 별 의미가 없는 거죠.

이제, (1, 0)에 있는 점을 사용해보기로 하죠.

$y = ax^2 + bx + 1$ 이니까, $(1, 0) \Rightarrow 0 = a + b + 1 \Rightarrow a + b = -1$.

그럼 이제, 다른 거 뭘 사용할 수 있나요?
그 포물선의 꼭지점에 관한 한계를 사용할 수 있죠.
구체적으로 이야기하면 그 꼭지점의 x 좌표는 양 그리고 y 좌표는 2 이어야 하는 겁니다.
그러면, 이제 그 꼭지점을 구해보도록 하죠.

$y = ax^2 + bx + 1 = a\{x^2 + bx/a + (b/2a)^2 - (b/2a)^2\} + 1 = a(x + b/2a)^2 - a(b/2a)^2 + 1$
$= a(x + b/2a)^2 - b^2/4a + 1 = a(x + b/2a)^2 - (b^2 - 4a)/4a$.

그러면, 그 꼭지점은 $\{-b/2a, -(b^2 - 4a)/4a\}$에 있네요.
그러니까, $-(b^2 - 4a)/4a = 2$ 이죠.

그러면 이제, a와 b에 관한 두개의 방정식으로 된 연립방정식을 갖는 거죠: $a + b = -1$ 과 $-(b^2 - 4a)/4a = 2$. 그러면:
$a + b = -1 \Rightarrow a = -(1 + b)$.
$-(b^2 - 4a)/4a = 2 \Rightarrow b^2 - 4a = -8a \Rightarrow b^2 + 4a = 0$.

그러면, $a = -(1 + b) \Rightarrow b^2 + 4a = b^2 - 4(1 + b) = b^2 - 4b - 4 = 0$.
$b^2 - 4b - 4 = (b - 2)^2 - 8 = 0 \Rightarrow b - 2 = \pm 8^{1/2} \Rightarrow b = 2 + 8^{1/2}$ 또는 $2 - 8^{1/2}$.

그러면:
$b = 2 + 8^{1/2} \Rightarrow a = -(1 + b) = -(3 + 8^{1/2})$.
$b = 2 - 8^{1/2} \Rightarrow a = -(1 + b) = -(3 - 8^{1/2})$.

자 이제, 위의 두 값들 중에서 a 의 값을 골라야 하는데 그 기울기가 음이라는 걸 알고있죠.
하지만, 두 값들이 모두 음이어서 이들 중 어느 것이 a 의 값인지 알 수가 없네요.
그런데, $-(3 - 8^{1/2})$이 음이라는 건 어떻게 아나요?
$3^2 = 9$ 이고 $(8^{1/2})^2 = 8$ 이죠. 그러면, $3 > 8^{1/2} \Rightarrow 3 - 8^{1/2} > 0 \Rightarrow -(3 - 8^{1/2}) < 0$.

그럼 이제, 그 꼭지점에서의 x 좌표에 관한 한계를 사용할 겁니다.
그 x 좌표 $= -b/2a > 0$ 라는 걸 알고있죠.
그러면, $-b/2a > 0 \Rightarrow b/2a < 0 \Rightarrow b/a < 0$. 그러면 이제:

$b = 2 + 8^{1/2} > 0$ 이고 $a = -(3 + 8^{1/2}) < 0$. 그러면, $b/a < 0 \Rightarrow$ 적합한 거죠.
$b = 2 - 8^{1/2} < 0$ 이고 $a = -(3 - 8^{1/2}) < 0$. 그러면, $b/a > 0 \Rightarrow$ 부적합한 거네요.

그러니까, 그 포물선의 방정식은 $y = -(3 + 8^{1/2})x^2 + (2 + 8^{1/2})x + 1$ 인 겁니다.

그럼 이제, 그 기울기가 양인 포물선으로 넘어가보도록 하죠.

그림. F.4

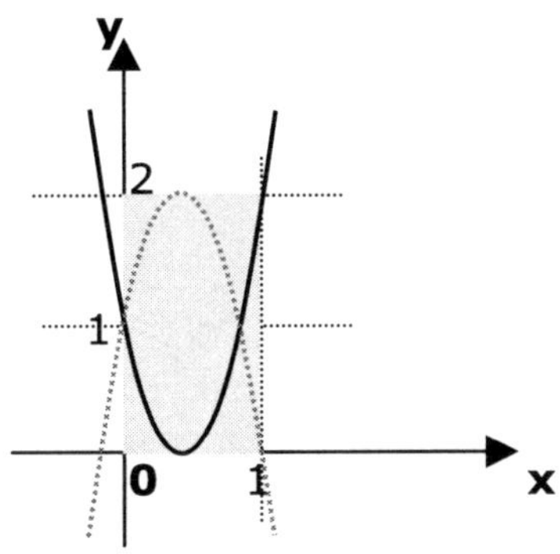

위의 커브들을 자세히 살펴보면 직선 $y = 1$ 에 관하여 서로 대칭이라는 걸 알 수 있죠.
그러면, 그 기울기들의 크기는 서로 같은 거네요. 즉, 방금 구한 포물선의 기울기의 부호를 바꾸면 지금 구하고자 하는 포물선의 기울기를 갖게 되는 거죠. 그러니까, $a = 3 + 8^{1/2}$.

그런데, b 의 값은 어쩌죠?
(1, 2)에 있는 그 점을 사용해서 b 의 값을 구하면 되죠. 이제, (1, 2)에 있는 그 점을 사용해보기로 하죠.
$y = ax^2 + bx + 1$ 이니까 $(1, 2) \Rightarrow 2 = a + b + 1 \Rightarrow a + b = 1 \Rightarrow b = 1 - a$.
그러면, $a = 3 + 8^{1/2} \Rightarrow b = 1 - a = 1 - (3 + 8^{1/2}) = -2 - 8^{1/2} = -(2 + 8^{1/2})$.
그러니까, 그 방정식은 $y = (3 + 8^{1/2})x^2 - (2 + 8^{1/2})x + 1$ 이네요.

그런데, 연습 삼아서 처음에 a 부터 직접 구해서 이 포물선을 다시 구해보도록 하죠.
그러면, 그 점 (1, 2)와 그 꼭지점을 사용하는 겁니다.

먼저, $y = ax^2 + bx + 1$ 이니까 $(1, 2) \Rightarrow 2 = a + b + 1 \Rightarrow a + b = 1 \Rightarrow a = 1 - b$.
다음으로, 그 꼭지점의 다음과 같은 구체적인 내용을 사용하겠어요:
그 꼭지점의 x 좌표는 양이고 y 좌표는 0 이다.

그 꼭지점은 $\{-b/2a, -(b^2 - 4a)/4a\}$에 있다는 걸 알고있죠. 그러니까, $-(b^2 - 4a)/4a = 0$.

그러면 이제, a와 b에 관한 두개로 된 연립방정식을 갖는 거죠: $a = 1 - b$와 $-(b^2 - 4a)/4a = 0$. 그러면:

$-(b^2 - 4a)/4a = 0 \Rightarrow b^2 - 4a = 0$.
$a = 1 - b \Rightarrow b^2 - 4a = b^2 - 4(1 - b) = b^2 + 4b - 4 = (b + 2)^2 - 8 = 0 \Rightarrow b + 2 = \pm 8^{1/2}$
$\Rightarrow b = -2 + 8^{1/2}$ 또는 $-2 - 8^{1/2}$.

그러면:
$b = -2 + 8^{1/2} \Rightarrow a = 1 - b = 1 - (-2 + 8^{1/2}) \Rightarrow a = 3 - 8^{1/2}$.
$b = -2 - 8^{1/2} \Rightarrow a = 1 - b = 1 - (-2 - 8^{1/2}) \Rightarrow a = 3 + 8^{1/2}$.

그러면 이제, 위의 두개의 값들 중에서 a의 값을 골라야 하는데 그 기울기가 양이라는 걸 알고있죠.
하지만, 둘 다 양이어서 a의 값으로 뭘 골라야 할지 모르는 거죠.

그럼 이제, 그 꼭지점에서의 x좌표에 관한 한계를 사용할 겁니다.
그 x좌표 $= -b/2a > 0$라는 걸 알고있죠. 그러면, $-b/2a > 0 \Rightarrow b/2a < 0 \Rightarrow b/a < 0$. 그러면 이제:
$b = -2 + 8^{1/2} > 0$이고 $a = 3 - 8^{1/2} > 0$. 그러면, $b/a > 0 \Rightarrow$ 부적합한 거죠.
$b = -2 - 8^{1/2} < 0$이고 $a = 3 + 8^{1/2} > 0$. 그러면, $b/a < 0 \Rightarrow$ 적합한 거네요.

그러니까, 그 포물선의 방정식은 $y = (3 + 8^{1/2})x^2 - (2 + 8^{1/2})x + 1$인 겁니다.

요약:

주어진 방정식은 $y = ax^2 + bx + 1$이고 $0 \le x \le 1$이며 $0 \le y \le 2$이죠.
그러면, 그 포물선은 다음 둘 중 어느 하나와 같아야 합니다.

그림. F.3

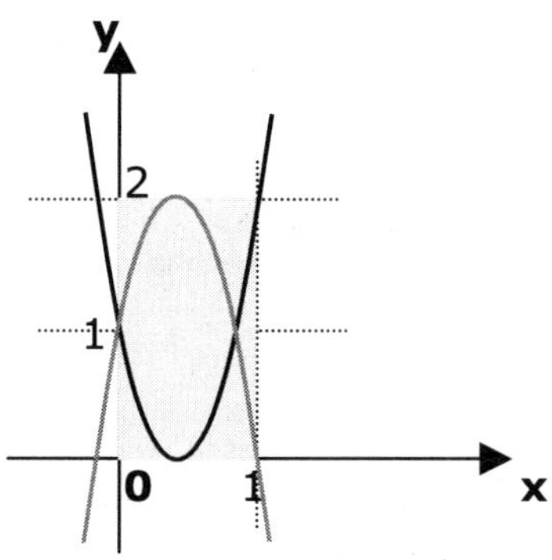

기울기가 음인 포물선부터 시작합니다.
그러면, 이는 점 (1, 0)을 지나고 그 꼭지점의 x좌표는 양 그리고 y좌표는 2이죠.

먼저, $y = ax^2 + bx + 1$이니까, $(1, 0) \Rightarrow 0 = a + b + 1 \Rightarrow a + b = -1$.
다음으로, 그 꼭지점을 구할 겁니다.
$y = ax^2 + bx + 1 = a\{x^2 + bx/a + (b/2a)^2 - (b/2a)^2\} + 1 = a(x + b/2a)^2 - a(b/2a)^2 + 1$
$= a(x + b/2a)^2 - b^2/4a + 1 = a(x + b/2a)^2 - (b^2 - 4a)/4a$.

그러면, 그 꼭지점은 $\{-b/2a, -(b^2 - 4a)/4a\}$. 그러므로, $-(b^2 - 4a)/4a = 2$.

그러면 이제, a와 b에 관한 두개로 된 연립방정식을 갖습니다: $a + b = -1$ 과 $-(b^2 - 4a)/4a = 2$. 그러면:

$a + b = -1 \Rightarrow a = -(1 + b)$.
$-(b^2 - 4a)/4a = 2 \Rightarrow b^2 - 4a = -8a \Rightarrow b^2 + 4a = 0$.

그러면, $a = -(1 + b) \Rightarrow b^2 + 4a = b^2 - 4(1 + b) = b^2 - 4b - 4 = 0$.
$b^2 - 4b - 4 = (b - 2)^2 - 8 = 0 \Rightarrow b - 2 = \pm 8^{1/2} \Rightarrow b = 2 + 8^{1/2}$ 또는 $2 - 8^{1/2}$.

그러면, $b = 2 + 8^{1/2} \Rightarrow a = -(1 + b) = -(3 + 8^{1/2})$이고 또한, $b = 2 - 8^{1/2} \Rightarrow a = -(1 + b) = -(3 - 8^{1/2})$.

그 꼭지점의 x 좌표 $= -b/2a > 0$ 라는 걸 알고있죠.
그러면, $-b/2a > 0 \Rightarrow b/2a < 0 \Rightarrow b/a < 0$.
그러면 이제:
$b = 2 + 8^{1/2} > 0$ 이고 $a = -(3 + 8^{1/2}) < 0$. 그러면, $b/a < 0 \Rightarrow$ 적합.
$b = 2 - 8^{1/2} < 0$ 이고 $a = -(3 - 8^{1/2}) < 0$. 그러면, $b/a > 0 \Rightarrow$ 부적합.

따라서, $a = -(3 + 8^{1/2})$이고 $b = 2 + 8^{1/2}$.

그럼 이제, 그 기울기가 양인 포물선으로 넘어가보도록 하죠.
그러면, 이는 (1, 2)을 지나고 그 꼭지점의 x 좌표는 양이고 y 좌표는 0 입니다.
먼저, $y = ax^2 + bx + 1$ 이니까 $(1, 2) \Rightarrow 2 = a + b + 1 \Rightarrow a + b = 1 \Rightarrow a = 1 - b$.
다음으로, 그 꼭지점은 $\{-b/2a, -(b^2 - 4a)/4a\}$에 있다는 걸 알고있죠. 그러니까, $-(b^2 - 4a)/4a = 0$.

그러면 이제, a와 b에 관한 두개로 된 연립방정식을 갖는 거죠: $a = 1 - b$와 $-(b^2 - 4a)/4a = 0$.

그러면:
$-(b^2 - 4a)/4a = 0 \Rightarrow b^2 - 4a = 0$.
$a = 1 - b \Rightarrow b^2 - 4a = b^2 - 4(1 - b) = b^2 + 4b - 4 = (b + 2)^2 - 8 = 0 \Rightarrow b + 2 = \pm 8^{1/2}$
$\Rightarrow b = -2 + 8^{1/2}$ 또는 $-2 - 8^{1/2}$.

그러면:
$b = -2 + 8^{1/2} \Rightarrow a = 1 - b = 1 - (-2 + 8^{1/2}) \Rightarrow a = 3 - 8^{1/2}$.
$b = -2 - 8^{1/2} \Rightarrow a = 1 - b = 1 - (-2 - 8^{1/2}) \Rightarrow a = 3 + 8^{1/2}$.

그 x 좌표 $= -b/2a > 0$ 라는 걸 알고있죠. 그러면, $-b/2a > 0 \Rightarrow b/2a < 0 \Rightarrow b/a < 0$. 그러면 이제:
$b = -2 + 8^{1/2} > 0$ 이고 $a = 3 - 8^{1/2} > 0$. 그러면, $b/a > 0 \Rightarrow$ 부적합.
$b = -2 - 8^{1/2} < 0$ 이고 $a = 3 + 8^{1/2} > 0$. 그러면, $b/a < 0 \Rightarrow$ 적합.

따라서, $a = 3 + 8^{1/2}$이고 $b = -(2 + 8^{1/2})$.

그러므로, $a = -(3 + 8^{1/2})$이고 $b = 2 + 8^{1/2}$ 또는 $a = 3 + 8^{1/2}$이고 $b = -(2 + 8^{1/2})$.

제안과 풀이
G 번 예제에서의 문제

$y = f(x) = ax^2 + bx + c$ 이고 $0 \le x \le 1$ 이며 $0 \le y \le 2$ 라고 하고 a와 b 그리고 c의 값들을 구하세요.

이 예제에서도 함수 f 가 $0 \le x \le 1$ 인 x 에 대해서 0 부터 2 까지의 값들을 만들어내게 해야 하니까 이 함수 f 가 $0 \le x \le 1$ 와 $0 \le y \le 2$ 로 명시된 활동영역에서 어떻게 활동해야 하는지 알아보아야 하죠. 함수의 활동은 그 커브로 알 수 있고 이 f 의 커브가 포물선이라는 걸 알고있죠. 그러면, 이 커브를 x-y 좌표계에서의 그래프에 놓고 그 활동영역에서의 커브를 보도록 하죠. 아래의 x-y 평면에서 옅은 회색 영역이 그 활동영역이죠.

그림. G.0

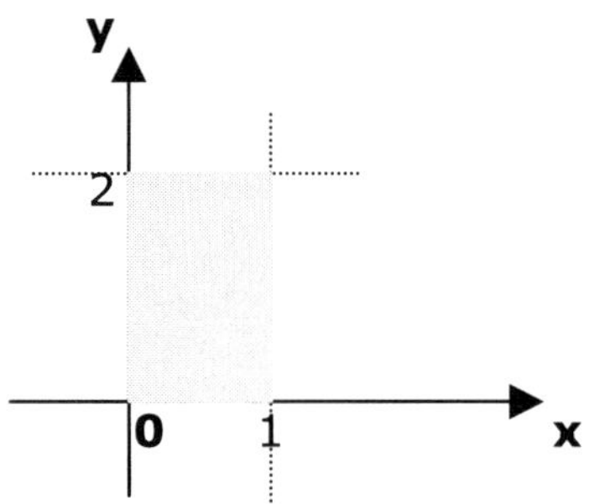

그럼 이제, 그 그래프에 커브를 놓는데 그 커브의 일부(포물선 선분)가 그 영역에 있도록 하는 겁니다. 하지만, 그 부분이 그 영역에 그저 놓이는 건 아니죠. 그 선분이 x 축을 따라 0 에서 1 까지 y 축을 따라서는 0 에서 2 까지 걸쳐서 그 영역을 채울 수 있도록 해야 하는 겁니다. 이제, 그래프에 커브들을 몇 개 놓아보도록 하죠.

그림. G.1

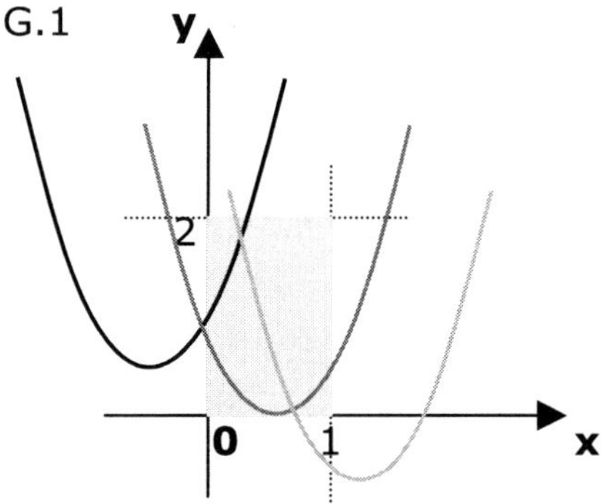

위와 같이 놓인 그 어느것도 맞지않는 거네요. 왜냐하면 이것들 그 어느 것도 위에서 명시한 양 만큼에 걸쳐서 그 활동영역을 채우지않기 때문이죠. 이제, 다시 시도해보기로 하죠.

그림. G.2

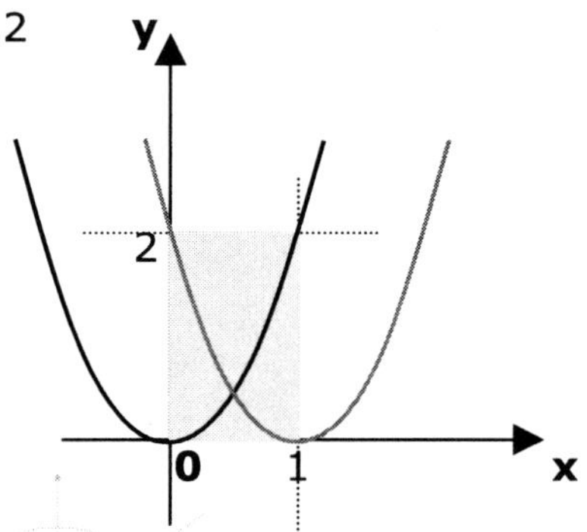

이것들은 되죠 하지만, 이것들만은 아닙니다. 그러니, 좀 더 해보기로 하죠.

그림. G.3

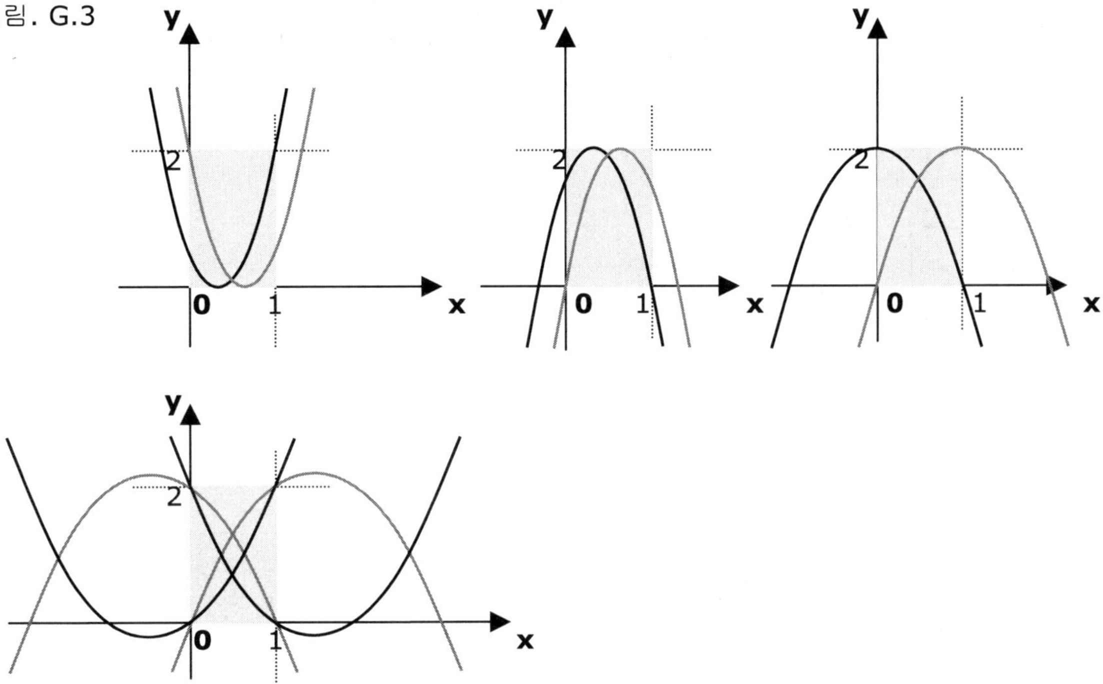

고려해봐야 할 경우들이 꽤 많군요. 그러니까, 그 모든 경우들을 두 부류들로 분류해보기로 하겠어요. 그 둘 중 한 부류에서는 포물선들이 양의 기울기들을 갖는 것이고 다른 한 부류에서는 그 포물선들이 음의 기울기들을 갖는 거죠. 그런데, 각 부류마다 답이 되는 포물선들이 많이 있다는 걸 알게 될 겁니다. 얼마나 많은데요? 무한정 많죠. 그러니까, $y = ax^2 + bx + c$에서의 상수들 a, b, c의 특정 값들이 아니라 이 상수들 각각의 범위를 구하게 되는 거죠. 그럼 이제, 포물선들이 양의 기울기를 갖는 부류부터 시작하죠.

그림. G.4

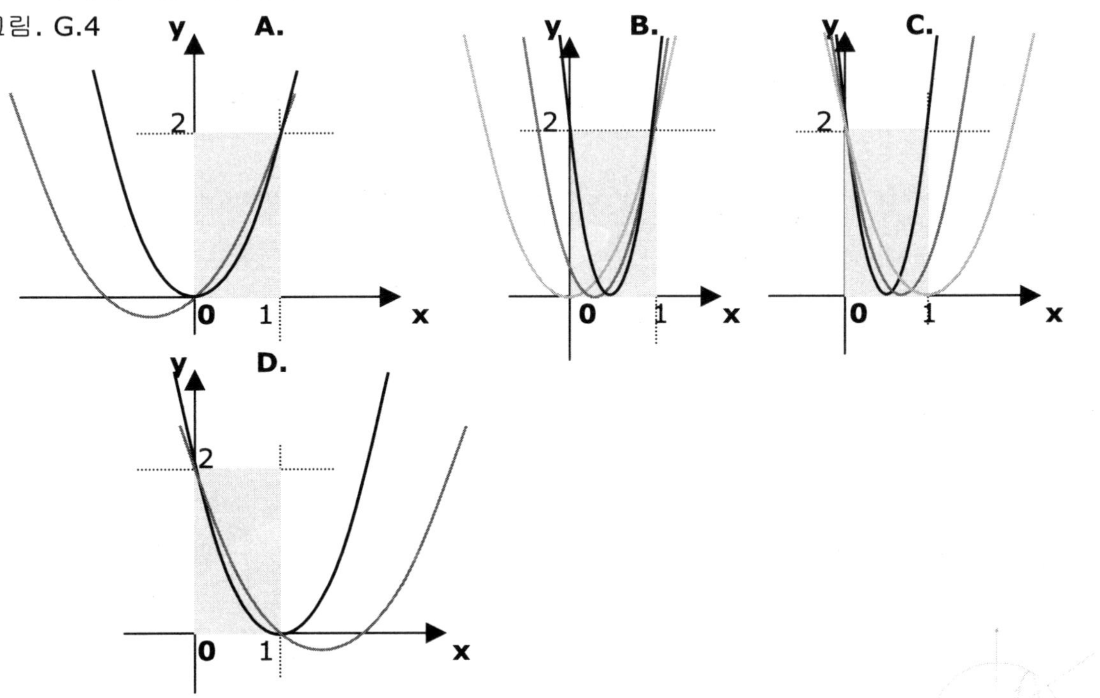

A의 경우, $f(0) = 0$이고 $f(1) = 2$; 또한, 꼭지점에서의 x 좌표 ≤ 0.
B의 경우, $0 \leq f(0) \leq 2$이고 $f(1) = 2$; 또한, 꼭지점에서는, $0 \leq$ x 좌표 $\leq 1/2$이며 y 좌표 $= 0$.
C의 경우, $f(0) = 2$이고 $0 \leq f(1) \leq 2$; 또한, 꼭지점에서는, $1/2 \leq$ x 좌표 ≤ 1이며 y 좌표 $= 0$.
D의 경우, $f(0) = 2$이고 $f(1) = 0$; 또한, 꼭지점에서는, x 좌표 ≥ 1.

자 이제, A의 경우부터 시작하죠.

그림. G.4

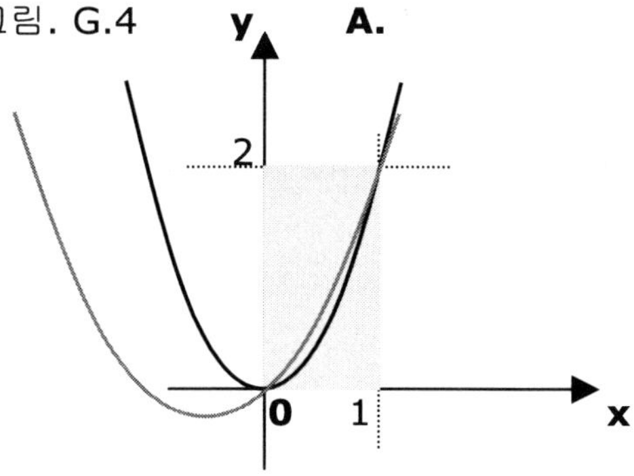

다음을 만족하는 a, b, c의 값들을 구해야 하죠:
$y = f(x) = ax^2 + bx + c$이고 $0 \leq x \leq 1$이며 $0 \leq y \leq 2$.

그러면, 다음과 같은 조건들을 만족시켜야 하죠: $f(0) = 0$이고 $f(1) = 2$; 또한, 그 꼭지점에서의 x 좌표 ≤ 0.

그런데, 그 꼭지점에서의 y 좌표는 어째서 고려하지 않는 거죠? 그것이 0보다 작거나 같아야 하는 것 아닌가요? 물론, 그건 0 보다 작거나 같아야 하죠. 하지만, 고려하지 않아도 되요. 고려하면 쓸데 없는 일을 하는 거죠. 위에서 명시한 조건들은 이제 더할 나위 없이 완벽한 것이고 y 좌표로 하여금 그렇게 되도록 강제할 겁니다. 다시 말해서, 그 모든 조건들이 완료되면 그 꼭지점에서의 y 좌표는 양이 될 수 없는 거죠. 그럼 이제, 그 조건들을 만족시켜 보기로 하죠.

$f(0) = 0 \Rightarrow 0 + 0 + c = 0 \Rightarrow c = 0 \Rightarrow f(x) = ax^2 + bx$. 그러면, $f(1) = 2 \Rightarrow a + b = 2$.

이제, 그 꼭지점을 구해보도록 하죠.
$y = ax^2 + bx + c = a\{x^2 + bx/a + (b/2a)^2 - (b/2a)^2\} + c = a(x + b/2a)^2 - a(b/2a)^2 + c$
$= a(x + b/2a)^2 - b^2/4a + c = a(x + b/2a)^2 - (b^2 - 4ac)/4a$.
그러면, 꼭지점 $= \{-b/2a, -(b^2 - 4ac)/4a\}$.

그런데, a는 기울기이고, 지금, 그 기울기 a는 양이죠.
그러면, 그 꼭지점의 x 좌표 $\leq 0 \Rightarrow -b/2a \leq 0 \Rightarrow b/a \geq 0 \Rightarrow b \geq 0$ 왜냐하면 $a > 0$.

그럼 이제, 중간 결과로서, 다음과 같죠: $c = 0$이고 $a + b = 2$이며 $b \geq 0$.

이제, a의 범위를 구해보기로 하죠.
$a + b = 2 \Rightarrow b = 2 - a$. 그러면, $b = 2 - a$이고 $b \geq 0$이니까 $2 - a \geq 0 \Rightarrow a \leq 2$.
그런데 또한, $a > 0$인 것을 알고있죠. 그러니까, $0 < a \leq 2$이네요.

따라서, 이제 다음과 같죠: $c = 0$, $a + b = 2$, $0 < a \leq 2$.

그러므로, A의 경우에는, 그 방정식이 $y = ax^2 + bx$이고 a와 b는 다음을 만족하죠:
$a + b = 2$ 이고 $0 < a \le 2$.

그럼 이제, B의 경우를 시작하기로 하죠.

그림. G.4

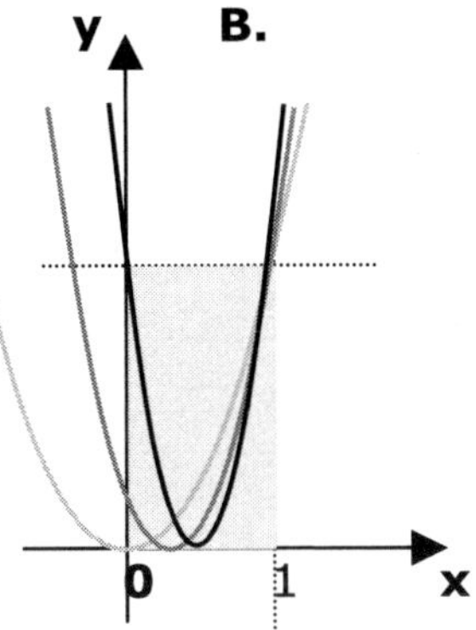

다음을 만족하는 상수 a, b, c의 값들을 구해야 하죠:
$y = f(x) = ax^2 + bx + c$ 이고 $0 \le x \le 1$ 이며 $0 \le y \le 2$.

그러면, 다음과 같은 조건들을 만족시켜야 하는 거죠:
$f(1) = 2$ 이고 또한, 그 꼭지점에서, $0 \le x$ 좌표 $\le 1/2$ 이고 y 좌표 $= 0$.

그런데, f(0)의 값은 어째서 고려하지 않는 거죠? 즉, $0 \le f(0) \le 2$ 이어야 하는 것 아닌가요?
물론, 그렇죠. 하지만, 그런 경우를 고려하지 않아도 되요. 고려하면 쓸데 없는 일을 하는 거죠. 위에서 명시한 조건들은 이제 더할 나위 없이 완벽한 것이고 f(0)로 하여금 그렇게 되도록 강제할 겁니다. 다시 말해서, 그 모든 조건들이 완료되면 $0 \le f(0) \le 2$가 될 겁니다. 그럼 이제, 그 조건들을 만족시켜보기로 하죠.

먼저, f(1)부터 시작합니다: $f(1) = 2 \Rightarrow 2 = a + b + c$.

다음으로, 그 꼭지점은 $\{-b/2a, -(b^2 - 4ac)/4a\}$에 있다는 걸 알고있죠. 그러면:

y좌표 $= 0 \Rightarrow -(b^2 - 4ac)/4a = 0 \Rightarrow b^2 - 4ac = 0$.
$0 \le x$ 좌표 $\le 1/2 \Rightarrow 0 \le -b/2a \le 1/2 \Rightarrow -1/2 \le b/2a \le 0 \Rightarrow -1 \le b/a \le 0$.

이제, 중간 결과로서, 다음과 같죠: $a + b + c = 2$, $b^2 - 4ac = 0$, $-1 \le b/a \le 0$.
그러면 이제, a, b, c의 범위들을 구해야 하는데 이용할 만한 것이 별로 없네요.
그런데, 위의 그래프를 자세히 보면, $2 \le a \le 8$ 이라는 걸 알 수 있죠. 어떻게 요?

위의 그래프에서 옅은 회색 포물선의 기울기가 2 인 것을 알 수 있죠. 어떻게 요?

꼭지점이 원점에 있으니까 그 방정식이 $y = ax^2$이라고 하세요. 그러면:
$y = ax^2$이고 그 포물선은 (1, 2)를 지나니까 $(1, 2) \Rightarrow 2 = a$.

또한, 검정 포물선의 기울기가 8 인 것도 알 수 있죠. 어떻게 요?

그 꼭지점이 (1/2, 0)에 있으니까 그 방정식이 $y = a(x - 1/2)^2$ 라고 하세요.
그러면, $y = a(x - 1/2)^2$이고 이는 (1, 2)를 지나니까 $(1, 2) \Rightarrow 2 = a(1 - 1/2)^2 = a(1/2)^2 = a/4 \Rightarrow a = 8$.

그러므로, 또 하나의 중간 결과로서, 다음과 같죠:
$2 \le a \le 8$, $a + b + c = 2$, $b^2 - 4ac = 0$, $-1 \le b/a \le 0$.

그러면 이제, b 의 가능한 값들을 구해보기로 하죠.
$-1 \le b/a \le 0 \Rightarrow -a \le b \le 0$. 그러면, $2 \le a \le 8$ 이니까 $-a \le b \le 0 \Rightarrow -8 \le b \le 0$ 이죠. 어떻게 요?

$2 \le a \le 8$ 는 a 가 2 에서 8 까지 어느 값이든 갖을 수 있다는 거죠. 그러면:
먼저, $a = 2$ 이면 $-a \le b \le 0 \Rightarrow -2 \le b \le 0$. 다음으로, $a = 8$ 이면 $-a \le b \le 0 \Rightarrow -8 \le b \le 0$.
그러니까, $-8 \le b \le 0$ 이네요.

그러면 이제, 다음과 같아요: $a + b + c = 2$, $b^2 - 4ac = 0$, $2 \le a \le 8$, $-8 \le b \le 0$.

그러면, c 는 어쩌고요?
그 c 는 소거해버리기로 하죠. 그러면, $a + b + c = 2 \Rightarrow c = 2 - (a + b)$.

또한, 두개의 관계식들 $a + b + c = 2$ 와 $b^2 - 4ac = 0$ 를 합쳐볼 수 있겠어요.

먼저, $a + b + c = 2 \Rightarrow c = 2 - (a + b) \Rightarrow -4ac = -8a + 4a(a + b) = -8a + 4a^2 + 4ab$
$= 4a^2 + 4ab + b^2 - b^2 - 8a = (2a + b)^2 - b^2 - 8a$.

다음으로, $b^2 - 4ac = b^2 + (2a + b)^2 - b^2 - 8a = (2a + b)^2 - 8a = 0$.

그러면, $(2a + b)^2 = 8a \Rightarrow 2a + b = \pm(8a)^{1/2} = \pm \quad \Rightarrow b = \pm\sqrt{8a} - 2a = \pm(8a) - 2a$이죠.

그런데, a 가 2 에서 8 까지 변할 때:
$\{(8a)^{1/2} - 2a\}$은 0 에서 -8 까지 변하죠 그러나 $\{-(8a)^{1/2} - 2a\}$은 -8 에서 -24 까지 변합니다.

$-8 \le b \le 0$ 이라는 것을 알고있죠.
그러니까, $b = (8a)^{1/2} - 2a = \sqrt{8a} - 2a$이네요.

그러면 이제, 다음과 같아요: $b = (8a)^{1/2} - 2a$, $2 \le a \le 8$, $c = 2 - (a + b)$.

그리하여, 구하고자 하는 방정식은 다음과 같이 나타낼 수 있어요:

$y = ax^2 + bx + 2 - (a + b)$이며 a, b는 다음을 만족한다: $b = (8a)^{1/2} - 2a$이고 $2 \le a \le 8$.

그런데, 다시 그래프를 조사해보면 $0 \le c \le 2$ 라는 것을 알 수 있죠.
이는 왜냐하면 $0 \le$ y절편 ≤ 2 이어야 하고 $y = ax^2 + bx + c$에서 c가 y절편이기 때문이죠.

또한, b 의 범위도 확인해보죠.

$0 \le c \le 2$ 이고 $2 \le a \le 8 \Rightarrow 2 \le a + c \le 10$.
$a + b + c = 2 \Rightarrow a + c = 2 - b \Rightarrow 2 \le 2 - b \le 10 \Rightarrow 0 \le -b \le 8 \Rightarrow -8 \le b \le 0$.

그럼 이제, C 의 경우를 가지고 계속하기로 하죠.

그림. G.4

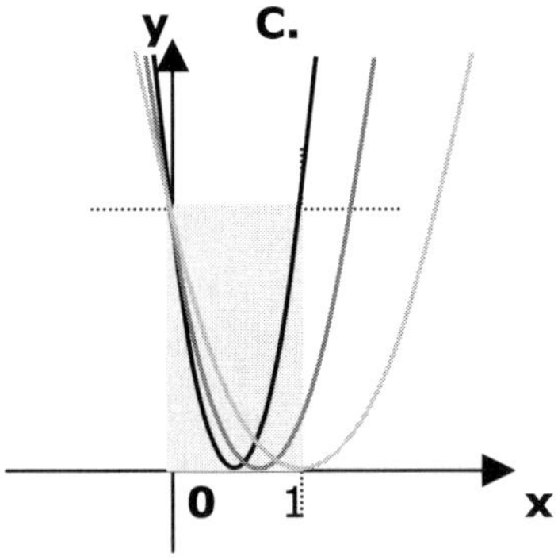

그러면 이제, 다음을 만족하는 상수들 a, b, c 의 값들을 구해야 하죠:
$y = f(x) = ax^2 + bx + c$이고 $0 \le x \le 1$ 이며 $0 \le y \le 2$.

그러면, 다음과 같은 조건들을 만족해야 합니다:
$f(0) = 2$ 이고 또한, 그 꼭지점에서, $1/2 \le$ x 좌표 ≤ 1 이고 y 좌표 $= 0$.

먼저, $f(0) = 2 \Rightarrow 2 = c \Rightarrow f(x) = ax^2 + bx + 2$.

다음으로, 그 꼭지점으로 넘어가보죠. 그 꼭지점은 $\{-b/2a, -(b^2 - 4ac)/4a\}$에 있는 거죠. 그러면:

먼저, y좌표 $= 0 \Rightarrow -(b^2 - 4ac)/4a = 0 \Rightarrow b^2 - 4ac = 0 \Rightarrow b^2 - 8a = 0$ 왜냐하면 $c = 2$.
다음으로, $1/2 \le$ x 좌표 $\le 1 \Rightarrow 1/2 \le -b/2a \le 1 \Rightarrow -1 \le b/2a \le -1/2 \Rightarrow -2 \le b/a \le -1$.

그러면, 중간 결과로서, 다음과 같죠: $c = 2$, $b^2 - 8a = 0$, $-2 \le b/a \le -1$.

그러면 이제, a와 b사이의 관계식 $b^2 - 8a = 0$ 이 있으니까 a나 b의 범위를 구해야겠어요.
이 관계식에 b^2항이 있는데 b의 부호를 모르죠.
그러면, 부등식 $-2 \le b/a \le -1$ 을 이용해서 b 의 범위를 구해보도록 하죠.
$b^2 - 8a = 0 \Rightarrow b^2 = 8a \Rightarrow b/a = 8/b$.
그러면, $-2 \le b/a \le -1$ 이니까 $-2 \le 8/b \le -1$. 따라서, $-1 \le b/8 \le -1/2 \Rightarrow -8 \le b \le -4$.

그럼 이제, 다음과 같습니다: $c = 2$, $-8 \le b \le -4$, $b^2 - 8a = 0$.
그러니까, C의 경우에는, 그 방정식이 $y = ax^2 + bx + 2$ 이고 a와 b는 다음을 만족하는 거죠:
$b^2 - 8a = 0$(또는, $a = b^2/8$)이고 $-8 \le b \le -4$.

그런데, 위의 B 의 경우에서와 같이, 위의 그래프만 봐도 $2 \le a \le 8$ 이라는 것을 알 수 있어요.
우선, 옅은 회색 포물선의 기울기가 2 인 것을 알 수 있죠.
그 꼭지점이 (1, 0)에 있으니까 방정식을 $y = a(x - 1)^2$라고 하세요. 그러면:
$y = a(x - 1)^2$이고 그 포물선은 (0, 2)를 지나니까 $(0, 2) \Rightarrow 2 = a$.
다음으로, 검정 포물선의 기울기는 8 인 것을 알 수 있죠.
그 꼭지점이 (1/2, 0)에 있으니까 방정식을 $y = a(x - 1/2)^2$이라고 하세요. 그러면:
$y = a(x - 1/2)^2$이고 그 포물선은 (1, 2)를 지나니까 $(1, 2) \Rightarrow 2 = a(1 - 1/2)^2 = a(1/2)^2 = a/4 \Rightarrow a = 8$.
이것이 맞는지 실제로 a 의 범위를 구해서 확인해보도록 하죠.
$-8 \le b \le -4 \Rightarrow 16 \le b^2 \le 64 \Rightarrow 2 \le b^2/8 \le 8$. 그러면, $a = b^2/8$ 이니까 $2 \le a \le 8$.

그럼 이제, D 의 경우로 넘어가보도록 하죠.

그림. G.4

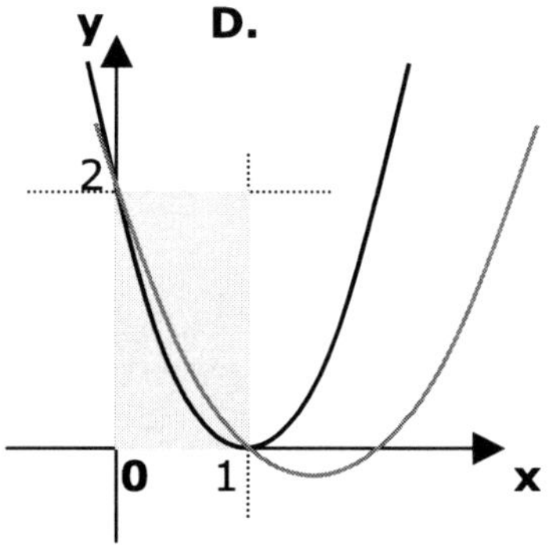

이제, 다음을 만족하는 상수들 a, b, c 의 값들을 구해야 하죠:
$y = f(x) = ax^2 + bx + c$이고 $0 \le x \le 1$ 이며 $0 \le y \le 2$.
그러면, 다음과 같은 조건들을 만족해야 합니다: $f(0) = 2$ 이고 $f(1) = 0$; 또한, 그 꼭지점에서, x 좌표 ≥ 1.
그러면, $f(0) = 2 \Rightarrow 2 = c \Rightarrow f(x) = ax^2 + bx + 2$.
따라서, $f(1) = 0 \Rightarrow 0 = a + b + 2$ 이네요. 그러니까, $a + b = -2$.
이제, 그 꼭지점으로 넘어가보죠. 그 꼭지점은 $\{-b/2a, -(b^2 - 4ac)/4a\}$에 있는 거죠.
또한, 그 기울기 a 가 양이라는 걸 알고있죠.
그러면, x 좌표 $\ge 1 \Rightarrow -b/2a \ge 1 \Rightarrow b/a \le -2 \Rightarrow b \le -2a$ 왜냐하면 $a > 0$.
그러면, 중간 결과로서, 다음과 같죠: $c = 2$, $a + b = -2$, $b \le -2a$.
이제, a 의 범위를 구해보도록 하죠.
$a + b = -2 \Rightarrow b = -2 - a$. 그러면, $b \le -2a \Rightarrow -2 - a \le -2a \Rightarrow a \le 2$.
그런데 또한, 지금 $a > 0$ 이죠. 그러니까, $0 < a \le 2$ 이네요.
그러니까, D의 경우, 방정식은 $y = ax^2 + bx + 2$ 이고 a와 b는 다음을 만족해요: $a + b = -2$ 이고 $0 < a \le 2$.

자 이제, 기울기가 음인 포물선들의 부류로 넘어가기로 하죠.

그림. G.5

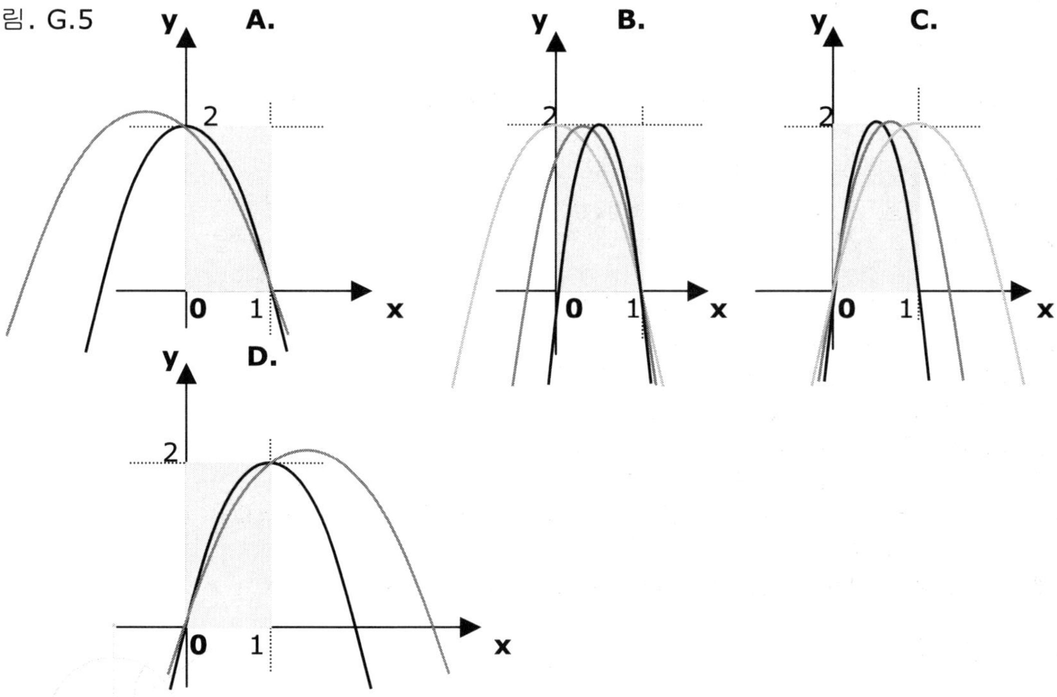

A 의 경우, f(0) = 2 이고 f(1) = 0; 또한, 꼭지점에서의 x 좌표 ≤ 0.
B 의 경우, 0 ≤ f(0) ≤ 2 이고 f(1) = 0; 또한, 꼭지점에서는, 0 ≤ x 좌표 ≤ 1/2 이며 y 좌표 = 2.
C 의 경우, f(0) = 0 이고 0 ≤ f(1) ≤ 2; 또한, 꼭지점에서는, 1/2 ≤ x 좌표 ≤ 1 이며 y 좌표 = 2.
D 의 경우, f(0) = 0 이고 f(1) = 2; 또한, 꼭지점에서는, x 좌표 ≥ 1.

그럼 이제, A 의 경우부터 시작하죠.

그림. G.5

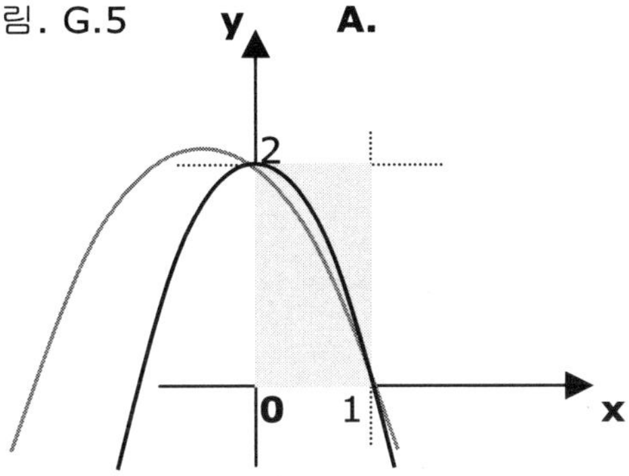

다음을 만족하는 상수 a, b, c 의 값들을 구해야 합니다:
$y = f(x) = ax^2 + bx + c$이고 $0 \le x \le 1$ 이며 $0 \le y \le 2$.

그러면, 다음과 같은 조건들을 만족시켜야 하죠: f(0) = 2 이고 f(1) = 0; 또한, 그 꼭지점에서의 x 좌표 ≤ 0.

먼저, $f(0) = 2 \Rightarrow 2 = 0 + 0 + c \Rightarrow c = 2 \Rightarrow f(x) = ax^2 + bx + 2$.
그러면, $f(1) = 0 \Rightarrow 0 = a + b + 2 \Rightarrow a + b = -2$.

다음으로, 그 꼭지점 = $\{-b/2a, -(b^2 - 4ac)/4a\}$이죠.
그런데, a 는 기울기이고, 지금, 그 기울기 a 는 음이죠.

그러면, 그 꼭지점의 x 좌표 $\le 0 \Rightarrow -b/2a \le 0 \Rightarrow b/a \ge 0 \Rightarrow b \le 0$ 왜냐하면 $a < 0$.

그럼 이제, 중간 결과로서, 다음과 같죠: c = 2 이고 a + b = -2 이며 b ≤ 0.

이제, a 의 범위를 구해보기로 하죠.

$a + b = -2 \Rightarrow b = -2 - a$. 그러면, b = -2 – a 이고 b ≤ 0 이니까 $-2 - a \le 0 \Rightarrow a \ge -2$.
그런데 또한, a < 0 인 것을 알고있죠. 그러니까, -2 ≤ a < 0 이네요.

따라서, 이제 다음과 같죠: c = 2, a + b = -2, -2 ≤ a < 0.

그러므로, A의 경우에는, 그 방정식이 $y = ax^2 + bx + 2$ 이고 a와 b는 다음을 만족하죠:
a + b = 2 이고 -2 ≤ a < 0.

그럼 이제, B 의 경우를 시작하기로 하죠.

그림. G.5

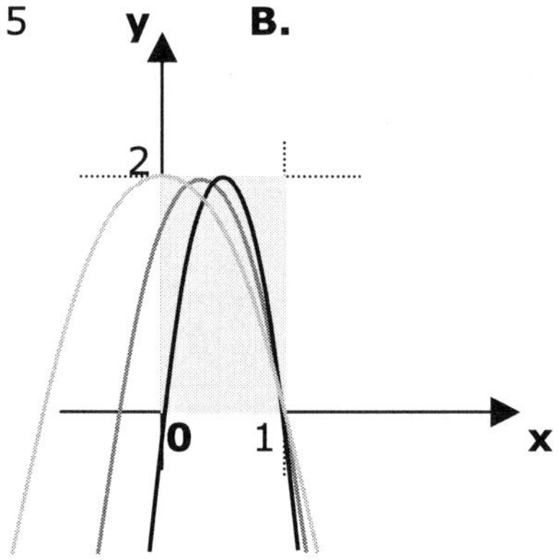

그러면 이제, 다음을 만족하는 상수 a, b, c 의 값들을 구해야 하죠:
$y = f(x) = ax^2 + bx + c$ 이고 $0 \le x \le 1$ 이며 $0 \le y \le 2$.

그러면, 다음과 같은 조건들을 만족시켜야 하는 거죠:
$f(1) = 0$ 이고 또한, 그 꼭지점에서, $0 \le x$ 좌표 $\le 1/2$ 이고 y 좌표 $= 2$.

먼저, f(1)부터 시작합니나: $f(1) = 0 \Rightarrow 0 = a + b + c$.
다음으로, 그 꼭지점은 $\{-b/2a, -(b^2 - 4ac)/4a\}$에 있다는 걸 알고있죠. 그러면:
y좌표 $= 2 \Rightarrow -(b^2 - 4ac)/4a = 2 \Rightarrow b^2 - 4ac = -8a \Rightarrow b^2 - 4ac + 8a = 0$.
$0 \le x$ 좌표 $\le 1/2 \Rightarrow 0 \le -b/2a \le 1/2 \Rightarrow -1/2 \le b/2a \le 0 \Rightarrow -1 \le b/a \le 0$.

그럼 이제, 중간 결과로서, 다음과 같죠: $a + b + c = 0$, $b^2 - 4ac + 8a = 0$, $-1 \le b/a \le 0$.

그러면 이제, a, b, c 의 범위들을 구해야 하는데 이용할 만한 것이 별로 없네요.
그런데, 위의 그래프를 자세히 보면, $-8 \le a \le -2$ 이라는 걸 알 수 있죠. 어떻게 요?

위의 그래프에서 옅은 회색 포물선의 기울기가 -2 인 것을 알 수 있죠. 어떻게 요?
꼭지점이 (0, 2)에 있으니까 그 방정식이 $y = ax^2 + 2$ 라고 하세요. 그러면:
$y = ax^2 + 2$ 이고 그 포물선은 (1, 0)을 지나니까 $(1, 0) \Rightarrow 0 = a + 2 \Rightarrow a = -2$.

또한, 검정 포물선의 기울기가 -8 인 것도 알 수 있죠. 어떻게 요?
그 꼭지점이 (1/2, 2)에 있으니까 $y = a(x - 1/2)^2 + 2$ 라고 하세요.
그러면, $y = a(x - 1/2)^2 + 2$ 이고 그 포물선은 (1, 0)을 지나니까:
$(1, 0) \Rightarrow 0 = a(1 - 1/2)^2 + 2 = a(1/2)^2 + 2 = a/4 + 2 \Rightarrow a = -8$.

그러므로, 또 하나의 중간 결과로서, 다음과 같죠:
$-8 \le a \le -2$, $a + b + c = 0$, $b^2 - 4ac + 8a = 0$, $-1 \le b/a \le 0$.

그러면 이제, b 의 범위를 구해보기로 하죠.
$-1 \le b/a \le 0 \Rightarrow 0 \le b \le -a$ 왜냐하면 $a < 0$. 그러면, $-8 \le a \le -2$ 이니까 $0 \le b \le -a \Rightarrow 0 \le b \le 8$.
어떻게 요?
$-8 \le a \le -2$ 는 a 가 -8 에서 -2 까지 어느 값이든 갖을 수 있다는 거죠. 그러면:
먼저, $a = -8$ 이면 $0 \le b \le -a \Rightarrow 0 \le b \le -(-8) = 8 \Rightarrow 0 \le b \le 8$.
다음으로, $a = -2$ 이면 $0 \le b \le -a \Rightarrow 0 \le b \le -(-2) = 2 \Rightarrow 0 \le b \le 2$.
그러니까, $0 \le b \le 8$ 이네요.

그러면 이제, 다음과 같아요: $a + b + c = 0$, $b^2 - 4ac + 8a = 0$, $-8 \le a \le -2$, $0 \le b \le 8$.

그 c는 소거해버리기로 하죠. 그러면, $a + b + c = 0 \Rightarrow c = -(a + b)$.
또한, 두개의 관계식들 $a + b + c = 0$과 $b^2 - 4ac + 8a = 0$을 합쳐볼 수 있겠어요.

먼저, $a + b + c = 0 \Rightarrow c = -(a + b)$
$\Rightarrow -4ac = 4a(a + b) = 4a^2 + 4ab = 4a^2 + 4ab + b^2 - b^2 = (2a + b)^2 - b^2$.

다음으로, $b^2 - 4ac + 8a = b^2 + (2a + b)^2 - b^2 + 8a = (2a + b)^2 + 8a = 0 \Rightarrow (2a + b)^2 = -8a$ 이는 양이죠 왜냐하면 a가 음이니까.
따라서, $2a + b = \pm(-8a)^{1/2} \Rightarrow b = -2a \pm (-8a)^{1/2} \Rightarrow b = -2a +$ 또는 $b = -2a - \sqrt{-8a}$.

그런데, a가 -8에서 -2까지 변할 때:
$\{-2a + \sqrt{-8a}\}$은 24에서 8까지 변하죠 그러나 $\{-2a - \sqrt{-8a}\}$은 8에서 0까지 변합니다.

$0 \le b \le 8$이라는 것을 알고있죠.
그러니까, $b = -2a - \sqrt{-8a}$이네요.

그러면 이제, 다음과 같아요: $b = -2a - \sqrt{-8a}$, $-8 \le a \le -2$, $c = -(a + b)$.

그리하여, 구하고자 하는 방정식은 다음과 같이 나타낼 수 있어요:

$y = ax^2 + bx - (a + b)$이며 a, b는 다음을 만족한다: $b = -2a - \sqrt{-8a}$이고 $-8 \le a \le -2$.

그런데, 다시 그래프를 조사해보면 $0 \le c \le 2$라는 것을 알 수 있죠.
이는 왜냐하면 $0 \le$ y절편 ≤ 2이어야 하고 $y = ax^2 + bx + c$에서 c가 y절편이기 때문이죠.

또한, b의 범위도 확인해보죠.

$(0 \le c \le 2$이고 $-8 \le a \le -2) \Rightarrow -8 \le a + c \le 0$.
$a + b + c = 0 \Rightarrow a + c = -b \Rightarrow -8 \le -b \le 0 \Rightarrow 0 \le b \le 8$.

그럼 이제, C의 경우를 가지고 계속하기로 하죠.

그림. G.5

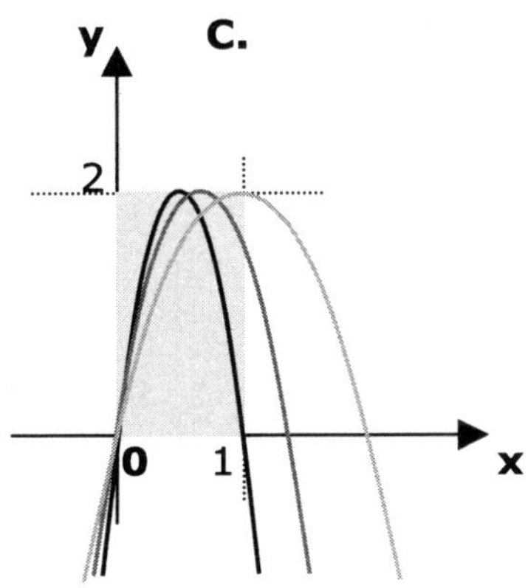

그러면 이제, 다음을 만족하는 상수들 a, b, c 의 값들을 구해야 하죠:
$y = f(x) = ax^2 + bx + c$이고 $0 \le x \le 1$ 이며 $0 \le y \le 2$.

그러면, 다음과 같은 조건들을 만족해야 합니다:
$f(0) = 0$ 이고 또한, 그 꼭지점에서, $1/2 \le x$좌표 ≤ 1 이고 y 좌표 $= 2$.

먼저, $f(0) = 0 \Rightarrow 0 = c \Rightarrow f(x) = ax^2 + bx$.
다음으로, 그 꼭지점으로 넘어가보죠. 그 꼭지점은 $\{-b/2a, -(b^2 - 4ac)/4a\}$에 있는 거죠. 그러면:

먼저, y좌표 $= 2 \Rightarrow -(b^2 - 4ac)/4a = 2 \Rightarrow b^2 - 4ac = -8a \Rightarrow b^2 + 8a = 0$ 왜냐하면 $c = 0$.
다음으로, $1/2 \le x$좌표 $\le 1 \Rightarrow 1/2 \le -b/2a \le 1 \Rightarrow -1 \le b/2a \le -1/2 \Rightarrow -2 \le b/a \le -1$.

그러면, 중간 결과로서, 다음과 같아요: $c = 0$, $b^2 + 8a = 0$, $-2 \le b/a \le -1$.

두개의 식들 $b^2 + 8a = 0$ 과 $-2 \le b/a \le -1$ 을 가지고 b의 범위를 구해보기로 하죠.

그러면, $b^2 + 8a = 0 \Rightarrow b^2 = -8a \Rightarrow a = -b^2/8$. 그러면, $a = -b^2/8$ 이니까:
$-2 \le b/a \le -1 \Rightarrow -2 \le -8/b \le -1 \Rightarrow 1 \le 8/b \le 2 \Rightarrow 1/2 \le b/8 \le 1 \Rightarrow 4 \le b \le 8$.

그러면 이제, 다음과 같아요: $c = 0$, $b^2 + 8a = 0$, $4 \le b \le 8$.

그리하여, 구하고자 하는 방정식은 다음과 같이 나타낼 수 있어요:

$y = ax^2 + bx$이며 a, b는 다음을 만족한다: $b^2 + 8a = 0$(또는, $a = -b^2/8$)이고 $4 \le b \le 8$.

그런데, 위의 B 의 경우에서와 같이, 위의 그래프만 봐도 $-8 \le a \le -2$ 라는 것을 알 수 있어요.

우선, 옅은 회색 포물선의 기울기가 -2 인 것을 알 수 있죠.
그 꼭지점이 (1, 2)에 있으니까 방정식을 $y = a(x - 1)^2 + 2$ 라고 하세요.

그러면, $y = a(x - 1)^2 + 2$ 이고 그 포물선은 (0, 0)을 지나니까:
$(0, 0) \Rightarrow 0 = a(0 - 1)^2 + 2 \Rightarrow 0 = a + 2 \Rightarrow a = -2$.

다음으로, 검정 포물선의 기울기는 -8 인 것을 알 수 있죠.
그 꼭지점이 (1/2, 2)에 있으니까 방정식을 $y = a(x - 1/2)^2 + 2$ 라고 하세요.

그러면, $y = a(x - 1/2)^2 + 2$ 이고 그 포물선은 (0, 0)을 지나니까:
$(0, 0) \Rightarrow 0 = a(0 - 1/2)^2 + 2 = a(1/2)^2 + 2 = a/4 + 2 \Rightarrow a/4 = -2 \Rightarrow a = -8$.

이것이 맞는지 실제로 a 의 범위를 구해서 확인해보도록 하죠.

$4 \le b \le 8 \Rightarrow 16 \le b^2 \le 64 \Rightarrow 2 \le b^2/8 \le 8 \Rightarrow -8 \le -b^2/8 \le -2$.
그러면, $a = -b^2/8$ 이니까 $-8 \le a \le -2$.

그럼 이제, D 의 경우로 넘어가보기로 하죠.

그림. G.5

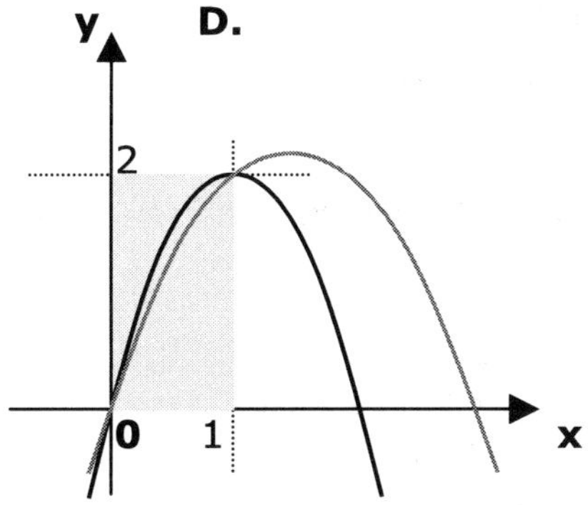

이제, 다음을 만족하는 상수들 a, b, c의 값들을 구해야 하죠:
$y = f(x) = ax^2 + bx + c$이고 $0 \leq x \leq 1$이며 $0 \leq y \leq 2$.

그러면, 다음과 같은 조건들을 만족해야 합니다: $f(0) = 0$이고 $f(1) = 2$; 또한, 그 꼭지점에서, x 좌표 ≥ 1.

그러면, $f(0) = 0 \Rightarrow 0 = c \Rightarrow f(x) = ax^2 + bx$.
따라서, $f(1) = 2 \Rightarrow 2 = a + b$네요. 그러니까, $a + b = 2$.

이제, 그 꼭지점으로 넘어가보죠. 그 꼭지점은 $\{-b/2a, -(b^2 - 4ac)/4a\}$에 있는 거죠.
또한, 그 기울기 a가 음이라는 걸 알고있죠.

그러면, x 좌표 $\geq 1 \Rightarrow -b/2a \geq 1 \Rightarrow b/a \leq -2 \Rightarrow b \geq -2a$ 왜냐하면 $a < 0$.

그러면, 중간 결과로서, 다음과 같죠: $c = 0$, $a + b = 2$, $b \geq -2a$.

이제, a의 범위를 구해보도록 하죠.
$a + b = 2 \Rightarrow b = 2 - a$. 그러면, $b \geq -2a \Rightarrow 2 - a \geq -2a \Rightarrow a \geq -2$.
그런데 또한, 지금 $a < 0$이죠. 그러니까, $-2 \leq a < 0$이네요.

그러니까, D의 경우, 방정식은 $y = ax^2 + bx$이고 a와 b는 다음을 만족합니다:
$a + b = 2$이고 $-2 \leq a < 0$.

제안과 풀이
H번 예제에서의 문제

$y = f(x) = (2x^2 + 3)/7$ 이고 $m \le x \le n$이며 $m \le y \le n$이고 여기에서 m 과 n은 상수라 하고 m과 n의 값들을 구하세요.

이 예제에서는, 함수 f의 출력의 범위가 입력의 범위와 동일하게 하는 거죠.
그러면, 우선 입력과 출력이 같아지는 두개의 특정 입력 값들을 찾아야 하는 거네요.
그러니까, 이 함수로 그 특정한 값들을 입력 했을 때, 그 출력들이 제각기 그 특정한 값들과 같아야 하는 거죠.
그런데, 입력과 출력이 제각기 동일한 함수가 하나 있는데 그 함수가 뭐죠?
그걸 항등 함수라고 하죠. 이 함수의 커브는 x-y 좌표계에서는 직선 $y = x$ 입니다.
그런데, 이 함수 f 의 커브는 포물선이죠. 그러면, 그 두 커브들 즉, 그 포물선과 그 직선이 서로 만나는 점들을 구하면 그 범위를 찾을 수 있는 거죠. 그 점들의 좌표들이 m 과 n 의 값들이 되거든요.
사실, 이는 포물선이 아닌 커브에 대하여도 마찬가집니다.
그러면 이제, 그 점들을 구해보도록 하죠.
그 포물선의 방정식은 $y = (2x^2 + 3)/7$ 이고 그 직선의 방정식은 $y = x$이죠. 그러면:

$x = (2x^2 + 3)/7 \Rightarrow 7x = 2x^2 + 3 \Rightarrow 2x^2 - 7x + 3 = 0 \Rightarrow (2x - 1)(x - 3) = 0 \Rightarrow x = 1/2$ 또는 3.

그러면, 그 커브들은 x가 1/2과 3일 때 서로 만나는 거죠.
그러니까, $m = 1/2$ 이고 $n = 3$인 거네요. 그런데, 꼭 이렇게 되는 것만은 아닙니다. 다음예제를 보면 알게 되죠.

그림. H.0

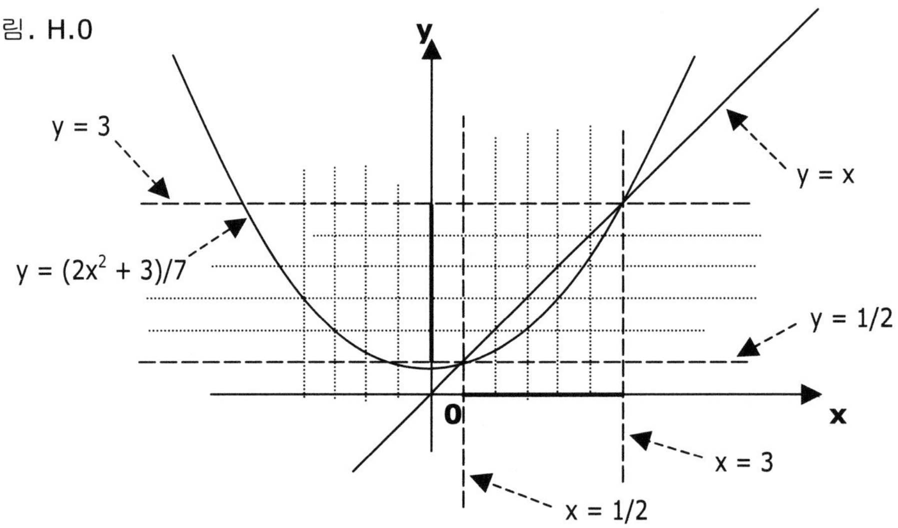

요약:

f의 커브와 그 직선 즉, 포물선 $y = (2x^2 + 3)/7$ 와 직선 $y = x$가 만나는 점들에서의 x 좌표의 값들이 m과 n.

$x = (2x^2 + 3)/7 \Rightarrow 7x = 2x^2 + 3 \Rightarrow 2x^2 - 7x + 3 = 0 \Rightarrow (2x - 1)(x - 3) = 0 \Rightarrow x = 1/2$ 또는 3.

그러므로, $m = 1/2$ 이고 $n = 3$.

제안과 풀이
I번 예제에서의 문제

$y = f(x) = (2x^2 - 3)/5$ 이고 $m \le x \le n$이며 $m \le y \le n$이고 여기에서 m과 n은 상수라 하고 m과 n의 값들을 구하세요.

이전 예제에서와 마찬가지로, 함수의 출력의 범위가 입력의 범위와 동일하게 하는 거죠.
그러면, 우선 입력과 출력이 같아지는 두개의 특정 입력 값들을 찾아야 하는 거네요.
그러니까, 이 함수로 그 특정한 값들을 입력 했을 때, 그 출력들이 제각기 그 특정한 값들과 같아야 하는 거죠.
그런데, 입력과 출력이 동일한 함수가 하나 있는데 그걸 항등 함수라고 하고 이 함수의 커브는 x-y 좌표계에서는 직선 $y = x$ 입니다. 이 함수 f 의 커브는 포물선이죠. 그러면, 그 두 커브들 즉, 그 포물선과 그 직선이 서로 만나는 점들을 구하면 그 범위를 찾을 수 있죠. 그러면 이제, 그 점들을 구해보도록 하죠.

그 포물선의 방정식은 $y = (2x^2 - 3)/5$ 이고 그 직선의 방정식은 $y = x$이죠. 그러면:

$x = (2x^2 - 3)/5 \Rightarrow 5x = 2x^2 - 3 \Rightarrow 2x^2 - 5x - 3 = 0 \Rightarrow (2x + 1)(x - 3) = 0 \Rightarrow x = -1/2$ 또는 3.

그러면, $m = -1/2$ 이고 $n = 3$ 인 것 같네요. 그러니까, $-1/2 \le x \le 3$ 이면 $-1/2 \le y \le 3$ 이어야 하는 거죠.
그런데, 이 y 의 범위에 문제가 좀 있군요. 즉, $-1/2 \le y \le 3$ 이 아닌 거죠. 어째서 요?

이 함수 f의 커브가 포물선인데 기울기가 양이고 그 대칭축이 직선 $x = 0$ 즉, y축이죠. 그러면, 이는 아래로 볼록하네요. 그러니까, 그 꼭지점의 y좌표는 $-1/2$ 보다 작은 거죠. 따라서, $-1/2 \le y \le 3$ 이 아닌 겁니다. 사실, 그 y 좌표는 $-3/5$ 이죠: $y = (2x^2 - 3)/5 = 2x^2/5 - 3/5$. 그러니까, $-1/2 \le x \le 3$ 이면 $-3/5 \le y \le 3$ 이 되는 것이고 이는 답이 될 수 없는 겁니다. 즉, 이 문제에 대한 답은 없는 거죠.

포물선의 모양을 항상 고려하는 것이 좋습니다. 이는 다른 커브들에 대하여도 마찬가집니다.

그림. I.0

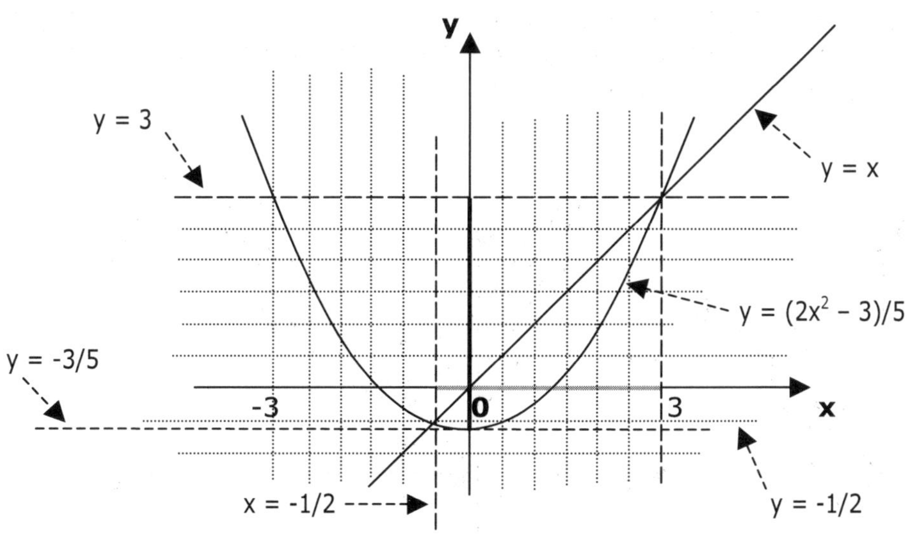

이제, 비슷한 예제를 하나 더 다루어 보죠.

제안과 풀이
J 번 예제에서의 문제

$y = f(x) = (2x^2 + c)/5$ 이고 **$m \le x \le n$**이며 **$m \le y \le n$**이고 여기에서 **c**와 **m** 그리고 **n**은 상수라 하고 **c**와 **m** 그리고 **n**의 값들을 구하세요.

앞서 나온 두 예제들에서와 마찬가지로, 함수의 출력의 범위가 입력의 범위와 동일하게 하는 거죠.
그러면, 우선 입력과 출력이 같아지는 두개의 특정 입력 값들을 찾아야 하는 거네요.
그러니까, 이 함수로 그 특정한 값들을 입력 했을 때, 그 출력들이 제각기 그 특정한 값들과 같아야 하는 거죠.
그런데, 입력과 출력이 동일한 함수가 하나 있는데 그걸 항등 함수라고 하고 이 함수의 커브는 x-y좌표계에서는 직선 $y = x$입니다. 이 함수 f의 커브는 포물선이죠. 그러면, 그 두 커브들 즉, 그 포물선과 그 직선이 만나는 점들을 구하면 그 범위를 찾을 수 있죠. 그런데, 그 점들만 구했다고 해서 답이 완성되는 건 아니죠. 그 꼭지점의 위치에 따라 답이 있거나 없을 수도 있으니까. 그러면, 어떤 상황에서 답이 존재할지 약식으로 그래프를 작성해서 알아보기로 하죠. 먼저, 그 포물선의 오목한 상태를 확인해보도록 하죠. 그 포물선 $y = (2x^2 + c)/5$ 의 기울기는 양이고 대칭축은 y축이죠. 그러면, 이는 아래로 볼록한 겁니다. 그러면, 그 꼭지점들이 y축에 있도록 포물선들을 좀 놓아보도록 하죠.

그림 J.0

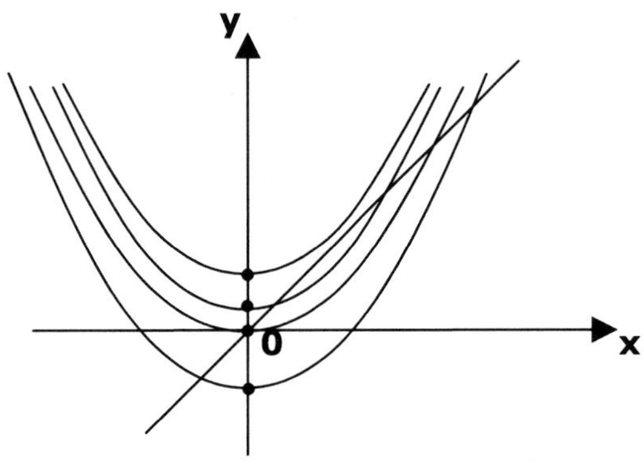

자 이제 우선, 그 포물선과 직선이 두 점들에서 만나야 하죠. 다음으로, 만나더라도 그 꼭지점이 원점이나 직선 아래쪽에 있으면 안 되는 거네요. 즉, 그 꼭지점의 y 좌표가 x 좌표보다 크거나 같고 포물선과 직선은 두 점에서 서로 만나야 하는 거죠. 그러면, 대칭축이 y 축은 아니지만 y 축과 평행한 다른 경우들을 보도록 하죠.

그림. J.1

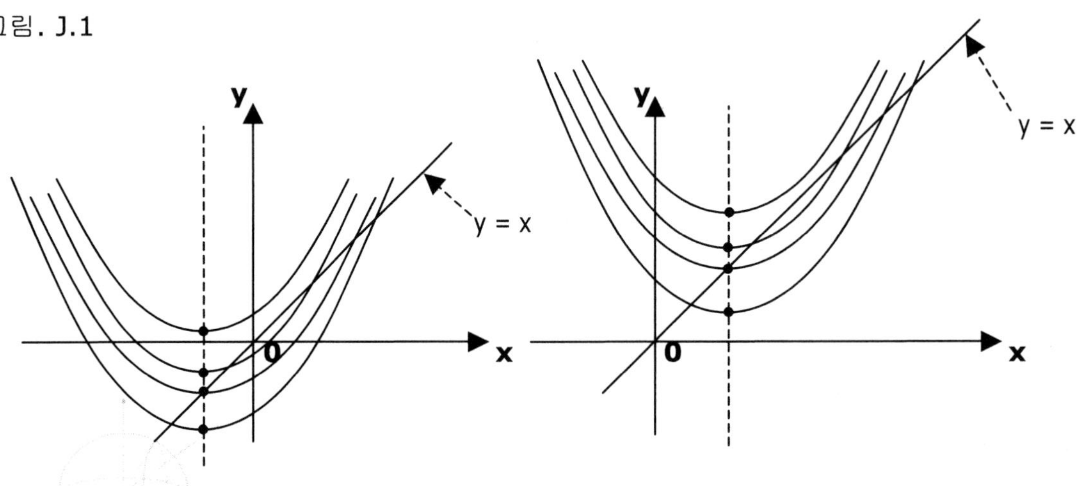

그러면 이제 우선, 포물선이 아래로 볼록하면 대칭축의 위치에 상관없이 꼭지점에서의 y 좌표가 x 좌표보다 작으면 안 되는 거네요. 즉, 그 y 좌표가 x 좌표보다 크거나 같아야 하는 겁니다. 이에 더하여, 그 포물선과 그 직선은 두 점들에서 서로 만나야 하는 거죠.

그러니까, 이 문제가 요구하는 그런 구간이 존재하려면, 꼭지점의 y 좌표가 x 좌표보다 크거나 같아야 하고 또한, 그 커브들 즉, 그 포물선과 직선이 두 점들에서 서로 만나야 하는 겁니다.

그러면, 다른 방법으로 그런 구간이 있는지 알 수 없을까요?
있죠. 위의 그래프들을 조사해보면 다음과 같은 사실도 알 수 있어요.

우선, 그 커브들 즉, 그 포물선과 직선이 두 점들에서 서로 만나야 하죠.
그 포물선의 방정식은 $y = (2x^2 + c)/5$ 이고 그 직선의 방정식은 $y = x$이죠.

그러면, $x = (2x^2 + c)/5 \Rightarrow 5x = 2x^2 + c \Rightarrow 2x^2 - 5x + c = 0$.

자 이제, 위의 방정식의 근들을 s_1과 s_2이라고 하세요; 즉, $2x^2 - 5x + c = (x - s_1)(x - s_2) = 0$.
또한, $s_1 < s_2$라고 해보죠.

그러면, 그 꼭지점에서의 x좌표가 s_1보다 작거나 같아야 하는 겁니다.

그러면 이제, 두 가지 방법 모두에서 우선, 그 포물선과 직선이 두 점들에서 서로 만나야 하죠.
이는 그 두 점들을 구하는 방정식의 판별식으로 확인할 수 있죠. 그 판별식이 0 보가 크면 그 방정식은 두개의 서로 다른 실근을 갖는 겁니다. 그러면, 방정식 $2x^2 - 5x + c = 0$ 의 판별식이 0 보다 커야 하는 겁니다.

이제, 방정식이 $ax^2 + bx + c = 0$ 이라 하고 판별식을 D라 하세요.
그러면, D는 다음과 같아요: $D = b^2 - 4ac$.

그러면 이제, $2x^2 - 5x + c = 0$ 의 판별식을 구해보죠.
그 판별식 $D = (-5)^2 - 4(2)c = 25 - 8c$이네요.
그러면, $D > 0$ 이어야 하죠. 즉, $25 - 8c > 0 \Rightarrow c < 25/8$ 이어야 하는 거죠.

자 이제, 그 꼭지점과 그 근들을 구하기로 하죠:

$y = (2x^2 + c)/5 = 2x^2/5 + c/5 \Rightarrow$ 그 꼭지점 $= (0, c/5)$.

$2x^2 - 5x + c = 2(x^2 - 5x/2) + c = 2(x^2 - 5x/2 + 25/16 - 25/16) + c = 2(x - 5/4)^2 - 25/8 + c = 0$
$\Rightarrow 2(x - 5/4)^2 = 25/8 - c \Rightarrow (x - 5/4)^2 = 25/16 - c/2 \Rightarrow x - 5/4 = \pm(25/16 - c/2)^{1/2}$
$\Rightarrow x = 5/4 + (25/16 - c/2)^{1/2}$ 또는 $5/4 - (25/16 - c/2)^{1/2}$.
그러면, $5/4 + (25/16 - c/2)^{1/2} > 5/4 - (25/16 - c/2)^{1/2}$이니까:
$m = 5/4 - (25/16 - c/2)^{1/2}$이고 $n = 5/4 + (25/16 - c/2)^{1/2}$.

그러면 이제, 먼저 첫번째 방법을 사용해보기로 하죠.
그러면, 그 꼭지점의 y 좌표가 x 좌표보다 크거나 같아야 하고 그 판별식 > 0 이어야 합니다.
먼저, 그 꼭지점 $= (0, c/5)$이니까 $c \geq 0$ 이죠 왜냐하면 y 좌표 $= c/5 \geq 0 =$ x 좌표이기 때문이죠.
다음으로, $D > 0 \Rightarrow c < 25/8$ 입니다.
따라서, $0 \leq c < 25/8$ 이어야 하네요.
그러므로, $m = 5/4 - (25/16 - c/2)^{1/2}$이고 $n = 5/4 + (25/16 - c/2)^{1/2}$이며 여기에서 $0 \leq c < 25/8$.

그러면 이제, 그 두 번째 방법을 확인해보도록 하죠.
그러면, 그 꼭지점의 x 좌표가 그 두 근들 중의 더 작은 근보다 작거나 같아야 하죠.

그 꼭지점 = (0, c/5)이니까 그 꼭지점에서의 x 좌표는 0 입니다.
그러면, x좌표 = $0 \le m = 5/4 - (25/16 - c/2)^{1/2}$이네요. 그러므로:

$5/4 - (25/16 - c/2)^{1/2} \ge 0 \Rightarrow (25/16 - c/2)^{1/2} \le 5/4 \Rightarrow 25/16 - c/2 \le 25/16 \Rightarrow -c/2 \le 0 \Rightarrow c/2 \ge 0 \Rightarrow c \ge 0$.

또한, 그 판별식 $D > 0$ 이니까 $c < 25/8$ 이죠.

따라서, $0 \le c < 25/8$ 이네요.

그러므로, $m = 5/4 - (25/16 - c/2)^{1/2}$이고 $n = 5/4 + (25/16 - c/2)^{1/2}$이며 여기에서 $0 \le c < 25/8$ 입니다.

요약:

함수 f 의 커브는 포물선인데 이는 직선 y = x 와 두 점들에서 만나야 하며 그 꼭지점의 y 좌표가 x 좌표보다 크거나 같아야 합니다.

먼저, 그 꼭지점을 구합니다.
$y = (2x^2 + c)/5 = 2x^2/5 + c/5 \Rightarrow$ 그 꼭지점 = (0, c/5).

다음으로, f 의 커브와 직선 y = x 가 서로 만나는 점들의 x 좌표들을 구합니다.
$2x^2 - 5x + c = 2(x^2 - 5x/2) + c = 2(x^2 - 5x/2 + 25/16 - 25/16) + c = 2(x - 5/4)^2 - 25/8 + c = 0$
$\Rightarrow 2(x - 5/4)^2 = 25/8 - c \Rightarrow (x - 5/4)^2 = 25/16 - c/2 \Rightarrow x - 5/4 = \pm(25/16 - c/2)^{1/2}$
$\Rightarrow x = 5/4 + (25/16 - c/2)^{1/2}$ 또는 $5/4 - (25/16 - c/2)^{1/2}$.

그러면, $5/4 + (25/16 - c/2)^{1/2} > 5/4 - (25/16 - c/2)^{1/2}$이니까:
$m = 5/4 - (25/16 - c/2)^{1/2}$이고 $n = 5/4 + (25/16 - c/2)^{1/2}$.

위의 방정식의 판별식 $D = (-5)^2 - 4(2)c = 25 - 8c$.

그러면, 그 꼭지점의 y 좌표가 x 좌표보다 크거나 같아야 하고 그 판별식 > 0.
먼저, 그 꼭지점 = (0, c/5) ⇒ y 좌표 = c/5 이고 x 좌표 = 0.
그러면, y 좌표 = $c/5 \ge 0 \Rightarrow c \ge 0$. 다음으로, $D > 0 \Rightarrow c < 25/8$.
따라서, $0 \le c < 25/8$.

그러므로, $m = 5/4 - (25/16 - c/2)^{1/2}$이고 $n = 5/4 + (25/16 - c/2)^{1/2}$이며 여기에서 $0 \le c < 25/8$.

제안과 풀이

K 번 예제에서의 문제

다음과 같은 포물선의 방정식이 있다고 하세요: $2x^2 + (2c + 1)x - (c + 1)y + c + 1 = 0$ 여기에서 c는 상수. 그러면, 그 포물선은 그 어느 c 값에 대하여도 특정한 두개의 점들을 지난다 하고 그 두 점들을 구하세요.

사실, 주어진 방정식은 하나의 특정 포물선의 방정식이 아니고 한 집단을 이루는 여러 포물선들을 나타내는 겁니다. 직선이 대하여 공부할 때, 두 개의 직선들이 하나의 특정한 점에서 만나면 그 두개의 직선들의 선형 조합은 그 특정한 점을 지난다는 사실을 배웠죠.
예를 들어, a, b, c, d, e, f, m, n 은 상수라 하고 두 직선들 $ax + by + c = 0$ 와 $dx + ey + f = 0$ 은 하나의 특정한 점에서 만난다 하세요. 그러면, 직선 $m(ax + by + c) + n(dx + ey + f) = 0$ 도 그 특정한 점을 지나는 겁니다. 그러면, $m(ax + by + c) + n(dx + ey + f) = 0$ 을 $ax + by + c = 0$ 과 $dx + ey + f = 0$ 의 선형 조합이라고 하는 거죠. 다시 말해서, 그 두 직선들에 어떤 수를 곱해서 합하더라도 그 합은 그 특정한 점을 지나죠. 즉, 두 상수들 m 과 n 에 대하여 그 어떤 수들을 선택하더라도 그 합이 나타내는 직선도 그 특정한 점을 지나는 겁니다.

그러면 이제, 이와 동일한 개념이 두개의 포물선들에도 적용되는 것이죠. 또한, 이는 하나의 포물선과 직선에 대하여도 마찬가집니다. 즉, 두 개의 포물선들이 특정한 점에서 서로 만나면 그 선형 조합들도 모두 그 특정한 점에서 서로 만나는 것이고 하나의 포물선과 직선이 특정한 점에서 서로 만나면 그 선형 조합들도 모두 그 특정한 점에서 서로 만나는 겁니다.
예를 들어, 포물선을 나타내는 두 개의 방정식들 A 와 B 를 합해서 하나의 새로운 포물선의 방정식 C 를 만든다 하세요. 또한, 그 포물선들 A 와 B 가 두 점들에서 서로 만난다 하세요. 그러면, 그 새로운 포물선 C 도 그 두 점들을 지나고 따라서, 그 세 포물선들이 모두 함께 그 두 점들에서 서로 만나는 거죠.

또 하나의 예를 들어, 먼저 아래의 방정식들에서 x 와 y 를 제외한 그 어느 문자도 상수라 하세요:
$A(x, y) = ax^2 + bx + c + dy = 0$
$B(x, y) = ex + fy + g = 0$
$C(x, y) = m(ax^2 + bx + c + dy) + n(ex + fy + g) = 0.$
다음으로, 포물선 A 와 직선 B 가 특정한 점들에서 서로 만난다 하세요.
그러면, 포물선 C 도 A 와 B 를 그 특정한 점들에서 만나는 겁니다.

그러면 이제, 하나 더 예를 들어서, 아래의 방정식들에서 x 와 y 를 제외한 그 어느 문자도 상수라 하세요:
$A(x, y) = ax^2 + bx + c + dy = 0$
$B(x, y) = ex^2 + fx + g + hy = 0$
$C(x, y) = m(ax^2 + bx + c + dy) + n(ex^2 + fx + g + hy) = 0.$
또한, 포물선들 A 와 B 가 특정한 점들에서 서로 만난다고 하세요.
그러면, 포물선 C 도 그 특정한 점들에서 그 포물선들 A 와 B 를 만나는 거죠.

사실, 이와 비슷한 개념이 다른 종류의 커브들에도 적용되는 겁니다.
예를 들어, m 과 n 이 상수라 하고 두 커브들 $f(x, y) = 0$ 과 $g(x, y) = 0$ 이 특정한 점에서 서로 만난다 하세요.
그러면, $mf(x, y) + ng(x, y) = 0$ 으로 나타내어지는 커브들도 모두 그 특정한 점을 지나는 거죠.

Math 에서 일반적으로 커브라고 하면 선을 곡선뿐만 아니라 직선도 의미하는 겁니다. 또한, 직선의 방정식이란 어느 직선이든 의미할 수도 있는 겁니다. 예를 들어, $x = 1$ 과 $x = -1/2$ 도 제각기 직선의 방정식이 될 수 있죠. 물론, $y = 2$ 도 직선의 방정식이 될 수 있죠. 방정식 $x = 0$ 도 직선의 방정식이 될 수 있고 이는 y 축을 나타내고 방정식 $y = 0$ 도 직선의 방정식이 될 수 있고 이는 x 축을 나타내죠. 또한, 직선 $x = 1$ 은 모든 y 의 값에 대하여 모든 점 (1, y)를 지나는 직선이죠. 즉, 직선 $x = 1$ 은 점 (1, 0), (1, -1/2), (1, 12) 등등을 지나는 겁니다.

이제, $f(x, y) = 0$과 $g(x, y) = 0$을 기본 커브들이라 하세요. 그러면, 그 기본 커브들이 특정한 점에서 서로 만나면 $mf(x, y) + ng(x, y) = 0$으로 나타내어지는 모든 커브들은 모두 함께 그 특정한 점에서 서로 만납니다.

그러면 이제, 이 예에서는 $2x^2 + (2c + 1)x - (c + 1)y + c + 1 = 0$이고 이는 포물선들의 집단을 나타내죠.
그러면, 이는 두개의 기본 커브들로 이루어져있고 그 기본 커브들이 그 두개의 점들에서 서로 만나는 겁니다.
그러니까, 그 두개의 기본 커브들을 추출해내서 그 커브들이 어디에서 서로 만나는지 알아내어야 하는 거죠.
즉, 그 커브들이 서로 만나는 그 두 점들을 구해야 하는 겁니다.

그러면 먼저, 그 두개의 기본 커브들부터 시작하기로 하죠.

$2x^2 + (2c + 1)x - (c + 1)y + c + 1 = 2cx - cy + c + 2x^2 + x - y + 1$
$= c(2x - y + 1) + 2x^2 + x - y + 1 = 0.$

그러면, 그 두개의 기본 커브들은 다음과 같은 거네요:
직선 $2x - y + 1 = 0$과 포물선 $2x^2 + x - y + 1 = 0.$

다음으로, 이 커브들이 서로 만나는 점들을 구해야 하죠.
그 x 좌표부터 시작하기로 하죠.
$2x - y + 1 = 0 \Rightarrow y = 2x + 1$ 이고 또한, $2x^2 + x - y + 1 = 0 \Rightarrow y = 2x^2 + x + 1.$
그러면, $2x + 1 = 2x^2 + x + 1$으로 놓고 이 방정식을 풀어야 하죠.
$2x + 1 = 2x^2 + x + 1 \Rightarrow 2x^2 - x = 0 \Rightarrow x(2x - 1) = 0 \Rightarrow x = 0$ 또는 $1/2.$
그러면: $x = 0 \Rightarrow y = 2x + 1 = 0 + 1 \Rightarrow y = 1$이고 또한, $x = 1/2 \Rightarrow y = 2x + 1 = 1 + 1 = 2 \Rightarrow y = 2.$
그러니까, 그 모든 포물선 $2x^2 + (2c + 1)x - (c + 1)y + c + 1 = 0$은 $(0, 1)$과 $(1/2, 2)$을 지나네요.

요약:

$2x^2 + (2c + 1)x - (c + 1)y + c + 1 = 0$은 포물선들의 집단을 나타내며 이 방정식은 두개의 기본 커브들의 방정식들로 이루어져있고 그 두 기본 커브들이 그 두개의 점들에서 서로 만납니다.
그러면 먼저, 그 두개의 기본 커브들을 구합니다.

$2x^2 + (2c + 1)x - (c + 1)y + c + 1 = 2cx - cy + c + 2x^2 + x - y + 1$
$= c(2x - y + 1) + 2x^2 + x - y + 1 = 0.$

그러면, 그 두개의 기본 커브들은 다음과 같죠: 직선 $2x - y + 1 = 0$과 포물선 $2x^2 + x - y + 1 = 0.$

다음으로, 이 커브들이 서로 만나는 점들을 구합니다.
$2x - y + 1 = 0 \Rightarrow y = 2x + 1$ 이고 또한, $2x^2 + x - y + 1 = 0 \Rightarrow y = 2x^2 + x + 1.$
$2x + 1 = 2x^2 + x + 1 \Rightarrow 2x^2 - x = 0 \Rightarrow x(2x - 1) = 0 \Rightarrow x = 0$ 또는 $1/2.$
그러면: $x = 0 \Rightarrow y = 2x + 1 = 0 + 1 \Rightarrow y = 1$이고 또한, $x = 1/2 \Rightarrow y = 2x + 1 = 1 + 1 = 2 \Rightarrow y = 2.$

그러므로, 그 점들은 $(0, 1)$과 $(1/2, 2)$.

제안과 풀이
L번 예제에서의 문제

다음과 같은 포물선의 방정식이 있다고 하세요: $(c + 2)x^2 + (2c + 1)x - (c + 1)y + 3c + 1 = 0$ 여기에서 c는 상수. 그러면, 그 포물선이 그 어느 c값에 대하여도 두개의 특정한 점들을 지난다 하고 그 두 점들을 구하세요.

앞의 예제에서와 같이, 이 방정식 $(c + 2)x^2 + (2c + 1)x - (c + 1)y + 3c + 1 = 0$도 포물선들의 모임을 나타내는 것 같네요. 그렇다면, 이 방정식은 몇 개의 기본 커브들의 선형 조합일 겁니다. 그러면, 이 방정식을 다음과 같이 c에 관해서 정리해보도록 하죠:

$(c + 2)x^2 + (2c + 1)x - (c + 1)y + 3c + 1 = cx^2 + 2cx - cy + 3c + 2x^2 + x - y + 1$
$= c(x^2 + 2x - y + 3) + 2x^2 + x - y + 1 = 0.$

그러면, 그 기본 커브들이 다음과 같은 두개의 포물선들이네요:
$x^2 + 2x - y + 3 = 0$과 $2x^2 + x - y + 1 = 0.$

그러면 이제, 이 포물선들이 서로 만나는 점들을 구해야 하죠.
그러면, 그 x 좌표들부터 시작하죠:
$x^2 + 2x - y + 3 = 0 \Rightarrow y = x^2 + 2x + 3$이고 또한, $2x^2 + x - y + 1 = 0 \Rightarrow y = 2x^2 + x + 1.$
그러면, 방정식 $x^2 + 2x + 3 = 2x^2 + x + 1$을 풀어야 하죠.
$x^2 + 2x + 3 = 2x^2 + x + 1 \Rightarrow x^2 - x - 2 = 0 \Rightarrow (x + 1)(x - 2) = 0 \Rightarrow x = -1$ 또는 2.
그러면, y 좌표들을 구합니다:
$x = -1 \Rightarrow y = x^2 + 2x + 3 = 1 - 2 + 3 = 2.$
$x = 2 \Rightarrow y = x^2 + 2x + 3 = 4 + 4 + 3 = 11.$

그러면, 그 기본 포물선들이 (-1, 2)와 (2, 11)에서 서로 만나죠.
그러므로, 각각의 c값에 대하여 모든 포물선 $(c + 2)x^2 + (2c + 1)x - (c + 1)y + 3c + 1 = 0$도 이 두 점들을 지나는 겁니다.

요약:

$(c + 2)x^2 + (2c + 1)x - (c + 1)y + 3c + 1 = cx^2 + 2cx - cy + 3c + 2x^2 + x - y + 1$
$= c(x^2 + 2x - y + 3) + 2x^2 + x - y + 1 = 0.$

그러면, 다음의 포물선들이 서로 만나는 점들을 그 포물선 $(c + 2)x^2 + (2c + 1)x - (c + 1)y + 3c + 1 = 0$이 지나는 겁니다: $x^2 + 2x - y + 3 = 0$과 $2x^2 + x - y + 1 = 0.$

그러면, 그 x 좌표들부터 시작하죠:
$x^2 + 2x - y + 3 = 0 \Rightarrow y = x^2 + 2x + 3$이고 또한, $2x^2 + x - y + 1 = 0 \Rightarrow y = 2x^2 + x + 1.$
그러면, $x^2 + 2x + 3 = 2x^2 + x + 1 \Rightarrow x^2 - x - 2 = 0 \Rightarrow (x + 1)(x - 2) = 0 \Rightarrow x = -1$ 또는 2.
그러면, y 좌표들을 구합니다:
$x = -1 \Rightarrow y = x^2 + 2x + 3 = 1 - 2 + 3 = 2.$
$x = 2 \Rightarrow y = x^2 + 2x + 3 = 4 + 4 + 3 = 11.$
그러므로, 그 두 점들은 (-1, 2)와 (2, 11).

그림. L.0

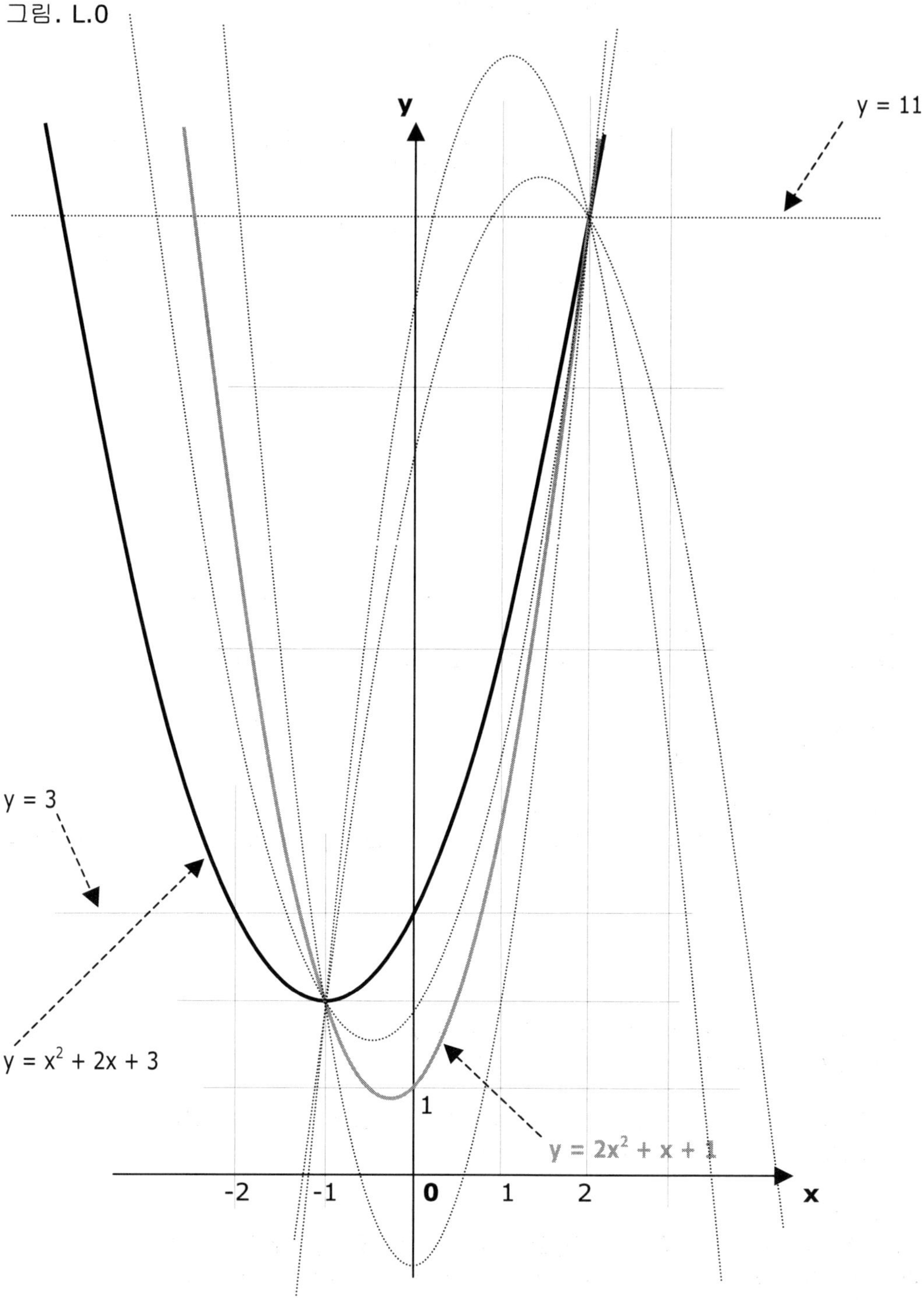

제안과 풀이
M번 예제에서의 문제

다음과 같은 포물선의 방정식이 있다고 하세요: $cx^2 + 2cx - (c + 1)y + 3c + 3 = 0$ 여기에서 c는 상수. 그러면, 그 포물선이 그 어느 c값에 대하여도 두개의 특정한 점들을 지난다 하고 그 두 점들을 구하세요.

앞의 예제에서와 같이, 이 방정식 $cx^2 + 2cx - (c + 1)y + 3c + 3 = 0$도 포물선들의 모임을 나타내는 것 같네요. 그러면, 이 방정식을 다음과 같이 c에 관해서 정리해보도록 하죠:
$cx^2 + 2cx - (c + 1)y + 3c + 3 = cx^2 + 2cx - cy + 3c - y + 3 = c(x^2 + 2x - y + 3) - y + 3 = 0$.
그러면, 그 기본 방정식들은 다음과 같아요: $x^2 + 2x - y + 3 = 0$과 $-y + 3 = 0$.
그러니까, 그 기본 커브들은 포물선 $x^2 + 2x - y + 3 = 0$과 직선 $y = 3$이네요. 그 직선은 x축에 평행이죠. 그러면 이제, 그 커브들이 서로 만나는 점들을 구해야 하죠. 그 직선 $y = 3$이 그 포물선과 만나니까 그 점들에서의 y좌표들은 3이네요. 그러면, 그 x좌표들을 구하기로 하죠.
$y = 3 \Rightarrow y = x^2 + 2x + 3 \Rightarrow 3 = x^2 + 2x + 3 \Rightarrow x^2 + 2x = 0 \Rightarrow x(x + 2) = 0 \Rightarrow x = 0$ 또는 -2.
그러니까, 그 커브들은 (0, 3)과 (-2, 3)에서 서로 만나죠.
그러므로, 각 c 값에 대해서 모든 포물선 $cx^2 + 2cx - (c + 1)y + 3c + 3 = 0$은 그 두 점들 (0, 3)과 (-2, 3)을 지나네요.

요약:
$cx^2 + 2cx - (c + 1)y + 3c + 3 = cx^2 + 2cx - cy + 3c - y + 3 = c(x^2 + 2x - y + 3) - y + 3 = 0$.
그러면, 그 기본 커브들은 포물선 $x^2 + 2x - y + 3 = 0$과 직선 $-y + 3 = 0$이고 그 직선 $y = 3$이 그 포물선과 만나니까 그 점들에서의 y좌표들은 3. 그러면:
$y = 3 \Rightarrow y = x^2 + 2x + 3 \Rightarrow 3 = x^2 + 2x + 3 \Rightarrow x^2 + 2x = 0 \Rightarrow x(x + 2) = 0 \Rightarrow x = 0$ 또는 -2.
그러므로, 그 점들은 (0, 3)과 (-2, 3).

그림. M.0

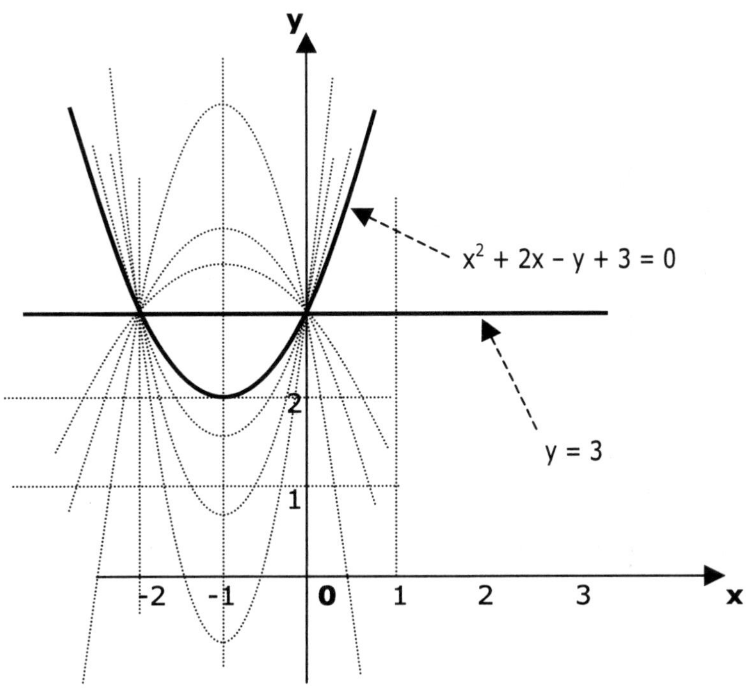

제안과 풀이
N 번 예제에서의 문제

다음과 같은 포물선의 방정식이 있다고 하세요: $cx^2 + (2c + 1)x - cy + 3c + 1 = 0$ 여기에서 c는 상수. 그러면, 그 포물선이 그 어느 c 값에 대하여도 하나의 특정한 점을 지난다 하고 그 특정한 점을 구하세요.

앞의 예제에서와 같이, 이 방정식 $cx^2 + (2c + 1)x - cy + 3c + 1 = 0$ 도 포물선들의 모임을 나타내는 것 같네요. 그러면, 이 방정식을 다음과 같이 c에 관해서 정리해보도록 하죠:
$cx^2 + (2c + 1)x - cy + 3c + 1 = cx^2 + 2cx - cy + 3c + x + 1 = c(x^2 + 2x - y + 3) + x + 1 = 0.$
그러면, 그 기본 방정식들은 다음과 같아요: $x^2 + 2x - y + 3 = 0$ 과 $x + 1 = 0$.
그러니까, 그 기본 커브들은 포물선 $x^2 + 2x - y + 3 = 0$ 과 직선 $x = -1$ 이네요. 이 직선은 y축에 평행이죠. 그러면 이제, 이 커브들이 서로 만나는 점들을 구해야 하는데 직선 $x = -1$ 이 그 포물선과 만나니까 그 특정 점에서의 x좌표는 -1 이네요. 그러면, 그 y좌표를 구하기로 하죠.
$x = -1 \Rightarrow y = x^2 + 2x + 3 = 1 - 2 + 3 \Rightarrow y = 2$. 따라서, 그 커브들은 (-1, 2)에서 서로 만나죠.
그러므로, 각 c 값에 대해서 모든 포물선 $cx^2 + (2c + 1)x - cy + 3c + 1 = 0$ 은 점 (-1, 2)를 지나네요.

요약:
$cx^2 + (2c + 1)x - cy + 3c + 1 = cx^2 + 2cx - cy + 3c + x + 1 = c(x^2 + 2x - y + 3) + x + 1 = 0.$
그러면, 그 기본 커브들은 포물선 $x^2 + 2x - y + 3 = 0$ 과 직선 $x = -1$ 이고 그 직선 $x = -1$ 이 그 포물선과 만나니까 그 특정 점에서의 x좌표는 -1. 그러면, $x = -1 \Rightarrow y = x^2 + 2x + 3 = 1 - 2 + 3 \Rightarrow y = 2$.
그러므로, 그 특정 점은 (-1, 2).

그림. N.0

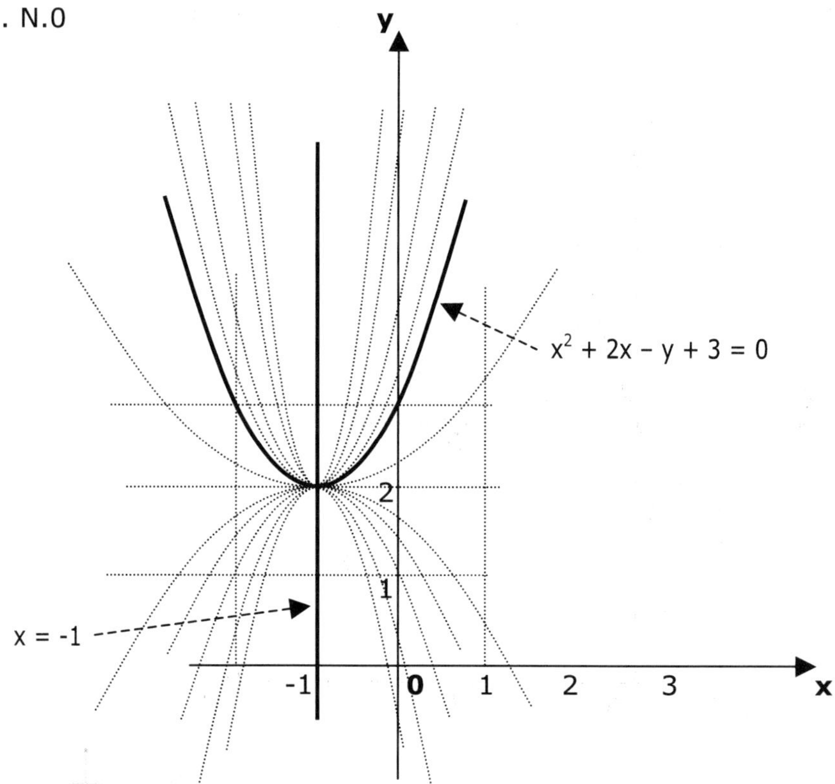

제안과 풀이
O 번 예제에서의 문제

다음과 같은 포물선의 방정식이 있다고 하세요: $y = cx^2 - (a - 2c)x - 2b(c - 1)$ 여기에서 c는 상수. 그러면, 그 포물선이 그 어느 c 값에 대하여도 두 점들 (1, 2)와 (s, t)를 지난다 하고 a 와 b 그리고 s 와 t 의 값들을 구하세요.

앞서 나온 예제에서와 마찬가지로, 이 방정식 $y = cx^2 - (a - 2c)x - 2b(c - 1)$도 포물선들의 모임을 나타내는 것 같네요. 그렇다면, 이 방정식은 기본 커브들의 선형조합이겠죠. 그러면, 그게 그런지를 확인해보아야 겠네요. 그러면, 이 방정식의 최고차 항이 x^2 항이니까 이 방정식을 다음과 같이 c에 관해서 정리해보도록 하죠:

$y = cx^2 - (a - 2c)x - 2b(c - 1) \Rightarrow cx^2 + 2cx - 2cb - ax + 2b - y = 0.$
$\Rightarrow c(x^2 + 2x - 2b) - (ax - 2b + y) = 0.$ 즉, $c(x^2 + 2x - 2b) + (-1)(ax - 2b + y) = 0.$

그러면, 그 기본 커브들은 $x^2 + 2x - 2b = 0$ 과 $ax - 2b + y = 0$ 이죠.
그런데, 방정식 $x^2 + 2x - 2b = 0$ 의 커브는 포물선이 아니네요. 그러면, 이거 뭐 잘못된 것 아닌가요?
잘못된 건 없어요. 이 방정식은 두개의 직선들을 나타내는 겁니다. 어떻게 요?
기본 커브들이 점 (1, 2)를 지나야 하죠. 이 점의 x 좌표는 1 이고 y 좌표는 2 이죠.
그러면, $x = 1$ 이면 $x^2 + 2x - 2b = 1 + 2 - 2b = 0 \Rightarrow b = 3/2$ 이네요.
따라서, $x^2 + 2x - 3 = 0$ 이 되는 거죠.
그러니까, $x^2 + 2x - 3 = 0 \Rightarrow (x - 1)(x + 3) = 0 \Rightarrow x = 1$ 또는 $x = -3$ 이네요.
따라서, 그 두개의 직선들은 $x = 1$ 과 $x = -3$ 인 것이고 이는 y 축에 평행한 거죠.

그런데, 그 기본 커브들은 두개의 점들 (1, 2)와 (s, t)를 지나죠.
그러면, 그 두 점들에서의 x 좌표들이 1 과 –3 인 겁니다.
그러니까, 점 (s, t)에서 s 가 –3 인 거죠.

그러면 이제, 그 나머지 기본 직선 $ax - 2b + y = 0$ 으로 넘어가보죠.
우선, $b = 3/2 \Rightarrow ax - 2b + y = ax - 3 + y = 0$ 이죠.
다음으로, 이 직선도 (1, 2)를 지나니까:
$(1, 2) \Rightarrow ax - 3 + y = a - 3 + 2 = 0 \Rightarrow a = 1$ 이네요.
그러니까, 그 나머지 기본 직선은 $x - 3 + y = 0$ 이죠.
그런데, 이 기본 직선도 점 (s, t) 즉, (-3, t)를 지나죠.
그러니까, 이 기본 직선의 방정식 $x - 3 + y = 0$ 에서 $x = -3$ 일 때, $y = t$ 가 되는 겁니다.
그러면, $x = -3 \Rightarrow x - 3 + y = -3 - 3 + t = 0 \Rightarrow t = 6$ 이네요.

그러면 이제, 그 기본 커브들은 $x^2 + 2x - 3 = (x - 1)(x + 3) = 0$ 으로부터 두 직선들 $x = 1$ 과 $x = -3$, 그리고 또 하나의 직선 $x - 3 + y = 0$ 이고 그 두 점들은 (1, 2)와 (-3, 6)입니다.
그런데, $y = cx^2 - (a - 2c)x - 2b(c - 1)$은 포물선의 방정식인데 어째서 그 기본 커브들 중에 포물선이 없나요?
비유적으로 설명하자면 포물선은 없지만 포물선의 뿌리는 있어요. 뿌리라니 요?
방정식 $x^2 + 2x - 3 = 0$ 은 포물선 $y = x^2 + 2x - 3$ 에서 온 것이죠. 포물선 $y = x^2 + 2x - 3$ 의 x절편들이 그 뿌리죠. 그러면, 방정식 $x^2 + 2x - 3 = 0$ 의 답을 포물선이라 하는 $y = x^2 + 2x - 3$ 의 커브의 뿌리라고 할 수 있어요. 그 답이 마치 나무의 뿌리와 같다고 하는 거죠. 그 x축을 지표면으로 보고 그 커브를 나무로 보면 그 답이 뿌리가 되는 거죠. 그러면 이제, 포물선의 뿌리인 두 직선들이 하나의 특정 직선과 합쳐져서 또 하나의 나무가 되는데 이것이 또 하나의 포물선이죠. 다른 포물선들도 그와 동일한 뿌리를 갖을 수 있고 그런 포물선들 각각의 방정식이 $y = c(x^2 + 2x - 3)$인 겁니다.

요약:

$y = cx^2 - (a - 2c)x - 2b(c - 1) \Rightarrow cx^2 + 2cx - 2cb - ax + 2b - y = 0.$
$\Rightarrow c(x^2 + 2x - 2b) - (ax - 2b + y) = 0.$ 즉, $c(x^2 + 2x - 2b) + (-1)(ax - 2b + y) = 0.$

그러면, 그 기본 커브들은 $x^2 + 2x - 2b = 0$ 과 $ax - 2b + y = 0.$
방정식 $x^2 + 2x - 2b = 0$ 은 두개의 직선들을 나타냅니다.
기본 커브들이 점 (1, 2)를 지나므로 $x = 1$ 이면 $x^2 + 2x - 2b = 1 + 2 - 2b = 0 \Rightarrow b = 3/2.$
따라서, $x^2 + 2x - 2b = x^2 + 2x - 3 = 0 \Rightarrow (x - 1)(x + 3) = 0 \Rightarrow x = 1$ 또는 $x = -3.$
그러므로, 그 두개의 직선들은 $x = 1$ 과 $x = -3$ 인 것이고 이는 y 축에 평행.

그 기본 커브들은 두개의 점들 (1, 2)와 (s, t)를 지나고 그 두 점들에서의 x 좌표들이 1 과 −3 이므로 점 (s, t)에서 s 가 −3.
$b = 3/2 \Rightarrow ax - 2b + y = ax - 3 + y = 0.$
이 직선도 (1, 2)를 지나니까 $(1, 2) \Rightarrow ax - 3 + y = a - 3 + 2 = 0 \Rightarrow a = 1 \Rightarrow x - 3 + y = 0.$
이 기본 직선도 점 (s, t) 즉, (-3, t)를 지나므로:
이 기본 직선의 방정식 $x - 3 + y = 0$ 에서 $x = -3$ 일 때, $y = t.$
그러면, $x = -3 \Rightarrow x - 3 + y = -3 - 3 + t = 0 \Rightarrow t = 6.$
그러므로, $a = 1, b = 3/2, s = -3, t = 6.$

그림. O.0

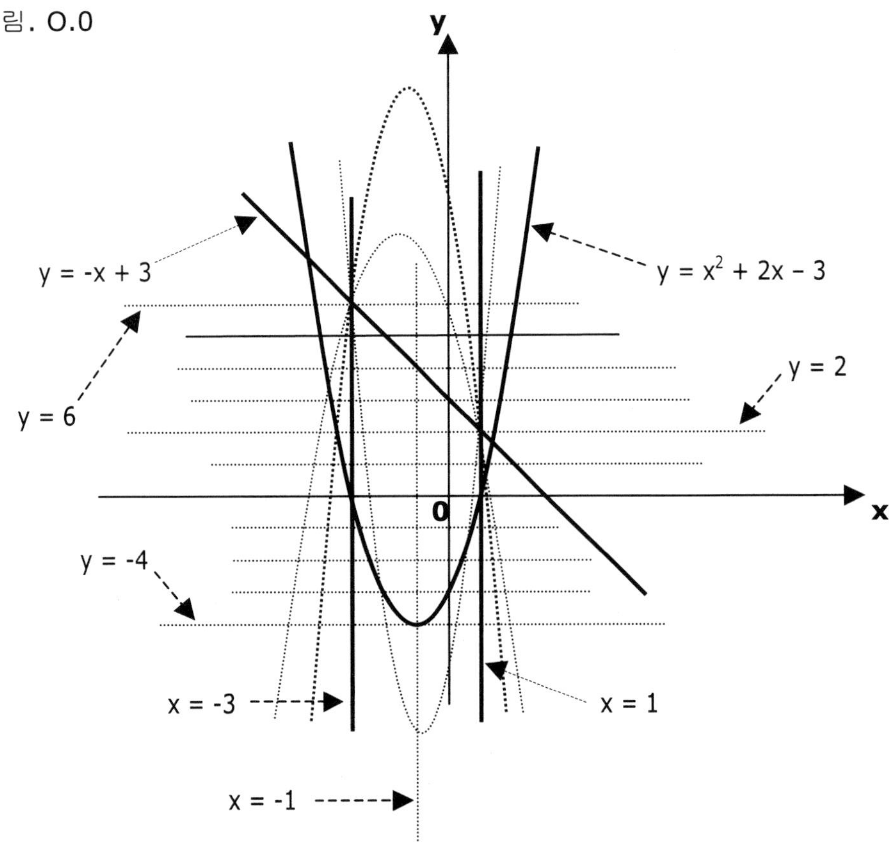

제안과 풀이
P 번 예제에서의 문제

다음과 같이 가정하세요: $y = f(x) = 2(x - 2)^2 + 5$ 이고 $f(x) = T_3 \bullet T_2 \bullet T_1 \bullet f$
T_1: (x, y) ⟶ (3 – x, -2 – y)
T_2: (x, y) ⟶ (-x, -y)
T_3: (x, y) ⟶ (x + u, y + v) 여기에서 u와 v는 상수.

그러면, u 와 v 의 값들을 구하세요.

함수 f 를 수리적 또는 대수적으로 변환시킬 수 있어요. 함수의 변환을 통해서 새로운 함수를 만들 때는 실제로는 그 원래함수가 갖는 커브에서의 모든 점들을 변환시키는 거죠. 그러니까, 그 커브의 일반점을 만들어서 그 일반점에 그 변환을 적용하죠. 하지만, 그래프 해법으로 그 변환을 시도해보도록 하겠어요. 이를테면, 주먹 구구식이라고 할 수도 있는데, 최종 답만을 빨리 구하고자 할 때 유용하죠. 그러면 이제, f 의 커브에 있는 하나의 점 P 를 이동시켜보도록 하죠. 그러면, 나머지 점들도 그 점 P 가 이동하는 것과 마찬가지로 이동하는 겁니다. 다음과 같이 x-y 평면에 그 점 P 가 있다고 하세요.

그림. P.0

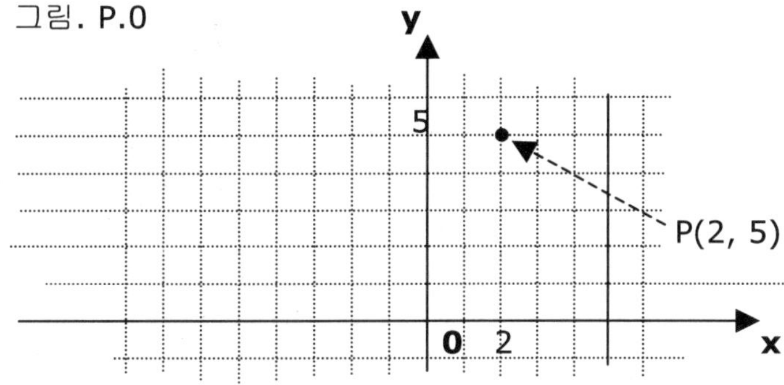

그러면, 점 P에 변환 T_1을 적용하는 것부터 시작해보기로 하죠:
T_1: (x, y) ⟶ (3 – x, 2 – y).

이 T_1은 사실, 다음과 같은 두개의 변환들의 합성 변환이에요:
A: (x, y) ⟶ (-x, -y)
B: (x, y) ⟶ (x + 3, y + 2)

그림. P.1

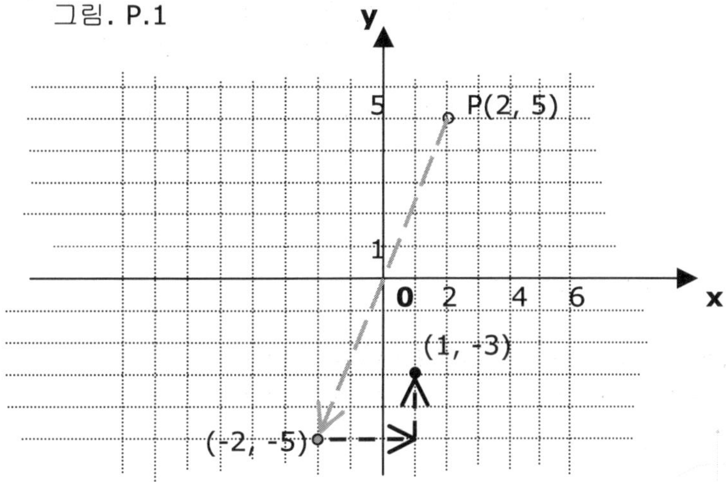

이 합성 변환 $T_1 = B \bullet A$는 변환 A부터 시작해서 변환 B로 끝나는 거라고 할 수 있겠어요.
그러니까, $T_1 = B \bullet A: (x, y) \longrightarrow (3 - x, 2 - y)$.
이 합성 변환 T_1은 다음과 같이 작동하죠. 아래의 그림에서와 같이 변환 A에서는, (2, 5)에 있는 점 P가 원점에 대칭으로 변환되어 (-2, -5)에 있는 점이 되죠. 변환 B에서는, (-2, -5)에 있는 그 점이 x축 방향으로 3 만큼 y축 방향으로 2 만큼 평행 이동되어 (1, -3)에 있는 점이 되요.

다음으로, T_2는 다음과 같아요. $T_2: (x, y) \longrightarrow (-x, -y)$.

변환 T_2에서는 (1, -3)에 있는 점이 원점에 대칭으로 변환되어 (-1, 3)에 있는 점이 됩니다.

그림. P.2

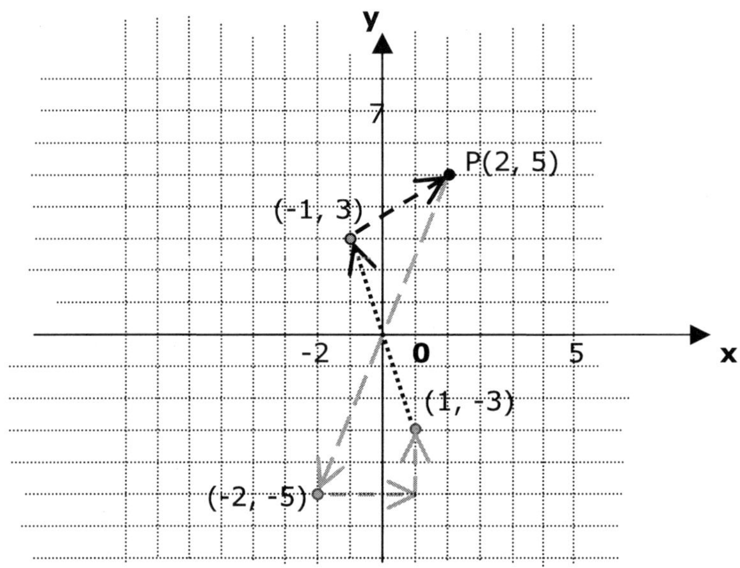

그러니까 간단히 하면, $T_2 \bullet T_1:(x, y) \longrightarrow \{-(3 - x), -(2 - y)\} = (x - 3, y - 2)$인 것이죠.

그러면 이제, T_3는 다음과 같아요. $T_3: (x, y) \longrightarrow (x + u, y + v)$ 여기에서 u와 v는 상수.

그러면, T_3에서는 평행 변환이 (-1, 3)에 있는 점에 적용되는 거죠.
그러면 이제, (-1, 3)에 있는 점을 원래의 위치 (2, 5)로 이동시키려면 u 와 v 의 값들이 무엇이어야 하나요?
그래프에 알 수 있는 바와 같이 (-1, 3)에 있는 그 점은 x 축 방향으로 3 만큼 y 축 방향으로 2 만큼 이동해야 하죠.
그러니까, u = 3 이고 v = 2 이네요.
이는 함수 f 의 커브에 있는 다른 모든 점들 각각에 대하여도 마찬가집니다.

요약:

수리적으로는 다음과 같이 하면 되죠:

$T_3 \bullet T_2 \bullet T_1:(x, y) \longrightarrow (x - 3 + u, y - 2 + v) = (x, y)$.
그러므로, $x - 3 + u = x \Rightarrow u = 3$ 이고 또한, $y - 2 + v = y \Rightarrow v = 2$.

제안과 풀이
Q번 예제에서의 문제

다음과 같이 가정하세요: $y = f(x) = 2x^2 - 2x + 3$ 이고 $g = T_2 \bullet T_1 \bullet f$
T_1: $(x, y) \longrightarrow (x - 3, y - 2)$
T_2: $(x, y) \longrightarrow (y, x)$

그러면, **g**를 구하세요.

물론, 함수 f를 수리적으로 변환시킬 수 있어요. 하지만 이번에도, 그래프 해법으로 이를 변환시켜보도록 하죠. 합성 변환 $T_2 \bullet T_1$에서는 변환 T_1으로 시작해서 T_2로 끝나죠. T_1은 평행 변환이죠. 이 T_1에서는 f의 커브에 있는 모든 점이 x축 방향으로 –3 만큼 y축 방향으로 –2 만큼 이동하죠. 그러면, 새로운 커브가 하나 만들어지는데 하나의 새로운 함수가 이 새로운 커브를 갖는 거죠. 그 새로운 함수가 $T_1 \bullet f$인 거죠.
그러면 이제, T_2에서는 그 새로운 함수 $T_1 \bullet f$의 커브가 직선 y = x에 관하여 대칭으로 이동할 겁니다. 그러면, 또 하나의 새로운 커브가 만들어질 텐데 하나의 새로운 방정식이 그 새로운 커브를 갖을 겁니다. 그 방정식은 $T_2 \bullet T_1 \bullet f$인데 이것이 바로 g인 겁니다.
그러면 이제, f 의 커브에 있는 하나의 특정한 점을 이동시켜보도록 하죠. 그러면, 그 나머지 모든 점들도 그 특정한 점을 이동시키는 것과 마찬가지로 이동시키면 되겠죠. 그러면, 그 꼭지점을 그 특정한 점으로 하죠. 그러면, 나머지 점들도 그 꼭지점을 이동시키는 것과 마찬가지로 이동시키는 겁니다.
먼저, 그 f 의 커브의 세부 사항을 보도록 하죠.
기울기가 2 이고 그 커브는 아래로 볼록하죠. 다음은, 그 꼭지점입니다:
$2x^2 - 2x + 3 = 2(x^2 - x) + 3 = 2(x - 1/2)^2 - 1/2 + 3 = 2(x - 1/2)^2 + 5/2 \Rightarrow$ 그 꼭지점 = (1/2, 5/2).
그러면 이제, 꼭지점부터 변환을 시작하는 겁니다.

그림. Q.0

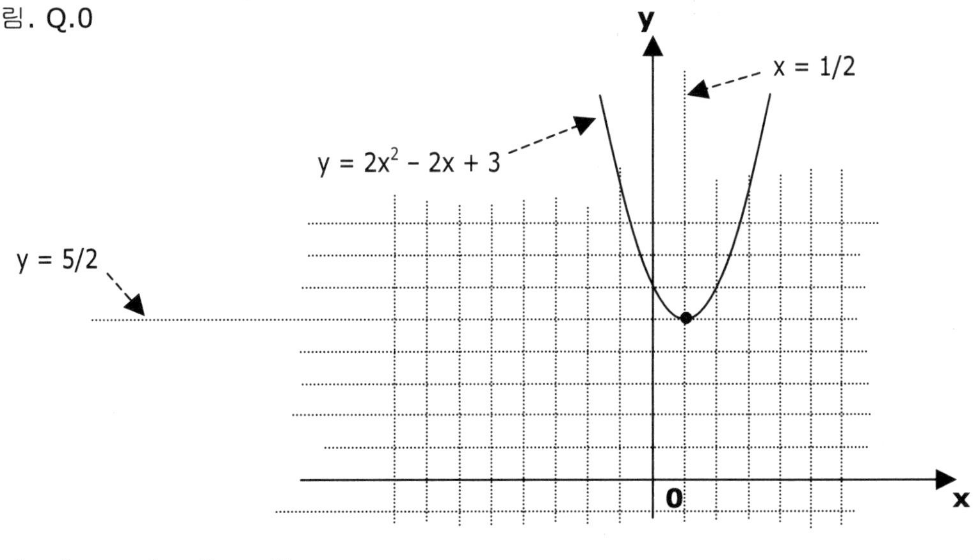

T_1: $(x, y) \longrightarrow (x - 3, y - 2)$.

T_1에서는 그 꼭지점이 x축 방향으로 –3 만큼 y축 방향으로 –2 만큼 이동하죠.
그럼 이제, 그 꼭지점을 이동하자마자 그 전체 커브도 이동하는 겁니다.
변환이 평행으로 이루어지므로 포물선의 오목한 상태와 기울기는 그대로 유지되는 거죠.

그림. Q.1

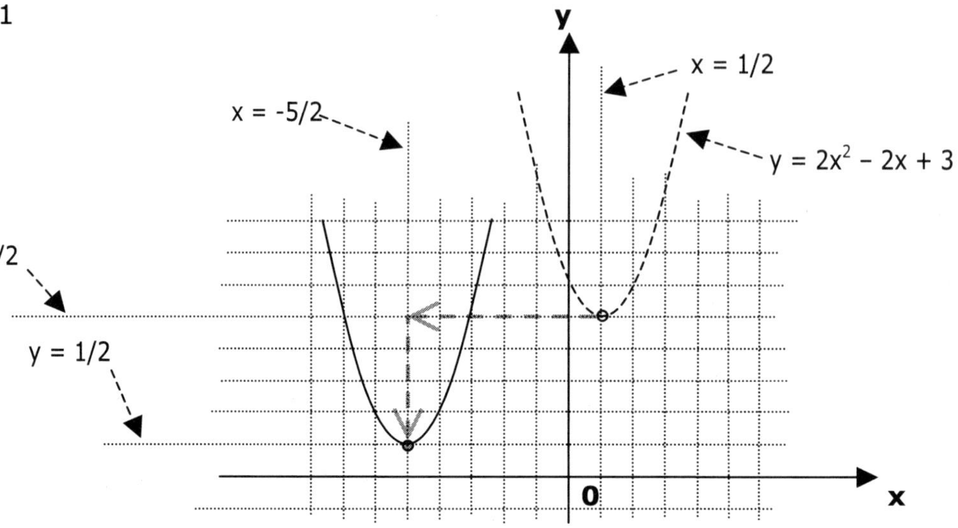

T_2: (x, y) ⟶ (y, x).

T_2에서는 그 꼭지점이 직선 y = x에 관하여 대칭으로 이동하죠.
그러면, 그 오목한 상태가 변환을 거치면서 변할 것이고 왼쪽으로 볼록할 거에요. 즉, 그 포물선은 오른쪽으로 열릴 거죠. 하지만, 대칭변환은 그 커브의 기울기의 크기에 영향을 끼치진 않아요. 그러니까, 그 기울기의 크기는 2로 남아있을 겁니다.

그림. Q.2

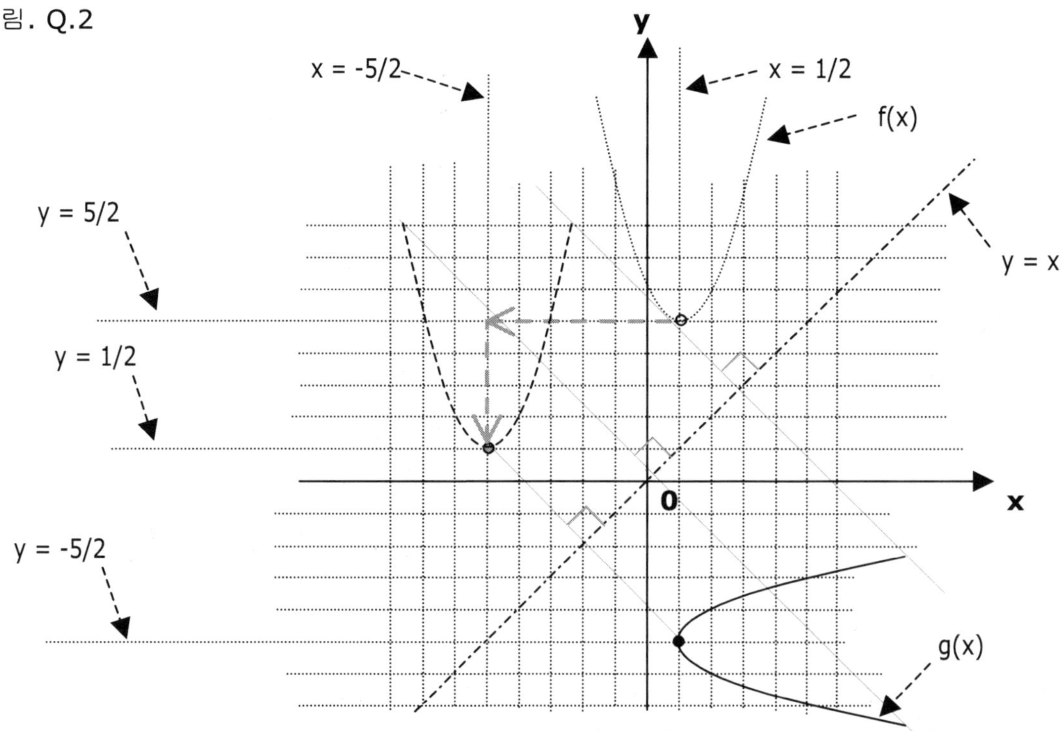

그러면, 좌로 볼록한 포물선이 T_1과 T_2의 합성 변환의 결과인 거죠.

그러면 이제, g 의 방정식을 구해보기로 하죠.

이런 방정식의 표준형은 $x - m = a(y - n)^2$ 이며 여기에서 (m, n)이 그 꼭지점이고 a가 기울기이죠.
그래프에서 그 새로운 꼭지점이 이제, (1/2, -5/2)이라는 걸 알 수 있죠.
그 기울기의 크기는 그대로 라는 것을 알고있죠. 그러면, 그 새로운 커브가 좌로 볼록하니까 그 기울기는 2 인 거죠. 따라서, 그 방정식은 다음과 같아요:

$x - 1/2 = 2\{y - (-5/2)\}^2 = 2(y + 5/2)^2 \Rightarrow x - 1/2 = 2(y + 5/2)^2$.

그러므로, g는 다음과 같죠: $g(x, y) = x - 2(y + 5/2)^2 - 1/2 = 0$.

요약:

수리적으로는 다음과 같이 하면 되죠:

다음과 같죠: $y = f(x) = 2x^2 - 2x + 3$ 이고 $g = T_2 \bullet T_1 \bullet f$
$T_1: (x, y) \longrightarrow (x - 3, y - 2)$
$T_2: (x, y) \longrightarrow (y, x)$

이제, (u, v)가 g 의 일반점이라고 하세요. 그러면:
$T_2 \bullet T_1: (x, y) \longrightarrow (u, v) = (y - 2, x - 3)$.

그러면, $u = y - 2 \Rightarrow y = u + 2$ 이고 또한, $v = x - 3 \Rightarrow x = v + 3$.

그런데, $y = 2x^2 - 2x + 3 = 2(x^2 - x) + 3 = 2(x - 1/2)^2 - 1/2 + 3 = 2(x - 1/2)^2 + 5/2$
$\Rightarrow y = 2(x - 1/2)^2 + 5/2$.

그러면, $y = u + 2$ 이고 $x = v + 3$ 이니까:
$y = 2(x - 1/2)^2 + 5/2 \Rightarrow u + 2 = 2\{(v + 3) - 1/2\}^2 + 5/2 = 2(v + 3 - 1/2)^2 + 5/2 = 2(v + 5/2)^2$
$\Rightarrow u + 2 - 5/2 = u - 1/2 = 2(v + 5/2)^2$.

그런데, 이 일반점(u, v)는 x-y 평면에 있는 거죠.
그러니까, 이 u 와 v 를 제각기 x 와 y 로 바꾸면 g 를 갖는 거죠.
그러면, $u - 1/2 = 2(v + 5/2)^2 \Rightarrow x - 1/2 = 2(y + 5/2)^2$.
그러므로, $g(x, y) = x - 2(y + 5/2)^2 - 1/2 = 0$.

제안과 풀이
R 번 예제에서의 문제

다음과 같이 가정하세요: $y = f(x) = -2x^2 + 8x - 5$ 이고 $y = g(x) = 2x^2 - 6x + 5/2$ 이며 $g = T \bullet f$
T: (x, y) ⟶ (u − x, v − y).

그러면, **u** 와 **v** 의 값들을 구하세요.

$u - x = -x + u$ 이고 $v - y = -y + v$ 이죠.
그러니까, 변환 T는 다음과 같은 두개의 변환들의 합성 변환이네요:
A: (x, y) ⟶ (-x, -y)
B: (x, y) ⟶ (x + u, y + v).

이 합성 변환 $T_1 = B \bullet A$에서는, A로 시작해서 B로 끝나요. 그러니까: $T_1 = B \bullet A$: (x, y) ⟶ (u − x, v − y).

이번에도 그래프 해법을 사용해보죠. 그럼 이제, A 에서는 f 의 커브에 있는 모든 점이 원점에 관하여 대칭으로 이동되어 새로운 커브의 점이 되는 거죠. B 에서는 그 새로운 커브에 있는 모든 점이 x 축 방향으로 u 만큼 y 축 방향으로 v 만큼 평행 이동되어 새로운 함수의 커브의 점이 되는데 이 새로운 함수가 g(x)입니다.
그러면 이제, 그 u 와 v 즉, x-y 평면에서 그 각각의 방향으로의 평행이동 변화량을 구해야 하는 거죠.
그러면, 그 f 와 g 의 커브들을 그래프에 놓아보도록 하죠.

$$y = f(x) = -2x^2 + 8x - 5 = -2(x^2 - 4x + 4 - 4) - 5 = -2(x - 2)^2 + 8 - 5 = -2(x - 2)^2 + 3.$$

$$y = g(x) = 2x^2 - 6x + 5/2 = 2(x^2 - 3x + 9/4 - 9/4) + 5/2 = 2(x - 3/2)^2 - 9/2 + 5/2 = 2(x - 3/2)^2 - 2.$$

그러면, 다음과 같습니다:
우선, f 의 커브의 꼭지점은 (2, 3)이고 그 기울기는 −2, 그러니까 그 커브는 위로 볼록하죠.
다음으로, g 의 커브의 꼭지점은 (3/2, -2)이고 그 기울기는 2, 그래서 그 커브는 아래로 볼록해요.

그림. R.0

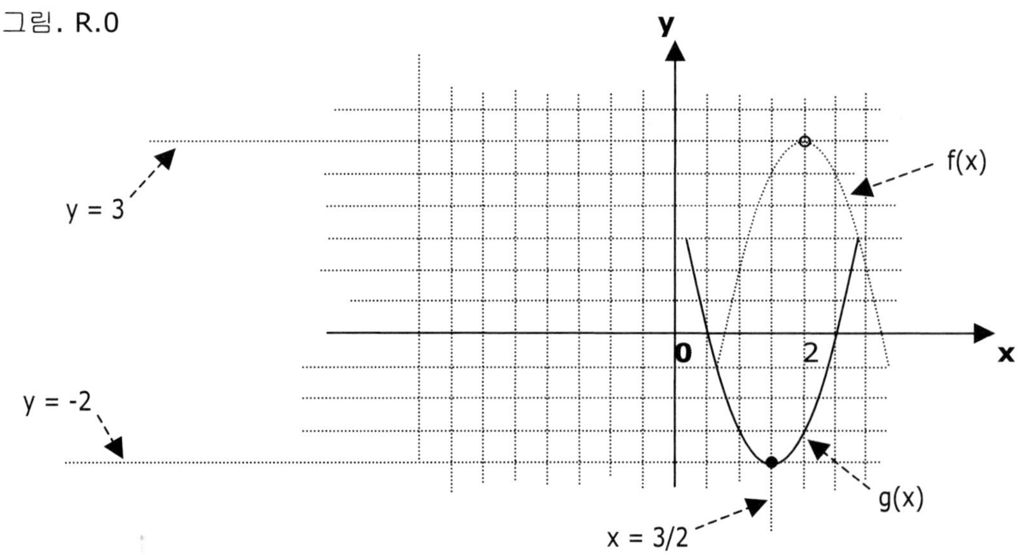

변환 T 에서는 두개의 변환들이 다음과 같이 일어나죠:
A: (x, y) ⟶ (-x, -y).
B: (x, y) ⟶ (x + u, y + v).

A 는 원점에 관하여 대칭인 변환이고 B 는 평행 변환이죠. 그러면 먼저, f 의 꼭지점은 A 에 의하여 원점에 관하여 대칭으로 변환되어 (-2, -3)의 위치로 이동하죠. 그러면, f 의 커브는 회색 점선 포물선으로 변환되는 거죠.

그림. R.1

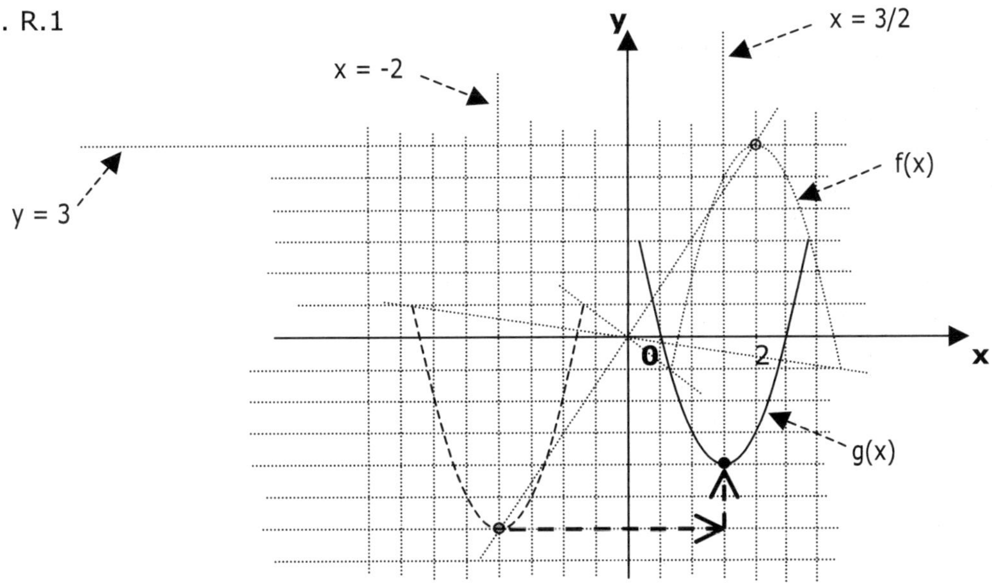

그리고 나서, 그 회색 점선 포물선은 B 에 의하여 x 축을 따라 u 만큼 y 축을 따라 v 만큼 이동되어 회색 포물선이 되는데 이는 g 의 커브인 겁니다. 위의 그래프에서, u = 7/2 이고 v = 1 이라는 것을 알 수 있죠.

요약:

수리적인 방법으로는 다음과 같이 해볼 수 있겠어요.

다음과 같죠: $y = f(x) = -2x^2 + 8x - 5$ 이고 $y = g(x) = 2x^2 - 6x + 5/2$ 이며 $g = T \bullet f$
T: (x, y) ⟶ (u − x, v − y).

그러면, f 와 g 의 기울기의 크기들은 동일하고 그 꼭지점들은 다음과 같아요:
$y = f(x) = -2x^2 + 8x - 5 = -2(x^2 - 4x + 4 - 4) - 5 = -2(x - 2)^2 + 8 - 5 = -2(x - 2)^2 + 3$
⇒ 꼭지점 = (2, 3).
$y = g(x) = 2x^2 - 6x + 5/2 = 2(x^2 - 3x + 9/4 - 9/4) + 5/2 = 2(x - 3/2)^2 - 9/2 + 5/2 = 2(x - 3/2)^2 - 2$
⇒ 꼭지점 = (3/2, -2).

그러면, 꼭지점 (2, 3)이 변환되어 꼭지점 (3/2, -2)가 되는 것이죠.
그런데, 변환되는 커브에 있는 모든 점은 동일한 방식으로 변환되죠.
변환 T 에서는 점 (x, y)가 점 (u − x, v − y)로 변환되는 거죠.
따라서, 그 커브에 있는 (2, 3)이 (3/2, -2)로 변환되니까 u − 2 = 3/2 이고 v − 3 = -2 이네요.
그러므로, u = 2 + 3/2 = 7/2 이고 v = -2 + 3 = 1 입니다.

제안과 풀이
S번 예제에서의 문제

다음과 같이 가정하세요: $x = f(y) = 2(y-1)^2 + 2$ 이고 $y = g(x) = (x - 3/2)^2 + 2$ 이며 $g = T_2 \bullet T_1 \bullet f$
T_1: $(x, y) \longrightarrow (cy, 2x)$ 여기에서 c는 상수
T_2: $(x, y) \longrightarrow (x + u, y + v)$ 여기에서 u와 v는 상수

그러면, c와 u 그리고 v의 값들을 구하세요.

T_2는 평행 변환이죠. 하지만, T_1은 직선 $y = x$에 관한 대칭 변환이 아닙니다.
이번에는 T_1때문에 그래프 해법은 부적합할 것 같으니까 수리적인 방법을 사용해보도록 하죠.
그러면 이제, 점 (s, t)를 새로운 함수 $T_1 \bullet f$ 의 일반점이라고 하세요. 그러면:
$T_1:(x, y) \longrightarrow (s, t) = (cy, 2x)$.

그러면, $s = cy$ 이고 $t = 2x$ 이죠. 그러니까, $y = s/c$ 이고 $x = t/2$ 이네요. 따라서:
$x = 2(y-1)^2 + 2 \Rightarrow t/2 = 2(s/c - 1)^2 + 2 = 2s^2/c^2 - 4s/c + 4 \Rightarrow t = 4s^2/c^2 - 8s/c + 8$.

그 일반점은 x-y 평면에 있으니까 s와 t를 제각기 x와 y로 교체하기로 하죠.
그러면, $T_1 \bullet f = y = 4x^2/c^2 - 8x/c + 8$ 이네요.

그러면 이제, T_2로 넘어가보기로 하죠.
그러면, $T_1 \bullet f$가 T_2로 입력되는 거고 그 출력 즉, $T_2 \bullet T_1 \bullet f$가 g입니다.
그런데, 포물선의 평행 변환에서는 그 포물선의 기울기는 변하지 않는 거죠.
그러면, 포물선 g의 기울기가 1 이니까 $T_1 \bullet f$의 기울기도 1 이어야 하네요.

그러니까, 포물선의 방정식 $y = 4x^2/c^2 - 8x/c + 8$ 에서 $4/c^2 = 1$ 이어야 하는 거죠.
그러면, $4/c^2 = 1 \Rightarrow c^2 = 4 \Rightarrow c = 2$ 또는 -2 입니다.

그러면 먼저, $c = 2 \Rightarrow y = 4x^2/c^2 - 8x/c + 8 = x^2 - 4x + 8 = (x-2)^2 + 4$ 이죠.
그러므로, $y = T_1 \bullet f = (x-2)^2 + 4$ 즉, $y - 4 = (x-2)^2$이네요.
그러면 이제, $T_1 \bullet f$와 g의 꼭지점들의 위치를 비교해서 u와 v의 값들을 구할 수 있겠죠.

하지만, 다음과 같이 다른 방법으로 해볼 수도 있어요. 이제, g의 일반점을 (s, t)라 하세요.
T_2는 다음과 같죠:
T_2: $(x, y) \longrightarrow (x + u, y + v)$ 여기에서 u와 v는 상수.
그러면, $(s, t) = (x + u, y + v) \Rightarrow s = x + u$ 이고 $t = y + v$ 이죠.
따라서, $x = s - u$ 이고 $y = t - v$ 이죠.
그러니까, $y - 4 = (x-2)^2 \Rightarrow (t - v) - 4 = \{(s - u) - 2\}^2$이죠.
이 일반점 (s, t)는 g의 일반점이고 g는 x-y 좌표계에 있으므로 s와 t를 제각기 x와 y로 교체해야 하죠.
그러니, $(y - v) - 4 = \{(x - u) - 2\}^2$이고 이는 g에 관한 겁니다.
그런데, $y = g(x) = (x - 3/2)^2 + 2 \Rightarrow y - 2 = (x - 3/2)^2$이죠.
그러면, $(y - v) - 4 = y - 2$ 이고 또한, $(x - u) - 2 = x - 3/2$ 이죠.
그러니까, $c = 2$ 이고 $v = -2$ 이며 $u = -1/2$.

다음으로, $c = -2 \Rightarrow y = 4x^2/c^2 - 8x/c + 8 \Rightarrow y = x^2 + 4x + 8 = (x+2)^2 + 4$ 이죠.
그러므로, $y = T_1 \bullet f = (x+2)^2 + 4$ 즉, $y - 4 = (x+2)^2$이네요.

그런데, (s, t)가 g의 일반점이고 (x, y)가 $T_1 \bullet f$의 일반점이면 $x = s - u$이고 $y = t - v$이죠.
그러니까, $y - 4 = (x + 2)^2 \Rightarrow (t - v) - 4 = \{(s - u) + 2\}^2$이죠.
이 일반점 (s, t)는 g의 일반점이고 g는 x-y 좌표계에 있으므로 s와 t를 제각기 x와 y로 교체해야 하죠.
그러니, $(y - v) - 4 = \{(x - u) + 2\}^2$이고 이는 g에 관한 겁니다.
그런데, $y = g(x) = (x - 3/2)^2 + 2 \Rightarrow y - 2 = (x - 3/2)^2$이죠.
그러면, $(y - v) - 4 = y - 2$ 이고 또한, $(x - u) + 2 = x - 3/2$ 이죠.
그러니까, $c = -2$ 이고 $v = -2$ 이며 $u = 7/2$ 입니다.

요약:

점 (s, t)를 함수 $T_1 \bullet f$ 의 일반점이라고 하세요. 그러면, $T_1:(x, y) \longrightarrow (s, t) = (cy, 2x)$.
그러면, $s = cy$ 이고 $t = 2x$. 그러니까, $y = s/c$ 이고 $x = t/2$. 따라서:
$x = 2(y - 1)^2 + 2 \Rightarrow t/2 = 2(s/c - 1)^2 + 2 = 2s^2/c^2 - 4s/c + 4 \Rightarrow t = 4s^2/c^2 - 8s/c + 8$.
그러면, $T_1 \bullet f = y = 4x^2/c^2 - 8x/c + 8$.

그러면 이제, $g = T_2 \bullet T_1 \bullet f$이고 $y = g(x) = (x - 3/2)^2 + 2 \Rightarrow y - 2 = (x - 3/2)^2$.
T_1은 평행 변환이고 포물선의 평행 변환에서는 그 기울기는 변하지 않고 포물선 g의 기울기가 1 이니까 $T_1 \bullet f$의 기울기도 1.
그러니까, 포물선의 방정식 $y = 4x^2/c^2 - 8x/c + 8$ 에서 $4/c^2 = 1$.
그러면, $4/c^2 = 1 \Rightarrow c^2 = 4 \Rightarrow c = 2$ 또는 -2.

이제, g의 일반점을 (s, t)라 하세요.
그러면, $(s, t) = (x + u, y + v) \Rightarrow s = x + u$ 이고 $t = y + v$.
따라서, $x = s - u$ 이고 $y = t - v$.

그러면 먼저, $c = 2 \Rightarrow y = 4x^2/c^2 - 8x/c + 8 = x^2 - 4x + 8 = (x - 2)^2 + 4$.
그러면, $y = T_1 \bullet f = (x - 2)^2 + 4$ 즉, $y - 4 = (x - 2)^2$.
그러므로, $y - 4 = (x - 2)^2 \Rightarrow (t - v) - 4 = \{(s - u) - 2\}^2$.
그러면 이제, $(y - v) - 4 = \{(x - u) - 2\}^2$이고 이는 g에 관한 것 $\Rightarrow y - 2 = (x - 3/2)^2$.
그러면, $(y - v) - 4 = y - 2$ 이고 또한, $(x - u) - 2 = x - 3/2$.
그러므로, $c = 2$ 이고 $v = -2$ 이며 $u = -1/2$.

다음으로, $c = -2 \Rightarrow y = 4x^2/c^2 - 8x/c + 8 \Rightarrow y = x^2 + 4x + 8 = (x + 2)^2 + 4$.
그러면, $y = T_1 \bullet f = (x + 2)^2 + 4$ 즉, $y - 4 = (x + 2)^2$.
그러니까, $y - 4 = (x + 2)^2 \Rightarrow (t - v) - 4 = \{(s - u) + 2\}^2$.
그러면 이제, $(y - v) - 4 = \{(x - u) + 2\}^2$이고 이는 g에 관한 것 $\Rightarrow y - 2 = (x - 3/2)^2$.
그러면, $(y - v) - 4 = y - 2$ 이고 또한, $(x - u) + 2 = x - 3/2$.
그러므로, $c = -2$ 이고 $v = -2$ 이며 $u = 7/2$.

유의사항:

함수 f(y)가 함수 g(x)로 변하는 두개의 경로들이 있죠. 그 둘 중의 하나는 다음과 같죠:
T_1: (x, y) ⟶ (2y, 2x).
T_2: (x, y) ⟶ (x − 1/2, y − 2).

간단히 하면:
$T_2 \bullet T_1$:(x, y) ⟶ (2y − 1/2, 2x − 2).

그 둘 중 나머지 하나는 다음과 같아요:
T_1: (x, y) ⟶ (-2y, 2x).
T_2: (x, y) ⟶ (x + 7/2, y − 2).

간단히 하면:
$T_2 \bullet T_1$:(x, y) ⟶ (-2y + 7/2, 2x − 2).

그림. S.0

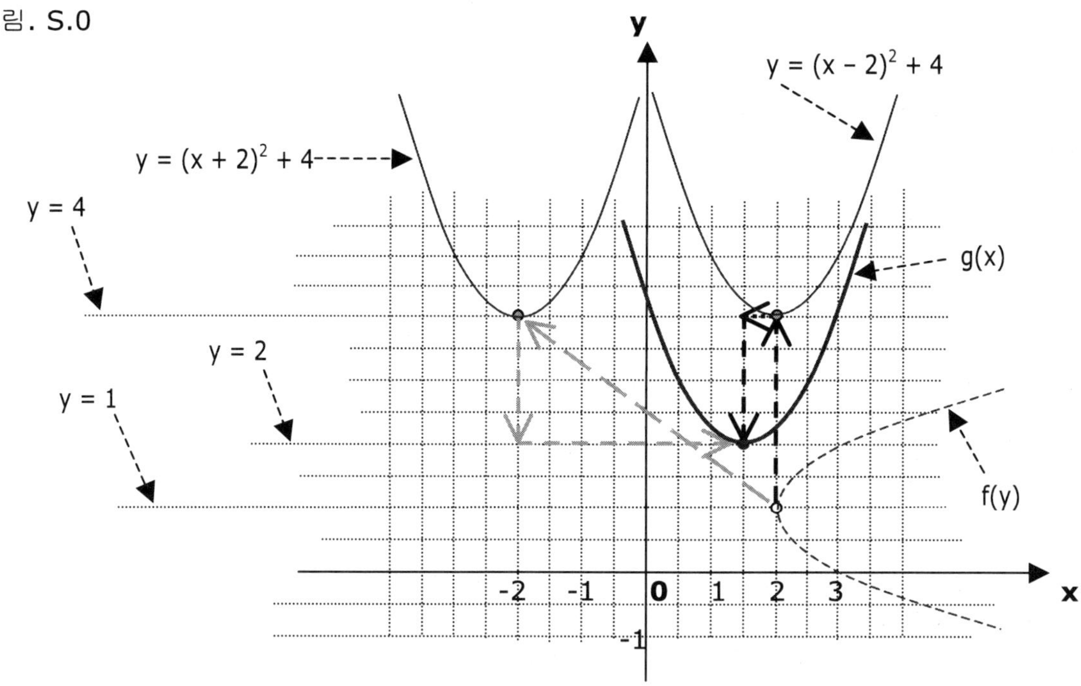

제안과 풀이
T번 예제에서의 문제

다음과 같이 가정하세요:
$f(x) = 2x^2 - 2cx + c$ 여기에서 c는 상수이고 $c \ge -1$. $y = f(x)$의 커브는 포물선인데 그 꼭지점은 (s, t)이다.
T: c ⟶ (s, t).
t = g(s).

그러면, g를 명시하세요.

이 f는 포물선의 집단을 나타내는 함수입니다. 즉, c의 값을 정하면 그 f가 정해지는 거죠. 물론, 그 값은 –1 보다 크거나 같아야 하죠. 포물선 $y = 2x^2 - 2cx + c$의 꼭지점은 (s, t)인데 여기에서 c는 상수이고 $c \ge -1$ 입니다.
그런데, T는 매개변수 변환이고 여기에서 그 매개변수는 c이죠.
T: c ⟶ (s, t)이고 여기에서 (s, t)가 그 포물선의 꼭지점이죠. 그러면, c가 s와 t를 연결해주네요. 즉, s와 t는 제각기 c의 식으로 나타내지는 거죠. 그러니까, s와 t는 제각기 c의 함수가 되는 겁니다. 매개변수 변환에 대한 자세한 내용은 **제 2권: 그래프 조작**을 참고하세요.

그러면 이제, s = u(c)이고 t = v(c)이며 $c \ge -1$ 라고 하세요.
이제, 그 꼭지점부터 시작하기로 하죠.
$y = 2x^2 - 2cx + c = 2(x^2 - cx + c^2/4 - c^2/4) + c = 2(x - c/2)^2 - c^2/2 + c$이고 $c \ge -1$.

그러니까, 그 꼭지점은 $(c/2, -c^2/2 + c)$이고$c \ge -1$ 이네요.
그러면, $s = u(c) = c/2$ 이고 $t = v(c) = -c^2/2 + c$이며 $c \ge -1$ 입니다.

이제, 다음과 같은 거죠:
T: c ⟶ $(s, t) = (c/2, -c^2/2 + c)$이고 $c \ge -1$.

그러면, $s = c/2 \Rightarrow c = 2s$. 따라서, $c \ge -1 \Rightarrow c/2 \ge -1/2 \Rightarrow s \ge -1/2$.
또한, $c = 2s \Rightarrow t = -c^2/2 + c \Rightarrow t = -4s^2/2 + 2s = -2s^2 + 2s$ 이고 $s \ge -1/2$.
그러므로, $t = g(s) = -2s^2 + 2s$ 이고 $s \ge -1/2$ 입니다.

함수 g를 x-y 좌표계에 놓으려면 s와 t를 제각기 x와 y로 교체해야 되죠.
그러면, $y = g(x) = -2x^2 + 2x$이고 $x \ge -1/2$ 입니다.

요약:

$y = 2x^2 - 2cx + c = 2(x^2 - cx + c^2/4 - c^2/4) + c = 2(x - c/2)^2 - c^2/2 + c$이고 $c \ge -1$.
그러므로, 그 꼭지점 $(s, t) = (c/2, -c^2/2 + c)$이고$c \ge -1$.

이제, s = u(c)이고 t = v(c)이며 $c \ge -1$ 라고 하세요.
그러면, $s = u(c) = c/2$ 이고 $t = v(c) = -c^2/2 + c$이며 $c \ge -1$.
그러면, $s = c/2 \Rightarrow c = 2s$. 따라서, $c \ge -1 \Rightarrow c/2 \ge -1/2 \Rightarrow s \ge -1/2$.
또한, $c = 2s \Rightarrow t = -c^2/2 + c \Rightarrow t = -4s^2/2 + 2s = -2s^2 + 2s$ 이고 $s \ge -1/2$.

그러므로, $t = g(s) = -2s^2 + 2s$이고 $s \ge -1/2$ 입니다.

이제, f와 g의 커브들을 x-y 평면에 놓아보기로 하죠.
$y = g(x) = -2x^2 + 2x = -2(x^2 - x) = -2(x^2 - x + 1/4 - 1/4) = -2(x - 1/2)^2 + 1/2$ 이고 $x \geq -1/2$.

그림. T.0

$y = g(x) = -2(x - 1/2)^2 + 1/2$ 이고 $x \geq -1/2$. **$y = f(x) = 2x^2 - 2cx + c$이고 $c \geq -1$**

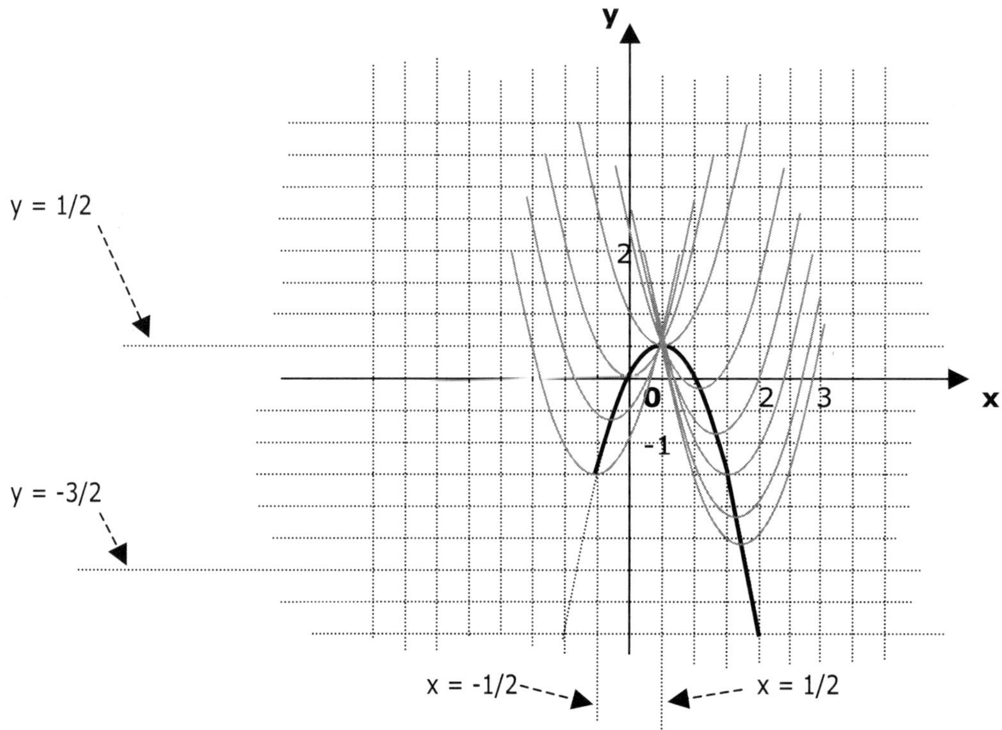

그러면 이제, g의 커브는 c가 변함에 따라 f의 꼭지점이 만드는 자취를 나타낸다고 할 수 있죠.

그런데, 직선의 방정식의 경우에서와 마찬가지로 여러 개의 포물선들이나 직선들 또는 이 두 가지 들의 조합 하나의 방정식으로 나타낼 수도 있어요. 예를 들면 다음과 같습니다.
먼저 $k = 0, 1, 2, \ldots$이고 a_k, b_k, 등등이 상수라고 하세요. 그러면:

두개의 포물선들: $(a_0y + b_0x^2 + c_0x + d_0)(a_1y + b_1x^2 + c_1x + d_1) = 0$.
다수의 포물선들: $(a_0y + b_0x^2 + c_0x + d_0)(a_1x + b_1y^2 + c_1y + d_1)\ldots(a_ny + b_nx^2 + c_nx + d_n) = 0$.
하나의 포물선과 하나의 직선: $(ay + bx^2 + cx + d)(ey + fx + g) = 0$.
하나의 포물선과 두개의 직선들: $(ay + bx^2 + cx + d)(ey + fx + g)(hy + ix + j) = 0$.
다수의 포물선들과 직선들: $(a_0y + b_0x^2 + c_0x + d_0)(a_1x + b_1y^2 + c_1y + d_1)\ldots(a_ny + b_nx^2 + c_nx + d_n)(a_0x + b_0y + c_0)(a_1x + b_1y + c_1)\ldots(a_nx + b_ny + c_n) = 0$.

어떻게 요?
예를 들면, $(ay + bx^2 + cx + d)(ey + fx + g) = 0 \Rightarrow ay + bx^2 + cx + d = 0$ 또는 $ey + fx + g = 0$.

또한, 이런 커브들을 점들의 집합으로 나타내면 다음과 같이 예를 들 수 있죠.
하나의 포물선: $\{(x, y) \mid ax + by^2 + cy + d = 0\}$이고 그 대칭축은 x축에 평행입니다.
두개의 포물선들과 하나의 직선: $\{(x, y) \mid (ay + bx^2 + cx + d)(px + qy^2 + ry + s)(ty + ux + v) = 0\}$.

제안과 풀이

U 번 예제에서의 문제

다음과 같이 가정하세요:
포물선 $x = y^2 - 2(c + 1)y + 2c^2$의 꼭지점은(s, t)인데 여기에서 c는 상수이고 $c \geq -1$.
T: c ⟶ (s, t).
t = g(s)

그러면, g 를 구하세요.

이전 예제에서와 마찬가지로 주어진 포물선의 방정식은 포물선들의 집단을 나타내죠.
포물선 $x = y^2 - 2(c + 1)y + 2c^2$의 꼭지점은 (s, t)인데 여기에서 c는 상수이고 $c \geq -1$ 입니다.
그런데, T 는 매개변수 변환이고 여기에서 그 매개변수는 c 이죠.
T: c ⟶ (s, t)이고 여기에서 (s, t)는 그 포물선의 꼭지점이죠. 그러면, c 가 s 와 t 를 연결해줍니다. 즉, s 와 t 는 제각기 c 의 식으로 나타내지는 거죠. 그러니까, s 와 t 는 제각기 c 의 함수가 되는 거죠.

그러면 이제, s = u(c)이고 t = v(c)이며 $c \geq -1$ 라고 하세요.
이제, 그 꼭지점부터 시작하기로 하죠.
$x = y^2 - 2(c + 1)y + 2c^2 = y^2 - 2(c + 1)y + (c + 1)^2 - (c + 1)^2 + 2c^2$
$= \{y - (c + 1)\}^2 - (c + 1)^2 + 2c^2 = \{y - (c + 1)\}^2 - c^2 - 2c - 1 + 2c^2$
$= \{y - (c + 1)\}^2 + c^2 - 2c - 1 = \{y - (c + 1)\}^2 + (c - 1)^2 - 2$ 이고 $c \geq -1$.

그러니까, 그 꼭지점은 $\{(c - 1)^2 - 2, c + 1\}$이고$c \geq -1$ 이네요.
그러면, $s = u(c) = (c - 1)^2 - 2$ 이고 $t = v(c) = c + 1$ 이며 $c \geq -1$ 입니다.

이제, 다음과 같은 거죠:
T: c ⟶ $(s, t) = \{(c - 1)^2 - 2, c + 1\}$이고 $c \geq -1$.

그러므로, $s = (c - 1)^2 - 2$ 이고 $t = c + 1$ 이네요.
그러면, $s = (c - 1)^2 - 2 \Rightarrow (c - 1)^2 = s + 2 \Rightarrow c = \pm(s + 2)^{1/2} + 1$.
따라서, $c = (s + 2)^{1/2} + 1$ 또는 $c = -(s + 2)^{1/2} + 1$ 이죠.
한편, $t = c + 1 \Rightarrow c = t - 1$ 입니다.
그러면 이제, $c = t - 1$ 이니까 $t - 1 = (s + 2)^{1/2} + 1$ 또는 $t - 1 = -(s + 2)^{1/2} + 1$ 이죠.
그러므로, $t = (s + 2)^{1/2} + 2$ 또는 $-(s + 2)^{1/2} + 2$ 이네요.

자 이제, $c \geq -1$ 이어야 합니다.
우선, $c = \pm(s + 2)^{1/2} + 1$ 은 s가 –2 보다 크거나 같을 때 성립하죠.
다음으로, $c = (s + 2)^{1/2} + 1 \geq 1$ 이죠. 하지만, 또한 $c = -(s + 2)^{1/2} + 1 \leq 1$ 입니다.
그러니까, c 가 항상 –1 보다 크거나 같을 수 있도록 s 의 범위를 구해야 하네요.
그러면, $c \geq -1 \Rightarrow -(s + 2)^{1/2} + 1 \geq -1 \Rightarrow (s + 2)^{1/2} \leq -2 \Rightarrow s + 2 \leq 4 \Rightarrow s \leq 2$.
하지만, s 는 –2 보다 크거나 같아야 하죠.
따라서, $-2 \leq s \leq 2$ 이죠.
그러므로, $t = g(s) = (s + 2)^{1/2} + 2$ 이고 $s \geq -2$ 이거나 또는 $t = g(s) = -(s + 2)^{1/2} + 2$ 이고 $-2 \leq s \leq 2$ 입니다.

함수 g 를 x-y 좌표계에 놓으려면 s 와 t 를 제각기 x 와 y 로 교체해야 되죠. 그러면:
$y = g(x) = (x + 2)^{1/2} + 2$ 이고 $x \geq -2$ 이거나 또는 $y = g(x) = -(x + 2)^{1/2} + 2$ 이고 $-2 \leq x \leq 2$ 입니다.

이제, 주어진 포물선들과 g 의 커브를 x-y 평면에 놓아보기로 하죠.
주어진 포물선들의 방정식은 $x = y^2 - 2(c + 1)y + 2c^2$이고 여기에서 c는 상수이며 $c \geq -1$.

그림. U.O.

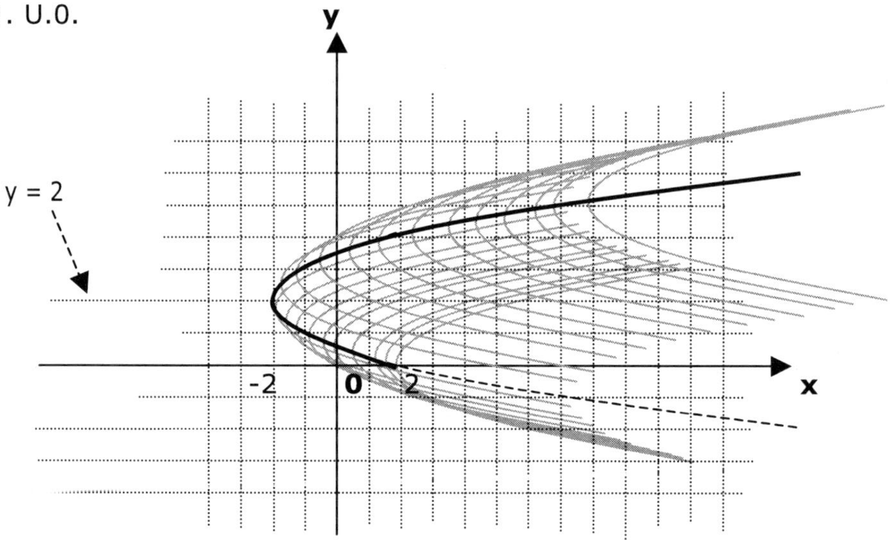

그러면 이제, g의 커브는 다음과 같이 주어진 방정식이 나타내는 그 모든 포물선들의 꼭지점들로 이루어져있다는 것을 알 수 있죠: $x = y^2 - 2(c + 1)y + 2c^2$이고 $c \geq -1$.
즉, g의 커브는 c가 변함에 따라 $x = y^2 - 2(c + 1)y + 2c^2$의 꼭지점이 만드는 자취인 거죠.

요약:

$x = y^2 - 2(c + 1)y + 2c^2 = y^2 - 2(c + 1)y + (c + 1)^2 - (c + 1)^2 + 2c^2$
$= \{y - (c + 1)\}^2 - (c + 1)^2 + 2c^2 = \{y - (c + 1)\}^2 - c^2 - 2c - 1 + 2c^2$
$= \{y - (c + 1)\}^2 + c^2 - 2c - 1 = \{y - (c + 1)\}^2 + (c - 1)^2 - 2$ 이고 $c \geq -1$.
그러므로, 그 꼭지점 $(s, t) = \{(c - 1)^2 - 2, c + 1\}$이고$c \geq -1$.
이제, $s = u(c)$이고 $t = v(c)$이며 $c \geq -1$ 라고 하세요.
그러면, $s = u(c) = (c - 1)^2 - 2$ 이고 $t = v(c) = c + 1$ 이며 $c \geq -1$.
그러면, $s = (c - 1)^2 - 2 \Rightarrow (c - 1)^2 = s + 2 \Rightarrow c = \pm(s + 2)^{1/2} + 1$.
따라서, $c = (s + 2)^{1/2} + 1$ 또는 $c = -(s + 2)^{1/2} + 1$.
한편, $t = c + 1 \Rightarrow c = t - 1$.
그러면 이제, $c = t - 1$ 이므로, $t - 1 = (s + 2)^{1/2} + 1$ 또는 $t - 1 = -(s + 2)^{1/2} + 1$.
그러므로, $t = (s + 2)^{1/2} + 2$ 또는 $-(s + 2)^{1/2} + 2$.

자 이제, $c \geq -1$ 이어야 합니다. 우선, $c = \pm(s + 2)^{1/2} + 1$은 s가 -2보다 크거나 같을 때 성립.
다음으로, $c = s + 2)^{1/2} + 1 \geq 1$. 하지만 또한, $c = -(s + 2)^{1/2} + 1 \leq 1$.
그러므로, c 가 항상 -1보다 크거나 같을 수 있도록 s 의 범위를 정해야 합니다.
그러면, $c \geq -1 \Rightarrow -(s + 2)^{1/2} + 1 \geq -1 \Rightarrow (s + 2)^{1/2} \leq -2 \Rightarrow s + 2 \leq 4 \Rightarrow s \leq 2$.
하지만, s 는 -2보다 크거나 같습니다. 따라서, $-2 \leq s \leq 2$.
그러므로, $t = g(s) = (s + 2)^{1/2} + 2$ 이고 $s \geq -2$ 이거나 또는 $t = g(s) = -(s + 2)^{1/2} + 2$ 이고 $-2 \leq s \leq 2$.
그런데, x-y 좌표계에서는:
$y = g(x) = (x + 2)^{1/2} + 2$ 이고 $x \geq -2$ 이거나 또는 $y = g(x) = -(x + 2)^{1/2} + 2$ 이고 $-2 \leq x \leq 2$.

함수와 그래프의 기초
– 제 3권 : 원뿔곡선 1 (직선과 포물선) –

2007년 12월 15일 초판 인쇄
2008년 1월 1일 초판 발행

저 자 / 김성렬
발행자 / 오판근
발행처 / 교우사

주 소 / 서울 동대문구 제기2동 148-19
전 화 / 925-2861(대), 편집부 / 925-2825
팩 스 / 925-2860, 953-2860
E-mail / kw2861 @ chollian. net
http: // www. kyowoo. co. kr
등록 / 1994. 3. 24 제6-0256호

값(한글 · 영문판) **25,000원**